CO_2 Free Ammonia as an Energy Carrier

Ken-ichi Aika · Hideaki Kobayashi
Editors

CO_2 Free Ammonia as an Energy Carrier

Japan's Insights

Springer

Editors
Ken-ichi Aika
Professor Emeritus, Tokyo Institute of Technology
Meguro-ku, Tokyo, Japan

Hideaki Kobayashi
Institute of Fluid Science
Tohoku University
Aoba-ku, Sendai, Japan

ISBN 978-981-19-4766-7   ISBN 978-981-19-4767-4 (eBook)
https://doi.org/10.1007/978-981-19-4767-4

This Springer imprint is published by the registered company Springer Nature Singapore Pte Ltd.
The registered company address is: 152 Beach Road, #21-01/04 Gateway East, Singapore 189721, Singapore

Preface

At present, the global effects of climate change, such as severe weather events, are a serious concern. Climate change is attributed to anthropogenic CO_2 emissions, leading to questions regarding how this issue can best be addressed by changes in industrial processes and societal lifestyles. One possible approach is the development of new technologies capable of mitigating global warming. Such technologies would be welcomed even by climate change skeptics, because they would result in energy savings and increased renewable energy usage that would conserve limited fossil fuel resources.

Hydrogen is a potential green alternative to fossil fuels that can also be produced using renewable energy sources, although the transport and storage of hydrogen are problematic. One means of addressing these challenges is to use ammonia as a hydrogen storage medium because ammonia can be decomposed to hydrogen for fuel cell use or can be burned directly as an energy source. There are drawbacks to this approach, including energy transition efficiency concerns related to the synthesis of ammonia and the generation of NO_x during combustion, but the use of ammonia as a fuel has seen revolutionary advances over the last several years in Japan.

The 2011 earthquake and tsunami and the associated incident at the Fukushima Daiichi nuclear power plant resulted in greater interest in the use of renewable energy. As noted, a major issue related to the use of hydrogen as a renewable energy source is the difficulty in identifying suitable carriers and in subsequently releasing hydrogen from these carriers. On this basis, "Energy Carriers" Project was initiated with the support of the Ministry of Education, Culture, Sports, Science and Technology (MEXT) and the Ministry of Economy, Trade and Industry (METI). The support from MEXT was in the form of an Advanced Low Carbon Technology (ALCA) program scheduled from July 2013 to June 2014. At the same period, the Japanese government has proposed a new Cross-ministerial Strategic Innovation Program (SIP), into which the Energy Carrier Project was traversed, running from July 2014 to March 2019 under the control of the Council for Science, Technology and Innovation Cabinet Office. The ALCA team expanded as a result, and this new program will be administered according to the principles of plan-do-check-action. The research targets were divided among 11 teams. These research goals comprise solar heat capture (Part II of this book), advanced hydrogen production (Part III), CO_2-free ammonia synthesis (Part IV), ammonia source hydrogen station (Part V), ammonia fuel cells (Part VI), ammonia direct combustion (Parts VII–IX), organic hydride, liquified hydrogen processing, hydrogen combustion, hydrogen station risk management, and separation process. But the project had instead been reconfigured to have ten teams during the second two-year interval, transitioning to five teams as industry partnerships were realized during the final year.

Initially, our intent was to record all results from the Energy Carrier Project in book form, but the amount of data was determined to be overly large. We have therefore focused solely on the results related to the use of ammonia as an energy carrier, because of the uniqueness of this aspect of the work, and have also included some related subjects. The direct combustion of ammonia as examined in this study is recognized as an immediate stage technology at the moment and could have a significant impact on industrial processes. Liquid hydrogen processing and organic hydrides also have potential industrial-scale applications, but these subjects have also been considered in other hydrogen-based projects funded by METI, and so results related to these topics are likely to be published elsewhere. This publication also contains several invited chapters that are not related to the SIP project but describe projects performed based on funding by other grants. These chapters were considered to provide a balance to the present subject matter.

Parts II to IX provide technical information obtained from related projects, while Parts I and X review the use of ammonia as a fuel and discuss future prospects. As noted above, the world is rapidly transitioning to a low-carbon society. Thus, we have invited two guest authors from the United States Ammonia Energy Association, with whom we exchanged information during the course of the SIP program, to contribute to Parts I and X.

As a consequence of the notable success of ammonia-based fuel technology in Japan, other countries have started to produce blue and green ammonia, so there has been a progress toward a worldwide ammonia chain. Thus, the present work has been titled *CO_2 Free Ammonia as an Energy Carrier: Japan's Insights*.

We appreciate the contributions of all authors to this publication, especially the lead editors for each part who devoted significant time to completing this book. We also sincerely thank the government officials and subject matter experts who supported the SIP project and related programs, as detailed in Chap. 1 of Part I. Finally, we are much obliged to our wives (Chigusa Aika and Shukuko Kobayashi) for supporting our research activities and the writing of this book.

Tokyo, Japan Ken-ichi Aika
Sendai, Japan Hideaki Kobayashi
May 2022

Contents

About the Editors

Professor Emeritus Ken-ichi Aika received his Ph.D. from the Tokyo Institute of Technology (Tokyo Tech) in 1969. He started his research career there in the group headed by the late Professor Atsumu Ozaki. He spent 3 years in the USA (Princeton University and Texas A&M University, from 1974 to 1977) and later was a professor at Tokyo Tech from 1981 to 2005. He worked as a professor from 2007 to 2012 for the Open University of Japan, where he devoted himself to science communication. After that, he was involved in the energy carrier research program from 2013 to 2019 for his first year as the program officer of the Japan Science and Technology Agency (JST) and for the following 5 years as the deputy program director of the Energy Carrier Cross-Ministerial Strategic Innovation Promotion Program (SIP). He is currently a visiting professor of Green Ammonia Research Center at KOSEN Numazu. He has published approximately 300 articles and contributed to some ten books. His research covers catalysis, reaction engineering, and the environmental chemical process, with the largest number of his works being related to ammonia synthesis catalysis. He invented many effective ruthenium catalysts.

Professor Hideaki Kobayashi received his Ph.D. from the Graduate School of Engineering, Tohoku University, in 1991. He is currently a professor at the Institute of Fluid Science, Tohoku University. He started his research career at the School of Engineering, Tohoku University, as a research associate in 1984 before moving to the Institute of Fluid Science at Tohoku University. His research focuses on combustion science and technology, covering laminar flame, turbulent flame, microgravity combustion, high-speed combustion for propulsion, and gas turbine combustion. He has published over 220 articles and contributed four books. He served as a team leader in ammonia direct combustion in the Energy Carrier Cross-Ministerial Strategic Innovation Promotion Program (SIP) from 2014 to 2019. He is a fellow of the Combustion Institute.

Part I
Green Ammonia as Energy Carrier

Chapter 1
Brief Review of the Japanese Energy Carrier Program and an Energy Science View of Fuel Ammonia

Ken-ichi Aika

Abstract The activity data of Japanese Energy Carrier Program (2013–2019) are reviewed briefly. The 2011 Tohoku earthquake was the impetus for proposing a new energy carrier program. Its relation to the existing long-term energy development program is described. Among the "SIP (Strategic Innovative Program) Energy Carrier Program," an ammonia study received a large budget allocation in the last stage because of its importance. The standard ammonia process from natural gas, which has the advantage of using air (mixture of nitrogen and oxygen) as a resource, is illustrated. The pure CO_2 by-product is easily passed to carbon capture and storage (CCS) to obtain blue ammonia. The standard ammonia process using renewable power through water electrolysis (green ammonia) also is illustrated. The characteristic feature of both processes is reviewed, and analyzed thermodynamically, and rough carbon footprints are proposed.

Keywords Japanese research · Blue ammonia · Green ammonia · Energy efficiency · Carbon footprint

1.1 Relation to the SIP Energy Carrier Program

The background of this book is briefly summarized in the preface. Here, some of the SIP program record and its relation to the contents of this book are described. Through the worldwide energy program [1], the Ministry of Economy, Trade and Industry (METI) organized hydrogen programs as are shown in Fig. 1.1. The 2011 Tohoku earthquake triggered industries and academia to seek development of energy carrier technology. The Ministry of Education, Culture, Sports, Science and Technology (MEXT) started an energy carrier program under the framework of the Advanced Low Carbon Research and Technology and Development Program (ALCA) in 2013. The following year this program was succeeded by the cross-ministerial Strategic Innovation Program (SIP) "Energy Carrier" supported by the Cabinet Office (CAO).

K. Aika (✉)
Professor Emeritus, Tokyo Institute of Technology, 2-12-1 Ohokayama, Meguro-ku, Tokyo 152-8550, Japan
e-mail: kenaika@c01.itscom.net

K. Aika and H. Kobayashi (eds.), *CO_2 Free Ammonia as an Energy Carrier*,
https://doi.org/10.1007/978-981-19-4767-4_1

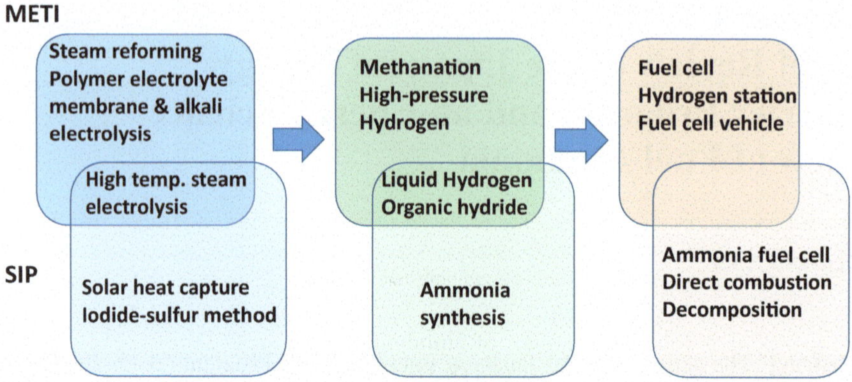

Fig. 1.1 METI hydrogen program and SIP "Energy Carrier" program (2014–19) including hydrogen production, carrier, and utilization technologies

While METI had been concerned about the practical technologies at that time, SIP was developing future technologies. Examples are solar heat capture (part II of this book [2]) and advanced hydrogen conversion (part III [3]). As is shown in part II, solar heat can be collected into fluid salts that can store the energy, even at night. This energy can be used for decomposing water to hydrogen through the iodide-sulfur (IS) method or electrolysis-steam turbine power. Neither solar power capture to fluid salt nor the IS method is economically feasible at present; however, many new technologies were developed under the SIP program. Since those technologies will be utilized in the future, we included them in this book (parts II and III). We also included work on high-temperature steam electrolysis, which is partly supported by METI and almost practical, in part III (together with the IS method).

The liquid hydrogen and organic hydride programs had been started before the SIP program supported by METI and several firms. Some branch programs also had been conducted under SIP and supplied results such as those shown in Figs. 1.2 and 1.3. However, that content has been released in other publications. Since ammonia-related research is unique and has not been reported in work other than that from the SIP program, these are described in the greatest detail in this book. The ammonia research is well summarized and discussed in Chap. 3 by Shiozawa [4], a deputy director of the SIP "Energy Carrier" program.

The activity and size of the SIP "Energy Carrier" program (2014–19) are shown in brief. Thirty-two university laboratories, 16 national research institute laboratories, and 59 industrial groups took part in this program through 11 categories, as shown in Fig. 1.2. Ammonia direct combustion teams were the greatest in number, because these teams were expanded in the last stage. The policy of the SIP program was "innovation" to change the current technology, which Kazuo Kyuma (SIP program representative, CISTI, CAO) proposed. The director of the SIP "Energy Carrier" program, Muraki, and others managed the project strategically by the plan-do-check-action method. The framework of the governing structure is shown in Fig. 1.3. The

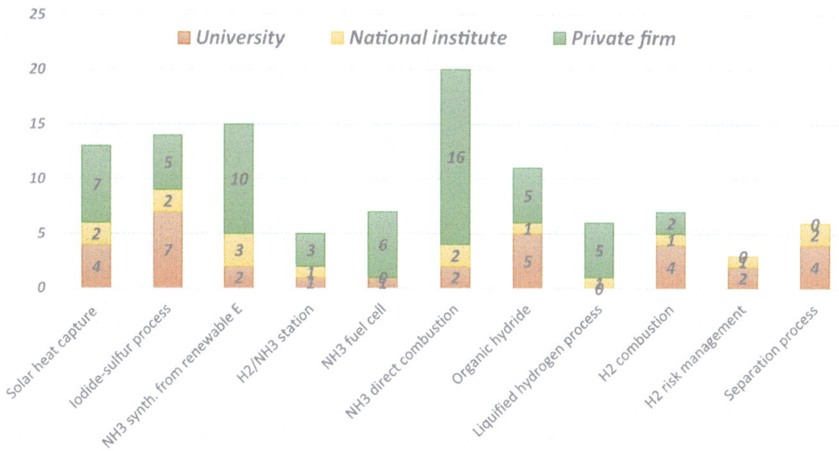

Fig. 1.2 Number of research units in each unit in the SIP "Energy Carrier" program (2014–19)

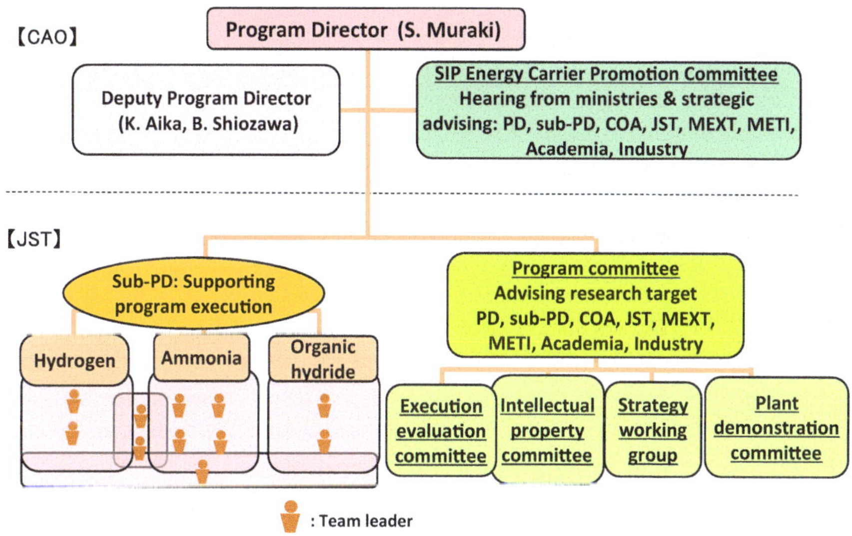

Fig. 1.3 Framework of the governing structure of the SIP "Energy Carrier" program (2014/7–19/3)

top half of the structure was supported by the CAO [5, 6] and the bottom half was supported by the Japan Science and Technology Institute (JSP) [7, 8]. We fully appreciate the meeting organizers and logistic support members and celebrate the success of the SIP "Energy Carrier" program.

The annual budgets for all the teams are shown in Fig. 1.4. Every year, about 3 billion yen was supplied for work toward the target. Since the committee evaluated their results through the principle of plan-do-check-action every 2 years, the

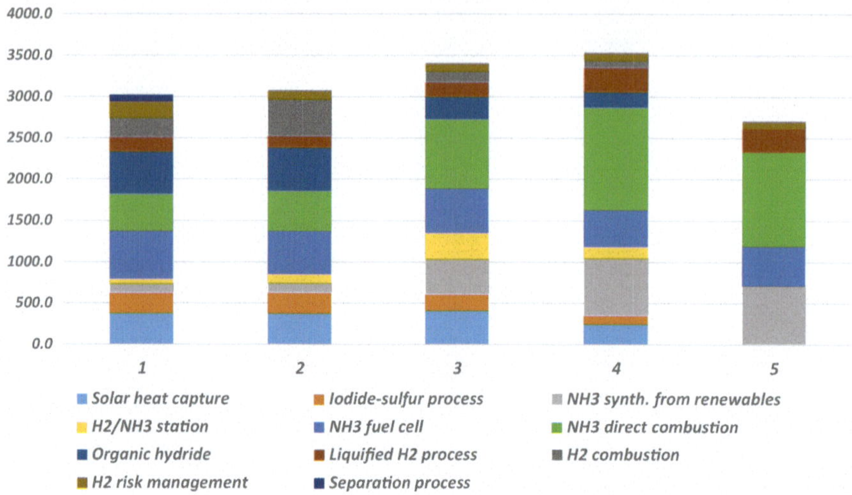

Fig. 1.4 Annual money input (M¥) for target research

budget of the next stage changed greatly. The biggest changes were for the "ammonia direct combustion" and "ammonia synthesis from renewable" teams, which had their budgets nearly doubled in the last 2 years. The money was used to install a mini pilot plant or to conduct the large practical experiments needed during the last stage and the results are included in this book. The amount of money input changed drastically throughout the 5 year program.

Numbers of papers published, and patents released for each team are shown in Fig. 1.5 for from 2014 (7) to 2018 (10). This book covers only the activity done by the six teams shown in Fig. 1.5. The main original papers from those teams are reviewed in parts II to IX of this book.

1.2 Evaluation of Blue and Green Ammonia from an Energy Science View

1.2.1 Energy View of Ammonia Synthesis from Natural Gas (Blue Ammonia)

When we think of ammonia as a fuel, the ammonia produced must be treated as energy. First, the most popular method of the ammonia synthesis process starts from natural gas. An interesting feature of this process is the use of air (mixture of nitrogen and oxygen) as a raw material. Oxygen and nitrogen are added in the partial oxidation unit. The nitrogen passes through the shift reactor and CO_2 separator, making the synthesis gas ($N_2 + 3H_2$). To achieve a stoichiometric ratio of synthesis gas, the

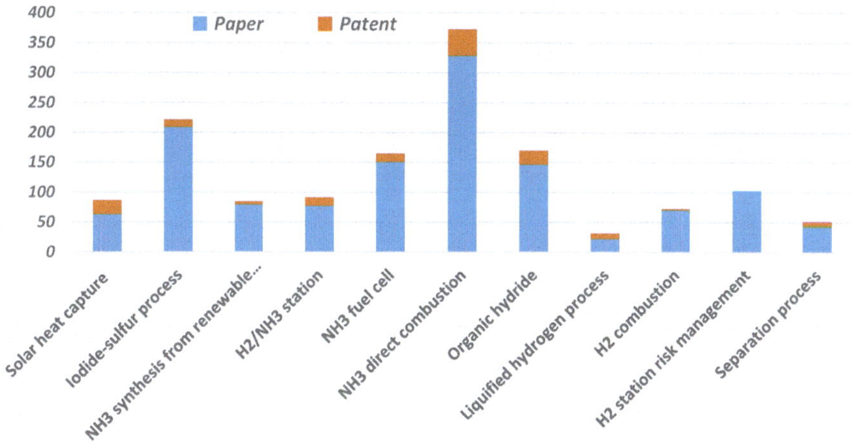

Fig. 1.5 Number of publications/patents by SIP team from July 2014 to October 2018

methane/air feed ratio is controlled. This is simpler than the way used to separate nitrogen at an earlier stage and add (and compress) it before the synthesis reactor. Another interesting feature is the production of CO_2 as a pure by-product, which may be used for urea production [$2NH_3 + CO_2 \rightarrow (NH_2)_2CO + H_2O$], carbon capture and storage (CCS), or enhanced oil recovery (EOR). This is an important feature of ammonia produced from natural gas; it is called CO_2-free (reduced) fuel or "blue ammonia".

To understand the process structure, valuable data can be obtained from documents [9–12] that can be presented as a simple model. The ammonia synthesis process is advanced each year by improving total energy efficiency; however, the chemical reactions do not change. An excellent review by Dybkjaer [12] covers the thermodynamic aspects of this process in detail. Those data are cited along with research of ours [13].

In this model, impurities (higher hydrocarbons, sulfur compounds in natural gas, and argon in air) are neglected and air is assumed to contain 21% O_2 and 78% N_2 (1% Ar). For thermodynamic calculations, water is assumed to be a liquid for convenience. Table 1.1 contains an example of the working conditions of every unit cited [12] and the chemical reactions reported by the author. Reactions from the primary reformer to shift reactors may be mixed up to some extent. The steam carbon ratio (*e.g.*, 2.5) supplied is greater than that in the reaction formula (1.39) shown afterward. The excess water is available for shifting CO to CO_2 and is corrected for after the shift converter. The primary reformer is located in a furnace to obtain heat for the endothermic reaction (Eq. 1.8).

To obtain one mole of ammonia, Eqs. (1.1), (1.2), (1.3), and (1.6) in Table 1.1 are summed to create material balance: (1.1) * 0.173 + (1.2) * 0.269 + (1.3) * 0.442 + (1.6) * 0.5. This gives the following equation and enthalpy change.

Table 1.1 Details of the main unit of the methane feed ammonia synthesis process. Some data[a)] are cited from example 7 (1700 t/d unit) of Fig. 6.24 in a reference [12] (original reference [14])

Unit	Press/bar-g[a)]	Temp/C[a)]	COx[a)]	Main reaction	ΔH/kJ 25C 1 bar	
Primary reformer	31.7–33.5	790–870		$CH_4 + H_2O \rightarrow CO + 3H_2$ (1.1)	250.1	In fuel furnace
Secondary reformer	30–32	980–1010		$CH_4 + 0.5O_2 + 1.859N_2 \rightarrow CO + 2H_2 + 1.859N_2$ (1.2)	−35.7	Partial oxidation
CO shift (HTS/LTS)	29–30		To 0.5% CO	$CO + H_2O(L) \rightarrow CO_2 + H_2$ (1.3)	2.8	Excess steam
Water removal						Condensation
CO_2 remover	26–28		To 0.01–0.1% CO_2	$CO_2 + MEA \leftrightarrow MEA - CO_2$ (1.4)		2-column MEA cycle
Methanation			5 ppm	$CO + 3H_2 \rightarrow CH_4 + H_2O(L)$ (1.5)		Impurity purged
Compressor	150–330					Steam driven
NH_3 synthesis	330			$N_2 + 3H_2 \rightarrow 2NH_3$ (1.6)	−92.2	Gas recycled
NH_3 separation		−33		$NH_3 \rightarrow NH_3(L)$ (1.7)	−34.1	Refrigeration

$$0.442\ CH_4 + 0.615\ H_2O(L) + 0.1345\ O_2 + 0.5\ N_2 \rightarrow 0.442\ CO_2 + NH_3 \cdots \Delta H = -11.2\ kJ \tag{1.8}$$

Methane, water, and air travel through the reactors (under different conditions) are converted into the final products, CO_2 and ammonia. Excess steam is added at the reformer and works as a driving force to shift the reaction and recovery after that. The argon and methane impurities (Eq. 1.5) accumulated in the synthesis loop are purged periodically. For thermodynamic reasons, ammonia conversion is limited to 20–30%, depending on the conditions. Ammonia produced is separated from the reactant gas through refrigeration (Eq. 1.7). For thermodynamic convenience, ammonia is treated as a gas here.

From an enthalpy point of view (Eq. 1.8), 393.6 kJ of methane (0.442 * 890.4 kJ/mol: combustion energy) is converted to 382.6 kJ of ammonia (combustion energy) and 11.2 kJ of heat on a mol-NH_3 basis. For one ton ammonia production, 23.15 GJ of methane is turned into 22.51 GJ of ammonia and 0.64 GJ of heat. However, the practical process consumes more methane than that predicted by Eq. (1.8) as fuel for the furnace. The high-pressure steam produced is used for driving turbines and compressors, and for heat exchange at the reactors. An example of the system, including a steam line and compressors, is illustrated in Fig. 1.6 along with a figure in a literature [15].

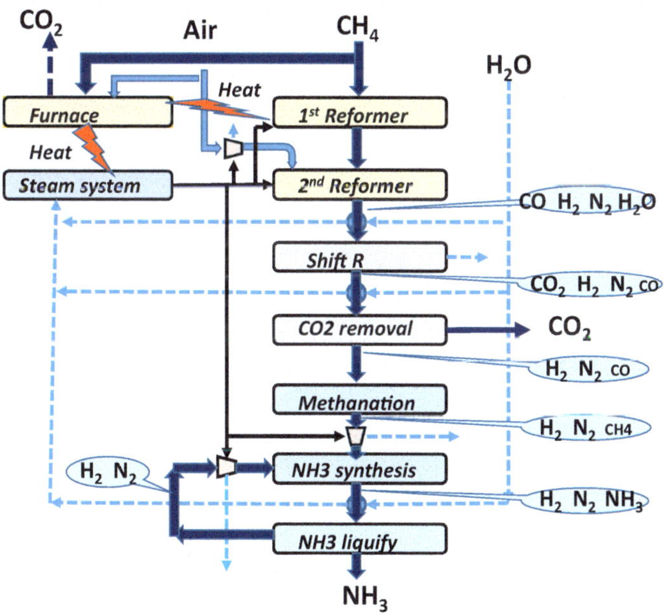

Fig. 1.6 Process diagram for ammonia synthesis from natural gas. Only 3 major heat exchangers and 3 compressors are shown. The data in Table 1.1 are from this process

The steam line energy is abandoned after process driving. Dybkjaer [12] reported a waste energy of 11.54 GJ/t-NH_3 (196.1 kJ/mol-NH_3). The heat loss comes from reforming (3.3%), steam generation (2.9%), shift-CO_2 removal-methanator (11.2%), synthesis (14.9%), turbines-compressor (56.5%), stack (8.3%), and miscellaneous loses (2.9%). This can help determine how the turbine-compressor work loses energy during this process. The major work involves compressing the gas to synthesis pressure and the recycling of unreacted synthesis-gas after separation of ammonia. The value of the NH_3 product enthalpy in this review [12] is 20.94 GJ/t-NH_3 (356.0 kJ/mol-NH_3), which is a little less than the calculation here (22.51 GJ/t-NH_3 or 382.6 kJ/mol-NH_3). One reason is the difference in the final state of NH_3; in the former, it is a liquid (−33 °C) and in the latter it is a standard gas. Thus, enthalpy-based efficiency is 66.1% (11.54/34.05). Interestingly, the exergy (Gibb's energy)-based efficiency is reported as 66% [12]. This situation is illustrated in Fig. 1.7a.

As is illustrated in Fig. 1.7a and Eq. (1.8), 0.442 mol methane (23.2 GJ/t-NH_3) is converted to 1 mol ammonia (22.5 GJ/t-NH_3) with an energy loss of 11.5 GJ/t-NH_3, which is scarified by methane burning. Burned gas is reported to contain 5–25% of CO_2, 3–10% of O_2, 0–500 ppm of SOx, and 0–100 ppm of NOx, with the rest being N_2 (60–80%). However, the pressure of CO_2 in the burned gas is so low, about 0.1 bar, that capturing it is too costly. In contrast, the process of CO_2 is captured by the amine absorption unit as a pure chemical product, which may be used as a raw material for urea or dry ice production or for enhanced oil recovery. The CO_2 pressure at the amine unit is estimated to be 4.9 bar (27 bar * 0.442/2.442). Such high pressure enables the amine absorption process to be practical, while burned CO_2 pressure is just 1/50 of the process CO_2 pressure. Of course, the burned CO_2 can be captured if

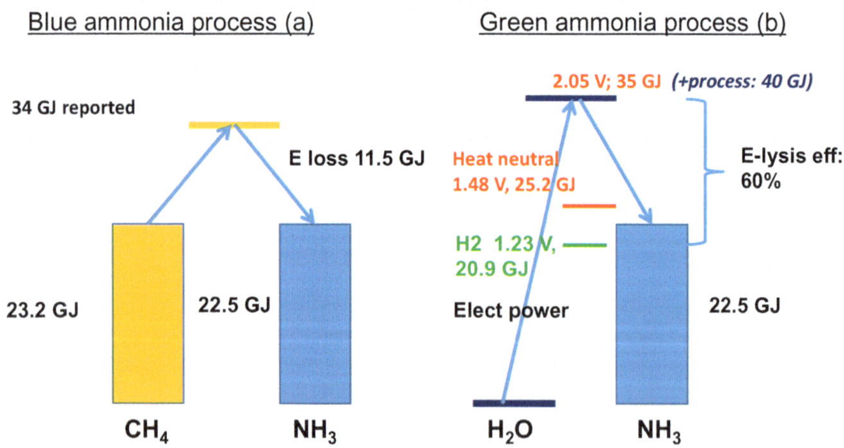

Fig. 1.7 Energy conversion of ammonia synthesis from **a** natural gas and **b** renewable energy. Numerical values indicate fuel enthalpy per ton ammonia at 25 °C under 1 bar (electrical voltage is related to Gibb's energy). Modified from a previous report [16]

the high cost is not an obstacle. Therefore, the process CO_2 can be delivered easily from methane-based ammonia to CCS or EOR use (blue ammonia), while burned CO_2 is abundant. Ammonia production from natural gas is economically feasible for CCS or EOR [17, 18].

Using this model, the carbon intensity of blue ammonia as a fuel was investigated. Energy loss included about 50% of the methane feed as process material (Fig. 1.7) but was partially compensated by the exotherm of the total reaction. As shown in Fig. 1.8, the amount of burned CO_2 was not 0.211 mol (50% of 0.422), but 0.198 mol/mol-NH_3. The carbon footprint (CFP) at the production site was calculated as 22.7 g-CO_2/MJ-NH_3 [19]. Here, a total process efficiency of 66.7% was set as a benchmark. If efficiency could be increased to 70 or 75%, the CFP might be decreased to 19.8 or 14.7 g-CO_2/MJ-NH_3.

Thus, this value was compared to that of octane combustion: $C_8H_{18} + 12.5O_2 \rightarrow 8CO_2 + 9H_2O$ (L) (5511.7 kJ exotherm). The CFP of octane combustion was 63.9 g-CO_2/MJ. However, the CFP of blue ammonia is about 1/3 that of a fossil fuel, so the CFP of ammonia from other fossil source through a similar production process was examined. For coal or pitch, the CFP is around 60 g-CO_2/MJ-NH_3 [15], because of the high endotherm of hydrogen production. This is nearly identical to the value of liquid hydrocarbon fuel, meaning it to be no CO_2 free ammonia. Thus, some different processes must produce hydrogen or ammonia from coal or pitch. To capture burned CO_2, nitrogen in air is troublesome. The N_2 is separated out before introducing O_2 to the coal gasification unit. In this way, the exhaust gas contains just water vapor and CO_2, which are easily separated [20]. As discussed above, ammonia synthesis from natural gas appears to be best with CCS or EOR [21], while the CFP here covers just the production process. As a fuel, however, the proper CFP in the certificate must be a "well to wheel" or "well to gate" value. If we are allowed to

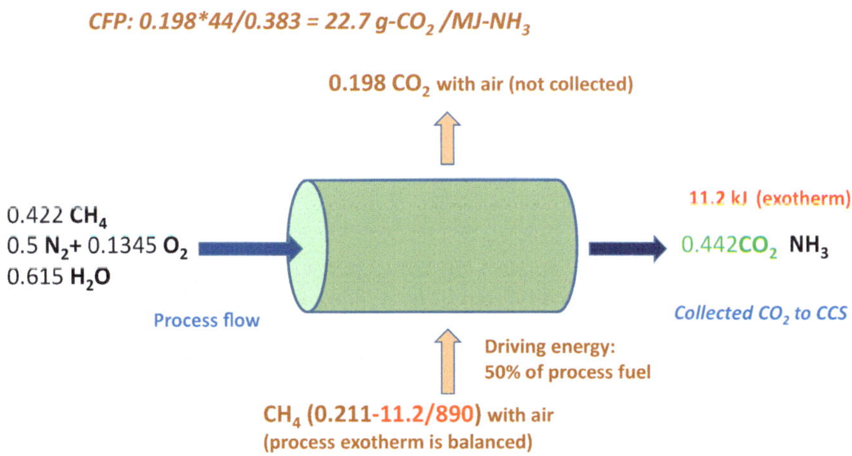

Fig. 1.8 Simplified ammonia synthesis process: comparing process mass/energy flow with furnace mass/energy flow (modified from [15])

use CFP data of Californian gasoline [22] where fossil fuel mining (11 g), refining (13 g), transportation (2 g), and fuel combustion (73 g-CO_2/MJ) are summed up (99 g-CO_2/MJ), the "well to wheel" value for blue ammonia is estimated as 49 g-CO_2/MJ (CFP of CCS is not added). The CFP of blue ammonia appears to be around half that of gasoline in California [15]. The detailed study is described in Chap. 46 [23] of this book.

1.2.2 Energy View of Ammonia Synthesis from Renewable Energy (Green Ammonia)

Several ways are available for converting the renewable energy to hydrogen. Characteristic features are summarized in Fig. 1.9. A type of molten salt is used as the energy storage material at high temperature (350–450 °C). Material heated through a mirror system generates steam and drives a turbine, which has been commercialized for dry regions. One SIP energy carrier group tried to improve the heating system operated at 650 °C by developing a new molten salt and sunlight capturing system, which are reviewed in chaps. 5–8 [2]. Another SIP group tried to improve the iodine–sulfur (IS) method (chemical water splitting) by decreasing the working temperature to around 600 °C (usually operated at 800–900 °C) to utilize the heat from sunlight. Those are reviewed in chaps. 11–15 [3]. The molten salt group succeeded in building a sample unit, and the new IS method group demonstrated their unit process parts within their site. This system provides the possibility of establishing a 100 MW-size hydrogen production factory in the future.

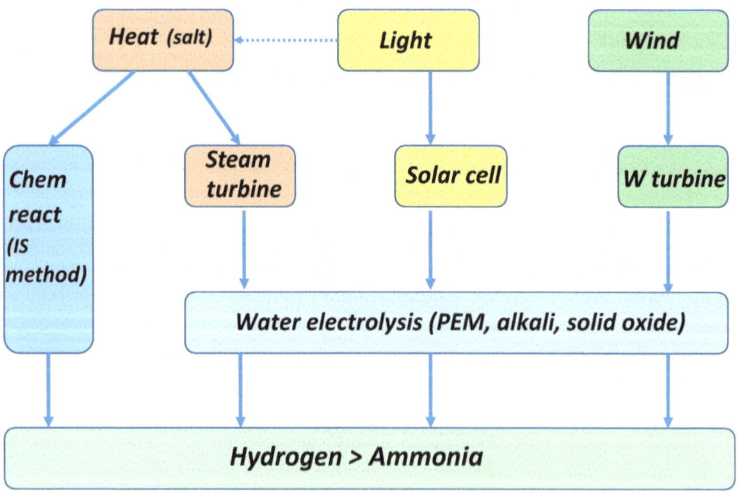

Fig. 1.9 Hydrogen production method from renewable energy

Solar cells and wind turbines (Fig. 1.9) are the most popular methods for generating renewable electric power. The renewable power generation sites usually are located remotely from big cities and are usually not very large. A standard wind power cluster or solar power village may supply a full power of around 100 MW. Another feature of renewable energy is the low rate of operation: approximately 25% of full spec. Therefore, a 100-MW renewable power station can supply only 25 MW, meaning 600 MWh (*i.e.*, 2169 MJ) in a day. If this is applied to ammonia synthesis with an energy efficiency of 56% (40 GJ/t-NH_3), as shown in Fig. 1.7b, a 54 t-NH_3/d plant can be practical [24]. The size of a renewable power-driven plant is just 1/37 of the present 2000 t/d plant. However, capital expenses would be about 3 times greater if a scale factor of 0.7 order is applied, assuming the process structure is the same. However, engineers believe the process structure may be different in a small plant, even for a natural gas-driven plant [25]. The complicated high-pressure steam process (Fig. 1.6) must be replaced by an electric motor to drive compressors in a small plant. It is natural to use electricity to drive compressors for renewable ammonia production based on size and power source. Another feature of a renewable ammonia plant is the large capital cost of water electrolysis, which is the greatest among the process units. Since the electrolysis unit is composed of 2-dimensional electrodes, the capital cost scale factor may be negligible. Large scale is not always economic. Three types of electrolysis units (Fig. 1.9) are expected to be developed [26]. Alkaline electrolysis cells (AEC) are the most economical at present; however, its advantage is small compared with polymer membrane electrolysis cells (PEMEC). A cost reduction might be possible in the future for solid oxide electrolysis cells (SOEC). A new type of solid oxide electrolysis system also is under development (chaps. 9 and 10) [3]. Renewable energy costs must be considered as a whole and a new system model has been proposed [27–30].

A proposed standard green ammonia process is shown in Fig. 1.10 and discussed on the basis of its energy and carbon footprint. The system structure is simple: Hydrogen is produced through water electrolysis and compressors are driven by electric motors. However, one obstacle to reliable renewable energy systems is its unreliability (because the power supply depends on the weather and season). The systems must be backed up by batteries and grid power and by adding a hydrogen reservoir in front of the ammonia synthesis line. The energy diagram of this process is shown in Fig. 1.7b. To obtain 1.5 mol of H_2 (3 electrons) by water electrolysis, 355.6 kJ/mol-NH_3 (237.1 kJ/mol-H_2 $\times$ 1.5 mol-H_2/mol-NH_3 or 20.9 GJ/t-NH_3) of Gibb's energy is necessary. This corresponds to 1.23 V for 1 electron activation. However, the enthalpy value of this state is 428.7 kJ/mol-NH_3 (285.8 kJ/mol-H_2 $\times$ 1.5 mol-H_2/mol-NH_3 or 25.2 GJ/t-NH_3), meaning that the difference of 73.1 kJ/mol-NH_3 is supplied as heat from the surroundings during electrolysis. This heat is supplied from the over voltage (1.48 V). Additional voltage is needed to drive the electrons at a proper rate, which is reported to be about 2.05 V (excess voltage of 0.8 V: efficiency of 60% or 35 GJ/t-NH_3) to decompose water [31].

As is shown in Fig. 1.10, electricity is used for the compressor and other devices, but its value is estimated to be about half of the steam system (5 GJ/t-NH_3) because

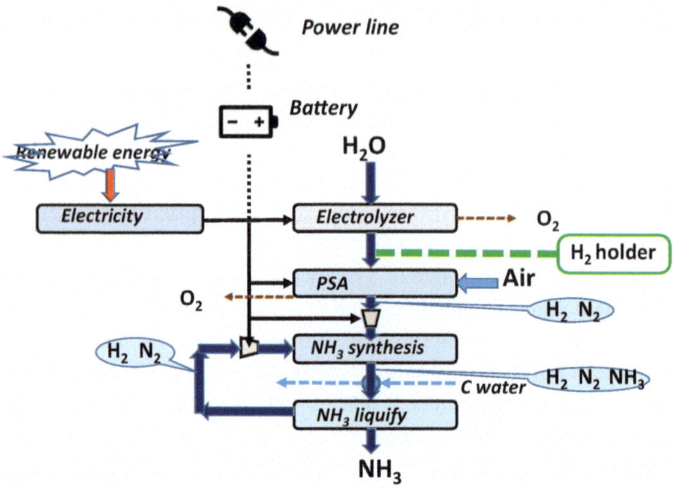

Fig. 1.10 Diagram of the ammonia synthesis process from renewable energy (electricity)

of the simplicity of the renewable resource systems and improvement in motor efficiency. Thus, the total energy amounts to about 40 GJ/t-NH_3, which is 1.2 times greater than that for the natural gas ammonia system. This disadvantage is not serious if the cost of renewable power is less than that of natural gas. The efficiency of water electrolysis, synthesis catalysts, and separation systems is being investigated [28–30]. Chapters 16–22 [32] discuss a new ammonia catalyst and a new process with absorption separation.

Finally, the CFP of green ammonia synthesis is discussed. As an example, CO_2 production from a 40-MW wind power plant is estimated as 5.2 g-CO_2/MJ-electricity. This gives an in situ CFP of 9.25 (5.2 * 40/22.5) g-CO_2/MJ-NH_3, which is 1/2.35 of that for blue ammonia. The well-to-wheel CFP is the same value (9.25 g/MJ: less than 10% of gasoline); however, some value may be added afterward. Details of this value will be discussed in chap. 46 [23]. It is to be noted that the concept of green ammonia system and the methods have been proposed and patented by Nakamura [33].

References

1. Example: World energy network program (WE-NET) from 1993
2. Kato Y, organizer (2022) Part II Solar heat storage (chapter 5–8). In: Aika K, Kobayashi H (eds) CO_2 free ammonia as an energy carrier- Japan's insights. Springer
3. Inagaki Y, Sakaba N, organizer (2022) Part III. Advanced hydrogen production through electrolysis and thermochemical reaction (chapter 9–15). In: Aika K, Kobayashi H (eds) CO_2 free ammonia as an energy carrier—Japan's insights. Springer

4. Shiozawa B (2022) The potential of ammonia as CO_2 free-fuel and hydrogen carrier (chapter 3). In: Aika K, Kobayashi H (eds) CO_2 free ammonia as an energy carrier—Japan's insights. Springer
5. SIP energy carrier promotion committee on Jan. 16th· 2017: from CAO (Kyuma K, Muraki S, Tanaka N, Matsumoto E, Takegami I, O-hora T, Shiozawa B, Aika K), from MEXT (Aizawa M, Ono M, Koga A, Miyazaki S), from METI (Kuriyama N, Okada T, Murayama M, Katayama H, Ohira H), from other ministries (Suzuki T, Tabuchi K, Ikemoto T, Matsuoka S), Industry & academy (Eguchi K, Kawai T, Kikukawa T, Saito K)
6. Miyake A, Taniguchi S, Nakanoh A, Takanohashi T, Matsubara T, Deguchi S CAO logistics members
7. Aika K, Uedono N, Takasu Y, Fuchigami T, Matsukata M, Shimada H, Makino H, Ishii S, Akai Y, Iijima M, Matsumoto S, Yakabe H Execution & evaluation committee
8. Koga A, Kaibe K, Ohya M, Miyazaki S, Umezawa J, Aoyama K, Edera M, Ishimura Y, Iwatake N, Katsumura A, Ishizuki M, Minami T, Fukuyama Y, Kon F JST logistics members
9. Bridger GW, Gadsby RE, Ridler DE (1979) High-pressure ammonia synthesis and plant operation. In: Hardy RWF, Bottomley F, Burns RC (eds) A TREATISE ON DINITROGEN FIXATION, sections 1 and 2, Inorganic and Physical Chemistry and Biochemistry. John Wiley & Sons Inc., New York, pp 291–332
10. Finneran JA, Czuppon TA (1979) Economics of production and consumption of ammonia. In: Hardy RWF, Bottomley F, Burns RC (eds) A TREATISE ON DINITROGEN FIXATION, sections 1 and 2, Inorganic and physical chemistry and biochemistry. John Wiley & Sons, Inc., New York, pp 333–382
11. Hooper CW (1991) Ammonia synthesis: commercial practice. In: Jennings JR (ed) Catalytic ammonia synthesis. Fundamentals and practice, Plenum Press 7:253–283
12. Dybkjaer I (1995) Ammonia production processes. In: Nielsen A (ed) Ammonia, catalysis and manufacture. Springer Verlag, Berlin: 199–327 (ISBN 3–540–58335–1)
13. Aika K (2013) Expecting new technology of ammonia synthesis catalysts: utilizing renewable energy. PETROTECH 36:271–277
14. Quartulli J, Wagener D (1973) Erdoel Kohle, Erdgas. Petrochem Brenns Chem 26(4):192
15. Smith C, Hill AK, Torrente-Murciano L (2020) Current and future role of Haber-Bosch ammonia in a carbon-free energy landscape. Energy Env Sci 13:331–344
16. Aika K (2016) Expecting LCA data of energy carrier: CO_2 free hydrogen and ammonia as examples. J LCA Japan 12(3):1–9
17. Anderson H (2002) Norsk Hydro, World hydrogen energy conference, Montreal
18. McFarlan A (2012) Canmet ENERGY. 9th annual ammonia fuel conference, San Antonio, TX, Oct 1st
19. Aika K, Inazu K, Sugimoto M (2021) Manufacture of CO_2 free ammonia: challenge and prospect. 127th meeting of Cat Soc Japan 2P15
20. Onozaki M, Makino H, Mochida I (2015) Significance and outline of the project: development of next-generation IGCC with CO_2 capture; oxy-fuel IGCC. J Japan Inst Energ 94(1):218–223
21. Aika K (2016) Background of SIP energy carrier program and technology of synthesis and utilization of ammonia. Chem Eng 80(7):402–405
22. Aguila J (2013) The 10th annual ammonia fuel conference, Sacramento, CA, USA
23. Kudoh Y, Ozawa A (2022) Life cycle carbon dioxide emissions from ammonia-based power generation technology (chapter 46). In: Aika K, Kobayashi H (eds) CO_2 free ammonia as an energy carrier—Japan's insights, Springer
24. Aika K (2013) Challenges and prospect of ammonia synthesis process through renewable energy. 45th Fall Meeting Chem Eng Soc, Sep 18th Okayama
25. Ennis R, Lesur PF (1977) How small NH_3 plants compete. Hydrocarbon Processing. Dec, pp 121–124
26. Schmidt O, Gambhir A, Staffell I, Hawkes A, Nelson J, Few S (2017) Future cost and performance of water electrolysis: an expert elicitation study. Intern J Hydrogen Energ 42:30470–30492

27. Aika K, Kakegawa T (1991) On-site ammonia synthesis in De-NOx process. Catal Today 10:73–80
28. Liu CY, Aika K (2004) Effect of the Cl/Br molar ratio of a $CaCl_2$-$CaBr_2$ mixture used as an ammonia storage material. Ind Eng Chem Res 43(22):6994–7000
29. Malmali M, Reese M, McCormick AV, Cussler EL (2018) Converting wind energy to ammonia at lower pressure. ACS Sustain Chem Eng 6(1):827–834
30. Smith C, Hill AK, Torrente-Murciano L (2020) Current and future role of Haber-Bosch ammonia in a carbon-free energy landscape. Energ Envir Sci 13:331–344
31. Grundt T, Christiansen K (1982) Hydrogen by water electrolysis as bases for small scale ammonia production—a comparison with hydrocarbon based technologies. Intern J Hydrogen Energ 7(3):247–257
32. Aika K, Fujimura Y (2022) Part IV CO_2 free ammonia synthesis under industrial conditions (chapter 16–21). In: Aika K, Kobayashi H (eds) CO_2 free ammonia as an energy carrier—Japan's insights. Springer
33. Nakamura N (2008) Method of storing solar heat energy. JP 2009197733A; Method of converting solar heat energy, WO 2009104813A1

Chapter 2
Physical and Chemical Properties of Ammonia as Energy and Hydrogen Carriers

Yoshitsugu Kojima

Abstract Ammonia has been expected as hydrogen and energy carriers because it has high gravimetric and volumetric H_2 densities with 17.8 wt% and 10.7kgH_2/100L, respectively. The volumetric hydrogen density is above 1.5 times of liquid hydrogen, and it is easily liquefied under 1 MPa at 298 K. Direct combustion of ammonia is also possible without emission of CO_2. The heat of combustion of ammonia is above 1.3 times of liquid hydrogen. Ammonia is a safest substance based on the flammability because of the high flash point. However, health hazard of ammonia is large. This correlated with the large solubility in water.

Keywords Ammonia · Hydrogen carrier · Energy carrier · Hydrogen density · Heat of combustion · Flash point · Solubility

2.1 Background

Based on energy and environmental issues using fossil fuels, we have an energy security crisis and negative effects of global warming such as sea level rise and abnormal weather, etc. In addition, air pollution by PM 2.5 (fine inhalable particles with diameters those are generally 2.5 μm and smaller) released from the combustion of fossil fuels also caused serious health damage. Global CO_2 emissions pathways reported by Intergovernmental Panel on Climate Change (IPCC) showed that it is necessary to reduce CO_2 emissions around 40–70% by 2050 compared to the level in 2010, and almost zero emission by 2100 to achieve 1.5–2 °C global temperature change limit target above industrial revolution level. Recently, China, EU and Japan are aiming toward a decarbonized society by 2050–2060.

Renewable energy such as solar power, solar heat, wind power, geothermal power, hydropower, tidal power, wave power and biomass generate electric energy. Among renewable energies, the ultimate sustainable primary energy source is inexhaustible solar energy. The solar energy that shines on the surface of the earth is about 1 kW/m^2,

Y. Kojima (✉)
Natural Science Center for Basic Research and Development, Hiroshima University, 1-3-1, Kagamiyama, Higashi-Hiroshima, Hiroshima 739-8530, Japan
e-mail: kojimay@hiroshima-u.ac.jp

K. Aika and H. Kobayashi (eds.), *CO_2 Free Ammonia as an Energy Carrier*,
https://doi.org/10.1007/978-981-19-4767-4_2

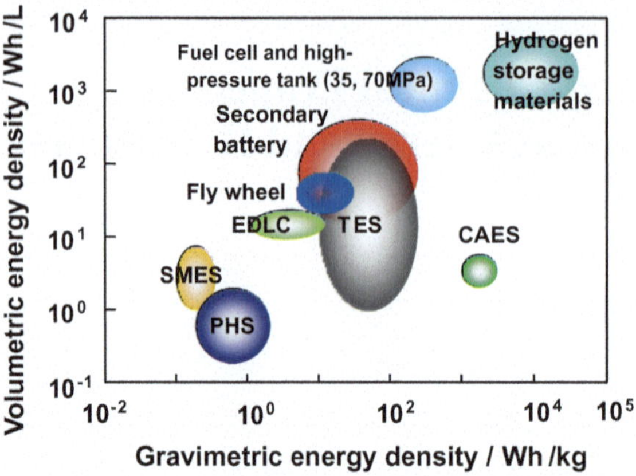

Fig. 2.1 Gravimetric and volumetric energy densities of energy storage systems and hydrogen storage materials [2, 3]

but when the sun is tilted, the energy becomes weaker, so it averages 250 W/m^2. The energy increases as the area increases, which is equivalent to 1 GW of electricity at 4 km^2 [1]. Besides solar energy, wind power has a simple structure and can convert wind energy into electricity at a relatively lower cost. Although renewable energy is used by converting it into electric energy, it fluctuates temporally and spatially [1]. It is also difficult to store and transport a large amount of electricity. Energy storage technology is required for renewable energy uptake in the world. Figure 2.1 shows gravimetric and volumetric energy densities of energy storage systems and hydrogen storage materials.

Developments of electric double-layer capacitors (EDLC), superconducting magnetic energy storage systems (SMES), fly wheels, fuel cell systems, secondary batteries, pumped hydro storage systems (PHS), thermal energy storage systems (TES), compressed air energy storage systems (CAES) and hydrogen storage materials including liquid hydrogen have been carried out [2, 3]. Hydrogen is produced from renewable energy by electrolysis of water and thermochemical water splitting, and it has been considered as a secondary energy that is clean and can be stored and transported. But, it is a gas at room temperature and atmospheric pressure. It requires extremely low temperature or high pressure of several tens MPa or more to achieve high density. Ammonia can store hydrogen more compactly than liquid hydrogen because it stores hydrogen in atomic form.

This paper describes the physical and chemical properties of ammonia as hydrogen and energy carriers.

2.2 Physical Properties of Ammonia

In 1909, Haber–Bosh process was developed to produce ammonia by reaction of N_2 and H_2 from natural gas or coal at high pressure and temperature (20–40 MPa, 673–873 K) using iron-based catalysts. Ammonia is synthesized by the following Eq. (2.1) [4, 5].

$$N_2 + 3H_2 \rightarrow 2NH_3 \tag{2.1}$$

The global production of ammonia is 150 million metric tonnes/year. Ammonia is a compound in which nitrogen has a valence number of −3. Ammonia is transparent in the visible and near-ultraviolet range, but has a wide continuous absorption band in the infrared range from 750 to1250 cm^{-1}, 1400 to 1800 cm^{-1} and 3200 to 3500 cm^{-1} [6, 7]. Figure 2.2 shows the molecular structure of ammonia (IUPAC name: Azan) [8]. The nitrogen atom (N) has one unshared electron pair and three shared electron pairs in the outer shell. Therefore, the hydrogen storage capacity reaches 17.8wt%. N is in the sp^3 hybrid state and forms a tetrahedral array by forming covalent bonds with H using three hybrid orbitals. Because ammonia is repulsive with the unshared electron pair, the angles formed by H–N–H are 106.6 ° and smaller than the sp^3 hybrid orbital angle of CH_4 109.5 ° [6]. Further, hydrogen in ammonia is positively charged and stably exists even in water which is different from metal hydrides.

Ammonia has several characteristics that makes it a desirable medium for storing hydrogen. The pressure required to liquefy ammonia is shown in Fig. 2.3 as a vapor–liquid equilibrium curve [6, 9]. When ammonia is compressed, it liquefies at 293 K and 0.86 MPa [6]. The physical properties are similar to those of propane (C_3H_8) (boiling point of ammonia: 240 K at 0.1 MPa, vapor pressure of ammonia: 1.0 MPa, at 298 K, boiling point of propane: 231 K at 0.1 MPa, vapor pressure of propane: 1.0 MPa at 298 K). This indicates that ammonia can be stored in relatively low pressure vessels below 1 MPa.

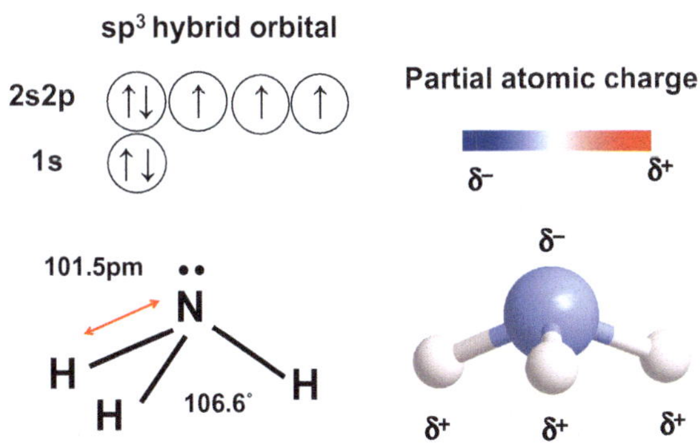

Fig. 2.2 Molecular structure of ammonia [8]

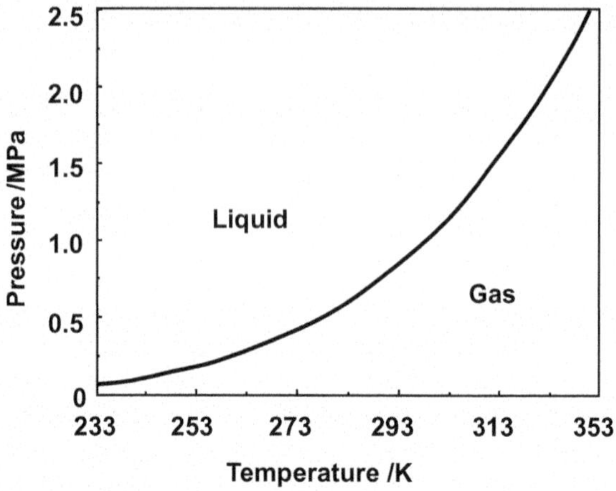

Fig. 2.3 Vapor–liquid equilibrium curve of ammonia

Tables 2.1 shows the physical properties of ammonia, together with the values of liquid hydrogen and organic hydrides [methylcyclohexane (MCH), toluene]. Hydrogen desorption from ammonia can be expressed by Eq. (2.2).

$$2NH_3 \rightarrow N_2 + 3H_2 \tag{2.2}$$

Ammonia has a hydrogen content of 17.8 wt%. The volumetric hydrogen density of liquid ammonia is 10.7–12.1kgH_2/100L (1 MPa, 298 K–0.1 MPa, 240 K) and 1.5–2.5 times as large as that of liquid hydrogen [4.9 kgH_2/100L (1 MPa, 31 K)—7.1kgH_2/100L (−0.1 MPa, 20 K)] [10]. Ammonia has large hydrogen content compared to organic hydrides (methylcyclohexane, toluene) in terms of gravimetric and volumetric hydrogen densities [8] and has a top-level gravimetric hydrogen density and a highest volumetric hydrogen density in the hydrogen storage materials [10]. The enthalpy of vaporization from liquid ammonia is 20–23 kJ/molNH_3. It is possible to vaporize ammonia at normal temperature. Thus, the theoretical energy required to decompose ammonia to release hydrogen is 30.6 kJ/molH_2 and is about 10% of the hydrogen combustion energy [Higher Heating Value (HHV): 286 kJ/molH_2] [10].

The volume expansion between hydrides and other elements by 1 mol of hydrogen molecule ΔV was calculated by Eq. (2.3) [11].

$$\Delta V = 2/x \cdot M_{\mathrm{MHx}}/d_{\mathrm{MHx}} - 2/x \times M_{\mathrm{M}}/d_{\mathrm{M}} \quad x : 1 - 3 \tag{2.3}$$

where x is a valence number. M_{MHx} ia a molecular weights, and d_{MHx} is a density of a hydride (liquid ammonia, MCH). M_{M} is a molecular weight, and d_{M} is a density of other element (liquid N_2, toluene). The standard entropy differences between hydrides (liquid ammonia, MCH) and other elements (liquid N_2, toluene), ΔS were

Table 2.1 Physical properties of ammonia, liquid hydrogen and organic hydrides [4, 7]

Physical properties	Ammonia	Liquid Hydrogen	Organic hydrides	
			Methylcyclo hexane (MCH)	Toluene
Gravimetric H_2 density/wt%	17.8	100	6.15	
Volumetric H_2 density/kgH_2/100L	10.7 (1 MPa, 298 K) 12.1 (0.1 MPa,240 K)	4.9 (1 MPa. 31 K) 7.1 (0.1 MPa. 20 K)	4.73	
Molecular weight	17.0	2.02	98.2	92.1
Boiling point/K (0.1 MPa)	240	20.3	374	384
Density/gcm^{-3}	0.682 (240 K, 0.1 MPa), 0.603 (298 K, 1 MPa)	0.0708 (20 K, 0.1 MPa)	0.769 (293 K)	0.862 (298 K)
Critical temperature/K	406	33.0	572	592
Critical pressure/MPa	11.4	1.29	3.48	4.13
Enthalpy of vaporization /kJmol^{-1}(0.1 MPa)	23.3 20.0 (298 K)	0.9	31.3 35.4 (298 K)	33.2 38.0 (298 K)
Standard heat of formation ΔH^0 /kJmol$H_2$$^{-1}$	−30.6 (gas)	−0.9 at 20 K −1.10 (van't Hoff Plot, NIST) [4]	−68.4 (gas), −67.5 (liquid)	
Standard entropy change ΔS^0/J/KmolH_2	−129 (gas)	−44.8 at 20 K −53.9 (van't Hoff Plot, NIST) [4]	−122 (gas) [4]	
Volume difference by hydrogenation ΔV/cm^3/molH_2	10.9, 13.0 (liquid) [9]	–	6.92 (liquid) [9]	
Standard entropy difference by hydrogenation ΔS/J/KmolH_2	23.7 (liquid) [9]		8.7 (liquid) [9]	

calculated by the Eq. (2.4) [11].

$$\Delta S = 2/x \cdot S_{\mathrm{MHx}} - 2/x \times S_{\mathrm{M}} \quad x : 1 - 3 \tag{2.4}$$

where S_{MHx} and S_{M} are standard entropies of hydrides and other elements (liquid N_2, toluene), respectively. The volume expansion and entropy difference of liquid NH_3 by hydrogenation of liquid N_2 is large compared with those of methylcyclohexane

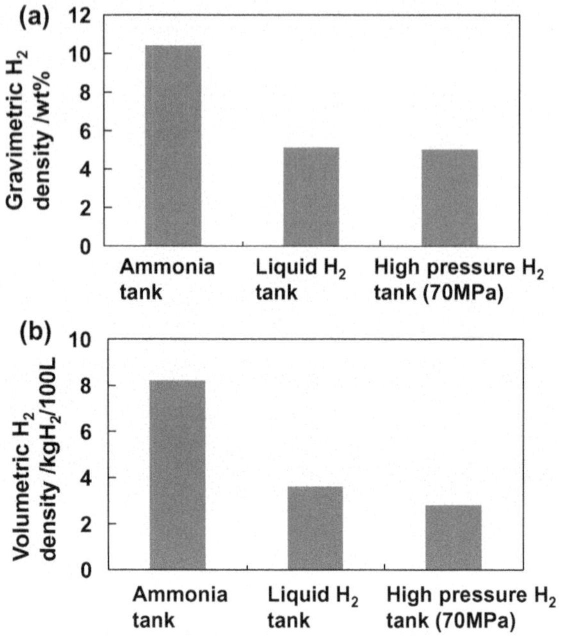

Fig. 2.4 **a** Gravimetric hydrogen densities of hydrogen storage systems [12, 13], **b** Volumetric hydrogen densities of hydrogen storage systems [12, 13]

(MCH) as shown in Table 2.1[11]. This suggests that hydrogen in ammonia has larger entropy and volume.

Figure 2.4 shows the gravimetric and the volumetric hydrogen densities of the three kinds of hydrogen storage systems. The gravimetric hydrogen density in the ammonia tank is 10.4 wt%, and the volumetric hydrogen density is 8.2 kgH_2/100L [12, 13], which are above twice compared to the liquid H_2 tank and 70 MPa high-pressure hydrogen tank as shown in Fig. 2.4 [12, 13]. Hydrogen carriers are certain kinds of hydrogen storage materials that transport large amount of hydrogen from one place to another. Ammonia has advantages as a hydrogen carrier for stable supply.

2.3 Chemical Properties of Ammonia

Figure 2.5 shows the solubility of ammonia in water as a function of temperature [14]. The solubility of ammonia is high and 46 g/100gH_2O [32 wt%, $(H_2O)_2NH_3$] at the ammonia vapor pressure of 0.1 MPa and 298 K. The solubility of ammonia in water decreases with increase in the temperature as shown in the figure. The solubility is large compared with those of hydrogen, organic hydrides (MCH and toluene) as shown in Table 2.2. The vapor concentration (ppm) of ammonia in ammonia water is

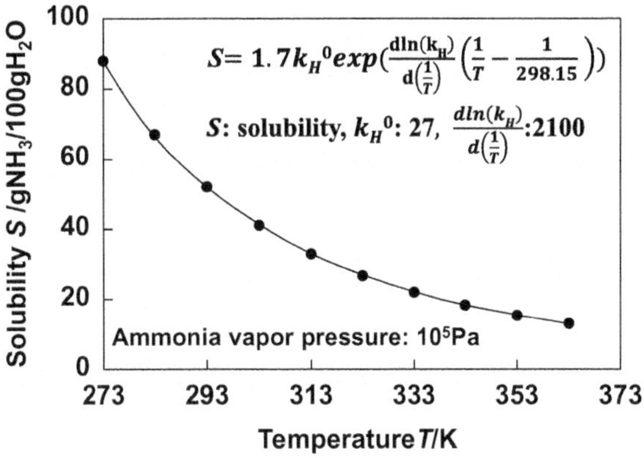

Fig. 2.5 Ammonia vapor concentration as a function of ammonia concentration in ammonia water [6, 14]

obtained by 10 times of the vapor pressure (Pa). The ammonia vapor concentration in the equilibrium state with ammonia water at 298 K is similar to the ammonia concentration in the water [14].

In the ammonia water, NH_3 and H_2O form hydrogen bond. It is well known that the large amount of water is an ammonia absorbent in the plant facilities to suppress the leaked ammonia because of the large solubility and the large diffusion coefficient in water (1.5×10^{-5} cm^2/sec at 293 K) [9]. The large diffusion coefficient will be based on the low viscosity of water 1 mPa s at 293 K. Direct combustion of NH_3 and H_2 is possible. The combustion of ammonia can be expressed by Eq. (2.5).

$$4NH_3 + 3O_2 \rightarrow 2N_2 + 6H_2O \tag{2.5}$$

The standard enthalpy change associated with the ammonia combustion is -383 kJ/molNH_3 (22.5 MJ/kg, 13.6 MJ/L) (higher heating value). The standard heat of combustion for liquid H_2 is 142 MJ/kg (10.1 MJ/L) and has lower volumetric value. Direct combustion of organic hydrides is difficult without emission of CO_2. According to narrow sense of ISO13600, energy carriers are substances those can be used to produce heat. Ammonia is also used as an energy carrier for direct combustion.

Tables 2.2 also shows another kinds of chemical properties of ammonia, together with the values of liquid hydrogen and organic hydrides (MCH, toluene) [9, 15]. The flash point and the ignition temperature of ammonia are large compared with another energy carriers such as liquid hydrogen, methylcyclohexane and toluene. This suggests that ammonia does not burn easily. The LC50 of ammonia, MCH and toluene have similar values of 7679 ppm, 7082 ppm and 7460 ppm, respectively.

The detected concentration of ammonia (NH_3) with nose is 2–5 ppm, which is a specific malodorous substance based on the Odor Control Law in Japan. The water solution having an ammonia concentration of 2–4% is marketed as a drug for insect bites, and its vapor pressure is about 30,000 ppm. Ammonia water with

Table 2.2 Chemical properties of ammonia, liquid hydrogen and organic hydrides [9, 15]

Chemical properties	Ammonia	Liquid hydrogen	Organic hydrides	
			Methylcyclohexane (MCH)	Toluene
Heat of combustion/MJ/kg [7]	22.5	142	–	–
Heat of combustion/MJ/L [7]	13.6	10.1	–	–
Heat of combustion/kWh/kg [7]	6.26	39.5	–	–
Heat of combustion/kWh/L [7]	3.78	2.81	–	–
Flash point/K [7]	405 (critical temperature)	33.0 (critical temperature)	269 [7]	277 [7]
Ignition temperature/K [7, 12]	924 [10]	773–844 [10]	523 [7]	753 [7]
Explosion limit/vol% [7]	16–25	4–74	1.2–6.7	1.1–7.1
Solubility in water/g/100 gwater [7]	46.1(298 K) 52.0(293 K) Vapor pressure 0.1 MPa	1.57×10^{-4} (298 K) Vapor pressure 0.1 MPa	1.51×10^{-3} (298 K)	51.9×10^{-3} (298 K)
LC50/ppm (50% lethal concentration) [12]	7679(4h rat inhalation)	>7500(4h rat inhalation)	7082(4h mouse inhalation)	7460(4h rat inhalation)

a concentration of 10% is also used as a stimulant. In USA, household ammonia solution (concentration 5–10%) is commonly used as a glass cleaner, a bath cleaner and a kitchen cleaner. Aqueous ammonia solution with an ammonia concentration of 10% or less is also used daily as a hair color. Thus, ammonia with a concentration of 10% or less already has social acceptability.

However, anhydrous NH_3 is designated as a deleterious substance. Therefore, anhydrous ammonia is used in the controlled area as raw materials for fertilizer, NOx treatment agent, refrigerant, etc. According to the American Conference of Governmental Industrial Hygienists (ACGIH) and Japan Association of Industrial Health, the allowable concentration by exposure to ammonia is 25 ppm (17 mg/m^3) under working environment of 8 h/day.

The safety of ammonia is classified into fire hazard and health hazard. The National Fire Protection Association NFPA in USA has established NFPA 704 as a "standard system for identifying material hazards in an emergency response." This code is a diamond-shaped sign commonly known as fire diamond and comes in four types: red "flammable", blue "health hazard", yellow "unstable" and white "special notes".

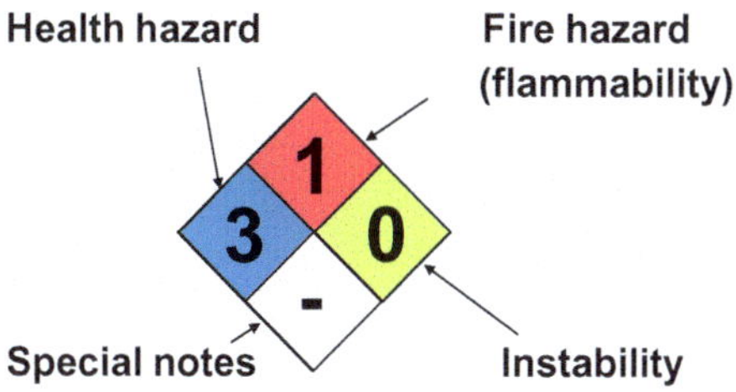

Fig. 2.6 Fire diamond of ammonia

It shows the quality characteristics, and each has a numerical value in 5 grades (0–4). Figure 2.6 shows a fire diamond of ammonia. The fire hazard and the health hazard of ammonia are 1 and 3, respectively. The Globally Harmonized System of Classification and Labeling of Chemicals (GHS) was reviewed by the United Nations GHS Subcommittee and published in 2003. In Japan, the GHS-related ministries liaison conference was established in 2001 to promote GHS classification projects. CLP (Regulation on Classification, Labeling and Packaging of substances mixtures) is a GHS-based regulation on the classification and labeling of chemicals in EU, which came into effect on January 20, 2009.

Figure 2.7 shows the relationship between the NFPA flammability (regulation categories) and flash points of various fuels according to NFPA704 [9, 15, 16]. The flammability of fuels becomes higher as the flash point becomes lower, and it becomes a constant value below 200 K. The flash point of ammonia is 405 K (the lowest temperature at which a combustible mixture with air can be formed), which is higher than any other fuel (methane: 86 K, hydrogen: 33 K, propane: 169 K, hexane: 251 K). The ignition point (the temperature at which combustion spontaneously starts even without an ignition source) showed that the ammonia has a maximum value of 924 K (ignition temperature of ammonia: 924 K, hydrogen: 773–844 K, MCH: 523 K, toluene: 753 K, methane: 810 K, propane: 723 K, hexane: 498 K) [9, 15]. From this, ammonia is the safest substance based on the flammability. Unlike Europe and USA, ammonia has the highest flammability (regulation category) in Japan. According to GHS, flammability is defined by the combustion range of a mixed gas of ammonia and air, and the range is 15–28% (GHS classification) and 12% or more.

Figure 2.8 shows the relation between fuel health hazard (regulation category) and water solubility according to NFPA704. The NFPA health hazard increases with the solubility in water. Ammonia has the highest value of 3, and hydrogen gas has the lowest value of 0. However, the health hazard of liquid hydrogen was as high as 3 [14]. The classification of GHS and CLP for the acute aqueous environmental hazard of anhydrous ammonia is 1 (environmental hazard: large). Anhydrous ammonia has a negative effect on the environment. However, ammonia water, MCH, and toluene have an acute aqueous environmental hazard of 2 (environmental hazard: medium) based on GHS. Although anhydrous ammonia is changed to ammonia water when dissolved in water, the classification of acute aqueous environmental toxicity is different from ammonia water [16].

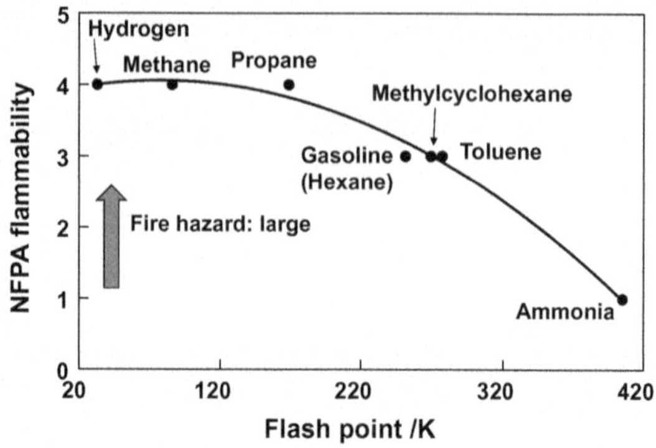

Fig. 2.7 Relation between NFPA flammability and flash point [16]

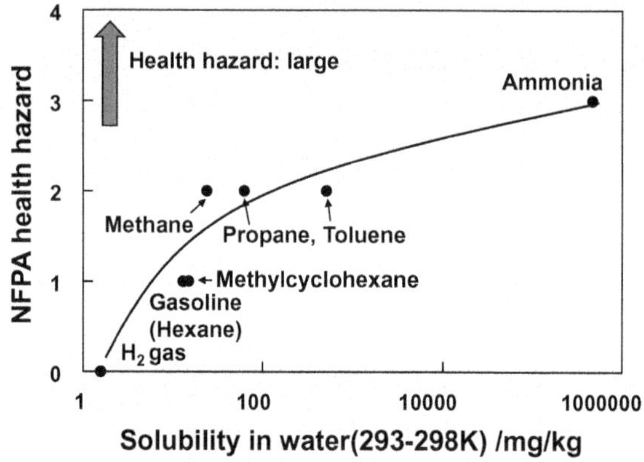

Fig. 2.8 Relation between NFPA health hazard and solubility in water [16]

Ammonia accidents occurred in Japan 2014 are only in the plant facilities. The ratio of ammonia accidents divided by the high-pressure gas accidents is about 10%. All accidents with ammonia are caused by leaks from a valve, a pipe and a flange in the plant facilities. The maximum ammonia vapor concentration is 2 ppm (vapor pressure: 0.2 Pa at atmospheric pressure) based on an offensive odor control law at industrial area in Japan. The new ammonia removal system combined water and insoluble proton-based materials will be a candidate for the leaked ammonia concentration lowering in the atmosphere by suppression of water pollution [16]. Zirconium phosphate has a low ammonia vapor pressure below 1 Pa (0.1 ppm) and a high capacity of 10.2 wt% as an insoluble proton-based materials [17–20]. The ammonia vapor pressure P_{NH3} of ammonia storage materials was expressed by the following equation, experimentally [21].

$$ln\left(\frac{P_{\mathrm{NH3}}}{\mathrm{P_0}}\right) = \frac{-3.1(93\ln\Delta V - 414)^2 - 49.3(93\ln\Delta V - 414)}{\mathrm{R}T} - \frac{(93\ln\Delta V - 414)}{\mathrm{R}} \quad (2.6)$$

where ΔV is a volume difference of ammonia storage materials before and after ammonia absorption, R is gas constant (8.314 J/K mol), T is an absolute temperature, and P_0 is a standard pressure (0.101 MPa). The volume difference was obtained from the densities and the molecular weights of the ammonia storage materials before and after ammonia absorption. ΔV for the zirconium phosphate absorbed ammonia is changed from 0 to 15.8cm^3/molNH$_3$. Therefore, the low P_{NH3} is based on the small ΔV by Eq. 2.3.

According to IEA report, the cost of importing ammonia from electrolytic hydrogen produced in Australia into Japan 2030 as a hydrogen carrier is low compared with liquid hydrogen and organic hydrides (methyl cyclohexane, toluene) [22]. If ammonia could be used by the end user without the need for reconversion, the cost would be even lower.

2.3.1 *Summary*

Ammonia was recognized as an attractive hydrogen and energy carriers because it has a high hydrogen storage density of 17.8 wt% and 10.7 kgH$_2$/100L, and it is easily liquefied under about 1 MPa at room temperature. Ammonia is the only mass-produced hydride that does not have carbon atoms. Ammonia will be produced from renewable energy such as solar power, wind power and geothermal. Ammonia is also burnable without emission of CO_2. It is expected that ammonia will be utilized as hydrogen and energy carriers for renewable energy in future.

Acknowledgements This work was partially supported by Council for Science, Technology and Innovation (CSTI), Cross-ministerial Strategic Innovation Promotion Program (SIP), "energy carrier" (funding agency: JST). The author is greatly indebted to Professor T. Ichikawa and Associate Professor H. Miyaoka of Hiroshima University. And this work was carried out as a study in Research Center for Nitrogen Recycling Energy Carrier of Hiroshima University.

References

1. Muller RA (2012) Energy for future presidents: the science behind the headlines, WW Norton & Company, New York, London
2. Guney MS, Tepe Y (2017) Classification and assessment of energy storage systems. Renew Sustain Energ Rev 75:1187–1197
3. Kojima Y (2020) Hybrid nickel-metal hydride/hydrogen gas battery (hybrid hydrogen battery). J Japan Inst Energ 99:214–220
4. Smith C, Hill AK, Torrente-Murciano L (2020) Current and future role of Haber-Bosch ammonia in a carbon-free energy landscape. Energ Environ Sci 13:331–344
5. Kojima Y (2015) Hydrogen storage and transportation using ammonia. J Surf Sci Soc Japan 36:583–588
6. NIST Chemistry WebBook, SRD 69. https://webbook.nist.gov/chemistry/; https://webbook.nist.gov/cgi/cbook.cgi?ID=C7664417&Mask=4&Type=ANTOINE&Plot=on; https://webbook.nist.gov/cgi/cbook.cgi?ID=C7664417&Mask=10

7. Ammonia, National Research Council, Committee on medical and biologic effects on environmental pollutions, 1979/11/1
8. Atkins P, Overton T, Rourke J, Weller M, Armstrong F (2010) Shriver & Atkins Inorganic Chemistry, 5th edn
9. Haynes WM (2016) Editor-in-Chief: CRC handbook of chemistry and physics, 97th edn, CRC Press 2016–2017
10. Kojima Y (2019) Hydrogen storage materials for hydrogen and energy carriers. Int J Hydrogen Energ 44:18179–18192
11. Kojima Y, Yamaguchi M (2022) Entropy differences between hydrides and other elements. Chem Commun, in press
12. Kojima Y (2015) Trends and issues in hydrogen storage systems. J Inst Electrostat Japan 39:192–197
13. Kojima Y (2013) A green ammonia economy. In: 10th annual NH_3 fuel conference, embassy suits riverfront promnade, Sacramento, CA, 22–25 of September
14. Kojima Y, Yamaguchi M (2020) Ammonia storage materials for nitrogen recycling hydrogen and energy carriers. Int J Hydrogen Energ 45:10233–10246
15. Kojima Y (2014) Hiroshima University, Final Report. SIP (Cross-ministreal strategic innovation promotion program) Subject "Energy Carrier" R&D Theme "Ammonia Hydrogen Station Basic Technology" Research Subject "Development of Ammonia Removal", July 1, 2014–March 31, 2018
16. Kojima Y, Ikeda H (2019) 4–5–3 Safety of ammonia as hydrogen and energy carriers. In: Abstracts of the 28th annual meeting of the Japan institute of energy, pp 156–157
17. Kojima Y (2019) Hiroshima University, Final Report, SIP (Cross-ministreal Strategic Innovation Promotion Program) Subject "Energy Carrier" R&D Theme "Development of technologies related to ammonia production, storage and transportation using CO2-free hydrogen", Research Subject "Emergency ammonia adsorption method", April 1, 2019-March 31, 2020
18. Yamaguchi M, Ichikawa TH, Miyaoka H, Xhang T, Miyaoka H, Kojima Y (2020) Proton-based solid acids for ammonia absorption in ammonia water. Int J Hydrogen Energ 45:22189–22194
19. Yamaguchi M, Miyaoka H, Kojima Y (2020) Concentration-composition-isotherm for the ammonia absorption process of zirconium phosphate. RSC Adv 10:20882–20885
20. Yamaguchi M, Miyaoka H, Kojima Y (2021) Thermodynamic and spectroscopic analyses of zirconium phosphate-absorbed ammonia. J Phys Chem C 125:3758–3763
21. Kojima Y, Yamaguchi M (2021) Thermodynamic analysis of ammonia storage materials. Int J Hydrogen Energ 46:11756–11760
22. Birol F (2019) The future of hydrogen. Seizing today's opportunities, Report prepared by the IEA for the G20, pp 82–83, Japan

Chapter 3
Potential of Ammonia as CO_2-Free Fuel and Hydrogen Carrier

Bunro Shiozawa

Abstract This chapter shows significance of hydrogen energy in decarbonizing the Japan's energy system and identifies areas to which hydrogen energy needs to be introduced as a matter of priority, after reviewing the present status of the Japanese energy demand and supply structure. Among varieties of forms of hydrogen energy, this article explains that ammonia (NH_3) has many advantages in transferring and using hydrogen energy in a large amount. Furthermore, this article points out that NH_3 is found to be usable as a CO_2-free fuel in power generation and heating in industrial processes, by explaining research findings of NH_3 combustion mechanism and technological developments in the NH_3 combustion technology which were realized by the SIP "Energy Carriers" (2014–18): a government-funded innovation promotion project. In addition to those technological aspects, cost advantages of NH_3 as hydrogen energy is also explained by referring to cost examination studies conducted by independent public organizations. Lastly, recent public and private sector movements both in Japan and foreign countries which aim at social implementation of above-mentioned research outcomes; that is introduction of NH_3 as a CO_2-free fuel are described. Furthermore, future possibility of use of NH_3 as hydrogen carrier (not CO_2-free fuel) is briefly discussed.

Keywords NH_3 · Ammonia · Ammonia combustion · CO_2-free fuel · Hydrogen energy · Energy carrier

3.1 Japan's Energy Supply and Demand Structure

The energy supply and demand structure of Japan is shown in Fig. 3.1. To achieve the goal of "carbon neutrality" by 2050 as Prime Minister Suga declared in October

The former Deputy Program Director of SIP "Energy Carriers" [1]

B. Shiozawa (✉)
Senior Researcher, International Environment and Economy Institute, 3-21-1273 Kanda-Nishikicho, Chiyoda-ku, Tokyo 101-0054, Japan
e-mail: bunro@n07.itscom.net

K. Aika and H. Kobayashi (eds.), *CO_2 Free Ammonia as an Energy Carrier*,
https://doi.org/10.1007/978-981-19-4767-4_3

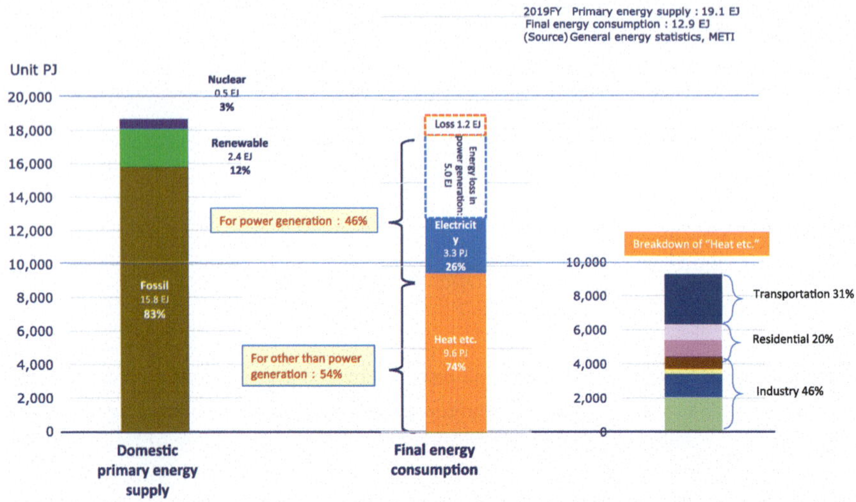

Fig. 3.1 Energy supply and demand structure of Japan (2019 Fy)

2020, we must substantially reduce emissions, and therefore this energy supply and demand structure must be fundamentally changed.

Presently, Japan relies on fossil energy for 83% of its primary energy supply. This fossil energy is converted into electricity, gas, and petroleum products, or consumed directly in the industry sector (fuels for production processes), transportation sector (automobile fuel, etc.), and residential sector (home and office heating and cooling, cooking, etc.), and it emits CO_2 in these conversion and consumption processes. Other primary energy sources which do not emit CO_2 in their use, such as nuclear and renewable energies only account for 17%. The weight of nuclear energy is only 3% due to slow progress in restarting nuclear power plants after the Fukushima nuclear power accidents caused by the huge Tsunami in 2011, and renewable energy remains at 12%.

Then, 46% of the primary energy supply is used for power generation; however, only about 40% of that energy is ultimately converted to electricity due to the efficiency of power generation. And this electricity is finally consumed by the demand sectors, such as the industry, transportation, and residential sectors. The remaining 54% of the primary energy supply is consumed as non-electric energy, which is mostly heat [2], in these sectors.

As a result, 26% of energy is ultimately consumed as electricity and 74% as non-electric energy which is mostly heat in Japan. Non-electric energy consumption is divided by sector as follows: 46% for the industry, 31% for the transportation, and 20% for the residential.

To achieve the carbon neutrality goal, dependence on fossil energy must be substantially reduced to the extent necessary to achieve the goal. For this, the present energy demand structure as described above needs to be drastically changed. As a matter of course, it is not a solution to the change only the supply side of energy, since the type of energy needed depends on how it is used.

3.2 Why Are Electrification and Hydrogen Energy Important in Achieving the Carbon Neutrality Goal?

Decarbonization of energy consumed as electricity can be achieved by decarbonizing the source of electricity. Of the energy sources, nuclear power and renewable energy are the two that do not emit CO_2, but unless nuclear power plants can newly be built and/or replaced, the power generation capacity of existing nuclear power plants will be greatly reduced in 2050 as they reach the end of their legally set service life (at the maximum, 60 years). The availability of renewable energies in Japan is limited in terms of both quantity and quality, and it seems difficult to obtain the necessary amount of renewable energy within Japan while ensuring a certain level of economic efficiency. A combination of fossil energy and carbon dioxide storage (CCS) could be a possible option, but there is no prospect yet of constructing and operating CCS in an economically viable manner in Japan. Besides, Japan cannot exchange energy with neighboring countries or regions via power transmission lines and gas pipelines like European countries.

Because of those difficulties, Japan needs to introduce cheap renewable energy in the form of hydrogen energy, which is available in overseas regions where rich renewable energy resources exist, as fuel for power generation to decarbonize the power supply.

Regarding the energy being consumed as heat, decarbonization of heat energy consumed in the transportation and residential sectors is progressing in the form of electrification. Electrification of automobiles, heating, and cooling, and cooking will continue to advance.

On the other hand, it is not easy to decarbonize heat used in the industrial sector. Some of heat being used in the industry can be changed to heat generated by electric energy such as heaters. Other heat sources that are difficult to electrify must be converted to CO_2-free fuels. The fossil energy used not only as heat source but as raw materials in the industrial sector, such as coking coal and naphtha, needs to be decarbonized by replacing the present production processes with innovative ones which do not emit CO_2 or by recycling carbons incorporated into the products. This will also require a large amount of hydrogen.

As can be seen from the above, decarbonization of Japan's energy system will require:

- Promotion of electrification;
- Introduction of hydrogen energy to decarbonize electricity;
- Decarbonization of some heat sources; and
- Introduction of hydrogen to decarbonize production processes and recycle carbons.

In other words, Japan needs to promote electrification and introduce a large amount of hydrogen energy for the decarbonization of its energy system. Given this reality, Japan needs to secure a large amount of hydrogen energy supply.

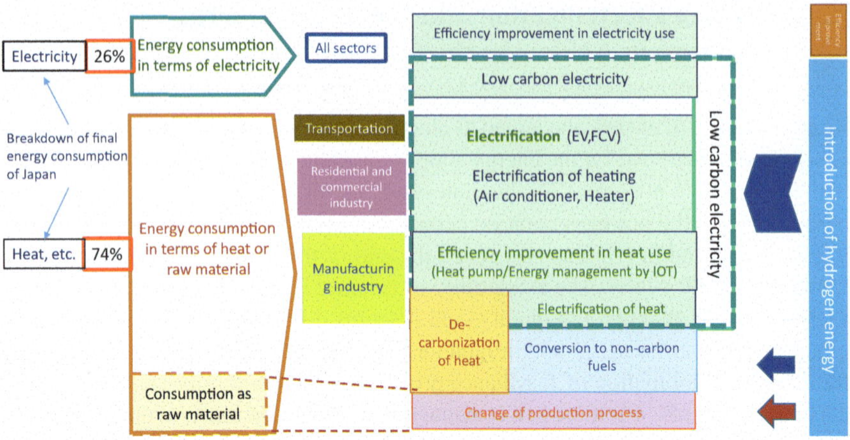

Fig. 3.2 Basic direction of decarbonization of the Japan's energy system

But there are limited options for Japan to fulfill this need. One of the options is to convert inexpensive renewable energy into hydrogen in overseas regions where rich renewable energy resources exist and introduce it to Japan using hydrogen energy carriers that make transportation of a large amount of hydrogen in a practically easy manner.

A schematic figure depicting this idea is shown in Fig. 3.2.

3.3 NH_3 is a Means of Mass Introduction of Hydrogen Energy

However, hydrogen, a typical form of hydrogen energy, is not easy to transport and store a large amount due to its physical–chemical property including its extremely low volumetric energy density. For increasing volumetric energy density, hydrogen needs to be liquefied by cooling up to an extremely low temperature of minus 253 °C. But the liquefaction of hydrogen requires a considerable amount of energy at the current level of technology, and transportation and storage technologies, including a carrier vessel transporting large volume liquefied hydrogen, are still under development and have not reached an implementation stage.

On the other hand, NH_3 is a substance with high volmetric hydrogen density, and infrastructure and technology for production, mass transportation, and storage of NH_3 already exist. About 200 million tons of NH_3 are produced annually in the world, and about 20 million tons of NH_3 are commercially traded internationally every year.

This means that there are no technical issues related to the mass transportation and storage of NH_3. Furthermore, since liquefaction conditions of NH_3 are almost the same as those for liquefied petroleum gas (LPG), it is possible to use LPG infrastructure for transportation and storage of NH_3 if necessary.

NH_3 can be also used as a hydrogen carrier in the same way as other carriers, i.e., NH_3 can be produced by reacting hydrogen with nitrogen (N_2) which is extracted from the air and transporting NH_3 to the point of use. Then NH_3 is decomposed (cracked) to obtain hydrogen for use (Fig. 3.3). However, the research and development conducted in the SIP "Energy Carriers" [3] revealed the fact that NH_3 can be used as a CO_2-free fuel as it is, without cracking NH_3 to convert to hydrogen ("direct use" in the figure). In other words, NH_3 can be used not only as a hydrogen carrier but as a CO_2-free fuel.

This is a great advantage of NH_3 because this eliminates the need for the process of cracking NH_3 to reconvert to hydrogen and energy input required for the process [4], thus significantly reducing the cost required for using hydrogen energy. In addition, it frees us from the difficulty of handling hydrogen produced after cracking.

But some issues needed to be clarified for the actual use of NH_3 as a fuel. These issues are:

(1) Stability of NH_3 combustion; and
(2) Containment of NO_X emissions in NH_3 combustion.

How these issues were overcome in the research and development of the SIP "Energy Carriers" and the results of the technological development regarding direct NH_3 utilization technologies including NH_3 combustion will be described in the next section.

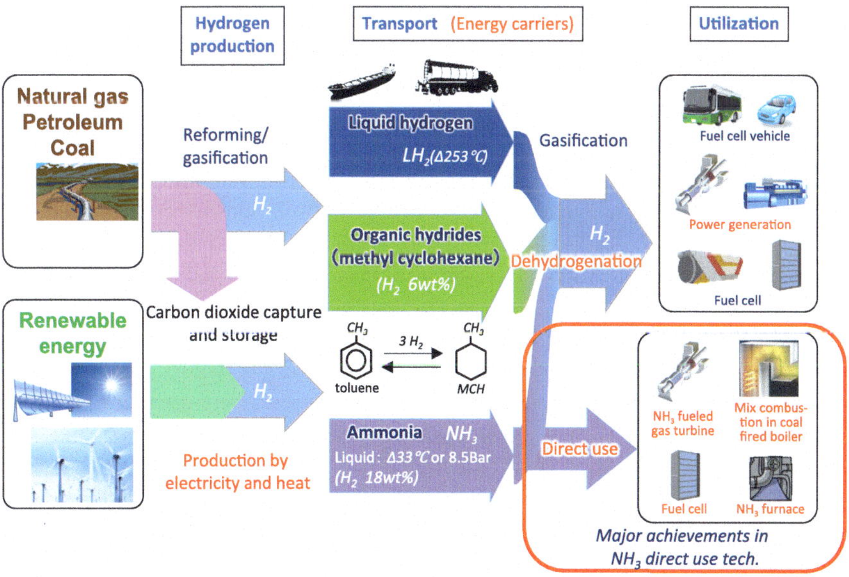

Fig. 3.3 Energy carriers

More detailed technical reports on the development of individual direct NH_3 utilization technologies will be provided in separate chapters in this book, which are written by the researchers who have actually engaged in the development of those technologies in the SIP "Energy Carriers" initiative.

3.4 Summary of Achievements in the Development of Combustion and Use Technology Conducted in the Sip "Energy Carriers"

As mentioned above, there were two concerns regarding use of NH_3 as a fuel; that is if:

(1) NH_3 can be combusted stably; and
(2) NO_X emission can be contained in the combustion of NH_3.

3.4.1 Research Findings on the Combustion Mechanisms of NH_3

The flame velocity of NH_3 is only about one-fifth of that of methane (CH_4), the flame temperature is lower, and a narrower combustible range is exhibited. Therefore, maintaining a stable flame was an issue for using NH_3 as fuel. In addition, there was another concern that a large amount of "fuel NO_X" produced from nitrogen atoms in the fuel would be generated when NH_3 is used as fuel. (In the case of the combustion of CH_4, "thermal NO_X" is produced from N_2 in the air.)

These fundamental issues related to the NH_3 combustion mechanism were explored in detail for the first time by a Tohoku University research team, led by Dr. Hideaki Kobayashi. The research findings elucidated a scientific understanding of NH_3 combustion [5] and helped devise designs and solutions to problems that had arisen in the course of development of NH_3 combustion equipment.

The problem of NH_3 combustion stability was solved by using a swirl burner [6] to mix the gases inside the combustor. It was also found that formation of NO_X could be contained by burning NH_3 in the combustor under the condition in which a slight surplus of NH_3 was present in the combustion gas. Under this condition, NO_X that happened to be produced in the combustion was converted to N_2, with the surplus NH_3 acting as a reductant. In other words, NH_3 worked both as fuel and reductant in NH_3 combustion. This reduction effect is not surprising, given that NH_3 is used as a reductant in exhaust gas denitration systems in thermal power plants and diesel trucks. It was also found that these effects were enhanced under high-pressure combustion. This is advantageous for the containment of NO_X emissions in larger gas turbines that burn fuel at higher pressures.

Since the conditions for NO_X emission containment were identified, a new combustion method called "rich-lean two-stage combustion [7]" was developed [8] to enable a further reduction of NO_X emissions.

3.4.2 *Results of the Development of NH_3 Combustion Technology*

These research findings on the combustion mechanism of NH_3 were the basis for the works supported by the SIP "Energy Carriers" initiative on combustion technologies, including gas turbines, coal-NH_3 co-firing boilers, and industrial furnaces.

3.4.2.1 Small Gas Turbines

In the field of small gas turbines, Toyota Energy Solutions Corp. developed 50 and 300 kW class 100% NH_3-fired micro gas turbines. This development was carried out in close cooperation with the research team at Tohoku University and combustion technology researchers at the National Institute of Advanced Industrial Science and Technology (AIST).

At AIST, a power generation test was conducted at the Fukushima Renewable Energy Institute (FREA) of AIST using a 50 kW micro gas turbine generator. The tests proceeded first with kerosene/NH_3 co-firing, then with CH_4/NH_3 co-firing, and finally with 100% NH_3 firing. It was confirmed that the gas turbine could generate electricity stably at the expected efficiency rate even during 100% NH_3 firing and that NO_X emissions were reduced to a level that could be removed by conventional denitration equipment [9]. Based on these results, Toyota Energy Solutions has developed a 300 kW class 100% NH_3-fired micro gas turbine generator.

3.4.2.2 Medium-Size Gas Turbine for Power Generation

IHI Corporation's contribution to the SIP "Energy Carriers" work involved devel opment of a low-NO_X combustor for the co-firing of CH_4/NH_3 in a medium-sized power generation turbine. The company was able to show that a 2 MW-class gas turbine could generate power by co-firing CH_4 and NH_3 [10] with a co-firing rate [11] of 20%. This demonstration verified that.

- NH_3 co-firing would reduce CO_2 emissions in proportion with NH_3 co-firing rate;
- NO_X emissions could be controlled below the environmental standard using conventional denitration equipment; and
- The essentially the NH_3 fuel was completely burned [12].

3.4.2.3 Large Gas Turbine for Power Generation

For large gas turbines of several hundred megawatts, Mitsubishi Heavy Industries Engineering Corporation (MHI) and Mitsubishi Power, Ltd. (formerly Mitsubishi-Hitachi Power Systems, Inc. (MHPS)) worked on development of a fuel supply system that used NH_3 as a hydrogen carrier to supply hydrogen to a CH_4/H_2 co-fired gas turbine which had already been developed by MHPS in a separate R & D project.

The reason MHI/MHPS did not use NH_3 directly as fuel in this technology development is that operating temperatures tend to go up with an increase in combustor size, and higher temperatures tend to produce higher levels of NO_X.

The MHI/MHPS fuel supply system involved decomposition of NH_3 using the gas turbine's exhaust heat and a catalyst to produce hydrogen. Here a key development target was the optimal distribution of exhaust heat from the gas turbine to the NH_3 decomposer without sacrificing overall energy efficiency of the power generation process. Another technological challenge was the development of an NH_3 decomposition system that can operate efficiently and stably. Detailed design studies supported by the SIP "Energy Carriers" initiative found that there may even be a possibility of increasing overall power generation efficiency beyond that of a natural gas (CH_4)-fired combined cycle gas turbine power generation system [13].

Development of this technology is continuing after the end of the SIP "Energy Carriers" initiative with financial support from the New Energy and Industrial Technology Development Organization (NEDO).

3.4.2.4 Ammonia Co-firing in a Pulverized Coal-Fired Boiler

It was thought that one of the promising applications of NH_3 as a fuel would be to co-fire it with coal at coal-fired power plants. The slow-burning speed of NH_3 was thought to be suitable for co-firing with pulverized coal, and co-firing NH_3 would reduce CO_2 emissions from coal-fired power generation in proportion to the amount of co-fired NH_3. But there was a big concern with co-firing in this way: emission of a large amount of NO_X from the combustion process.

In the SIP "Energy Carriers" initiative, the combustion behavior of pulverized coal and NH_3 during co-firing was studied in detail by a research team at Osaka University [14], while co-firing tests with an NH_3 ratio of 20% were conducted by the Central Research Institute of Electric Power Industry (CRIEPI) using both single and multi-burner furnaces. The results of these tests showed that by adjusting the injection point and feed rate of NH_3 into the furnaces, NO_X emission could be contained and controlled [15], and no N_2O which might be formed in the combustion process was detected in the exhaust gas.

Based on these basic study results, the Chugoku Electric Power Co., Ltd. conducted a demonstration test of coal-NH_3 co-firing at its commercial power plant, Mizushima Thermal Power Station Unit 2 (156 MW rated output). The demonstration test confirmed that CO_2 emissions were reduced by the amount of NH_3 co-firing; the NH_3 used as fuel was completely burned and not emitted outside the plant; NO_X emissions were not significantly different from those of 100% coal-fired power plants; and the plant met environmental standards without any problems [16, 17]. Although the ratio of NH_3 was only 0.6–0.8% due to the limited capacity of the existing NH_3 vaporizer, the most significant achievement was the confirmation that the coal-NH_3 co-firing technology can be applied to power plants in commercial operation without causing any problems as a CO_2 emissions reduction measure.

Based on the results of this demonstration test, the Chugoku Electric Power Co., Ltd. evaluated the coal-NH_3 co-firing technology as "a low-cost CO_2 reduction technology which does not require a large remodeling of coal-fired power generation facility including denitration equipment, and therefore maximizes the use of existing coal-fired power generation facilities." [18]

Following this demonstration, IHI Corporation developed a coal-NH_3 co-firing burner that can be attached to an existing coal-fired boiler in a way that minimizes incremental NO_X emissions. After installing its coal-NH_3 co-firing burner, IHI analyzed changes in the boiler's heat collection characteristics when NH_3 is mixed into the fuel. It was confirmed that at 20% NH_3 co-firing, CO_2 emissions decreased 20%; NO_X emissions could be reduced to the level of 100% coal combustion; and the heat collection performance of the boiler did not change significantly. This study was conducted because there was a concern that the heat distribution on the furnace wall of a boiler may change, as the flame temperature of NH_3 is generally lower than that of coal and co-firing with NH_3 reduces the amount of soot and coal powder particles in the furnace which produce radiative heat [19].

This coal-NH_3 co-firing technology for coal-fired power generation boilers does not require major equipment modification. As a result, this technology is attracting a great deal of attention from the power industry as a means of reducing CO_2 emissions from existing coal-fired power plants which have had no economic CO_2 emission reduction options other than introduction of biofuels.

For the commercial implementation of this technology, NEDO is planning to provide support for a long-term technical demonstration at an actual coal-fired power plant. Companies involved in this implementation plan are currently working on detailed design and feasibility analysis for the demonstration.

3.4.2.5 Industrial Furnaces

Japan's industrial sector accounts for more than 20% of the country's fossil fuel consumption. Research supported by the SIP "Energy Carriers" initiative found that NH_3 could also be used as a fuel for industrial furnaces even though such furnaces come in various sizes and types.

One of the challenges of using NH_3 as a fuel in industrial furnaces is reduced radiative heat transfer from the flame since NH_3 in combustion does not have the radiative heat transfer effect generated by soot combustion. In the course of a study conducted at Osaka University, it was found that enhanced flame radiation from oxygen-enriched combustion and multi-stage combustion to achieve flame temperature uniformity could overcome these problems in both 100% NH_3 combustion and

CH_4-NH_3 mixed combustion (co-firing ratio: 30%) in a 10 kW class model furnace. In addition, the same results were confirmed by a demonstration study using a 100 kW class model furnace, which is close to the practical scale of industrial furnaces being used in industry [20].

In a project supported by the SIP "Energy Carriers" initiative, this technology was actually applied to a degreasing furnace used in the pretreatment process for hot-dip galvanized steel sheet production. Taiyo Nippon Sanso Co., Ltd. and Nippon Steel Nisshin Co., Ltd. conducted a verification test to confirm applicability of this technology at one of the latter's production lines. By mixing conventional CH_4 fuel with NH_3 (co-firing ratio: 30%) to heat the furnace, they confirmed that such co-firing could be applied to this process and reduce CO_2 emission by 30% [21].

3.4.2.6 NH_3 Fuel Cell

Direct use of NH_3 in forms other than combustion has also yielded important results, specifically in the use of NH_3 as a fuel for fuel cells.

Solid oxide fuel cells (SOFCs) operate at temperatures of 700–1000 °C. NH_3 decomposes into hydrogen and nitrogen (N_2) at temperatures above 500 °C. Therefore, it was thought that NH_3 itself could be used as a fuel for SOFCs. Thus, the idea was to supply NH_3 instead of hydrogen as a fuel for SOFCs to avoid the difficulties of transporting and storing hydrogen in large amounts.

At present in Japan, SOFCs are supplied with city gas and LP gas. These commodities are inexpensive and easy to transport and store. To obtain hydrogen from these gases to fuel SOFCs, it is necessary to integrate a gas reformer within the SOFC's balance of plant. The reformer emits CO_2 during the reforming process. One way to solve these problems is with a pure hydrogen supply system. However, if NH_3 can be used as a fuel, the hydrogen transportation and storage problems can be avoided. And, of course, no reformer is needed. Thus, if NH_3 can be used directly as a fuel for SOFCs, it will greatly improve the convenience of SOFCs. Since SOFCs are expected to play a key role as a distributed CO_2-free power source in coming years, the SIP "Energy Carriers" initiative explored the option of NH_3 direct-use technology for SOFCs.

In the course of this research and development, a research team at Kyoto University confirmed that NH_3 could be used as a fuel for SOFCs and that the power generation performance of NH_3-fueled SOFCs was equivalent to that of pure hydrogen (53% efficiency (LHV) at 255 W of direct current power generation) [22]. Based on these results, Noritake Company, Ltd. fabricated a 1 kW-class stack, and IHI Corporation developed a 1 kW NH_3-fueled SOFC system [23]. IHI Corporation is continuing the development of a larger commercial-use NH_3-fueled SOFC system.

3.4.3 Significance of the Results of the Development of NH_3 Direct Utilization Technology

The International Energy Agency (IEA) has recognized the significance of these achievements in the development of NH_3 direct utilization technology as follows, in its "The Future of Hydrogen" [24] report:

- Coal-NH_3 co-firing technology in coal-fired boilers will be an important means to reduce CO_2 emissions from existing coal-fired power plants;
- Direct use of NH_3 in gas turbine power generation will be an important means of reducing CO_2 emission of thermal power generation—which is expected to play a vital role in ensuring the "adjustability" of the electricity grid once a large amount of solar and wind energy has come online.

Although not described in detail in this article, the results of the development of these technologies for direct use of NH_3 are also expected to play a significant role in the decarbonization of thermal energy, (except fot the case of industrial furnaces mentioned in 3.4.2.5) which accounts for about 75% of Japan's final energy consumption.

3.4.4 Summary and Present Status of NH_3 Direct Utilization Technology After SIP "Energy Carriers" Initiative

As described above, the SIP "Energy Carriers" initiative has scientifically proven and demonstrated that NH_3 can be directly used as a CO_2-free fuel in gas turbines and coal-fired boilers for power generation, industrial furnaces, etc., without decomposing NH_3 into hydrogen. Figure 3.4 summarizes the achievements of the NH_3 direct-use technology development in the SIP "Energy Carriers" initiative and their present status after the initiative ended.

As mentioned above, technical issues for using NH_3 as a CO_2-free fuel for power generation have almost been cleared by the R & D conducted in the SIP "Energy Carriers" initiative, and the phase of technological development has moved into the stage of the implementation using actual power generation equipment and facilities. The remaining important development theme is the development of a large (500–600 MW class) combined cycle gas turbine (CCGT) generator that uses NH_3 as a hydrogen carrier as described in 3.4.2.3 above.

For other power generation equipment and facilities, the focus of development activities has shifted to the development of peripheral technologies, such as technologies to further improve power generation efficiency and reduce power generation costs.

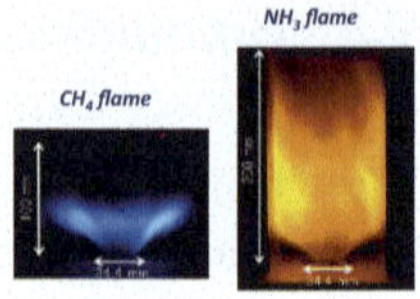

- Elucidation of NH_3 combustion mechanism
  By the scientific research on NH_3 combustion mechanism, it was confirmed that:
  - stable combustion of NH_3 is possible;
  - formation of NO_x can be controlled and emission of other air pollutants, such as N_2O and NH_3 can be contained by adjusting combustion conditions.

- Development of 100% NH_3 or Coal-NH_3, CH_4-NH_3 mixed combustion equipment

(Gas turbine)		
- Micro (50-300kW) gas turbine	(NH_3 100% fuel) :	Developed and in a commercialization stage
- MW class gas turbine	(20% NH_3 mixed combustion):	Developed and in a commercialization stage
	(~70% NH_3 mixed combustion):	Under development
- 500-600 MW class gas turbine	(NH_3 used as H_2 carrier):	Under development
(Coal fired boiler)		
- 10 MW pulverized coal fired boiler	(20% NH_3 mixed combustion):	Developed and in a commercialization stage
	(~60% NH_3 mixed combustion):	Under development
- 1 GW pulverized coal fired boiler	(20% NH_3 mixed combustion):	Plan to implement in 2025-26
(Industrial furnace)		
- 100 kW furnace:	(20% NH_3 - Coal mixed combustion):	Demonstrated.
- Degreasing furnace:	(30% NH_3- CH_4 mixed combustion):	Verified applicability of the mixed combustion.

- NH_3 fueled SOFC (Solid oxide fuel cell)

- 1 kW NH_3 fueled solid oxide fuel cell system:	Developed.
- 300 kW NH_3 fueled solid oxide fuel cell system:	Under development

Fig. 3.4 Developments of NH_3 direct utilization technologies achieved in the SIP "Energy Carriers" and their progress afterward

3.5 Cost of CO_2-Free Ammonia

If ammonia is to be introduced into the energy system as a CO_2-free fuel, its cost must be at least competitive with that of other CO_2-free fuels such as CO_2-free hydrogen. In the discussion below, the cost aspect of CO_2-free ammonia will be considered.

To state a conclusion at the beginning, the cost of CO_2-free NH_3 can be less than 30 yen/Nm^3-H_2, which is the cost target of hydrogen energy for 2030, set out by the Japanese government in its "Basic Hydrogen Strategy" [25] for introducing hydrogen energy into Japan. Moreover, it would even almost reach the level of 20 yen/Nm^3-H_2, which is the target for 2050. For ease of reference, Table 3.1 shows the cost per ton of NH_3 that corresponds with the government's 2030 and 2050 hydrogen cost targets.

Table 3.1 Corresponding cost of calorific equivalent hydrogen and NH_3 1$ = ¥110

Hydrogen cost value	Equivalent NH_3 cost value
20 (Yen/Nm^3-H_2) (≈2 $/kg-$H_2$) (The cost target of hydrogen for 2050 as stated in the "Basic Hydrogen Strategy")	320 ($/t-$NH_3$)
30 (Yen/Nm^3-H_2) (≈3 $/kg-$H_2$) (The cost target of hydrogen for 2030 as stated in the "Basic Hydrogen Strategy")	480 ($/t-$NH_3$)

3.5.1 Estimation of the Production Cost of NH_3

The production process of NH_3 was first commercialized by Haber and Bosch more than 100 years ago, and since then various improvements have been made to the process to establish it as an industrial process. The energy efficiency of ammonia production using a natural gas feedstock has almost reached the theoretical limit. A study by the International Fertilizer Industry Association in 2008 on energy efficiency of 93 NH_3 plants in 33 countries [26] found that the top decile of energy-efficient plants using natural gas as a feedstock achieves an efficiency of 32 GJ/ton-NH_3. The most modern plants come even closer to the theoretical limit with efficiencies of 28–29 GJ/ton-NH_3.

Under such circumstances, knowing the price of feedstock natural gas, it is possible to estimate the variable production cost of NH_3 for each production plant. Specifically, the variable cost of NH_3 (\$/ton-NH_3) can be approximated by simply multiplying the cost of natural gas (\$/MMBtu) by 30. If the cost of natural gas is 3 \$/MMBtu, for example, the variable production cost of NH_3 from this feedstock would be about 90 \$/ton-NH_3. To estimate the total production cost, the fixed cost of the plant (plant capital, labor, etc.), which varies with location and age of the plant's facilities (and often other factors), must be added to the variable cost. A typical value for fixes cost would be \$100–120/ton-NH_3. Thus, the pre-shipment cost of NH_3 produced in overseas regions where the cost of natural gas is around 3 \$/MMBtu can be estimated as around 200 \$/ton-NH_3.

The ammonia produced with this cost structure is of course the conventional variety, with the full greenhouse gas footprint that originates in the carbon embodied in the natural gas feedstock. To estimate the cost of carbon-free ammonia, the cost of removing CO_2 by carbon capture and sequestration (CCS), with or without enhanced oil recovery (EOR), needs to be added. Such estimations have been made by independent public organizations, both domestic and foreign, as discussed below.

3.5.2 Estimation of the Production Cost of CO_2-Free NH_3

There are two ways to produce CO_2-free NH_3:

(a) NH_3 produced using natural gas feedstock with the CO_2 generated in the production process removed by CCS/EOR; and
(b) NH_3 produced from renewable hydrogen.

Since slightly more than 1.6 tons of CO_2 is emitted per ton of NH_3 production [24], the cost of producing CO_2-free NH_3 by production method (a) can be estimated by adding the cost of CO_2 removal to the NH_3 production cost.

The cost of producing CO_2-free NH_3 from renewable hydrogen is determined by the cost of the renewable hydrogen feedstock, which is largely dependent on the cost of the renewable electricity and facility cost and capacity factor of electrolyzer to produce renewable hydrogen through water electrolysis [27].

3.5.2.1 CO_2-Free NH_3 Production Cost Estimated by the IEA

The International Energy Agency estimated production costs of CO_2-free NH_3 in its 2019 report "The Future of hydrogen" [24] as shown in Fig. 42 of the report.

The analysis represented in the figure indicates the following:

(1) If natural gas is used as feedstock and CO_2 emitted from the production process is removed by CCS, the production cost of CO_2-free NH_3 is about 300 \$/ton-NH_3. (Assuming a natural gas price of 3 \$/MMBtu) At a raw natural gas price of 10 \$/MMBtu, the cost of producing CO_2-free NH_3 is about 530 \$/ton-NH_3.
(2) If the cost of renewable electricity becomes cheaper than about 50 \$/MWh, the cost of producing CO_2-free NH_3 from renewable hydrogen may become lower than that from natural gas with CCS.

3.5.2.2 CO_2-Free NH_3 Production Cost Estimated by the Institute of Energy Economics, Japan (IEEJ)

Completely independent estimation of the CO_2-free NH_3 cost was conducted under the auspices of the SIP "Energy Carriers" initiative by the Institute of Energy Economics, Japan (IEEJ). It was based on the following cases and assumptions:

- Construction of a new NH_3 production plant with an annual capacity of 1.1 million tons in either Saudi Arabia, another Middle Eastern country, or North America;
- Natural gas feedstock priced at 3 \$/MMBtu;
- The CO_2 generated in the production process removed by carbon capture at the production location;
- The resulting CO_2-free NH_3 delivered to Japan by ocean freight; and
- Realization of a 10% equity internal rate of return (EIRR) as supplier profit [28].

The analysis shows that if CO_2-free NH_3 can be sold at prices between 276 \$/ton-NH_3 (for Saudi Arabia) or 300 \$/ton-NH_3 (for other locations), the business will be profitable on the production side.

The EOR Effect

If an enhanced oil recovery (EOR) field exists near the NH_3 plant, high concentration of CO_2 emitted from the NH_3 production process can be sold almost intact for EOR. In this case, the revenue from the process can be used to offset the production costs. Since CO_2 for EOR can be sold at about 20 \$/t-CO_2, this would reduce the production cost of CO_2-free NH_3 by about 35 \$/ton-NH_3. The IEEJ analysis took this into account in its estimation of the cost of CO_2-free NH_3 produced in the case of the US. This analysis makes clear that the availability of EOR can meaningfully influence the economics of CO_2-free NH_3 from natural gas.

3.5.2.3 Cost of CO_2-Free NH_3 Arriving in Japan

The cost of transporting CO_2-free NH_3 to Japan and then to its point of use within the country needs to be added to the cost of production. This cost was estimated to be around 40–80 \$/ton-NH_3 by the IEEJ study. Logistics costs vary depending on the point of origin, the size of the ship, and the number of times in which the product is loaded and unloaded. For the Middle East to Japan, logistics costs are 40 \$ton-NH_3; for North America to Japan: 80 \$/ton-NH_3. In both cases, it was assumed that very large gas carrier (VLGC) vessels with approximate capacity of 80,000 m^3 are used.

According to the results from these analyses, the cheapest CO_2-free NH_3 would be the one that was produced in the Middle East. The cost of this CO_2-free NH_3 after logistics costs was taken into account by the IEEJ at 320 \$/ton-NH_3, or about 340 \$/t-NH_3 according to the IEA's estimated production cost of CO_2-free NH_3 including the cost of delivery to Japan estimated by IEEJ.

The validity of this cost estimate is also supported by the IEA analysis shown in Figure 31 of "The Future of Hydrogen", which shows the cost estimate for importing hydrogen energy in the form of CO_2-free NH_3 from Australia and the Middle East to Japan, and from Russia to the EU.

From this analysis, the cost of CO_2-free NH_3 from the Middle East to Japan is almost in line with the results estimated by the IEEJ above. More specifically, the cost of importing NH_3 from the Middle East to Japan in the form of CO_2-free NH_3, produced from natural gas with subsequent CCS, is estimated at 2.3 \$/kg-H_2 (about 360 \$/t-NH_3). (This cost includes 50 km inland transportation cost in Japan.) Here, it is worth recalling the fact that CO_2-free NH_3 can be used as fuel as it is, so the cost of CO_2-free NH_3 is the "cost of hydrogen" presented in the figure minus "reconversion cost."

Two conclusions can be derived from these analyses:

First, these estimates suggest that the cost of CO_2-free NH_3 arriving in Japan is likely to be around 350 \$/t-NH_3. Since the actual price of CO_2-free NH_3 available to Japanese users will be determined by negotiations between buyers and sellers, there is little point in discussing the validity of the individual cost estimates in more detail. What is important is to assess the social implementation potential of CO_2-free NH_3 from this cost level of roughly 350 \$/t-NH_3 derived from these analyses. This cost level corresponds to about 22 yen/Nm^3-H_2, as can be seen from Table 3.1, which almost meets the 2050 cost target for hydrogen set by the "Basic Hydrogen Strategy."

Second, until around the year 2030, the cost of CO_2-free NH_3 from a combination of natural gas and CCS will be considered considerably cheaper than that of renewable hydrogen. This is shown in Fig. 31 of the IEA analysis, where the estimated cost of importing CO_2-free NH_3 from renewable hydrogen produced in Australia into Japan is estimated to be 4.4 \$/kg-H_2 (equivalent to about 700 \$/t-NH_3, just about double the cost of CO_2-free NH_3 produced from natural gas with CCS). After 2030, however, the IEA expects that the cost of producing NH_3 from renewable hydrogen will eventually fall to less than half its 2030 level [24], as the cost of electrolyzers and renewable electricity decreases further.

3.6 Toward Introduction of NH_3 as CO_2 Fuel

As described above, the results of the research and development in the SIP "Energy Carrier" initiative have revealed the great potential of NH_3 that NH_3 can play a key role in decarbonizing the power generation sector, which is the largest source of the CO_2 emissions in Japan. Remaining issue for societal implementation of NH_3 is the establishment of a supply chain for CO_2-free NH_3. Specifically, securing a cost-competitive and stable supply source of CO_2-free NH_3 and developing and expanding infrastructure necessary to handle large quantities of CO_2-free NH_3 for power generation.

Here, the reason for the development of a supply chain of NH_3 has become a prominent issue - although a supply chain of NH_3 is already existing -, is that volume of NH_3 fuel needed for power generation is very large. At present, Japan's annual consumption of NH_3 is about 1 million tons, with imports accounting for about 20% of that amount. But amount of NH_3 required for reducing CO_2 emission from power generation would reach to 3 million tons per year in 2030 and 30 million tons per year in 2050.

Because of the importance of developing this scale of a supply chain of NH_3, Ministry of Economy, Trade, and Industry (METI) led to launch the Public–Private Council on Fuel Ammonia Introduction in October 2020. The members of the Council include the Ministry of Land, Infrastructure, Transport and Tourism, electric power companies, trading companies, power generation equipment and machinery manufacturers, shipping companies, government-affiliated resource development organizations, and public financial institutions, in addition to METI itself. And in February 2021, the Council developed a report which sets out the 2030 and 50's importation targets of NH_3 mentioned above and lists policy measures and private company activities which will be necessary to construct a supply chain.

Before the establishment of the Council, the Clean Fuel Ammonia Association (CFAA) [formerly Green Ammonia Consortium (GAC)] was established in April 2019 by the initiative of private companies which are interested in CO_2-free NH_3. Presently, the CFAA consists of more than 100 government agencies, other public institutions and private companies from both domestic and foreign, which would make up a value chain for CO_2-free NH_3. Using the CFAA as an information platform, various activities are underway to develop a value chain of NH_3 fuel in both supply and user sides of countries.

In parallel with these movements, JERA, Japan's largest thermal power generation company, announced "JERA Zero Emissions-2050" in October 2020 as a roadmap to decarbonize its power generation. The roadmap includes a plan to start using NH_3 as a co-fired fuel in its 1GW coal-fired power generation facility around 2025 and to expand such operation to a full scale by 2030, and the start of 100% NH_3-fired power generation in the 2040s. Currently, JERA is working on securing stable and cost-competitive supply sources of NH_3 and engaging in the necessary modification and construction of related facilities in that power generation site for the introduction of NH_3, while getting financial support from NEDO.

In addition to the use of NH_3 as a CO_2-free fuel, some trials to use NH_3 as a hydrogen carrier for transporting hydrogen have also started in the world. The project announced in July 2020 by Saudi Arabia's NEOM [29], ACWA Power [30], and Air Products, the world's largest industrial gas manufacturer, would be a typical example of this kind. The plan is to produce 650 tons of renewable hydrogen per day from a 4 GW solar and wind power generation facility in Saudi Arabia and to use that hydrogen to produce 2.5 million tons of CO_2-free NH_3 per year for export. According to the announcement, Air Products will take all this CO_2-free NH_3 produced and sell it around the world as CO_2-free hydrogen for transportation fuel.

This is an interesting business plan because NH_3 needs to be cracked and purified after the delivery to a site of use in order to use NH_3 as a hydrogen carrier, thus it requires additional cost to convert NH_3 to hydrogen for producing transportation fuel. Nonetheless, the decision to use NH_3 as a hydrogen carrier in the above-mentioned case is presumably based on the judgment that the benefits of transporting NH_3, which is easier to transport than hydrogen, outweigh the additional costs incurred by that cracking and pulification processes [31].

In this regard, it is worth recalling the IEA's cost analysis, as shown in Figs. 30 and 32 of the "The Future of Hydrogen" report, which shows that transporting hydrogen using NH_3 as a hydrogen carrier would be the cheapest method for transporting hydrogen energy, among others.

As discussed above, mass introduction of hydrogen will be inevitable to decarbonize electricity and fuels of transportation. Furthermore, in the industry sector, demand for hydrogen is also expected to increase as discussed in Sect 3.2 for its decarbonization. Since practical and economic means of transporting large amounts of hydrogen are limited, the potential of NH_3 as a hydrogen carrier would be likely to attract more attention in future.

In this way, social implementation of NH_3 has already begun and a potential role of NH_3 in constructing a carbon–neutral economy may further expand in coming future.

References

1. SIP "Energy Carriers" is an innovation promotion project carried out in 2014–18 under the Cross Ministerial Strategic Innovation Promotion Program (SIP) of the Cabinet Office of the Japanese government
2. The non-electric energy other than heat includes fossil energy (or resource) which is used for raw materials, such as coking coal for blast furnaces in the steel industry and naphtha and other hydro-carbon raw materials for the petrochemical industry
3. Cross Ministerial Startegic Innovation Promotion Program (SIP) run by the Cabinet Office of the Japanese government took up the development and promotion of production, transportation, and use technologies of Hydrogen energy as one of the themes of the program. The title of the theme was SIP "Energy Carriers" and was conducted from 2014–18 with 150 million dollars funded by the government
4. The decomposition reaction of NH_3 is an endothermic reaction, which means that about 13% of the hydrogen energy that NH_3 can carry needs to be used in the decomposition process

5. Kobayashi H et al (2018) Science and technology of ammonia combustion. In: International symposium of the Combustion Institute, Dublin [Proc Combust Inst 37(1):109 (2019)]
6. A swirl burner uses a set of guiding vanes to give a swirling motion to the combustion air
7. A combustion method in which NH_3 is combusted in slightly NH_3 rich condition in the primary combustion zone, then in the secondary combustion zone diluting combusted gas by air and burning H_2 generated by the decomposition of NH_3 occurred in the primary combustion zone
8. Kobayashi H, Hayakawa A (2019) Significance and impact of ammonia combustion research. J Combust Soc Jpn 61(198):277–282
9. Iki N (2019) Demonstration of ammonia fuel with small gas turbines. J Combust Soc Jpn 61(198):283–288
10. IHI (2019) The world first ammonia mixed combustion at 2000kW class gas turbine realized (in Japanese) https://www.ihi.co.jp/ihi/all_news/2018/technology/1190442_1624.html. Accessed 13 Sept 2021
11. In terms of heat value. In the following, co-firing rates described in this article are this base
12. Ito S, Uchida M, Suda T, Fujimori T (2019) Demonstration test of ammonia/natural gas co-firing power generation with 2MW class gas turbine. J Combust Soc Jpn 61(198):289–292
13. Nose M, Araki H, Senba FH, Tanimura S (2019) Development of ammonia utilization technology for large gas turbines for power generation. J Combust Soc Jpn 61(198):291–298
14. Fukui J, Nakatsuka N, Hatanaka K, Higashino H, Hayashi J, Akamatsu F (2016) Co-combustion behavior of pulverized coal particles with ammonia held by hydrogen/air diffusion flame in the laminar counterflow field (in Japanese). Report No. C334 at 54th combustion symposium of the Combustion Society of Japan, Sendai, Japan
15. Kimoto M, Yamamoto A, Ozawa Y, Hara S (2019) Ammonia co-firing technology applicable to pulverized coal fired boilers. J Combust Soc Jpn 61(198):299–303
16. Tanigawa H, Ouchi M (2018) Demonstration Test of Ammonia-Coal Co-Firing at No. 2 Plant of Mizushima Power Generation Plant (in Japanese), Electrical Review April 2018, pp 52–55
17. Yoshizaki T (2019) Test of the co-firing of ammonia and coal at Mizushima power station. J Combust Soc Jpn 61(198):309–312
18. Sakoya A (2018) History of innovations achieved in The Chugoku Electric Power Co. Inc. (in Japanese), Electrical Review, Jan 2018, pp 138–152
19. Ito T, Zhang J, Ishihara S, Ishii H, Suda T, Fujimori T (2019) Development of the coal co-firing technology with ammonia and numerical evaluation of the boiler performance. J Combust Soc Jpn 61(198):304–308
20. Murai R, Nakatsuka N, Higashino H, Akamatsu F (2019) Review of fundamental study on ammonia direct combustion in industrial furnaces. J Combust Soc Jpn 61(198):320–325
21. Numata M, Matsuda T, Hagiwara Y, Yamamoto Y (2019) Development of impinging jet burner using ammonia fuel for degreasing steel sheets. J Combust Soc Jpn 61(198):326–330
22. JST (2015) Power generation by SOFC fueled by NH_3 (in Japanese) http://www.jst.go.jp/pr/announce/20150722-6/index.html. Accessed 13 Sept 2021
23. IHI (2018) IHI succeeded in power generation using 1kW SOFC power generation system fueled by NH_3 (in Japanese) http://www.ihi.co.jp/ihi/all_news/2018/technology/2018-5-16/index.html Accessed Sept 13, 2021
24. The Future of Hydrogen International Energy Agency (IEA) (2019)
25. Basic Hydrogen Strategy (2017) Decided by relevant Ministers Council on renewable energies and hydrogen energy of the Japanese government
26. Energy efficiency and CO_2 emissions in ammonia production 2008–09 Summary Report (2009) International Fertilizer Industry Association
27. The NH_3 production cost of this method can be estimated by the chemical reaction formula of NH_3 synthesis ($1/2N_2 + 3/2H_2 \rightarrow NH_3$) and the cost of hydrogen production by electrolysis
28. A Feasibility Study on the Supply Chain of CO_2-Free Ammonia with CCS and EOR (2019) The Institute of Energy Economics, Japan
29. NEOM is a new industrial city development project promoted by the Kingdom of Saudi Arabia, which aims to create a new way of life through cutting-edge technology. It is working to build a new city based on 100% reliance on renewable energy sources such as solar and wind for

electricity consumption, full e-governance, and construction of net zero-carbon houses as the standard. A project entity has been established with the Public Investment Fund (PIF), a Saudi Arabian government fund, as the project fund

30. An electric power company headquartered in Riyadh, Saudi Arabia
31. The reason behind this business decision is believed to be the judgment that the price of hydrogen in the automobile fuel market is at a high level at this stage, and therefore, sufficient profit can be secured even with the effort and cost required for the reconversion of NH_3 to produce hydrogen

Chapter 4
Ammonia's Role in the Hydrogen Society

Stephen H. Crolius

Abstract By the start of the 2020s, it had become clear that a number of hydrogen-rich fuels would power the hydrogen society, not just elemental hydrogen itself. Ammonia stands out among the alternatives because, like hydrogen, it is an entirely carbon-free molecule. The fact that hydrogen and ammonia have emerged as the only two carbon-free molecules with mainstream recognition inevitably casts them as rivals who will compete for adoption by developers of sustainable energy systems. The relative importance that each achieves will be determined by the accelerating flow of decisions being made by actors in business, governmental, and institutional settings. For those with an interest in the energy economy of the future, it can be instructive to see the decision-making dynamic at play in real-world settings. This chapter presents three case studies in which hydrogen and ammonia are both seen as leading fuel options. In the first, involving fuel cell vehicles, it appears that hydrogen has locked up the dominant role, largely by virtue of its multi-decade head start on ammonia. In the second, involving utility-scale combustion turbines, current indications are that the two fuels will each secure substantial shares of the market. Selection is likely to occur on a site-by-site basis, based on project-specific considerations. In the third, involving maritime shipping, all current signs suggest that ammonia applications will far outstrip those of hydrogen. This outcome is based on the ammonia's superior fit with the design requirements and constraints of maritime fuel systems.

Keywords Sustainable energy economy · Low-carbon fuel selection · Fuel cell vehicles · Combustion turbines · Maritime fuel

4.1 Introduction

Human society has started to build a sustainable energy economy, i.e., an energy economy that does not destabilize the earth's climate, produce pollutants that are

---

S. H. Crolius (✉)
Carbon-Neutral Consulting LLC, Warwick, RI, USA
e-mail: scrolius@carbonneutralconsulting.com

Ammonia Energy Association, New York City, USA

K. Aika and H. Kobayashi (eds.), *CO_2 Free Ammonia as an Energy Carrier*,
https://doi.org/10.1007/978-981-19-4767-4_4

harmful to life, or depend on non-renewable resources. We are fortunate to have many options to draw on as we create this economy. We are not faced with the question of how to find enough energy to meet the needs of society. Instead, two other questions present themselves. First, what form of a sustainable energy economy can be achieved most readily, given our economic, technological, and political starting points? Attempts to answer this question focus on the shortest, most easily implemented paths to sustainability. Second, what architecture will ultimately impose the lowest economic burden on energy consumers? Attempts to answer this question gravitate toward solutions that embody techno-economic optimization. Both questions are legitimate and important. The right answer will be the one that responds to both while causing the least compromise to either.

Navigating the two questions will involve choices in dozens of areas. One relates to the role that energy stored in the form of fuel will play. Hydrogen and ammonia, both widely seen as important fuels for the sustainable energy economy, are caught up in a particular version of the implementability/optimization dynamic. Many hydrogen advocates begin with the belief that hydrogen can be readily implemented as a fuel, not least because most of the invention needed for a functioning production-use cycle is already in place. They argue that technological refinement can and will continue, but that should not slow down the process of implementation. Ammonia advocates make the case that for many use cases, ammonia-based systems offer significant techno-economic advantages. They argue that these advantages far outweigh delays that might be associated with ammonia implementation.

This chapter presents three case studies involving at-scale, real-world systems that illuminate the selection dynamics between hydrogen and ammonia. In the first one, centered on light-duty fuel cell vehicles, hydrogen seems to have secured its position as the dominant fuel. In the second one, centered on combustion turbines for electricity generation, it is too early to say which fuel will prevail while technologists and project developers in both camps seek to advance their respective systems. In the third one, maritime freight transport, it is ammonia that appears destined to become the dominant fuel. The forces operating within these case studies will go far in determining the ultimate design of the sustainable energy economy.

4.2 Case Study One: Fuel Cell Vehicles

On December 15, 2014, after twenty-plus years of development, the Toyota Motor Company introduced its Mirai hydrogen fuel cell vehicle (HFCV) to the Japanese market, and shortly thereafter to California. While Toyota and a variety of other automotive OEMs had been offering HFCVs on limited bases since the early 2000s, this was the first instance of an HFCV being offered for sale on an unlimited commercial basis.

Although a key impetus for HFCV development in the early 1990s was early recognition of the climate crisis, the automakers were returning to a technology whose roots substantially predated awareness of the impacts of anthropogenic greenhouse gases. Interest in automotive fuel cells originated first in the United States' space program of the 1950s and 1960s. In that period, the U.S. National Aeronautics and Space Administration oversaw wide-ranging efforts to develop hydrogen-fueled proton-exchange membrane (PEM) fuel cells. Picking up on the spirit of technological possibility that characterized that era, General Motors developed the first HFCV in 1966 with a stated goal of near-term commercialization. However, a combination of factors, including the high cost of the technology, the number and nature of potential hazards it created, and the lack of interest in developing hydrogen fueling infrastructure, caused GM to abandon the program shortly after the initial vehicle was demonstrated [1].

Interest in FCVs was kept alive in the 1970s by the emergence of two new societal concerns: the world's dependence for a substantial share of its energy supply on oil-rich autocratic nations; and the impacts of unregulated vehicle emissions on human health [2]. Together these concerns stoked interest in hydrogen as an endlessly abundant renewable energy vector that could be produced by the indigenous solar resources of virtually every country on earth; and the fuel cell as an energy conversion device whose efficiency and emissions cleanliness far surpass those of the internal combustion engine. As a consequence, when awareness of the climate crisis took hold in the early 1990s, it seemed natural for the leading automotive companies to focus their efforts on HFCV technology. Daimler took an initial lead with the unveiling of a prototype vehicle in 1994. In the second half of the decade, General Motors, Toyota, Mazda, Ford, Honda, Nissan, and Volkswagen all brought fuel cell vehicle projects to fruition whose economic and safety profiles promised early commercialization. And indeed, commercialization seemed imminent in the early years of the 2000s when these companies all launched demonstration vehicles, with Toyota becoming the first company to offer an HFCV (albeit with a hybrid propulsion system) to the public on a limited leasing basis [3].

The "chicken-and-egg" fueling system impasse (no hydrogen fueling stations without hydrogen-fueled vehicles; no vehicles without fueling stations) was a part of the public discourse by the latter half of the 1990s. However, the first serious effort to resolve the impasse via fueling station development did not come together until the Japan Hydrogen and Fuel Cell Demonstration Project was formed in 2002. This body, initially consisting of 22 automotive OEMs, energy companies, industrial gas companies, and miscellaneous industrial players pledged itself to addressing both HFCV production and fuel supply. By the end of 2005, the group had opened or started construction on 13 fueling stations [4]. The State of California followed suit in 2004 when it conceived and launched its Hydrogen Highway initiative, which four years later led to the opening of the state's first public hydrogen fueling station [5]. A decade-and-a-half later, in 2021, Japan has 135 hydrogen fueling stations in operation [6]. California has a similar number in operation or in development [7]. Station developers have also been active in Europe, contributing to a global total of almost 500 stations [8] (While this number represents a tiny fraction of the global population of petrol stations, estimated at approximately 500,000, it shows over-representation relative to the number of HFCVs: approximately 25,000 [9] out of a global automotive fleet of 1.4 billion [10]).

There are three reasons why ammonia may never displace hydrogen as a terrestrial transportation fuel. The first is the technological lead held by hydrogen. The automotive-grade PEM fuel cell has been the target of billions of dollars of investment over decades by the world's leading automotive companies and the governments of its largest economies. While development of ammonia-fueled analogs to PEM fuel cells, i.e., fuel cells whose alkaline chemistry matches the reaction conditions of electrolytic ammonia oxidation, is ongoing at a number of research laboratories [11], it seems unlikely that alkaline fuel cells will ever reach the scale of investment that flowed into hydrogen fuel cell development.

This unlikelihood is compounded by the second factor working in hydrogen's favor: any party who might be inclined to develop an automotive-grade alkaline fuel cell must take serious account of the slow progress being made in the development of the HFCV ecosystem. A potential investor must ask, is the development process inherently this slow, or is it being retarded by the success of the other sustainable light-duty automotive option: electric cars, whose current global population of 10 million is on a rapid growth trajectory [12]?

The third reason working in hydrogen's favor relates to the appearance of technologies that can be deployed at the fueling-station level to reform ("crack") ammonia and produce hydrogen at the level of purity demanded by HFCVs. These technologies make it possible to "get the best of both worlds", i.e., to use hydrogen fuel cells as onboard power plants while taking advantage of ammonia's lower-cost fuel storage and distribution economics. This is indeed a plausible vision for the development of the HFCV ecosystem, although it should be noted that the on-site hydrogen-cracking-and-purification units will add significant capital cost to the fueling stations, and the energy penalty associated with the cracking process is non-trivial.

It can thus be concluded that in the hydrogen versus ammonia selection dynamics, hydrogen's position in the terrestrial transportation domain appears secure. It achieved this position, however, based on a unique set of historical circumstances. It does not appear that techno-economic comparison of hydrogen's attributes versus those of other sustainable fuel options was conducted to justify the major investments that have flowed into HFCV technology. In fact, the 60-year history of HFCV development, which focused first on the PEM fuel cell, and much later on hydrogen fueling-station networks, and, much later still, on the design of fuel-supply logistics systems, shows a process wholly lacking in techno-economic optimization.

4.3 Case Study Two: Combustion Turbines

Combustion turbines are used to convert natural gas to electricity at generating stations around the world. In 2015, natural gas was used as a primary energy source to generate 23% of the world's electricity. This share has been rising over the last

25 years, at first at the expense of nuclear energy (which fell from a high of 18% in 1996 to 8% in 2015) and more recently at the expense of coal (which fell from a high of 41% in 2007 to 39% in 2015) [13].

The expansion has been facilitated by two considerations. First and most important is the cost competitiveness of electricity produced from natural gas. The United States Energy Information Administration projects that the most efficient natural gas plants commissioned in the first years of this decade will have a levelized cost (LCOE) per megawatt-hour (MWh) of USD$37. The most efficient coal-fired plants, by contrast, will have an LCOE of USD$73, and "advanced nuclear" plants will have an LCOE of USD$69 [14]. The second consideration is that natural gas plants have a carbon footprint that is about half that of coal-fired plants [15].

The prevalence of combustion turbines in the electricity sector is proving to be a helpful factor in the energy transition. One reason for this is that combustion turbines are a versatile technology that can be adapted to run on a variety of liquid and gaseous fuels. Combustion turbine manufacturers have traditionally exploited fuel versatility for use cases outside the electricity sector. Oil refining, petrochemical processing, and steel production all generate byproduct gases whose constituents include molecular hydrogen. Plant operators in these industries employ combustion turbines to convert these gases into electricity that can be used on-site or exported to the electric grid. For combustion turbine manufacturers who supply machines to these industries, the challenge of developing a utility-scale (>100 MW) combustion turbine capable of running on 100% hydrogen is more incremental than monumental.

Another fortuitous circumstance for the energy transition is that a change from one fuel to another can be accomplished without the disruption and expense involved in replacing complete machines (much less complete generating stations). Combustion turbine design is such that some parts, such as blades, vanes, and nozzles, are subject to regular wear and require periodic replacement or refurbishment. These parts also happen to be the most prominent components that must be changed out when a combustion turbine is being converted from one fuel to another. Converting a machine in this way involves less expense than acquiring and commissioning a completely new unit [16].

It was only as the current decade approached that combustion turbine manufacturers, sensing the imminence of a new market, committed to the development of machines that could run on pure hydrogen. By 2019, all of the major players, including Mitsubishi Power, GE Power, and Siemens Energy, were well along in their hydrogen programs [17]. At this point, the situation was broadly similar to the one that pertained for light-duty vehicles in the mid-1990s: a technologically sophisticated cohort of original equipment manufacturers saw the emergence of interest in hydrogen-fueled machines and mobilized to seize the opportunity. The question of techno-economic optimization across a range of alternative fuels does not appear to have retarded their initiation of hydrogen programs. In this case, though, there was one critical difference: the forward-looking stance of the Government of Japan.

The Great East Japan Earthquake of 2012 led, among many other consequences, to a national reconsideration of nuclear power. In 2010, approximately 26% of the country's primary energy was produced by nuclear-generating stations [18]. A pull-back from nuclear would clearly involve scaling up another source of primary energy, much of which would need to be imported. It was also clear that fossil fuels could not be a part of the long-term solution and that hydrogen needed to be.

In 2014, the Cabinet Office of the Government of Japan launched the Strategic Innovation Promotion Program (SIP). One of the SIP's initiatives, Energy Carriers, was tasked with investigating options for bringing large quantities of hydrogen into the country. The Energy Carriers team evaluated the technical feasibility and economic implications of the alternatives and chose to focus on three methods that would allow hydrogen to be transported to Japan in the requisite quantities. One method involves transporting elemental hydrogen in liquid form. A second method involves transporting it as a liquid organic hydride. (The specific LOH chosen for study by the Energy Carriers is methylcyclohexane.) Ammonia is the third method. Once these methods had been selected, the Energy Carriers mission was to map out development pathways for each option.

Ammonia was selected in the expectation but not the certainty that it could be used to fuel mainstream combustion processes, not least in compliance with twenty-first century emissions standards. Previous instances of ammonia combustion had occurred before modern environmental regulations had been enacted (one example was the use of ammonia as fuel for transit buses during the Second World War) or where environmental standards do not apply (such as a fuel for the United States' X-15 "rocket plane" in the late 1950s). The Energy Carriers team turned to combustion technology experts at Tohoku University to investigate the essential questions. The Tohoku group, led by Hideaki Kobayashi, worked throughout the years 2015–2017. Through the application of novel fuel mixing techniques and adjustment of parameters such as fuel-to-air ratio, they were able to establish that ammonia could viably function as a combustion fuel, including in the use case of combustion turbines [19].

Development of ammonia combustion turbines is now being taken forward by Toyota Energy Solutions (turbines of 300 kW power capacity), IHI (turbines of 2 MW power capacity), and Mitsubishi Power ("utility-scale" turbines of >100 MW power capacity), all in collaboration with the Tohoku University researchers [20]. Japan's development roadmap calls for a utility-scale combustion turbine that runs on pure ammonia to be ready for commercialization by 2030 [21]. In the meantime, several developments have materialized outside Japan. The Euro-centric industry association ETN Global has launched an Ammonia Working Group to conduct its own research on ammonia combustion turbines [22]. A number of other European players have come together in a multi-pillar project that will include the evaluation of hydrogen and ammonia as combustion turbine fuels [23]. And General Electric and IHI have signed an agreement to develop an Ammonia Gas Turbine Business Roadmap aimed at developing technology and markets in Asia [24].

Two leading-edge projects involving hydrogen-fueled combustion turbines provide perspective on how hydrogen and ammonia are likely to compete as combustion turbine fuels going forward. In one, Mitsubishi Power will install two combustion turbines with a combined power capacity of 840 MW at the Intermountain Power Plant (IPP) in Utah in the United States. When the units are commissioned in 2025, their fuel will consist of 30% hydrogen and 70% natural gas. The turbines will be converted to 100% hydrogen in a phased process that will run through 2045 [25, 26].

The existing plant, with a power capacity of 1900 MW, is currently fired with coal. In the new configuration, the plant will be interconnected with at least 370 MW of wind power (with the possibility of as much as 1900 MW) and 2300 MW of solar generation (and possibly an additional quantity of geothermal generation). The plant will meet daily load profiles with a combination of renewably generated electricity and the output of the combined-cycle turbines. Hydrogen will be produced from the wind- and solar-based electricity that is not needed for real-time demand. It will be stored in an existing underground facility that was originally created from an on-site salt dome to store liquid fuels [25, 27].

In the other leading-edge project, Mitsubishi Power is converting a combustion turbine at Nuon's Magnum Generating Station in the Netherlands to run on pure hydrogen. The Magnum facility was originally conceived in the mid-aughts as an integrated gasification combined-cycle (IGCC) plant that would generate electricity from a range of fuel materials including coal, petroleum, and biomass. Phase one of the project came to fruition in 2014 with the initiation of power generation using natural gas as fuel. The gasification aspect of the project was scrapped in 2016 [28]. In 2017 a new team came together around a new concept: use of blue hydrogen as a generating fuel. Nuon will continue to serve as project host, while Equinor will produce the hydrogen via conversion of natural gas from one of its Norwegian gas fields with sequestration of the CO_2 in a nearby storage reservoir. The Dutch natural gas utility Gas Unie will transport the hydrogen to the Netherlands.

Different possibilities are under consideration once the hydrogen unit is commissioned in 2023. (The next two units are to be converted by 2030.) The long-term vision is to move to green hydrogen produced from renewably generated electricity [29]. The questions of where that production might occur and what methods of transport might be employed are still under consideration, although Gas Unie is exploring transport via existing natural gas pipelines and storage in caverns [30]. Another scenario involving ammonia as the hydrogen carrier was floated early in the project conceptualization process. Mitsubishi Power has studied the conversion of ammonia to hydrogen for a utility-scale combustion turbine using the turbine's waste heat and a catalyst [31].

Examined from a techno-economic perspective, hydrogen appears to be the optimal fuel for storing renewably generated energy at the Intermountain Power Plant. The key factor is the on-site presence of the existing geologic storage facility. This means that the costs of hydrogen storage and transport will be minimal. In this setting, there does not appear to be a scenario in which the costs and energy losses of hydrogen-to-ammonia conversion would make sense.

By contrast, the best approach for the Nuon Magnum facility is yet to be determined. The project stakeholders will no doubt model transportation and storage costs for both elemental hydrogen and ammonia. This analysis will influence which fuel the second and third turbine units will be converted to and if the first unit should be eventually converted from hydrogen to ammonia.

In summarizing the selection dynamics that will apply as system developers consider hydrogen and ammonia as potential combustion turbine fuels, four major conclusions are already clear. First, efforts are well underway to develop utility-scale combustion turbines for both hydrogen and ammonia. Hydrogen may have a small lead currently but ammonia will catch up in short order. Second, both fuels will find a place in the electricity sector of the future. For a situation like the IPP in Utah, hydrogen is clearly the right choice. For other situations, where transport challenges are significant and/or on-site storage facilities must be built from scratch, ammonia is likely to be the better choice. Third, it cannot be projected at this time if one fuel will ultimately emerge as the dominant solution. And fourth, while the outcome of the hydrogen-ammonia selection process will matter to the companies involved in the respective fuel value chains, society should feel confident that the site-by-site selection process will lead to an optimal techno-economic result.

4.4 Case Study Three: Maritime Propulsion

The maritime sector features an extraordinary range of working vessels, from 6-m water taxis to 400-m cargo vessels. Not surprisingly, the vast majority of sectoral GHG emissions come from the category of large ships engaged in international trade [32]. Tangible action on the climate impact of the fuels used to power these vessels was initiated in the latter half of the 2010s. Although the International Maritime Organization's (IMO's) earliest evaluation of the maritime sector's GHG footprint occurred in 1997, its initial regulatory steps, aimed at increasing the energy efficiency of ships, were not taken until 2011. By 2017 the IMO was at work on a comprehensive framework for regulating GHG emissions, and in 2018 the organization adopted its *Initial IMO Strategy on reduction of GHG emissions from ships*, which included hard targets for GHG emission reductions [33].

The *Initial Strategy* spawned numerous techno-economic analyses of alternative maritime fuels. One of the first, issued in 2017 by maritime classification society Lloyds Register and commercial advisory service UMAS, included hydrogen and ammonia in a comparison of cost and profitability impacts of various fuels across multiple vessel types and fuel production scenarios. The profitability of ammonia-fueled vessels was found to be higher than that of hydrogen-fueled vessels in every scenario [34].

Figure 4.1 details relative profitability, expressed on an indexed scale of 0–1, under three future energy scenarios. (Note that the scenario names were chosen based on the main "green" fuel available in each, although the naming does not necessarily indicate the most preferable fuel.) Biofuel is the most profitable option because it can be deployed with minimal modification to ship machinery and storage. The ammonia internal combustion vessel is the second-most-profitable option in all cases. The profitability of the hydrogen-based options ranks behind the ammonia-based options in all cases.

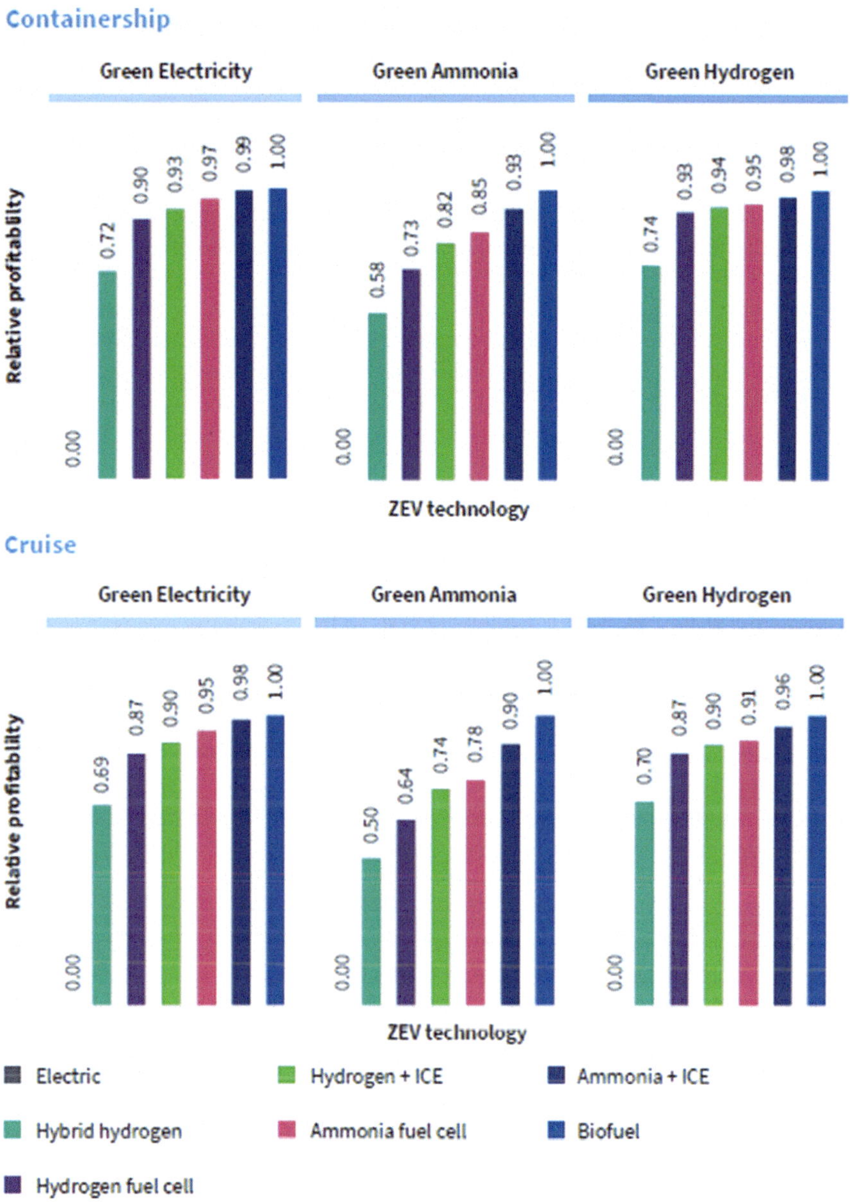

Fig. 4.1 The relative profitability of two different types of ship powered by seven different zero-emission vehicles (ZEV) technologies [34]

Ammonia's key advantages versus hydrogen relate to the economics of onboard fuel storage. One underlying driver is volumetric energy density. Lower volumetric density necessitates larger tanks to store the same quantity of energy, which in turn means higher capital expense for storage, and less cargo-carrying capacity. Heavy fuel oil (HFO), the traditional mainstay for maritime shipping, has a volumetric energy density of approximately 41 MJ/L [35]. While ammonia's volumetric energy density is 16 MJ/L—only 39% of HFO's—liquid hydrogen's is just 9 MJ/L—22% of HFO's. At 70 MPa, gaseous hydrogen's value is 6 MJ/L—just 15% of HFO's [36]. When it comes to fuel storage, vessel owners will incur higher costs when switching to ammonia or hydrogen, but the impact will be more severe for hydrogen.

Another driver of economic disparity between hydrogen and ammonia is storage conditions. Onboard storage of commodities such as ammonia and hydrogen is regulated by the IMO's International Code of the Construction and Equipment of Ships Carrying Liquefied Gases in Bulk (IGC Code). Ammonia, per the IGC Code, must be stored in refrigerated tanks at a temperature below its boiling point (−33 °C). The corresponding value for liquid hydrogen is −253 °C. The cryogenic systems required in this case are extensive, and also have operating expense implications [37].

A third driver is the measures needed to manage the hazards of stored hydrogen and ammonia. One challenge comes from the small size of the hydrogen molecule, which increases the risk of hydrogen escape via tank-wall permeability and microscopic leaks [38]. Another challenge comes from hydrogen's wide range of flammability (4–75% gas-in-air) and the low minimum ignition energy of hydrogen/air mixtures (0.02 mJ). Ammonia, by contrast, is readily contained in insulated steel tanks. Its range of flammability is much narrower than hydrogen's (15–28% gas-in-air) while the minimum ignition energy of ammonia/air mixtures is much higher (8 mJ) [39]. It must be said that ammonia is a hazardous chemical in its own right and that the cost of hazard management systems needed on ships fueled with ammonia will also be significant. The Lloyd's/UMAS conclusion, though, is that the costs for hydrogen will be higher.

Forward-looking maritime players started announcing ammonia-oriented technology development efforts shortly after the IMO's *Initial Strategy* was announced in April 2018. In October 2018, maritime engine manufacturer MAN Energy Systems announced that the company had an ammonia-fueled engine under development with the goal of completing development within three years [40]. In June 2019, engine and energy equipment manufacturer Wärtsilä unveiled a concept for producing and distributing hydrogen and/or ammonia from purpose-built facilities in the waters off northern Europe [41]. By the end of 2019, three different shipyards in China were working on designs for ammonia-fueled vessels [42]. The pace accelerated throughout 2020 and by the middle of 2021, dozens of maritime companies, including shipbuilders, carriers, classification societies, and fuel suppliers, were engaged in developing an ammonia fuel value chain for the maritime sector.

The surge of ammonia-related activity has been accompanied by a more modest set of hydrogen-oriented initiatives. The Lloyds/UMAS report noted that hydrogen's competitive disadvantage would tend to decrease as vessel range decreases—for the simple reason that disadvantages related to onboard fuel storage tend to shrink as range, and therefore the amount of onboard fuel storage, decreases. And indeed, many of the early maritime hydrogen projects are centered on vessel types such as ferry boats and near-shore workboats whose duty schedules allow them to be refueled on a frequent basis. In 2018, a team in California started work on a ferry boat (21 m in length and fueled by gaseous hydrogen) that will carry passengers and freight primarily within San Francisco Bay. In 2019 a team in Norway started work on a ferry boat (82 m in length and fueled by liquid hydrogen) that will carry passengers and cars between Norway and Germany. Both of these projects employ power plants based on PEM fuel cells, presumably because of the maturity the technology has achieved in the automotive sector.

Two sets of considerations mean that hydrogen's competitiveness even in smaller vessels is not a foregone conclusion. The first is that duty cycles in the maritime sector, even for small vessels, differ substantially from those of terrestrial light-duty vehicles. The rapid on–off cycling of light-duty vehicles puts a premium on the low operating temperatures of PEM fuel cells. By contrast, most types of maritime vessels, even those of smaller size, operate continuously or have a single daily on–off cycle. This circumstance facilitates deployment of solid oxide fuel cells (SOFCs) by allowing their high operating temperatures to be maintained continuously. Fuel-to-electricity conversion efficiencies for SOFCs are typically in the 60–70% range [43], while those of PEM fuel cells are in the 50–60% range [44]. To the degree that SOFCs become the fuel cell of choice for maritime applications, one argument in favor of hydrogen over ammonia goes away. This is because ammonia is as readily accepted as a fuel by SOFCs as hydrogen [45]. The ShipFC consortium, based in Norway, plans to demonstrate an offshore supply ship powered by a 2 MW ammonia-fueled SOFC starting in 2024 [46].

The second set of considerations relates to the case for moving to fuel cell propulsion on any basis. The main pillar supporting that case is the quest for higher fuel-conversion efficiencies. In the automotive world, where petrol-fueled cars typically have efficiencies in the 15–20% range, the merit of a 50%-efficient fuel cell is compelling. However, the internal combustion engines (ICEs) used to power ships already have the ability to reach 50% efficiency. Few maritime stakeholders are likely to embrace a wholly unfamiliar propulsion technology (PEM fuel cells) if there is no economic or compliance advantage in doing so. Thus, if the only way to dispense with fossil fuels would be to change to an energy conversion device that runs on hydrogen, then the case for adopting PEM fuel cells would be clear. But, courtesy of ammonia, the maritime sector has a path to a climate-friendly fuel that will allow retention of ICE technology.

Several companies are now developing hydrogen-fueled ICEs for maritime applications. One prominent effort is the BeHydro joint venture between Belgian shipping company CMB and ICE manufacturer Anglo Belgian Corporation. In 2020 the venture received an order, its first, for two 2 MW engines that will power a tugboat for the Port of Antwerp [47]. BeHydro indicates that its technology platform could be used for engines of up to 10 MW in power capacity. Transoceanic vessels are powered by engines with ratings of up to 80 MW.

It is possible that hydrogen, whether in elemental form or chemically packaged as ammonia, will never become an important maritime fuel. But in the event that the maritime sector does turn to hydrogen, there seems little doubt that the preferred form in most applications will be ammonia. This is not because elemental hydrogen will not be an option, but because in most applications, ammonia will have the better techno-economic case.

References

1. Hydrogen cars now (1966). GM Electrovan. Web page consulted May 2021
2. Fuelcellworks.com. History. Web page consulted May 2021
3. Qin N et al (2014) Analysis of fuel cell vehicle developments. U.S. Department of Transportation Electric Vehicle Transportation Center, FSEC Report Number: FSEC-CR-1987-14, Sept 2014
4. Japan Hydrogen and Fuel Cell Development Project. JHFC Phase 1: FY2002–FY2005. Web page consulted May 2021
5. California Fuel Cell Partnership. Los Angeles—West LA 1 ("2008-06-26 involved organizations: Shell Hydrogen, DOE, GM; Station type: Electrolyzer (Green H2); Description: First open access retail like station in California"). Web page consulted May 2021
6. Nikkei Asia. Be water: Japan's big, lonely bet on hydrogen, Dec 23, 2020
7. California Fuel Cell Partnership. FCEV sales, FCEB, and hydrogen station data. Web page consulted May 2021
8. International Energy Agency. Hydrogen more efforts needed tracking report, June 2020
9. Ibid.
10. Urban Guide. How many cars are there in the world? Web page consulted May 2021
11. See for example Zhao et al. An efficient direct ammonia fuel cell for affordable carbon-neutral transportation. Joule, July 30, 2019
12. CNBC. Global electric vehicle numbers set to hit 145 million by end of the decade. IEA Says, April 29, 2021
13. International Energy Agency. Electricity production from natural gas sources. World Bank Web page consulted June 2021
14. United States Energy Information Agency. Levelized costs of new generation resources in the annual energy outlook 2021, Feb 2021
15. United States Energy Information Agency. How much carbon dioxide is produced when different fuels are burned? Web page consulted June 2021
16. General Electric. A second life is possible for old turbines—and lucrative, Dec 14, 2018. Web page consulted June 2021
17. Power. High-volume hydrogen gas turbines take shape, May 1, 2019. Web page consulted June 2021
18. Statista. How Fukushima changed Japan's energy mix, Mar 11, 2021
19. Kobayashi H et al (2018) Science and technology of ammonia combustion. In: Proceedings of the combustion institute, Nov 9, 2018
20. Shiozawa B (2020) A deep dive into SIP "energy carriers" ammonia combustion research. Ammonia Energy, Sept 23, 2020
21. Muraki S (2019) Ammonia, key green energy for decarbonization, p 12. Presented at ammonia = hydrogen 2.0 conference, Melbourne, Australia, Aug 22, 2019
22. ETN Global. Ammonia one-pager, April 2019
23. FLEXnCONFU. Concept. Web page consulted June 2021
24. General Electric. GE and IHI sign agreement to develop ammonia fuels roadmap across Asia, June 22, 2021

25. Intermountain Power Agency. IPP renewed. Web page consulted June 2021
26. Note that the five-year intervals specified in the phasing correspond with the recommended maintenance intervals for utility-scale combustion turbines. See TWI, Factors influencing maintenance intervals for gas turbines. Web page consulted June 2021
27. Morehouse C (2019) Utility dive. Natural gas plant replacing Los Angeles coal power to be 100% hydrogen by 2045: LADWP, Dec 12, 2019
28. NS Energy. Nuon Magnum Power Plant. Web page consulted June 2021
29. Vattenfall. Vattenfall aims for carbon-free gas power. Blog post, April 23, 2018
30. Gas Unie. Magnum Power Station. Web page consulted June 2021
31. Shiozawa B, op. cit.
32. Olmer N et al (2017) Greenhouse gas emissions from global shipping, 2013–2015 (figure 3). ICCT, Oct 2017
33. IMO. Historic background. Web page consulted May 2021
34. Lloyd's Register. UMAS, Zero-emission vessels 2030. How do we get there? (2017)
35. IOR Energy. Fuel energy density. Web page consulted May 2021
36. Valera-Medina A et al (2018) Ammonia for power. Prog Energy Combust Sci 69:63–102
37. Lloyd's Register. World first for liquid hydrogen transportation. Web page consulted May 2021
38. Ibid.
39. Verkamp FJ et al (1967) Ammonia combustion properties and performance in gas-turbine burners. In: Symposium (International) on combustion, vol 11, issue 1
40. Laursen RS, MAN Energy Solutions. Ship operation using LPG and ammonia as fuel on MAN B&W dual fuel ME-LGIP engines, delivered at the NH3 fuel conference, Oct 2018
41. Crolius S. New coalition plans to build offshore green fueling hubs. Ammonia Energy, June 13, 2019
42. Brown T. Ammonia-fueled ships: entering the design phase. Ammonia Energy, Dec 12, 2019
43. For a representative data point see Elcogen, How SOFC can change the world. Web page consulted May 2021
44. For a representative data point see Green Car Congress, Ballard launches high-power density fuel cell stack for vehicle propulsion; 4.3 kW/L; Audi partner, Sept 15, 2020. Web page consulted May 2021
45. Barelli L et al (2020) Operation of a solid oxide fuel cell based power system with ammonia as a fuel: experimental test and system design. Energies, Nov 24, 2020
46. CMR Prototech. Prototech awarded contract to supply 2 MW zero-emission ammonia fuel cell module, Jan 23, 2020
47. Fuel Cells Works. CMB and Anglo Belgian Corporation (ABC) launch latest hydrogen-powered engine, Sept 18, 2020

Part II
Solar Heat Storage

Chapter 5
Solar Thermal Energy Conversion System

Kazuo Yoshida and Minoru Yuasa

Abstract High-temperature solar thermal energy will be the most promising energy source for hydrogen production by pyrolysis of water. It is well known that such high-temperature energy is currently widely used as solar thermal power generation. High-temperature heat can be obtained from sunlight by concentrating and converting the light coming straight from the sun, which is called direct normal irradiation. In general, the higher the light concentration, the higher the temperature. There are four types of collector technology currently in widespread use, but the parabolic trough and power tower are typical. It is very important to note that the characteristic of the solar thermal power generation is a decoupling of the time zone for obtaining the energy of the sun and the use of the energy. The decoupling can be accomplished by means of thermal energy storage system at a relatively low cost. In the daytime, electricity is generated by using solar thermal energy from the solar collector, and surplus thermal energy is stored in the storage system. And the electricity can be supplied by using the stored thermal energy during nighttime. It is important that the storage system of sufficient capacity enables an around-the-clock generation. In hydrogen production, stable operation is crucial due to the heat supply around the clock. Therefore, it is considered that the technology of solar heat acquisition and its storage can greatly contribute to hydrogen production.

Keywords Solar thermal energy · Concentrating solar power (CSP) · Solar energy concentration · Thermal storage

K. Yoshida (✉)
The Institute of Applied Energy, Shimbashi SY Building, 14-2 Nishi-Shimbashi 1-Chome, Minato-ku, Tokyo 105-0003, Japan
e-mail: yoshida@iae.or.jp

M. Yuasa
Chiyoda Corporation, Minato Mirai Grand Central Tower, 4-6-2, Minatomirai, Nishi-ku, Yokohama 220-8765, Japan

K. Aika and H. Kobayashi (eds.), *CO_2 Free Ammonia as an Energy Carrier*,
https://doi.org/10.1007/978-981-19-4767-4_5

5.1 Introduction

There exist multiple hydrogen production processes required to synthesize ammonia, but using high-temperature heat such as thermal decomposition of water is also one of the important routes. As we aim to reduce carbon dioxide, high-temperature heat needs to be obtained from renewable energy. Considering the enormous amount of energy and sustainability, the most important high-temperature heat source derived from renewable energy is probably solar heat. In this chapter, we would like to explain the solar thermal power generation technologies that have already been put into practical use as a technology related to high-temperature solar heat supply.

5.2 Technological Bases of Solar Thermal Energy

Solar heat has been widely used by humankind since ancient times, but the required technology differs depending on the operating temperature. The most common form of using low-temperature solar heat supply is probably a solar water heater placed on the roof of a house or apartment. In the case of this system, it consists of a flat plate that converts sunlight into heat and a tube that flows water in combination with it. Hot flat plate illuminated by sunlight transfer heat to water flowing in tube attached to the plate. The temperature of the water easily attained from 50 to 70 °C. On the other hand, if a high temperature is required that causes something to burn up by using sunlight, another type of a device is required. An interesting example is the Olympic torch, which is ignited by sunlight collected by a parabolic dish mirror. To give a more familiar example, we think everyone did it as a kid, but when you put the paper in the focus of the magnifying glass, it burns immediately.

The difference between supplying hot water and getting a high temperature that burns paper is a "concentration" that collects the sunlight into a small area. It can be seen that high-temperature heat can be obtained by increasing the light flux intensity using something like a parabolic mirror or magnifying glass. There are many solar thermal power plants in the world that convert sunlight into high-temperature heat to generate electricity. Many reflectors are used at these power plants to collect sunlight and convert it into heat to generate electricity. When the concentration is high, a large amount of energy is given to a small area, and small heat loss resulting from the small area, in turn, high temperature can be obtained. The following equation shows the efficiency of the collector that collects the sunlight and converts it into heat.

$$\eta_{\mathrm{col}} = \eta_{\mathrm{opt}} - \frac{\left[h_c(T_r - T_a) + \varepsilon\sigma T_r^4\right]}{C I_a}$$

where,

η_{opt} Optical efficiency
h_{c} Heat transfer coefficient

T_r Temperature on receiver surface in Kelvin
T_a Ambient temperature in Kelvin
ε Emissivity
σ Stefan-Boltzmann constant
C Concentration
I_a Solar flux intensity on receiver surface.

The first term of the equation expresses the optical efficiency which is the ratio of the amount of absorbed light onto receiver surface reflected from the mirrors of collector to the amount of sunlight entering the collector. The second term is related to heat loss, which is obtained from the ratio of the heat loss caused by radiation and convection from the receiver to the amount of sunlight entering the collector. As is clear from the equation, the efficiency of the collector can be increased by reducing the term related to heat loss, but for that purpose, the emissivity or the condensing degree should be increased. As is clear from the equation, the collector efficiency increased by reducing the term related to heat loss, but for that purpose, the emissivity and/or the concentration should be increased.

Figure 5.1 shows the relationship between collector efficiency and operating temperature using the above equation with different concentrations. In this figure, for the sake of simplicity, only the heat loss caused by radiation, which has a large effect at high temperatures, is considered. Since the heat loss due to radiation increases in proportion to the fourth power of the absolute temperature of the receiver, the higher the temperature of the receiver, the greater the radiant heat loss compared to the convective heat loss. As is clear from the figure, in the region where the concentration is low, the efficiency decreases immediately as the temperature rises. However, when the concentration is high, the decrease in efficiency is small even if the temperature rises.

Figure 5.2 depicts the system efficiency by multiplying the temperature change of the collector efficiency and the Carnot efficiency shown in Fig. 5.1. Under the condition of concentration being unity, the system efficiency peaks at low temperature, but the value of the efficiency is low. As the concentration increases, the peak efficiency shifts to the higher temperature side, and the value of efficiency at the peak also increases. As the temperature rises, the system efficiency improves, but it is not necessarily a good idea to raise the operating temperature. This is because higher temperatures require more expensive materials to improve efficiency, which can lead to a sharp increase in CAPEX in turn generation cost rise. For example, considering the power generation by a steam turbine generally used in solar thermal power generation, if the steam temperature is used under subcritical conditions of 530–540 °C, generally inexpensive metal materials can be used, so the equipment cost is low and the power generation cost is low. In sharp contrast supercritical steam conditions give rise to higher efficiency but need expensive high-temperature alloy, therefore generation cost should be increased. Therefore, it is necessary to select the optimum value for the concentration according to the application. A solar furnace using a parabolic mirror is used to evaluate highly heat-resistant materials used in

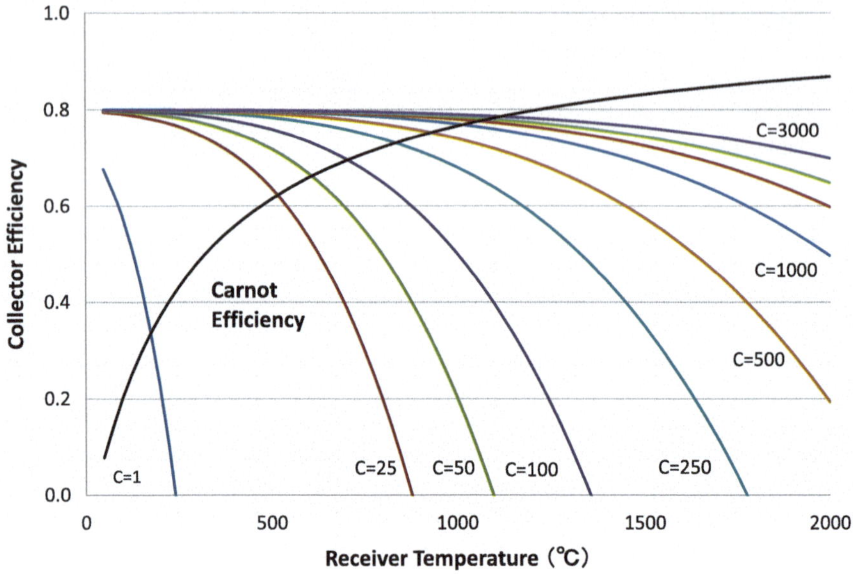

Fig. 5.1 The effect of concentration on collector efficiency as a function of receiver temperature

space, but it has a very high concentration of light and can evaluate the materials at high temperatures such as several thousand degrees.

When high-temperature heat is required, such as solar thermal power generation and high-temperature solar heat utilization, it is not possible to seek a plant efficiently at low cost anywhere in the world. The area suitable for solar thermal power generation is the area called a Sunbelt, as shown in Fig. 5.3 [1]. A characteristic of these areas is that the air is dry and hot, resulting in high *Direct Normal Irradiation (DNI)*. DNI is the sunlight that comes directly from the sun, and the vectors of the sunlight are almost aligned. The amount of DNI is an important index for efficiently collecting a large amount of solar thermal energy at low cost because it can only be focused by light with a uniform vector. On the other hand, the "scattered light" in which the light emitted from the sun is scattered by water vapor or its very small particles in the atmosphere and/or aerosols has a random vector, but such light cannot be collected. DNI is low in areas with high levels of atmospheric moisture and fine particles such as aerosols, and even if a solar thermal power plant is constructed, the cost of power generation will be higher than that of the Sun Belt. The Sunbelt is located in an area called Horse latitude in meteorology, and deserts or arid zone are likely to form in dry, high-temperature areas. The reason why the sun belt occurs in the area shown in Fig. 5.3 is due to the general circulation of the atmosphere. Sunbelt is a region where dry air comes down from the sky and becomes hot due to adiabatic compression as it approaches the ground. Therefore, the high DNI region results from clean atmosphere having less "dirt" such as moisture and aerosols.

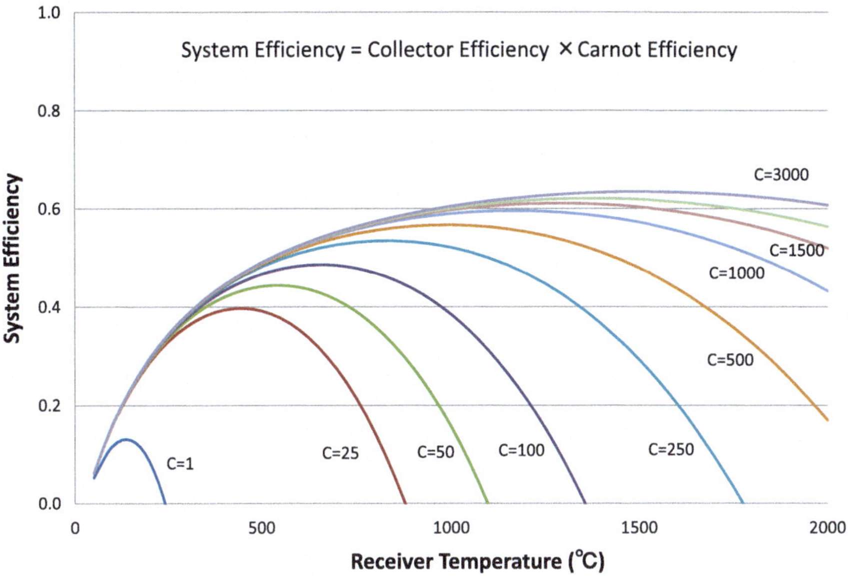

Fig. 5.2 The effect of concentration on system efficiency as a function of receiver temperature

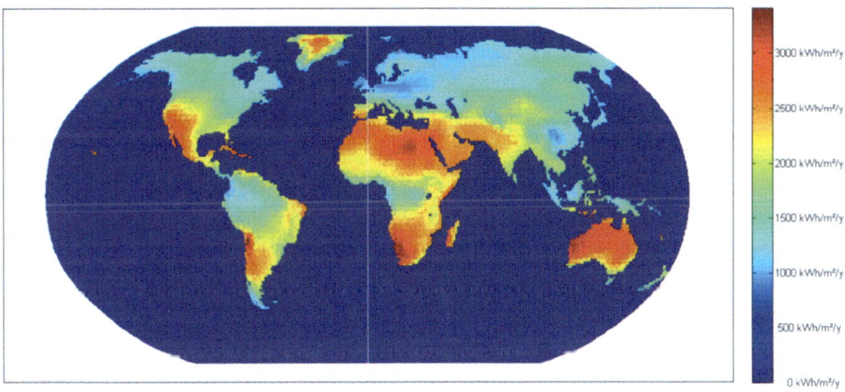

Fig. 5.3 Direct normal insolation map (kWh/m^2/annual) [1]

5.3 Technology

There are four main types of collectors in concentrating solar power and high-temperature solar thermal utilization, namely parabolic trough and parabolic dish using curved mirrors in the shape of parabola, and a linear Fresnel and power tower system using off-axis reflection as shown in Fig. 5.4 [2]. A shape of the reflected image gives another classification, namely line focus and point focus.

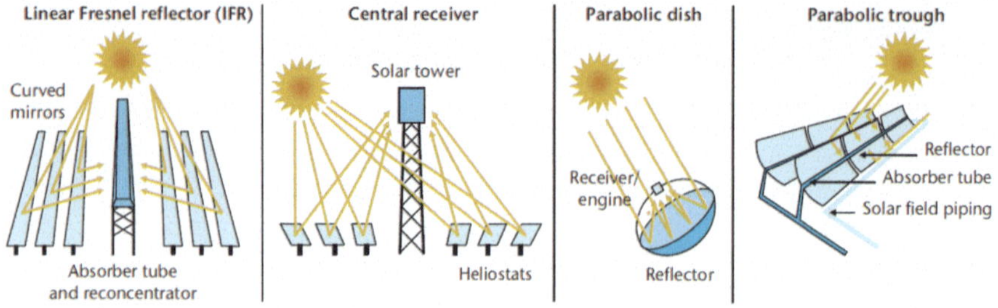

Fig. 5.4 Main CSP technologies [2]

Another collector classification is based on the number of axes that track the sun. Line focus collectors that track the sun on one axis include the parabolic trough and the linear Fresnel. The uniaxial tracking collector tracks only the diurnal motion of the sun and does not track the changes in the altitude of the sun. Therefore, it is easily affected by seasonal fluctuations and the latitude of the installation location. Point focus systems that track two axes include the power towers (or central receiver system) and the parabolic dish. These systems can track both diurnal motion and altitude, so the effects of seasonal fluctuations and latitude are small.

Here, the parabolic trough and the power tower will be described as typical examples of line focus systems and point focus systems, respectively because they are common technologies in commercial plants.

5.3.1 Parabolic Trough

The parabolic trough is a collector consisting of a reflector having a parabolic cross-section and a receiver tube at its focal point as shown in Fig. 5.5 [3]. A principle of the concentration of light using a paraboloid is that reflected incoming light parallel to the central axis of the paraboloid on its surface gathers at the focal point definitely. Therefore, it is easily possible to collect solar heat constantly by moving the central axis of the parabolic collector from east to west following the diurnal motion of the Sun. Obtained thermal energy by the collector elevates the temperature of a *heat transfer fluid (HTF)* flowing in the receiver tube. In commercial solar thermal power plants using the parabolic-trough collector, the HTF is heated by a large number of collectors installed on a large site, collected in one place, then a steam turbine generates electricity by using high temperature and high-pressure steam generated by a steam generator.

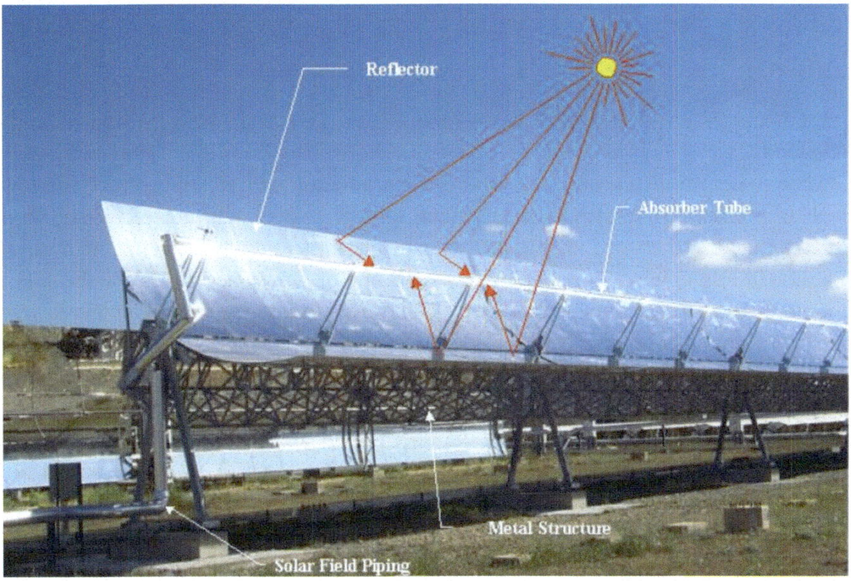

Fig. 5.5 The parabolic-trough collector principle [3]

5.3.1.1 Collector

Parabolic-trough collectors in actual plants are aligned in the north–south direction and track the sun moving from east to west in a single axis. However, since the change in the altitude of the sun cannot be tracked, the change in the solar energy obtained by the collector will be significantly different between summer and winter. Also, at high latitudes, the amount of energy that collectors can obtain is smaller than at low latitudes.

It is apparent that the parabolic-trough collector must always gather reflected sunlight onto the receiver tube regardless of collector rotation in accordance with the diurnal motion of the sun. Therefore, the rigidity of the collector frame is quite important, and both lightweight and highly rigid frame structure is required simultaneously.

The parabolic-trough collectors used in solar thermal power plants generally have apertures from 5.77 to 7.51 m, but they have been enlarged to reduce capital costs. The standard receiver used in common collectors incorporates metal tubes of 70 mm in diameter, and some are bigger than that. Therefore, the concentration of solar light is as low as about 80–100, and as described above, it is suitable for operation at a relatively low temperature when considering only the magnitude of the concentration. However, the receiver contains multiple technologies that reduce heat loss, enabling efficient high-temperature heat supply.

5.3.1.2 Receiver

The receiver used in the large trough used for power generation is a vacuum tube, a metal tube is covered with a glass tube having high transparency, and annulus between these tubes is evacuated. As a result, heat conduction from the metal tube, convective heat transfer, and heat loss due to radiation can be suppressed to a low level. In addition, the surface of the metal tube is coated with a selective coating which enhances to absorb visible and near-infrared in sunlight but reduces infrared radiation of longer wavelength than around 2500-μm simultaneously. With the aid of the selective coatings heat loss caused by radiation decreases as well as an increase in the absorption of sunlight. The detailed structure of the receiver and the structure and performance of the selective coating will be described in detail in another chapter. The receiver, which consists of a metal tube and a glass tube, repeats thermal expansion and contraction every day depending on whether the sun is shining or not. Since the coefficient of thermal expansion differs between metal and glass, both ends of the receiver have a bellows structure to absorb thermal expansion and contraction.

5.3.1.3 Heat Transfer Fluid

A common HTF used in parabolic-trough solar thermal power plants is a eutectic mixture of biphenyl ether and diphenyl. The advantages of this organic fluid have a low viscosity and low pour point. In contrast, the drawback of this HTF is that it is flammable and has a high vapor pressure. In addition, another important disadvantage is that it gradually pyrolysis and releases hydrogen during use at high temperatures. Some of the released hydrogen penetrates metal tubes and enter to the annulus between the glass tube and the metal receiver tube. Since hydrogen has high thermal conductivity, and when it enters the vacuum annulus, the heat loss increases considerably. To suppress this problem, there is a hydrogen getter in the annulus in order to absorb the hydrogen.

Another very important disadvantage of this HTF should be relatively low operating temperature of about 400 °C at most. Resulting steam temperature is not enough steam condition to operate steam turbine at high efficiency. Therefore, endeavor to develop alternative HTFs has been done in order to increase operating temperature of the plants. Water was one of candidate of HTFs considered in the past. Water is directly heated in the receiver tube to generate superheated steam, called a *Direct Steam Generation (DSG)*. In the case of the DSG, unlike plants that use the synthetic oil, a steam generator is not required, therefore, plant costs can be reduced. However, since water evaporation process takes place in the receiver tube, water flows from liquid to gas phases via gas–liquid phase. The phase change of the water results in various problems to the receiver tube. The problems related to the receiver tube were solved by changing wall thickness of metal tube and controlling water flow conditions. At the same time, the thermal storage system incorporated in the DSG plant is complicated and costly because it involves an evaporation process of water, which is another reason why it is not widely used.

Currently, nitrate-based molten salts are considered to be a promising option for raising the operating temperature. Commonly used molten salt in solar thermal power plants is a solar salt, which is a two-component system of sodium nitrate (60 wt%) and potassium nitrate (40 wt%). The use of solar salt enables to high-temperature operation as high as 565 °C. Plants using the solar salt have already been commercialized as power tower plants. It is well known that the solar salt not only acts as an HTF but also as a thermal storage material. The solar salt heated by solar energy is stored directly in the tank and released to generate electricity whenever needed. The details of the storage system will be described later. The commercialization of the parabolic-trough plant using solar salt has been delayed compared to the power tower, but it is expected that its introduction will expand in the future.

5.3.2 *Power Tower*

In a power tower plant, a large number of heliostats placed around a tower reflect sunlight onto a receiver located atop of the tower. The heliostat is a device that tracks the sun on two axes and always reflects the sunlight on the surface of the receiver regardless of the movement of the sun. Since the reflected light can be collected from a large number of heliostats for one receiver, the concentration of solar light is as high as 500–1000, and it is inevitably easy to obtain high temperatures. The receiver converts concentrated sunlight to high-temperature heat and transfers it to HTF flowing in receiver tubes.

5.3.2.1 Heliostat

In recent large plants, heliostats are generally placed 360° around the tower as shown in Fig. 5.6 [4], but as the size of the plant increases, the distance between the receiver and the heliostat located at the fringe of the heliostat field attains more than 1500 m. Therefore, a high-precision heliostat should be required to reflect sunlight onto small receiver surface because of such a long-distance between them. Such heliostat needs precise shape, tracking accuracy, frame rigidity, and so on, but these requirements raise manufacturing costs. The amount of solar energy reflected by the heliostat entering the receiver must reach the amount of energy that meets the power generation capacity. Rather than using a large number of heliostats with low manufacturing costs and low accuracy, using a smaller number of heliostats with high accuracy at a certain cost may result in lower overall equipment costs. Therefore, the accuracy of the heliostat must be balanced with the overall capital cost.

Another important fact to diminishes the efficiency of the heliostat located far from the receiver is enlargement of image caused by the size of the sun. The sun has a viewing angle of about 32 min when viewed from the ground, from which sunlight enters the reflector in a conical shape and is reflected with the same shape.

Fig. 5.6 Aerial view of a typical power tower plant (Gemasolar, Spain) [4]

Therefore, the longer the distance from the heliostat to the receiver, the wider the reflected light image since the light reflected by the reflector goes out in a conical shape. The sunlight reflected by one point of the reflector spreads to approximately 8 m in diameter at a distance of 1 km. Furthermore, the magnitude of image becomes even larger when the surface roughness and waviness of mirror surface, and errors of frame. In addition to this, the larger the reflector area of the heliostat, the larger the spread of reflected sunlight image in the receiver portion. The smaller the reflected image is, the higher the light intensity is, which also reduces the heat loss from the receiver and increases the thermal efficiency. The reflective surface of the heliostat is concave to reduce the spread of reflected light on the receiver surface.

Heliostats vary in size widely from 1 to 170 m^2. Currently, heliostats used for large-scale power generation are relatively large, with a low cost of about 60–170 m^2. On the other hand, small heliostats are relatively expensive. However, since the spread of reflected light image on the receiver surface is small, it is suitable for obtaining high temperatures. It is often used in processes that carry out high-temperature thermochemical reactions such as fuel production.

5.3.2.2 HTF and Receiver

The receiver of the power tower has a different structure depending on the properties of HTF. The HTF used in the power tower includes molten salt, water/steam, and air, but recently liquid sodium has also been used. In the case of liquid HTF such as the molten salt, a bundle of metal tubes is generally arranged in a plane, and the HTF flows in parallel through the tubes. The surface of this receiver panel is exposed to the intensive reflected sunlight from a large number of heliostats. The generated heat on the receiver surface is transferred to the HTF flowing through the tube, but in order to prevent local heating of the receiver surface, the reflected flux from the heliostat should be uniform to some extent. The target of the heliostat has been adjusted. The smaller the receiver, the better in order to increase concentration and to reduce the heat loss. HTF being liquid phase under the operating condition such as solar salt has less difficulties related to heating in the receiver tube, unlike water which undergoes phase change.

In order to increase the efficiency of the receiver, it is better to have high concentration of solar light. However, the intensity of the reflected light flux that can be applied to the receiver tube is naturally determined by the thermal conductivity of the HTF flowing in the tube. Common molten salt used in solar thermal power plants called solar salt has thermal conductivity as low as about 0.5 W/mK. On the other hand, the thermal conductivity of liquid sodium is as high as 134 W/mK at room temperature. Therefore, when sodium is used as an HTF, it is possible to provide a reflected light flux having a higher concentration than the molten salt. Sodium is a liquid in a wide temperature range and has excellent physical properties as an HTF. In particular, the density is quite low and the power to pump up to the receiver atop of the tall tower should be small. In 1980s efforts had been made to use sodium as the HTF, but it was discontinued due to an accident. However, in recent years there has been a move in Australia to use liquid sodium again as the HTF for power towers, and the plant is actually in operation.

Plants that use water/steam as the HTF are still in operation, but new construction is not active. One of the reasons for this is thought to be the behavior of water/water vapor in the receiver tube. That is, the water is heated in the tube to raise the temperature, evaporate, and become superheated steam. The thermal conductivity of water and that of water vapor are quite different. When the water flowing in the tube is liquid phase, it can withstand intensive solar flux reflected by a number of heliostats, but after vaporizing the water it is not able to do because of thermal conductivity drop by an order of magnitude. Therefore, it is difficult to control the flux intensity and its distribution across the receiver surface. Another reason why there are few new plants that use water/steam as the HTF is that the phase change of the heat medium complicates the heat storage system and leads to an increase in cost.

Another type of receiver is applied when air is used as the HTF because of a low thermal conductivity of air. When air is heated by the tubular receiver, sufficient heat cannot be applied due to its low thermal conductivity. Since the strong flux hits only one side of the receiver surface, applying a strong flux causes a large temperature gradient in the tube and the tube breaks. On the other hand, if the flux is weak enough not to damage the tube, the air cannot be heated sufficiently. Therefore, as shown in Fig. 5.7, a volumetric receiver is used in which a strong flux is applied to a porous material made of ceramics or high-temperature alloys, and air is passed through the flux to heat the porous material.

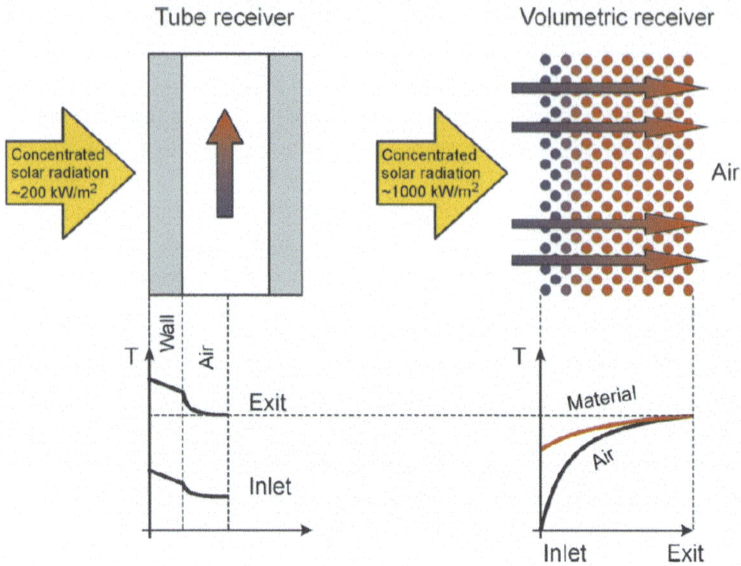

Fig. 5.7 The volumetric receiver principle compared to a tublar receiver [5]

5.3.3 *Thermal Energy Storage*

Thermal energy storage is one of the characteristics of solar thermal power generation. The thermal energy collected by various methods can be stored by a relatively inexpensive method, and stable/demand-based power can be generated. Initially, the amount of unstable renewable energy introduced was small, so the effect to the grid was small and it was not a problem. But the effect of the instability on the grid is due to the rapid spread of renewable energy, especially of photovoltaic power generation, in recent years. This feature has attracted attention when it can no longer be ignored, and much research has been conducted in recent years.

The simplest method of thermal energy storage used in solar thermal power generation is to store HTF as it is. This method is used in most cases when molten salt is used as the HTF. On the other hand, when synthetic oil is used as HTF, a method of exchanging heat from HTF to molten salt, which is a heat storage medium, is adopted. The reason is that synthetic oil requires a pressure vessel to store because its vapor pressure is high at 400 °C (heat storage temperature), and that synthetic oil is generally more expensive than molten salt.

The two-tank system, as shown in Fig. 5.8 is used in most cases as a method of storing molten salt. The two-tank system consists of a low-temperature molten salt tank and a high-temperature molten salt tank. The low-temperature molten salt pumped from the low-temperature molten salt tank is directly or indirectly heated by solar energy and stored in the high-temperature molten salt tank. The high-temperature molten salt stored in the high-temperature molten salt tank is pumped

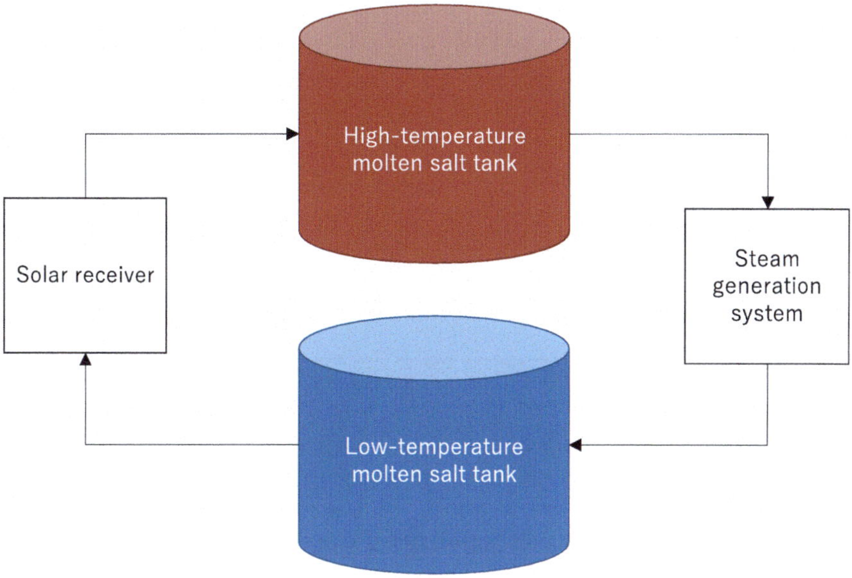

Fig. 5.8 Two-tank thermal energy storage system

to the steam generation system according to the electric power demand, and heat energy is given to water/steam, and after the temperature drops, it is returned to the low-temperature molten salt tank.

One of the simplest questions when storing heat energy in a tank is that the stored energy is reduced due to heat dissipation from the tank. The ratio of the amount of heat lost due to heat dissipation to the amount of heat storage can be easily calculated from the size (diameter, height) of the tank, heat insulation performance, tank temperature, and outside air temperature. For example, if the tank diameter is 20 m, the tank height is 15 m, the heat insulation performance is 100 W/m^2, the outside temperature is 20 °C, the high-temperature tank is 550 °C, and the low-temperature tank is 290 °C, the amount of heat lost in 24 h is less than 1% of the amount of heat storage.

There is a heat storage method using a thermocline as shown in Fig. 5.9 as research and development of heat storage technology. In a two-tank system, the molten salt moves back and forth between the two tanks, so the amount of molten salt that can be stored is equivalent to one tank after all. In other words, the equipment is not being used efficiently because two tanks cannot be filled at the same time. Focusing on this point, thermocline heat storage system is a method aimed at realizing a heat storage amount equivalent to that of a two-tank system in one heat storage tank. In this method, there is a high-temperature molten salt in the upper part and a low-temperature molten salt in the lower part in the tank, and the amount of heat storage can be changed by raising or lowering the boundary between the high-temperature molten salt and the low-temperature molten salt called a thermocline layer. Further,

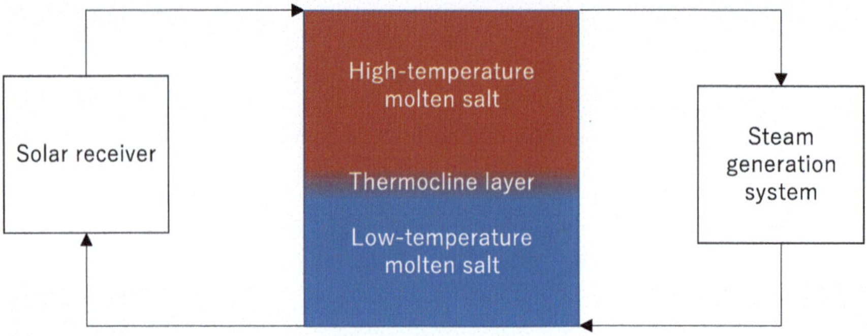

Fig. 5.9 Thermocline thermal energy storage system

it is also considered to further reduce the heat storage cost by filling the tank with a solid heat storage medium which is cheaper than the molten salt.

In addition to the thermocline system, heat storage technology that utilizes latent heat and chemical reaction heat is being researched and developed. Latent heat storage mainly utilizes thermal power generation by Gemasolar seems to be particularly important.

The latent heat of an object during a phase change between the solid phase and the liquid phase. As an example, there is a capsule of a solid metal or ceramic having a high melting point in which a salt having a melting point slightly lower than the heat collection temperature is encapsulated. By using this capsule as a substitute for the filler of the thermocline system, it is expected that the amount of heat storage will increase and the heat output will be stabilized. Chemical heat storage utilizes a thermochemical reaction and is a method that can be expected to have the same effect as latent heat storage.

References

1. Breyer C, Knies G (2009) Global energy supply potential of concentrating solar power. SolarPACES
2. International Energy Association (2014) Technology roadmap, solar thermal electricity
3. Plataforma Solar de Allmerríía (2001) Annual report
4. Burgaleta JI, Arias S, Ramirez D (2011) Gemasolar, the first tower thermosolar commercial plant with molten salt storage. SolarPACES
5. Fend T (2010) High porosity materials as volumetric receivers for solar energetics. Optica Appl XL(2):271–284

Chapter 6
Solar Selective Absorbers Based on Semiconducting Thin Films

Yoshiki Okuhara, Masasuke Takata, Takuhito Tsutsui, and Kazuto Noritake

Abstract In order to produce CO_2 free hydrogen using thermal energy at high temperatures, concentrated solar power systems are one of the most promising candidates. For mass production of hydrogen, linear focusing systems consisting of parabolic trough mirrors are suitable because of the high scalability of the power plant capacity. Although most of the conventional linear focusing systems have been operated under 400 °C, higher temperatures are desirable for more efficient solar-thermal energy conversion and hence for lower cost of hydrogen production. A key component in the linear focusing systems is solar receiver tubes with "solar selective absorbers" by which concentrated solar radiation is absorbed and thermal infrared radiation is inhibited at the operating high temperatures. This chapter introduces novel solar selective absorbers for efficient and durable solar-thermal energy conversion operating at 650 °C. The solar absorptance has been enhanced by band-edge absorption of semiconducting β-$FeSi_2$, and the infrared emissivity at high temperatures has been reduced by thermally stabilized silver (Ag). The optimally designed multilayers achieved the averaged solar-thermal conversion efficiency of 87% between 300 and 650 °C, and ensured superior thermal durability over 25 years under high temperatures.

Keywords Solar · Absorber coatings · Silicide · Silver · Semiconductor

Y. Okuhara (✉) · M. Takata
Japan Fine Ceramics Center, 2-4-1 Mutsuno, Atsuta-ku, Nagoya 456-8587, Aichi, Japan
e-mail: okuhara@jfcc.or.jp

T. Tsutsui · K. Noritake
Toyota Industries Corporation, Kyowa Plant, 8 Chaya, Kyowa-cho, Obu 474-9180, Aichi, Japan

K. Aika and H. Kobayashi (eds.), *CO_2 Free Ammonia as an Energy Carrier*,
https://doi.org/10.1007/978-981-19-4767-4_6

6.1 Background

Sustainable solar energy is expected to be the best primary energy resource to produce CO_2 free hydrogen and ammonia. There are various methods to split H_2O using solar energy such as artificial photosynthesis, water electrolysis combined with photovoltaic cells, and so on [1]. The authors believe that the solar-thermal energy conversion using concentrated solar power (CSP) systems is one of the promising ways to produce hydrogen because of the expected highest energy conversion efficiency and the lowest cost including energy storage systems for 24 h operation [2]. Based on the thermal energy, hydrogen can be produced by not only water electrolysis combined with efficient steam turbine generators but also thermochemical reactions. An iodine–sulfur process is a thermochemical cycle to produce hydrogen from H_2O using thermal energy at a high temperature above 850 °C. If considerable efforts to reduce the reaction temperature below 650 °C using membrane techniques come to fruition [3], efficient production of CO_2 free hydrogen can be expected using the CSP systems combined with the thermochemical cycle.

The CSP systems have been mainly classified according to their solar-concentrating methods, i.e. point focusing CSP like a solar tower or parabolic dish systems, and linear focusing CSP like linear Fresnel or parabolic trough collectors as shown in Fig. 6.1 [4]. Although the point focus CSP can produce higher temperatures because of their higher concentrating ratio, the power plant capacity is restricted due to the solar-concentrating limitation. In contrast, the linear focus CSP has no limitation of the solar field capacity, but the temperature range is relatively lower such as under 400 °C. The recent development of the linear focusing CSP systems has been focused on enhancing the operation temperature to 550 °C [5]. Although increasing the operating temperature is a possible approach to enhance the thermodynamic efficiency and to combine with the thermochemical cycle for hydrogen production, ensuring the solar-thermal conversion efficiency and the thermal durability in a harsh environment is a challenge for some components in the linear focusing CSP plants. The highest temperature in the system is produced at the solar receiver tubes under concentrated solar radiation. Thus, the receivers must achieve high conversion efficiency and long-term stability at high temperatures.

As shown in Figs. 6.1 and 6.2, the evacuated receiver tubes used in proven linear focusing systems require solar selective absorber coatings on stainless steel tubes, through which both high and low absorptance in the visible-near-infrared (NIR) and NIR-infrared (IR) regions, i.e., spectrally selective absorption [6–18], should be achieved for efficient solar-thermal conversion. Assuming the modeled absorptance spectra with various gradients of transition at the cut-off wavelength $\lambda_{\text{cut-off}}$, it was confirmed that the selective absorptance curve should have a steeper transition to achieve lower infrared emissivity ε as well as higher solar absorptance α, for enhancing the efficiency η especially at higher temperature [14]. In addition to the steepness, the cut-off position strongly influences the α and ε. If the $\lambda_{\text{cut-off}}$ was defined as the wavelength above which the absorptance became less than 50%,

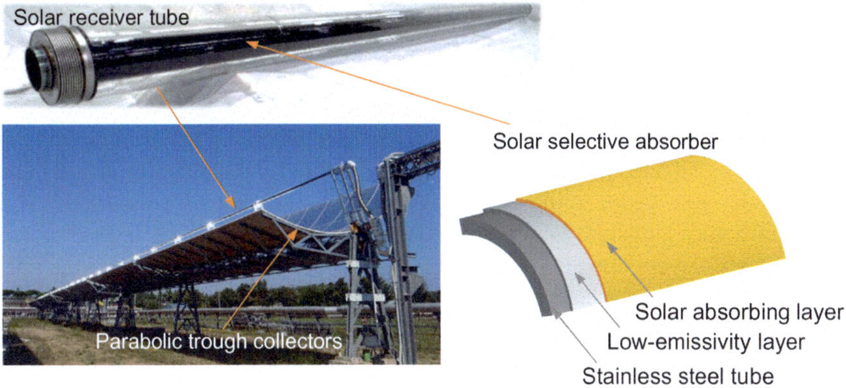

Fig. 6.1 Solar receiver tubes in linear focusing solar power systems with parabolic trough mirrors and a schematic diagram of solar selective absorbers

the ideal $\lambda_{cut\text{-}off}$ to provide the highest efficiency at 650 °C is around 1760 nm in wavelength.

Solar selective absorbers are composed of solar absorbing layers and low emissivity layers on stainless steel tubes as depicted in Fig. 6.1. Conventional solar absorbing layers mostly consist of cermet materials composed of SiO_2 or Al_2O_3 matrix and metallic Mo or copper (Cu) particles, stacked on a low emissive metallic surface with high infrared reflectance [6–13]. Great efforts have been made to improve the solar selective absorbing behavior and hence to enhance conversion

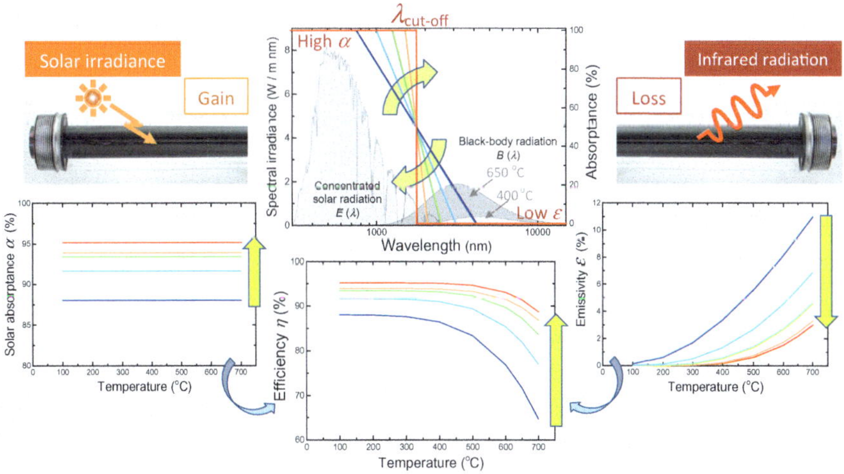

Fig. 6.2 Modeled selective absorbing spectra with various gradients of cut-off behavior. Steeper gradients of the cut-off spectra provide higher solar absorptance α and lower emissivity ε, and hence higher conversion efficiency η. https://doi.org/10.1063/1.5067227

efficiency, but novel solar absorbing layers instead of the conventional cermet layers are necessary to break through the plateaued selective absorbing performance.

In terms of low-emissivity layers, innovative material design is required to overcome the existing metallic base, because infrared radiation losses become more pronounced at higher temperatures. It is well known that the lowest emissivity can be achieved by silver (Ag) surface, but the utilization of the Ag thin films has been limited under 400 °C due to the thermal instability at a higher temperature. Instead of Ag, refractory tungsten (W) or molybdenum (Mo) surfaces have been used for low-emissivity layers operating above 500 °C, but their infrared optical properties have serious problems at high temperatures as described in Sect. 6.2.1.

In this chapter, for integrating the solar-thermal energy with CO_2 free hydrogen production, the target temperature generated by solar selective absorbers was set at 650 °C with the prospect of future progress in thermal water splitting technologies such as thermochemical cycles at the temperature. In order to achieve more efficient and durable solar-thermal conversion at the unexplored temperature of 650 °C, a novel approach to design advanced solar selective absorbers has been developed.

6.2 Design of Solar Selective Absorbing Multilayer

6.2.1 Optimum Low-Emissivity Layer

At high temperatures, the solar-thermal conversion efficiency of receiver tubes is dominated by the thermal emittance of their metallic bases, but only a few reports have discussed the infrared emittance measured at actual high temperatures [1, 12, 13]. The measured temperature dependences of refractive index, *n,* and extinction coefficient, *k,* spectra for W, Mo and thermally stabilized Ag films are shown in Figs. 6.3a–c, respectively [15, 17]. The stabilization method used for the Ag film is described later in Fig. 6.4. Based on the optical constants, their absorptance (= emittance = 100%—reflectance) spectra were calculated at each temperature, as indicated in Fig. 6.3. With an increase in the temperature, the *n* and *k* spectra of the W and Mo layers were found to change more drastically than those of the thermally stabilized Ag layer. The temperature dependence of the *n* and *k* spectra could be interpreted by a decrease in the electron relaxation time resulting from the enhanced phonon scattering in the Drude model, suggesting that the influence of temperature on the relaxation time was more pronounced in W and Mo than in Ag. Because of the change in the *n* and *k* spectra, the calculated absorptance spectra of the W and Mo layer also showed a strong temperature dependence. The higher temperature caused an emittance rise of about up to 4.5% in the infrared region and an absorptance drop in the solar spectrum region, as shown in Figs. 6.3a–b. When both the selective absorptance and low emissivity of the W or Mo base were used, a decrease in the cut-off gradient and an increase in the emissivity deteriorates the solar-thermal conversion efficiencies at high temperatures. In addition, the selective absorbing behavior of cermet

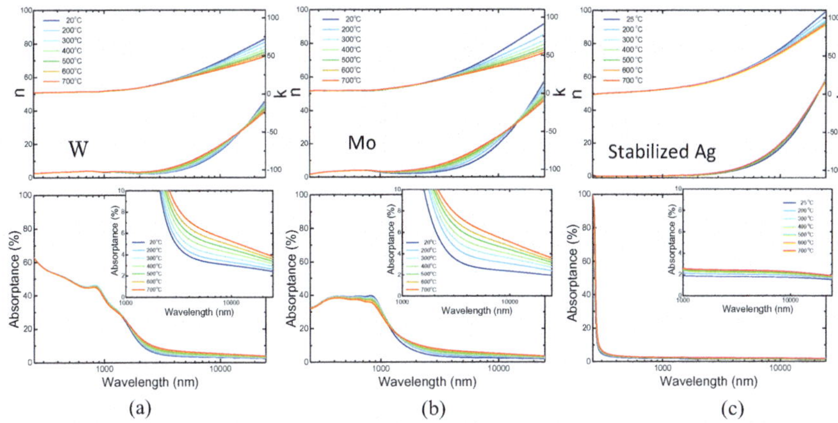

Fig. 6.3 Refractive index *n* and extinction coefficient *k* spectra of **a** tungsten (W), **b** molybdenum (Mo) and **c** thermally stabilized silver (Ag) films. Their optical absorptance spectra were calculated based on these optical constants. https://doi.org/10.1063/1.5117631

layers containing Mo particles is believed to be deteriorated at high temperatures. In contrast, the Ag film exhibited remarkably more stable *n* and *k* spectra (Fig. 6.3c), and the change in the absorptance in the infrared region was limited to 0.6% even at 700 °C. Even though Ag base layers have no solar absorbing effect because of their characteristic absorption at wavelengths shorter than 400 nm, the lowest emissivity even at high temperatures is more significant for efficient photo-thermal conversion.

In order to utilize the lowest stable infrared emissivity of Ag, serious issues of thermal instability of Ag films must be solved. It is well known that the Ag thin films drastically agglomerated and vaporized even below the melting point at 961.8 °C. Although much effort has been dedicated to suppressing the agglomeration of Ag films by adding various kinds of additives into them or through stacking with some layers [19–22], most of their target temperatures were under 500 °C for the manufacturing process of semiconductors.

Thus, more effective additives and stacking structures at higher temperatures must be determined. We have attempted to co-sputter Ag layers with certain additives, such as silicon (Si), aluminum (Al), tungsten (W), molybdenum (Mo), palladium (Pd), niobium (Nb), and tantalum (Ta), on quartz substrates, and to investigate the stability after annealing at 700 °C for 1 h in a vacuum. The formation of pinholes was detected from an increase in the transmittance spectra, and the influence of these impurities on the NIR-IR reflectance spectra of Ag was evaluated. The comprehensive experimental investigation implied that an Ag layer including the Ta additive was the best combination. For co-sputtering of Ag DC 60 W with Ta DC 2 W, which was the lowest power to maintain the Ta plasma, the Ag layers included approximately 0.9 at. % of Ta. The Ag + Ta layers were found to be thermally stabilized, to suppress the agglomeration of Ag layers, as shown in Fig. 6.4a, and maintaining almost zero percent transmittance implied no formation of pinholes after annealing at 700 °C.

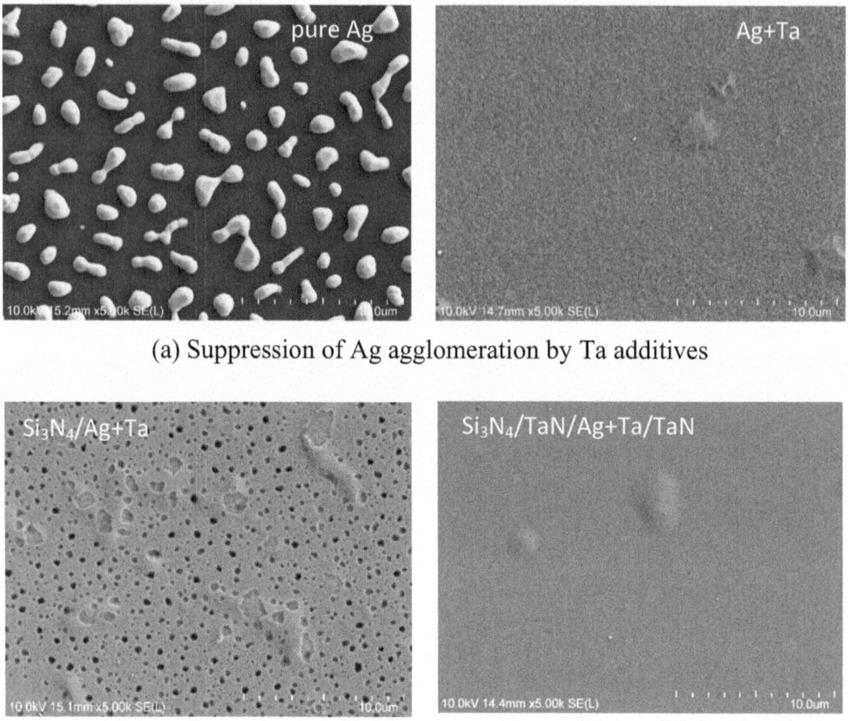

(a) Suppression of Ag agglomeration by Ta additives

(b) Suppression of Ag vaporization by TaN adhesive layers

Fig. 6.4 Surface SEM images of **a** Ag films with and without Ta additives, and **b** Si_3N_4 (50 nm)/Ag + Ta (300 nm) multilayers and Si_3N_4 (50 nm)/TaN (15 nm)/Ag + Ta (300 nm)/TaN (50 nm) multilayers after heating at 700 °C in a vacuum. https://doi.org/10.1016/j.solmat.2019.110304

Moreover, it was confirmed that the small amount of Ta additives had no influence on the NIR-IR reflectance of pure Ag.

Even after suppressing the agglomeration of Ag, vaporization of Ag occurred as shown in Fig. 6.4b. If the Ag + Ta surface was covered with an Si_3N_4 capping layer, then blister formation was observed after heating at 700 °C for 51 h. These voids, observed as dark dots at the V_{acc} of 10 kV in the SEM images, disappeared upon decreasing the accelerating voltage V_{acc} to 3 kV, suggesting that these voids formed not at the topmost Si_3N_4 surface, but rather at the interface between the Si_3N_4 cap and vaporized Ag + Ta layer. We attempted to prevent the vaporization of Ag by stacking with Ta, $TaSi_2$ or TaN layers. These Ta-based compounds were chosen to share the Ta target for stacked layers as well as additives, and to utilize these compounds with a metallic nature. Because metallic bonding is a type of chemical bonding that arises from the electrostatic attractive force between free electrons and positively charged metal ions, it is important to suppress the vaporization or sublimation of Ag atoms from the surface by stacking with refractory metallic layers. Stacking with Ta and $TaSi_2$ layers was also effective in suppressing blister formation, but the TaN layers had the strongest effect of stabilizing the Ag + Ta layer. Although Mardani et al. reported that Ag films on the Ta layer were more stable than those on the TaN layer at 700 °C for 1 h [23], their free Ag surfaces in a vacuum offered no insight into the vaporization. In contrast, the thermal stability, shown in Fig. 6.4b, provides evidence of the suppression of vaporization through the upper and lower TaN layers at both sides of the Ag layer. Furthermore, the utilization of TaN layers has a more crucial advantage than Ta or $TaSi_2$ layers in terms of not only the stabilizing Ag in Fig. 6.4b, but also yielding a diffusion barrier effect, as described in Sect. 6.2.4. Consequently, the basic stacking structure of low-emissive layers could be established as "TaN/Ag + Ta/TaN".

6.2.2 Optimum Solar-Absorbing Layer

As described in Sect. 6.1, progressive studies of solar selective absorbers using cermet layers have been performed to improve the conversion efficiency and thermal stability at 550 °C or lower [6–13], but there have not been many studies on optimizing the absorptance spectra as shown in Fig. 6.2 in terms of the spectral position and steepness of the transition curve suitable for higher temperature. Moreover, the authors have found that duplicated conventional solar absorber "SiO_2/Mo + SiO_2 composites/Mo-base" cannot achieve the ideal selective absorbing behavior expected from the effective medium approximation of the Mo + SiO_2 composites [8], because the Mo nanoparticles dispersed in the SiO_2 matrix formed silicide Mo_3Si after annealing [16]. One of the potential methods for providing steeper cut-off transition is to utilize interband absorption of semiconductors [6], where high absorption only occurs at wavelength ranges shorter than the band edge. The band-edge absorption consistent with the intended cut-off wavelength is expected to offer the highest conversion efficiency, for example, the cut-off wavelength should be approximately 1760 nm for an absorbing operation at 650 °C. Although it has been reported that absorbers using semiconductors such as Si, Ge, PdS, or silicides exhibit lower solar absorptance α and higher thermal emissivity ε than cermet-based absorbers [6–13], the potential of semiconducting absorbers should be reconsidered through more accurate evaluations. The previous issue of silicide formation in the SiO_2 matrix gave us an important clue to select the semiconducting materials [16]. As a topmost layer ensuring low reflectance loss, SiO_2 is the best candidate due to the low refractive index $n = 1.5$. Thus, the semiconducting materials directly contact SiO_2 at the interlayer or in the SiO_2 matrix to form the composite layers, so the semiconducting materials must have superior chemical compatibility with SiO_2 at high temperatures. Therefore, "semiconducting silicides" are the optimum material combined with SiO_2 to ensure stable band-edge absorption with high thermal durability. Harding investigated the selective reflectance of iron silicide and other silicides deposited on Cu surfaces [24]. Unfortunately, their semiconducting nature had not been revealed at that time; thus, the unoptimized chemical compositions of these silicides spoiled the intrinsic selective absorbing behavior. Moreover, because of an increase in emissivity resulting from thermal deterioration under 500 °C, the scientific reason for which was not identified in this paper, absorber coatings based on a series of silicides have been classified in the mid-temperature range under 400 °C in the reviewed article [6]. Since the improperly determined thermal instability of the iron silicide is attributed to the thermal reaction with Cu, the intrinsic thermal stability of iron silicide is expected to be achieved at higher temperatures by the appropriate design of a multi-layered structure. Moreover, based on the subsequent studies on silicide materials thus far, the iron silicide β-$FeSi_2$ has been clarified as a semiconductor with a direct band gap ranging from 0.78 to 0.85 eV at room temperature and a large absorption coefficient of over 10^5 cm^{-1} at 1 eV [25–29]. The band gap, which is slightly higher than ideal cut-off energy (e.g., 0.70 eV at 650 °C), is expected to be lowered toward the ideal energy with increasing temperature [30, 31], suggesting that β-$FeSi_2$ is a promising candidate for spectrally selective absorbers.

As the first step, band-edge absorbing behavior of the monolithic β-$FeSi_2$ layer was evaluated at high temperatures. Figure 6.5 shows the temperature dependency of the n and k spectra for the monolithic β-$FeSi_2$, in which a rise in k resulting from the interband absorption was observed at wavelengths shorter than 2,000 nm, and the band-edge absorption was found to be kept even at high temperatures of 600 °C. It should be noted that the gradient of the rise in k depended on the temperature, indicating that the higher temperature provided a narrower band gap. This phenomenon can be explained by Einstein's model based on the variation of band gap with temperature: thermal dilation of the lattice and electron–phonon interactions [30, 31]. A disadvantage of the β-$FeSi_2$ is the high refractive index of $n = 4 \sim 6$ in the solar wavelength. In order to suppress the reflective loss at the interface with the SiO_2 top layer, middle layers with intermediate refractive indices are necessary using β-$FeSi_2 + SiO_2$ composites. The optical constants of the composites were controlled by adjusting the β-$FeSi_2$ content [16], as indicated in Fig. 6.6. The refractive indices n of the β-$FeSi_2$ were decreased to those of SiO_2 by decreasing the content of β-$FeSi_2$ in the composites. The high k of β-$FeSi_2$ below the cut-off wavelength declined monotonically, whereas it decreased after increasing from zero to 0.3 above the cut-off wavelength by decreasing the β-$FeSi_2$ content. These n and k spectra can be used to design layered structures of the absorbers and to simulate the selective absorbing properties. The existing condition of the β-$FeSi_2$ in the SiO_2 matrix was investigated by X-ray photoelectron spectroscopy (XPS). The XPS spectra support the idea that the β-$FeSi_2$ phase was formed and thermally stable in the SiO_2 matrix during the deposition and annealing. Consequently, the basic structure of optimum solar-absorbing layers was determined as "SiO_2/SiO_2 + β-$FeSi_2$ composite/monolithic β-$FeSi_2$".

6.2.3 Solar-Thermal Energy Conversion Performance

Solar-selective absorbing multilayers have been designed using the "TaN/Ag + Ta/TaN" low-emissivity bases in Sect. 6.2.1 and the "semiconducting β-$FeSi_2$" solar

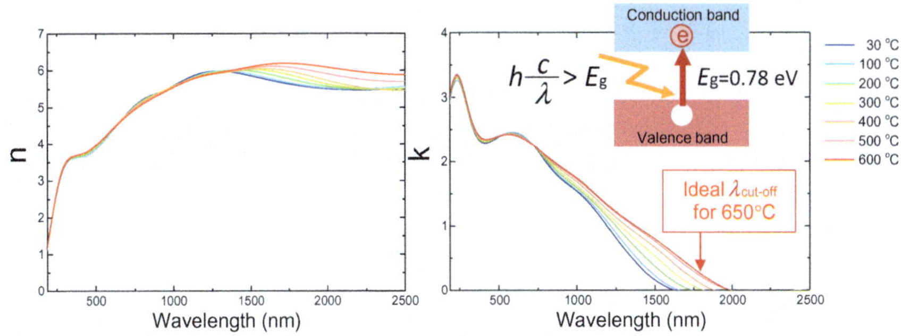

Fig. 6.5 Spectral changes in refractive index n and extinction coefficient k of β-$FeSi_2$ films with increasing temperature

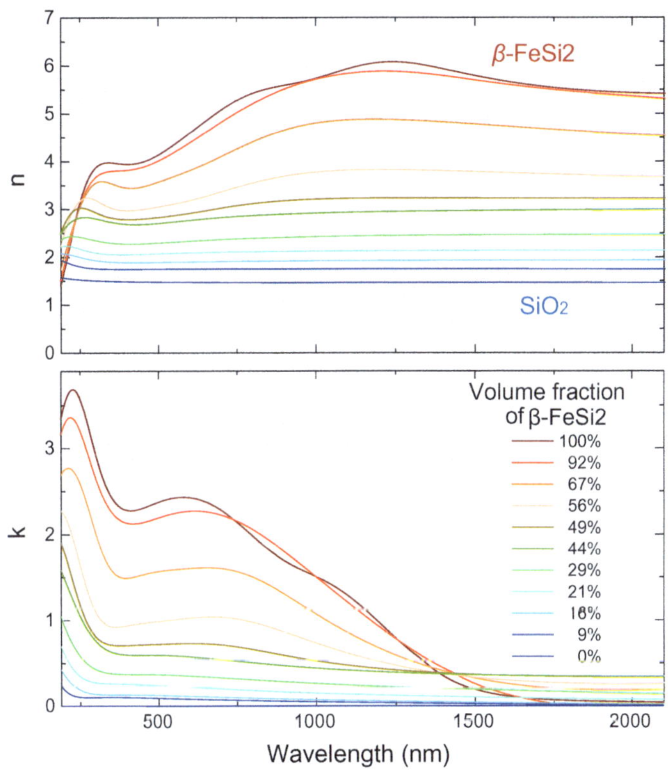

Fig. 6.6 Refractive index n and extinction coefficient k spectra of the composite layers containing various β-$FeSi_2$ nanoparticles in SiO_2 matrix. https://doi.org/10.1016/j.solmat.2016.12.005

absorbing layers in Sect. 6.2.2. Figure 6.7 shows the n and k spectra for each layer composed in the multilayer. The SiO_2 film was chosen to be the top layer because of its low refractive index. The n of around 1.5 was insensitive to temperature below 700 °C, thus Fig. 6.7a shows the n and k spectra only at room temperature. Two oscillation behaviors were observed around 9400 (1060 cm^{-1}) and 21,000 nm (470 cm^{-1}), which are attributed to the Si–O-Si stretching and Si–O-Si rocking vibrations, respectively [32]. The $SiO_2 + \beta$-$FeSi_2$ composite layer with a composition of β-$FeSi_2$ of approximately 20 vol.% was chosen as the middle layer. The intermediate n and k spectra between SiO_2 and β-$FeSi_2$ is also independent of temperature, thus the spectra at room temperature are shown in Fig. 6.7b. The n and k spectra for the monolithic β-$FeSi_{2\text{-}x}$ shown in Fig. 6.7c indicated a slight difference in spectral shape around the band-edge absorption shown in Fig. 6.5. The k spectrum around the infrared region did not reach zero even at wavelengths longer than the cut-off wavelength $\lambda_{cut\text{-}off}$ (smaller energy than the band edge). This is attributed to the compositional control of Si vacancies in β-$FeSi_{2\text{-}x}$ by adjustment of sputtering condition in order to enhance the solar absorptance α. The n and k spectra for the β-$FeSi_{2\text{-}x}$ shown in Fig. 6.7c exhibited another characteristic increase in the infrared region. Similar increases in the n and k spectra at the infrared region were also observed for the TaN film in Fig. 6.7d. These temperature dependencies were caused by increases in carrier density of thermally excited free electrons in the β-$FeSi_{2\text{-}x}$ and TaN film with an increase in the temperature. The increase in the carrier density resulted in a plasma wavelength shift to shorter wavelengths in the Drude model and hence induced the increases in n and k in the infrared region. Although the temperature dependency of the infrared n and k spectra for the β-$FeSi_{2\text{-}x}$ and the TaN layers were obvious, their influence on the solar selective absorbing performance was negligible after calculating the absorptance spectra using their temperature dependency. As shown in Fig. 6.7e, a very few Si was added into the Ag + Ta base for enhancing the thermal durability of the multilayers, and the details will be discussed in the later Sect. 6.2.4.

All these n and k spectra including their temperature dependencies were considered design of the multilayer to achieve the highest solar selective absorbing performance. And the result of optical calculation induced the best stacking structure as follows,

SiO_2 (70 nm)/$SiO_2 + \beta$-$FeSi_2$ (55 nm)/β-$FeSi_2$ (10 nm)/SiO_2 (10 nm)/TaN (20 nm)/Ag + Ta + Si (200 nm).

As shown in the TEM image in Fig. 6.8, the designed multilayers were prepared on the SS310S substrates with "TaN (100 nm)/SiO_2 (100 nm)" layers for ensuring thermal stability. Each layer had good homogeneity without any voids, and these continuous films had uniform thicknesses and good contacts at each interface, with no delamination after initial annealing at 750 °C for 1 h. Because the bright field TEM observation mode shows a darker contrast for heavier elements, many dark dots in the $SiO_2 + \beta$-$FeSi_2$ composite layer denote β-$FeSi_2$ nano-particles smaller than 5 nm in diameter.

As depicted in Fig. 6.9, the absorptance spectrum measured for the prepared multilayer exhibited superior solar selective absorbing behavior, accompanied with

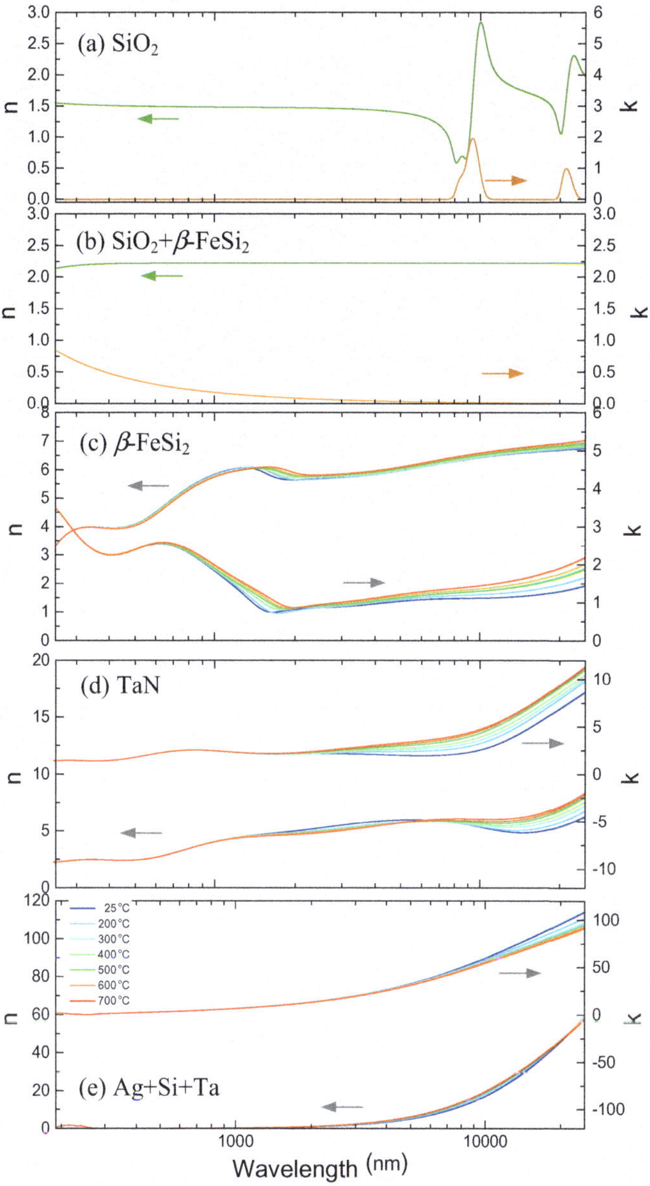

Fig. 6.7 Refractive index n and extinction coefficient k spectra of **a** SiO_2, **b** β-$FeSi_2$ + SiO_2, **c** β-$FeSi_2$, **d** TaN, and **e** Ag + Ta + Si layers. The n and k spectra of the **a** SiO_2 and **b** β-$FeSi_2$ + SiO_2 layers were temperature-independent, while those of the **c** β-$FeSi_2$, **d** TaN, and **e** Ag + Ta + Si layers depended on the temperature up to 700 °C. https://doi.org/10.1016/j.solmat.2019.110304

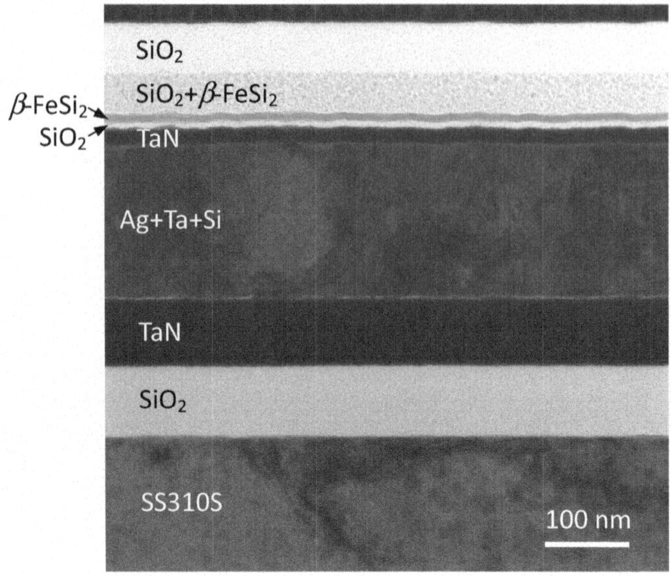

Fig. 6.8 Cross-sectional TEM image of the developed multilayer "SiO_2/SiO_2 + β-$FeSi_2/\beta$-$FeSi_2/SiO_2$/TaN/Ag + Ta + Si/TaN/SiO_2" on SS310S after annealing at 750 °C for 1 h in a vacuum. https://doi.org/10.1016/j.solmat.2019.110304

extremely low infrared absorptance (i.e., low emissivity). Moreover, the experimentally measured absorptance spectrum was in good agreement with the expected spectrum based on the optical calculation using the n and k spectra at room temperature. Comparing the absorptance spectrum with that of the conventional multilayer "SiO_2/Mo + SiO_2 composites/Mo-base" [8, 16], the developed multilayer with band-edge absorption by semiconducting β-$FeSi_2$ layer was found to have a steeper cut-off transition, and lower emittance in the infrared region resulting from thermally stabilized Ag. It should be noted that the conventional multilayer "SiO_2/Mo + SiO_2 composites/Mo-base" will induce an increase in emissivity and decline in the gradient of the cut-off transition at a higher temperature because of the deterioration in optical properties of Mo as shown in Fig. 6.3b. In contrast, the absorptance spectra for the developed multilayer will shift the cut-off wavelength $\lambda_{\text{cut-off}}$ toward the ideal position due to the narrowing of band gap energy of β-$FeSi_2$, which induces enhancement of solar absorptance α at higher temperatures.

If all the temperature dependencies of the n and k spectra in Fig. 6.7 are considered, then the calculated absorptance spectra at an operating temperature ranging from 300 to 650 °C provided the temperature dependency of solar absorptance α, thermal emittance ε and conversion efficiency η as shown in Fig. 6.10. The detailed calculations have been described elsewhere [16]. The higher solar absorptance α owing to the interbond absorption of β-$FeSi_2$ and the lower infrared emissivity ε

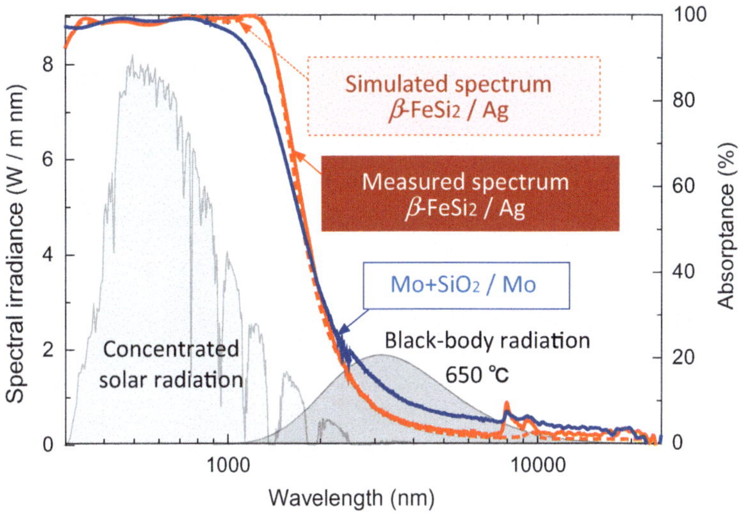

Fig. 6.9 Measured (solid red line) and simulated (dashed red line) absorptance spectra of the developed multilayer "$SiO_2/SiO_2 + \beta\text{-}FeSi_2/\beta\text{-}FeSi_2/SiO_2$/TaN/Ag + Ta + Si/TaN/$SiO_2$" on SS310S at room temperature. For comparison, measured absorptance spectrum of the conventional multilayer "SiO_2/Mo + SiO_2 composites/Mo-base" is shown as the solid blue line. https://doi.org/10.1016/j.solmat.2019.110304

resulting from the thermally stabilized Ag provided drastic enhancement in the averaged solar thermal conversion efficiency η of 87% in the temperature range from 300 to 650 °C.

6.2.4 *Thermal Durability*

As shown in the TEM image of the prepared solar selective absorbers in Fig. 6.8, the Ag + Ta base layer included a small amount of Si, and the TaN and SiO_2 layers of 100 nm in thickness were inserted at the interface between the Ag + Ta + Si layer and the stainless-steel substrate (SS310S). These additionally designed structures are for suppressing diffusion of Si from the β-$FeSi_2$ to Ag layers and SS310S substrates. The deterioration mechanism was found by considerable investigations of structural changes in the multilayers consisting of the β-$FeSi_2$ and Ag layers at high temperatures. Cross-sectional SEM and TEM observations of the thermally deteriorated multilayers, with $TaSi_2$ layers instead of TaN layers for suppressing Ag vaporization as described in Sect. 6.2.1, showed a lot of void formation in the β-$FeSi_2$ layers. X-ray diffraction patterns revealed changes in the crystal phase from the semiconducting β-$FeSi_2$ to metallic FeSi phase, and X-ray photoelectron spectroscopy showed the Ag base included a small amount of diffused Si after heating. These deteriorations of the β-$FeSi_2$ layers were able to be prevented on the Si substrate, which

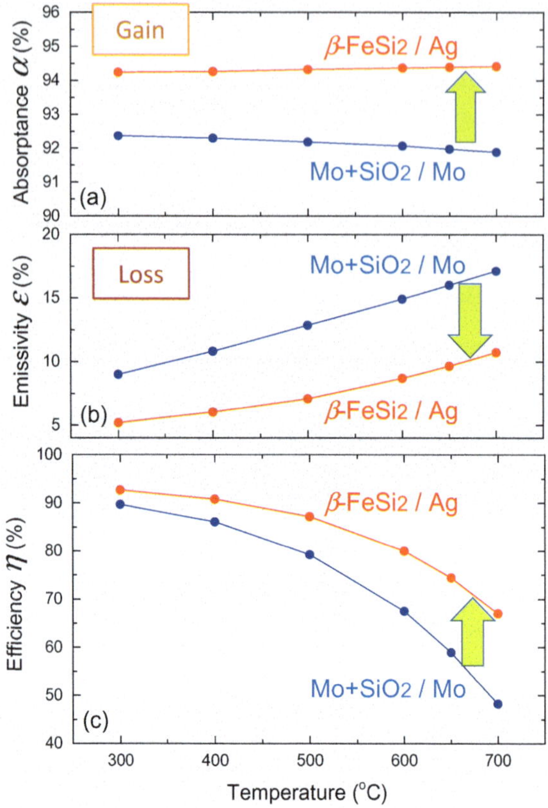

Fig. 6.10 **a** Solar absorptance α, **b** infrared emissivity ε, and **c** solar-thermal conversion efficiency η of the multilayers "$SiO_2/SiO_2 + \beta\text{-}FeSi_2/\beta\text{-}FeSi_2/SiO_2/TaN/Ag + Ta + Si/TaN/SiO_2$" on SS310S as a function of temperature. The temperature dependence of absorptance spectra was taken into consideration in their results. For comparison with the conventional solar selective absorbers, the blue lines indicated the results based on the absorptance spectra of "$SiO_2/Mo + SiO_2$ composites/Mo-base"

was the firmest evidence that the diffusion of Si is the main issue to be suppressed for ensuring thermal durability [18]. As reported in some studies [33–36], TaN in transition metal nitrides is one of the best candidates as a diffusion barrier for Si migration. In the developed multilayers as shown in Fig. 6.8, the Si diffusion was successfully suppressed by the TaN layers inserted between the Ag layers and SS310S substrates as expected. This result means that the TaN layers upper and lower the Ag layer were effective not only for suppressing the Ag vaporization but also for inhibiting the Si diffusion. The doping of Si into the Ag + Ta base layer was also the answer to prevent the diffusion of Si from the β-$FeSi_2$ layers to the Ag + Ta layers. The bottom SiO_2 layer inserted between the lower TaN layer and SS310S was required to avoid reactive deterioration at the interface.

To estimate the long-term lifetimes of the prepared solar selective absorbers below 650 °C, thermal aging tests were conducted under accelerated heating temperatures of 750, 775, and 800 °C. Changes in the absorptance spectra during the thermal degradation at each high temperature were observed, and utilized to calculate the solar absorptance α, emissivity ε, and conversion efficiency η as a function of the elapsed time. Although only representative spectral changes in the absorptance at 750 °C are shown in Fig. 6.11a, it was confirmed that at higher temperatures of 775 and 800 °C the changes in the spectral shape exhibited almost the same behavior, excluding their faster deterioration. This result suggests that the deterioration progressed consistently by the same mechanism under these temperatures, and this consistency was necessary to utilize Arrhenius plots based on these optical degradation data to estimate the lifetime at 650 °C. As shown in Fig. 6.11d, the conversion efficiencies η at each accelerated temperature declined after maintaining a constant level for a certain period, and the time until the decline in efficiency decreased with an increasing accelerated temperature. The solar absorptance α in Fig. 6.11b began to gradually decline from the initial stage of degradation, which reduced the conversion efficiency. In contrast, the emissivity ε at these temperatures appeared to slightly decrease at the initial stage, as shown in Fig. 6.11c, which enhanced the conversion efficiency. Therefore, the changes in both α and ε led to a balance of the conversion efficiency, which remained constant. However, at the late stage the emissivity ε began to increase and the decline in the solar absorptance α accelerated. Thus, the conversion efficiencies η began to decline drastically.

In this study, the lifetime of a solar selective absorber was defined as the time above which the conversion efficiency decreased by 1% from the initial constant level, and the lifetimes were used to determine the deterioration rates (%/h) at these accelerated temperatures. The natural logs of the deterioration rates are shown as functions of the reciprocal of the absolute temperature (1/K) in the Arrhenius plot in Fig. 6.12. Because these three degradation rates at 750, 775, and 800 °C were arranged linearly in the plot, the extrapolated line illustrates the degradation rates at lower temperatures. Assuming that the solar absorbers are under high operating temperatures for 8 h per day, the natural log of the deterioration rate must be lower than -11.2 to provide a lifetime of over 25 years. Figure 6.12 reveals that the extrapolated degradation rate at 650 °C was in this range, and the prepared solar selective absorbers can be thermally durable at 650 °C for 42.6 years. The Arrhenius plot indicates that the activation energy for the degradation of our absorbers consisting of β-$FeSi_2$ with an Ag base on SS310S was 427.1 kJ/mol, which is higher than the previously reported value (325 kJ/mol) for traditional solar absorbers based on cermet and W layers for solar-thermal conversion below 550 °C [11].

6.3 Summary

This chapter introduced the advanced solar selective absorbers for providing efficient and durable solar-thermal conversion at a high temperature of 650 °C. The high

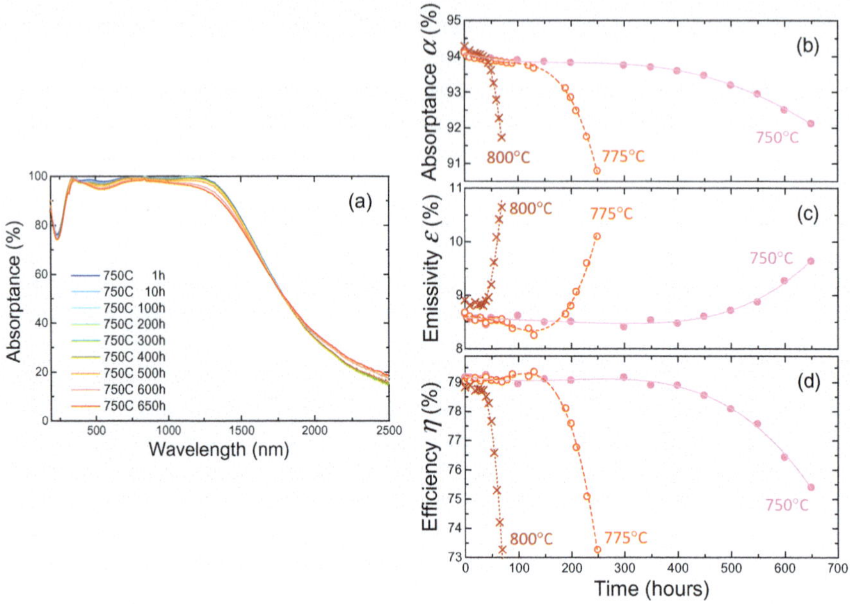

Fig. 6.11 **a** Spectral changes in the absorptance of the developed multilayer "$SiO_2/SiO_2 + \beta$-$FeSi_2/\beta$-$FeSi_2/SiO_2$/TaN/Ag + Ta + Si/TaN/SiO_2" on SS310S under accelerated heating at 750 °C for 650 h in a vacuum. The degradation of the **b** solar absorptance α, **c** emissivity ε, and **d** solar-thermal conversion efficiency η for the multilayers at 750, 775, and 800 °C as a function of the accelerated heating time. https://doi.org/10.1016/j.solmat.2019.110304

energy conversion efficiency was able to be achieved by the high solar absorptance α due to the steeper cut-off of the band-edge absorption of semiconducting β-$FeSi_2$ and the low infrared emittance ε owing to the thermally stabilized Ag. Moreover, the superior thermal durability over 25 years at 650 °C was realized resulting from suppressing the Si diffusion from the β-$FeSi_2$ to SS310S substrates by the TaN layers and stabilizing the agglomeration and vaporization of Ag layers by the Ta additives and the TaN laminations.

For realizing the production of CO_2 free hydrogen and ammonia based on the developed solar-thermal energy conversion, there are still a lot of issues to be solved, for example, novel heat transfer fluids useful at 650 °C, a drastic decrease in operating temperature of the thermochemical iodine–sulfur process to 650 °C or lower. Nonetheless, the progress in efficient thermal energy generation from sustainable solar power is expected to be a significant step to actualize the CO_2 free hydrogen production.

From another perspective, the developed design concept of the solar selective absorbers can be applied to proven CSP systems operating at practical temperature ranges such as 400 and 550 °C for enhancing their conversion efficiencies. The cut-off wavelength $\lambda_{\text{cut-off}}$ of the absorptance spectra can be adjusted to the appropriate shorter wavelength by selecting the semiconducting layers with a narrower band

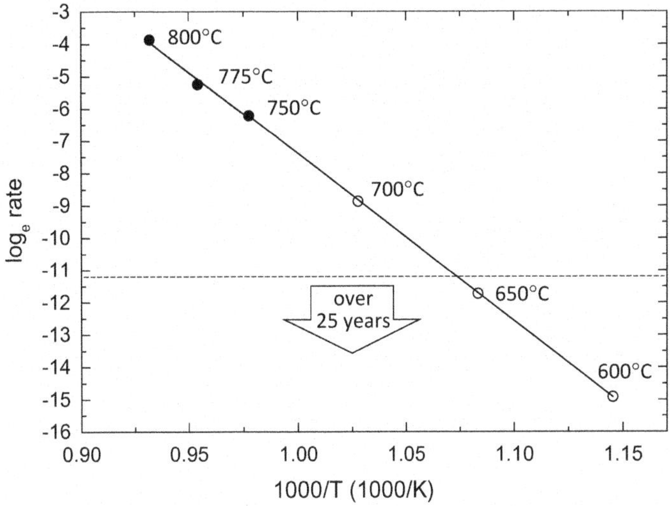

Fig. 6.12 Arrhenius plot of the deterioration rates as a function of reciprocal of the absolute temperature (1/K). The deterioration rates (%/h) were defined by the time above which the conversion efficiency decreased by 1% from the initial constant level in Fig. 6.11d. Based on the three linearly arranged degradation rates at 750, 775, and 800 °C, the extrapolated line illustrates the degradation rates at lower temperatures. https://doi.org/10.1016/j.solmat.2019.110304

gap than that of the β-$FeSi_2$ [37–39]. And a lot of knowledge about improving the thermal durability can also be applied to the conventional solar absorbers operating at lower temperatures for expanding their lifetimes or for simplified their multi-layered structures, both of which contribute to reducing the cost of solar receiver tubes.

Acknowledgements This study was supported by the Council for Science, Technology and Innovation (CSTI), the Cross-ministerial Strategic Innovation Promotion Program (SIP), "Energy carrier" (Funding agency: Japan Science and Technology Agency (JST)).

References

1. Seraphin BO (1979) Solar energy conversion: solid state physics aspects. Springer, Berlin
2. Islam MT, Huda N, Abdullah AB, Saidurcd R (2018) A comprehensive review of state-of-the-art concentrating solar power (CSP) technologies: current status and research trends. Renew Sustain Energy Rev 91:987–1018
3. Myagmarjava O, Iwatsuki J, Tanaka N, Noguchi H (2019) Research and development on membrane IS process for hydrogen production using solar heat. Int J Hydrogen Energy 44(35):19141–19152
4. Price H, Lüpfert E, Kearney D, Zarza E, Cohen G, Mahoney R (2002) Advances in parabolic trough solar power technology. J Sol Energy Eng 124:109–125

5. Giaconia A, Iaquaniello G, Metwally AA, Caputo G, Balog I (2020) Experimental demonstration and analysis of a CSP plant with molten salt heat transfer fluid in parabolic troughs. Sol Energy 211(15):622–632
6. Kennedy CE (2002) Review of mid- to high- temperature solar selective absorber materials. NREL/TP-520–31267 Golden CO 2002
7. Selvakumar N, Barshilia HC (2012) Review of physical vapor deposited (PVD) spectrally selective coatings for mid-and high-temperature solar thermal applications. Solar Energy Mater Solar Cells 98:1–23
8. Carlo R, Alessandro A (2004) Surface coating of the collector tube of a linear parabolic solar concentrator. European Patent EP1397622B1, 1 Dec 2004
9. Esposito S, Antonaia A, Addonizio ML, Aprea S (2009) Fabrication and optimization of highly efficient cermet-based spectrally selective coatings for high operating temperature. Thin Solid Films 517:6000–6006
10. Antonaia A, Addonizio ML, Esposito S, Ferrara M, Castaldo A, Guglielmo A, D'Angelo A (2014) Adhesion and structural stability enhancement for Ag layers deposited on steel in selective solar coatings technology. Surf Coat Technol 225(25):96–101
11. Antonaia A, D'Angelo A, Esposito S, Addonizio ML, Castaldo A, Ferrara M, Guglielmo A, Maccari A (2016) Accelerated aging tests on ENEA-ASE solar coating for receiver tube suitable to operate up to 550 °C. AIP conf proc 1734(030003):1–8
12. Seraphin BO (1976) Chemical vapour deposition of semiconductor films for solar energy conversion. Thin Solid Films 39:87–94
13. Seraphin BO (1979) Chemical vapour deposition of spectrally selective surfaces for high temperature photothemal conversion. Thin Solid Films 57:293–297
14. Okuhara Y, Kuroyama T, Takata M, Tsutsui T, Noritake K (2018) Solar selective absorbers based on semiconducting β-$FeSi_2$ for high temperature solar-thermal conversion. AIP conf proc 2033(220005):1–8
15. Okuhara Y, Kuroyama T, Yokoe D, Kato T, Takata M, Tsutsui T, Noritake K (2019) Solar selective absorbers consisting of semiconducting silicide absorbing layers with thermally stabilized Ag base. AIP Conf Proc 2126(120013):1–8
16. Okuhara Y, Yokoe D, Kato T, Suda S, Takata M, Noritake K, Sato A (2017) Solar-selective absorbers based on semiconducting β-$FeSi_2$ for efficient photothermal conversion at high temperature. Solar Energy Mater Solar Cells 161:240–246
17. Okuhara Y, Kuroyama T, Yokoe D, Kato T, Takata M, Tsutsui T, Noritake K (2018) High-temperature solar-thermal conversion by semiconducting β-$FeSi_2$ absorbers with thermally stabilized silver layers. Solar Energy Mater Solar Cells 174:351–358
18. Okuhara Y, Kuroyama T, Yokoe D, Kato T, Takata M, Tsutsui T, Noritake K (2020) Thermal durability of solar selective absorbers consisting of β-$FeSi_2$ with low emissive Ag layers on stainless steel. Solar Energy Mater Solar Cells 206(11304):1–8
19. Sugawara K, Kawamura M, Abe Y, Sasaki K (2007) Comparison of the agglomeration behavior of Ag(Al) films and Ag(Au) films. Microelectron Eng 84(11):2476–2480
20. Minamide Y, Kawamura M, Abe Y, Sasaki K (2009) Agglomeration suppression behavior and mechanisms of Ag–Cu and Ag–Nb thin films. Vacuum 84(5):657–662
21. Kim HC, Alford TL (2003) Improvement of the thermal stability of silver metallization. J Appl Phys 94(8):5393–5395
22. Glickman EE, Bogush V, Inberg A, Diamand YS, Croitoru N (2003) Electrical resistivity of thin electroless Ag–W films for metallization. Microelectron Eng 70(2–4):495–500
23. Mardani S, Primetzhofer D, Liljeholm L, Vallin Ö, Norstöm H (2014) Electrical properties of Ag/Ta and Ag/TaN thin films. Microelectron Eng 120:257–261
24. Harding GL (1978) Sputtered metal silicide solar selective absorbing surfaces. J Vac Sci Technol 15(1):65–69
25. Borisenko VB (2000) Semiconductor silicides. Springer, Berlin
26. Bost MC, Mahan JE (1985) Optical properties of semiconducting iron disilicide thin films. J Appl Phys 58(7):2696–2703

27. Bost MC, Mahan JE (1988) A clarification of the index of refraction of beta-ion disilliside. J Appl Phys 64(4):2034–2037
28. Christensen NE (1990) Electric structure of β-$FeSi_2$. Phys Rev B 42(11):7148–7153
29. Dimitriadis CA, Werner JH, Logothetidis S, Stutzmann M, Weber J, Nesper R (1990) Electronic properties of semiconducting $FeSi_2$ films. J Appl Phys 68(4):1726–1734
30. Mckinty CN, Kewell AK, Sharpe JS, Lourenco MA, Butler TM, Valizadeh R, Colligon JS, Reeson Kirkby KJ, Homewood KP (2000) The optical properties of β-FeSi2 fabricated by ion beam assisted sputtering. Nucl Instr Meth B 161-163:922–925
31. Yang Z, Homewood KP, Finney MS, Harry MA, Reeson KJ (1995) Optical absorption study of ion beam synthesized polycrystalline semiconducting $FeSi_2$. J Appl Phys 78(3):1958–1963
32. Hoex B, Peeters FJJ, Creatore M, Blauw MA, Kessels WMM, Sanden MCM (2006) High-rate plasma-deposited SiO_2 films for surface passivation of crystalline silicon. J Vac Sci Technol A 24(5):1823–1830
33. Nicolet MA (1978) Diffusion barriers in thin films. Thin Solid Films 52:415–443
34. Wittmer M (1984) Barrier layers: principles and applications in microelectronics. J Vac Sci Technol A 2(2):273–280
35. Tsai MH, Sun SC, Tsai CE, Chuang SH, Chui HT (1996) Comparison of the diffusion barrier properties of chemical-vapor-deposited TaN and sputtered TaN between Cu and Si. J Appl Phys 79(9):6932–6938
36. Mehrotra B, Stimmell J (1987) Properties of direct current magnetron reactively sputtered TaN. J Vac Sci Technol B 5(6):1736–1740
37. Panknin D, Wieser SEW, Henrion W, Lange H (1996) Buried $(Fe_{1\text{-}x}Co_x)Si$ layers with variable band gap formed by ion beam synthesis. Appl Phys A 62:155–162
38. Dalapati GK, Liew SL, Wong ASW, Chai Y, Chiam SY, Chi DZ (2011) Photovoltaic characteristics of β-$FeSi_2$(Al)/n-Si(100) heterojunction solar cells and the effects of interfacial engineering. Appl Phys Lett 98(013507):1–3
39. Kishimoto K, Nagatomo Y, Miki T, Koyanagi T (2002) Microstructure and thermoelectric properties of Cr-doped β-$FeSi_2$ sintered with micrograins treated in radio frequency plasmas of SiH_4 and GeH_4 gases. J Appl Phys 92(8):4393–4401

Chapter 7
Enhancing Corrosion Resistance of Stainless Steel by Hot-Dip Aluminizing for High-Temperature Solar Thermal Application

Santosh Prasad Sah and Atsushi Nishikata

Abstract In this chapter, an overview has been made from the recent researches on molten salt corrosion of popular alloys like ferritic–martensitic steels and stainless steels with a particular focus on failure of the corrosion resistance in terms of dissolution of the corrosion products. Dissolution of corrosion products undermines the corrosion resistance. Compared to ferritic substrates, aluminizing of austenitic substrates can reduce the porosity of coatings. The corrosion and interdiffusion resistance of hot-dip aluminized 310S stainless steel, a study carried out under the SIP project, has been presented. Aluminized coatings are largely damaged by interdiffusion, so understanding the microstructure of coatings is important. The heat pretreatment at 800 °C for 24 h made the coating highly interdiffusion resistant at 650 °C. The dense aluminized coating sufficiently alloyed with Cr showed a corrosion rate of ~2 μm $year^{-1}$ in the carbonates melt at 650 °C. The achievement of thicker protective aluminized coatings for commercial purposes still requires optimization in coating process, heat pretreatment and final-stage polishing procedure.

Keywords Molten salt · Stainless steel · Coating · Corrosion · Aluminizing · Diffusion

7.1 Introduction

Generally, molten salts are deemed as better heat storage media at high temperature because of high specific heat capacity, conductivity and low volatility. However, high-temperature molten salts are very aggressive, so they easily corrode containment metal or alloys. Therefore, corrosion is a challenging issue for the development of high-temperature solar–thermal systems. Corrosion is of course not a newly noticed problem of molten salts, but the current thrust for developing high-temperature solar–thermal systems has renewed the interest in corrosion problems. A lot of studies

S. P. Sah (✉) · A. Nishikata
Faculty of Materials Science and Engineering, Tokyo Institute of Technology, Meguro-ku, Tokyo, Japan
e-mail: santoshpsah@gmail.com

K. Aika and H. Kobayashi (eds.), *CO_2 Free Ammonia as an Energy Carrier*,
https://doi.org/10.1007/978-981-19-4767-4_7

were carried out in molten salts for nuclear reactors [1], biomass incinerators [2], molten carbonate fuel cells [3] and concentrated solar power (CSP) systems [4, 5] to understand the process and the extent of corrosion. Many alloys and coatings were observed to have significant corrosion resistance depending on the type and the temperature of molten salts.

Ammonia has received a significant attention as the hydrogen carrier, and its mass production through a clean energy process is on high demand. For efficient production of ammonia, a thermal system operating at 650 °C is preferable. However, a brief review of recent corrosion studies in molten salts can point out the potential corrosion resistant alloys for the given application. For heat storage, metal nitrate/nitrite, metal carbonate and metal halide salts are usually chosen as the candidate. Thermochemically, an inert or passivated metal/alloy can show high corrosion resistance in molten salts. Active corrosion occurs in the melt of halides, while passivation of the alloys is likely to occur in nitrate, nitrite or carbonate melts. Besides corrosion resistance, containment alloys also require to have high creep resistance at high temperature.

7.2 Corrosion Resistance of Steels and Stainless Steels

Ferritic–martensitic (FM) steels and stainless steels have generally high creep and corrosion/oxidation resistances [6, 7]. FM steels and austenitic stainless steels were often tested in molten salts at 500 °C or higher temperature. Chemical compositions of some commercial FM steels are shown in Table 7.1, and the Cr composition of FM steels is very important for controlling the corrosion damage. Commercial FM steels can contain Cr as high as 12 wt.%. When Cr content was 5 wt.% or less, defective and porous corrosion scales were formed. Figure 7.1 shows the cross-sectional morphology of corrosion scale formed by 1000 h corrosion of T5 steel in $LiNO_3$–$NaNO_3$–KNO_3 eutectic melt at 550 °C. There are pores and voids in the scale which might be resulted from dissolution of corrosion products. However, such corrosion formed a void-free duplex corrosion scale on the higher Cr containing steels like in T9 (8.25% Cr) or X20 (11.08% Cr) as shown in Fig. 7.2. In addition to the dense and duplex morphology of corrosion scales, Fig. 7.2 also shows elemental analysis of the duplex corrosion scale. The outer layer (O) of the scale is Fe-rich ($LiFeO_2$), and the inner layer (I) is Cr-rich $(Fe, Cr)_3O_4$. Both the scale thickness and the mass gain due to corrosion decreased with increasing Cr content of the steels. The corrosion rate decreased from 139 to 5.5 μm $year^{-1}$ upon increasing Cr from 2.34 to 11.08 wt.% in the steels. The thickness of scales formed by 1000 h corrosion decreased from ~22 μm (5.25% Cr) to ~2 μm (11.08% Cr) [8].

On increasing the temperature of $NaNO_3$–KNO_3 melt to 580 °C, the corrosion resistance of P91 FM steel (8.75% Cr) was undermined, and a higher corrosion rate of 0.3 mm $year^{-1}$ was observed from EIS spectra at 1000 h of corrosion. This corrosion rate observed by EIS corresponds well with that from gravimetric analysis. Corrosion of P91 steel also showed the presence of diffusion impedance [9]. In the same experimental conditions, VM12 FM steel showed the corrosion rate of 50 μm $year^{-1}$

Table 7.1 Composition (wt.%) of ferritic-martensitic steels [8, 12, 34]

	Alloy					
Element	T5	T9	X20	P91	VM12	X20CrMoV11-1
Cr	5.25	8.25	11.08	8.75	11–12	10.37
Ni	–	–	0.79	0.17	0.1–0.4	0.693
Mo	0.55	0.93	0.87	0.944	0.2–0.4	0.852
Mn	0.34	0.40	0.49	0.412	0.15–0.45	0.58
Si	0.48	0.36	0.25	0.356	0.4–0.6	0.278
P	0.03	0.02	0.02	0.014	–	0.022
C	0.14	0.09	0.21	0.102	0.1–0.14	0.208
V	–	–	0.26	0.203	0.2–0.3	0.247
Fe	Bal	Bal	Bal	Bal	Bal	Bal
Others	–	–	–	Nb, Cu, W, Co, Ti, Al, S	N, Cu, W, Co, Ti, Al	Nb, Cu, W, Co, Ti, Al

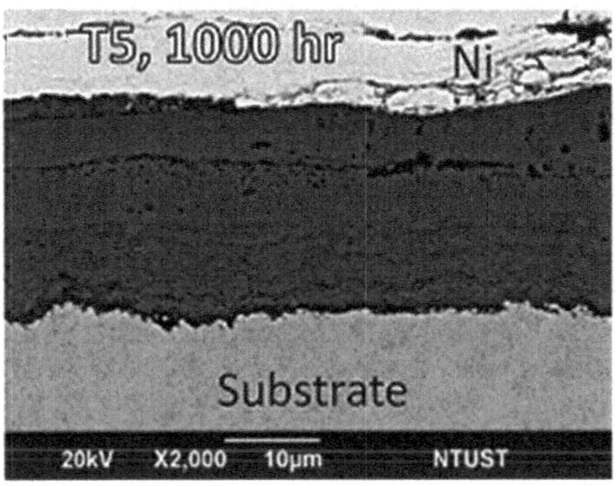

Fig. 7.1 Cross-sectional BSE micrograph of T5 following exposure to molten salt at 550 °C for 1000 h [8]

[10], a value 1/6th of the corrosion rate of P91. The enhanced corrosion resistance of VM12 steel must be due to its higher Cr content. However, corrosion of VM12 steel proceeded with appearance of the diffusion impedance and formation of a duplex corrosion scale with voids. Therefore, VM12 steel suffers from a severe corrosion. Increasing temperature of the melt has devastating effects on corrosion. At 600 °C, 1000 h corrosion of P91 steel formed a multilayered corrosion scale thicker than 1 mm in $NaNO_3$–KNO_3 melt [11]. However, a 400 μm thick corrosion scale was formed on the higher Cr (10.37%) containing X20CrMoV11-1 steel by 1000 h corrosion in $NaNO_3$–KNO_3 melt. This scale was also porous and multilayered indicating aggressive nature of the corrosion associated with the dissolution of corrosion products [12]. Although the increased Cr content of FM steels is effective to suppress corrosion, corrosion does turn very aggressive on elevating the temperature of melts

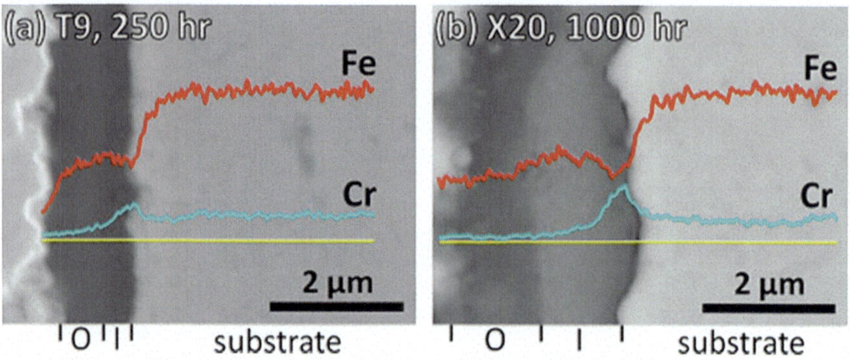

Fig. 7.2 Element distribution determined by EDS line scan analysis: **a** T9 alloy following exposure to molten salt at 550 °C for 250 h and **b** X20 alloy following exposure to molten salt corrosion at 550 °C for 1000 h. O and I represent outer layer and inner layer of corrosion scales, respectively [8]

to 600 °C. Therefore, FM steels are no longer the choices for making containments of the melts which have the temperature of 600 °C or higher.

However, stainless steels are corrosion resistant materials in the nitrates melt, and 321H stainless steel showed the corrosion rate of ~7 or 9 μm year^{-1} [13, 14] at 500 °C and ~18 μm year^{-1} at 530 °C [15]. 316 & 347H stainless steels grew thinner scales of ~20 μm by 600 °C/1000 h corrosion in the nitrate melt. The corrosion product was a duplex layer, the outer part was enriched with Fe, and the inner part was enriched with Cr. Extending corrosion of 347H to 5000 h virtually did not thicken the corrosion scale, showing the excellent corrosion resistance of 347H. But the test temperature of the nitrates melt being further elevated to 680 °C (i.e., beyond salt decomposition temperature), the corrosion rate of 321 or 347 stainless steels was observed ~0.5 mm year^{-1}, and this rate is ~30–40 times high to that at 600 °C [13]. Similarly, when temperature of carbonates melt was 700 °C, much thicker and porous corrosion products (>150 μm thick by 500 h, Fig. 7.3) were formed. The higher temperature of 700 °C must have accelerated the corrosion of 304 and 316 stainless steels [16]. Stainless steels suffer from a severe corrosion if the temperature of the melt is close to 700 °C. Severe corrosion by the melts quite often creates porosity of the corrosion scales. In many studies, the thickness of corrosion scales was considered as an index of corrosion rate, without considering the dissolution of corrosion products. If the dissolution of corrosion products is significant, the thickness of corrosion scales cannot be considered as the index of corrosion rate in molten salts. Encinas-Sanchez et al. concluded that dissolution of corrosion products undermines the corrosion resistance [14].

The corrosion rate of 304 stainless steel was observed 3.5 mm year^{-1} in carbonates melt with CO_2–2%O_2 environment at 650 °C [17], and generally stainless steels have corrosion rates in the order of mm year^{-1} in the melts at 600 °C or above [4, 18]. These corrosion rates can vary to some extent depending on the chemistry of melts. The significant acceleration of corrosion of stainless steels in the melts of temperature

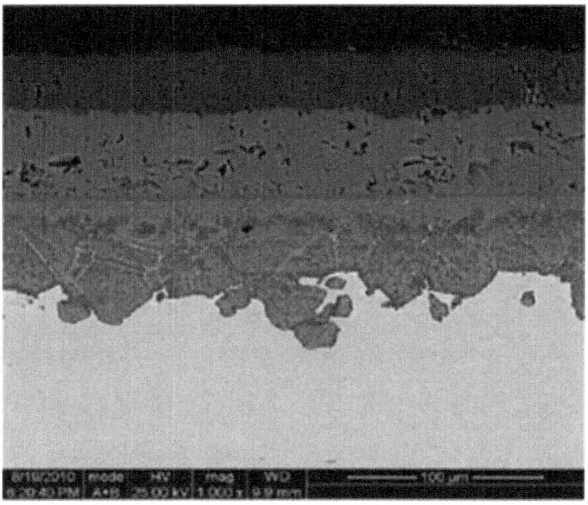

Fig. 7.3 Cross-sectional morphology of 304 SS corroded at 973 K in $(Li–Na–K)_2CO_3$ for 500 h [16]

above 600 °C is likely due to accelerated kinetics of metal oxidation and accelerated dissolution of corrosion products. Although the features of high frequency diffusion and the consequent dissolution of corrosion products of 321H stainless steel are obvious from EIS spectra sampled at 7000 h in the low temperature (500 °C) nitrate melt [19], no detailed explanation was made. The effect of increased kinetics was well known from high-temperature oxidation [20]; the change of rate constant with temperature follows the Arrhenius-type equation. On the other hand, the effect of dissolution was recently demonstrated from the corrosion of 304 stainless steel in carbonates melt kept under different gas environments [21]. Mass loss increased linearly with time as shown in Fig. 7.4, 6.2 mg cm^{-2} as a constant mass loss for every 24 h of corrosion. This showed that corrosion products were not protective at all. In case the protective corrosion products were formed, mass loss kinetics could slow down with time.

Severe dissolution of oxides at oxide/melt interface drives corrosion control kinetics through the diffusion of cations through oxide scales. Such a solid-state diffusion was noticed through the high frequency Warburg (diffusion) impedance [21]. This type of diffusion impedance was often observed from solid-state electrolytes where ions diffuse through the bulk of electrolytes [22]. It is likely that diffusion became the corrosion controlling step soon after the initial loss of protective properties of the oxidation film of 304 stainless steel. This phenomenon was obvious from the transition of semicircle to straight line of the Nyquist plots and shortening of the length of straight lines with the progress of corrosion. These straight lines of Nyquist plots were considered to represent ionic diffusion through the oxide scales and were associated with the subsequent dissolution of oxides into the melt [21]. The generation of Fe-oxide layer on Cr-oxide layer by a pretreatment led to disappearance of the Warburg impedance and to the emergence of large semicircles from

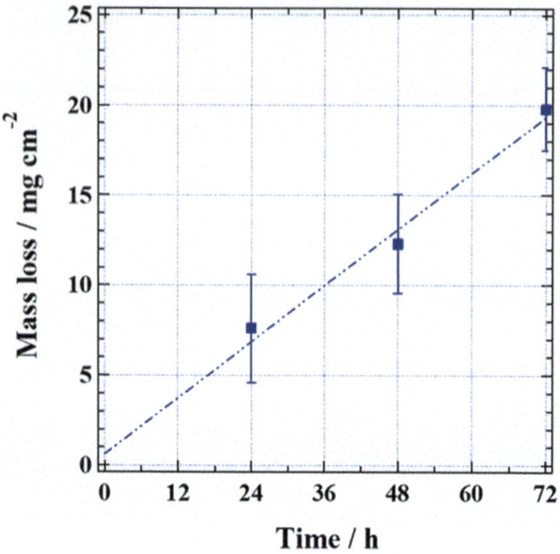

Fig. 7.4 Mass loss observed for 304 SS corroded for different duration in $0.43Li_2CO_3$–$0.32Na_2CO_3$–$0.25K_2CO_3$ with flowing CO_2–2%O_2 environment at 650 °C [21]

the Nyquist plot (Fig. 7.5). Large semicircles represent enhanced interfacial charge transfer resistance, R_{ct}. The disappearance of Warburg impedance is the signature of attainment of the high corrosion resistance. R_{ct} increased by 3 orders of magnitude by the pretreatment [21].

Selective as well as synergistic dissolution of oxides of Fe and Cr in the melt resulted into different corrosion kinetics. The basicity of carbonates melt was crucial to cause (differential) dissolution of corrosion products of 304 stainless steel. In the acidic melt obtained under CO_2–2%O_2 gas environment, facile synergistic dissolution of the mixture of $LiCrO_2$ and $LiFeO_2$ undermined the corrosion resistance. In the basic melt obtained under the environment of N_2–20%O_2, mainly $LiCrO_2$ dissolved during the pretreatment, and $LiFeO_2$ remained on the surface [21]. Therefore, a very poor corrosion resistance was also observed (Fig. 7.5c). The stacking of layers of $LiFeO_2$/$LiCrO_2$–$LiFeO_2$ obtained by the pretreatment offered a high interfacial charge transfer resistance in the acidic melt. At the same time, dissolution of oxides was significantly suppressed [21]. Remarkably, the microstructure of corrosion products was highly instrumental to suppress metal oxidation and to enhance corrosion resistance. The dissolution of corrosion products was noticed even in the low temperature nitrates melt [19], although to a different extent. Therefore, corrosion products of stainless steels have the inherent properties to dissolve out into the melts. In summary, mixed corrosion products of $LiCrO_2$–$LiFeO_2$ are usually resulted from molten salt corrosion of stainless steels, so stainless steels are subject to an accelerated corrosion. Therefore, a secondary method of corrosion protection is required for stainless steels.

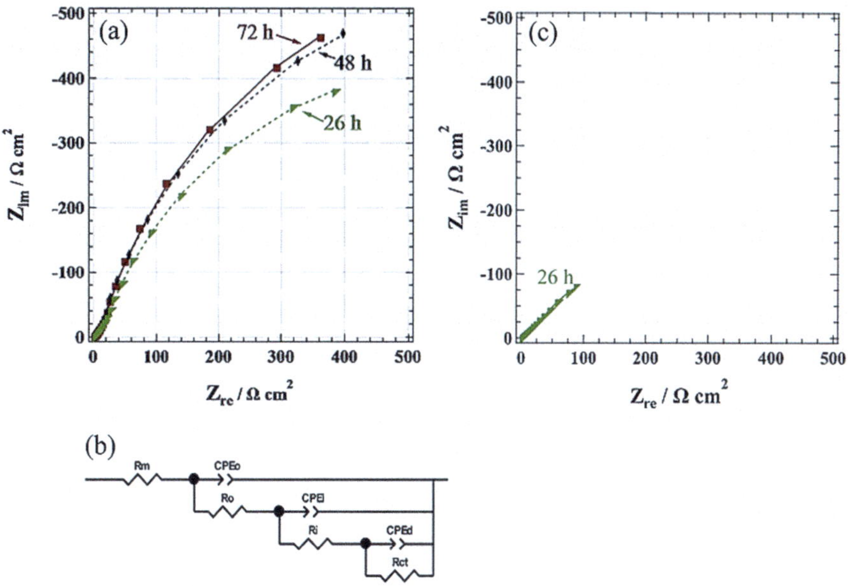

Fig. 7.5 **a** Nyquist plots observed during corrosion of the 24 h-pretreated 304 SS in $0.43Li_2CO_3$–$0.32Na_2CO_3$–$0.25K_2CO_3$ with CO_2–2%O_2 environment at 650 °C, **b** equivalent circuit used for **a** and **c** Nyquist plot observed during extending the pretreatment to 26 h [21]

7.3 Aluminizing of FM Steel and Stainless Steel

In order to exploit reasonable cost and excellent mechanical properties of FM steels and stainless steels, coatings were often applied to enhance the corrosion resistance in high-temperature gaseous or molten salts environments. Coatings of Al metal are economic, and a very protective layer of alumina can be formed by corrosion or oxidation of Al. Addition of Fe and Cr to the alloys promoted the formation of α-alumina [23, 24], the most stable form of alumina, in high-temperature environments. FeAl intermetallic showed higher corrosion resistance in carbonates melt at 650 °C [25]. Therefore, aluminizing of FM steels and stainless steels is likely to result into very protective coatings. The coating of Al metal can be formed through a variety of ways, like electrodeposition [26], ion vapor deposition [27, 28], thermal spray [29], electrostatic spark discharge [30], slurry coating [11, 31–33], pack-cementation [34, 35] or hot-dip [36] for different applications. Out of these, slurry, pack-cementation and hot-dip coatings were so far studied for heat containers of solar–thermal systems. These coatings were often heat-treated at the end of fabrication process, and these coatings are simply 'diffusion coatings'. The microstructure and composition of substrates and the parameters of heat pretreatment always do matter to microstructures and properties of diffusion coatings. The slurry coating method can produce coatings as thick as 80 μm. Thick coatings were formed on ferritic steels (i.e., body-centered cubic, BCC, structure) in comparison with that on austenitic stainless steels

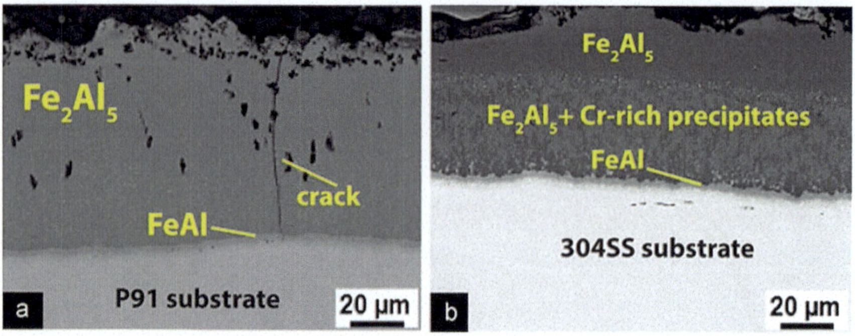

Fig. 7.6 Back-scattered SEM micrographs of **a** as-coated P91 and **b** as-coated 304 [11]

as shown in Fig. 7.6 [11]. Numerous pores and cracks were formed in the coatings of P91 steel, while Cr-rich precipitates were formed in the coating of 304 stainless steel. No voids were formed on 304 stainless steel. The cracks were not observed on the same substrate of P91 by others [33]. The cracks are likely to be formed by rapid cooling after the heat treatment. However, faster diffusion of Al into BCC structure of the ferrite [37] led to form numerous voids, popularly known as Kirkendal's voids, in the coatings. As a general, these voids are detrimental to the integrity of coatings. Diffusion of Al into FCC structure of austenite is slower [37, 38]; therefore, austenitic alloys are better substrates for aluminizing.

Hot-dip aluminizing is an established method to coat steels of various grades in order to enhance oxidation, corrosion and sulfidation resistances [39]. As an industrial practice, hot-dip samples are subsequently heat pretreated to enhance bonding of the coating with the substrate. The hot-dip aluminized coating studied under SIP project was formed by dipping coupons of 310S stainless steel into Al melt (purity >99%) at 714 °C for 90 s. The hot-dip coatings were subsequently heat pretreated at 700 °C/ 40 min to enhance adhesion and shot-blasted (for 60 s) to remove the brittle outer Al-rich parts of the coating [36]. This coating is termed as 'aluminized' in this chapter.

7.4 Microstructure of Hot-Dip Aluminized Coatings and Interdiffusion Resistance

Sah et al. explored the microstructure of hot-dip aluminized coating [36]. Figure 7.7 shows the surface and cross-sectional morphology of aluminized 310S stainless steel before and after immersion into the carbonates melt at 650 °C. The surface contains numerous pores of 5–53 μm sizes that penetrate beyond the center of the coating (Fig. 7.7a, b). These pores are most likely formed by the heat treatment (700 °C/40 min) and/or by the shot blasting. The coating has a ~23 μm thick continuous inner layer. At the very coating/substrate interface, there is a thin intermetallic layer as shown by an arrow in Fig. 7.7c. After 72 h of immersion into the carbonates melt at 650 °C, the morphology of the aluminized coating was changed possibly by the interdiffusion. The aluminized coating thickened from 50 to 64–70 μm. The porosity of the outer layer (Fig. 7.7e) was not changed noticeably. However, a different level of porosity of the outer layer was observed from another sample (Fig. 7.7h). Therefore, the porosity of the aluminized coating must be the result of the shot blasting carried out at the end of coating process.

The thin intermetallic layer at coating/substrate interface grew significantly from 1 μm to 7 μm in both the samples (Fig. 7.7 f, i) when the aluminized coatings were exposed to the molten carbonates at 650 °C for 72 h. Chemical analysis by EDS showed the composition of the intermetallic layer of 53 at% Al–10 at% Cr–26 at% Fe–8 at% Ni before the immersion and of 50 at% Al–9 at% Cr–31 at% Fe–11 at% Ni after the immersion. The composition of intermetallic layer remained intact, and this showed that the exposure to the melt caused only the thickening of the intermetallic layer. Therefore, the interfacial microstructure/composition of the aluminized coating is unstable and is likely to undergo significant changes upon a long-term exposure to the hot molten salts. Such an interdiffusion must be driven by the high concentration gradient of Al, Fe, Cr and Ni across the coating/substrate interface. As it is well known that the interdiffusion damages both substrate and coating, its countermeasure is

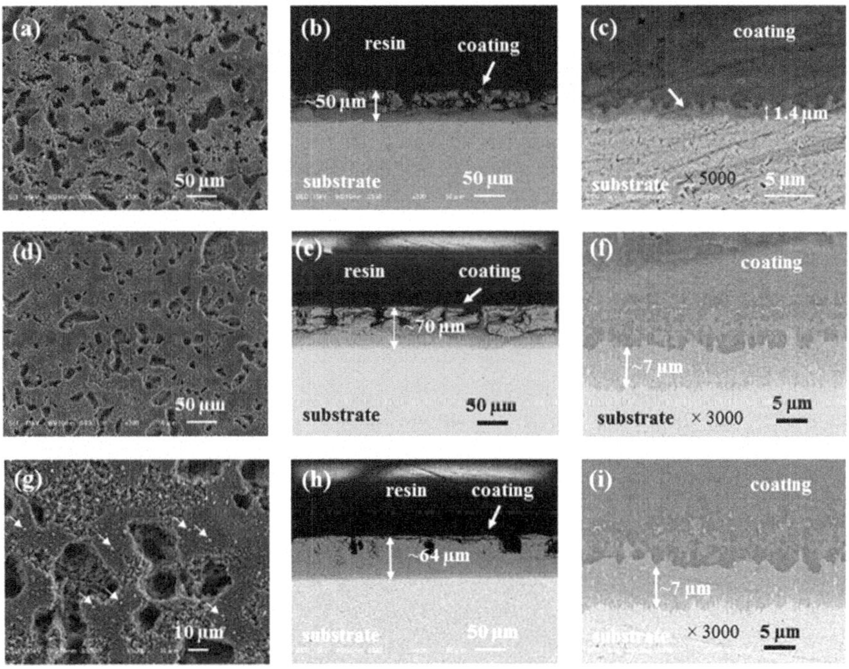

Fig. 7.7 Morphology of the aluminized 310S stainless steel before **a**, **b**, **c** and after **d**–**g** 72 h of corrosion in $0.43Li_2CO_3$–$0.32Na_2CO_3$–$0.25K_2CO_3$ kept at 650 °C under flowing CO_2–2%O_2 mixed gas. **a**, **d**, **g** are secondary electron images of the surface; **b**, **c**, **e**, **f** are back-scattered electron images of the cross section. **h**, **i** are the cross-sectional images of another sample [36]

direly required. Generally, metallic and ceramic interlayers were preferred [40–50]; however, introduction of such interlayers does matter to the cost and feasibility of the coating process. A cheaper method for mitigation of the interdiffusion is required for solar–thermal applications.

As an effective method, an additional heat pretreatment of 800 °C/ 24 h was used to enhance the interdiffusion resistance [36]. The microstructure of aluminized coating was drastically changed by the heat pretreatment. All the layers of coating thickened. Roughly, the cross-sectional morphology of the coating was demarcated into three layers. The outer layer was highly porous and was a heterogeneous mixture of light gray (Fe-rich) and dark gray (Al-rich) phases. Contrarily, the middle and inner layers were compact. Unlike the middle layer, the inner layer was mostly homogeneous. The coating/substrate interface was heterogeneous; it must be evolved from the differential alloying capability of Al with changing microstructure at the alloying front of the stainless steel [51]. Based on atomic ratio observed from EPMA, the dark gray phase of the outer layer was $FeAl_3$-type aluminide while the light gray phase was Fe–Cr–Ni–aluminide (MAl_{1+}-type; M: Fe, Ni, Cr. ' + ' sign indicated Al: M > 1) provided C content was neglected. ~50 at% of C was also detected in the coating. The middle layer was compact but heterogeneous. The light gray phase of the middle layer was composed of ~53 at% Al–31 at% Fe–8 at% Cr–7 at% Ni–1 at% Si, while the dark gray phase was composed of ~57 at% Al–20 at% Fe–21 at% Cr–2 at% Si. The inner part of the middle layer was a thin continuous Al/Cr-rich (dark gray) sublayer which was Fe–Cr-aluminide (MAl_{1+}- type). The inner layer of the coating constituted a large share of the dense part of the coating. It was Fe-rich Fe–Cr–Ni-aluminide (MAl_{1+}-type) with a nominal composition of 53 a% Al–31 at% Fe–8 at% Cr–7 at% Ni–<1 at% Si, a composition very similar to that of the intermetallic layer as shown in Fig. 7.7. The inner layer also contained some Fe- and Cr-rich precipitates [36]. Hence, the coating composition and microstructure evolved after the additional heat pretreatment was rather complicated. The porous layer was removed by wet-polishing before the corrosion test.

When the heat pretreated coating was exposed to the molten salt at 650 °C, the thickness of inner layer was retained intact even after 9 days of the exposure. The morphology of inner layer showed that the coating/substrate interface was quite stable after the secondary heat pretreatment (Fig. 7.8) [36].

The importance of heat pretreatment of the coating was also realized from the morphology of a poorly heat pretreated aluminized coating exposed to the carbonates melt at 650 °C as shown in Fig. 7.9 [52]. The small pits observed at 3 days of exposure increased in number and dimensions upon extending the exposure time to 9 days. Some very large and deep pits were also formed. The pits must be formed by inhomogeneous composition of the outer layer, and the pits might grow inwards down to the band of Kirkendal's voids. The growing pit damaged severely the inner homogeneous layer as shown in Fig. 7.10 [27]. Therefore, a proper heat pretreatment is required.

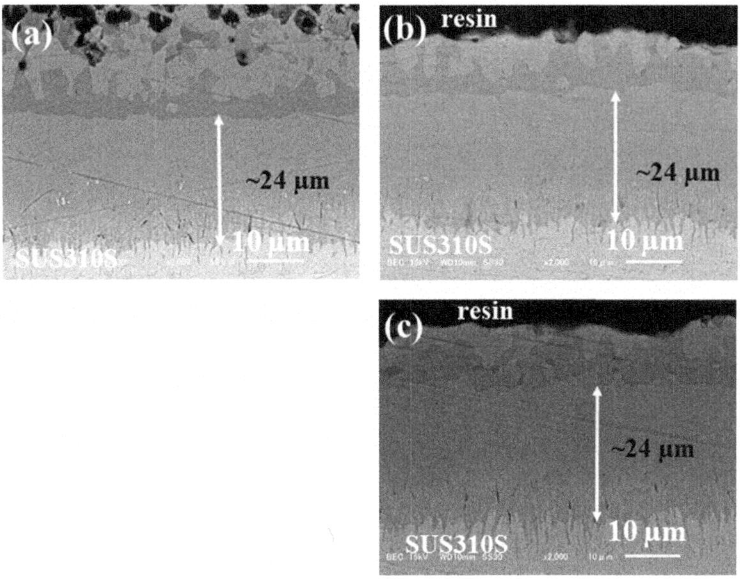

Fig. 7.8 Back-scattered electron image of the cross section of the heat pretreated aluminized 310S stainless steel before **a**, after 72 h **b** and 216 h **c** of immersion in $0.43Li_2CO_3$–$0.32Na_2CO_3$–$0.25K_2CO_3$ kept at 650 °C under flowing CO_2–2%O_2 mixed gas. The pores on the top of **a** must be a part of the porous outer layer [36]

7.5 Corrosion Resistance of Hot-Dip Aluminized Coatings

For evaluating corrosion performance of the interdiffusion–resistant aluminized coating, the porous outer layer was removed by wet-polishing [36]. Basically Al-rich phases of the coatings were removed, and mostly Cr, Fe-enriched phases ($FeCr_{0.5}Al_{0.5}$, Al_3Cr_7, $Al_{16}Cr_{9.5}$, FeAl, FeNi, Fe_3Al and $Al_{0.5}Fe_{1.5}$) were retained. For corrosion test by mass loss, the polished samples were immersed into the carbonates melt at 650 °C for 72 h under flowing CO_2–2%O_2. After chemical etching of the corrosion products, it was found that the mass loss due to corrosion was 7.5–11.8 µg cm^{-2} which is equivalent to a corrosion rate of ~2 µm $year^{-1}$. This corrosion rate was at least two orders of magnitude lower than that of the bare 310S stainless steel [17]. The higher corrosion resistance of the heat pretreated aluminized 310S stainless steel was also confirmed from electrochemical measurements. Anodic polarization gave a comparison of anodic current of the heat pretreated aluminized sample with that of the bare substrate (Fig. 7.11) [36]. The anodic current of the heat pretreated aluminized sample was highly suppressed over a large overpotential (~900 mV); this must be due to the formation of a protective anodic oxide layer. However, such anodic passivity was not observed from the bare substrate. For further evidences of high corrosion resistance of the heat pretreated aluminized samples, EIS spectra were observed during corrosion (Fig. 7.12) [36]. Bode impedance plots

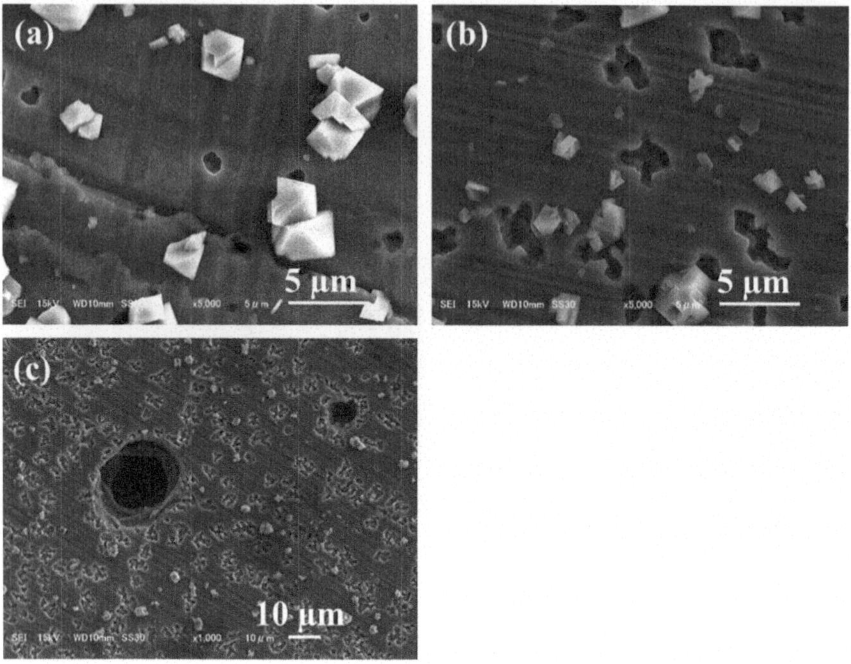

Fig. 7.9 Surface morphology of heat pretreated (800 °C/8 h) aluminized SUS310S after **a** 3 days **b** 6 days and **c** 9 days corrosion in 0.43Li_2CO_3–0.32Na_2CO_3–0.25K_2CO_3 kept at 650 °C under flowing CO_2–2%O_2 mixed gas [52]

showed that the modulus of impedance at 10 mHz of the heat pretreated aluminized sample was ~1.1 × 10^4 Ω cm^2, which is ~550 times higher than that of the bare substrate. The equivalent circuit fitting revealed that R_{ct} of the heat pretreated aluminized 310S stainless steel (sample 1) was 1.3 × 10^4 Ω cm^2. Since the surface of the polished coating was compositionally and microstructurally heterogeneous, manual wet abrasion of the porous layer must have exposed a surface of different composition to the melt. This difference was obvious from the spectra of sample 1 and sample 2 [36]. The modulus of impedance of sample 2 at 10 mHz was 2.2 × 10^4 Ω cm^2. This value is higher than that of sample 1. Moreover, the heat pretreated aluminized coating was observed highly corrosion resistant in the given environment. In addition, EIS was monitored (for sample 1) for 15 days (Fig. 7.13), and it was found that the modulus of impedance (10 mHz) increased slowly after several hours (e.g., 8 h) of corrosion; later the impedance remained constant around the maximum value (~1.4 × 10^4 Ω cm^2) [36]. This retention of high impedance is due to high durability of the passive film in the given corrosive environment.

The high corrosion resistance observed as above is the revelation of protective and durable properties of the passive film in the melt. Figure 7.14 shows the cross-sectional morphology of corrosion layer formed on the heat pretreated aluminized 310S stainless steel by 72 h corrosion [36]. The passive film was dense and continuous

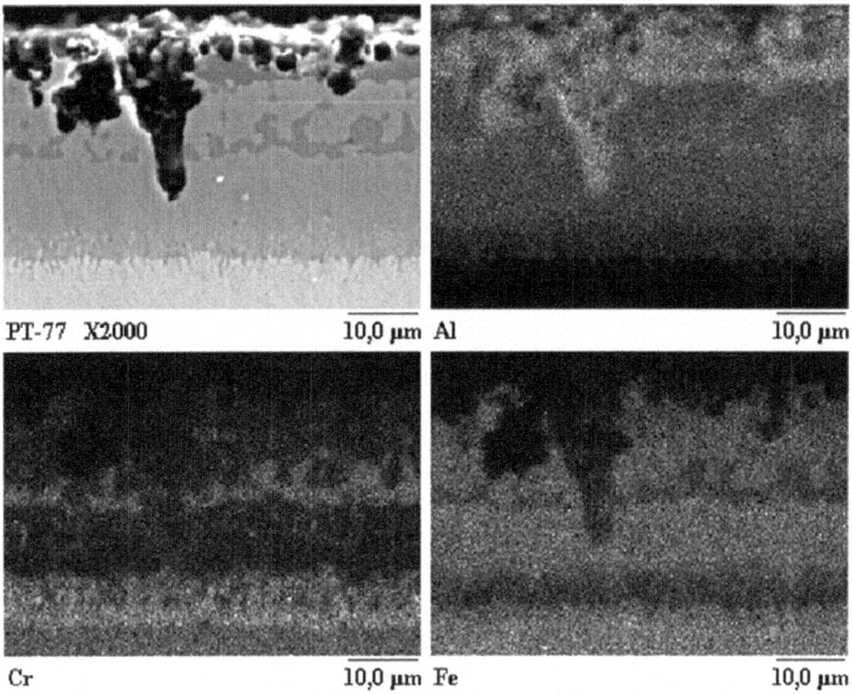

Fig. 7.10 SEM cross-sectional micrograph of a COAT1 (Al-slurry coating) sample after the diffusion treatment and 1000 h of immersion in molten carbonate. Some crevices appear, but most parts of the sample are still protected by the $LiAlO_2$ [27]

over the surface. The passive film was ~54 nm thick and was a composition of 54 at% Al–43 at% O–3 at% Fe. The corrosion layer must be formed by pure alumina (i.e., not containing Cr, Fe and Ni). Grazing X-ray diffraction at a very low angle of 0.2° explored further that the passive film was composed of α-$LiAlO_2$ [36]. The α-$LiAlO_2$ was also detected elsewhere from the corrosion layer of a similar coating [28]. The inner oxide layer did not grow significantly over an extended exposure of 15 days, and the composition of the inner oxide layer formed by the 15 days of corrosion was 58 at% Al–31 at% O–6 at% Fe–3 at% Cr–2 at% Ni which is quite similar to that observed after 72 h of corrosion [36]. Hence, the composition of oxide layer must be retained as α-$LiAlO_2$. Therefore, morphology and composition of the passive film remained almost the same irrespective of corrosion duration.

Formation and stability of the passive film formed on the aluminized/aluminide coating is crucial to protect the underlying coating and the substrate in high-temperature environments [36]. Basically, the composition of coatings determines formation and growth of the protective oxide layer. A passive film composed of α-$LiAlO_2$ is formed on the aluminized coating by the carbonate molten salt corrosion. The stability of the oxide film is very crucial to maintain the passivity over an extended period of corrosion. Generally, the third element effect of Cr is responsible

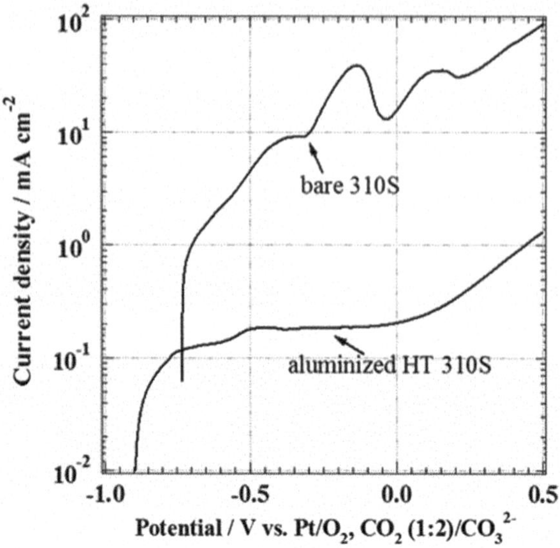

Fig. 7.11 Anodic polarization curves of the aluminized HT 310S stainless steel and bare 310S stainless steel observed after 30 min of immersion in $0.43Li_2CO_3$–$0.32Na_2CO_3$–$0.25K_2CO_3$ kept at 650 °C under flowing CO_2–2%O_2 mixed gas. Scan rate was 10 mV s^{-1} [36]

for formation and regeneration of a continuous and thin alumina layer. Cr comes from the substrate, and it is required in a minimal quantity of 4–5 at% [53] to show the effect. In the study by Sah et al. [36], the 800 °C/24 h heat treatment brought at least 6 at% Cr just below the porous layer of the coating. Therefore, a uniform and protective alumina-based α-$LiAlO_2$ layer was formed. Retention of high impedance (10 mHz) for at least 15 days showed the high protective nature of the α-$LiAlO_2$ passive film. The higher degree of passivation of the sample 2 might be caused by higher surface content of Cr exposed by the polishing. The importance of chemical composition and microstructure of the above hot-dip aluminized coating can be realized from the cross-correlation with other reports. Plasma sprayed FeCrAl coatings [29] showed excellent corrosion behavior; such a high corrosion resistance must be due to formation of the passive film composed of α-$LiAlO_2$. The FeCrAl coatings contained a significant amount of Cr; hence, formation of the protective passive film was likely. The polarization resistance (i.e., R_{ct}) of the hot-dip coating was at least one order of magnitude higher than that of the FeCrAl coating. The high corrosion resistance of the polished heat pretreated hot-dip coating must be due to its beneficial composition and the dense microstructure [36]. The heat pretreated (820 °C, 10 h; Fig. 7.10) Al-slurry coating [27] suffered from a significant localized corrosion that should be caused by the coating's failure to regenerate the passive film. The insufficient alloying with Cr throughout the coating is the reason for the failure in regeneration and maintenance of the passive film. Therefore, coating composition and morphology are very important to provide high corrosion resistance in the given melt environment. The morphology and composition of the polished heat pretreated aluminized coatings

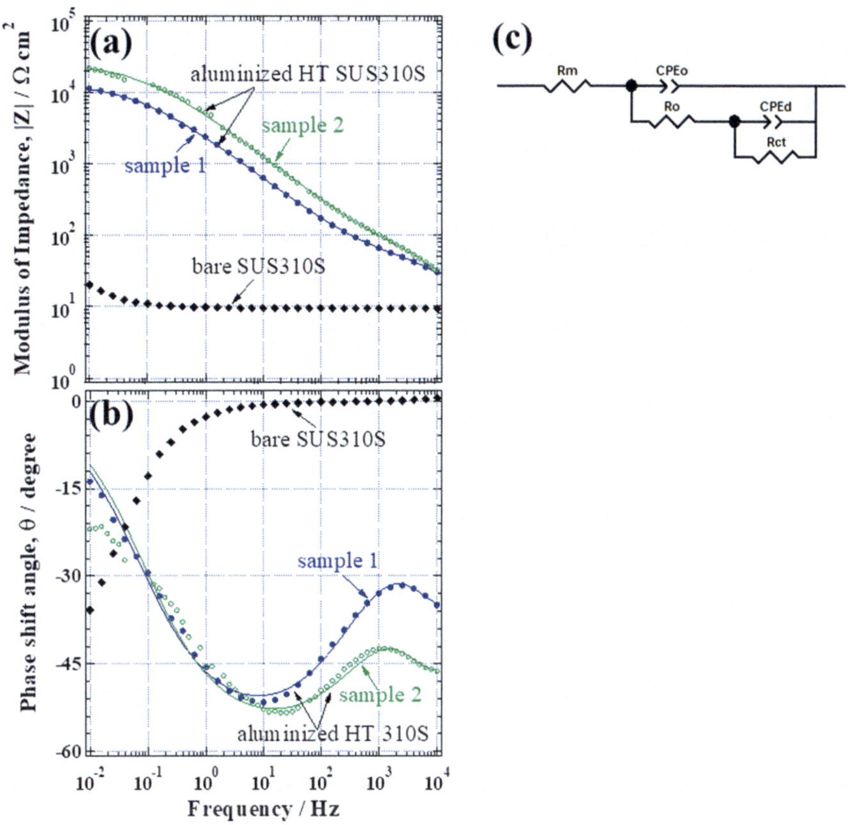

Fig. 7.12 Bode impedance plots **a**, **b** of the aluminized HT 310S stainless steel and the bare 310S stainless steel observed at 72 h of immersion in $0.43Li_2CO_3$–$0.32Na_2CO_3$–$0.25K_2CO_3$ kept at 650 °C under flowing CO_2–2%O_2 mixed gas. Markers and lines are the observed and the fitting data, respectively. **c** is the equivalent circuit for the aluminized HT 310S stainless steel [36]

can be the suitable requirements for commercial coatings. However, commercial applications require the service life of coatings of few decades. Therefore, a further optimization in coating process, heat treatment and polishing is required to produce thicker protective coatings. Although a very thick coating of NiCoCrAlHfSiY was produced by HVOF, the coating contained a lot of fissures [54]. A critical evaluation is required to observe the corrosion rate of such defective coatings.

7.6 Conclusions and Challenges

The temperature of melts and the composition of steels and stainless steels are very crucial parameters to define corrosion resistance in the given molten salt. Increasing

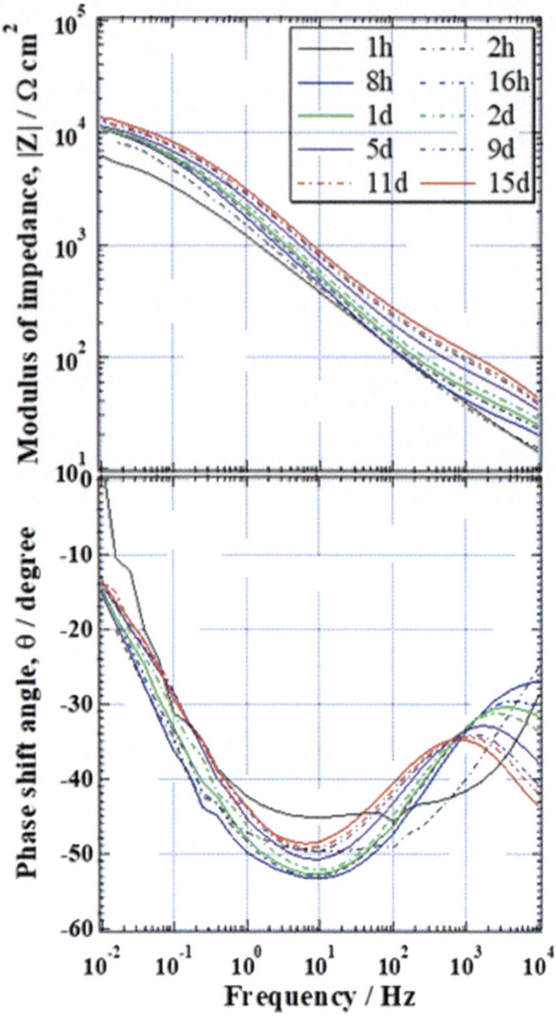

Fig. 7.13 Bode impedance plot of the aluminized HT 310S stainless steel (sample 1) at different immersion time in $0.43Li_2CO_3$–$0.32Na_2CO_3$–$0.25K_2CO_3$ kept at 650 °C under flowing CO_2–2%O_2 mixed gas. '*h*' and '*d*' represent hours and days of the corrosion, respectively [36]

Cr content of the FM steels improves the corrosion resistance in nitrates melt. However, increasing temperature of the melt undermines corrosion resistance, and commercial FM steels cannot be choices for the making containment vessels for molten salts at 600 °C or above. Stainless steels of 18–8 grade like 347, 321 or 304 can provide excellent corrosion resistance at 600 °C, but corrosion resistance is significantly undermined at 650 °C or above. Melt basicity is one of the crucial parameters that affect the corrosion resistance. It was found that corrosion products of 304 stainless steel dissolved significantly into carbonates melt under CO_2–2%O_2

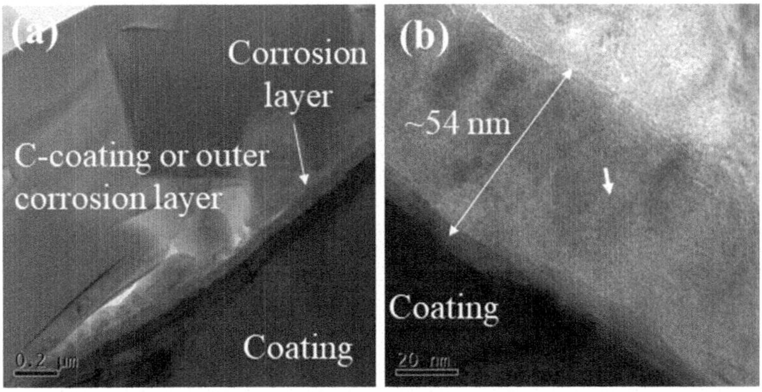

Fig. 7.14 Bright field TEM images of the cross section of corrosion layer formed on the aluminized HT 310S stainless steel after 72 h of immersion in 0.43Li_2CO_3–0.32Na_2CO_3–0.25K_2CO_3 kept at 650 °C under flowing CO_2–2%O_2 mixed gas [36]

gas at 650 °C, and such dissolution undermined the corrosion resistance by 3 orders of magnitude. Mixed oxides of Fe and Cr dissolved with an accelerated rate in the acidic melt of carbonates. A protective coating is required to suppress dissolution of oxides and to enhance corrosion resistance of stainless steels in carbonates melt 650 °C.

Aluminized coatings are economically attractive for enhancing corrosion resistance of promising FM (ferritic–martensitic) steels and stainless steels. Aluminized coatings produce protective passive films due to corrosion. However, interdiffusion can cause the loss of Al into the substrates and can create voids in the coating at high temperature. Such an interdiffusion is unfavorable for Al coating to provide promising corrosion resistance over an extended period. Interdiffusion is slower on austenitic stainless steels than on ferritic steels. A heat pretreatment of 800 °C/24 h is required to produce dense intermetallic diffusion layers on the hot-dip aluminized 310S stainless steel, and such heat pretreated coating is interdiffusion-resistant at 650 °C. The porous outer layer must be ablated by polishing.

The polished heat pretreated coating showed the corrosion rate of ~2 μm $year^{-1}$ in carbonate melts at 650 °C. This low corrosion rate is the consequence of formation of ~50 nm thick and protective passive films of α-$LiAlO_2$. The durability of the passive film comes from the beneficial composition and microstructure of the coating. The passive film can maintain a corrosion rate of 2 μm $year^{-1}$; hence, the expected service life of 25 years for a solar–thermal system requires a coating thickness at least of 50 μm. Unfortunately, the dense layer of the presented coating is ~35 μm. Therefore, coating fabrication and heat pretreatment method should be modified to produce thicker coatings. An easy-and-effective polishing method should be also developed to remove the porous outer layer from the coatings tailored on pipes and tubes. The practical corrosion is dynamic, so corrosion rate in the dynamic system is usually higher than that in the static one. Therefore, further modification in hot-dip, heat pretreatment and polishing procedures is required to produce thicker protective coatings.

Acknowledgements The research work of hot-dip aluminizing of 310 stainless steel was supported by Council for Science, Technology and Innovation (CSTI), Cross-ministerial Strategic Innovation Promotion Program (SIP), 'energy carrier' (Funding agency: Japan Science and Technology Agency, JST).

References

1. Guo S, Zhang J, Wu W, Zhou W et al (2018) Corrosion in the molten fluoride and chloride salts and materials development for nuclear applications. Prog Mater Sci 97:448–487
2. Mudgal D, Singh S, Prakash S et al (2014) Corrosion problems in incinerators and biomass-fuel-fired boilers. Int J Corros 505306
3. Donado RA, Marianowski LG, Maru HC, Selman JR et al (1984) Corrosion of the wet-seal area in molten carbonate fuel cells: I. analysis. J Electrochem Soc 131:2535–2540
4. Walczak M, Pineda F, Fernández ÁG, Mata-Torres C, Escobar RA et al (2018) Materials corrosion for thermal energy storage systems in concentrated solar power plants. Renew Sust Energ Rev 86:22–44
5. Bell S, Steinberg T, Will G et al (2019) Corrosion mechanisms in molten salt thermal energy storage for concentrating solar power. Renew Sust Energ Rev 114:109328
6. Maruyama K, Sawada K, Koike J-i et al (2001) Strengthening mechanisms of creep resistant tempered martensitic steel. ISIJ Int 41:641–653
7. Singh Raman RK, Al-Mazrouee A et al (2007) High-temperature oxidation of Cr-Mo steels and its relevance to accelerated rupture testing and life assessment of in-service components. Metall Mater Trans A 38:1750–1759
8. Cheng W-J, Chen D-J et al (2015) High-temperature corrosion of Cr-Mo steel in molten $LiNO_3$–$NaNO_3$–KNO_3 eutectic salt for thermal energy storage. Sol Energ Mat Sol C 132:563–569
9. Encinas-Sánchez V, de Miguel MT, Lasanta MI, García-Martín G, Pérez FJ et al (2019) Electrochemical impedance spectroscopy (EIS): an efficient technique for monitoring corrosion processes in molten salt environments in CSP applications. Sol Energ Mat Sol C 191:157–163
10. Pérez FJ, Encinas-Sánchez V, Lasanta MI, de Migue MT, García-Martín G et al (2019) Corrosion monitoring of ferritic-martensitic steels in molten salt environments for CSP applications. AIP Conf Proc 2126:030040
11. Soleimani Dorcheh A, Galetz MC et al et al (2016) Slurry aluminizing: a solution for molten nitrate salt corrosion in concentrated solar power plants. Sol Energ Mat Sol C 146:8–15
12. Soleimani Dorcheh A, Durham RN, Galetz MC et al (2016) Corrosion behavior of stainless and low-chromium steels and IN625 in molten nitrate salts at 600 °C. Sol Energ Mat Sol C 144:109–116
13. Kruizenga A, Gill D et al (2014) Corrosion of iron stainless steels in molten nitrate salt. Energ Procedia 49:878–887
14. Encinas-Sánchez V, Lasanta MI, de Miguel MT, García-Martín G, Pérez FJ et al (2021) Corrosion monitoring of 321H in contact with a quaternary molten salt for parabolic trough CSP plants. Corros Sci 178:109070
15. Federsel K, Wortmann J, Ladenberger M et al (2015) High-temperature and corrosion behavior of nitrate nitrite molten salt mixtures regarding their application in concentrating solar power plants. Energ Procedia 69:618–625
16. Zeng CL, Liu Y et al (2011) A comparative study of the corrosion behavior of three stainless steels in an eutectic (Li, Na, $K)_2CO_3$ melt with and without (Na, K)Cl additives at 973 K in air. High Temp Mater Proc 30:161–169

17. Sah SP, Tada E, Nishikata A et al (2018) Corrosion behaviour of austenitic stainless steels in carbonate melt at 923 K under controlled CO_2-O_2 environment. Corros Sci 133:310–317
18. Gallardo-González J, Martínez M, Barreneche C, Fernández AI, Liu M, Steven Tay NH, Bruno F, Segarra M et al (2018) Corrosion of AISI316 as containment material for latent heat thermal energy storage systems based on carbonates. Sol Energ Mat Sol C 186:1–8
19. Pérez FJ, de Miguel MT, Encinas-Sánchez V, Lasanta MI, Illana A, García-Martín G et al (2020) Online corrosion monitoring system for thermal storage tanks using molten salts. Laboratory scale-up to demonstration parabolic-trough plant. AIP Conf Proc 2303:110006
20. Chen RY, Yuen WYD et al (2003) Review of the high-temperature oxidation of iron and carbon steels in air or oxygen. Oxid Met 59:433–468
21. Sah SP (2020) Corrosion of 304 stainless steel in carbonates melt– a state of enhanced dissolution of corrosion products. Corros Sci 169:108535
22. Touhami S, Mainka J, Dillet J, Ait Hammou Taleb S, Lottin O et al (2019) Transmission line impedance models considering oxygen transport limitations in polymer electrolyte membrane fuel cells. J Electrochem Soc 166:F1209–F1217
23. Kitajima Y, Hayashi S, Nishimoto T, Narita T, Ukai S et al (2010) Rapid formation of α-Al_2O_3 scale on an Fe–Al alloy by pure-metal coatings at 900 °C. Oxid Met 73:375–388
24. Airiskallio E, Nurmi E, Heinonen MH, Vayrynen IJ, Kokko K, Ropo M, Punkkinen MPJ, Pitkanen H, Alatalo M, Kollar J, Johansson B, Vitos L et al (2010) High temperature oxidation of Fe–Al and Fe–Cr–Al alloys: The role of Cr as a chemically active element. Corros Sci 52:3394–3404
25. Zeng CL, Wang W, Wu WT et al (2000) Electrochemical-impedance study of the corrosion of Ni and FeAl intermetallic alloy in molten $(0.62Li, 0.38K)_2CO_3$ at 650 °C. Oxid Met 53:289–302
26. Fujimoto N, Yamamoto M, Nagoya T et al (1998) Estimation of the lifetime of Al/Ni-plated material for wet-seal area in molten carbonate fuel cells. J Power Sources 71:231–238
27. Pérez FJ, Duday D, Hierro MP, Gomez C, Aguero A, Garcıa MC, Muela R, Sanchez Pascual A, Martinez L et al (2002) Hot corrosion study of coated separator plates of molten carbonate fuel cells by slurry aluminides. Surf Coat Technol 161:293–301
28. Ni CS, Lu LY, Zeng CL, Niu Y et al (2014) Evaluation of corrosion resistance of aluminium coating with and without annealing against molten carbonate using electrochemical impedance spectroscopy. J Power Sour 261:162–169
29. Aguero A, Garcıa de Blas FJ, Garcıa MC, Muelas R, Roman A et al (2001) Thermal spray coatings for molten carbonate fuel cells separator plates. Surf Coat Technol 146–147:578–585
30. Frangini S, Loreti S, Mascı A et al (2005) Development of new aluminizing method for molten carbonate fuel cell wet seals. Trans ASME 2:60–64
31. Pedraza F, Proy M, Boulesteix C, Krukovskyi P, Metel M et al (2016) Slurry aluminizing of IN 800HT austenitic stainless steel and pure nickel. Correlations between experimental results and modelling of diffusion. Mater Corros 67:1059–1067
32. Audigié P, Bizien N, Baráibar I, Rodríguez S, Pastor A, Hernández M, Agüero A et al (2017) Aluminide slurry coatings for protection of ferritic steel in molten nitrate corrosion for concentrated solar power technology. AIP Conf Proc 1850:070002
33. Audigié P, Encinas-Sánchez V, Juez-Lorenzo M, Rodríguez S, Gutiérrez M, Pérez FJ, Agüero et al (2018) High temperature molten salt corrosion behavior of aluminide and nickel-aluminide coatings for heat storage in concentrated solar power plants. Surf Coat Technol 349:1148–1157
34. Fähsing D, Oskay C, Meißner TM, Galetz MC et al (2018) Corrosion testing of diffusion-coated steel in molten salt for concentrated solar power tower systems. Surf Coat Technol 354:46–65
35. Oskay C, Meißner TM, Dobler C, Grégoire B, Galetz MC et al (2019) Scale formation and degradation of diffusion coatings deposited on 9% Cr steel in molten solar salt. Coatings 9:687
36. Sah SP, Tada E, Nishikata A et al (2018) Enhancing corrosion resistance of type 310S stainless steel in carbonate melt by hot-dip aluminizing. J Electrochem Soc 165:C403–C411
37. Pint BA, Zhang Y et al (2011) Performance of Al-rich oxidation resistant coatings for Fe-base alloys. Mater Corros 62:549–560
38. Lopez C, Kvryan A, Kasnakjian S, Coronado A, Sujittosakul S, Villalpando O, Ravi VA et al (2015) Effect of austenite stability on pack aluminizing of austenitic stainless steels. JOM 67:61–67

39. Shinko-kinzoku co, Aluminizing treatment (in Japanese): http://www.shinko-kinzoku.co.jp/works/aluminais.html Accessed 1 Sep 2022
40. Haynes JA, Zhang Y, Colley KM, Walker L, Reeves KS, Pint BA et al (2004) High-temperature diffusion barriers for protective coatings. Surf Coat Technol 188–189:153–157
41. Narita T (2011) Diffusion barrier coating system concept for high temperature applications. Can Metall Quart 50:278–290
42. Narita T, Izumi T, Nishimoto T, Shibata Y, Thosin KZ, Hayashi S et al (2005) Advanced coatings on high temperature applications. Mater Sci Forum 522–523:1–14
43. Matsumura Y, Fukumoto M, Hayashi S, Kasama A, Iwanaga I, Tanaka R, Narita T et al (2004) Oxidation behavior of a Re-base diffusion barrier/β-NiAl coating on Nb-5Mo-15W at high temperatures. Oxid Met 61:105–124
44. Müller J, Neuschütz D et al (2003) Efficiency of α-alumina as diffusion barrier between bond coat and bulk material of gas turbine blades. Vacuum 71:247–251
45. Kuranishi T, Habazaki H, Konno H et al (2005) Oxidation-resistant multilayer coatings using an anodic alumina layer as a diffusion barrier on γ-TiAl substrates. Surf Coat Technol 200:2438–2444
46. Müller J, Schierling M, Zimmermann E, Neuschütz D et al (1999) Chemical vapor deposition of smooth α-Al_2O_3 films on nickel base superalloys as diffusion barriers. Surf Coat Technol 120–121:16–21
47. Wang QM, Wu YN, Guo MH, Ke PL, Gong J, Sun C, Wen LS et al (2005) Ion-plates Al-O-N films on Ni-base superalloys as diffusion barriers. Surf Coat Technol 197:68–76
48. Knotek O, Lugscheider E, Löffer F, Beele W et al (1994) Diffusion barrier coatings with active bonding, designed for gas turbine blades. Surf Coat Technol 68(69):22–26
49. Knotek O, Löffer F, Beele W et al (1993) Diffusion barrier design against rapid interdiffusion of MCRAlY and Ni-base material. Surf Coat Technol 61:6–13
50. Coad JP, Rickerby DS et al (1985) The use of titanium nitride as a diffusion barrier for M-Cr-Al-Y coatings. Mater Sci Eng 74:93–103
51. Awan GH, ul Hasan F et al (2008) The morphology of coating/substrate interface in hot–dip aluminized steel. Mater Sci Eng A 472:157–165
52. Nishikata A (2018) Mechanism of metallic material corrosion by molten salts and the solution of protecting it, SIP Final Report: https://www.jst.go.jp/sip/dl/k04/end/team1-9.pdf Accessed 1 Sep 2022
53. Brady MP, Smialek JL, Humphrey DL, Smith J et al (1997) The role of Cr in promoting protective alumina scale formation by γ-based Ti-Al-Cr alloys: 2. Oxidation behavior in air. Act Mater 45:2371–2382
54. Gomez-Vidal JC, Noel J, Weber J et al (2016) Corrosion evaluation of alloys and MCrAlX coatings in molten carbonates for thermal solar applications. Sol Energ Mat Sol C 157:517–525

Chapter 8
Solar-Thermal Energy Conversion System: Design and Practice

Seon Tae Kim, Hiroki Takasu, and Yukitaka Kato

Abstract In this chapter, the solar-thermal energy conversion system is investigated with a particular focus on the characterization of raw materials and the feasibility of the thermal driving system. Lithium orthosilicate (Li_4SiO_4) was selected as a suitable material for storing thermal energy at approximately 700 °C, and advanced pelletizing methods were proposed for practical applications. Additionally, special lithium orthosilicate-packed bed reactors (LPRs) and zeolite-packed bed reactors (ZPRs) were designed and developed for thermal driving demonstrations at the laboratory scale. The developed Li_4SiO_4 tablet (K-tablet) with a thermal driving demonstration system showed sufficient potential to be considered for the solar-thermal energy conversion system instead of the existing thermal energy storage (TES) methods.

Keywords Thermochemical energy storage · Chemical heat pump · Lithium orthosilicate · Carbon dioxide · Thermal driving demonstration

8.1 Background

Among the various thermal energy storage (TES) methods, thermochemical energy storage (TCES), which uses reversible endothermic and exothermic chemical reactions for heat storage and output, is proposed as a suitable thermal energy storage technology for solar-thermal conversion systems. TCES technology has several advantages, such as less energy loss during long-term storage, higher thermal energy storage density, and constant temperature output, owing to its chemical equilibrium. TCES technology can also be used in chemical heat pumps (CHPs), which can increase the temperature level by exothermic reaction [1–3].

S. T. Kim
Institute of Low-Carbon Industrial Processes, German Aerospace Center (DLR), Zittau, Germany
e-mail: seon.kim@dlr.de

S. T. Kim · H. Takasu · Y. Kato (✉)
Laboratory for Zero-Carbon Energy, Institute of Innovative Research, Tokyo Institute of Technology, 2-12-1-N1-22, Ookayama, Meguro-ku Tokyo 152-8550, Japan
e-mail: kato.y.ae@m.titech.ac.jp

K. Aika and H. Kobayashi (eds.), *CO_2 Free Ammonia as an Energy Carrier*,
https://doi.org/10.1007/978-981-19-4767-4_8

Since the available temperature range of the TCES method depends on the chemical reaction, each chemical reaction has to be investigated in various temperature ranges. Various chemical reactions, including inorganic chlorides/ammonia, inorganic chlorides, bromides, sulfide/water, metal/hydrogen, and metal oxide/water, have been reported in different temperature ranges [4–11]. However, most of these investigations focused on temperature ranges lower than 400 °C; therefore, there are few reports on the carbonation of metallic oxides, such as CaO and PbO, at relatively high-temperature TCES [12]. These reactions can be described as the gas–solid reactions (Eq. 8.1).

$$M_xO(s) + CO_2(g) \leftrightarrow M_xCO_3(s) \tag{8.1}$$

According to this equation, the equilibrium conditions are determined only by the reaction temperature and CO_2 pressure. Therefore, the endo-/exothermic reaction temperature could be controlled by varying the CO_2 pressure. The CaO/CO_2 system has been reported to be a TCES and CHP material and is used for heat storage at 830 °C with energy output at ~900 °C from the CO_2 storage [12]. However, there is no suitable TCES material at approximately 700 °C even though this temperature level is very important for heat utilization systems, such as solar-thermal power plants, high-temperature gas-cooled reactors, steam reforming processes for hydrogen production, and some high-temperature industrial processes. Thus, studies on TCES at approximately 700 °C are strongly required.

The challenges in achieving high-temperature TCES materials are attributed to their peculiar reaction temperatures. The pressure equilibrium constant for the forward reaction, K_p, is defined by Eq. 8.2 and was calculated for the carbonations of several alkali oxides using a simulation software (HSC chemistry, Outotec), as shown in Fig. 8.1 [13]. The metallic oxide with $-\ln K_p = 0$ at the target temperature is suitable for the temperature range; however, there is no available material at ~700 °C.

$$K_p = \frac{1}{P_{CO_2}/P_0} \tag{8.2}$$

P_{CO_2} CO_2 pressure (kPa)
P_0 standard pressure (kPa)

Carbonation of metallic oxides has also been investigated for the separation of CO_2 gas from the exhaust gas of thermal power plants and high-temperature industrial processes, with the objective of CO_2 capture and storage. Nakagawa and Ohashi (1998) reported a novel method of CO_2 separation at 680 °C using lithium zirconate (Li_2ZrO_3) [14]. Kato and Nakagawa (2001) also found that lithium-containing complex oxides can effectively adsorb CO_2 and show high durability [15]. According to these studies, the reaction temperature can be reduced by developing a composite material containing metallic oxides. Consequently, various types of metal-containing complex oxides, especially those based on lithium, have been extensively studied for CO_2 separation. These oxides are expected to be used not only as CO_2 separators

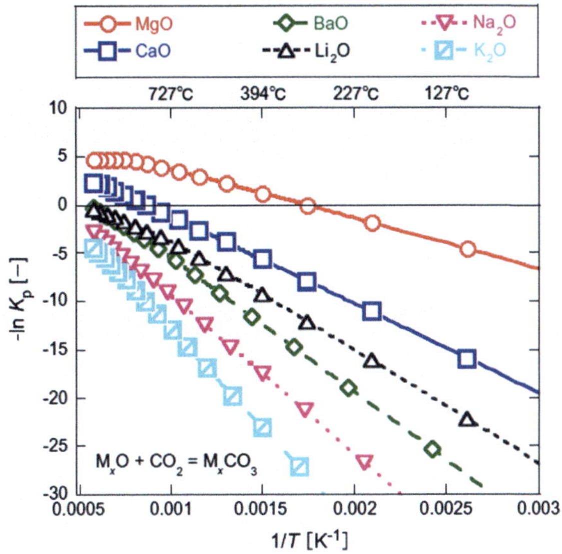

Fig. 8.1 Equilibrium diagram for the carbonation of several alkali and alkaline earth metal oxides [13]

but also as TCES materials at relatively high temperatures. However, there are few studies on metal-containing complexes for TCES materials.

Unlike the abovementioned carbon capture and storage studies that were conducted in a flow reactor or under low CO_2 concentrations, we considered a 100% CO_2 batch reactor for the TCES system. A lithium orthosilicate (Li_4SiO_4), which is one of the best materials for CO_2 separation, was investigated for application as a TCES material at approximately 700 °C.

8.2 Solar Heat Storage Material and Potential Evaluation

8.2.1 Preparation and Carbonation of Lithium Orthosilicate

Among the various TCES reactions, a lithium orthosilicate/carbon dioxide (Li_4SiO_4/CO_2) reaction system has been proposed for high-temperature thermal energy storage at approximately 700 °C.

$$Li_4SiO_4(s) + CO_2(g) \leftrightarrow Li_2CO_3(s) + Li_2SiO_3(s) \quad \Delta H_r = -94\,\text{kJ/mol} \tag{8.3}$$

The carbonation of Li_4SiO_4 is an exothermic reversible reaction. In this reaction, Li_2CO_3 reacts with Li_2SiO_3 to produce Li_4SiO_4 and CO_2 in the reverse endothermic decarbonation reaction. Li_4SiO_4 also has an excellent CO_2 storage

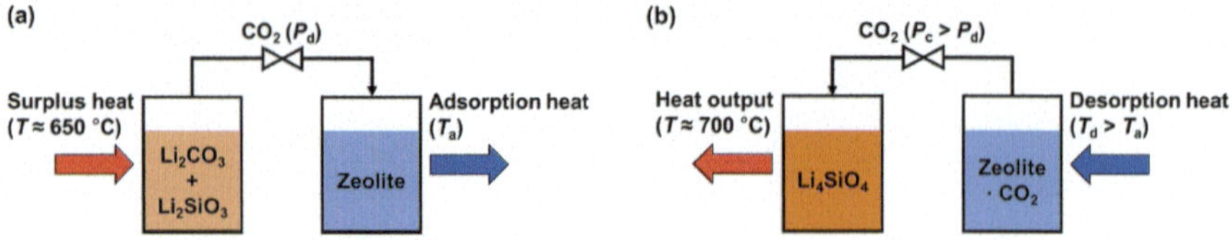

Fig. 8.2 Principle of Li_4SiO_4/CO_2/zeolite chemical heat pump system: **a** thermal energy storage process by Li_2CO_3 decarbonation and zeolite CO_2 adsorption, **b** thermal energy output process by Li_4SiO_4 carbonation and zeolite CO_2 desorption

capacity (36.7 wt%) and a high durability for cyclic experiments. Owing to these advantages, the Li_4SiO_4 and Li_4SiO_4/CO_2/zeolite system was proposed as a practical TCES system operating at approximately 700 °C. Figure 8.2 shows a schematic diagram of the Li_4SiO_4/CO_2/zeolite TCES system. Decarbonation (an endothermic reaction) is a thermal energy storage process (Fig. 8.2a), while carbonation (an exothermic reaction) is a process that releases thermal energy (Fig. 8.2b). This TCES system is expected to store the surplus thermal energy and release it when needed using low-grade heat.

As a first step, the reaction behavior and potential of Li_4SiO_4 as a thermochemical energy storage material were studied in a raw powder state; Li_4SiO_4 was prepared using a solid-state reaction as follows.

$$2Li_2CO_3(s) + SiO_2(s) \rightarrow Li_4SiO_4(s) + CO_2(g) \quad (8.4)$$

First, the starting materials, lithium carbonate Li_2CO_3 (99.9%) and SiO_2 (99.0%) (both obtained from Wako Pure Chemical Industries, Ltd.), were thoroughly mixed using mortar and pestle in the molar ratio of 2.0:1.0. Next, the mixture was heated at 700 °C for 15 h under an argon atmosphere to form CO_2 and Li_4SiO_4. Furthermore, mortar and pestle were used to grind the Li_4SiO_4 sample into a powder with mean particle diameter of approximately 20 μm. Finally, X-ray diffraction (XRD, MiniFlex 600, Rigaku) and scanning electron microscopy (SEM, SM-200, Topcon Corp.) were used to characterize the synthesized Li_4SiO_4.

8.2.2 Closed Thermogravimetric Experiments

The potential of Li_4SiO_4 as a TCES material was investigated using a thermogravimetric analysis (TG) system (TG-9600S, Ulvac Shinku-Riko Inc.) under various conditions (Fig. 8.3).

The CO_2 buffer tanks were connected to the thermobalance to avoid pressure changes during the reaction. The absolute CO_2 pressure in the reactor was measured using a pressure gauge, and the pressure in the reactor was controlled by the CO_2 supply and evacuation system. Each sample was pretreated to remove the gas initially adsorbed at room temperature. After obtaining the vacuum state, the reactor was set

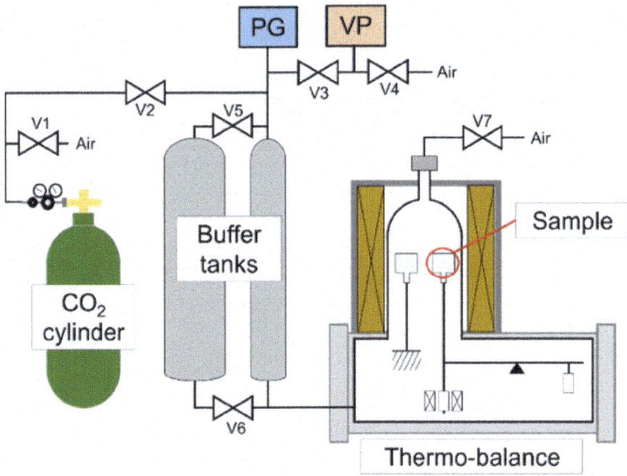

Fig. 8.3 Schematic diagram of the experimental apparatus for closed thermogravimetric analysis (*PG* pressure gauge, *V* stop valve, *VP* vacuum pump)

to the experimental temperature at a constant heating rate of 5 °C/min, and CO_2 was introduced at 400 °C to avoid side reactions.

The cyclic experiments, temperature swing (TS), and pressure swing (PS) were also conducted to study the material durability, and the reacted mole fraction, *x*, was calculated from the change in mass of the sample during the experiment as follows:

$$x = \frac{\Delta m / M_{CO_2}}{m_0 / M_{Li_4SiO_4}} \tag{8.5}$$

Δm	change in mass of the sample (g)
M_{CO_2}	molecular weight of CO_2 (g/mol)
$M_{Li_4SiO_4}$	molecular weight of Li_4SiO_4 (g/mol)
m_0	initial mass of Li_4SiO_4 in the sample or system (g)

8.2.3 *Non-isothermal Experimental Results and Discussion (Li_4SiO_4 Powder)*

Non-isothermal carbonation was conducted at various CO_2 pressures (10, 40, 80, and 100 kPa) to determine the reaction behavior of Li_4SiO_4. Figure 8.4 shows the pressure dependence of the carbonation reactivity of Li_4SiO_4 between 400 and 900 °C. All carbonation started over 500 °C, and the reactivity increased suddenly at approximately 700 °C, especially under 100 kPa. Conversely, decarbonation started at different temperatures depending on the CO_2 pressure.

The peaks of the *x* profiles are assumed to be the temperatures at which decarbonation starts under CO_2 pressures of 10, 40, 80, and 100 kPa, and the peak temperature increases with increasing CO_2 pressure. From Fig. 8.4, the relationship between

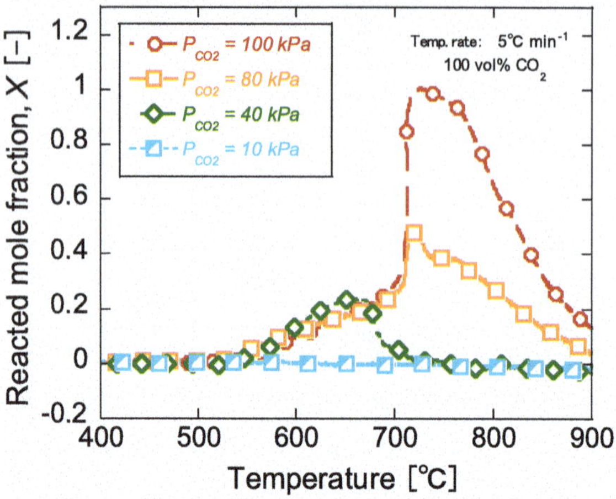

Fig. 8.4 Pressure dependence of Li_4SiO_4 reactivity between 400 and 900 °C

the CO_2 pressure and equilibrium temperature can be obtained. The Gibbs energy change, ΔG, and enthalpy change, ΔH, can be defined by Eqs. 8.6 and 8.7.

$$\Delta G = -\mathrm{R}T \ln K_P \tag{8.6}$$

ΔG Gibbs energy change of Eq. 8.3 (kJ/mol)
R Gas constant [kJ/(K mol)]
T Equilibrium temperature (K)

$$\ln\left(P_{\mathrm{CO}_2}/P_0\right) = \frac{\Delta G}{\mathrm{R}T} = \frac{\Delta H}{\mathrm{R}T} - \frac{\Delta S}{\mathrm{R}} \tag{8.7}$$

ΔH Enthalpy change of Eq. 8.3 (kJ/mol)
ΔS Entropy change of Eq. 8.3 [kJ/(K mol)]

The obtained ΔH value was −109 kJ/mol, which is in line with the value in literature (−94 kJ/mol at 800–1000 K) [16]. This means that the carbonation of Li_4SiO_4 is a dominant reaction under the given conditions, and Li_4SiO_4 is a promising candidate material for use in TCES systems at a working temperature of approximately 700 °C.

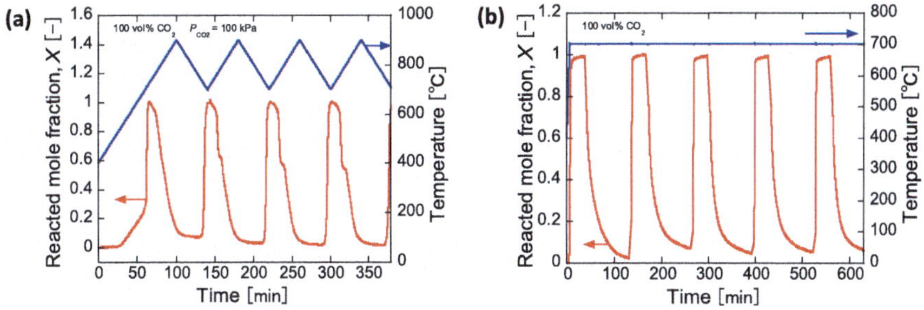

Fig. 8.5 Durability of Li_4SiO_4 reactivity in **a** TS experiment at 400 °C under a CO_2 pressure of 100 kPa and **b** PS experiment at 700 °C (carbonation: $P_{CO_2} = 80$ kPa for 30 min and decarbonation: $P_{CO_2} = 10$ kPa for 90 min)

8.2.4 *Cyclic Experimental Results and Discussion (Li_4SiO_4 Powder)*

Cyclic experiments were performed using the TS and PS methods to investigate the cyclic durability of Li_4SiO_4 at high temperatures. For the TS experiment, the temperature was varied four times between 700 and 900 °C under 100 kPa of CO_2. Except for the first cycle in the TS experiment, Li_4SiO_4 showed a maximum reacted mole fraction for carbonation of 1.0, and the minimum mole fraction for decarbonation was 0.04, as shown in Fig. 8.5a. The material durability was also investigated through a PS experiment $\left(T_{CO_2} = 700^{\circ}\text{C}, P_{CO_2} = 10 - 80\text{kPa}\right)$, shown in Fig. 8.5b. In the PS experiment, the maximum value of x reached almost 1.0, and the minimum value of x was less than 0.07.

The carbonation process was very rapid, finishing within the first 5 min of each cycle. This was due to a considerable enhancement of the CO_2 diffusivity resulting from Li_2CO_3 melting at 700 °C. The decarbonation process also proceeded rapidly, and x decreased to less than 0.07 after 90 min. In summary, Li_4SiO_4 showed sufficient cyclic durability and high potential for use as a high-temperature TCES material at around 700 °C.

8.2.5 *Kinetic Analysis (Li_4SiO_4 Powder)*

The volumetric thermal energy, q_v (kJ/L), and gravimetric thermal energy density, q_g (kJ/kg), of Li_4SiO_4 can be calculated using Eqs. 8.8 and 8.9.

$$q_v = \frac{-\Delta H \rho (1 - \varepsilon)}{M_{Li_4SiO_4} \times 100} \tag{8.8}$$

$-\Delta H$ heat of reaction (kJ/mol)

$M_{Li_4SiO_4}$ molecular weight of Li_4SiO_4 (g/mol)
ρ density of Li_4SiO_4 (g/cm^3)
ε porosity of Li_4SiO_4 (–)

$$q_g = \frac{-\Delta H}{M_{Li_4SiO_4}} \tag{8.9}$$

The porosity of Li_4SiO_4, ε, was 0.59, measured by the gravimetric method, and the heat of reaction, $-\Delta H$, was 94 kJ/mol in the range of 800–1100 K. The molecular weight M and density ρ of Li_4SiO_4 is 120 g/mol and 2.33 g/cm.

The calculated thermal energy densities of Li_4SiO_4 powder were 750 kJ/L and 780 kJ/kg, respectively. Additionally, an estimation on the performance of Li_4SiO_4 CHP, which can store thermal energy at 650 °C and release it at 700 °C, showed that 14% of $-\Delta H$ was used for self-heating from 650 to 700 °C, decreasing the thermal energy densities to 650 kJ/L and 670 kJ/kg.

8.3 Lithium Orthosilicate Pellets for High-Temperature Chemical Heat Pump

8.3.1 Lithium Orthosilicate Tablet (K-Tablet)

To date, the potential of Li_4SiO_4 as a high-temperature TCES material has been investigated, and the Li_4SiO_4/CO_2/zeolite CHP system integrated into the thermal energy grid is expected to play an important role in the stable supply of renewable energy, as shown in Fig. 8.6 [17]. The heat source heats the cold heat transfer fluid (HTF), which in turn charges the CHP system through the decarbonation of Li_2CO_3 (Fig. 8.6a). When needed, the stored thermal energy of the HTF is released to the power plant either directly or indirectly through the heat exchanger, Fig. 8.6b.

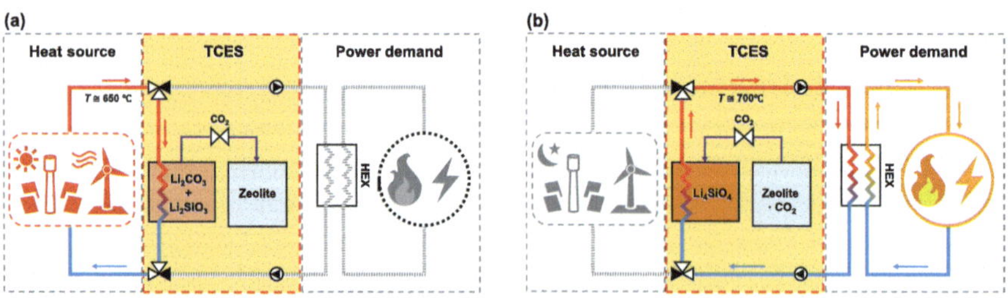

Fig. 8.6 Schematic diagram of Li_4SiO_4/CO_2/zeolite chemical heat pump system integrated into the thermal energy grid: **a** thermal energy storage process and **b** thermal energy release process

Although CHP technology has many advantages, it is still in the research and development stage. Some of the main obstacles in the actual deployment of CHP systems are the low reactivity and low effective thermal conductivity of natural raw materials [18]. Therefore, most CHP working pairs have been investigated in powder form. However, it has been difficult to solve the problem of performance degradation of CHP materials in practical shapes, such as spheres, tablets, and pellets. [19].

In this section, a Li_4SiO_4 tablet (K-tablet) was developed for practical application in a CHP system with a Li_4SiO_4/CO_2 working pair, and the potential of the K-tablet was evaluated through thermogravimetric experiments. The developed K-tablet constitutes a lithium orthosilicate-packed bed reactor (LPR) to pair with the zeolite-packed bed reactor (ZPR), as shown in Fig. 8.2.

8.3.2 Making K-Tablet and Open Thermogravimetric Experiment

Lithium carbonate (Li_2CO_3, 99.0%), silicon dioxide (SiO_2, 99.0%), and potassium carbonate (K_2CO_3, 99.0%), acquired from Wako Pure Chemical Industries, Ltd., were prepared as precursors. K_2CO_3 was added to enhance the CO_2 sorption property; chemical sorbents doped with other alkali metals (Na and/or K) are expected to enhance the CO_2 uptake rate and sorption capacity [20, 21]. Potassium titanate ($K_2O_8TiO_2$, Otsuka Chemical, TISMO D) and carbomethoxymethyl (CMC, AS ONE) were also added as additives to form and sustain the tablet shape. The Li_4SiO_4 tablet (K-tablet) molding process is as follows:

1. Li_2CO_3, SiO_2, and K_2CO_3 were mixed in a molar ratio of 2.0:1.0:0.1, using a pestle and mortar.
2. The other components, 10 wt% $K_2O{\cdot}8\ TiO_2$ and 5 wt% CMC, were added to the powder mixture as a binder, and then mixed with purified water (20 wt%).
3. The powder mixture was then compressed into tablets ($\phi 7.0 \times 4.0$ mm^2, $\rho = 0.84$ g/cm^3) at 10 MPa using a stainless-steel molding set.
4. The tablets were sintered and precast under an argon atmosphere at 850 °C for 12 h in a tubular electric furnace.

Unlike other pelletizing methods, this process has only one sintering stage; hence, the build time is significantly reduced [22, 23]. The morphology of the K-tablet produced is shown in Fig. 8.7. The scanning electron microscopy images showed that all the materials were mixed homogeneously, and a porous structure was well-formed, augmenting CO_2 diffusivity in the tablet.

The characterization of the produced K-tablet was carried out using an open thermogravimetric system (TG-9600, Ulvac Shinku-Riko, Inc.) (Fig. 8.8). Isothermal carbonation, decarbonation, and 10 cyclic experiments were conducted in argon and carbon dioxide atmospheres in succession, with the temperature set to 600, 650, and 700 °C during each round. Pure Li_4SiO_4 and Li_2CO_3/Li_2SiO_3 powders were also evaluated under the same conditions for comparison. Before the experiment,

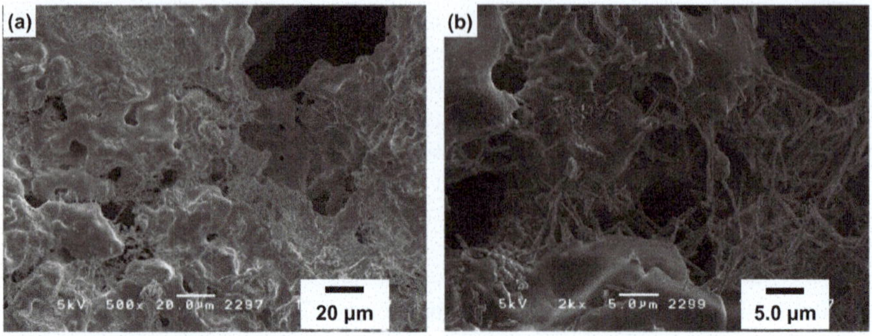

Fig. 8.7 Scanning electron microscopy (SEM) images of the K-tablet

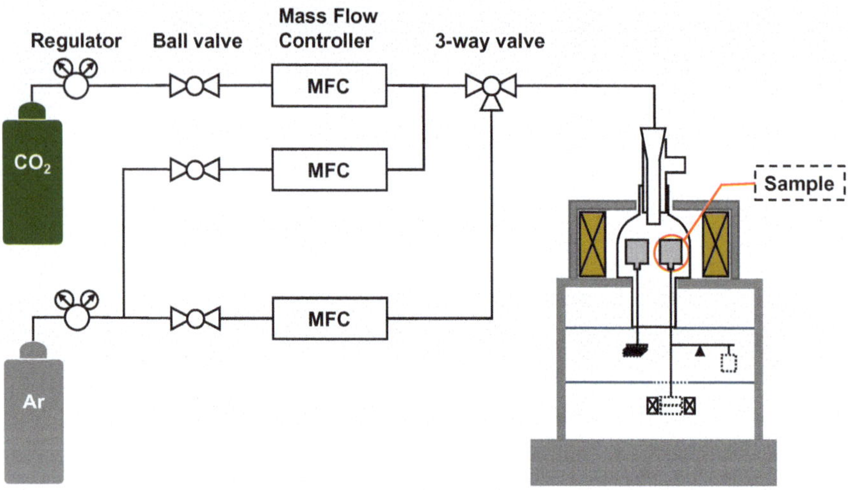

Fig. 8.8 Schematic diagram of the opened thermogravimetric analysis system

a pretreatment was performed at 700 °C for 90 min under an argon atmosphere to remove the CO_2 absorbed by Li_2SiO_3 at room temperature. After pretreatment, the reactor was set to the target temperature, and the temperature was increased at a constant rate of 50 °C/min. To evaluate the reactivity of the produced K-tablet, the reacted mole fraction (Eq. 8.5) was used.

8.3.3 Reaction Mechanism (Double-Shell Model)

The improvement of CO_2 uptake by forming a molten carbonate shell was verified, and a carbonation mechanism is proposed in this section. Nakagawa (2004) pointed out that an additional amount of molten alkali carbonate may improve the CO_2

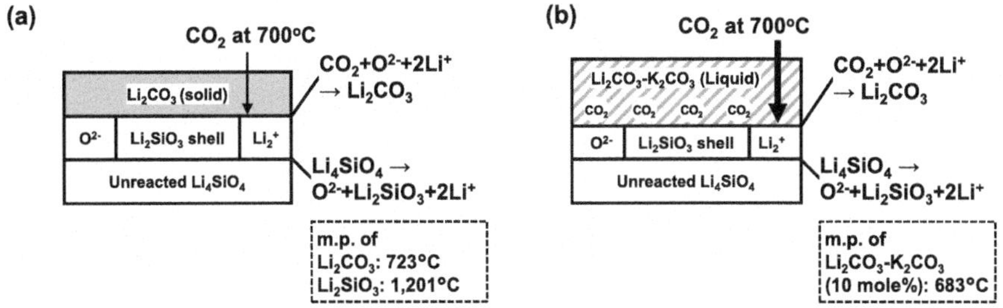

Fig. 8.9 Schematic illustration of carbonation mechanism of **a** pure Li_4SiO_4 and **b** modified Li_4SiO_4

diffusion rate in the carbonate shell and enhance the CO_2 sorption capacity above what is observed with the solid carbonate shell alone [24]. Olivares-Marin et al. (2010) investigated the rheological properties of alkaline elements in Li_2ZrO_3 under a CO_2 atmosphere [25]. In addition, Pfeiffer et al. (2007) found that the solid solution of $Li_{2-x}Na_xZrO_3$ first chemisorbed CO_2 through the formation of a carbonate shell on the surface at low temperatures (473–573 K). Then, Li and/or Na atoms diffused from the core to the surface of the particles through the external carbonate shell when the temperature reached 673 K or higher [26].

The double-shell reaction model was proposed by integrating the above studies, as shown in Fig. 8.9 [27]. The carbonation of Li_4SiO_4 is accelerated through a carbonation mechanism with the formation of a solid Li_2CO_3-Li_2SiO_3 layer surrounding the unreacted core of Li_4SiO_4. The formation and growth of the external shell limit the diffusion of gases and ions (Li^+ or O^{2-}), leading to a decreased rate of CO_2 sorption (Fig. 8.9a). Desorption of CO_2 occurs accompanied by the regeneration of Li_4SiO_4 through the decomposition of Li_2CO_3 [28]. As explained above, it is generally accepted that the introduction of alkaline elements (K_2CO_3 in this study) into Li_4SiO_4 changes the melting points of the system, and a liquid eutectic (mixed-salt in molten form) shell is formed on the outer surface of Li_4SiO_4, thereby significantly improving the CO_2 diffusion rate of the samples (Fig. 8.9b [29]).

8.3.4 *Experimental Results and Discussion (K-Tablet)*

The temperature dependency on carbonation and decarbonation of the K-tablet and of the powder state of Li_4SiO_4 and $Li_2CO_3 + Li_2SiO_3$ is shown in Fig. 8.10. The K-tablet exhibited superior carbonation performance compared to the Li_4SiO_4 powder at all temperatures except 700 °C (Fig. 8.10a). The fine particle state of Li_4SiO_4 powder, which has a high effective surface area, exhibited better carbonation reactivity at the initial temperature of 700 °C; the equilibrium temperature of Li_4SiO_4 carbonation at $P_{CO_2} = 100$ kPa is approximately 700 °C. However, the introduction of K_2CO_3 induced a melting point decrease in Li_2CO_3 and enhanced carbonation reactivity of the K-tablet at 650 °C and 600 °C, although both Li_4SiO_4 and K_2CO_3 have higher melting points than the experimental temperatures of 1255 °C and 891 °C, respectively.

According to Fig. 8.10b, although the initial decarbonation reactivity of Li_2CO_3 + Li_2SiO_3 powder was lower than that of the K-tablet, both materials showed sufficient reactivity at 700 °C; however, the decarbonation reactivity of the Li_2CO_3 + Li_2SiO_3 powder significantly decreased at 690 °C. In contrast, the superior thermal energy storage capacity of the K-tablet at low temperatures was confirmed by its high decarbonation reactivity at 650 and 600 °C. According to the double-shell reaction model described in the previous section, the diffusivity of decarbonized CO_2 is limited by the internal Li_2SiO_3 shell at low temperature, whereas at 700 °C, Li_2CO_3 (T_m = 723 °C) is slightly in a molten state and accelerates the decarbonation process. This liquid layer also results in a sharply decreased decarbonation rate of the K-tablet after the initial 10 min.

The durability of the produced K-tablet was also investigated through a cyclic experiment. The initial carbonation reactivity on the 10th cycle was higher than that of the 1st cycle, and the reacted conversion value, Δx, stabilized with increasing number of cycles and converged to 0.75, as shown in Fig. 8.11a. The initial decarbonation reactivity also slightly increased after 10 cycles (Fig. 8.11b). Both carbonation and decarbonation reacted mole fraction values converged to a similar value at the 10th cycle, regardless of the reaction temperature. This means that even if the reaction temperature influences the reaction rate, there is not much difference in the total thermal energy capacity at temperatures >600 °C. The exterior and morphological structure of the K-tablet after the 10-cycle experiment at 700 °C was confirmed using a digital microscope (VHX-100, KEYENCE) and a scanning electron microscopy (Fig. 8.12). The K-tablet not only maintained its external tablet shape without any deformation, as shown in Fig. 8.12a, but also preserved its porous structure on the surface efficiently (Fig. 8.12b), even though the surface became smooth (Fig. 8.12c).

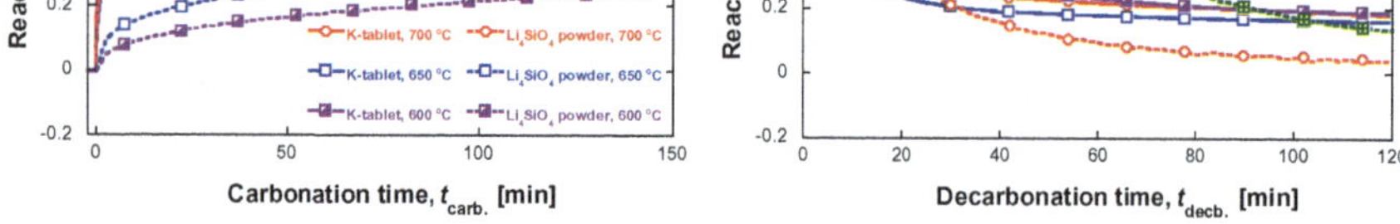

Fig. 8.10 Isothermal **a** carbonation and **b** decarbonation under various temperature

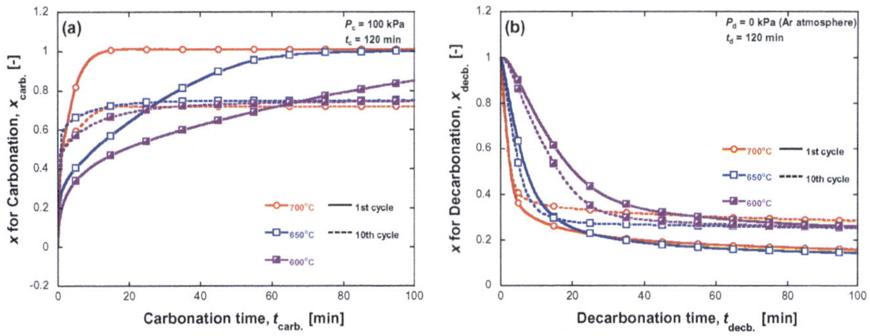

Fig. 8.11 Comparison of 1st and 10th cycles of the cyclic experiment: **a** carbonation and **b** decarbonation

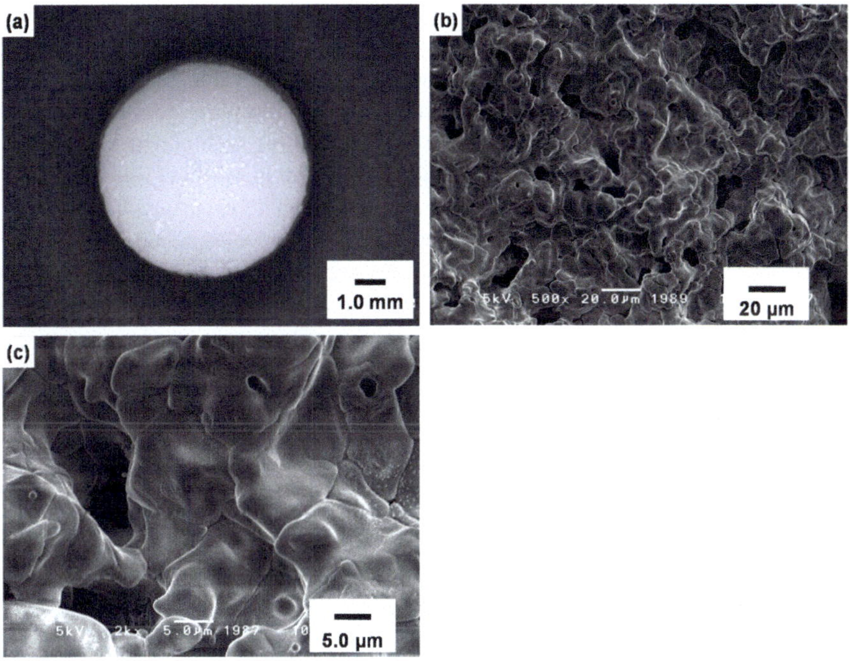

Fig. 8.12 Image of K-tablet after the 10-cycle experiment at 700 °C: **a** exterior, and **b**, **c** morphology structure

8.3.5 *Kinetic Analysis (K-Tablet)*

The total thermal energy output capacities, Q (kJ/kg), and storage and output rate, W (kW/kg), of the K-tablet were estimated from the measured reacted mole fraction, and defined by the following equations:

$$Q = q_g \cdot \Delta x \cdot \alpha \tag{8.10}$$

$$W = \frac{dQ}{dt} \tag{8.11}$$

where t [s] is the time interval during the experiment, and α is the ratio of contained Li_4SiO_4 in the K-tablet.

Comparing the K-tablet and the Li_4SiO_4 powder (Fig. 8.13), the Li_4SiO_4 powder exhibited better thermal energy output performance than the K-tablet at 700 °C owing to the low Li_4SiO_4 content (~77%) of the K-tablet. However, a superior performance of the K-tablet can be achieved at low temperatures; the estimated performance of the K-tablet is at least three times higher than that of the Li_4SiO_4 powder at low temperatures (650 and 600 °C), despite its low Li_4SiO_4 content and tablet shape.

In the cyclic experiment, all Q_{output} values converged to ~450 kJ/kg after the 10th cycle, and the thermal energy output rate, W_{output}, at the 10th cycle (Fig. 8.14a) showed higher values compared to those of the 1st cycle. Specifically, carbonation at temperatures below 650 °C exhibited the highest W_{output} value, 9.82 kW/kg, and the values at 600 °C and 700 °C were 8.43 kW/kg and 7.58 kW/kg, respectively. The W_{output} values over 7.00 kW/kg at the 10th cycle prove the high potential of the K-tablet as a CHP material for practical applications. Conversely, there was no change in the thermal energy storage rate, W_{store}, during the cyclic experiment Fig. 8.14b. The W_{store} values increased with increasing reaction temperature, and these values slightly increased after 10 cycles of CO_2 diffusivity enhancement [30].

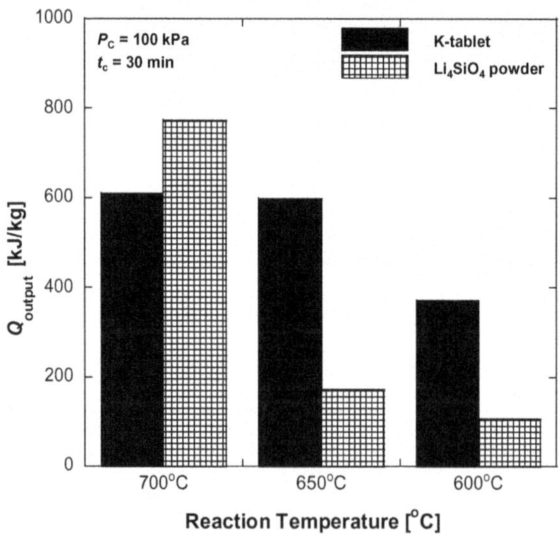

Fig. 8.13 Changes in the thermal energy output capacity, Q_{output}, during the initial 30 min of the isothermal experiment

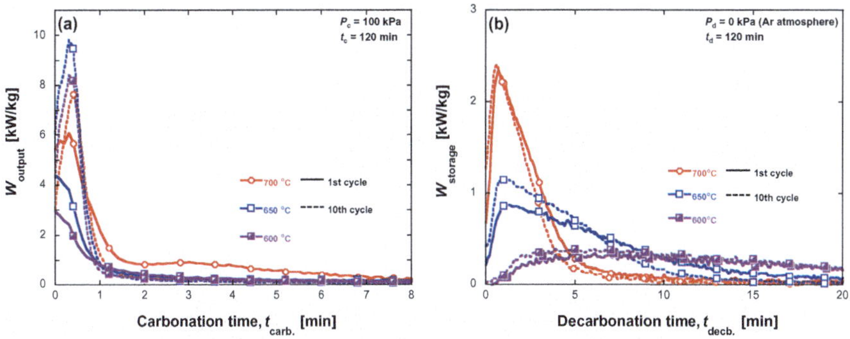

Fig. 8.14 Comparison of thermal energy capacities of the 1st and 10th cycles: **a** thermal energy output rate, W_{output}, and **b** thermal energy storage rate, W_{store}

8.4 Thermal Driving Demonstration of Li_4SiO_4/CO_2/Zeolite Thermochemical Energy Storage System

8.4.1 *Introduction of Thermal Driving Demonstration*

In this section, two different large-scale reactors, a lithium orthosilicate-packed bed reactor (LPR) and a zeolite-packed bed reactor (ZPR), were designed, and the thermal driving feasibility of a high-temperature thermochemical energy storage system was demonstrated without any mechanical power.

The advanced Li_4SiO_4 tablet (K-tablet) was mass-produced (Fig. 8.15) according to the method described in Sect. 8.3.2, and 329 g of the K-tablet was used for the LPR in this study.

8.4.2 *Packed Bed Reactors and Experimental Procedure*

The two reactors, LPR and ZPR (SUS 304), were designed considering the actual operating conditions. LPR operated under high temperatures (around 700 °C) in a vacuum state, and ZPR implemented rapid temperature changes during a specific time. In addition, thermocouples were set in each reactor to measure the temperature changes during the experiment.

Figure 8.16 shows the cross-section of the vacuum chamber and the top view of the LPR. Three alumina crucibles (SSA-H-T8, OD × ID × L = 50.0 × 40.0 × 180 mm, volume = 225 mL) with a sheath heater were used as the electric furnaces. The electric furnaces were fixed in a vacuum chamber, which was encased by ribbon and jacket heaters. The ZPR was designed for effective heating and cooling of the zeolite bed as a CO_2 reservoir. The ZPR consisted of two layers of the cooling tube

Fig. 8.15 Exterior image of mass-produced K-tablet

and sheath heater on the inner and outer sides of the vacuum chamber as shown in Fig. 8.17. The volume of the vacuum chamber was 4.9 L, and a metal mesh tube was positioned at the center of the reactor to enhance the CO_2 distribution.

A schematic diagram of the thermal driving demonstration system for the Li_4SiO_4/CO_2/zeolite TCES system is shown in Fig. 8.18. The two reactors were connected by a flexible and rigid pipe with a solenoid valve and were mounted on an electric balance to measure the weight change during the experiment. The solenoid valve controlled the CO_2 supply pressure by receiving the current value from the pressure gauge setting. All data relating to temperature, pressure, and mass were collected using a data acquisition system.

The experimental procedure can be divided into two processes for each reactor: carbonation of LPR/desorption of ZPR (CD experiment) corresponding to the thermal

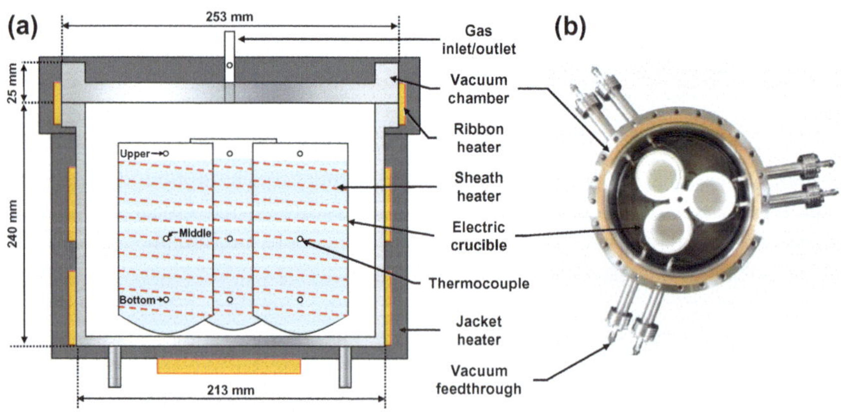

Fig. 8.16 Image of **a** the cross-section of the vacuum chamber and **b** the top view of the LPR

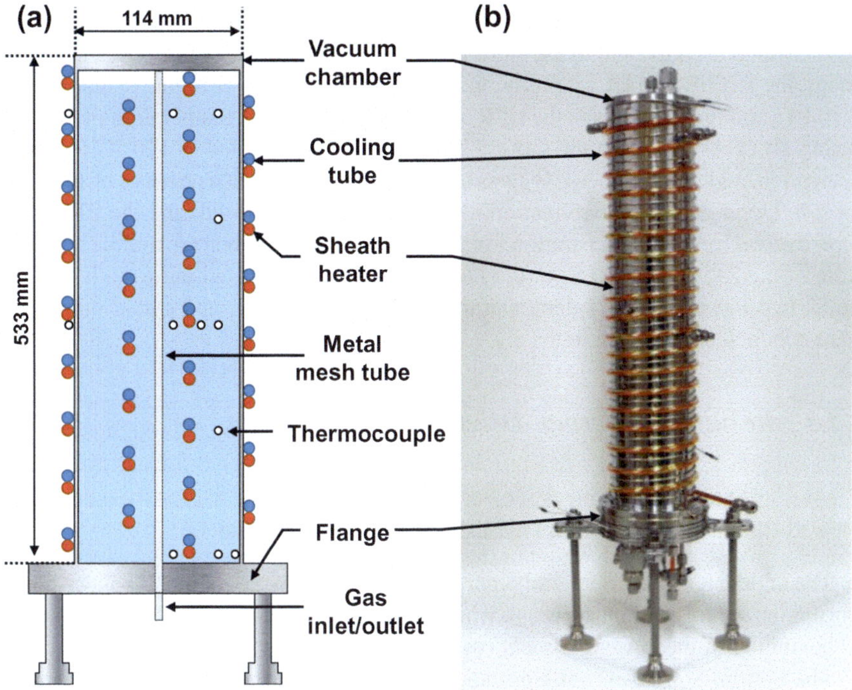

Fig. 8.17 Image of **a** the cross-section of the vacuum chamber and **b** the side view of the ZPR

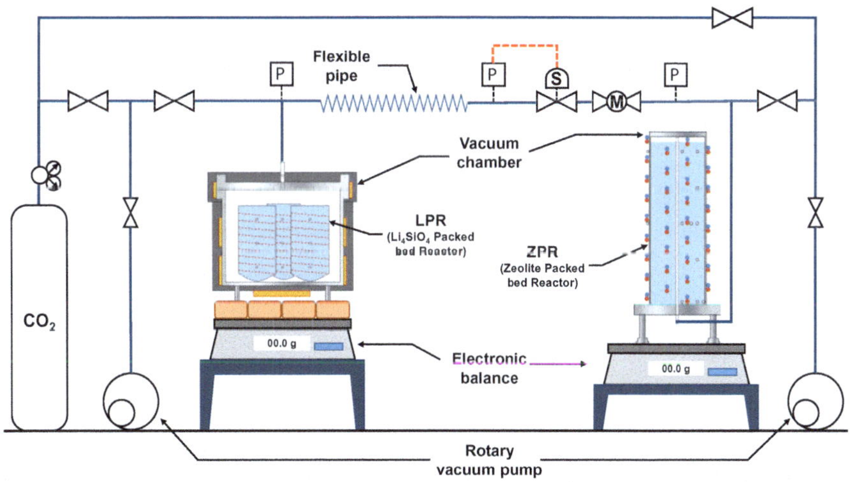

Fig. 8.18 Schematic diagram of the thermal driving demonstration system

energy output process and decarbonation of LPR/adsorption of ZPR (DA experiment), which is equivalent to the thermal energy storage process. Before each experiment, the Li_4SiO_4 tablet was fully decarbonated by heating it in the vacuum state (~10 Pa), and the crucibles in the LPR were kept at a constant temperature of 715 °C during all experiments. In the CD experiment, ZPR was heated to 120 °C for CO_2 desorption, and the desorbed CO_2 was introduced to the LPR at a pressure of 150 kPa for 4 h. Upon complete carbonation, the main valve was closed, and the ZPR was cooled down for 2 h, in preparation for the DA experiment. When the middle temperature of the ZPR reached 15 °C, the DA experiment was carried out for 6 h. In this study, two thermal driving demonstrations, single operation, and cyclic operation were conducted according to the above procedure.

8.4.3 Results of Thermal Driving Demonstration

The results of thermal driving demonstration are shown in Fig. 8.19. All experiments showed similar temperature and mass change trends as follows:

- The middle temperature sharply increased at the initial stage of the CD experiment (thermal energy output process), while the temperature suddenly decreased at the beginning of the DA experiment (thermal energy storage process).
- The bottom temperature did not show any fluctuations during the entire experiment. This could be attributed to the insufficient CO_2 diffusion in the electric crucible to cause a small temperature change as well as insufficient reacted conversion of LPR.

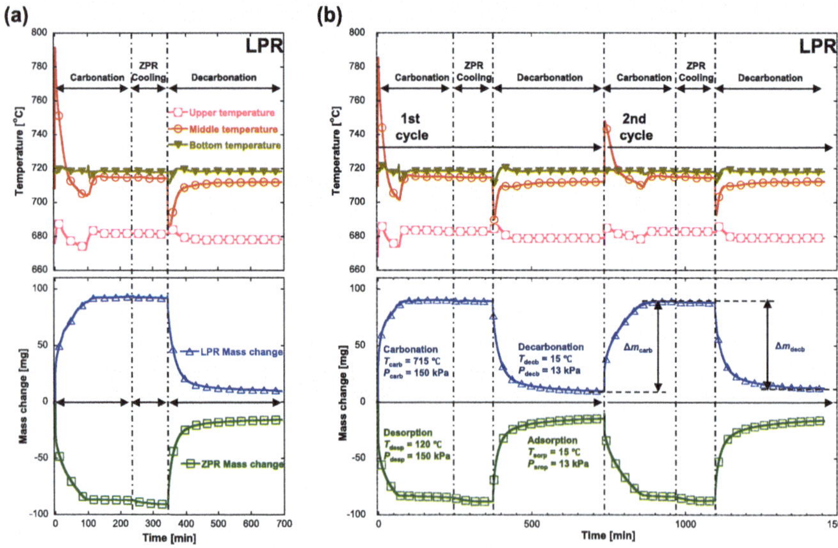

Fig. 8.19 Thermal driving demonstrations results: **a** single operation and **b** cyclic operation

- The upper temperature was measured above the K-tablet-packed bed and represented the atmosphere of the electric crucible. In the CD experiment, the temperature was increased by the exothermic reaction, which was also increased in the DA experiment by high-temperature desorption of CO_2 gas from the K-tablet.

The overall weight change of the LPR and ZPR corresponded closely to each other, and these results were used for the calculation of the reacted mole fraction and thermal energy capacity.

The single operation (SO) (Fig. 8.19a) and the first cycle of the cyclic operation (1st CO) (Fig. 8.19b) showed almost same results. The highest temperatures were 792 °C and 786 °C, and mass conversions in carbonation, Δm_{carb}, were 93.7 g and 91.4 g, respectively. However, the highest temperature and the mass conversion of the carbonation process decreased to 748 °C and 80.3 g, respectively, during the second cycle of cyclic operation (2nd CO). The poor decarbonation on the 1st CO may have induced a decrease in the highest temperature and the mass conversion of the carbonation process during the 2nd CO.

8.4.4 Kinetic Analysis of Results (Thermal Driving Demonstration)

The ZPR was expected to measure the change in mass more precisely than the LPR because of its lighter weight and lower operating temperature conditions. Therefore, the reacted mole fraction was calculated based on the mass change of the ZPR. The calculated reacted conversion, Δx, and the mass conversion in Sect. 8.4.3 have similar trends. The SO and 1st CO showed similar Δx values of 0.79 and 0.76, respectively, but the value of Δx dropped to 0.66 in the 2nd CO. However, all values of Δx for decarbonation were within the range of 0.62–0.68.

The thermal energy output capacity, Q_{LPR}, which is the released thermal energy per unit volume of LPR, was estimated from the calculated reacted mole reaction. Figure 8.20 shows the graphs of the thermal energy output and storage capacities against time. The total thermal energy output capacities of the SO and 1st CO were estimated as 395 kJ/L-bed and 381 kJ/L-bed, respectively, with the latter decreasing to 331 kJ/L-bed in 2nd CO, as shown in Fig. 8.20a. Specifically, the 1st CO showed a higher thermal energy output capacity than the others at 50 min, but after 150 min, SO showed the highest thermal energy output capacity value. On the contrary, there is no distinct change in the thermal storage process, as shown in Fig. 8.20b. Additionally, the thermal energy recovery ratio, which is the ratio of thermal energy storage capacity to thermal energy output capacity ($Q_{\text{storage}}/Q_{\text{ouput}}$), was used to estimate the performance degradation for further cycles. The SO and 1st CO showed the same value, 0.87, and this value increased to 0.96 in the 2nd CO. From this result, it is expected that even if the number of cycles is increased, the thermal energy output and storage capacities do not decrease significantly. The total heat capacities of the produced Li_4SiO_4 tablets were also evaluated from a material perspective; the thermal energy output and storage processes were estimated to be over 700 kJ/L of material.

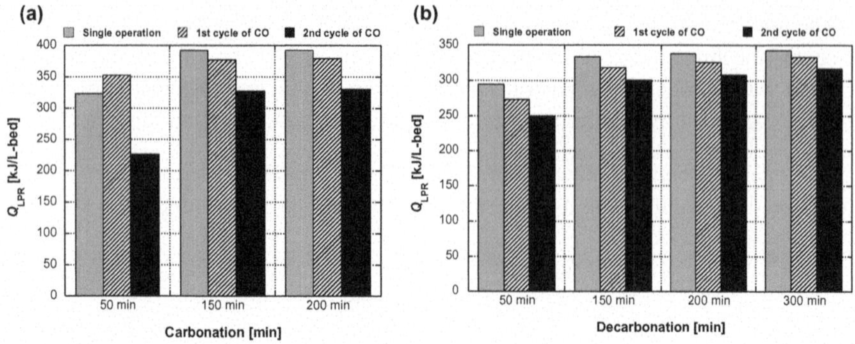

Fig. 8.20 Thermal energy output and storage capacity changes: **a** thermal energy output process, **b** thermal energy storage process

These values are meaningful because the energy density of the TCES method is over 20 times higher than that of the sensible and latent TES methods [31].

8.5 Conclusions

A lithium orthosilicate (Li_4SiO_4) powder was prepared by a solid-state method, and its potential was evaluated by non-isothermal and cyclic experiments under various conditions. The Li_4SiO_4 powder showed high reactivity at approximately 700 °C and sufficient durability in the cyclic experiments. In particular, the volumetric and gravimetric thermal energy densities of Li_4SiO_4 powder were estimated to be 750 kJ/L and 780 kJ/kg, respectively, when the porosity was 0.59.

A tablet form of Li_4SiO_4 (K-tablet) was produced using a new pelletizing method for practical applications. Compared to the pure Li_4SiO_4 and $Li_2CO_3 + Li_2SiO_3$ powders, the produced K-tablet exhibited superior carbonation and decarbonation performances. The double-shell model was also suggested to understand the carbonation and decarbonation mechanisms of the K-tablet, and it clearly explained the experimental results. The produced K-tablet exhibited sufficient durability during the cyclic experiments, and all reacted conversions, Δx, converged to 0.75. In particular, the K-tablet exhibited better thermal energy output and storage capacity at low temperatures, despite containing less Li_4SiO_4 and its formation into a tablet shape. The produced K-tablet had considerable potential as a practical CHP material for thermal energy transformation at temperatures from 600–650 °C to over 700 °C.

The Li_4SiO_4/CO_2/Zeolite TCES system was developed on a laboratory scale, and a thermal driving demonstration was conducted. Specially designed reactors, LPR and ZPR, were developed, and the experimental system was configured for efficient thermal driving demonstrations. Single and cyclic thermal driving demonstrations were carried out successfully, and the thermal energy output density of the LPR was estimated to be 331–395 kJ/L-packed bed (759–904 kJ/L-material).

The developed TCES system shows a high possibility of enhancing the efficiency of high-temperature industrial processes, such as a solar-thermal energy system, and its high energy density is expected to help downsize the existing sensible and latent TES systems [32].

References

1. Nevau P, Castaing J (1993) Solid–gas chemical heat pumps: fields of application and performance of the internal heat of reaction recovery process. Heat Recov Syst CHP 13(3):233–251
2. Goetz V, Elie F, Spinner B (1993) The structure and performance of single effect solid/gas chemical heat pumps. Heat Recov Syst CHP 13(1):79–96
3. Li TX, Wang RZ, Kiplagat JK, Chen H, Wang LW (2011) A new target-oriented methodology of decreasing the regeneration temperature of solid–gas thermochemical sorption refrigeration system driven by low-grade thermal energy. Int J Heat Mass Tran 54(21–22):4719–4729
4. Li TX, Wang RZ, Kiplagat JK, Wang LW (2009) Performance study of a consolidated manganese chloride-expanded graphite compound for sorption deep-freezing processes. Appl Energ 86(7–8):1201–1209
5. Bao HS, Wang RZ, Wang LW (2011) A resorption refrigerator driven by low grade thermal energy. Energ Convers Manage 52(6):2339–2344
6. Xu J, Oliveira RG, Wang RZ (2011) Resorption system with simultaneous heat and cold production. Int J Refrig 34(5):1262–1267
7. Oliveira RG, Wang RZ (2007) A consolidated calcium chloride-expanded graphite compound for use in sorption refrigeration systems. Carbon 45(2):390–396
8. N'Tsoukpoe KE, Schmidt T, Rammelberg HU, Watts BA, Ruck WKL (2014) A systematic multi-step screening of numerous salt hydrates for low temperature thermochemical energy storage. Appl Energ 124:1–16
9. Linder M, Mertz R, Laurien E (2010) Experimental results of a compact thermally driven cooling system based on metal hydrides. Int J Hydrogen Energ 35(14):7623–7632
10. Kato Y, Sasaki Y, Yoshizawa Y (2005) Magnesium oxide/water chemicals heat pump to enhance energy utilization of a cogeneration system. Energy 30(11–12):2144–2155
11. Ogura H, Ishida H, Yokooji R, Kage H, Matsuno Y, Mujumdar AS (2001) Experimental studies on a novel chemical heat pump dryer using a solid–gas reaction. Dry Technol 19(7):1461–1477
12. Kato Y, Yamada M, Kanie T, Yoshizawa Y (2011) Calcium oxide/carbon dioxide reactivity in a packed bed reactor of a chemicals heat pump for high temperature gas reactors. Nucl Eng Des 210(1–3):1–8
13. HSC Chemistry ver. 7.0 (2011) Outotec, Finland
14. Nakagawa K, Ohashi T (1998) A novel method of CO_2 capture from high temperature gases. J Electrochem Soc 145(4):1344–1346
15. Kato M, Nakagawa K (2001) New series of lithium containing complex oxides, lithium silicates, for application as a high temperature CO_2 absorbent. J Ceram Soc Jpn 109(1275):911–914
16. Barin I, Platzki G (1995) Thermochemical data of pure substances, 3rd edn. Wiley, New York
17. Takasu H, Ryu J, Kato Y (2017) Application of lithium orthosilicate for high-temperature thermochemical energy storage. Appl Energ 193:74–83
18. Kim ST, Ryu J, Kato Y (2011) Reactivity enhancement of chemical materials used in packed bed reactor of chemical heat pump. Prog Nucl Energ 53(7):1027–1033
19. Kousksou T, Bruel P, Jamil A, Rhafiki TE, Zeraouli Y (2014) Energy storage: application and challenges. Sol Energ Mat Sol C 120:59–80
20. Seggiani M, Puccini M, Vitolo S (2013) Alkali promoted lithium orthosilicate for CO_2 capture at high temperature and low concentration. Int J Greenh Gas Con 17:25–31
21. Wang S, An C, Zhang QH (2013) Syntheses and structures of lithium zirconates for high-temperature CO_2 absorption. J Mater Chem A 11:3540–3550

22. Toshiba Corporation (2005) Carbonic acid gas absorbent, and method and apparatus for separating carbonic acid gas. Japan Patent 2005075139, 16 Mar 2005
23. Puccini M, Seggiani M, Vitolo S (2013) Lithium silicate pellets for CO_2 capture at high temperature. Chem Engineer Trans 35:373–378
24. Nair BN, Yamaguchi T, Kawamura H, Nakao SI, Nakagawa K (2004) Processing of lithium zirconate for applications in carbon dioxide separations: structure and properties of the powders. J Am Ceram Soc 87(1):68–74
25. Olivares-Marin M, Castro-Diaz M, Drage TC, Maroto-Valer MM (2010) Use of small-amplitude oscillatory shear rheometry to study the flow properties of pure and potassium-doped Li_2ZrO_3 sorbents during the sorption of CO_2 at high temperatures. Sep Purif Technol 73(3):415–420
26. Pfeiffer H, Vazquez C, Lara VH, Bosch P (2007) Thermal behavior and CO_2 absorption of $Li_{2-x}Na_xZrO_3$ solid solutions. Chem Mater 19:922–926
27. Ida J, Lin YS (2003) Mechanism of high temperature CO_2 sorption on lithium zirconate. Environ Sci Technol 37:1999–2004
28. Wang Q, Luo J, Zhong Z, Borgna A (2011) CO_2 capture by solid adsorbents and their applications: current status and new trends. Energy Environ Sci 4:42–55
29. Ochoa-Fernandez E, Ronning M, Yu X, Grande T, Chen D (2008) Composition effects of nanocrystalline lithium zirconate on its CO_2 capture properties. Ind Eng Chem Res 47:434–442
30. Kim ST, Nihei T, Kurahashi C, Hoshino H, Takasu H, Kato Y (2019) Kinetic study of lithium orthosilicate pellets for high-temperature chemical heat pumps. Energ Storage 1(4):e72
31. N'Tsoukpoe KE, Liu H, Pierres NL, Luo L (2009) A review on long-term sorption solar energy storage. Renew Sust Energ Rev 13(9):2385–2396
32. Kim ST, Kurahashi C, Hoshino H, Takahashi C, Tamura Y, Takasu H, Saito S, Kurihara M, Kato Y (2019) Thermal driving demonstration of Li_4SiO_4/CO_2/Zeolite thermochemical energy storage system for efficient high-temperature heat utilizations. ISIJ Int 59(4):721–726

Part III
Advanced Hydrogen Production Through Electrolysis and Thermochemical Reaction

Chapter 9
Steam Electrolysis Cells that Can Operate at 600 °C

Hiroshige Matsumoto, Kwati Leonard, Takaya Fujisaki, and Kuninori Miyazaki

Abstract Steam electrolysis using proton-conducting metal oxides can potentially increase electrolyzer efficiency while reducing electrical energy demand and lowering hydrogen production costs. The present work evaluated and selected baseline cell materials and demonstrates the feasibility and operation of proton-conducting solid oxide electrolysis cells at intermediate temperature. A series of electrolyte compositions and steam electrode materials were examined using conductivity, steam stability, and electrocatalytic activity. Several cell configurations have also been evaluated. Some demonstrate excellent performance and durability ($\leq$2.5%/1000 h in terminal voltage rise and 2.1%/1000 current efficiency reduction rate). Three and ten cell stacks were built and tested. The three cell stack reached a cell terminal voltage of 4.7 V, and the hydrogen generation rate was 42 $^{N}Lh^{-1}$, with a current efficiency of 83%. The cell performance stability was improved by inserting an active NiO/SZCY541 layer between the electrolyte and the cathode substrate. After 1500 h or more of durability testing, very minimal Co and Ni defused to the electrolyte layer and are not the leading cause of the minor degradation observed. The present proton-conducting solid oxide electrolysis cell design reveals a more feasible approach for realizing both highly efficient and economically viable systems.

Keywords Hydrogen production · Steam electrolysis · Proton conductor · Cell stacks · Carbon neutrality

---

H. Matsumoto (✉) · K. Leonard
Kyushu University, Ohaza-motooka, Nishi-ku, Fukuoka 819-0395, Japan
e-mail: matsumoto@i2cner.kyushu-u.ac.jp

K. Miyazaki
Nippon Shokubai Co. Ltd, Suita-shi, Osaka 564-8512, Japan

T. Fujisaki
Tohoku University, Aoba-ku, Sendai, Miyagi 980-8579, Japan

K. Aika and H. Kobayashi (eds.), *CO_2 Free Ammonia as an Energy Carrier*,
https://doi.org/10.1007/978-981-19-4767-4_9

9.1 Introduction

Proton-conducting metal oxides and their applications to hydrogen-related electrochemical devices have been studied since the first report of appreciable proton conduction found in ytterbium-doped strontium cerate [1]. Materials with high proton conductivity have been found mainly in perovskite-type structures. Since the material conducts protons (hydrogen ion, H^+), it is possible to construct a fuel cell using hydrogen as a fuel. It can be used to produce hydrogen and oxygen by electrolyzing steam. It can also be used for hydrogen sensors, electrochemical hydrogen separation (hydrogen pumps), and electrochemical membrane reactors [2].

For solid oxide fuel cells and steam electrolyzers, oxide ion-conducting materials such as stabilized zirconia can also be used. Stabilized zirconia has a long history, and this is rather the mainstream. Compared with the case where an oxide ion-conducting material is used, proton-conducting oxide has the following advantages.

First, the operating temperature can be lowered. Proton conductors show relatively high proton conductivity even in the intermediate temperature region of 600 °C or lower. The phenomenon of proton conduction can be explained by the proton hopping mechanism: Protons stay in the metal oxide lattice by covalently bonding with lattice oxygens and move by interchanging the covalent bonding and hydrogen-bonding oxygens. This mechanism leads to lower activation energy compared with the oxide ion conduction in which the lattice oxygen itself is conducted by the vacancy mechanism. Solid oxide fuel cells using stabilized zirconia operate at 700 °C to 900 °C, and lowering the operation temperature is beneficial for cost, durability, etc. As shown later, some proton conductors exhibit higher electrical conductivity than oxide ion conductors in the temperature range below 600 °C.

Second, hydrogen is not diluted with water vapor in fuel cells and steam electrolysis as is the case with oxide ion conductors. Taking steam electrolysis as an example, the electrolyte cases using protons and oxide ions as the conducting species are compared in Fig. 9.1. When the electrolyte is an oxide ion conductor, the product hydrogen is generated in steam, in which case separation is required. In the proton conductor case, hydrogen is generated at the counter electrode separately from steam, although some humidification is required to suppress the extremely reducing atmosphere [3]. When the cells are operated in the fuel cell mode, hydrogen fuel is diluted with generated steam when the electrolyte is oxide ion conducting. However, the steam is generated separately from hydrogen at the counter electrode when the electrolyte is proton proton-conducting. As a result, the voltage can be kept high at the same fuel utilization rate [4].

Focusing on electrolysis to produce hydrogen from electrical energy, we have studied steam electrolysis using proton conductors. To reduce the use of fossil fuels as primary energy and increase the use of renewable energy, it is necessary to convert electricity into storable secondary energy, and hydrogen production by electrolysis meets this purpose [5, 6]. Taking advantage of the above merits, it has become possible to operate steam electrolysis at 600 °C or lower. This section summarizes

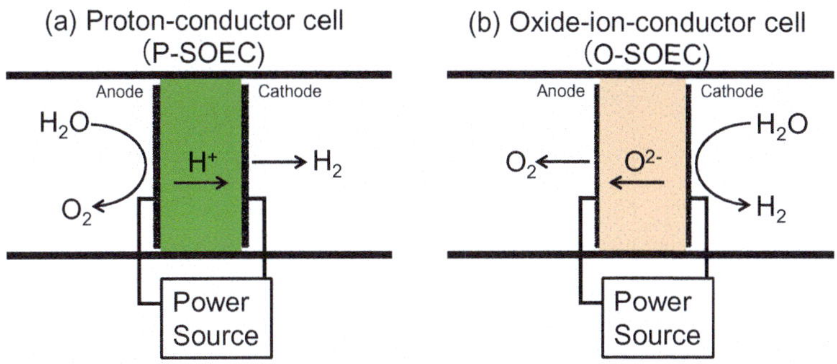

Fig. 9.1 Schematic illustration of steam electrolysis comparing **a** proton conductor cell and **b** oxide-ion-conductor cell

the development of suitable proton-conducting electrolytes, electrode materials, and steam electrolysis cells from the viewpoint of intermediate temperature application.

9.2 Development of Proton Conductive Electrolyte

Proton-conducting oxides used as electrolyte materials are required to have high proton conductivity, chemical stability, and mechanical strength. When used for steam electrolysis, stability in steam is particularly important. There is a trade-off between the materials reported so far. In designing a proton-conducting oxide, a perovskite-type oxide having a chemical formula represented by ABO_3 is most suitable as a crystal structure. Also, it is the basis of the material design to develop proton conductivity by substituting a part of the cations, mostly *B*-site ones, with a lower valence cation (so-called dopant) [1, 7]. Therefore, it is essential to select the A, B site cations and dopants appropriately. In the previous study on the $SrZr_{0.9-x}Ce_xY_{0.1}O_{3-\delta}$ system ($x = 0$–0.9), $SrZr_{0.5}Ce_{0.4}Y_{0.1}O_{3-\delta}$ (from now on, SZCY541) has relatively high proton conductivity with sufficient chemical stability for steam electrolysis [7]. In this study, the SZCY541 composition was set as the starting composition, and two directions of changing the dopant concentration and changing the *A*-site element were examined (Fig. 9.2). Before the investigation, the solid-state reaction method (SSR) and the chemical solution method (CS) were compared for the sample preparation. SSR is a method of mixing, firing, crushing, molding, and using carbonate of alkaline earth metals and oxide. The CS method uses mainly metal nitrate for each constituent metal: After mixing the aqueous solution with citric acid and EDTA, evaporation and heat treatment are conducted, followed by crushing, molding, and the main firing. As shown in Fig. 9.3, denser sintered bodies can be obtained using the CS method and were subsequently used for this work.

Figure 9.4 shows the conductivity of each composition in wet hydrogen. When Zr/Ce was kept at 5/4, it was advantageous in terms of conductivity to select Ba for the *A* site, and a high conductivity was obtained when the amount of Y was

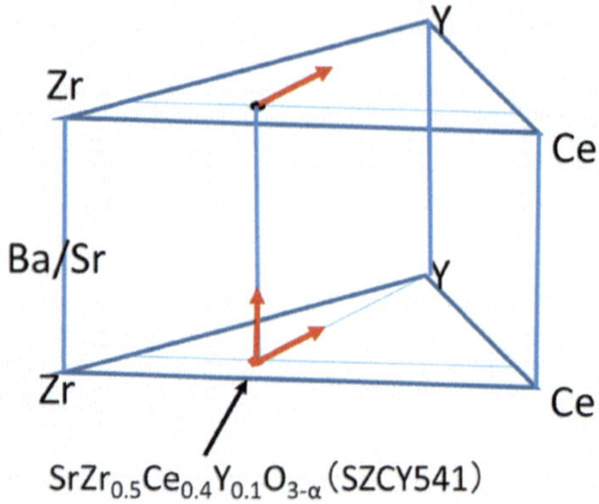

Fig. 9.2 (Ba, Sr)(Zr,Ce,Y)O_3 system for the optimization of the electrolyte composition

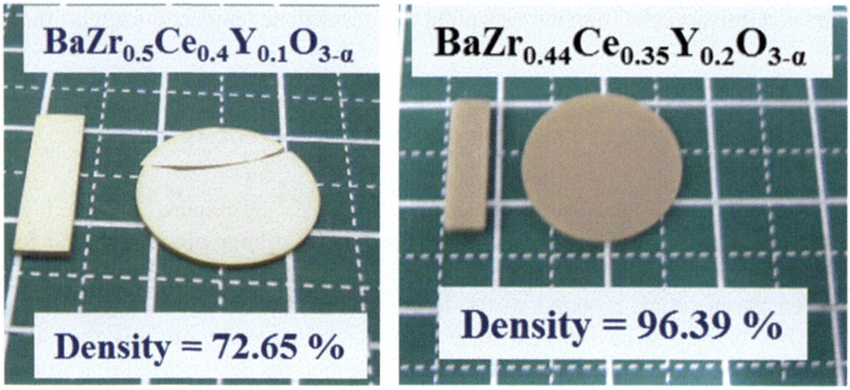

Fig. 9.3 Electrolyte specimen prepared via **a** solid-state reaction (SSR) and **b** chemical solution (CS) methods

20 mol%. (Fig. 9.4a). On the other hand, when the ratio of Zr and Ce was changed, it was found that the maximum conductivity was reached at Zr/Ce = 5/4. Therefore, it is concluded that $BaZr_{0.44}Ce_{0.36}Y_{0.2}O_{3-\delta}$ (BZCY $(54)_{8/9}2$) has high conductivity [8].

The stability of $BaZr_{0.44}Ce_{0.36}Y_{0.2}O_{3-\delta}$ (BZCY $(54)_{8/9}2$) in steam was investigated. The sintered sample was exposed to an atmosphere containing 80% steam in either air or 1%H_2-99%Ar at at 600 °C for 100 h. The changes in conductivity over time and crystal phase before and after the exposure were investigated. The results are shown in Fig. 9.5. No differences were observed in both instances upon exposure to water vapor and the coexisting gases. In addition, the X-ray diffraction pattern after steam exposure was the same as that before exposure with the same single perovskite phase. Therefore, it was concluded that BZCY$(54)_{8/9}2$ has both high conductivity and stability in 80% steam.

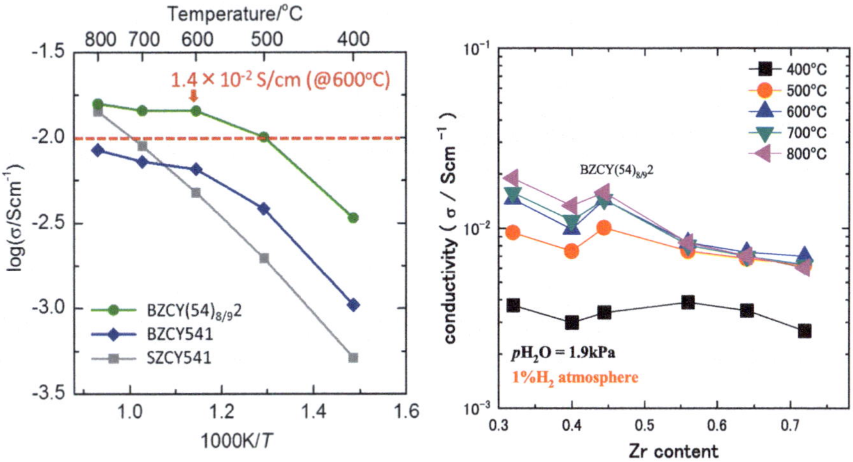

Fig. 9.4 **a** Electrical conductivity of (Ba,Sr)(Zr,Ce,Y)$O_{3-\delta}$ measured in wet hydrogen (1.9 kPa water vapor in 1% hydrogen)

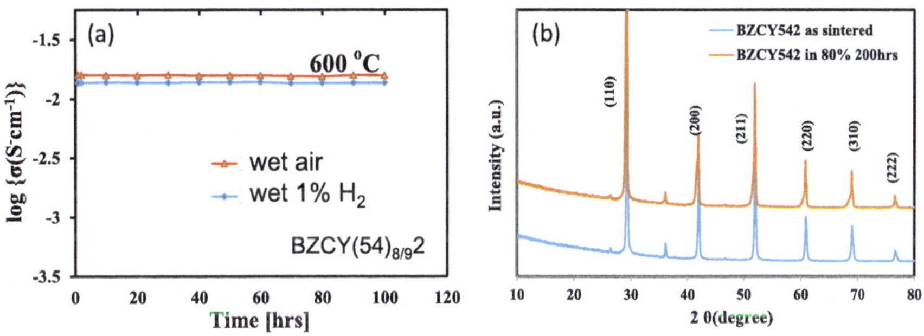

Fig. 9.5 Stability test of BZCY$(54)_{8/9}$2 in 80% water vapor atmosphere (600 °C). **a** Conductivity as a function of time, **b** X-ray diffraction pattern before and after exposure. Coexisting gas is air or 1% hydrogen diluted with argon, and measurement is performed for 100 h under each condition

9.3 Steam Electrode

The oxygen-evolving electrode (anode) is required to have both stability against steam and electrode catalytic activity. The steam stability and electrode activity of the candidate materials listed in Table 9.1 were investigated.

The thermodynamic stability in a gas contains 80% steam and 20% air at 600 °C for 100 h. The synthesized anode electrode material paste was screen-printed on SZCY541 electrolyte disk and baked at 1000–1100 °C. The crystal phases before and after exposure were measured by XRD. BLC55 and SSC55 showed some changes in the XRD pattern, and the formation of decomposition products was also observed in

Table 9.1 List of anode candidate materials

Single-phase powder	Precursors	Stability in steam
$Ba_{0.5}La_{0.5}CoO_{3-\delta}$, (BLC55)	From metal nitrates (CSS)	Fair
$Sm_{0.5}Sr_{0.5}CoO_{3-\delta}$ (SSC55)	From metal acetates	Fair
$Ba_{0.5}La_{0.5}Co_{0.9}Ti_{0.1}O_{3-\delta}$	$LaCO_3$,$BaCO_3$,TiO_2, and Co_3O_4	Good
$Ba_{0.5}La_{0.5}Co_{0.9}Zr_{0.1}O_{3-\delta}$	$LaCO_3$,$BaCO_3$, ZrO_2, and Co_3O_4	Good
$PrBaCo_2O_{5+\delta}$ (PBCO)	Pr_6O_{11}, $BaCO_3$, and Co_3O_4	Good

SSC55. Whereas, the main phases of BLCT, BLCZ, and PBCO remained unchanged after exposure.

The electrode activity was measured by an impedance method using an electrolyte support cell. Ni-SZCY541 cermet was used as the counter electrode. Since proton conductors have partial hole conduction in the anode atmosphere, measurement in a symmetric atmosphere was avoided. Measurements were thus carried out using the fuel cell configuration. Wet (2% H_2O) hydrogen diluted with argon in the cathode side, and 80% water vapor and 20% air introduced into the anode side. The results are shown in Fig. 9.6. Among the studied material, the electrode resistance of PBCO was the lowest, followed by the electrode resistance of BLC55. Based on these results, $Ba_{0.5}La_{0.5}CoO_{3-\delta}$ (BLC55) and $PrBaCo_2O_{5+\delta}$ (PBCO) were considered for steam electrolysis evaluation.

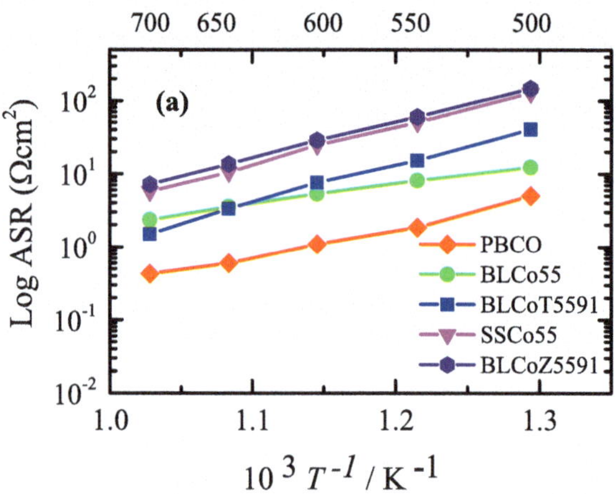

Fig. 9.6 Temperature dependence of electrode resistance values of the studied electrodes

9.4 Steam Electrolysis Evaluation with Bottom Cells

An anode support cell was prepared using the electrolyte and electrode material optimized as described above, and its characteristics were examined. The cell preparation procedure is as follows: NiO and the powder of the electrolyte sample were wet-mixed at a volume ratio of 1: 1 and then molded by a hydrostatic press (250 MPa). An electrolyte paste was applied to this by screen printing and fired at 1450–1500 °C to obtain half cells, and the anode paste was applied to prepare an electrolytic cell. A mixed gas of 80% steam and 1% O_2–99% argon was used as the anode gas, and moistened 1% H_2 was used as the cathode gas, respectively. The polarization characteristics of the electrodes were measured by a current interrupt method. Hydrogen generated at the cathode by steam electrolysis was quantified by gas chromatography.

Figure 9.7 shows the cross section of the produced electrolytic cell. It can be seen that an electrolyte membrane with a thickness of 10–20 μm is densely formed on the cermet substrate of NiO and the electrolyte. The electrolysis test equipment is shown in Fig. 9.8. It has a double-tube structure in which an alumina tube with a diameter of 13 mm is covered with a quartz glass outer tube. With the electrolytic cell sandwiched between the alumina tubes via a glass seal, the anode gas containing steam is flown inside, and the sweep gas is 1% outside. An electrolysis test was carried out by applying an electric current while flowing hydrogen (Table 9.2).

When PBCO was used as the steam electrode material, steam electrolysis test showed a decrease in current efficiency (that is, an increase in electron leakage) at the hydrogen generation rate during operation for several tens of hours although the initial activity was high. No further studies have been conducted; however, it was suggested that PBCO undergoes some changes during the operation of steam electrolysis.

Figure 9.9 shows steam electrolysis test (600 °C) results using BLC55 as the anode material and $BZCY(54)_{8/9}2$ as the electrolytes. Two different cases on the electrolyte components in the anode substrate to be $BZCY(54)_{8/9}2$ (a–c in Fig. 9.9) and SZCY541 (d-f in Fig. 9.9) are compared. It should be reminded that $BZCY(54)_{8/9}2$ has higher

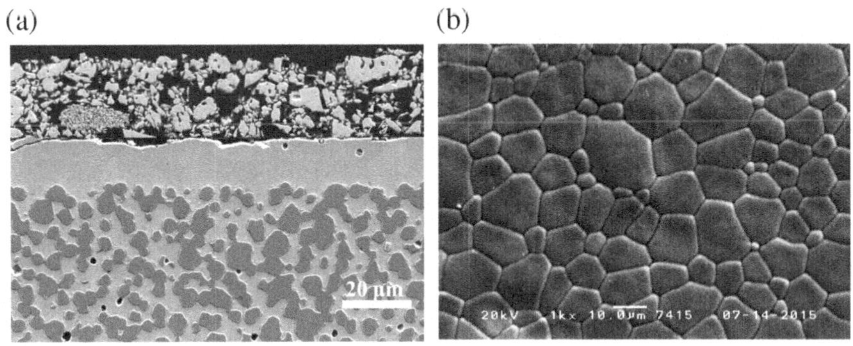

Fig. 9.7 SEM of half cell **a** cross section, **b** top view

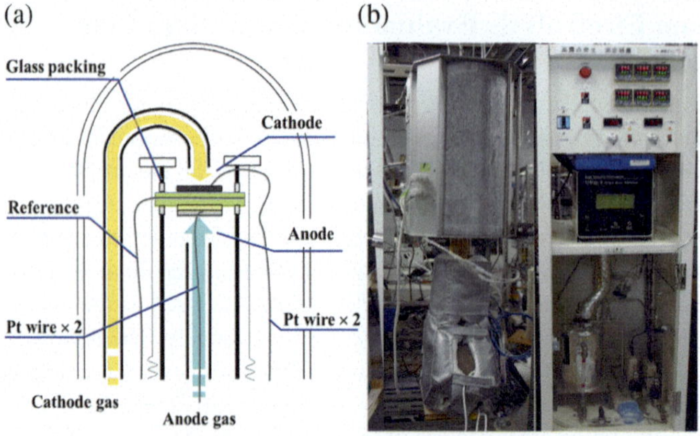

Fig. 9.8 Steam electrolysis equipment: **a** schematic electrochemical cell, **b** furnace and steam generator

Table 9.2 Summary of steam electrolysis test

Temperature/°C	OCV	Appl. V (@0.5 A/cm^2)	η (%)
600	0.79	1.56	83
550	0.81	1.68	68
500	0.80	1.80	46

conductivities than SZCY541; however, for the substrate components, the SZCY541 works better than the other in the current–voltage characteristics (by comparing Fig. 9.9 a, d). The separated overpotentials in figures b and d showed the ohmic loss to be the main difference between the two cases. In the substrate consisting of BZCY(54)$_{8/9}$2, the Ba-Ni oxide phase formation can be identified, suggesting the reaction of the Ba-based electrolyte with NiO. From the above, it was concluded that the combination of BZCY(54)$_{8/9}$2 for the electrolyte and NiO/SZCY541 for the cathode is advantageous for both the electrolytic voltage and the hydrogen generation efficiency.

Figure 9.10 shows steam electrolysis test results using BLC55 as the anode material and BZCY(54)$_{8/9}$2 as the electrolyte. The data shows that the open-circuit voltage rises as the temperature drops. In contrast, the hydrogen generation efficiency decreased as the temperature decreased (83% @ 600 °C → 46% @ 500 °C). In the studies so far, it has been found that the hydrogen generation efficiency correlates with the overvoltage of the anode (oxygen generating side electrode). Therefore, when the electrode impedance in the open-circuit state was measured (Fig. 9.10), it was found that the electrode overvoltage increased about four times by lowering the operating temperature from 600 to 500 °C. It is highly possible that the electrode catalytic activity of the anode is insufficient and is one of the causes of the decrease in hydrogen generation efficiency. Therefore, further improvement in the performance of the anode is essential for reducing the operating temperature.

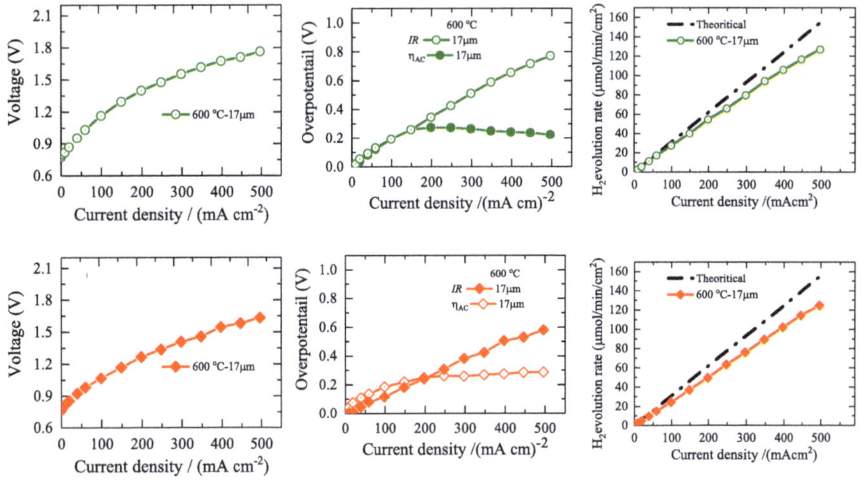

Fig. 9.9 Water vapor electrolysis test results using BLC55 anode, BZCY(54)$_{8/9}$2 electrolyte and NiO cermet substrate containing either BZCY(54)$_{8/9}$2 (**a**–**c**) or SZCY541 (**d**–**f**) (600° C). Left: current–voltage characteristics, middle: overvoltage characteristics, right: hydrogen generation rate

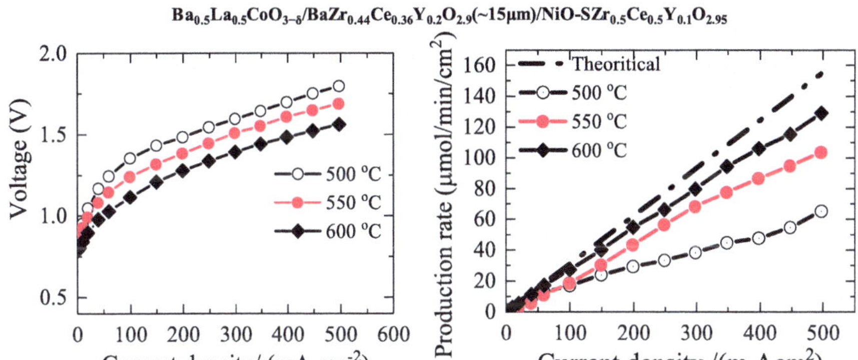

Fig. 9.10 Steam electrolysis test results evaluated at 600 °C or less (electrolyte: BZCY(54)$_{8/9}$2, cathode: NiO / SZCY541). Left: current–voltage characteristics, Right: hydrogen generation rate

9.5 Proton-Conducting Electrolysis Cell Production

The above section describes the development of proton-conducting electrolytes for practical use. This section focuses on the production of highly efficient and durable large area electrolysis cells for practical application.

The developed proton-conducting oxide electrolysis cells consist of a hydrogen electrode-supported Ni cermet for the substrate and hydrogen electrode, Sr $Zr_{0.5}Ce_{0.4}Y_{0.1}O_{3-\delta}$ (SZCY541), $Ba_{0.9}Sr_{0.1}$ $Zr_{0.44}Ce_{0.36}Y_{0.2}O_{3-\delta}$ (BSZCY91(5)$_{8/9}$2) for the electrolyte material, and $Ba_xLa_{1-x}CoO_{3-\delta}$ $(0.3 < x < 0.9)$ (BLC) for the oxygen electrode. The proton-conducting oxide electrolysis cells are prepared, using a tape-casting, screen-printing, and co-sintering method, which are mass-producible techniques.

An example of the electrolysis cell process is schematically illustrated, in Fig. 9.11. The NiO/SCZY451 green sheet was prepared from a mixture of NiO and SZCY541 with a solvent, binder, plasticizer, dispersant, and tape casted. Subsequently SZCY541and BSZCY91(5)$_{8/9}$2 pastes were laminated on the sheet by a screen-printing process. After which, the laminate of NiO/SCZY451 and the electrolyte were co-sintered at about 1300 °C. BLC paste was screen-printed onto the resulting half cell and fired to form the electrolytic cells. The deformation and warpage of the cells could be suppressed by optimizing the amount of binder, plasticizer, dispersant, and calcination temperature. Electrolysis cells withsizes from 25 mm × 25 mm to 100 mm × 100 mm could be manufactured (Fig. 9.12).

Half cells consisting of BZCY(54)$_{8/9}$2 lNiO/BCZY (45)$_{8/9}$2 by our company's tape-casting technique had a 10% or less yield. In contrast, half cell consisting of SZCY541 (Half Cell-A) after co-sintering were all flat, and the yield was 95% or

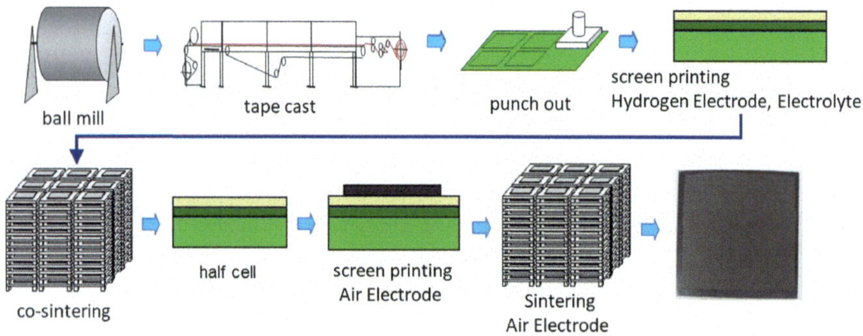

Fig. 9.11 Manufacturing process of proton-conducting oxide cells

Fig. 9.12 Planar large proton-conducting oxide cells (Cell size: 100 mm × 100 mm, electrode size: 90 mm × 90 mm)

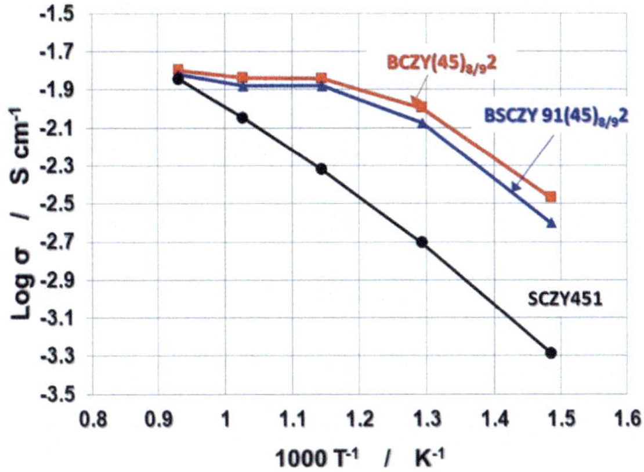

Fig. 9.13 Conductivity of proton-conducting oxides (BZCY(54)$_{8/9}$2, BSZCY91(54)$_{8/9}$2, and, SCZY 451)

more.. In the half cell using BZCY(54)$_{8/9}$2, the deviation between the shrinkage ratio of the substrate and that of the electrolyte layer was significant, in which case deformation and warpage occurred, resulting in a lower yield. This fact reveals another benefit of SZCY541 as the component in the Ni-cermet substrate. Therefore, optimization of the composition based on BZCY(54)$_{8/9}$2 electrolyte material and design of the half cell were examined.

SCZY 451 with Sr in the A-site produces flat half cells, however, with lower conductivity. On the other hand, BZCY(54)$_{8/9}$2 with A-site Ba has high conductivity but a meager yield. Therefore, some amount of Sr was added to the BZCY(54)$_{8/9}$2-based electrolyte to improve its yield. The degree of conductivity decrease could also be minimized by optimizing the doping amount. The results show that the conductivity of BZCY(54)$_{8/9}$2 doped with 10 mol% Sr in $B_{0.9}Sr_{0.1}Zr_{0.44}Ce_{0.36}Y_{0.2}$ $O_{3-\delta}$ (BSZCY91(54)$_{8/9}$2 at 873 K was approximately equivalent to that of BZCY(54)$_{8/9}$2 1.44×10^{-2} S cm^{-1}, and 1.32×10^{-2} S cm^{-1} for the former (Fig. 9.13).

The structure of the half cell using BSZCY91(54)$_{8/9}$2 and the half cell after co-sintering in a size of 20 mm × 20 mm are shown in Fig. 9.14. Half Cell-B, in which the electrolyte layer was BSZCY91(54)$_{8/9}$2 and the substrate was NiO/BSZCY91(45)$_{8/9}$2, had a yield of 30–40%. Half Cell C, in which the substrate was changed from NiO/ BSZCY91(54)$_{8/9}$2 to NiO/ SZCY541, had a high frequency of deformation of the half cell after firing in the range of 50–65%. In half Cell-D of BSZCY91(54)$_{8/9}$2 and NiO/ SZCY541, in which the SCZY 451 electrolyte layer was inserted between them, it was found that there was no generation of deformation and warpage and that the yield was improved to 80–90%. Even when the size of the half cell is increased to 100 mm × 100 mm, a flat half cell could be obtained.

Although a relatively high yield was obtained by improving the cell configuration, it is necessary to reduce the manufacturing cost for practical application. The examination of the manufacturing process for the cost reduction of the electrolytic cell will be investigated in the future.

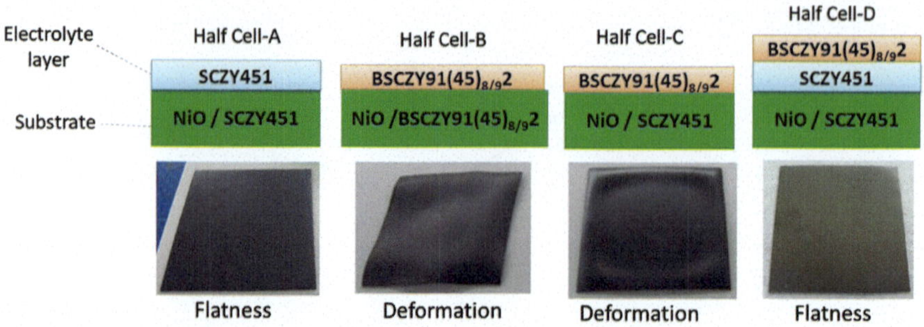

Fig. 9.14 Composition of half Cell-A, B, C, D and Appearance after co-sintering

On the electrolyte layers of half Cell-A using SZCY541 electrolyte material and half Cell-D using BSZCY91(54)$_{8/9}$2 electrolyte, electrolysis cells (Cell-A, B) using BLC as the oxygen electrode was fabricated and evaluated. The results o are shown in Fig. 9.15. The cell terminal voltage at 0.5 A cm^{-2} of Cell-B was improved from 1.71 V of the Cell-A to 1.45 V. The current efficiency of the Cell-B was 80% and that of Cell-A 81%. We found a cell configuration in which the yield of the electrolysis cell and the electrolysis performance were compatible (Table 9.3).

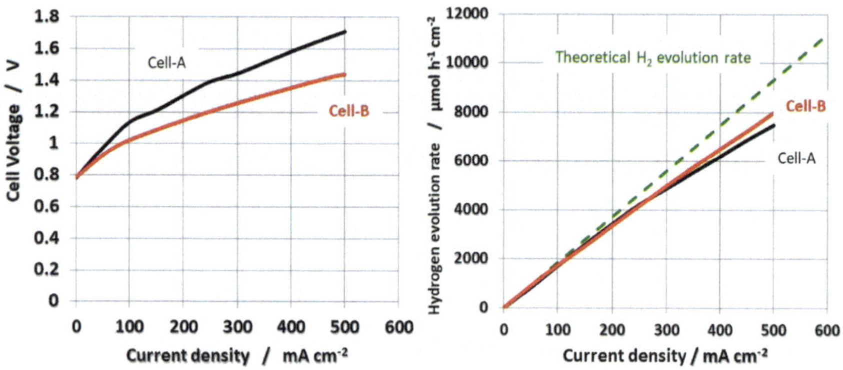

Fig. 9.15 Single-cell performance in electrolysis mode (Cell-A, B). Cell size: φ 20 mm Electrode size: φ 10 mm

Table 9.3 Cell A, B specifications

	Cell-A	Cell-B
Anode	BLC	BLC
Electrolyte	SCZY 451	BSCZY 91 (45)$_{8/9}$2SCZY 451
Cathode	Non	Non
Substrate	NiO/SCZY 451	NiO/SCZY 451

9.6 Durability of Proton-Conducting Elecrolysis Cells

9.6.1 Cell Durability Evaluation

Since the present steam electrolysis's operating temperature is relatively high at 873 K, its durability is a vital issue [9]. Thus, the long-term tests performed on Cell-B, which showed relatively good electrolysis characteristics with a 20 mm diameter cell, are shown in Fig. 9.16. The durability test was performed for Cell-B for 140 h. The cell terminal voltage change with time was relatively small, however, the current efficiency decreases significantly.

The configuration of the improved cell (Cell-C, D, E) based on Cell-B is shown in Table 9.4, and the long-term test results of Cell-C, Cell-D, and Cell-E are demonstrated in Fig. 9.16. Cell-B is an electrode-supported cell consisting of three layers: an oxygen electrode, electrolyte, and a substrate. Generally, in the oxide ion-conducting oxide cells for SOECs, an active hydrogen electrode for promoting the reaction is installed between the electrolyte layer and the substrate. Therefore, in Cell-C in which Ni/SZCY541 (active hydrogen electrode) was installed between the electrolyte layer and the cathode substrate, the long-term tests were carried out for more than 1300 h. In Cell-C, the rate of increase in cell terminal voltage per 1000 h was 1.4%, and the rate of decrease in current efficiency was 6%. This is influenced by the cathode layer's material design, the optimization of the cell structure, and the physical properties of NiO used.

Then, Cell-D, in which the thickness of the electrolyte layer was reduced from 15 μm of Cell-C to 8 μm, and Cell-E, in which the half cells' calcination conditions were optimized, was also fabricated. Durability test was also conductor for the later cells for 1500 h or longer. The initial cell terminal voltage of Cell-D, E was 1.37 V, an improvement from the formal Cell-C type of 1.46 V. The rate of rise of the terminal voltage was 2.5%/1000 h. The current efficiency reduction rate of Cell-D was 5.4%/1000 h, and that of Cell-E was 2.1%/1000 h. The rate of increase in

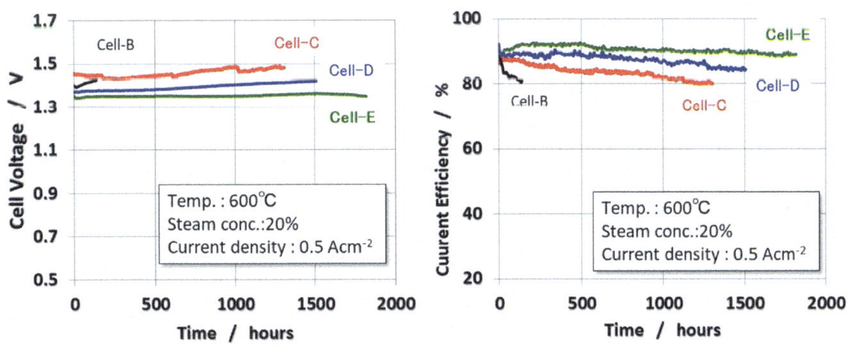

Fig. 9.16 Long-term performance at 873 K for Cell-B, C, D, E

Table 9.4 Cell-B, C, D, E, F specifications

	Cell-B	Cell-C	Cell-D	Cell-E
Anode	BLC	BLC	BLC	BLC
Electrolyte	BSCZY 91 $(45)_{8/9}2$ SCZY 451 thickness: 15 μm	BSCZY 91 $(45)_{8/9}2$ SCZY 451 thickness: 15 μm	BSCZY 91 $(45)_{8/9}2$ SCZY 451 thickness: 8 μm	BSCZY 91 $(45)_{8/9}2$ SCZY 451 thickness: 8 μm
Cathode	Non	NiO/SCZY 451	NiO/SCZY 451	NiO/SCZY 451
Substrate	NiO/SCZY 451	NiO/SCZY 451	NiO/SCZY 451	NiO/SCZY 451

hydrogen power consumption rate at the cell level of Cell-E was 4.7%/1000 h of Cell-D and 2.3%/1000 h of Cell-D. By optimizing the cell configuration and the half cell's calcination conditions, the durability of the electrolysis cell could be significantly improved.

On the other hand, since the current efficiency tends to decrease with time, it is necessary to clarify the degradation factors to improve further the electrolysis cell.

9.6.2 Investigation of Deterioration Factors for Electrolysis Cells

To clarify the degradation factors of the electrolysis cells, Cell-D, which was subjected to the long-term test for more than 1300 h, was characterized. Proton-conducting oxides are reported to be affected by transition metals, particularly Ni and Co [10–12].

Since the NiO content in the substrate of Cell-D is as high as 50 Vol% or more, it is assumed that Ni and Co diffused into the electrolyte layer during half cell fabrication or long-term testing. The defused components change the composition and physical properties of the electrolyte material and decrease the current efficiency with time. The analytical results of FE-EPMA for Co and Ni in Cell-D after the long-term test are shown in Fig. 9.17. The distribution of Co and Ni after the long-term test was smaller than that before the test. Co and Ni diffuse into the half cell during fabrication. Still, the change with time after the durability test is small, so we concluded that the diffusion of Co and Ni is not the main cause of the performance degradation in the long-term test.

Oxygen electrode (Anode): In steam electrolysis, the anode is an important electrode that splits water into oxygen and protons, and it is necessary to understand its state after the long-term test. Therefore, an X-ray diffraction (XRD) analysis was performed on BLC, the anode material of Cell-D that was subjected to the long-term test for more than 1300 h. From the X-ray diffraction of the anode layer in Cell-D, the diffraction peak of Co_3O_4 was confirmed (Fig. 9.18).

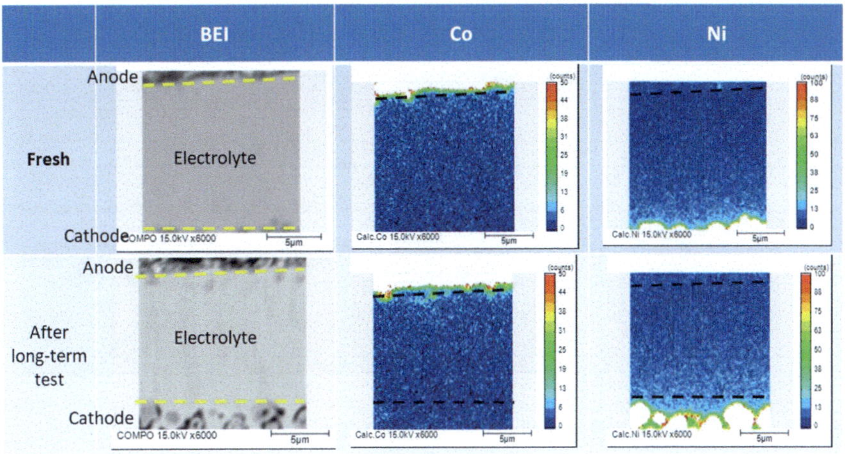

Fig. 9.17 FE-EPMA analysis of Cell-F section after durability evaluation (Co, Ni)

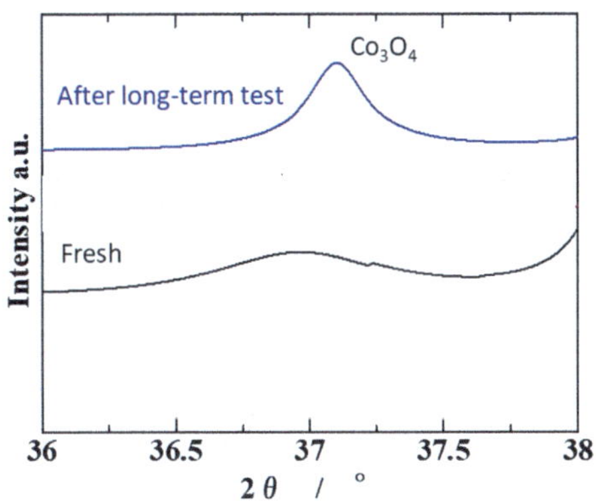

Fig. 9.18 Result of XRD analysis of anode in Cell-D after long-term test

Next, to grasp the component distribution of the anode material, SEM–EDS analysis was performed on the anode cross section (Fig. 9.19). Thus, although the distribution of Ba and La in BLC was almost the same as that in the initial stage, Co was sintered, and particles grew up to about 1 μm. Based on these results, it is possible that the current efficiency decreased over time due to the Co separation from BLC during the durability test and the decrease in the active site.

Electrolyte layer: To confirm the dispersion state of each component in the electrolyte layer in Cell-D subjected to the long-term test, FE-EPMA analysis was performed, and the results are shown in Fig. 9.20. Accordingly, after co-sintering, Ba, which is a component of the electrolyte material, diffused to the substrate, and its diffusion further progresses in the long-term test. Tong et al. [13] and Yoo et al. [12]. reported

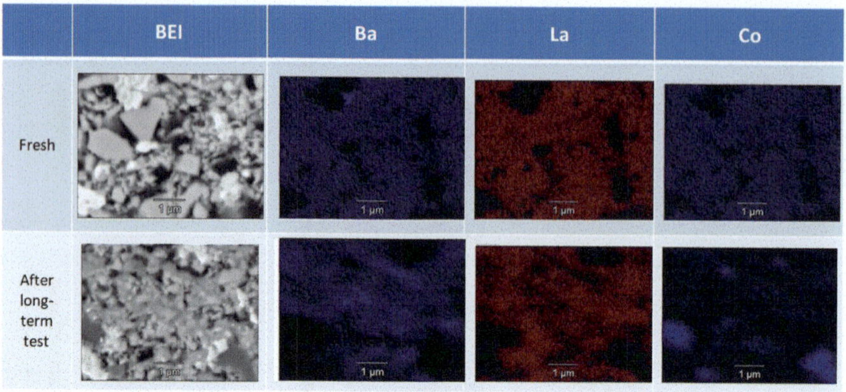

Fig. 9.19 Result of SEM–EDS of Cell-D after long-term test (cross section)

that Ba in $BaCe_{0.6}Zr_{0.2}Y_{0.2}O_{3-\delta}$ reacted with NiO above 1073 K (Eq. (1)–(3)). Since the half cells preparation temperature of Cell-D is 1573 K or higher, it is considered that Ba diffused to the cathode substrate.

$$BaCe_{0.6}Zr_{0.2}Y_{0.2}O_{3-\delta} + yNiO \rightarrow yBaNiO_x + Ba_{1-y}Ce_{0.6}Zr_{0.2}Y_{0.2}O_{3-\delta} \quad (9.1)$$

$$BaCe_{0.6}Zr_{0.2}Y_{0.2}O_{3-\delta} \rightarrow BaO + 0.8(Ce_{0.75}Zr_{0.25}O_2) + Y_2O_3 \quad (9.2)$$

$$BaNiO_2 + Y_2O_3 \rightarrow BaY_2NiO_5 \quad (9.3)$$

The expected electrolytic characteristics may not have been obtained because the features of the electrolyte material are changed by the inter-defused component. Further improvement in durability can be expected by suppressing the component transfer of the electrolyte material. Studies are currently underway to solve problems such as the component transfer of the anode material and electrolyte.

9.7 Development of Large Area Proton-Conducting Electrolysis Cells

An electrolysis cell with a size of 100 mm × 100 mm was fabricated in the Cell-E configuration. Electrolysis test was subsequently performed at 873 K with a steam concentration of 20 Vol% (steam utilization: 70%). The results are shown in Fig. 9.21. At 0.5 A cm $^{-2}$, the cell terminal voltage was 1.39 V, and current efficiency 82% (hydrogen evolution rate: 14.6 $^{N}Lh^{-1}$), showing relatively good characteristics.

Even though electrolysis was carried out with a 20% steam concentration, it is assumed that the water vapor concentration will be over 20% in the existing system.

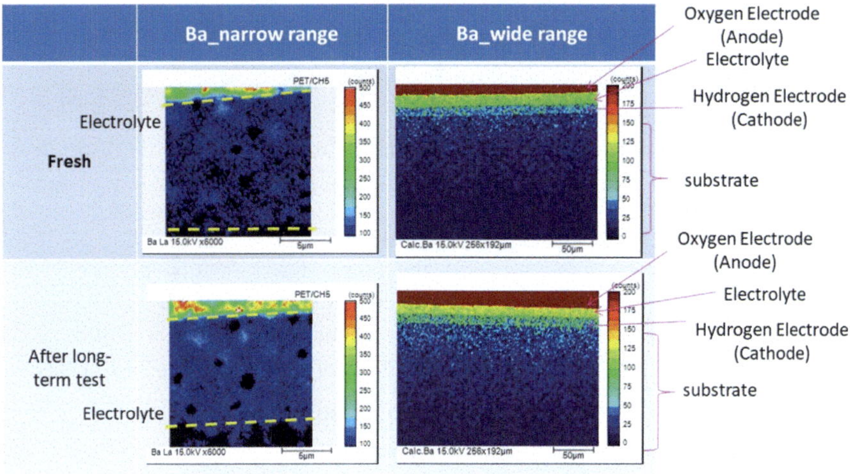

Fig. 9.20 Result of FE-EPMA mapping (Ba) of Cell-D after the long-term test (cross section)

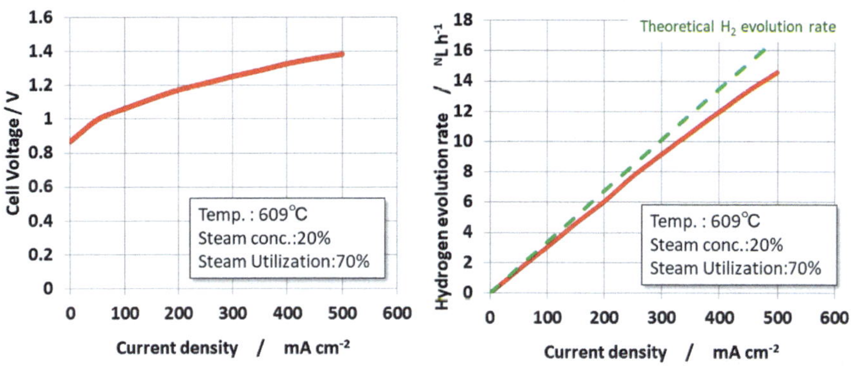

Fig. 9.21 Single-cell performance in electrolysis mode (Cell-E)

T. Electrolysis was also evaluated using 50%, 70%, and 90% steam; the results are shown in Fig. 9.22. The cell terminal voltage at 0.5 A cm $^{-2}$ decreased with increasing water vapor concentration, and the change in the cell terminal voltage was small when the steam amount was 50% or more. In addition, the current efficiency at 0.5 A cm $^{-2}$ tends to be higher than that of the 20% steam concentration, which was 82%, as the steam concentration increased.

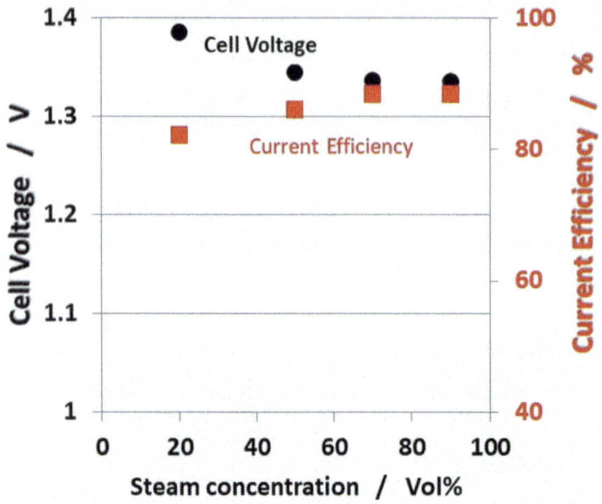

Fig. 9.22 Water vapor concentration dependence in Cell-E. Cell size: 100 mm × 100 mm, electrode size: 90 mm × 90 mm

9.8 Stack of Proton-Conducting Oxide Electrolysis Cells

We designed and assembled cell stacks based on prior literature, using the Cell-E configuration of 100 mm × 100 mm size, (3 and 10 cells stack). The assembled 3 and 10 cells stack are shown in Fig. 9.23, and the steam electrolysis test results with the 3 cells stack are shown in Fig. 9.24. The cell terminal voltage of the 3 cells stack was 4.7 V, the hydrogen generation rate was 42 $^{N}Lh^{-1}$, and the current efficiency was 83%. The hydrogen production rate was about 3 times that of a 100 mm × 100 mm single cell. There are various issues for practical application such as improving stack characteristics and cost reduction of stack manufacturing. To resolve these issues, we are studying the stack structure, etc.

Fig. 9.23 Short stack of proton-conducting oxide cell (Cell-E). Cell size: 100 mm × 100 mm, electrode size: 90 mm × 90 mm. Left: 3-cell stack; right: 10-cell stack

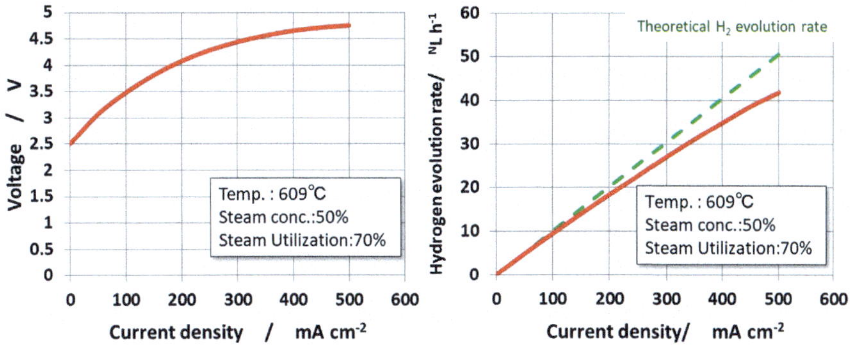

Fig. 9.24 A 3-cell-stack performance in electrolysis mode (Cell-E). Cell size: 100 mm × 100 mm, electrode size: 90 mm × 90 mm

9.9 Summary

For electrolyte development, $BaZr_{0.44}Ce_{0.36}Y_{0.2}O_{3-\delta}$ (BZCY $(54)_{8/9}2$) has high conductivity (1.4×10^{-2} S/cm @ 600 °C) and chemical stability in water vapor. This material is suitable as an electrolyte for steam electrolysis. Regarding the development of the electrode, it was concluded that $Ba_{0.5}Co_{0.5}O_{3-\delta}$ (BLC55) is a suitable anode material from the viewpoint of stability in water vapor, electrode activity, and hydrogen generation efficiency in steam electrolysis. In terms of steam electrolysis, the electrolytic voltage is lowest when NiO/SZCY541 (where SZCY541 is $SrZr_{0.5}Ce_{0.4}Y_{0.1}O_{3-\delta}$) is used as the cathode substrate, BZCY$(54)_{8/9}2$ as the electrolyte, and BLC55 the anode. When operated at 600 °C using a 12 μm thick BZCY$(54)_{8/9}2$ electrolyte, the electrolysis voltage is 1.48 V at 0.5 Acm^2, and a hydrogen generation efficiency exceeding 80% was obtained. The electrode overvoltage was about 0.3 V. In contrast, the electrolyte overvoltage was dominant, implying the thinner the electrolyte, the smaller the electrolyte overvoltage. In addition, a decrease in hydrogen generation efficiency was observed during operation at 550 and 500 °C, suggesting that the lack of anode performance was one of the causes.

Proton-conducting solid oxide electrolysis cells for large-scale hydrogen production from steam at the intermediate temperature of 600 °C have been fabricated using a tape casting, screen-printing, and co-sintering approach. A laminated combination of BSZCY91$(54)_{8/9}2$ and SZCY541 electrolytes on a NiO/SZCY541 active and substrate support produced 100 mm × 100 mm flat and warpage free half cells. Steam electrolysis voltage of the optimized Cell-E type configuration using BLC as anode and 20% steam concentration reached 1.39 V at 0.5Acm^{-2} at 600 °C and current efficiency of 82%. Inserting an active NiO/SZCY541 layer between the laminated electrolyte and cathode substrate layer significantly improves cell durability during long-term testing. After 1500 h of steam electrolysis operation, the terminal voltage rate (rising) was 2.5%/1000 h. At the same time, the current efficiency reduction rate was 2.1%/1000 h (Cell-E). Transition metal (Co and Ni) diffusion into the electrolyte component is pretty minimal and is not the leading cause of the performance degradation during the long-term test. However, noticeable Ba interdiffusion into the substrate layer and the formation Co_2O_3 oxide particles in the BLC anode material possibly affect the cells' durability during long-term testing. The cell terminal voltage at 0.5 A cm^{-2} decreased with increasing steam concentration, and the change was small 50% steam or more. Three and ten cell stacks were built and tested under electrolysis mode. The cell terminal voltage of the three cell stack was 4.7 V, the hydrogen generation rate was 42 $^{N}Lh^{-1}$, and the current efficiency was 83%. This study's technology advancements form a basis for future work to move proton-conducting solid oxide electrolysis cell technology toward practical application.

References

1. Iwahara H, Esaka T, Uchida H, Maeda N (1981) Proton conduction in sintered oxides and its application to steam electrolysis for hydrogen production. Solid State Ionics 3–4:359–363. https://doi.org/10.1016/0167-2738(81)90113-2
2. Iwahara H (1995) Technological challenges in the application of proton conducting ceramics. Solid State Ionics 77:289–298. https://doi.org/10.1016/0167-2738(95)00051-7
3. Matsumoto H, Hamajima S, Yajima T, Iwahara H (2001) Electrochemical hydrogen pump using SrCeO[sub 3]-based proton conductor: effect of water vapor at the cathode on the pumping capacity. J Electrochem Soc 148:D121. https://doi.org/10.1149/1.1400121
4. Matsuzaki Y, Tachikawa Y, Somekawa T et al (2015) Effect of proton-conduction in electrolyte on electric efficiency of multi-stage solid oxide fuel cells. Sci Rep 5:1–10. https://doi.org/10.1038/srep12640
5. Leonard K, Deibert W, Ivanova ME et al (2020) Processing ceramic proton conductor membranes for use in steam electrolysis. Membranes (Basel) 10:1–18. https://doi.org/10.3390/membranes10110339
6. Vøllestad E, Strandbakke R, Tarach M et al (2019) Mixed proton and electron conducting double perovskite anodes for stable and efficient tubular proton ceramic electrolysers. Nat Mater 18:752–759. https://doi.org/10.1038/s41563-019-0388-2
7. Matsumoto H, Sakai T, Okuyama Y (2013) Proton-conducting oxide and applications to hydrogen energy devices. Pure Appl Chem 85:427–435. https://doi.org/10.1351/PAC-CON-12-07-11
8. Leonard K, Lee YS, Okuyama Y et al (2017) Influence of dopant levels on the hydration properties of SZCY and BZCY proton conducting ceramics for hydrogen production. Int J Hydrogen Energy 42:3926–3937. https://doi.org/10.1016/j.ijhydene.2016.10.120
9. Anti-Corrosion Methods Mater 47:30–34
10. Leonard K, Okuyama Y, Takamura Y et al (2018) Efficient intermediate-temperature steam electrolysis with Y:$SrZrO_3$-$SrCeO_3$ and Y:$BaZrO_3$-$BaCeO_3$ proton conducting perovskites. J Mater Chem A 6:19113–19124. https://doi.org/10.1039/c8ta04019b
11. Han D, Shinoda K, Tsukimoto S et al (2014) Origins of structural and electrochemical influence on Y-doped $BaZrO_3$ heat-treated with NiO additive. J Mater Chem A 2:12552–12560. https://doi.org/10.1039/c4ta01689k
12. Yoo Y, Lim N (2013) Performance and stability of proton conducting solid oxide fuel cells based on yttrium-doped barium cerate-zirconate thin-film electrolyte. J Power Sources 229:48–57. https://doi.org/10.1016/j.jpowsour.2012.11.094
13. Tong J, Clark D, Bernau L et al (2010) Solid-state reactive sintering mechanism for large-grained yttrium-doped barium zirconate proton conducting ceramics. J Mater Chem 20:6333–6341. https://doi.org/10.1039/c0jm00381f

Chapter 10
Advances in High Temperature Electrolysis Using Solid Oxide Electrolysis Cells

Norikazu Osada

Abstract Hydrogen is the most attractive regenerative energy media and can be used for fuel cells, direct combustion, and so on. One of the key technologies for the hydrogen energy system is the production of large amounts of hydrogen with high efficiency. At present steam reforming reaction (SRR) of hydrocarbons is used worldwide, however, water electrolysis from renewable energies will be developed and used near future from the viewpoint of reducing greenhouse effects as well as saving fossil fuels. Here, several kinds of water electrolysis technologies are reviewed and finally high temperature electrolysis using solid oxide is shown in detail. Alkali electrolysis (AE), solid polymer electrolysis (SPE), and high temperature electrolysis (HTE) of water vapor are illustrated first in view of thermodynamics then of practical technology. As for the THE advanced technologies are illustrated from the standpoint of cell material (cathode, anode, and interconnector/separator) design and stack structure design. Lastly, Toshiba Energy Systems & Solutions Corp., 10 kW-class test system installed multiple SOEC stacks for hydrogen production recently developed are shown.

Keywords Hydrogen production · High temperature electrolysis · Solid oxide electrolyte · Multi-cell tube · Planar cell

10.1 Hydrogen Energy Systems

The development of sustainable energy systems is very important from the view point of reducing greenhouse effects as well as saving fossil fuels. Hydrogen is the most attractive regenerative energy media due to its cleanliness and potential applications to various energy conversion processes. Hydrogen can be used for fuel cells, direct combustion, and so on. Fuel cells exhibit higher energy conversion efficiency than the conventional internal combustion engine system because the chemical energy of fuel is directly converted into electric energy while that of the combustion engines

N. Osada (✉)
Toshiba Energy Systems & Solutions Corp., Saiwai-ku, Kawasaki, Japan
e-mail: norikazu1.osada@toshiba.co.jp

K. Aika and H. Kobayashi (eds.), *CO_2 Free Ammonia as an Energy Carrier*,
https://doi.org/10.1007/978-981-19-4767-4_10

are limited by Carnot efficiency. For example, polymer electrolyte fuel cells (PEFCs) have attracted much attention as clean and efficient power sources for electric vehicles (EVs) and residential co-generation systems. One of the key technologies for the hydrogen energy system is the production of large amounts of hydrogen with high efficiency.

10.2 Hydrogen Production Methods

Hydrogen has been produced by various methods as shown in Fig. 10.1 [1]. Industrial processes are steam reforming of hydrocarbons (naphtha, city gas, and methanol) and water electrolysis. Besides such direct productions, an appreciable amount of hydrogen has been obtained as a by-product in chlor-alkali electrolysis, off gas from a petroleum refinery (catalytic reforming or ethylene decomposition) or a coke oven gas in steelworks, etc. The amounts of H_2 production and the purity levels in these processes are summarized in Table 10.1 [1].

In the steam reforming reaction (SRR) of hydrocarbons, hydrogen and carbon monoxide are produced, followed by water gas shift reaction (WGSR) in the catalytic reactors,

$$\text{SSR: } CxHy + xH_2O \rightarrow (x + y/2)H_2 + xCO \tag{10.1}$$

$$\text{WGSR: } xCO + xH_2O \rightarrow xCO_2 + xH_2 \tag{10.2}$$

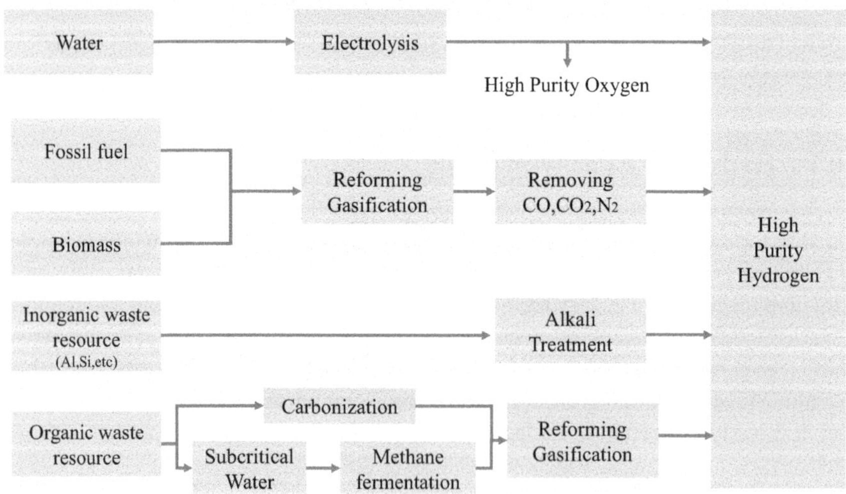

Fig. 10.1 Processes of hydrogen production [1]

Table 10.1 Characteristic of hydrogen production methods [1]

Type	Process	Amount of hydrogen production in Japan (2000)	Purification method (%)	Purity (%)
Direct production	Steam reforming of naphtha (oil factory)	7000	PSA: 95, methanation: 5	95–98
	Steam reforming of manufactured gas/methanol	120	PAS:100	99.99–99.999
	Water electrolysis	–	Unnecessary	>99.99
By-production	Chlor-alkali electrolysis (caustic soda)	1252	PSA9.6, raw: 90.4	>99.9
	Off gas oil refinery	6704	PSA:25, membrane separation 35, raw: 40	70–80 (raw) 95–97 (after purification)
	Off gas for petrochemistry (enthylene decomposition)	2100	PSA:5, Cryogenic separation:35–50, raw: 45–60	90
	Coke oven gas	9591	PSA: little, raw: 100	54–60(raw)

$$\text{Overall: } CxHy + 2xH_2O \rightarrow (2x + y/2)H_2 + xCO_2 \tag{10.3}$$

For the use of PEFCs in FCVs, the purity of H_2 must be >99.97% (ISO14687-2), and CO level must be reduced below 0.2 ppm to prevent poisoning of the Pt-based anode. Therefore, the reformed gas is purified further by a preferential oxidation (PROX) of CO, methanation, pressure swing adsorption, membrane separation, etc.

In contrast, water electrolysis is an attractive option to generate hydrogen by a single step without any purification. Using commercial electricity and water, on-site H_2 production is possible. As an advantage of electrochemical devices, the efficiency of the cell itself is almost independent of the scale or size. When combined with solar cells or wind power generation, pure H_2 and O_2 can be produced without any CO_2 emission. However, water electrolysis plays only a minor role so far because the cost of electricity required for the conventional water electrolysis is much higher than other methods; less than 1% of worldwide hydrogen demand is produced electrolytically.

10.3 Water Electrolysis

Water electrolysis is classified by the electrolyte employed in the cell; Alkali Electrolysis (AE), Solid Polymer Electrolysis (SPE), and High Temperature Electrolysis of water vapor (HTE). The efficiency of water electrolysis, ε is defined by the following equation,

$$\varepsilon = (\Delta H^0/2FE_{\mathrm{app}})\varepsilon_F \tag{10.4}$$

where ΔH^0 is the standard enthalpy of water decomposition reaction (285.8 kJ/mol at 25 °C), F is the Faraday constant (96,485 Coul/mol), E_{app} is the applied cell voltage (V), ε_F is the current efficiency for hydrogen production. The lower the E_{app}, the higher the efficiency is obtained. With decreasing the current density (the hydrogen production rate), the E_{app} can be reduced but a large electrode area is necessary, resulting in high capital cost. Thus, appropriate current density is required for the practical operation with a reasonable electrode area.

In the following sections, some characteristics of AE, SPE, and HTE are described.

10.3.1 Alkali Electrolysis (AE)

AE is a conventional and commercialized method. The principle of AE is shown in Fig. 10.2, and the electrode reactions are given as follows,

$$\text{Anode: } 2\mathrm{OH}^- \rightarrow 1/2\,\mathrm{O_2} + \mathrm{H_2O} + 2\mathrm{e}^- \tag{10.5}$$

$$\text{Cathode: } 2\mathrm{H_2O} + 2\mathrm{e}^- \rightarrow \mathrm{H_2} + 2\mathrm{OH}^- \tag{10.6}$$

$$\text{Overall: } \mathrm{H_2O} \rightarrow \mathrm{H_2} + 1/2\,\mathrm{O_2} \tag{10.7}$$

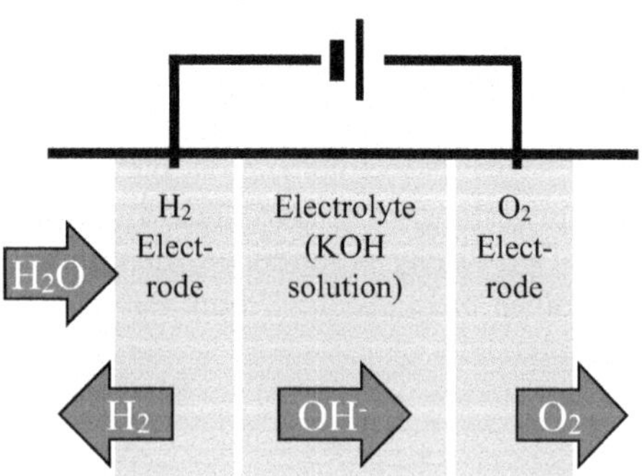

Fig. 10.2 Scheme for alkali electrolysis

Nickel plated Fe as the anode, Ni or Fe as the cathode, and 25–30% KOH as the electrolyte are employed in the conventional alkali electrolysis cell. For stacking the cells, monopolar and bipolar type have been adopted. The former type can reduce leak current but requires a larger floor space, compared with the later type. The later type is, however, dominant due to high capacity and compactness. Typical performances of these AE cells are shown in Table 10.2 [1]. The operating temperature is 70–90 °C. The current densities are limited to <0.2 A cm^{-2} to avoid an ohmic resistance by bubble formation. The energy conversion efficiencies range from 70 to 80%.

Since 1970s, advanced AE cells have been developed in Europe, USA, Canada, and Japan. As shown in Table 10.3 [1], high performance electrodes such as Raney Ni and a diaphragm material (potassium titanate) were typical advanced materials. High conversion efficiency was achieved at relatively high current density and high operating temperature with narrow-zero gap configuration. In the Sunshine Project of Japan, 100 kW class pilot plant was developed and the applied voltage of 1.75 Vat 0.25 A cm^{-2} was achieved under 120 °C and 20 atm condition. However, the operating temperature was lowered below 100 °C, due to its material cost from 1980s.

Since 1990, larger size AE systems have been developed and larger size AE plant is being set to a commercial level. NEL in Norway and Asahi KASEI in Japan were major companies for larger-scale AE plant. They developed kW to MW class AE stacks and systems [2, 3].

10.3.2 Solid Polymer Electrolysis (SPE)

The most characteristic material in the SPE is a proton conducting polymer membrane such as DuPont's Nafion, in place of a conventional liquid electrolyte. Alloys of PtIr and PtRh are employed as the anode and cathode catalysts in this cell. The principle of SPE and the electrode reactions are given as follows,

$$\text{Anode: } H_2O \rightarrow 2H^+ + 1/2O_2 + 2e^- \quad (10.8)$$

$$\text{Cathode: } 2H^+ + 2e^- \rightarrow H_2 \quad (10.9)$$

$$\text{Overall: } H_2O \rightarrow H_2 + 1/2\,O_2 \quad (10.10)$$

The SPE was first developed by General Electric Company (GE), USA, in 1970s as a derivative technology for PEFCs. Pure water is supplied to the anode compartment to generate O_2 and protons. Protons are discharged at the cathode to evolve H_2. Because of the nature of the polymer electrolyte, the SPE cell has following three major advantages. The cell is free from any problem regarding the maintenance of the concentration and volume of the electrolyte solution. The SPE allows an operation with differential pressure between the anode and the cathode. As shown in Fig. 10.3, H_2 and O_2 evolve at the porous catalyst layers without formation of bubbles in

Table 10.2 The typical results for alkali electrolysis [1]

Company (country)	Ability of hydrogen production/$Nm^3\ h^{-1}\ unit^{-1}$	Electrode size/m	Number of cells/unit	Operating temperature/°C	Pressure/atm	Current density/A cm^{-2}	Cell voltage/V
Lurgi (Germany)	750	Φ1.5	556	90	30	0.2	1.86
Delano (Italy)	540	2*4	108	80	Ordinary pressure	0.15	1.85
Norsk Hydro (Norway)	340	Φ1.5	235	80	Ordinary pressure	0.175	1.75
Brown Boveri (Swizerland)	300	1.5*3	80	80	Ordinary pressure	0.2	2.04
Elevtrolyzer (Canada)	12.5	0.6*0.8	***	70	Ordinary pressure	0.134	1.90

Table 10.3 Construction materials for advanced alkali electrolysis [1]

Company (country)	Diaphragm	Cathode materials	Anode materials	Current density/A cm^{-2}	Cell voltage/V
CEN, Mol (Belgium)	Sb_2O_3/polysulfone	(Ni, Co)S	Co_3O_4	0.3–1.0	1.65–1.80
Marcoussis (France)	K-hexatitanate/PTFE	MoNi	FeMoNi	0.3	1.85
SODETEG (France)	Stabilized asbesto	Raney Ni/Co	Co oxide-raney Ni	0.3	1.8–1.9
Sunshine project (Japan)	Potassium titanate impregnated PTFE	Surface area extended Ni	Surface area extended Ni	0.4	1.73–1.78

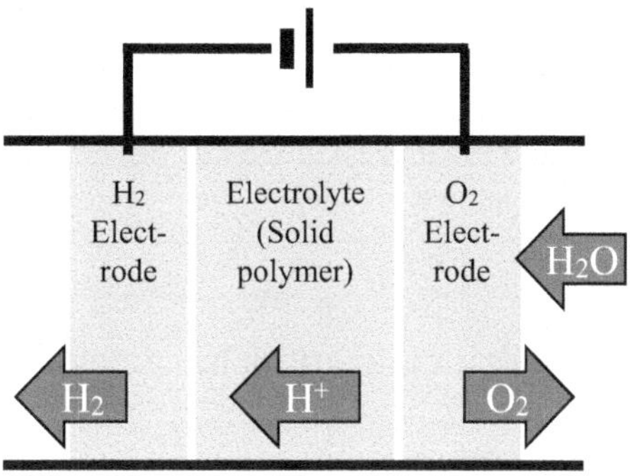

Fig. 10.3 Scheme for solid polymer electrolysis

the electrolyte. This results in low ohmic resistance and allows high current density operation. However, the gas crossover (permeation) of H_2 an O_2 through the polymer membrane to the counter electrode reduces the current efficiency slightly.

The typical performances of the SPE cells are shown in Table 10.4 [1]. The applied voltage of 1.5–1.7 V at 1.0–1.3 A cm^{-2} was achieved in the WE-NET project in Japan. However, the largest drawback of the SPE is the high cost of both the electrolyte' membrane and the electrocatalysts based on Pt.

10.3.3 *High Temperature Electrolysis (HTE) of Water Vapor*

The HTE is also known as Solid Oxide Electrolysis Cells (SOEC) because solid oxide materials such as yttria stabilized zirconia (YSZ) are used as the solid electrolyte. The concept of HTE was first proposed in 1960s by GE [4, 5]. The applied voltage is expected to be reduced due to favorable thermodynamic and kinetics conditions,

Table 10.4 The typical results for solid polymer electrolysis [1]

Company (country)	Year	Electrode area/cm^2	Number of cells/unit	Operating temperature/°C	Pressure/kg cm^2	Current density/A cm^{-2}	Cell voltage/V
General electric (United States)	1981	233	60	96	7	1.08	1.09
General electric (Unites States)	1985	233	60	60	3.5	1.08	1.9–2.0
B.B. (Swiss)	1986	400	18	80	***	1.00	1.70
RITE (Japan)	1999	1000	5	0–140	1–5	1.0–1.3	1.48–1.72
WE-NET (Japan)	2001	1000	50	***	5	1.0–1.5	***

providing the highest efficiency of electrolytic hydrogen production. The details of HTE are shown in the following section.

10.4 Principle and Characteristics of HTE

10.4.1 Principle of HTE

The principle of the HTE is shown in Fig. 10.4. The electrode reactions in the HTE are given as follows,

$$\text{Anode: } O^{2-} \rightarrow 1/2\,O_2 + 2e^- \tag{10.11}$$

$$\text{Cathode: } H_2O + 2e^- \rightarrow H_2 + O^{2-} \tag{10.12}$$

$$\text{Overall: } H_2O \rightarrow H_2 + 1/2\,O_2 \tag{10.13}$$

The HTE cell consists of a solid-oxide electrolyte and two electrodes (the anode and cathode). Steam is supplied to the cathode compartment to evolve H_2 and oxide ions (O^{2-}); (10.11). The oxide ions are transported through the solid electrolyte from the cathode to the anode, where O^{2-} is discharged to O_2 gas; (10.12). The overall reaction is written by (10.13).

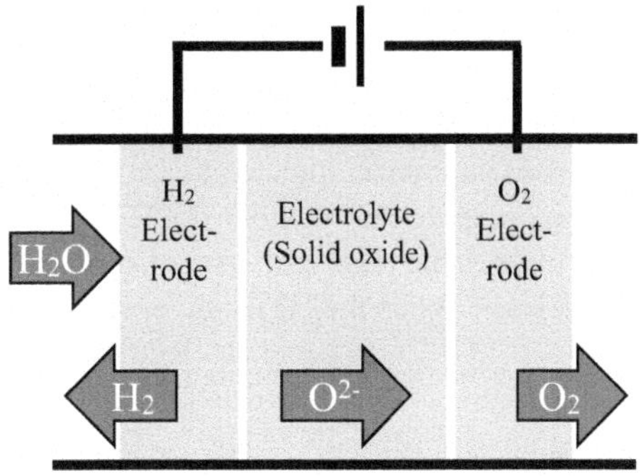

Fig. 10.4 Scheme for high temperature electrolysis

Figure 10.5 shows an energy diagram for the steam electrolysis. The applied voltage E_{app} required is given by Nernst equation.

$$E_{\text{app}} = \Delta G/2\text{F} \tag{10.14}$$

where ΔG is Gibbs free energy for the decomposition of water (steam) at the operating temperature. As shown by the dashed line, ΔG decreases with temperature, which is the favorable thermodynamic condition for the HTE. In contrast, the total energy required for the reaction is the enthalpy change ΔH at the operating temperature.

$$\Delta H = \Delta G + T\Delta S \tag{10.15}$$

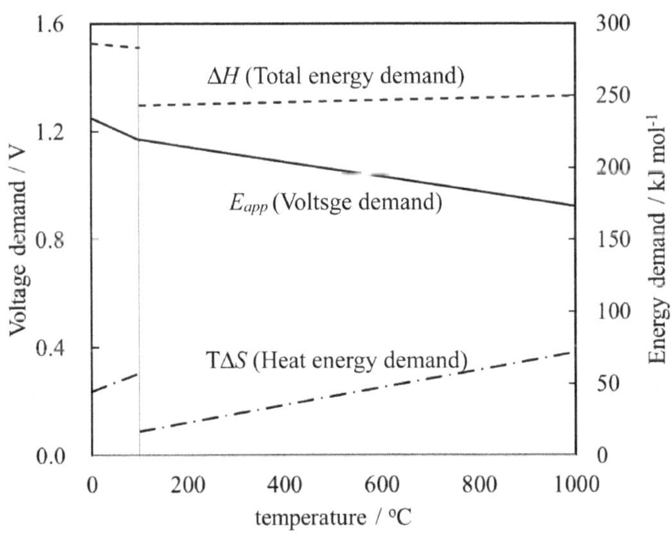

Fig. 10.5 Voltage and energy demand for water and steam electrolysis

Compared with ΔG, the temperature-coefficient of ΔH is relatively small. For example, the theoretical E_{app} for water electrolysis at 25 °C is 1.23 V but the value at 1000 °C is reduced to 0.95 V. Furthermore, the reaction rate (current density) becomes faster under high temperature conditions, resulting in lowering cathodic and anodic overpotential without noble metal electrocatalysts such as Pt. The ceramic electrolyte in the HTE is also free from the corrosion problem seen in acid or alkaline electrolytes. When the HTEs are operated reversely, they can work as the solid oxide fuel cells (SOFCs) to generate electricity with a high efficiency by consuming natural gas, coal gas, or stored hydrogen. Such solid electrolyte devices are also known as regenerative or reversible SOFCs. Thus, such cells are regarded as a reciprocal direct energy converter between hydrogen and electricity.

10.4.2 Typical Reports

The HTE was intensively investigated in 1970–1980s [6–14]. Typical results for the HTE cells are summarized in Table 10.5. In Germany, the "HOT ELLY" (High Operating Temperature Electrolysis) project was performed under the grant by German Government. In the HOT ELLY project, Dornier-Systems GmbH and Lurgi GmbH were major companies since the middle of 1970s [15–24]. They developed kW to MW class stacks. In their electrolysis tube with 10 cells, the performance of 13.3 V (1.33 V per cell) was achieved at 0.37 A cm and 997 °C. Their single cell showed a stable performance of 1.07 V at 995 °C and 0.3 A cm^{-2} for a long term of 1000 h with 100% Faradaic efficiency [24–26].

In United States, Westinghouse Electric. Co. mainly conducted the HTE development in the same term of HOT ELLY project [27, 28]. They developed original type HTE cells which had 75 cm^2 electrode area per cell. The performance of 1.31 V was achieved at 0.4 A cm^{-2} and 1000 °C. The applied voltage of 1.23 V at approximately 0.3 A cm^{-2} and 1000 °C was maintained for 450 h.

In Japan, Mitsubishi Heavy Industries, Ltd. (MHI) [29, 30] and Japan Atomic Energy Research Institute (JAERI) [31–33] conducted HTE development in the late 1980–90s. Especially, in MHI, the performance of their planar type cell was higher than those of HOT ELLY and Westinghouse, i.e., the applied voltages of 1.12 and 1.3 V were achieved at 0.45 and 0.94 A cm^{-2}, respectively, for a single cell operated at 1000 °C.

Since 2000, Technical University of Denmark in Denmark, University of Yamanashi, Kyushu University, National Institute of Advanced Industrial Science and Technology and Toshiba Corporation were started to research/develop high performance electrodes for HTE cells and HTE systems [34–41]. And Fuel Cell Energy in USA, Sunfire and Fraunhofer in Germany also have developed HTE systems.

Table 10.5 The typical results for high temperature electrolysis [15–33]

Company (country)	Year	Operating temperature/°C	Electrolyte	Cathode	Anode	Current density/A cm^{-2}	Cell voltage/V
General Electric (United States)	1969	1000	YSZ	Ni-YSZ	$PrCoO_3$	***	***
Dornier (Garmany)	1985	997	YSZ	Ni-YSZ	$LaMnO_3$	0.37	1.33
Westinghouse (United States)	1986	1000	YSZ	Ni-YSZ	$La(Sr)MnO_3$	0.4	1.31
Mitsubishi Heavy Industries (Japan)	1993	1000	YSZ	Ni-YSZ	$La(Sr)MnO_3$	0.45	1.12
Kyushu University (Japan)	1996	1000	YSZ	Ni-YSZ	$La(Sr)MnO_3$	0.4	1.60
JAERI	1995–2004	950	YSZ	Ni-YSZ	$La(Sr)MnO_3$	0.13	1.31

10.5 Cell Material Design of SOEC for THE

The materials of the solid oxide electrolysis cells as well as the stack design are described in this section. Similar to SOFCs, the materials for conventional HTE cells have been limited to ceramics which are stable at high temperatures.

10.5.1 Solid Oxide Electrolyte

The function of the solid electrolyte is to transport the oxide ions from the cathode to the anode and to separate gases between both compartments. Therefore, the solid electrolyte must have high oxide-ion conductivity and be sintered well to prevent gas permeation. The electrolyte should also have a negligible electronic conductivity to avoid a short-circuit within the cell. A chemical stability is also required in a large oxygen partial pressure gradient from highly reducing to oxidizing conditions. Furthermore, it should not react with the adjacent components, such as electrodes or interconnector. Stabilized zirconia, such as yttria-stabilized zirconia [$(ZrO_2)_x(Y_2O_3)_{1-x}$, $x = 0.08$ to 0.10 denoted as YSZ], is the most common material.

Figure 10.6 shows specific ionic conductivities as a function of reciprocal temperature, i.e., Arrhenius plots [42]. The conductivity of 8 mol% YSZ is slightly higher than 0.1 S cm^{-1} at 950–1000 °C and 0.03–0.003 cm^{-1} at reduced temperature of 600–800 °C. The mechanically preferred tetragonal phase containing 3 or 4 mol% Y_2O_3 (partially stabilized) is a less conductive alternative. Therefore, for the SOEC operated at lower temperatures, the YSZ should be thinner, and supported on the porous substrate of the anode or cathode to keep sufficient mechanical strength. Otherwise, an alternative electrolyte material of higher ionic conductivity should be used.

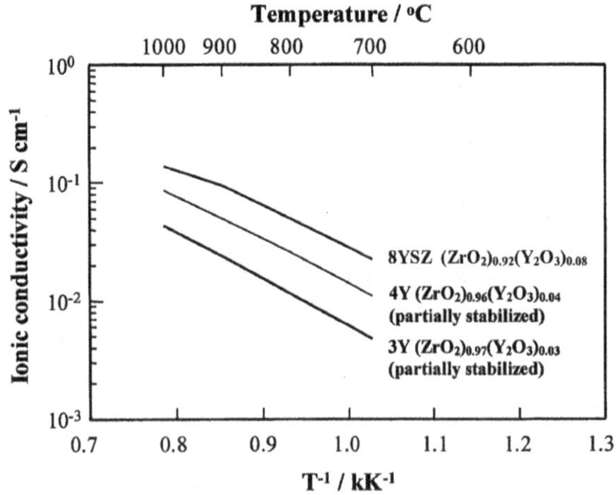

Fig. 10.6 Arrhenius plots of ionic conductivities of 3Y, 4Y, 8YSZ [42]

10.5.2 Electrodes

The electrode reaction proceeds in a triple phase boundary (TPB) where gases, electrons, and oxide ions are associated. To minimize the overpotential loss, the electrode must have not only high electrocatalytic activity but also an appropriate microstructure with high gas diffusivity and large TPB.

Cathode: At the cathode, electrons are transferred to water vapor, generating hydrogen gas and oxide ions. It is essential that the cathode is stable under reducing atmosphere and has electrocatalytic activity for the hydrogen evolution reaction. It must be compatible with the other cell components, such as the solid electrolyte and the interconnector. Typically, Ni-YSZ cermet (ceramic–metal composite) has been used as the cathode material for the SOEC [43–46]. Nickel has a high electronic conductivity as well as appropriate catalytic activity for the reaction. The nickel content in the cermet is typically 50–60 vol%. The YSZ acts as the ionic conductive path and matrix that prevents Ni from agglomeration.

University of Yamanashi has developed a mixed conduction ceria-based cathode for SOEC. The samaria-doped ceria (SDC) cathode containing with less than 20% Ni catalysts exhibited a higher performance than conventional Ni-YSZ cathode. Toshiba Corporation has also developed ceria-based cathode for SOEC. The gadolinia-doped ceria (GDC) with Ni catalysts exhibited high performance and long-term durability as a SOEC cathode.

Anode: At the anode, oxygen gas evolves from oxide ions passing through the electrolyte. The conventional SOEC anode material is an electronically conducting oxide ceramic based on the perovskite-type lanthanum manganite, $LaMnO_3$ in which La is partly substituted with Sr or Ca ($La(Sr)MnO_3$, LSM/$La(Ca)MnO_3$, LCM). The LSM or LCM are *p*-type electronic conductor with low oxide ion conductivity. While the electronic conductivity is high enough over wide temperature range (600–1000 °C), the oxide ion conductivity appreciably decreases at lower temperatures. On this account, the LSM and LCM can be only used in elevated temperature SOEC operated at around 1000 °C.

The lanthanum cobaltite, $LaCoO_3$ type materials were employed as the anode for lower operation temperature SOEC. These materials could be reacted with conventional electrolyte materials for example YSZ same as SOFCs. Toshiba Corp. has developed a high performance and long-term durability anode using $LaCoO_3$ type materials in which La/Co are partly substituted with Sr/Fe ($La(Sr)Co(Fe)O_3$, LSCF) with GDC interlayer. The SOEC, comprised of the Ni-GDC cathode, YSZ electrolyte, and LSCF anode with GDC interlayer, exhibited fairly high stability performance. The degradation rate was reduced to 0.3%/kh.

10.5.3 Interconnector/Separator

The interconnector connects cells in series, and, therefore, is exposed to both highly reducing and highly oxidizing atmospheres at the anode and the cathode, respectively. For some cell designs (planar type), the gas flow fields are formed on the interconnector by machining. The material should be, in principle, a pure electronic conductor without any ionic conduction, because the mixed conduction of ions and electrons causes a chemical short circuit reaction. Additional requirements for the interconnector are the dimensional stability in the $p[O_2]$ gradient and compatible thermal expansion coefficient with the electrodes and the electrolyte.

The choice of the interconnector material is strongly dependent upon the operating temperature. The conventional interconnecting material for the SOEC operated at 900 to 1000 °C is Ca- or Sr-doped $LaCrO_3$ same as SOFC [47, 48]. However, the demerits of the $LaCrO_3$-based interconnector are its difficulty to maintain the gas tightness and its high cost. If the operating temperature can be reduced below 800 °C, inexpensive heat-resistant alloys such as Fe-based stainless steels can be used as the interconnector.

10.6 SOEC Design

Three types of SOEC stacking design have been proposed so far: two tubular-types and one planar type. Tubular structures are single-cell per tube and multi-cells on a support tube. The most important advantage of the tubular-type design is the easy separation of gases. However, the H_2 production rate per unit volume is low and its manufacturing cost is relatively high. In contrast, the planar structure has a higher volumetric production rate (high power density) with short current pass, but the gas sealing is relatively difficult. The details of these types are described below.

Multi-cell tube design: This structure was first proposed by GE and adopted by Dornier System GmbH. Porous supporting-tube type and electrolyte-supporting type were examined as shown in Fig. 10.7. In the porous supporting-tube type, the porous support tube with a porosity of 34–38% was used. The cathode, gastight electrolyte, and anode were accumulated on that tube. These cells were connected with the interconnector in series as shown in Fig. 10.7a.

In contrast, the electrolyte-supporting tube type (Fig. 10.7b) without a porous support tube was also examined at Dornier System GmbH. Small cell segments were connected with the interconnector in series. The cathode and the anode were formed inside and outside the gastight tubular YSZ electrolyte (YSZ, about 300 μm thickness), respectively. Dornier System fabricated 10 cells on a tube and bundled 10 tubes into a module (10×10 cells).

Single-cell per tube design: Similar to the SOFCs, the cell structure shown in Fig. 10.8 had been developed by Westinghouse Electric Corp. The support tube

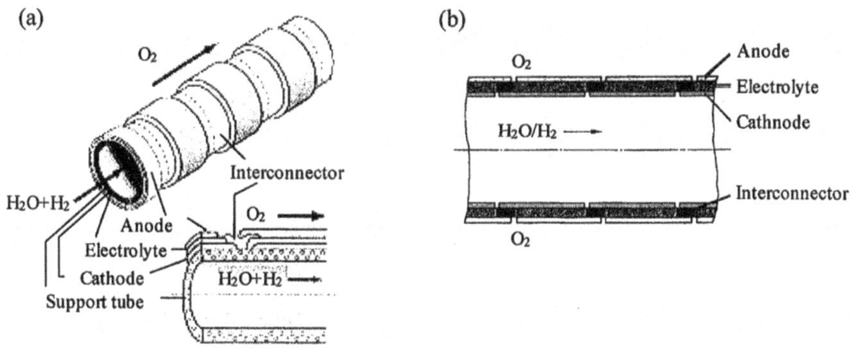

Fig. 10.7 The structure of multi-cell tube design for porous support tube type (**a**) and electrolyte support type (**b**) [18, 28]

was a porous calcia stabilized zirconia (CSZ) tube with porosity of around 35%, The inner diameter was 12–13 mm with a wall thickness of 1–1.5 mm and maximum length of 1000 mm. The support tube was overlaid with porous anode such as LSM. A gastight electrolyte layer such as YSZ was coated on the support tube with the thickness of around 40 μm. After that, the cathode such as Ni-YSZ cermet (100 μm thickness) was formed on the electrolyte surface. A band of the interconnector was formed on the anode with the YSZ insulating from the cathode.

Planar cell design: The electrolyte is sandwiched between two planar electrodes as shown in Fig. 10.9. Two types of this design have been examined: electrolyte-support and electrode substrate-support type. In the electrolyte-support type, porous

Fig. 10.8 The structure of single-cell per tube design [27, 28]

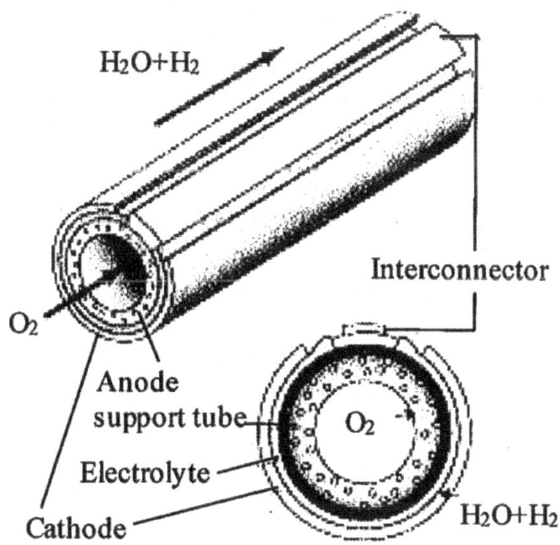

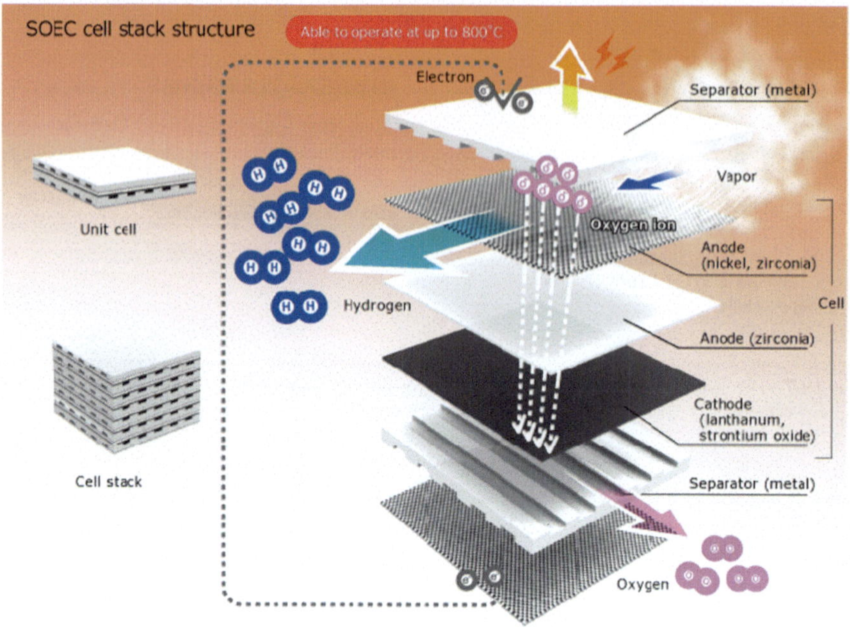

Fig. 10.9 The structure of planar cell design [41]

electrodes with 30–50 μm were coated on 100–300 μm thick electrolyte with sufficient mechanical strength. A combination of the various sintering processes was anticipated to save production costs and to guarantee proper contacts. On the other hand, in the electrode substrate-support type, the cell was supported by Ni-YSZ or metallic porous layer with the thickness of 0.3–2 mm. The cell had a 10–45 μm thick Ni-YSZ cathode, a 10–150 μm thick YSZ electrolyte, and a 10–50 μm LSM anode. They were produced by tape-casting, calendaring or pressing with a thickness of several hundred micrometers. In the planar cell, the interconnector also acts as a gas separation and flow field, and occupies a very large portion compared with any other designs. $LaCrO_3$-based materials are mainly used but its cost is high. The gas sealing material is important in fabricating the cell stacks. Compression ceramic seal, glass seal, and glass–ceramic seals are all proposed.

10.7 System Design [49, 50]

In order for the SOEC system to operate at high temperature (600–1000 °C), it needs to heat the water to make steam. The SOEC systems, generally comprised of SOEC stacks, steam generator, gas feed pump, heat exchanger for anode and cathode inlet and outlet gas, and so on. The other parts of the system, except for the SOEC stacks are known as BOP (balance of power). In order to increase the efficiency for SOEC systems, it is necessary to reduce not only electrical energy for SOEC stacks but the energy for BOP.

In Toshiba Corp., 10 kW-class test system installed multiple SOEC stacks for hydrogen production was developed. They evaluated scaling-up while maintaining the performance of each stack, system efficiency with heat exchangers, and optimization of system operation (Fig. 10.10). The Hydrogen production rate has achieved a maximum of about 3.5 Nm^3/h at 700 °C. And they conducted a hydrogen production test with fluctuating input power in order to ascertain whether hydrogen production can be carried out by supplying electricity generated by renewable energy directly to the system. The hydrogen production rate for the SOEC test system was in good agreement with its theoretical rate estimated from fluctuating input power (Fig. 10.11).

Based on the result of the 10 kW-class test system, the heat and mass balance analysis and the calculation of electric needs for the 200 kW-class SOEC system was

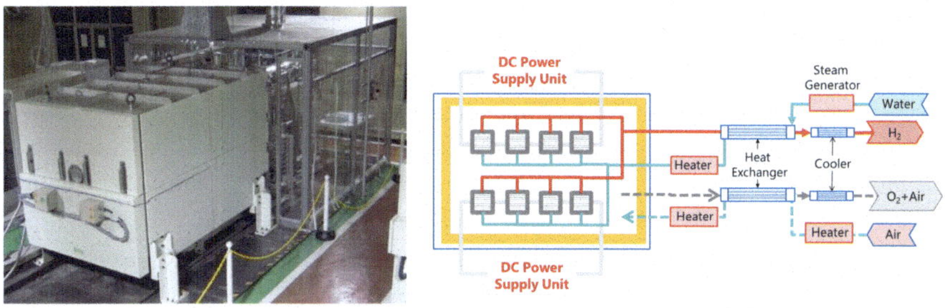

a b

Fig. 10.10 The exterior photograph (**a**) and schematic diagram of 10 kW-class SOEC test system (**b**) [49]

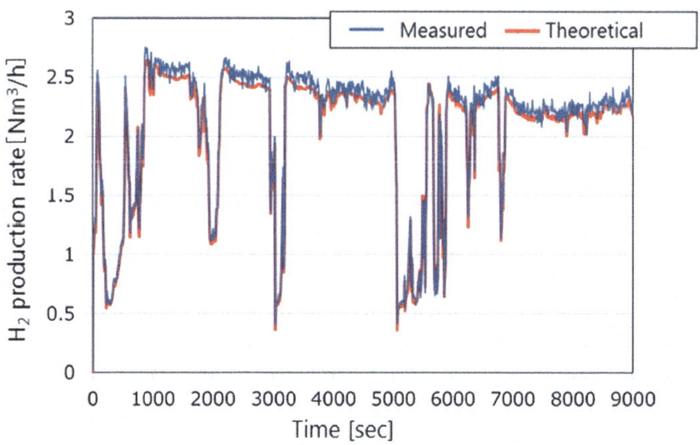

Fig. 10.11 The hydrogen production rate for 10 kW-class SOEC test system with fluctuating input power [49, 50]

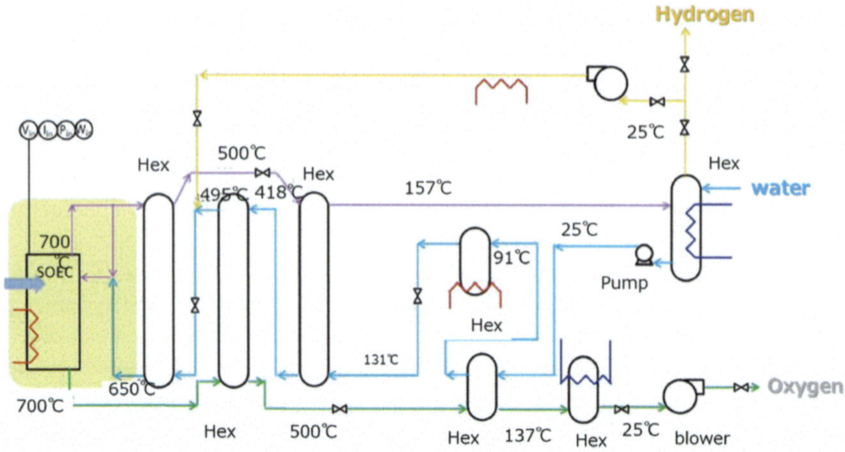

Fig. 10.12 The diagram, of 200 kW-class SOEC test system [49, 50]

Table 10.6 Specification of 200 kW (50 Nm^3/h)-class SOEC system [49, 50]

Item	Specification
H_2 production rate	50 Nm^3/h
SOEC electric power	151.6 kW
Operating temperature	700 °C
Steam feed amount	46.2 kg/h
Steam utilization	85%
H_2 feed amount	0.27 kg/h
H_2 concentration for cathode inlet	5%
Target of energy demand	4.0 kWh/Nm^3

carried out. The results of heat and mass balance analysis and specifications for the SOEC system were shown in Fig. 10.12 and Table 10.6. The system efficiency better than 4 kWh/Nm^3 was confirmed in the design study of the 50 Nm^3/h-class with the 200 kW-class system.

Acknowledgements A part of this work is based on results obtained from a project commissioned by the New Energy and Industrial Technology Development Organization (NEDO) in Japan.

References

1. Tamura H, Uchida H, Ikeda H, Iwakura C, Takasu Y (2003) Functional chemical series for electricity and ions 4

2. Technical development and strengthening of hydrogen energy storage and utilization; Fukushima-NEDO hydrogen system program (in Japanese), https://www.asahi-kasei.com/jp/news/2020/ze200915.html Accessed 1 Sep 2022
3. Water electrolysers / hydrogen generators, https://nelhydrogen.com/ Accessed 1 Sep 2022
4. Spacil H, Tedmon C (1969) Electrochemical dissociation of water vapor in solid oxide electrolyte cells: I . thermodynamics and cell characteristics. J Electrochem Soc 116:1618
5. Spacil H, Tedmon C (1969) Electrochemical dissociation of water vapor in solid oxide electrolyte cells: II. Materials, fabrication, and properties. J Electrochem Soc 116:1627
6. Badwal S, Bevan D, Bockris J (1980) The electrode kinetics of the evolution and dissolution of oxygen at the urania-zirconia interfaces. J Electrochim Acta 25:1115
7. Pound B, Bevan D, Bockris J (1981) The electrolysis of steam using uranium oxide electrodes. Int J Hydrogen Energy 6:473
8. Accorsi R (1982) Hydrogen production by high temperature electrolysis of steam. EUR Rep Commun EUR 8054:367
9. Olmer L, Viguie J, Schouler E (1982) An increase in the water vapor reduction rate by using an yttria-stabilized zirconia electrolyte with ceria-doped surface. Solid State Ionics 7:23
10. Barbi G, Mari C (1985) High temperature electrochemical reduction of the water molecule at cerium dioxide electrodes. Solid State Ionics 15:335
11. McElroy J (1983) Status of the development of solid polymer electrolyte water electrolysis for large scale hydrogen generation. US DOE Rep CONF 830974:127
12. Salzano F, Skaperdas G, Mezzina A (1985) Water vapor electrolysis at high temperature: systems considerations and benefits. Int J Hydrogen Energy 10:801
13. Liepa M, Borhan A (1986) High-temperature steam electrolysis: Technical and economic evaluation of alternative process designs. Int J Hydrogen Energy 11:435
14. Perfiliev M (1994) Problems of high-temperature electrolysis of water vapour. Int J Hydrogen Energy 19:227
15. Doenitz W, Schmidberger R, Steinheil E, Streicher R (1980) Hydrogen production by high temperature electrolysis of water vapour. Int J Hydrogen Energy 5:55
16. Doenitz W, Schmidberger R (1982) Concepts and design for scaling up high temperature water vapour electrolysis. Int J Hydrogen Energy 7:321
17. Dietrich G, Schafer W (1984) Advances in the development of thin-film cells for high temperature electrolysis. Int J Hydrogen Energy 9:747
18. Doenitz W, Erdle E (1985) High-temperature electrolysis of water vapor—status of development and perspectives for application. Int J Hydrogen Energy 10:291
19. Doenitz W, Schmidberger R (1988) Electrochemical high temperature technology for hydrogen production or direct electricity generation. J Hydrogen Energy 13:283
20. Erdle E, Donitz W, Schamn R, Koch A (1992) Reversibility and polarization behavior of high temperature solid oxide electrochemical cells. Int J Hydrogen Energy 17:817
21. Dietrich G, Schafer W (1982) A preparation technique of thin-film ZrO_2-Y_2O_3 solid electrolyte on porous support structures for high temperature electrolysis application. Hydrogen Energy Progress-IV:419
22. Dietrich G, Schafer W, Doenitz W (1983) Development of thin film cells for high temperature electrolysis. Hydrogen Energy Carrier 237
23. Dietrich G, Donitz W, Hermeking H, Oehme G (1982) Development of a technique for thin-film preparation of ZrO_2-Y_2O_3 solid electrolyte on porous support structures for high temperature electrolysis application EUR Rep. EUR 8054:371
24. Donitz W, Erdle E (1991) Electrochemical cell design and optimization procedures. Dechema-Monographs 123:21
25. Quantdt K, Streicher R (1986) Concept and design of a 3.5 MW pilot plant for high temperature electrolysis of water vapor. Int J Hydrogen Energy 11:309
26. Streicher R, Quantdt K (1984) Hydrogen Energy Progress-V:777
27. Maskalick N, Buzzelli E, Zuckerbrod D, MBonner M, Mezzina A, Salzano F (1984) Hydrogen production employing high temperature solid oxide cells. Proceedings of intersociety energy conversion engineering conference, vol 19, p 1415

28. Maskalick N (1986) High temperature electrolysis cell performance characterization. Int J Hydrogen Energy 11:563
29. Nanjo F, Takenobu K, Watanabe K, Miyamoto H, Sumi M, Tokunaga S, Koshiro I (1993) Development of the Mitsubishi planar reversible cell. Sci Technol Zirconia V 5:705
30. Koshiro I, Miyamoto H, Sumi M, Mori K (1995) The improvements of the solid oxide steam electrolysis cell components. Hydrogen Energy Progress-X:695
31. Hino R, Aita H, Sekita K, Haga K, Iwata T (1997) Study on hydrogen production by high temperature electrolysis of steam. JAERI-Res 97–064:1
32. Konishi S, Ohno H, Yoshida H, Katsuta H, Naruse Y (1986) Solid oxide electrolysis cell for decomposition of tritiated water. Int J Hydrogen Energy 11:507
33. Momma A, Kato T, Kaga Y, Nagata S (1997) Polarization behavior of high temperature solid oxide electrolysis cells. J Ceram Soc Jpn 105:369
34. Uchida H, Osada N, Watanabe M (2004) High-performance electrode for steam electrolysis. Mixed conducting Ceria-based cathode with highly-dispersed Ni electrocatalysts. Electrochem Solid-State Lett 7:A500
35. Osada N, Uchida H, Watanabe M (2006) Polarization behavior of SDC cathode with highly dispersed Ni catalysts for solid oxide electrolysis cells. J Electrochem Soc 153:A816
36. Uchida H, Nishino H, Puengjinda P, Kakinuma K (2020) Remarkably improved durability of Ni–Co dispersed Samaria-doped ceria hydrogen electrodes by reversible cycling operation of solid oxide cells. J Electrochem Soc 167:134516
37. Uchda H, Brito M, Nishino H (2021) D epth-direction analysis of nickel depletion in a Ni-gadolinia-doped ceria hydrogen electrode after steam electrolysis operation. J Ceramic Soc Japan 129:111
38. Hauch A, Jensen S, Ramousse S, Mogensen M (2006) Performance and durability of solid oxide electrolysis cells. J Electrochem Soc 153:A1741
39. Hauch A, Jensen S, Ramousse S, Mogensen M (2007) Silica segregation in the Ni/YSZ electrode. J Electrochem Soc 154:A619
40. Toshiba hydrogen energy research and development (in English), https://www.toshiba-energy.com/en/hydrogen/rd/index.htm Accessed 1 Sep 2022
41. Inuzuka R, Kameda T, Watanabe H, Yoshino M (2015) Development of hydrogen production and power storage systems using solid oxide electrolysis cell. Fuel Cell Seminar and Energy Exposition 2015
42. Tagawa H (1998) Solid oxide fuel cells and global environment
43. Dees D, Claar T, Easler T, Fee D, Mrazek F (1987) Conductivity of Porous Ni/ZrO_2-Y_2O_3 Cermets. J Electrochem Soc 134:2141
44. Kawada T, Sakai N, Yokokawa H, Dokiya M, Mori M, Iwata T (1990) Characteristics of Slurry-coated nickel zirconia cermet anodes for solid oxide fuel cell. J Electrochem Soc 137:3042
45. Iwata T (1996) Characterization of Ni-YSZ anode degradation for substrate-type solid oxide fuel cells. J Electrochem Soc 143:1521
46. Eguchi K, Kunisa Y, Adachi K, Arai H (1996) Effect of anodic concentration overvoltage on power generation characteristics of solid oxide fuel cells. J Electrochem Soc 143:3699
47. Sakai N, Horita T, Yokokawa H, Dokiya M, Kojima I (1993) Liquid-phase-assisted sintering of calcium-doped lanthanum chromites. J Am Ceram Soc 76:609
48. Yasuda I, Ogiwara F, Yakabe H (2001) Proceedings of the 7th international symposium on solid oxide fuel cells (SOFC-VII): 783
49. Tsuchiya N, Ohmura H, Kanamura S, Yoshino M, Fujiwara S, Kameda T, Yamada M, Yamada K (2017) Demonstration of hydrogen production by multi-typed SOEC system. Proceedings of 2017 AIChE annual meeting: 221d
50. Tsuchiya N, Ohmura H, Yoshino M, Osada N, Fujiwara S, Kameda T, Yamada K (2019) Hydrogen production system using SOEC—evaluation of the energy by long-term operation—proceedings of 2019 AIChE annual meeting: 526i

Chapter 11
Hydrogen Production by Thermochemical Water Splitting; Membrane IS Process

Shinji Kubo, Nariaki Sakaba, Yoshiyuki Inagaki, and Yasuo Hosono

Abstract The concept of the membrane iodine–sulfur (IS) process to produce H_2 by the thermochemical decomposition of water with solar heat of approximately 650 °C was proposed. In the conventional IS process, the decomposition of SO_3 in the H_2SO_4 decomposition process requires a high temperature heat of approximately 900 °C. As SO_3 will be decomposed at 650 °C when using solar heat, a technology to improve the conversion rate in the SO_3 decomposition reaction is required. In addition, the conventional IS process has technical problems such as excessive amount of I_2 required in the Bunsen reaction (approx. nine times the amount required for the reaction) and a low conversion rate in the HI decomposition reaction. To solve these problems, research was conducted to apply membrane technology to the three main reactions of the IS process, namely the Bunsen reaction, SO_3 decomposition, and HI decomposition. The thermal mass balance calculation of the membrane IS process showed that the H_2 production efficiency of 40% or more can be obtained at 650 °C.

Keywords Hydrogen production · Water decomposition · Iodine–sulfur process · Membrane · Solar heat

11.1 Introduction

Limited H_2 exists in nature with the main source of H_2 existing in chemical compounds such as water and fossil resources, etc. Therefore, it is necessary to decompose these compounds using heat and electricity obtained from fossil resources, nuclear power, and renewable energy to produce H_2 which is a raw material in the synthesis of ammonia. Currently, H_2 production from water using solar energy is mainly performed through electrolysis using solar photovoltaic power generation. However, the use of solar photovoltaic power generation is limited to the daytime,

S. Kubo (✉) · N. Sakaba · Y. Inagaki
Japan Atomic Energy Agency, Tokai, Oarai, Ibaraki 311-1393, Japan
e-mail: kubo.shinji@jaea.go.jp

Y. Hosono
CHIYODA Corporation, Minatomirai, Nishi-ku, Yokohama 220-8765, Japan

K. Aika and H. Kobayashi (eds.), *CO_2 Free Ammonia as an Energy Carrier*,
https://doi.org/10.1007/978-981-19-4767-4_11

and it is difficult to achieve economies of scale. However, the heat utilization system can be operated even at night by storing solar heat, and the economies of scale make it suitable for upsizing. By directly converting heat into H_2 is used instead of converting heat into electricity with a steam turbine, etc. and then converting into H_2, H_2 can be produced with higher efficiency than electrolysis. As H_2 production methods by the decomposition of water using heat directly, there are some methods such as thermochemical process, steam electrolysis, etc. for H_2 production through the decomposition of water using heat directly.

This chapter describes the outline of the membrane IS process (Iodine–Sulfur) to thermochemically decompose water with solar heat as a method for producing H_2. So far the IS process has been developed for a nuclear reactor, High Temperature Gas-cooled Reactor (HTGR), which can supply high temperature heat of approximately 900 °C. To use solar heat of approximately 650 °C in the IS process, it is necessary to accelerate the reaction at temperatures of approximately 600 °C. Therefore, a new process, called the membrane IS process, which applies membrane technology has been developed. This development includes component technologies on membranes, high-activity catalysts, and a corrosion resistant material.

11.2 Outline of IS Process

To decompose water to obtain H_2, it is necessary to provide the work corresponding to the free energy of water by inputting energy for the enthalpy of formation. Electrolysis of water is a method to supply the free energy with electricity. On the other hand, direct thermal decomposition decomposes water using heat without inputting work in the temperature range where the free energy of water decomposition is negative. However, this method requires an extremely high temperature of several thousand degrees, which is severe condition for component materials.

The thermochemical H_2 production process decomposes water by combining multiple chemical reactions and uses thermal energy in the lower temperature range in comparison to direct thermal decomposition. This process was considered thermodynamically in 1964 by Funk et al. [1] in the United States, and screening has been performed for many processes from the viewpoint of a maximum temperature of 1000 °C or less, feasibility in component materials, the reaction rates of elemental reactions, product separation, toxicity, and corrosiveness of working substances, H_2 production efficiency, etc. At present, the most actively researched thermochemical process is the iodine–sulfur (IS) process.

Figure 11.1 shows the reaction diagram of the IS process, which consists of the following three chemical reactions.

$$\text{Bunsen reaction: } I_2 + SO_2 + 2H_2O \rightarrow 2HI + H_2SO_4 \tag{11.1}$$

$$\text{Hydrogen iodine decomposition reaction: } 2HI \rightarrow H_2 + I_2 \tag{11.2}$$

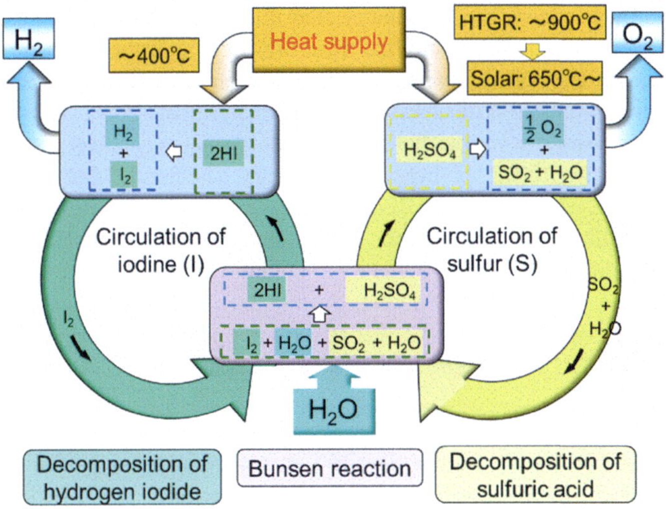

Fig. 11.1 Reaction diagram of thermochemical IS process

$$\text{Sulfuric acid decomposition reaction: } H_2SO_4 \rightarrow 1/2O_2 + SO_2 + H_2O \quad (11.3)$$

As shown in Eq. 11.1, HI and H_2SO_4 are produced by dissolving I_2 and SO_2 in water. HI and H_2SO_4 produced by the Bunsen reaction are separated, and HI is thermally decomposed at approximately 400 °C by the decomposition of HI (Eq. 11.2) to generate H_2 and I_2. The decomposition of H_2SO_4 (Eq. 11.3) proceeds in two steps (Eqs. 11.4 and 11.5). H_2SO_4 is concentrated and evaporated at approximately 300 °C or higher to produce SO_3, and then decomposed into SO_2 and oxygen O_2 at approximately 800 °C or higher.

$$H_2SO_4 \rightarrow H_2O + SO_3 \quad (11.4)$$

$$SO_3 \rightarrow 1/2O_2 + SO_2 \quad (11.5)$$

I_2 and SO_2 are returned to the Bunsen reaction after each reaction. Thus, materials other than water, O_2 and H_2 are circulated in a closed cycle and the IS process has the advantage of excellent reaction continuity and mixability of the reactants as only liquids and gases are used. However, issues in the IS process are as follows: (i) development of an operation method for stable H_2 production using a special chemical process such as the closed cycle; (ii) improvement in H_2 production efficiency converting thermal energy into chemical energy; and (iii) development of corrosion-resistant components for handling process fluids such H_2SO_4 and HI.

The Japan Atomic Energy Agency (JAEA) has been developing an IS process using heat supplied from HTGR. Since a light water reactor (LWR) mainly used for nuclear power generation uses metal clad tubes for fuel in the core and light water for the coolant, the heat supplied from the LWR is limited to approximately 300 °C. However, as HTGR can supply very high temperature heat of approximately 1000 °C by using graphite as the main component material of a core and helium gas as a coolant, it can be used in various fields such as high-efficiency power generation and H_2 production, as well as supplying heat to industry. The first HTGR with a thermal output of 30 MW, named the High Temperature Engineering Test Reactor (HTTR), was installed at the Oarai Research and Development Institute in JAEA. The HTTR achieved criticality in 1998 and a coolant temperature of 950 °C at the reactor outlet for the first time in 2004, a world first. Furthermore, JAEA conducted long-term high-temperature operation (950 °C over 50 days of operation), which demonstrated the capability of high temperature heat supply to heat utilization systems such as the IS process [2].

Regarding the operation method of the IS process, JAEA developed the measurement and control technology of the plant and succeeded in stably producing approximately 30 L/h of H_2 for one week in 2004 using a test apparatus with glass components heated by electric heaters [3]. The H_2 production efficiency of approximately 40% is possible at present, the target is approximately 50% for practical applications in the future. As for the corrosion-resistant components, JAEA fabricated a H_2SO_4 decomposer using a reaction tube made of silicon carbide (SiC), and reactors and piping in which the insides were coated with corrosion-resistant materials such as a glass lining. A H_2 production test facility with maximum H_2 production rate of 100 L/h using components made of industrial materials such as metals, ceramics, etc. [4]. JAEA has been developing plant technology to verify the structural integrity of components, operability, and controllability of the plant using this test facility. In 2019, continuous H_2 production for 150 h was successfully performed at production rate of 30 L/h. In the future, a H_2/electricity cogeneration system will be demonstrated by connecting power generation system with a helium gas turbine and a H_2 production system, such as the IS process, to the HTTR [5].

11.3 Concept of Membrane IS Process

Regarding H_2 production using renewable energy, thermochemical methods using solar heat are being investigated in the European research project Hycycle S, etc. and a solar reactor for H_2SO_4 decomposition at approximately 800 °C through the collection of solar heat has been developed [6]. In Japan, research to operate the IS process using stored solar heat at approximately 650 °C was performed with the support of the Council for Science, Technology and Innovation, Cross-ministerial Strategic Innovation Promotion Program, "Energy Carrier". Although the operation time is limited in the solar reactor, this method has the following major advantages: (1) 24 h operation regardless of whether day or night is possible; (2) economical stainless-steel materials can be used instead of expensive heat resistant materials.

When solar heat of approximately 650 °C is used, the conversion rate of SO_3 decomposition (Eq. 11.5) is greatly reduced owing to the decrease in reaction temperature. As a result, an increase in energy loss causes a decrease in H_2 production efficiency. Therefore, a goal was set to achieve a higher H_2 production efficiency at approximately 40% than that of electrolysis of water. For SO_3 decomposition, a method was selected to improve the conversion rate by extracting O_2 during the decomposition process using an O_2 permselective membrane. To improve of the conversion rate in the HI decomposition reaction and reduce of the amount of I_2 circulating in the Bunsen reaction, which is also technical issues in the conventional IS process using high-temperature heat sources such as HTGR, the application of a H_2 permselective membrane and a cation exchange membrane (CEM) was considered. In this way, the new process in which membranes were applied to the three main reactions, named the membrane IS process, was proposed [7].

11.3.1 Bunsen Reaction

In the Bunsen reaction, the produced HI exists as a mixture with unreacted I_2. The mixture of HI and I_2 (a heavy liquid) and H_2SO_4 (a light liquid) are separated naturally (liquid-liquid phase separation) by addition of excess amount of I_2 ($I_2/SO_2 = 9$) and are then transported to each decomposition process. The excess amount of I_2 causes the component to increase in size, and the energy required to circulated I_2 causes a reduction in the efficiency of H_2 production. Therefore, the membrane Bunsen reaction was investigated. In the reaction, an excess amount of I_2 for separation can be basically eliminated by advancing the production reaction of H_2SO_4 and HI on both sides of the CEM (Fig. 11.2).

A CEM, which could be used in a strong acid environment such as H_2SO_4 and HI, was developed by the National Institute for Quantum and Radiological Science and Technology, and the electrode catalyst was developed by the Shibaura Institute of Technology (SIT), and the reduction of I_2 by the CEM was experimentally verified

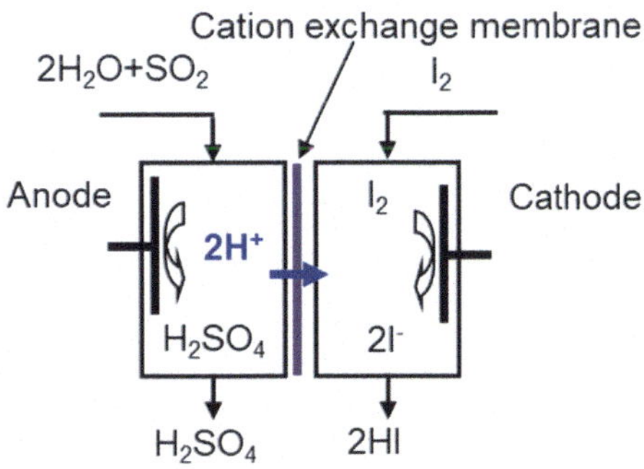

Fig. 11.2 Conceptual reaction diagram of Bunsen reaction associated with cation exchange membrane

by JAEA. Further details of the research and development process are described in Chap. 12.

11.3.2 Sulfuric Acid Decomposition Reaction

As shown in Fig. 11.3, the SO_3 decomposition reaction has a conversion rate of 80% or higher at 800 °C, but the rate decreases to approximately 30% at 600 °C. For this reason, increase of the circulation amount of undecomposed SO_3 and H_2SO_4 becomes one of the causes of reducing the H_2 production efficiency. Figure 11.4 is a conceptual diagram of SO_3 decomposition reaction with the O_2 permselective membrane. By extracting O_2 from the reaction products, the reaction equilibrium can be shifted to improve the conversion rate.

The development of the O_2 permselective membrane was performed at Hiroshima University, and improvement of the support material of membrane was performed at Noritake Co., Ltd. Improvement of the conversion rate by the membrane was experimentally verified by Hiroshima University. The catalyst, which can decompose SO_3 stably in a strong acid environment for a long period, was developed at Kumamoto

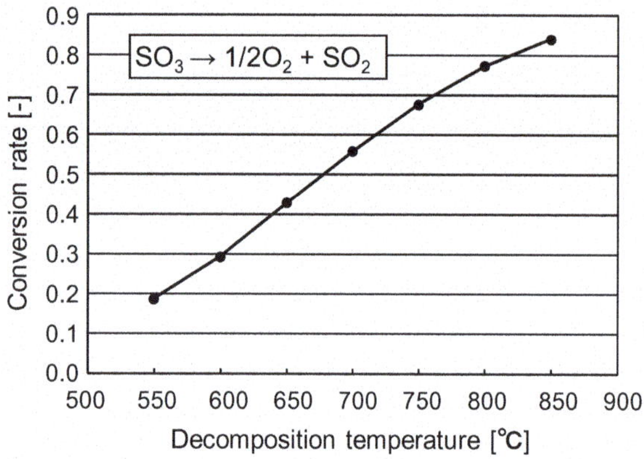

Fig. 11.3 Conversion rate of SO_3 decomposition

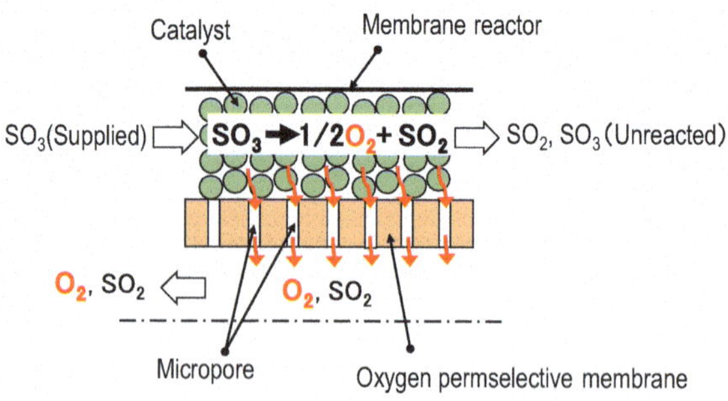

Fig. 11.4 Conceptual reaction diagram of SO_3 decomposition associated with O_2 permselective membrane

University. Further details of research and development process are described in Chap. 13.

11.3.3 Hydrogen Iodine Decomposition Reaction

The HI decomposition reaction is performed at approximately 400 °C, but the conversion rate remains low at approximately 20% owing to the restriction of chemical equilibrium in which there is a low-temperature dependence (Fig. 11.5). Therefore, it is difficult to increase the conversion rate even at high-temperatures. Therefore, the circulation of a large amount of HI (approx. five times the amount required for H_2 production) in the HI decomposition reactor is a cause of energy loss and enlargement of the HI decomposition components. Figure 11.6 is a conceptual diagram of the HI decomposition reaction with the H_2 permselective membrane. By extracting H_2 from the reaction products, the reaction equilibrium can be shifted to improve the conversion rate.

The development of the H_2 permselective membrane was performed at SIT, and the improvement of the conversion rate by the membrane was experimentally verified by JAEA. In addition, the catalyst, which can decompose HI stably in a strong acid

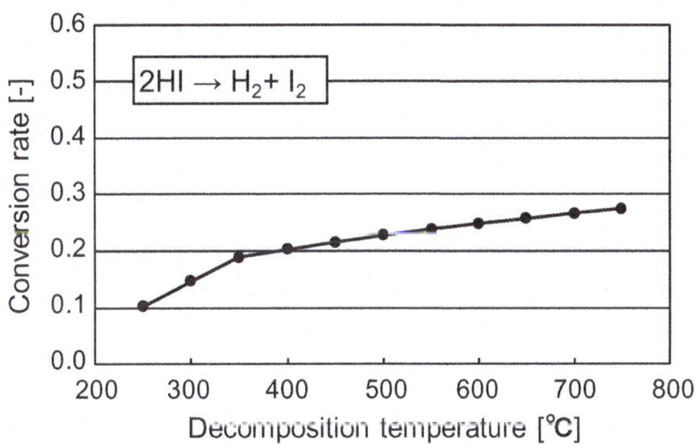

Fig. 11.5 Conversion rate of HI decomposition

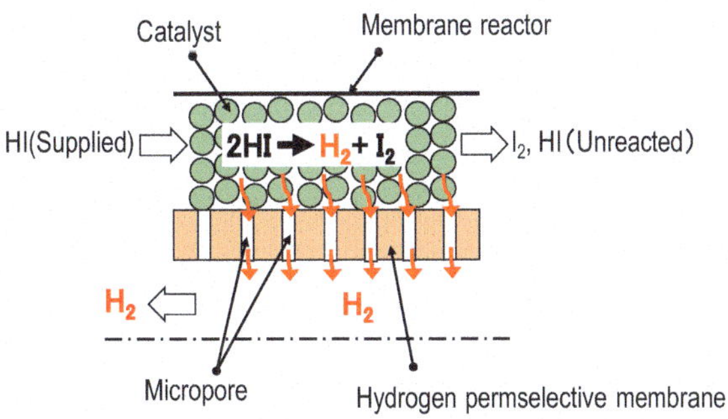

Fig. 11.6 Conceptual reaction diagram of HI decomposition associated with H_2 permselective membrane

environment for a long period, was developed at Kyushu University. Further details of R&D are described in Chap. 14.

11.3.4 Corrosion-Resistant Material

In the conventional IS process, the reactor material must be resistant not only to corrosion resistance but also creep damage as SO_3 is decomposed at 800 °C or higher. For this reason, research for the reactor material has been mainly conducted on ceramics (SiC). However, in the membrane IS process, a metal can be used because it is not necessary to consider creep damage at 600 °C.

The corrosion-resistant technology by the thermal spray of ceramics on the metal surface was developed by JGC CORPORATION and JAEA. Further details of R&D are described in Chap. 15.

11.3.5 Thermal Mass Balance and Hydrogen Production Efficiency

Evaluation of the H_2 production efficiency and heat and thermal mass balance of the membrane IS process plant (H_2 production of 14.7 t/h corresponding to a plant with ammonia production of 2000 t/d) was conducted by the CHIYODA Corporation. Figure 11.7 shows a schematic flow sheet and the amount of circulating materials in the membrane IS process plant calculated using the simulation code, Aspen HYSYS [8].

In the membrane Bunsen reactor, the CEM separates the H_2SO_4 production reaction ($SO_2 + 2H_2O \rightarrow H_2SO_4 + 2H^+ + 2e^-$) on the anode side and the HI production reaction ($I_2 + 2H^+ + 2e^- \rightarrow 2HI$) on the cathode side. By moving protons from the

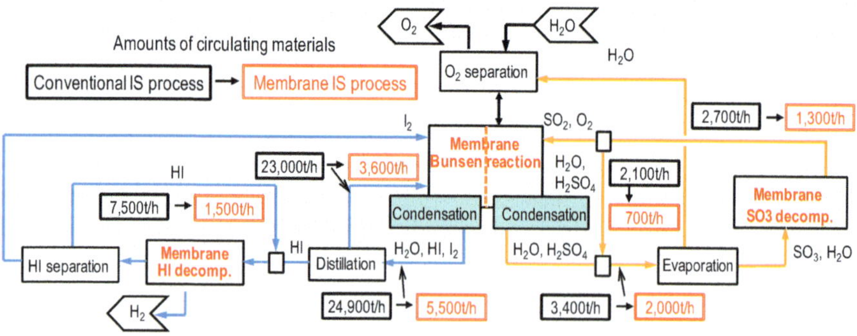

Fig. 11.7 Schematic flow sheet and amounts of circulating materials of membrane IS process plant [8]

anode side to the cathode side through the membrane, the Bunsen reaction proceeds as a whole. H_2SO_4 produced on the anode side is evaporated according to Eq. (11.4) and SO_3 is decomposed according to Eq. (11.5) using a membrane reactor in the H_2SO_4 decomposition process. SO_2 and undecomposed SO_3 after the decomposition reaction returns to the anode side of the membrane Bunsen reactor. HI produced on the cathode side is distilled to remove water, and HI is decomposed according to Eq. (11.2) using a membrane reactor in the HI decomposition process. I_2 and undecomposed HI after the decomposition reaction returns to the cathode side of the membrane Bunsen reactor. The amounts of circulation materials in comparison with the conventional IS process (Fig. 11.7) are significantly reduced to approximately 50% in the membrane SO_3 decomposer, approximately 20% in the membrane HI decomposer, and approximately 16% in the membrane Bunsen reactor, respectively, in the membrane IS process.

Table 11.1 shows the thermal energy in each reaction of the conventional and the membrane IS processes. The consumed electric power for pumps, etc. was converted to thermal energy, assuming that the power generation efficiency was 40%. The conversion rates of the membrane SO_3 and HI decomposers depend on the membrane areas. Using a parametric analysis of each decomposer, it was shown that the membrane SO_3 decomposer had a conversion rate of approximately 62% at 600 °C and the HI decomposer had a conversion rate of approximately 59% at 400 °C. The H_2 production efficiency η_H (%) was calculated from the combustion thermal energy Q_C (HHV) of the produced H_2 and the total input thermal energy Q_{in} of the process as follows.

$$\eta_H = \frac{Q_C}{Q_{in}} \times 100 \tag{11.6}$$

In the conventional IS process, the H_2 production efficiency is reduced by approximately 27% owing to the low conversion rate of the SO_3 decomposition, but the thermal energy required for H_2SO_4 decomposition is reduced by approximately 40% owing to the membrane SO_3 decomposition. In addition, the thermal energy used in the concentration and distillation process of HI decomposition by the membrane Bunsen reaction and in the decomposition and separation process by the membrane HI decomposition can be reduced. Owing to the reduction in thermal energy, the H_2 production efficiency of the membrane IS process was estimated to be approximately 42%.

Table 11.1 Thermal energy in each reaction of conventional and the membrane IS processes [8]

		Conventional	Membrane	Remarks
H_2SO_4 decomposition		1120 MWt	676 MWt	Reduced by the membrane SO_3 decomposition
HI decomposition	Condensation/distillation	740 MWt	570 MWt	Reduced by the membrane Bunsen reaction
	Decomposition/separation	140 MWt	30 MWt	Reduced by the membrane HI decomposition
Electricity		128 MWt	128 MWt	Calculated to thermal energy with the power generation efficiency of 40%
Total input thermal energy		2128 MWt	1404 MWt	
H_2 production efficiency		27%	42%	

References

1. Funk JE, Reinstorm RE (1964) System study of hydrogen generation by thermal energy. Gen Motors Rept EDR3714 II(SupplA):3.1–3.25
2. Goto M et al (2012) Long-term high-temperature operation of the HTTR. Nucl Engrg Des 251:181–190
3. Kubo S et al (2004) A demonstration study on a closed-cycle hydrogen production by thermochemical water-splitting iodine-sulfur process. Nucl Engrg Des 233:347–354
4. Takegami H et al (2020) Development of strength evaluation method of ceramic reactor for iodine-sulfur process and hydrogen production test in Japan Atomic Energy Agency. Nucl Engrg Des 360:110498
5. Yan X et al (2017) Design of HTTR-GT/H_2 test plant. Nucl Engrg Des 329:223–233
6. Thomey D et al (2013) Development and test of a solar reactor for decomposition of sulphuric acid in thermochemical hydrogen production. Int J Hydrogen Energy 37:16615–16622
7. Myagmarjav O et al (2019) Research and development on membrane IS process for hydrogen production using solar heat. Int J Hydrogen Energy 44:19141–19152
8. Final report "Membrane IS Process (CHIYODA Corporation)", Cross-ministerial strategic innovation promotion program. Japan Science and Technology Agency. https://www.jst.go.jp/sip/dl/k04/end/team2-9.pdf

Chapter 12
Production of Sulfuric Acid and Hydrogen Iodine by Bunsen Reaction Assisted with Membrane

Mikihiro Nomura, Tetsuya Yamaki, Shinichi Sawada, and Nobuyuki Tanaka

Abstract The Bunsen reaction assisted with membrane was carried out to skip the separation processes after the reaction. In order to improve the reaction efficiency, the radiation-grafted cation exchange membrane (CEM) with a high ion exchange capacity (IEC) and the highly-porous Au-electroplated electrode were developed. The divinylbenzen-crosslinked radiation-grafted CEMs showed the high proton conductivity and low water permeation flux. The porous electrodes with large electrochemically active surface area reduced the effective overvoltage for SO_2 oxidation at the anode side. The membrane Bunsen reaction was carried out using the CEM and the porous Au electrode. The overpotential largely decreased to 0.21 V by using the CEM (IEC = 2.9 mmol/g) at 50 °C.

Keywords Cation exchange membrane · Radiation grafting · Crosslinking · Porous metal electrode · Overpotential

12.1 Introduction

Dokiya et al. proposed an idea to efficiently proceed the Bunsen reaction by using a two-compartment electrolysis cell equipped with an activated carbon anode, a platinum cathode, and a CEM (cation exchange membrane) [1]. In this so-called membrane Bunsen reaction, the anode and cathode compartments separated by a CEM were filled with SO_2 and I_2 aqueous solutions, respectively. During the electrolysis, the following electrode reactions take place, and simultaneously H^+ is transported through a CEM from the anode to the cathode side:

N. Tanaka
Japan Atomic Energy Agency, Tokai, Ibaraki 311-1393, Japan

M. Nomura (✉)
Shibaura Institute of Technology, Toyosu, Koto-ku, Tokyo 135-8548, Japan
e-mail: Lascathy@shibaura-it.ac.jp

T. Yamaki · S. Sawada
National Institutes for Quantum Science and Technology, Anagawa, Inage-ku, Chiba 263-0024, Japan

K. Aika and H. Kobayashi (eds.), *CO_2 Free Ammonia as an Energy Carrier*,
https://doi.org/10.1007/978-981-19-4767-4_12

$$\text{Anode}: SO_2 + 2H_2O \rightarrow H_2SO_4 + 2H^+ + 2e^- \tag{12.1}$$

$$\text{Cathode}: I_2 + 2H^+ + 2e^- \rightarrow 2HI \tag{12.2}$$

The cell voltage for electrolysis, ΔE_{cell}, is expressed by

$$\Delta E_{cell} = \Delta E_{eq} + \eta_{anode} + \eta_{cathode} + IR_{cell} \tag{12.3}$$

where ΔE_{eq} is the equilibrated voltage at the current of 0, IR_{cell} is the ohmic overpotential associated with the proton transport resistance of the CEM and the electron conduction resistance of the cell circuit, η_{anode} is the activation overpotential for the sluggish anodic SO_2 oxidation, and $\eta_{cathode}$ is the activation overpotential for the cathodic I_2 reduction. Because the $\eta_{cathode}$ value is negligible due to its fast kinetics [2], reduction of the total cell overvoltage ($=\eta_{anode} + IR_{cell}$) is desirable from the viewpoint of low electric power consumption.

The CEM for the membrane Bunsen reaction is required to have as high a proton conductivity as possible. A highly-conductive CEM results in the overvoltage reduction in the reaction (IR_{cell}), thereby increasing the energy efficiency of the IS process. In addition, water permeation can occur in parallel through a CEM from the anode to the cathode side or vice versa, and so its lowering is required for the CEM. An ideally-zero or very low water permeation flux enables one to estimate the concentration of HI or H_2SO_4 from the electrolytic current, leading to quantitative and efficient production of both the acids for the IS process.

The proton conductivity of CEMs is easily increased by increasing their ion exchange capacity (IEC), which is the density of cation exchange groups. However, the CEM with a high IEC is strongly hydrophilic and, therefore, tends to absorb and permeate a large amount of water. Thus, it is challenging to prepare the CEM satisfying both the high proton conductivity and low water permeation flux. As described in Sect. 12.2, we have recently developed novel CEMs for the first time for applications to the membrane Bunsen reaction.

The electrochemical evaluation of catalysts for the SO_2 oxidation reaction has been carried out on several electrode materials in the researches on Hybrid S (HyS) process [3, 4]. It has been reported that carbon materials including graphite, activated carbon, and carbon black have no or intermediate activity for the SO_2 oxidation reaction. Among the noble metal catalysts tested previously, platinum (Pt) was consistently found to have high activity [5]. Palladium (Pd) exhibited the best performance in several works, although the gradual decrease in the SO_2 oxidation current at Pd was observed at high potentials in 3.5 mol/l H_2SO_4 due to electro-corrosion [6]. Quijada and Vazquez suggested that inactivation of electrode surface caused by strong adsorption of SO_2 did not occur on Au and perform well kinetically [7]. Regarding the anode material for the membrane Bunsen reaction, graphite and glassy carbon were used in the early study [8–10], and Pt electrodes including Pt/C were employed later [11, 12]. In our previous membrane, Bunsen reaction tests, Pt or Au plated on Ti plate, were employed as an anode due to its superior catalytic ability, but the cell overvoltage was unfortunately too high to be acceptable for industrial implementation. As described in Sect. 12.3, we have developed new electrode to reduce cell overvoltage of the membrane Bunsen reaction.

Finally, the experiments of membrane Bunsen reaction were carried out by using new electric cell assembled with the developed CEM and electrode. As described in Sect. 12.4, the experiments will demonstrate that the new cell showed low cell voltage such an extent to effectively improve the efficiency of H_2 production.

12.2 Development of Cation Exchange Membranes

A radiation-induced grafting method is one of the promising techniques to prepare functional polymer materials. This method involves irradiation of a polymer substrate with γ-rays or electron beams to produce radicals in the main chains and graft polymerization of vinyl monomers initiating from the radicals. The graft chains, which are post-functionalized if needed, provide the substrate with desired new properties. Advantageously, there is a high degree of freedom in molecular design: Various types of substrates and monomers can be used, and graft contents can be widely controlled by experimental conditions such as irradiation dose, temperature, and grafting time. To date, Takasaki Advanced Radiation Research Institute (TARRI) in QST (National Institutes for Quantum Science and Technology) has exploited the radiation grafting method in order to find wide-ranging applications in battery separators, fragrance-free deodorants, air cleaning filters, polymer electrolyte fuel cells, fibrous metal adsorbents, etc.

The most well-known procedure for preparation of CEMs consists of the following three steps: radical generation in a base thin polymer film by irradiation, grafting of styrene and styrene derivatives, and sulfonation of aromatic rings in the graft chains [13–15]. The resulting CEM contains a base polymer matrix to maintain mechanical strength and dimensional stability and poly (styrene sulfonic acid) graft chains to give cation conductivity. In recent two decades, much effort has been worldwide devoted to develop the radiation-grafted CEMs for application of polymer electrolyte fuel cells. Some research groups successfully synthesized the novel CEMs with chemical stability high enough for long-time cell operation and demonstrated their fuel-cell output power exceeding that of the commercially available benchmark fluoropolymer-based CEMs [16–19].

Following these research achievements due to a wide choice of materials and high controllability of experimental conditions, we have been using the radiation grafting technique to develop the CEMs for the membrane Bunsen reaction [20]. The base films and graft monomers can properly be selected, while the graft contents corresponding to the IECs of the CEM can be optimized. This is expected to result in the creation of the CEMs endowed with both the high proton conductivity and low water permeation flux.

Figure 12.1 shows a procedure of the radiation grafting method to prepare a CEM for the membrane Bunsen reaction. A 50-μm-thick ethylene-*co*-tetrafluoroethylene

(ETFE) film was used as a base polymer film due to its excellent chemical stability. At first, an ETFE film was irradiated with 10-kGy ^{60}Co γ-rays (Cell No. 6 at TARRI, QST, Japan) in Ar at room temperature to produce radicals. For graft polymerization, an irradiated film was immersed in a toluene solution of styrene and divinylbenzene (DVB) with volume ratios of styrene/DVB/toluene = 12.5/5/82.5, 25/5/70, 50/5/45, and 95/5/0, and then kept at 60 °C for 1–10 h. DVB was used to crosslink polystyrene graft chains to improve the mechanical strength and dimensional stability of the prepared CEM. After the grafting reaction, the film was taken out and extracted with toluene to remove any excess styrene monomers and homopolymers. The grafted film was then dried in vacuum at 50 °C for 12 h. The degree of grafting (DOG), corresponding to the ratio of grafted styrene and DVB, was calculated as follows:

$$\mathrm{DOG} = 100 \times (W_\mathrm{G} - W_\mathrm{O})/W_\mathrm{O}, \tag{12.4}$$

where W_O and W_G are the weights of original and grafted films, respectively. Figure 12.2 shows the DOG as a function of the grafting time. The DOG was controlled in a wide range of 3.6–112% by changing the grafting time and the monomer concentration.

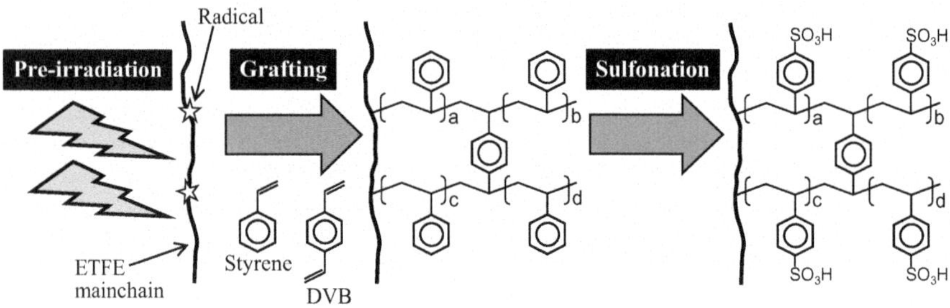

Fig. 12.1 Representative procedure of radiation grafting method to prepare CEM

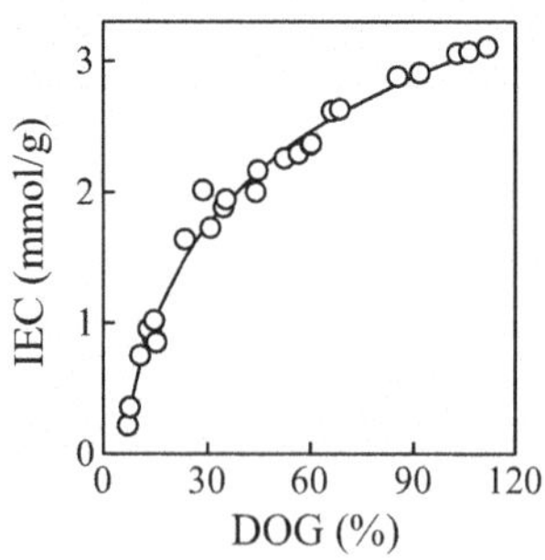

Fig. 12.2 DOG of grafted films as function of grafting time. The volume ratio of styrene/DVB/toluene in reaction solutions is (black circle) 12.5/5/82.5, (white circle) 25/5/70, (black triangle) 50/5/45, and (triangle) 95/5/0. The solid lines are guides for the eye

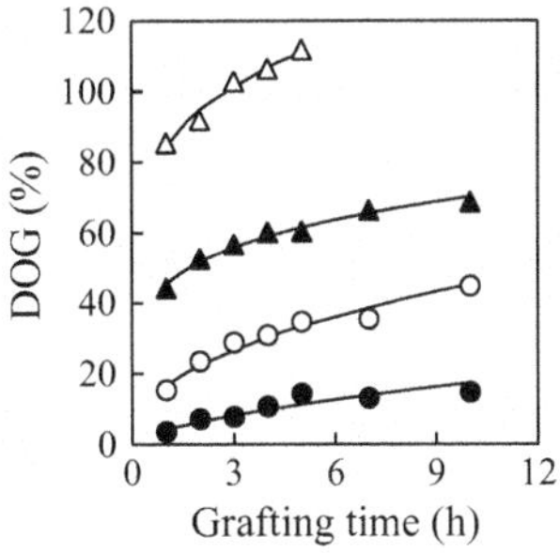

Fig. 12.3 The IEC of DVB-crosslinked radiation-grafted CEMs as function of DOG. The solid line is a guide for the eye

For sulfonation, the styrene/DVB-grafted film was immersed in a 1,2-dichloroethane solution of 0.2 mol/L chlorosulfonic acid at 50 °C for 6 h. Finally, the membrane was immersed in pure water at 50 °C for 24 h to prepare the DVB-crosslinked radiation-grafted CEM. The IEC of the CEMs was measured by acid–base titration. A CEM in an acid form was equilibrated with a 3 mol/L NaCl aqueous solution at 50 °C for 6 h. The amount of protons liberated from a CEM, N_{H+} (mol), was measured with a 0.01 mol/L NaOH aqueous solution by using an automatic neutralization titration apparatus. The IEC was calculated as follows:

$$\mathrm{IEC} = 1000\mathrm{N}_{\mathrm{H+}}/W_{\mathrm{dry}}, \tag{12.5}$$

where W_{dry} (g) is the weight of the dried CEM. Figure 12.3 shows the IEC of the CEMs as a function of the DOG. The IEC increased with increasing the DOG and reached the maximum of 3.1 mmol/g. This value was far larger than that of Nafion 212 (0.90 mmol/g), the benchmark commercially available CEM.

For the water uptake test, a CEM was stored in pure water for 12 h to ensure equilibration. After a CEM was taken out and the surface was wiped off to remove water droplets, its weight, W_{wet}, was measured. The water uptake was calculated as follows:

$$\text{Water uptake} = 100 \times (W_{\mathrm{wet}} - W_{\mathrm{dry}})/W_{\mathrm{dry}}. \tag{12.6}$$

Figure 12.4 shows the relationship between the IEC and water uptake for the DVB-crosslinked radiation-grafted CEMs [21]. As references, this figure includes the results of Nafion 212 and the non-crosslinked CEMs prepared by the grafting of only styrene (DVB was not used). The CEM with a higher IEC became more hydrophilic and exhibited higher water uptake. In the high IEC range (>2.2 mmol/g), the water uptake of the crosslinked CEMs was rather lower than that of the non-crosslinked ones as reported earlier [20]. Suppression of the water uptake by the DVB crosslinks was expected to restrict water permeation in the membrane Bunsen reaction.

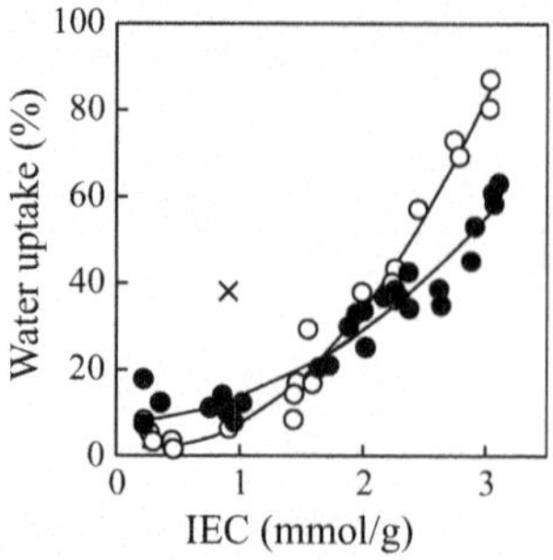

Fig. 12.4 Water uptake of DVB-crosslinked radiation-grafted CEMs with different IECs. The results of the non-crosslinked radiation-grafted CEMs (open circle) and Nafion 212 (cross) are also included as references. Modified and reprinted from reference [21]

A proton conductivity of the DVB-crosslinked radiation-grafted CEM was measured by an AC impedance technique. A stripe-shape fully-hydrated membrane was clamped between the two platinum electrodes connected to an LCR meter. The in-plane membrane resistance, R (Ω), was measured at room temperature by applying a 100-mV AC voltage at 100 kHz. The proton conductivity was calculated as follows:

$$\text{Proton conductivity} = L/(RS), \tag{12.7}$$

where L is the distance between the two electrodes, and S is the cross-sectional area of a CEM. Figure 12.5 shows the proton conductivity of the grafted CEMs as a function of the IEC. The proton conductivity increased with increasing the IEC and, importantly, reached the maximum of 0.18 S/cm at 2.8 mmol/g, which was more than twice that of Nafion 212 (0.083 S/cm).

A water flux of the DVB-crosslinked radiation-grafted CEM was measured by an osmotic pressure-driven permeation experiment. A two-compartment glass cell was used, and a CEM was set at its center to separate the two compartments. To imitate the environment during the reaction, one compartment was filled with an aqueous

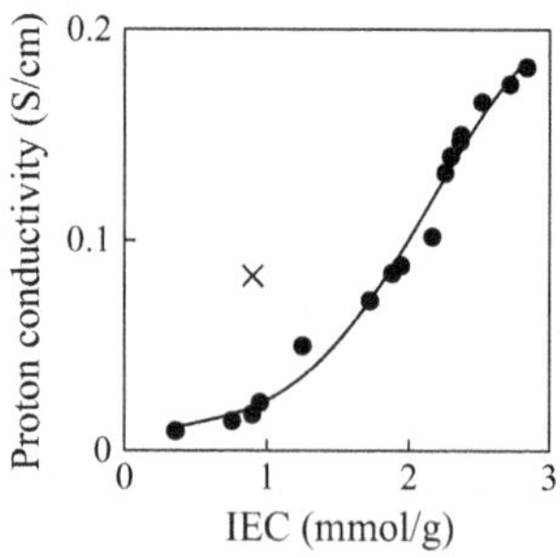

Fig. 12.5 Proton conductivity of DVB-crosslinked radiation-grafted CEMs with different IECs. The result of Nafion 212 (cross) is also included as a reference

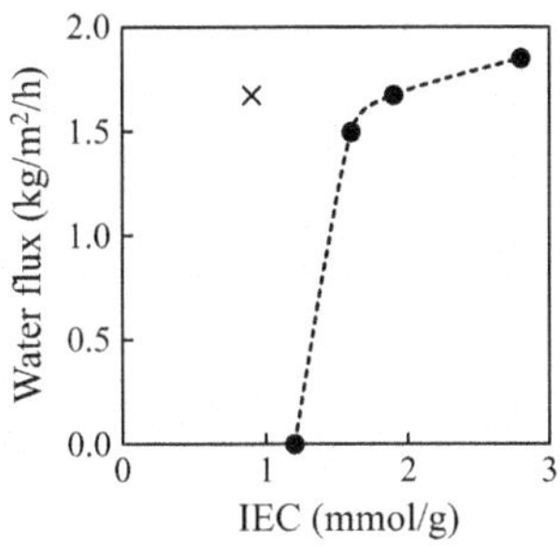

Fig. 12.6 Water permeation flux of DVB-crosslinked radiation-grafted CEMs with different IECs. The result of Nafion 212 (cross) is also included as a reference. Reproduced with permission from Ref. [23]. Copyright © 2020, The Society of Sea Water Science, Japan

HI/I_2 solution ($[HI] = [I_2] = 2.5$ mol/kg), while the other was filled with a 3.2 mol/kg H_2SO_4 aqueous solution. In this case, an osmotic pressure difference caused water permeation through a CEM from the HI/I_2 side to the H_2SO_4 side, and the volume of permeated water, ΔV, was measured by a graduated capillary. The water permeation flux was calculated as follows:

$$\text{Water permeation flux} = \Delta V/(S\Delta t), \quad (12.8)$$

where S is the permeation area (=7.1 cm^2), and Δt is the permeation time. Figure 12.6 shows the water permeation flux of the grafted CEMs as a function of the IEC [22]. The water flux steeply rose when the IEC increased from 1.2 to 1.6 mmol/g, and its increasing rate became slow in the higher IEC range. Even at the highest IEC of 2.8 mmol/g, the water flux was no more than 1.9 kg/m^2 h. This value was low enough and comparable with that of Nafion 212 (1.7 kg/m^2 h).

In conclusion, the DVB-crosslinked radiation-grafted CEMs with a high IEC satisfied the two key requirements: the high proton conductivity and low water permeation flux. This CEM promises the reduction of the cell overvoltage and efficient production of the acids in the membrane Bunsen reaction.

12.3 Development of Electrodes

First, to assess suitable electrode material, oxidation behavior of SO_2 was measured at Pt, Au, Pd, and glassy carbon disk electrodes in 1 mol/L H_2SO_4 containing 30 mmol/L SO_2 using cyclic voltammetry with multiple potential cycling. Based on the onset potential (E_{onset}), peak current (i_p), and Tafel slope in the part of initial current rise of cyclic voltammograms for the SO_2 oxidation, the three noble metal electrodes were superior to GC electrode, and Au revealed the best oxidation properties among them. The i_p value of Au and Pt electrodes decreased during the repeated potential cycling due probably to adsorption of SO_2.

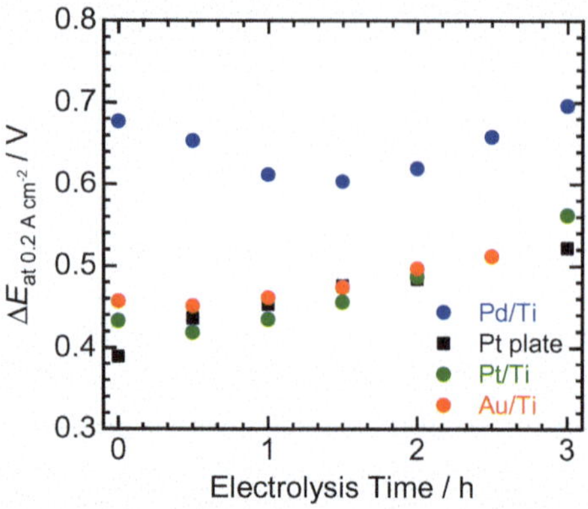

Fig. 12.7 Cell voltage at 0.2 A cm^{-2} as function of electrolysis time measured in test cell consisting of anolyte (1.2 mol/kg SO_2 in 9.3 mol/kg H_2SO_4), catholyte (5 mol/kg I_2 in 10 mol/kg HI), and Nafion 212. The data measured at Pt plate is also present

The above noble metals were also examined as an anode in the membrane Bunsen reaction cell. The noble metals were electrochemically plated with 2 μm thickness on Ti plate, because carbon substrates are unstable [23] and Ti has high corrosion resistance in highly acidic and oxidative atmosphere. Figure 12.7 represents the dependence of the cell voltage on the time of electrolysis at 0.2 A/cm^2 measured at Au/Ti, Pt/Ti, and Pd/Ti electrodes. The cell voltages tend to increase during the electrolysis due to deterioration of electrode and CEM, although the reason for the cell voltage drop observed at Pd/Ti in the initial one hour was not known. No significant difference in the cell voltage appeared between Au/Ti and Pt/Ti, both of which have the cell voltages comparable to that at Pt plate, and lower than that at Pd/Ti. The overvoltage for SO_2 oxidation after 3 h of electrolysis evaluated based on the cell voltage at the current of 0 was larger than 0.3 V, which was higher than the value acceptable for industrial implementation.

Next, we examined whether porous electrodes with large electrochemically active surface area (ECSA) can reduce effective overvoltage for SO_2 oxidation. Porous Au (Au/CF) and Pt (Pt/C/CF) electrodes were fabricated by Au plating and adsorption of Pt/C on carbon felt (CF, GF-20, porosity 93%) purchased from Nippon Carbon Co., Ltd. The SO_2 oxidation properties on those electrodes were examined in 1 mol/L H_2SO_4 containing 30 mmol/L SO_2 by cyclic voltammetry with the scan rate of 1 mV/s. Peak-shaped voltammograms were obtained for both electrodes, and a large current drop was observed from the first to second potential cycle, implying that supply of SO_2 to the inner surface of CF is insufficient. The i_p and E_{onset} at the first cycle tend to increase and decrease, respectively, with increasing the ECSA-projected surface area ratio (ECSA/PSA) as shown in Fig. 12.8, revealing that the use of porous structured electrodes is effective to reduce the overvoltage for SO_2 oxidation. However, the i_p was not in proportional to the ECSA.

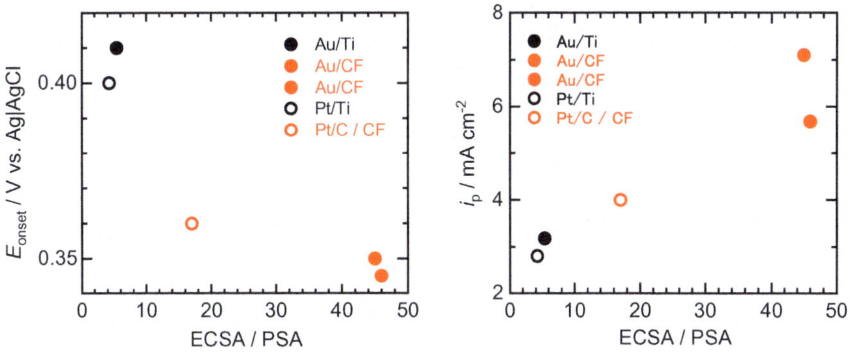

Fig. 12.8 Peak current and onset potential for SO_2 oxidation as a function of ECSA–projected area ratio measured in 1 mol/L H_2SO_4 containing 30 mmol/L SO_2

Another series of porous electrodes was made by electrodeposition of Au or Pt on two porous Ti substrates (MF-40-1t, porosity 87% and MF-80A-1t, porosity 70%) purchased from Nagamine Manufacturing Co., Ltd., Japan [20]. Sigmoidal polarization curves were measured for all porous electrodes, reflecting improvement in the transportation of reactant to the inner surface of porous electrode compared with the CF-based porous electrodes. Figure 12.9 represents the dependences of the steady-state current (i_s) and E_{onset} at the first potential cycling on the ECSA/PSA value for the porous Ti-based electrodes. Similar relationships to those observed for the CF-based porous electrodes as shown in Fig. 12.8 were obtained, although the comparable E_{onset} and higher i_s values were realized at smaller ECSA/PSA values. The dependences of the steady-state current (i_s) and E_{onset} at the first potential cycling on the ECSA/PSA value were similar to those observed for the porous electrodes with CF substrate. Despite the larger ECSA, the current values for Au(Pt)/MF-80A-1t and Au/CF were lower than that for Au(Pt)/MF-40-1t, suggesting importance of the enhancement in not only the ECSA but also the utilization of inner surface of porous electrodes. It was confirmed that the electrolyte circulation through the Au and Pt electrode with porous Ti was effective to improve the utilization of inner electrode surface.

Figure 12.10 shows the relationship between the cell voltage and the time of electrolysis during the membrane Bunsen reaction test for three Au or Pt electrodes with porous Ti substrates. For all electrode, the ΔE at 0.2 mA/cm^2 decreased in the initial 30 min, and started to increase after 2 h of electrolysis. The difference in the ΔE values before and after 3 h of electrolysis became small. Regarding the ΔE value at 0.2 A/cm^2, MF-40-1t substrate and Au brought about better performance than MF-80A-1t and Pt, respectively, which is similar tendency to that obtained in the fundamental examination described above. The overvoltage for SO_2 oxidation evaluated based on the cell voltage at the current of 0 after 3 h of electrolysis was 0.28 V for Au/MF-40-1t, 0.30 V for Pt/MF-40-1t, and 0.36 V for Au/MF-80A-1t.

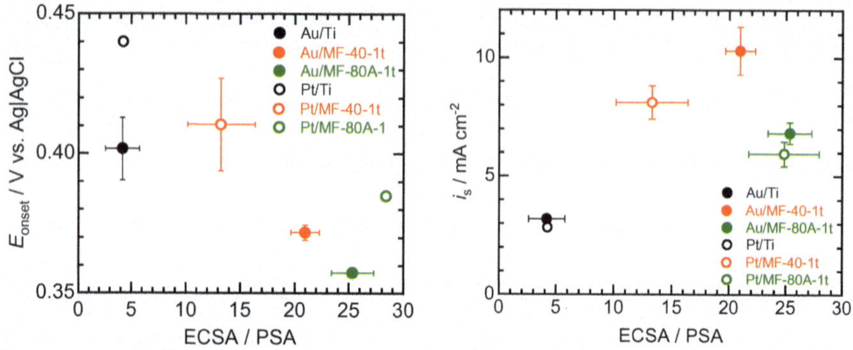

Fig. 12.9 Peak current and onset potential for SO_2 oxidation as function of ECSA–projected area ratio measured in 1 mol/L H_2SO_4 containing 30 mmol/L SO_2

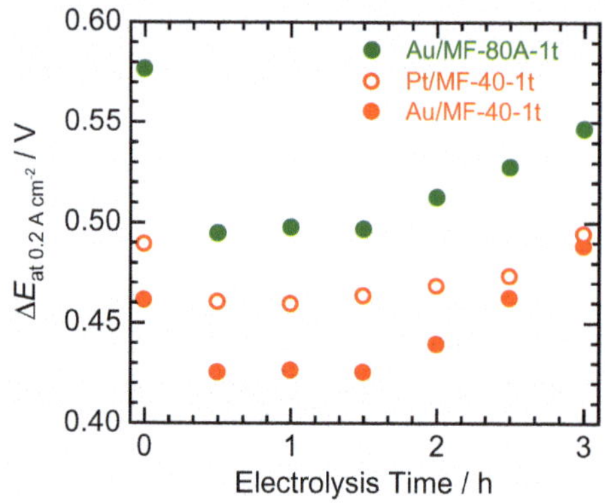

Fig. 12.10 Cell voltage at 0.2 A cm^{-2} as function of electrolysis time. The anolyte, catholyte, and CEM used in the test cell are the same as those in Fig. 12.7

The use of MF-40-1t porous Ti is able to reduce the ΔE at 0.2 mA/cm^2 by more than 0.07 V compared with the values for Ti plate-based electrodes.

12.4 Bunsen Reaction Assisted with Membrane

The membrane Bunsen reaction was carried out using the cation-exchange membranes and the porous Au electrode discussed in the previous section. The experimental procedure of the membrane Bunsen reaction in this section was described as follows: An electrochemical cell with the effective electrolysis area of 5 cm^2 was used for the test. A swollen CEM equilibrated with pure water was sandwiched between electrodes. The crosslinked membranes and Nafion, a commercial membrane, were tested. A graphite plate was put at cathode side, whereas platinum plate or the prepared porous Au anode was put at anode side. An anolyte was prepared by feeding SO_2 gas into 3.3 or 9.5 mol/kg H_2O H_2SO_4 aqueous solution at 30 mL/min for 3 h. A catholyte was prepared by dissolving HI and I_2 in purified water at the molality of $[HI] = [I_2] = 6.0$ or 2.5 mol/kg H_2O. 300 g of solutions were prepared for each side. An anolyte and catholyte kept at experimental temperature were cyclically supplied to the electrolysis cell by a tubing pump. The electrolysis was performed in the galvanostatic mode at the current of 1 A (current density: 200 mA/cm^2) for 3 h by using a potentiostat/galvanostat (HAL-3001, HOKUTO DENKO Co., Japan). The anolyte and catholyte solutions were sampled every half-hour, and their component was analyzed by an automatic titrator (Comite COM-900, Hiranuma Co., Japan). The amount of I^-, I_2, and total H^+ was titrated by using 0.02 mol/L KIO_3, 0.1 mol/L $Na_2S_2O_3$, and 0.1 mol/L NaOH aqueous solutions, respectively.

First of all, it was confirmed that the membrane Bunsen reaction can proceed by using the prepared membrane. The crosslinked membrane with IEC = 1.54 mmol/g was used for the experiment, and Nafion 212 membrane was used as a reference. The anode electrode was platinum plate. The operating temperate was a room temperature. Figure 12.11 showed concentrations change of anolyte and catholyte solution during the membrane Bunsen reaction [21]. Each component seemed to linearly change against the operation time. As for the crosslinked membrane, H_2SO_4 in anolyte increased from 3.3 to 3.8 mol/kg, and HI in catolyte increased from 6.0 to 6.7 mol/kg. On the other hands, SO_2 in anolyte and I_2 anolyte decreased. The solid lines in Fig. 12.11 show the theoretical concentration changes based on Eqs. (12.1) and (12.2) under ideal conditions for membrane and reaction performance.

The experimental values almost corresponded to the theoretical lines, suggesting that Bunsen reaction could properly proceed without side reactions and permeation of components except for H^+. The over potentials across the cation-exchange membranes were investigated. Figure 12.12 shows the cell voltages and currents for the Nafion membrane and the crosslinked radiation-grafted membrane [21]. The

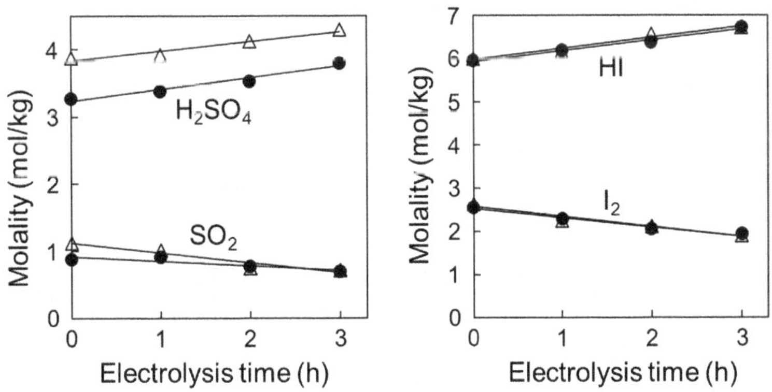

Fig. 12.11 Concentration change of reaction components at anode (left) and cathode (right) (black circle: crosslinked membrane triangle: Nafion 212) [21]

cell voltages were almost proportional to the currents and showed approximately 0.02 V at 0 mA/cm^2. The overpotential was defined by the difference between the cell voltages at 0 and 200 mA/cm^2. The cell voltage of the crosslinked membrane was 0.42 V. There was little difference compared with Nafion 212 (0.35 V). Consequently, it was confirmed that the crosslinked membrane could be applicable to the membrane Bunsen reaction.

Figure 12.13 shows photographs of the membrane surfaces after the membrane Bunsen reaction [24]. All the membranes were transparent before the reaction. The central effective membrane areas (5 cm^2) of the Nafion and the non-crosslinked membranes appear white. These white deposits were sulfur that were confirmed by the energy dispersive X-ray spectrometry measurements. The sulfur formation can be explained by the permeation of SO_2 through the cation-exchange membranes due to water permeation. The white color of the crosslinked membrane is pale in the figure, suggesting less sulfur deposition. Thus, the lower water permeation through the crosslinked membrane must be the reason for less sulfur deposition.

In order to improve the hydrogen production efficiency of the IS process, the overpotential of the cell voltage should be decreased. Considering that the cathode reaction would be sufficiently high [2], the overpotential can be regarded as the sum of the activation potential due to the anodic SO_2 oxidation (V_{anode}), and the ohmic loss associated with a cation exchange membrane (V_{ohm}). First, to reduce V_{anode},

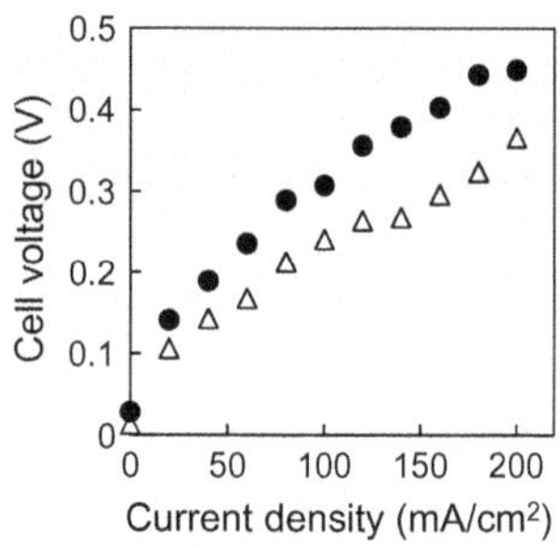

Fig. 12.12 Relationship between current density and cell voltage (black circle: crosslinked membrane triangle: Nafion 212) [21]

Fig. 12.13 Photographs of the membrane surfaces after the membrane Bunsen reaction for 3 h at a current density of 200 mA cm^{-2} [24]

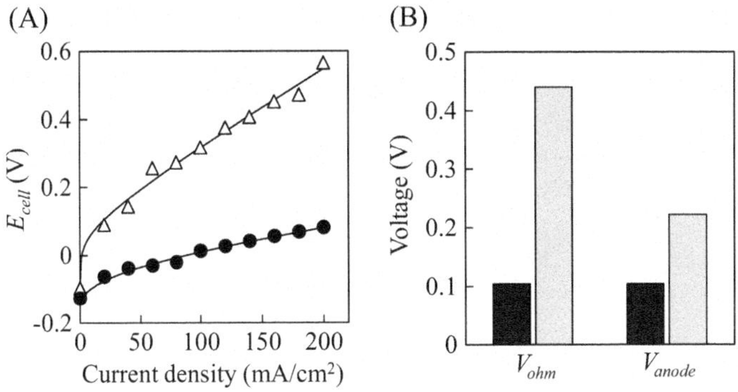

Fig. 12.14 **a** Relationship between current density and E_{cell} for Runs 1 (black circle) and 2 (triangle). The solid curves were obtained by fitting the experimental data points to Eq. (12.9). **b** The values of V_{ohm} and V_{anode} for Runs 1 (black) and 2 (grey) [20]

the porous Au electrode described in Sect. 2.3 was applied. Secondly, the cation exchange membrane whose IEC was enhanced (IEC = 2.9 mmol/g) was prepared to decrease the membrane resistance. In addition, the operating temperature increased to 50 °C (Run 1). Figure 12.14a showed the cell voltage as a function of the current density. In the graph, the result under conventional condition (membrane: Nafion 212, electrode: non-porous Pt, temperature: 20 °C) was also shown as Run 2. The overpotential largely decreased from 0.66 V (Run 2) to 0.21 V (Run 1). The cell voltage was theoretically given as following formula by using equilibrium potential (E_0), V_{ohm} and V_{anode} [20].

$$E_{\text{cell}} = E_0 + V_{\text{ohm}} + V_{\text{anode}} = E_0 + R_{\text{ohm}} I + \frac{RT}{F} \ln\left(\frac{I}{2I_0} + \sqrt{\frac{I^2}{4I_0^2} + 1}\right) \quad (12.9)$$

where I, I_0, R, F, and T represent the current density, the exchange current density, gas constant, Faraday constant, and temperature, respectively. According to Eq. (12.9), the breakdown of the overpotential was analyzed. The values of two parameters, R_{ohm} and I_0, obtained by the curve fitting are shown in Table 12.1 [20]. For Run 1 compared to Run 2, the value of R_{ohm} was only 28% and that of I_0 was 120-times higher. Lower R_{ohm} was due to high IEC of the radiation-grafted CEM, while higher I_0 was due to an enlarged effective surface area of the porous Au anode. Besides, higher operation temperature for Run 1 likely led to a further decrease of R_{ohm} and increase of I_0. Finally, V_{ohm} and V_{anode} at I of 200 mA/cm^2 were individually calculated from Eq. (12.9) and shown in Fig. 12.14b. V_{ohm} and V_{anode} for Run 1 was 1/4 and 1/2 for those of Run 2, respectively. It was found that employment of the low-resistive grafted CEM was more effective for the overvoltage reduction compared to the porous Au anode [20].

Table 12.1 Experimental conditions of membrane Bunsen reaction tests and constant values obtained by fitting of experimental plots by Eq. (12.9) [20]

Run	Membrane type	IEC (mmol/g)	Anode type	Temp. (°C)	Overvoltage (V)	R_{ohm} (Ωcm^2)	I_0 (mA/cm^2)
1	Grafted CEM	2.9	Porous Au	50	0.21	0.52	4900
2	Nafion 212	0.90	Non-porous Pt	20	0.66	2.1	39

References

1. Dokiya M (1977) The study of thermochemical hydrogen preparation (II). Electrochemical hydride cycle using the sulfur-iodine system. Denki Kagaku 45:139–143
2. Yoshida M (2008) Concentration of HIx solution by electro-electrodialysis using Nafion 117 for thermochemical water-splitting IS process. Int J Hydrogen Energy 33:6913–6920
3. Lu PWT, Ammon RL (1980) An investigation of electrode materials for the anodic oxidation of sulfur dioxide in concentrated sulfuric acid. J Electrochem Soc 127:2610–2616
4. O'Brien JA (2010) The electrochemical oxidation of aqueous sulfur dioxide: a critical review of work with respect to the hybrid sulfur cycle. Electrochim Acta 55:573–591
5. Xue L (2015) In-situ electrochemical impedance spectroscopy measurement of anodic reaction in SO_2 depolarized electrolysis process. Chem Eng Procces 89:70–74
6. Colon-Mercado HR, Hobbs DT (2007) Catalyst evaluation for a sulfur dioxide-depolarized electrolyzer. Electeochem Commun 9:2649–2653
7. Quajada C, Vazquez JL (2000) Electrochemical reactivity of aqueous sulphur dioxide at polycrystalline noble metal electrodes in acidic media. Recent Res Dev Electrochem 3:137–181
8. Nomura M (2004) Application of an electrochemical membrane reactor to the thermochemical water splitting IS process for hydrogen production. J Membra Sci 240:221–226
9. Nomura M (2004) Development of an electrochemical cell for efficient hydrogen production through the IS process. AIChE J 50:1991–1998
10. Ying Z (2013) Electrochemical investigation of the Bunsen reaction in the sulfureiodine cycle. Int J Hydrogen Energy 38:14391–14401
11. Sivasubramaniana P (2007) Electrochemical hydrogen production from thermochemical cycles using a proton exchange membrane electrolyzer. Int J Hydrogen Energy 32:463–468
12. Lee SK (2009) The effect of Pt loading amount on SO_2 oxidation reaction in an SO_2-depolarized electrolyzer used in the hybrid sulfur (HyS) process. Int J Hydrogen Energy 34:4701–4707
13. Sawada S (2006) Development of fluoropolymer electrolyte membranes for fuel cell applications by the γ-ray induced crosslinking and graft polymerization methods. Kobunshi Ronbunshu 63:149–159
14. Sawada S (2008) Water transport properties of crosslinked-PTFE based electrolyte membranes. Solid State Ionics 179:1611–1614
15. Yamaki T (2010) Quantum-beam technology: a versatile tool for developing polymer electrolyte fuel-cell membranes. J Power Sources 195:5848–5855
16. Chen J (2008) Fuel cell performance of polyetheretherketone-based polymer electrolyte membranes prepared by a two-step grafting method. J Membr Sci 319:1–4
17. Hamada T (2015) Poly(ether ether ketone) (PEEK)-based graft-type polymer electrolyte membranes having high crystallinity for high conducting and mechanical properties under various humidified conditions. J Mater Chem A 3:20983–20991
18. Gubler L (2005) Radiation grafted membranes for polymer electrolyte fuel cells. Fuel Cells 5:317–335
19. Gubler L (2006) Advanced monomer combinations for radiation grafted fuel cell membranes. Electrochem Commun 8:1215–1219

20. Sawada S (2020) Overvoltage reduction in membrane Bunsen reaction for hydrogen production by using a radiation-grafted cation exchange membrane and porous Au anode. Int J Hydrogen Energy 45:13814–13820
21. Yamaki T (2016) Development of cation exchange membranes for electrochemical Bunsen reactions by radiation-induced graft polymerization. Membrane (in Japanese) 41:114–120
22. Youcef HB (2008) The influence of crosslinker on the properties of radiation-grafted films and membranes based on ETFE. J Membr Sci 311:208–215
23. Inagaki Y (2019) Contribution of membrane technology to hydrogen society—development of membrane IS process. Bull Soc Sea Water Sci Jpn 73:194–202
24. Nomura M (2018) Development of ion-exchange membranes for the membrane Bunsen reaction in thermochemical hydrogen production by iodine-sulfur process. J Chem Eng Jpn 51:726–731

Chapter 13
Oxygen Production by Sulfuric Acid Decomposition Assisted with Membrane

Toshinori Tsuru and Masato Machida

Abstract The SO_3 decomposition reaction in the H_2SO_4 decomposition process has an equilibrium conversion rate of 80% or higher at a reaction temperature of 800 °C, but the conversion decreases to about 30% at 600 °C. SO_3 decomposition assisted with a membrane, i.e., SO_3 decomposition membrane reactor, has been developed in order to improve the conversion at moderate temperatures ($\leq$650 °C) by extracting O_2 with the O_2 permselective membrane from the reaction products and shifting the reaction equilibrium. To realize the concept, SO_3 decomposition catalysts operating at moderated temperatures and O_2 permselective membranes were developed. In addition to the catalytic activity, tolerance to the high-temperature corrosive environment, which severely damages most conventional catalytic materials within a short time, is an essential factor. In order to satisfy these conditions, the catalysts have been developed with active components such as platinum (Pt), vanadium (V), and support materials such as titanium dioxide (TiO_2), silicon dioxide (SiO_2), which have excellent acid resistance. Silica and bis(triethoxysilyl)ethane (BTESE)-derived organosilica membranes formed on a porous α-Al_2O_3 support has been developed as the O_2 permselective membrane showing O_2/SO_3 selectivity of 10–20 with thermal and chemical stability. In addition, a membrane reactor for SO_3 decomposition assisted with the O_2 permselective membrane was confirmed to increase SO_3 conversion.

Keywords Sulfur trioxide decomposition · Oxygen permselective membrane · Platinum catalyst · Vanadate catalyst · Silica/organosilica membrane · Catalytic membrane reactor

T. Tsuru (✉)
Hiroshima University, Kagamiyama, Higashi-Hiroshima, Hiroshima 739-8527, Japan
e-mail: tsuru@hiroshima-u.ac.jp

M. Machida
Kumamoto University, Kurokami, Chuo-ku, Kumamoto 860-8555, Japan

K. Aika and H. Kobayashi (eds.), *CO_2 Free Ammonia as an Energy Carrier*,
https://doi.org/10.1007/978-981-19-4767-4_13

13.1 Introduction

The thermal decomposition of H_2SO_4 is divided into the following two sub-reactions as described in Sect. 13.2:

$$H_2SO_4 \rightarrow H_2O + SO_3(573 - 873\ \mathrm{K}) \quad (13.1)$$

$$SO_3 \rightarrow SO_2 + 1/2O_2(1073 - 1173\ \mathrm{K}) \quad (13.2)$$

Both reactions are highly endothermic and smoothly achieve the equilibrium conversion without side reactions. Figure 13.1 shows the calculated equilibrium concentrations obtained from the decomposition of 1 mol of pure H_2SO_4 as a function of temperature [1]. The decomposition of H_2SO_4 to H_2O and SO_3 was predominant between 400 and 700 K while the reduction of SO_3 to SO_2 reached high conversions at temperatures higher than 1200 K. If it is possible to decompose SO_3 at medium temperatures (900–1000 K), the catalyst lifetime will be extended and hydrogen production will be both highly efficient and economical. According to theoretical evaluation, a membrane reactor (MR) enables SO_3 decomposition at intermediate-temperatures. Simulation results revealed that MR could effectively improve SO_3 conversion beyond the equilibrium status via selective removal of products, and could simultaneously reduce the reaction temperature less than 600 °C [1]. Importantly, the proof-of-concept was, for the first time, successfully confirmed from experimental viewpoints [2].

Another challenge to realize a membrane reactor for H_2SO_4 decomposition is the chemical and thermal stability, because SO_3 and SO_2 are strongly reactive and corrosive [3]. To realize the new concept, the chemical and thermal stability of the microporous membranes and membrane materials, including separation top layer (pure

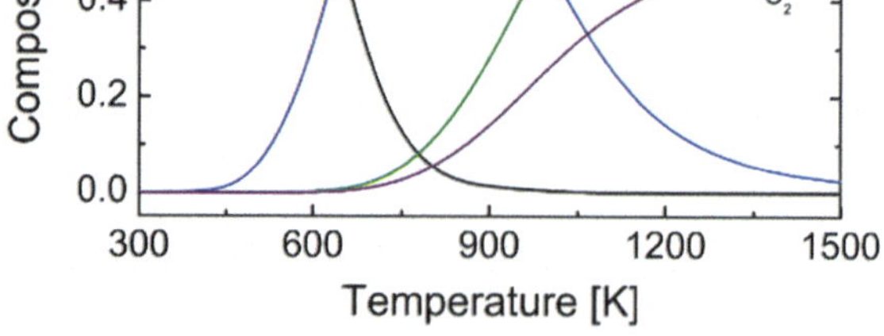

Fig. 13.1 Temperature dependence of equilibrium composition for the decomposition of 1 mol of pure H_2SO_4 at 1 atm [1]

SiO_2 or organosilica), intermediate layer (SiO_2, SiO_2–ZrO_2), and Al_2O_3 substrates, were evaluated and optimized for use in SO_3 decomposition membrane reactor as described in Sect. 13.2.

In the reaction (13.1), sulfuric acid dissociates nearly completely into mixtures of H_2O and SO_3 in the gas phase at above 400 °C, whereas the second step shown by Eq. (13.2) is a slow process requiring much higher temperatures (≥800 °C). A number of catalysts for the reaction (13.2) have been studied from 1970s [4–6], focusing on how they work at ≥800 °C because the conventional sulfur-based thermochemical water splitting cycles have been mainly designed to be installed on high-temperature nuclear reactors [7, 8]. Conversely, solar-driven SO_3 decomposition needs catalysts operating at moderated temperatures (≤650 °C), where the equilibrium conversion of SO_3 to SO_2 decreases below 50%. In addition to the catalytic activity, tolerance to the high-temperature corrosive environment, which severely damages most conventional catalytic materials within a short time, is an essential factor. In order to satisfy these conditions, innovations need to be introduced and explored concern novel catalyst materials (active components and supports), catalyst preparation technique, and catalytic reactor designs. These factors should also be in harmony with the optimal combination of improved performance, reusability, and process cost reduction. The catalysts have been developed with active components such as Pt, vanadium (V), and support materials such as titanium dioxide (TiO_2), silicon dioxide (SiO_2), which have excellent acid resistance, as described in Sect. 13.3.

The decomposition of SO_3 at 600 °C was, for the first time, examined using a SiO_2-based microporous catalytic membrane reactor with Pt-catalyst. Compared with a fixed-bed reactor, the SO_3 conversion increased from 0.09 (equilibrium conversion: 0.093) to 0.225 by extraction from the feed stream to the permeate stream via a membrane reactor as described in Sect. 13.4.

13.2 Development of Oxygen Permselective Membrane

The permeation mechanism of porous membranes is dominated by the molecular sieving, so that the molecular size of a permeating gas is crucial to selective extraction by a membrane reactor. The permeation of SO_2 through organosilica membranes revealed that 0.429 nm, Lennard–Jones (LJ) collision diameter, was more appropriate for the effective molecular size of SO_2 than the kinetic diameter [9]. Based on fitting SO_3 permeance onto a normalized Knudsen-based permeance (NKP) curve, the effective molecular size of SO_3 was estimated about 0.470 nm [2], which is similar to that of CF_4. To the best of our knowledge, this is the first report on the molecular size of SO_3, and this information should be helpful in the structural design of microporous membranes for SO_3 separation based on the molecular sieving mechanism.

Figure 13.2 shows the time course of permeances of single-gas (He, H_2, N_2, CF_4, SF_6) and binary-gas (O_2/SO_3) mixture thought a silica membrane, measured at 300 and 600 °C, respectively [2]. At 300 °C, the as-prepared silica membrane showed a H_2 permeance of 3.9×10^{-6} mol m^{-2} s^{-1} Pa^{-1} with selectivity for H_2/N_2 and

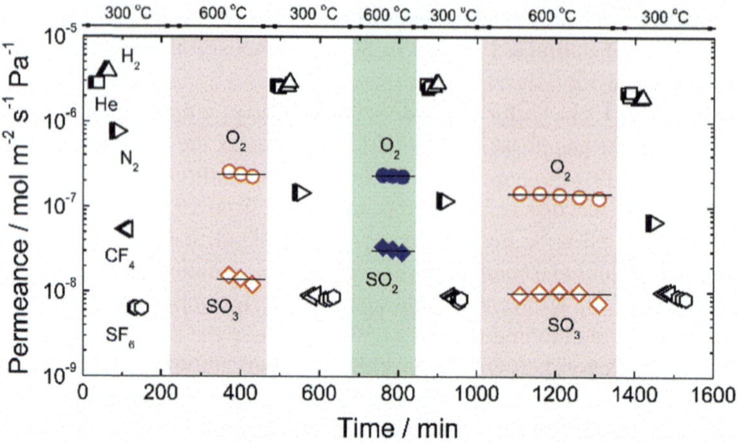

Fig. 13.2 Time course of gas permeance for a microporous silica membrane. Single-gas and binary-gas (95:5 O_2/SO_3 or 95:5 O_2/SO_2) permeation tests were conducted at 300 and 600 °C, respectively [2]. Reproduced from Ref. [2] with permission from the Royal Society of Chemistry

H_2/SF_6 of 5.1 and 610, respectively. After the first exposure to O_2/SO_3 at 600 °C during 210–450 min, the permeance of H_2 became lower than that of He, and the H_2/N_2 selectivity was increased to 28.2, indicating that the average pore size of the silica membrane was slightly narrowed. After additional exposure to O_2/SO_2 during 690–840 min and O_2/SO_3 during 1010–1350 min, approximately steady permeation properties were observed for He, H_2, N2, CF_4, and SF_6. A high O_2/SO_3 selectivity of 18.7 (Knudsen selectivity: 1.6) was achieved, which confirmed the high chemical and thermal stability of the microporous silica membrane consisting of pure SiO_2 top layer, a SiO_2 intermediate layer, and porous–silica, showed high chemical stability against acid and high thermal stability under the oxidizing atmosphere of a SO_3 or SO_2 flow.

In general, the network pore sizes of amorphous SiO_2 membranes is smaller than 0.35 nm [10], which made it difficult to obtain high permeance for O_2 (kinetic diameter 0.346 nm) and SO_2 (kinetic diameter 0.36 nm, LJ collision diameter 0.429). In addition, the SiO_2 membrane is not suitable for the IS process because the water vapor that was produced by H_2SO_4 decomposition may damage the network structure, which will result in a loss of gas permeation performance [10, 11]. Since SO_3 and SO_2 are similar in molecular size and both are strongly corrosive, MRs require a membrane with a well-defined pore size distribution for high levels of gas permeance and selectivity, as well as chemical and thermal stability under an O_2/SO_3 atmosphere.

Organosilica bis(triethoxysilyl)ethane (BTESE)-derived membranes with $-CH_2CH_2-$ linking units between two Si atoms (Spacer technique) reportedly showed higher gas permeance than pure silica membranes prepared from tetraethoxysilane (TEOS) [11]. BTESE-derived membranes with layered structures consisting of BTESE/SiO_2–ZrO_2/α-Al_2O_3 layers are generally prepared by coating BTESE-derived sol as a top layer onto a SiO_2–ZrO_2 intermediate layer, which is also formed on a porous α-Al_2O_3 support. These membranes demonstrated high permeance ratios for H_2/C_3H_8 (>10^3) and N_2/CF_4 (>60) due to the uniform pore sizes. In addition, BTESE-derived membranes prepared at 550 and 700 °C under N_2 were found to show high thermal stability and reach N_2/CF_4 permeance ratio as high as 80 even after heat treatment under air [10], indicating that BTESE-derived membranes is promising for O_2/SO_3 separation since the kinetic diameters of O_2 and SO_3 are similar to those of N_2 and CF_4, respectively. Regarding the intermediate layers, SiO_2–ZrO_2 have been utilized due to their great advantage in pore size control within a wide range from subnanometer to several nanometers, as well as excellent selectivity and stability. However, the chemical stability of either the intermediate layer (SiO_2–ZrO_2) or membrane supports (α-Al_2O_3) under a SO_3 atmosphere have never been reported.

Different types of powdered metal oxides used in membranes, including SiO_2, ZrO_2, Al_2O_3, and SiO_2–ZrO_2 were exposed to SO_3 and characterized by N_2 adsorption, Energy Dispersive X-ray Spectroscopy (EDS), and Scanning Electron Microscopy (SEM). As summarized in Table 13.1, the BET surface areas before and after SO_3 treatment remained similar for SiO_2, confirming the stability of SiO_2 powders under SO_3 exposure in addition to SiO_2 membranes shown in Fig. 13.2. On the other hand, the BET surface area was decreased for homemade Al_2O_3, SiO_2–ZrO_2, and ZrO_2 powders after SO_3 exposure. The BET areas of homemade Al_2O_3 powders fired at 400 and 600 °C decreased from 340 and 226 $m^2\ g^{-1}$ to 43 and 200 $m^2\ g^{-1}$ with high S/Al mole ratios of 66 and 22%, respectively, which indicated that more $Al_2(SO_4)_3$ had formed with firing at 400 °C. The homemade amorphous Al_2O_3 showed high S/Al ratios and a large decrease in surface area, suggesting the instability of α-Al_2O_3 under a SO_3 atmosphere. So, an innovative idea to coat α-Al_2O_3 porous substrates with amorphous SiO_2 was proposed, and it was experimentally confirmed that the stability of α-Al_2O_3 substates was dramatically improved due to reduced formation of $Al_2(SO_4)_3$ [12]. In a similar fashion, the BET area was decreased for SiO_2–ZrO_2 and ZrO_2 powders, suggesting that ZrO_2 contact with SO_3 at 550 °C might have resulted in the formation of $Zr(SO_4)_2$, while Si/Zr ratios of 9/1 and 7/3 retained relatively high values. The S ratios for SiO_2–ZrO_2 powders at Si/Zr ratios of 7/3 and 5/5 were 2.2 and 1.8%, respectively, both of which were smaller than that of ZrO_2 (2.7%). We suspected that SiO_2-rich SiO_2–ZrO_2 membranes might not be stable under a hydrothermal atmosphere when used for H_2SO_4 decomposition in the IS process, and, therefore, a Si/Zr ratio of 7/3 would match good balance between hydrothermal and chemical stability.

BTESE/SiO_2–ZrO_2 (Si/Zr = 7/3)/α-Al_2O_3 membrane prepared by coating BTESE-derived sols that were then calcined at 550 °C under N_2. SiO_2–ZrO_2 (Si/Zr = 7/3) was used as the intermediate layer. The permeance of fresh single gases (He, H_2, N_2, CH_4, CF_4) was measured after air treatment and after SO_3 exposure for 3 h. As shown in Fig. 13.3, H2 permeance of the fresh membrane was 6.1 × 10^{-7} mol $m^{-2}\ s^{-1}\ Pa^{-1}$ with H_2/CH_4 and H_2/C_3H_8 permeance ratios of more than 350 and 6000, respectively. Even after oxidation at 550 °C, BTESE-derived membranes

Table 13.1 BET area measured by N_2 adsorption, S/Al or S/(Si + Zr) mole ratio measured by EDS, and pH values detected after immersing powders into water [3]

Powders	Air-fired temperature (°C)*1	BET area ($m^2\ g^{-1}$)		S/Al or S/(Si + Zr) mole ratio (%)	pH after SO_3
		Before SO_3	After SO_3		
Home-made Al_2O_3	400	340	43*2	66	4.6
Home-made Al_2O_3	600	226	200	22	5.2
α-Al_2O_3	>1200	8.8	9.3	0.87	7.0
SiO_2	600	280	280	0.17	7.0
Si/Zr = 9/1	600	210	77	4.5	4.5
Si/Zr = 7/3	600	115	63	2.2	6.8
Si/Zr = 5/5	600	57	5	1.8	6.5
ZrO_2	600	21	15	2.7	6.0

*1 All the powders were fired under air for 24 h before SO_3 exposure
*2 The SO_3 exposure temperature was 400 °C

showed an increase in H_2 permeance with maintaining H_2/CH_4 and N_2/CF_4 permeance ratios probably due to stable SiC structure formed by the rearrangement of Si–C_2H_4–Si. Binary gas separation revealed that an O_2 permeance of (2.5–3.13) × 10^{-8} mol m^{-2} s^{-1} Pa^{-1} and O_2/SO_3 selectivity that ranged from 13 to 10 after 3 h of exposure to SO_3, which surpassed the Knudsen selectivity of 1.58. Single gases after O_2/SO_3 binary separation as summarized in Fig. 13.3 showed slightly decreased permeance after exposure to SO_3, while the permeance ratios for H_2/CH_4 and H_2/CF_4 remained still high, showing promise for use in O_2/SO_3 separation due to its thermal and chemical stability.

13.3 Development of Sulfur Trioxide Decomposition Catalysts

It has been reported that two classes of materials are candidates for active SO_3 decomposition catalysts that work at ≤650 °C. One is precious metals, the activity of which was reported to decrease in the following order: Pt > Pd > Rh > Ir > Ru [13, 14]. Supported Pt catalysts are considered to be the most promising for the required temperature range [5, 6, 14, 15]. The SO_3 decomposition reaction steps over Pt can be described as follows:

$$SO_{3s} \rightarrow SO_{2s} + O_s \tag{13.3}$$

$$SO_{2s} \rightarrow SO_2 \tag{13.4}$$

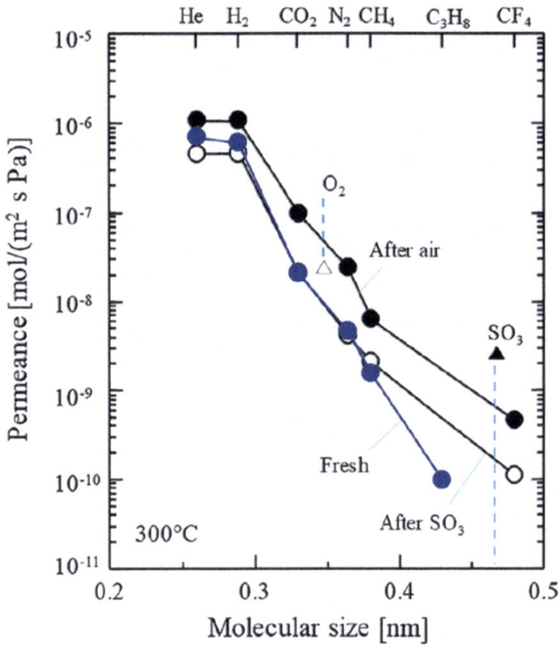

Fig. 13.3 Kinetic diameter dependency of gas permeance at 300 °C for the BTESE/SiO_2–ZrO_2 (Si/Zr = 7/3)/α-Al_2O_3 membrane prepared at 550 °C before (closed, after air treatment at 550 °C) and after exposure to SO_3 at 550 °C (open). O_2 and SO_3 permeance were measured at 550 °C in a binary system [3]

$$O_s + O_s \rightarrow O_2 \tag{13.5}$$

Herein, a subscript "s" means adsorbed species. The activity and deactivation of the catalysts can be determined by the rate of the dissociation of adsorbed SO_3 species of Eq. (13.3) and the removal rate of the decomposition products from the surface of the metal nanoparticles of Eqs. (13.4) and (13.5). Rashkeev et al. [14] reported based on density functional theory calculations that the rate for reaction of Eq. (13.3) is always faster than the rates for reactions of Eqs. (13.4) and (13.5). Pt is therefore more efficient when its metallic state is favored over its oxidized states. Metallic Pt surfaces should also be more efficient for the detachment of O atoms from SO_3 species of Eq. (13.3) compared to that from oxidized Pt surfaces.

The selection of support materials is a key to ensure the activity and long-term stability of supported Pt catalysts. Although γ-Al_2O_3 is the most widely used Pt support in industry, the instability in the presence of SO_3 leads to the formation of $Al_2(SO_4)_3$ overlayers at ~600 °C that block access to the active Pt sites [16]. Stable Pt catalysts can therefore be prepared by dispersing Pt onto SO_3-resistant support materials, including TiO_2, SiO_2, Ta_2O_5, and SiC [13, 17]. Among these candidates, TiO_2 has been studied most extensively, whereas most studies have focused only on the rutile phase, which is thermodynamically stable at high reaction temperatures of ~800 °C. Machida et al. [18] found that Pt supported on anatase is much more active and stable at 600 °C than on rutile as shown in Fig. 13.4. At this temperature Pt is mainly present in the active metallic state on anatase TiO_2, whereas less-active Pt oxides (PtO_2 and PtO) are dominant on rutile TiO_2. This stark contrast is rationalized by the greater interface formation energy, which stabilizes the oxidized states of Pt on rutile [18, 19]. During continuous catalytic testing at 600 °C for 1000 h, Pt/anatase exhibited activity loss of approximately 4% (of the initial activity). It was also reported that Ta_2O_5 showed most active and stable SO_3 decomposition performance at 600 °C among SO_3-resistant metal oxide supports (TiO_2, Nb_2O_5, Ta_2O_5, and WO_3) for Pt catalysts [17]. A long-term stability of Pt/Ta_2O_5 for the SO_3 decomposition reaction at 600 °C was demonstrated over a total of 1800 h, where the activity loss was less than 1.5% per 1000 h. Ramos-Fernandez et al. [20] studied the interaction between Pt metal and a Ta_2O_5 support and showed that the electronic density of Pt is increased via interaction with Ta_2O_5, which affects the dissociation ability of the S–O bond. These features are beneficial for catalytic activity and stability even under a strongly oxidizing atmosphere containing SO_3 and O_2.

Various metal oxides have been highlighted as economically viable catalysts. Their activities are lower, and the following activity order was obtained: Pt > Cr_2O_3 > Fe_2O_3 > CuO > CeO_2 > NiO > Al_2O_3 [5, 6]. Although various complex oxides containing Fe–Al, Fe–Ti, Fe–Cu, Fe–Cr, Cr–Cu, and Ni–Fe were also studied at high-temperatures (≥800 °C), none of these materials has efficiently satisfied the activity requirement below 650 °C [13]. As thermodynamics predicts, the deactivation of metal oxides due to sulfate formation under the reaction atmosphere is more preferable at lower reaction temperatures. According to the reaction scheme

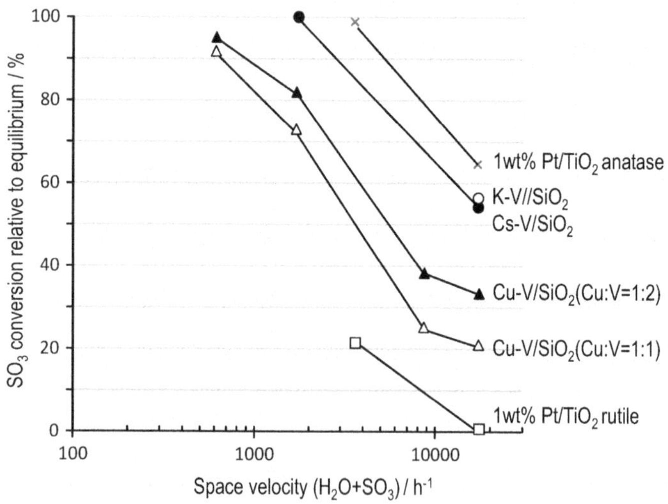

Fig. 13.4 Catalytic activity for SO_3 decomposition at 600 °C as a function of space velocity

firstly proposed by Tagawa and co-worker [6], SO_3 decomposition over metal oxide catalysts proceeds via a sulfate intermediate.

$$MO + SO_3 \rightarrow (MSO_4) \rightarrow MO_2 + SO_2 \tag{13.6}$$

$$MO_2 \rightarrow MO + 1/2O_2 \tag{13.7}$$

According to the reaction of Eq. (13.6), the metal oxides with a weak basicity interact so weakly with SO_3 that the rate of formation of surface sulfate (MSO_4) is low. By contrast, the metal oxides with a high basicity yield so stable sulfates that their decomposition becomes rate-determining. The subsequent step of Eq. (13.7) suggests that the oxygen removal after desorption of SO_2 is a key step for the SO_3 decomposition.

Another class of candidate catalysts for solar-driven SO_3 decomposition catalysts are vanadates (vanadium oxide) supported on porous SiO_2 supports, which show SO_3 decomposition activities approaching those of Pt catalysts at a moderate temperature range (Fig. 13.4). Unlike the solid metal oxide catalysts [13], vanadates can be regarded as molten-phase catalysts because their melting points are nearly equal to or lower than the reaction temperature for SO_3 decomposition. Machida et al. focused on vanadate-based catalyst systems including Cu−V [21–24], Ce−V [25], K−V, and Cs−V [26, 27]. Because vanadate forms the framework resistant to sulfate formation, redox of Cu provides an active site for SO_3 decomposition. The catalytic activity can be enhanced by supporting the catalyst on mesoporous SiO_2 followed by thermal aging above the melting point of vanadate, which allows the molten phase to penetrate the SiO_2 mesopores [22]. The simultaneous dissolution-reprecipitation of SiO_2 in the melt converts mesopores to macroporous cavities, the surface of which are completely covered with a thin layer of active vanadate.

A mechanism for the accelerated catalytic SO_3 decomposition over the supported molten copper vanadate catalyst was proposed [24]. On the surface of the molten vanadate, SO_3 adsorption to Cu oxide species produces $Cu-SO_4$ species, because it is thermodynamically favored around 600 °C. Here, one important contribution of the liquid surface to the catalytic reaction is the dissolution of $Cu-[SO_4]$ species in the molten phase, leaving the liquid surface with more vacant sites for further SO_3 adsorption. Because the molten phase consists of tetrahedral VO_4 and Cu^+, the dissolved $[SO_4]$ is immediately reduced to evolve SO_2 and then the reoxidized Cu^{2+} oxides, $Cu-[2O]$, release oxygen. The resulting free monovalent copper oxide species, $Cu-[O]$, defuses back to the liquid surface, where the reaction with SO_3 forms $Cu-[SO_4]$ again. Consequently, the molten vanadate enables fast cycles of SO_3 capture and decomposition to SO_2/O_2.

Among various metal vanadates, those containing alkaline metal elements (A) showed the highest SO_3 decomposition activity, which increases in the order Na < K < Rb < Cs in accordance with the basicity of the A element [26]. Considering that catalytic SO_3 decomposition over metal oxides proceeds via reaction intermediates in the form of metal sulfates as shown in reactions of Eqs. (13.6) and (13.7) [6], oxides with a higher basicity interact more strongly with SO_3. However, alkaline metal elements have negative effects on the pore structure of the support due to the corrosive effect of molten phase. As a result, the highest activities are exhibited at moderate A/V ratios (0.5–1.0). Although high activity is closely associated with molten phase, the most serious drawback of the catalyst is the lack of long-term stability under typical reaction condition. Stability testing of the Cs–V/SiO_2 catalyst over a 1000 h continuous reaction at 600 °C resulted in deactivation by ~20% of the initial activity. Kinetic analysis of the activity versus time-on-stream indicated that the observed deactivation behavior can be divided into an induction period ($\leq$100 h) and an acceleration period (>100 h). The vaporization loss of active components (Cs and V) from the molten phase is the primary reason for the observed deactivation. At the earliest stage, most vapor is generated in the upstream section of the catalyst bed and then redeposits inside the catalyst bed, thus moving slowly downstream. During this induction period, the deactivation is not obvious because the total Cs and V content of the catalyst bed remains almost unchanged. After this period, however, detachment of Cs and V from the downstream end of the catalyst bed induces accelerated deactivation. The vaporization loss was found to be significantly suppressed by inverting the catalyst bed every 100 h during the stability test [27]. It is necessary to develop a strategy to overcome the vaporization loss under harsh reaction condition.

13.4 Sulfur Trioxide Decomposition Reaction Assisted with Membrane

A membrane reactor (MR) is a combination of a permselective membrane and heterogeneous catalysts. MRs have the inherent capability and advantage of combing both a reaction and a separation in a single unit. The membrane provides selective removal of products, which shifts the equilibrium toward the product side and thus results in a higher reaction conversion even at lower temperatures [1, 28]. The modeling and simulation of MRs is beneficial to understanding and designing of MRs, including membrane performance (e.g., permeance and selectivity) and operation conditions such as temperature, pressure, and feed flow rate. The mathematical model of SO_3 decomposition in catalytic membrane reactors considers the following hypotheses similarly to dehydrogenation reaction [28]: (1) steady-state operation; (2) plug flow in retentate and permeate streams; (3) isothermal reaction, which only occurs over the catalyst on the retentate side; (4) negligible pressure drop along the axial direction of the membrane reactor; and (5) no concentration polarization effect for gas permeation through the membrane. For an alumina-palladium catalyst, the reaction rate of SO_3 decomposition (r_d) can be described as follows [1]:

$$r_d = \frac{\left(k_s K_{SO_3} p_{SO_3}\left(1 - \frac{p_{SO_2} p_{O_2}^{0.5}}{p_{SO_3}} \frac{1}{K_e}\right)\right)}{\left(1 + K_{SO_3} p_{SO_3} + K_{SO_2} p_{SO_2} + K_{O_2}^{0.5} p_{O_2}^{0.5}\right)^2} \quad (13.8)$$

where p_{SO_3}, p_{SO_2}, and p_{O_2} are the partial pressures of the components. The equilibrium constant for the SO_3 decomposition, K_e, the reaction rate constant, k_s, and the adsorption coefficients of the components, K_{SO_3}, K_{SO_2}, and K_{O_2}, are summarized in [1].

SO_3 decomposition was simulated in a co-current configuration. The membrane permeability and selectivity, reaction temperature (T), pressure in the feed and permeate streams, catalyst weight and SO_3 feed flow rate ($F_{SO_3,0}$) were adopted as parameters to study the effect on a catalytic membrane reactor (CMR) performance in terms of SO_3 conversion. The volume of the membranes was equated with the reactor volume to calculate the weight hourly space velocity (WHSV) and gas hourly space velocity (GHSV) since the catalysts were often loaded into the bore side of the tubular membranes in the catalytic membrane reactors.

SO_3 decomposition requires high-temperatures that make the use of polymer membranes impossible. One option is the use of dense inorganic membranes such as oxygen ion conducting membranes, which allow only the transport of oxygen ions. Another alternative is porous inorganic membranes (mainly zeolite, silica, and carbon) that can separate gas components based on molecular sizes. As shown in Fig. 13.5 [1], without a membrane-based extraction whereby the permeate stream is kept at atmospheric pressure, the SO_3 conversion was approximately equal to an equilibrium conversion of 0.28. When the MR was equipped with a porous inorganic membrane that showed a Knudsen selectivity for permeating O_2 over SO_2 ($\alpha(O_2/SO_2) = 1.414$) and a high selectivity for O_2 over SO_3 ($\alpha(O_2/SO_3) = 70$), the SO_3 conversion was significantly enhanced to 0.93, which was far beyond the equilibrium conversion of 0.28. On the other hand, only a slight increase was predicted for oxygen ion conducting membranes which were reportedly degraded once the SO_2 was introduced into the feed stream. These results suggest that porous inorganic membranes with excellent chemical stability have the potential to improve SO_3 conversion for SO_3 decomposition in a membrane reactor.

The first success to apply a membrane reactor to SO_3 decomposition at 600 °C was reported in 2015 [2]. As shown in Fig. 13.6, a fixed-bed reactor (FBR) showed the SO_3 conversion of approximately 0.090 (equilibrium conversion: 0.093), indicating that SO_3 decomposition on the feed side had reached a stable state. After 90 min, a portion of the reaction mixtures was extracted from the feed stream to the permeate stream by using O_2 as a sweep gas, leading to significant increase in SO_3 conversion to 0.225. Since the microporous silica membrane utilized partial pressure differences to selectively remove the produced O_2 and SO_2 from the reaction side, it was mainly a shift in the reaction equilibrium that contributed to the enhanced SO_3 conversion. It was clear that the achievement ratio of equilibrium conversion (φ) was increased about threefold. This is defined as the ratio of SO_3 conversion obtained in the MR to the equilibrium conversion at 600 °C and 140 kPa, confirming that the MR offers a

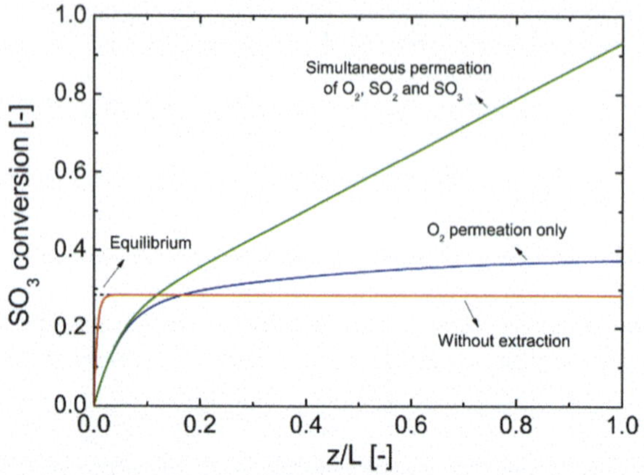

Fig. 13.5 Axial profiles of SO_3 conversion and O_2 partial pressure normalized by feed pressure in reactor by different membrane types (T = 900 K, $F_{SO_3,0} = 1.5 \times 10^{-5}$ mol s^{-1}, $P_{O_2} = 10^{-7}$ mol m^{-2} s^{-1} Pa^{-1}, $\alpha(O_2/SO_2) = 1.414$, α $(O_2/SO_3) = 70$, $p_R = 100$ kPa, $p_P = 5$ kPa, $w_{cat} = 1.5$ g) [1]

strong reduction in the reaction temperature for SO_3 decomposition and a significant increase in the SO_3 conversion.

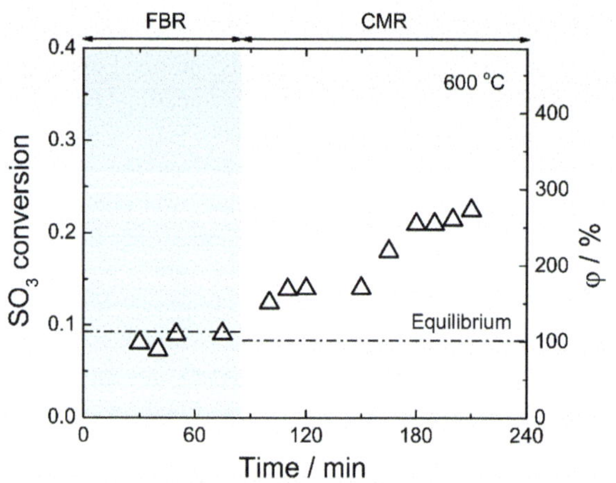

Fig. 13.6 SO_3 conversions and the achievement ratios of equilibrium conversion in fixed-bed reactor (FBR) and in CMR as function of time. Pt/Al_2O_3 particles (0.7 g) were used as catalysts. $T = 600$ °C; SO_3 feed flow rate = 2.5 ml/min, O_2 feed flow rate = 43.6 ml/min, space velocity = 500 h^{-1}; FBR: feed pressure = 110 kPa; CMR: feed pressure = 130–140 kPa, permeate pressure = 108 kPa, sweep gas flow rate = 8.75 ml/min [2]. Reproduced from Ref. [2] with permission from the Royal Society of Chemistry

References

1. Meng L et al (2015) Catalytic membrane reactors for SO_3 decomposition in Iodine-sulfur thermochemical cycle: a simulation study. Int J Hydrogen Energy 40:12687–12696
2. Meng L et al (2016) Enhanced decomposition of sulfur trioxide in the water-splitting iodine–sulfur process via a catalytic membrane reactor. J Mater Chem A 4:15316–15319
3. Yu X et al (2019) Evaluating the chemical stability of metal oxides in SO_3 and applications of SiO_2-based membranes to O_2/SO_3 separation. J Am Ceram Soc 102:6946–6956
4. Dokiya M et al (1977) Thermochemical hydrogen preparation-Part V. A feasibility study of the sulfur iodine cycle. Bull Chem Soc Jpn 50:2657–2660
5. O'Keefe DR et al (1980) Catalysis research in thermochemical water-splitting processes. Catal Rev Sci Eng 22:325–369
6. Tagawa H et al (1989) Catalytic decomposition of sulfuric acid using metal oxides as the oxygen generating reaction in thermochemical water splitting process. Int J Hydrogen Energy 14:11–17
7. Kasahara S et al (2007) Flowsheet study of the thermochemical water-splitting iodine-sulfur process for effective hydrogen production. Int J Hydrogen Energy 32:489–496
8. Sakaba N et al (2007) Conceptual design of hydrogen production system with thermochemical water-splitting iodine-sulphur process utilizing heat from the high-temperature gas-cooled reactor HTTR. Int J Hydrogen Energy 32:4160–4169
9. Meng L et al (2015) Permeation properties of BTESE–TEOS organosilica membranes and application to O_2/SO_2 gas separation. J Membr Sci 496:211–218
10. el Yu X (2018) Improved thermal and oxidation stability of bis (triethoxysilyl) ethane (BTESE)-derived membranes, and their gas-permeation properties. J Mater Chem A 6:23378–232387
11. Tsuru T (2018) Silica-based membranes with molecular-net-sieving properties: development and applications. J Chem Eng Jpn 51:713–725
12. Noritake Co Ltd. (2019) SIP finishing report
13. Brown NR et al (2012) A review of catalytic sulfur (VI) oxide decomposition experiments. Int J Hydrogen Energy 37:2685–2698
14. Rashkeev SN et al (2009) Catalytic activity of supported metal particles for sulfuric acid decomposition reaction. Catal Today 139:291–298
15. Petkovic LM et al (2008) Pt/TiO_2 (rutile) catalysts for sulfuric acid decomposition in sulfur-based thermochemical water-splitting cycles. Appl Catal A: Gen 338:27–36
16. Banerjee AM et al (2015) A comprehensive study on Pt/Al_2O_3 granular catalyst used for sulfuric acid decomposition step in sulfur–iodine thermochemical cycle: changes in catalyst structure, morphology and metal-support interaction. Appl Catal B: Environ 162:327–337
17. Nur ASM et al (2018) Pt supported on Ta_2O_5 as a stable SO_3 decomposition catalyst for solar thermochemical water splitting cycles. ACS Appl Energy Mater 1:744–750
18. Nur ASM et al (2017) Catalytic SO_3 decomposition activity and stability of Pt supported on anatase TiO_2 for solar thermochemical water splitting cycles. ACS Omega 2:7057–7065
19. Nur ASM et al (2018) Phase-dependent formation of coherent interface structure between PtO_2 and TiO_2 and its impact on thermal decomposition behavior. J Phys Chem C 122:662–669
20. Ramos-Fernández EV et al (2008) Pt/Ta_2O_5–ZrO_2 catalysts for vapour phase selective hydrogenation of crotonaldehyde. Appl Catal A: Gen 349:165–169
21. Machida M et al (2011) Efficient catalytic decomposition of sulfuric acid with copper vanadates as an oxygen-generating reaction for solar thermochemical water splitting cycles. Chem Commun 47:9591–9593
22. Machida M et al (2012) Macroporous supported Cu–V oxide as a promising substitute of the Pt catalyst for sulfuric acid decomposition in solar thermochemical hydrogen production. Chem Mater 24:557–561
23. Machida M et al (2013) Role of oxygen vacancies in catalytic SO_3 decomposition over $Cu_2V_2O_7$ in solar thermochemical water splitting cycles. J Phys Chem C 117:26710–26715
24. Kawada T et al (2014) Molten copper hexaoxodivanadate: an efficient catalyst for SO_3 decomposition in solar thermochemical water splitting cycles. Catal Sci Technol 4:780–785

25. Kawada T et al (2015) Structure and SO_3 decomposition activity of $CeVO_4/SiO_2$ catalysts for solar thermochemical water splitting cycles. Int J Hydrogen Energy 40:10726–10733
26. Kawada T et al (2016) Catalytic SO_3 decomposition activity and stability of A-V–O/SiO_2 (A = Na, K, Rb, and Cs) for solar thermochemical water-splitting cycles. Ind Eng Chem Res 55:11681–11688
27. Nur ASM et al (2018) Stability of molten-phase Cs–V–O catalysts for SO_3 decomposition in solar thermochemical water splitting. ACS Appl Energy Mater 1:2041–2047
28. Tsuru T et al (2004) Methane steam reforming by microporous catalytic membrane reactors. AIChE J 50:2794–2805

Chapter 14
Hydrogen Production by Hydrogen Iodine Decomposition Assisted with Membrane

Mikihiro Nomura, Tatsumi Ishihara, and Odtsetseg Myagmarjav

Abstract In order to improve the low equilibrium of the HI decomposition, a membrane reactor assisted with H_2 permselective membrane was developed. Silica membranes and stable HI decomposition catalysts have been investigated for the membrane reactor. Silica membranes were prepared on porous ceramic tubes by using a counter diffusion chemical vapor deposition. The membrane deposited at 450 °C showed the H_2 permeance of 5.0×10^{-7} mol m^{-2} s^{-1} Pa^{-1} with the H_2/HI permeance ratio of 6820. For HI decomposition catalysts, Pt–Rh/TiO_2/active carbon showed high conversion and stability. This catalyst shows HI decomposition conversion, ca.95% of the equilibrium conversion which was sustained up to 200 h. Finally, membrane reactor test was conducted to improve the HI decomposition reaction at 400 °C. The conversion was stable at about 0.48 by the assist with the H_2 permselective membrane and the conversion was higher than the equilibrium one (0.22).

Keywords Hydrogen permselective membrane · Chemical vapor deposition · Pt catalyst · Catalyst stability · Membrane reactor

14.1 Introduction

HI decomposition is performed at about 400 °C to produce H_2 and I_2. The problem is the low equilibrium conversion at about 20% [1]. The conversion can be improved by separating the H_2 from the reaction system. H_2 can be separated from I_2 and HI by a molecular sieving mechanism that utilizes the difference in molecular size. An H_2 permselective membrane that can be used at a high temperature at 400 °C should mainly consist with an inorganic material such as Pd [2] or silica [3]. Pd

M. Nomura
Shibaura Institute of Technology, Toyosu, Koto-ku, Tokyo 135-8548, Japan

T. Ishihara (✉)
Kyushu University, Ohaza-motooka, Nishi-ku, Fukuoka 819-0395, Japan
e-mail: ishihara@cstf.kyushu-u.ac.jp

O. Myagmarjav
Japan Atomic Energy Agency, Tokai, Oarai, Ibaraki 311-1393, Japan

K. Aika and H. Kobayashi (eds.), *CO_2 Free Ammonia as an Energy Carrier*,
https://doi.org/10.1007/978-981-19-4767-4_14

reacts with I_2, and stability of Pd membranes under I_2 atmosphere is not enough. Silica is stable under acid atmosphere. Silica-based membranes have been developed as H_2 permselective membranes as described in Sect. 14.2. Silica-based membranes have been developing by a sol–gel method [4] or a chemical vapor deposition (CVD) method [5]. The sol–gel method has been described in Chap. 13. Here, the CVD method is described. The membrane preparation by the CVD method can be classified in two types, a one-side geometry method [6] and a counter diffusion method [5], by depending on the supply method of the silica precursor(s). For the one-side geometry method, all sources are supplied from one side of the porous substrate. The opposite side of the substrate can be vacuumed to avoid the formation of the pinholes. Alkoxides such as tetraethoxysilane (TEOS) [6–8] have been used as silica precursors.

At present, there are many catalysts reported for decomposition of HI for large-scale H_2 production. The major research for the active catalyst to HI conversion is summarized by O'Keefe et al. [9]. It is reported that Pt group metal catalyst on support such as active carbon, γ-Al_2O_3, TiO_2, ZrO_2 shows the high catalytic performance in both wet and dry HI gas. Up to present, there are much more progress about development of Pt supporting catalyst for high activity and better stability. However, Pt particles will be aggregated and become larger particle size during HI decomposition reaction. This is the reason for decreasing catalytic activity and stability with reaction time. Therefore, the HI decomposition catalyst having stable and highly active performance for a long operation time has been developed as described in Sect. 14.3.

A silica ceramic membrane reactor to improve the efficiency of HI decomposition in the IS process has been explored at JAEA [10–12], and a study using silica-based membrane for HI decomposition has been reported in the early 2000s [1]. In the study, the HI decomposition conversion was improved to 76.4% from the equilibrium one (25%) at 600 °C with a utilization of the membrane reactor based on a silica H_2 permselective membrane that was prepared by a counter diffusion chemical vapor deposition (CVD) of tetraethoxysilane supported on γ-alumina-coated α-alumina tubes. For realization of industrial H_2 production by the IS process, it is desirable to construct a large-scale reactor using metallic materials. When the working temperature is decreased to approximately 400 °C, metallic materials could be used to replace high-cost structural materials (Ni-based alloys and Hastelloy used for HI decomposer). Also, the temperature dependence of the conversion rate is low as shown in Fig. 11.7, so there is little merit even if the reaction temperature is raised. However, no experimental work dealing with the investigation of an intermediate temperature (400 °C) HI decomposition in a catalytic membrane reactor of the IS process has so far been reported. The HI decomposition at an intermediate temperature (400 °C) using an H_2 permselective membrane was carried. The effect of the membrane based on a hexyltrimethoxysilane (HTMOS)-derived silica membranes was evaluated at 400 °C by varying the HI flow rates. The membrane was prepared by CVD method on γ-alumina-coated α-alumina tubes. H_2 was successfully extracted from the membrane and a significant acceleration in HI decomposition was experimentally verified [13], as described in Sect. 14.4.

14.2 Development of Hydrogen Permselective Membrane

Initially, the one-side geometry method has been investigated for an H_2 permselective membrane of the HI decomposition reaction. Hwang [14] had evaluated the stability of silica membranes in a HI–H_2O mixed vapor at 450 °C. The silica membrane deposited on a γ-alumina substrate with an H_2 permeance of 7×10^{-8} mol m^{-2} s^{-1} Pa^{-1} was stable for 150 h. Nomura [15] reported that the HI conversion had improved by applying the silica membrane. TEOS was used as the silica precursor. H_2 permeance through the silica membrane was 1.3×10^{-7} mol m^{-2} s^{-1} Pa^{-1}, and the HI conversion was evaluated at 51.6%. However, H_2 permeance through the silica membranes should be improved. Therefore, the silica-based membranes were developed by using the counter diffusion method [16, 17].

The important parameters for the counter diffusion method are silica sources and deposition temperature. Figure 14.1 shows the schematic diagram for the counter diffusion CVD method. Two kinds of reactants are supplied from the opposite sides of the porous substrate, and the deposition reaction is stopped by the deposited silica in the pores of the porous substrate to form the thin deposition layer. The pore sizes of the deposited membranes can be controlled in the order of Å by using the silica precursors containing the organic group(s) instead of the alkoxy groups [18, 19]. The molecular diameter of HI is larger than that of H_2. Thus, the pore size of the silica-based membrane should be smaller than the molecular diameter of HI to separate by a molecular sieve mechanism. If the effective thickness and number of the pores are the same, the permeance through a porous membrane is proportional to the pore diameter. Thus, it is important to control the pore size of the deposited silica-based membranes.

First, the pore size of the silica-based membranes prepared from six kinds of the silica precursors was evaluated. The pore sizes of the membranes were calculated from the single gas permeation tests [20]. A γ-alumina capillary with a pore size of 4 nm (O.D. 3 mm, effective membrane length 40 mm) was used as the substrate. The silica precursors were tetramethoxysilane (TMOS), ethyltrimethoxysilane (ETMOS), propyltrimethoxysilane (PrTMOS), hexyltrimethoxysilane (HTMOS),

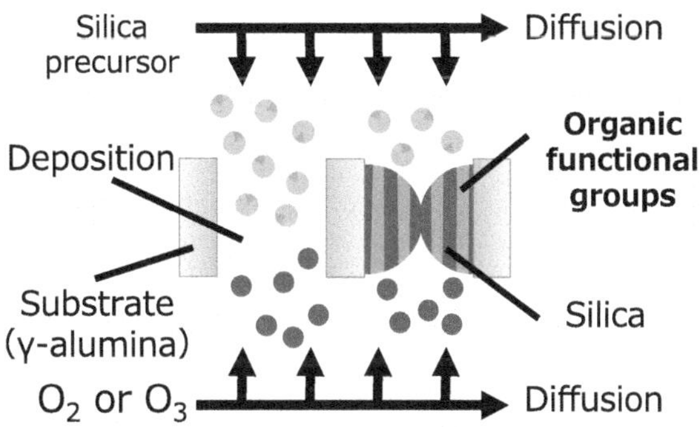

Fig. 14.1 Schematic diagram for counter diffusion CVD method

XTMOS

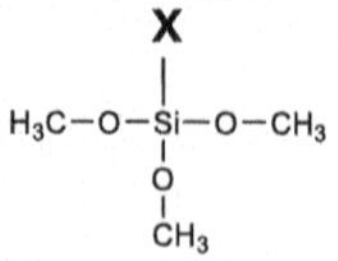

X= $-OCH_3$:	Tetramethoxysilane (TMOS)
X= $-C_2H_5$:	Ethyltrimethoxysilane (ETMOS)
X= $-C_3H_7$:	Propyltrimethoxysilane (PrTMOS)
X= $-C_6H_{13}$:	Hexyltrimethoxysilane (HTMOS)
X= $-C_6H_5$:	Phenyltrimethoxysilane (PhTMOS)
2X= $-C_6H_5$:	Diphenyldimethoxysilane (DPhDMOS)

Fig. 14.2 Structural formulas of silica precursors

Phenyltrimethoxysilane (PhTMOS) and diphenyldimethoxysilane (DPhDMOS). Figure 14.2 shows the molecular formula of the silica precursors. O_2 (1.0 L min^{-1}) was introduced into an ozone generator (ZOS-YB-20G or SOW-5000R, Shoken) to produce ozone. During membrane preparation, the ozone was introduced at the rate of 0.2 L min^{-1} to the inside of the substrate, and the silica precursor was introduced to the outside of the substrate with 0.2 L min^{-1} of N_2 bubbler (45–125 °C). The deposition was carried out at 150–500 °C by using O_2 or O_3 as a reactant, and the deposition period was 5 min or 90 min. The permeances of H_2 (0.28 nm), N_2 (0.36 nm), and SF_6 (0.55 nm) at 270 °C were measured with a soap film flow meter or a pressure change method. Figure 14.3 shows the apparatus for HI permeation. HI was supplied at the rate of 100 mL min^{-1} from the outside of the membrane, and N_2 was introduced at the rate of 100 mL min^{-1} as the sweep gas from the inside of the membrane. The permeation temperature was room temperature and 400 °C. The permeated HI gas was trapped with 100 mL of pure water in a Teflon bottle. The HI permeation rate was determined by neutralization titration of the trapped HI aqueous solution with 0.1 N NaOH aqueous solution. (COM-1700, Hiranuma Sangyo).

Figure 14.4 shows the pore sizes evaluation of the TMOS, ETMOS, PrTMOS, HTMOS, PhTMOS, and DPhDMOS-derived membranes deposited at 270 °C. The vertical axis shows the pore size calculated from the organic groups for the silica precursors based on the following two assumptions. The pore size of silica network (–Si–O–Si–) is assumed at 0.3 nm. The size of the organic group is calculated from the MM2 potential in a spherical shape. The silica-based membranes deposited from the silica precursors with alkyl groups (ETMOS, PrTMOS, HTMOS) showed larger pore sizes by increasing the number of carbons of the alkyl groups. The calculated pore size of the DPhDMOS (two organic groups in the precursor) was much larger than that of the PhTMOS (one organic group in the precursor). The pore sizes of the deposited membranes can be roughly predicted by the size of the organic groups in the silica precursors. However, the difference in the measured pore sizes was smaller than those calculated pore sizes for all membranes deposited from the silica precursors with alkyl groups (ETMOS, PrTMOS, HTMOS). The pore size of the HTMOS-derived membrane and that of the DPhDMOS-derived membrane showed difference of 0.4–1 nm from the value of the calculated pore sizes. The further investigation is required for the detailed discussions. Figure 14.5 shows the H_2 permeances plotted as a function of the measured pore sizes. The H_2 permeances increased by using the silica precursors with organic groups from that without the organic group (TMOS). The H_2 permeance through the membranes deposited from the silica precursors with

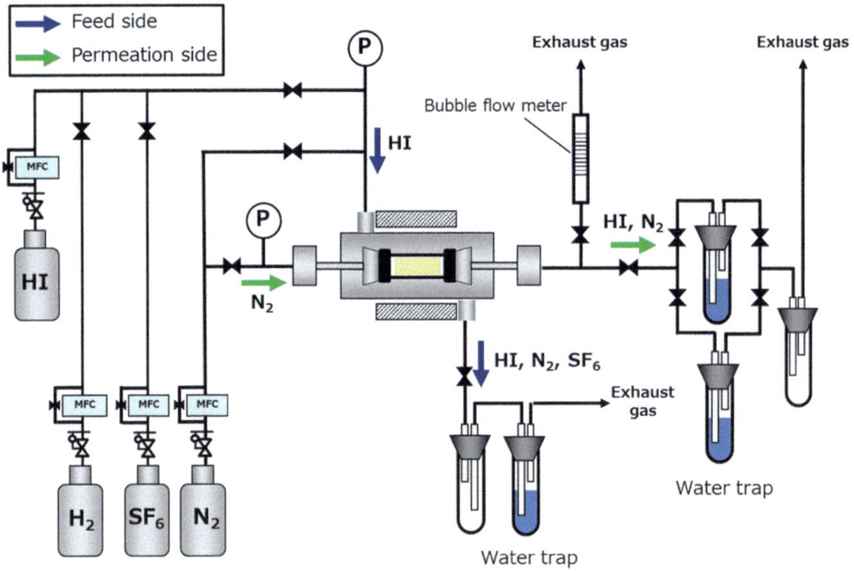

Fig. 14.3 Schematic diagram of apparatus of HI permeation test

a phenyl group is slightly higher than those from an alkyl group. However, the pore sizes for the PhTMOS or DPhDMOS-derived membranes are too large to separate HI from H_2.

Fig. 14.4 Pore sizes evaluation for membranes deposited at 270 °C

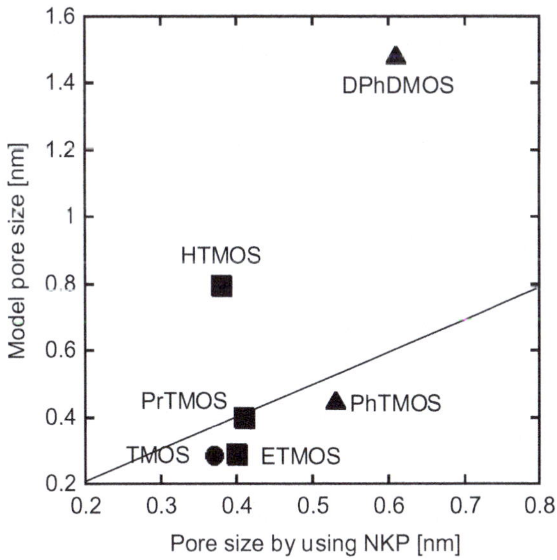

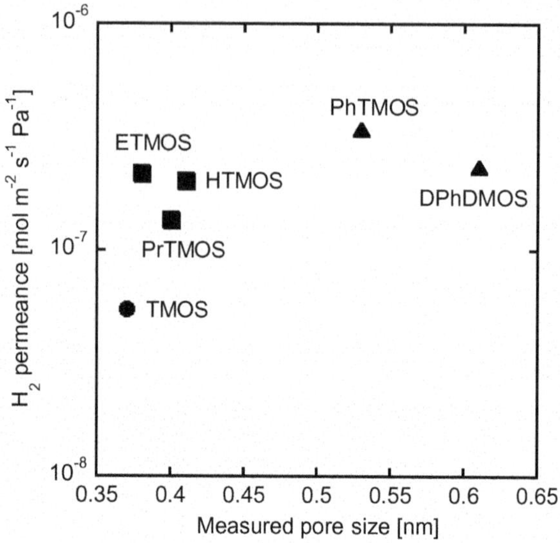

Fig. 14.5 H_2 permeances for membranes deposited at 270 °C

Figure 14.6 shows the measured pore sizes by changing the deposition temperatures for the HTMOS-derived membranes. The measured pore sizes can be classified into three patterns depending on the deposited temperature. The deposition temperature can be classified by the decomposition of organic groups and O_3. The O_3 concentration during the deposition is higher at the deposition temperature lower than 200 °C. Thus, all the deposited hexyl groups had decomposed during the deposition. As a result, the measured pore size was 0.76 nm which is the similar to that for the PrTMOS-derived membrane. On the other hand, all the deposited hexyl groups had thermally decomposed at the temperature higher than 500 °C. The measured pore size deposited at 500 °C was 0.76 nm which is the same to that deposited at 180 °C. The measured pore sizes were the similar at 0.47 nm for the deposition temperature between 270 and 450 °C.

The HTMOS-derived membrane was applied for H_2/HI separation tests at 400 °C. The deposition temperature was at 450 or 480 °C for 5 min. Figure 14.7 shows the HI permeance and the H_2/HI permeance ratio through the HTMOS-derived membrane. The HTMOS-derived membrane deposited at 450 °C showed the H_2 permeance of 5.0×10^{-7} mol m^{-2} s^{-1} Pa^{-1} with the H_2/HI permeance ratio of 6820. The H_2 and HI permeance at room temperature was 5.0×10^{-8} mol m^{-2} s^{-1} Pa^{-1} and 5.2×10^{-11} mol m^{-2} s^{-1} Pa^{-1}, respectively. Both permeances were lower than those measured at 400 °C. Figure 14.8 shows the relationship between the HI permeances and the SF_6 permeances. A positive correlation was obtained between the HI and SF_6 permeances, which indicates that SF_6 can be an index of HI permeation.

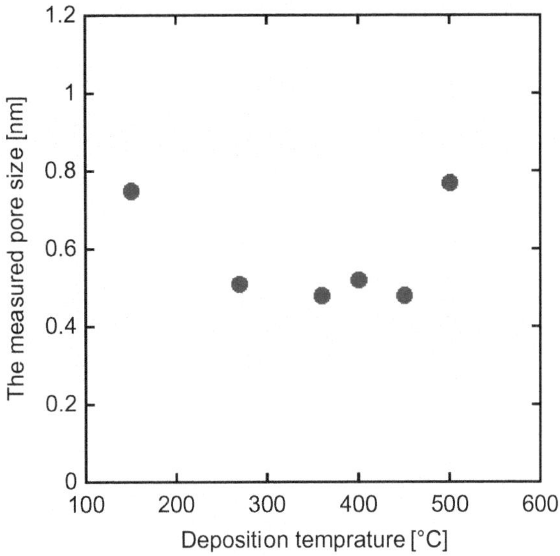

Fig. 14.6 Measured pore sizes by changing deposition temperatures for HTMOS-derived membranes

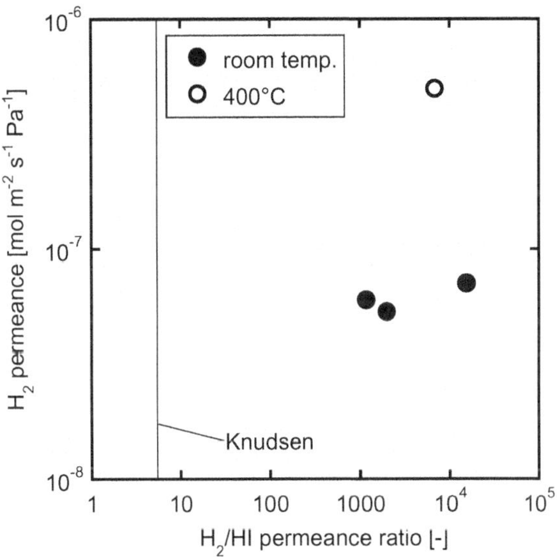

Fig. 14.7 HI permeance and H_2/HI permeance ratio through HTMOS-derived membrane

14.3 Development of Hydrogen Iodine Decomposition Catalyst

Active carbon has been the typical support due to its high surface area and low cost [21–23]. Other carbon materials such as carbon nanotube (CNT) and reduced

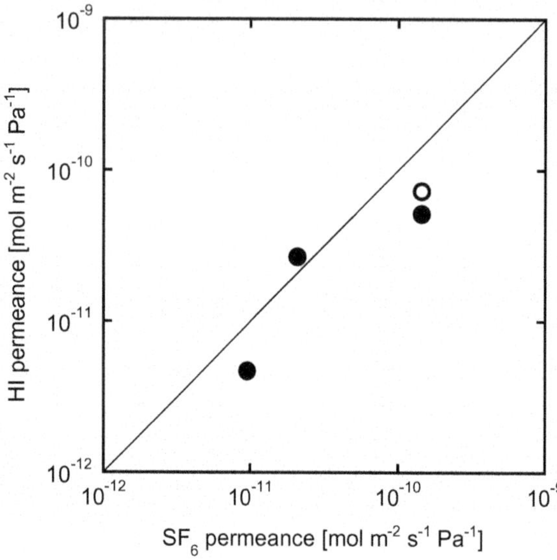

Fig. 14.8 The relationship between HI and SF_6 permeances

graphene oxide (rGO) are also studied. In CNT support, high Pt dispersion and microstructure of support are the reasons for the high HI conversion [24]. On the other hand, rGO support, which is active form of graphene oxide by increasing of active site, is attributed to the high dispersion and uniformity of Pt particles [25]. Furthermore, metal oxide such as CeO_2 and ZrO_2 are applied for supporting material [26, 27]. Pt dispersion on ceria shows strong metal-support interaction (SMSI) phenomena and increases O_2 vacancies and O_2 mobility, and electron transfer between Pt and Ce is also suggested [28]. Zr is doped to ceria lattice by sol–gel method in order to improve the thermal stability and reducibility [8]. Considering the cost for industrial development, Ni is one of an attractive choices for Pt replacement. However, during HI decomposition, Ni metal is changed to NiI_2 which resulted in the deactivation of Ni supporting catalyst [29].

Regarding the aggregation and increase in size of Pt particles, fortunately, supporting Pt–M bimetal catalyst can solve this problem. Among them, Pt–Ir supporting catalyst shows the highest catalytic activity, anti-sintering, and a high metal dispersion [30, 31]. In addition, Ni, Pd, or Rh added Pt bimetallic alloy are also studied [32, 33]. Although activity is lower than that of Pt–Ir alloy, the thermal stability, small particle size and well dispersion of metal particles are sustained on such Ni, Pd, or Rh added Pt.

CeO_2 is an attractive for the catalyst to HI decomposition because of inexpensive, abundant, nontoxic, act as both catalyst and support, and good physiochemical properties such as O_2 storage capacities and particular interaction with noble metal. In order to obtain more O_2 diffusivity, other transition metals are doped to ceria lattices such as Pd [34], Pt [35], and Zr [26]. Particularly, Pt doping ceria or solid solution is effective for increasing HI conversion. Therefore, electron can be transferred from

$Ce^{4+} + Pt^0$ to $Ce^{3+} + Pt^{2+}$ which improves the catalytic activity [28]. Moreover, other less expensive transition metals such as Co, Fe, and Ni are studied for the catalyst by sol–gel method [36]. Ni–CeO_2 had highest activity because of larger O_2 vacancies in Ce–Ni solid solution. In spite of efforts, HI decomposition activity on the reported catalyst is still low under a high space velocity (SV) and the stability is insufficient because of the deactivation by I_2 adsorption and reaction.

Advantage of metal oxide-supported Pt is high stability, however, the activity is still not high enough. In our previous study, we found that Pt/Ce(Cu)O_2/activated carbon is highly active to HI decomposition (almost the equilibrium conversion) as well as high stability over 100 h [37]. However, from the equilibrium, CeO_2 has reactivity with I_2 resulting in a decreased HI conversion. Therefore, the effects of metal oxide loading on an activated carbon from the activity and the stability are investigated for replacing CeO_2 [38].

Table 14.1 shows HI conversion on Pt/MO_x/M563 ($MO_x = CeO_2$, La_2O_3, Pr_2O_3, Nd_2O_3, Sm_2O_3, CuO, TiO_2, V_2O_5, Y_2O_3, MoO_3 and ZrO_2 at 10 wt%) catalyst at 300, 400 and 500 °C. Here, M563 is activated carbon (Shirasagi Co. Ltd, Japan, BET surface area of 1351 m^2/g) and obviously, activity of M563 catalyst to HI decomposition was strongly affected by the loaded oxide and in case of CuO, CeO_2, TiO_2, and La_2O_3, the high HI conversion was achieved and so these oxides are effective for increasing the HI decomposition activity. HI decomposition is strongly limited by the chemical equilibrium and it is found that CuO, CeO_2, and La_2O_3 loaded catalyst shows little higher HI conversion than that of equilibrium one. This could be explained by trapping of I_2 on the catalyst resulting in the shift of chemical equilibrium to the product site. On the other hand, TiO_2 loaded catalyst shows almost the same with the equilibrium conversion in all temperature ranges and so, Pt/TiO_2/M563 shows the reasonably high activity to HI decomposition and stability against formed I_2.

Figure 14.9 shows the temperature dependence of HI conversion of Rh and Ir added Pt/TiO_2/M563 catalyst (0.9 wt% Pt and 0.1 wt% Rh or Ir). It is reported that the addition of Ir [30] or Rh [34] is effective for increasing HI conversion, in particular, Ir addition showed significant positive effects on stability and so we studied the effects of Ir or Rh as additives. Since the HI conversion was close to the equilibrium conversion, the positive effects on HI conversion were not significantly

Table 14.1 HI conversion on Pt–MO_x/M563 added with metal oxide, where M = lanthanide (Ce, La, Pr, Nd, Sm) and transition metal (Cu, Ti, V, Y, Mo, Zr) [38]

Metal oxide	300 °C	400 °C	500 °C	Metal oxide	300 °C	400 °C	500 °C
Equilibrium	17.89	20.82	23.48	Sm	14.89	19.50	21.92
Cu	18.12	24.08	27.16	Y	14.00	19.33	21.85
Ce	16.67	23.63	26.33	Nd	12.91	18.70	21.69
Pr	16.26	23.33	25.37	V	12.55	17.90	20.06
Ti	16.01	20.25	22.92	Mo	11.37	17.57	20.03
La	15.61	20.06	22.08	Zr	9.81	17.01	19.32

observed as shown in Fig. 14.9. HI conversion at 480 °C was increased from 19.5 to 19.8% by Rh addition, however decreased to 18.2% by addition of Ir. Therefore, it was found that Rh addition shows positive effects on HI decomposition activity. In case of single Pt, TEM observation suggests that the average particle size was increased from 1.5 to 9.0 nm because the aggregation was significantly occurred by reaction with I_2, and further larger Pt size like 50 nm level was also observed. However, in case of Pt–Rh, large part of Pt particles was still sustained a smaller size than 4 nm and so, obviously, addition of Rh is highly effective for preventing Pt aggregation. This might be explained by low reactivity of Rh with I_2 and so, alloying of Pt with Rh may prevent the formation of metal I_2 compounds.

Further, longer stability in HI decomposition was measured with Pt–Rh/TiO_2/M563. Figure 14.10 shows the HI conversion at 400 °C as a function of reaction period up to 500 h. In the previous report, long-term stability of the activity like few 100 h is not reported, however, as shown in Fig. 14.10, Pt–Rh/TiO_2/M563 catalyst shows almost the equilibrium conversion, or slightly higher at initial period. This might be explained by the chemical shift to a product side by the adsorption of formed I_2 on the activated carbon. Although HI conversion decreased with reaction time, almost equilibrium HI conversion was sustained up to 200 h and slightly lower (ca.95%) than equilibrium conversion. Although HI conversion decreased with time on stream at an initial period, the activity was kept almost the constant after 200 h. This deactivation could be explained by aggregation of Pt and the change in TiO_2 crystal structure at initial period and also I_2 adsorption in activated carbon [38]. Therefore, Pt–Rh/TiO_2/activated carbon, M563, is highly active and stable to HI decomposition. Although the reaction condition is different, it is seen that the Pt–Rh/TiO_2/C catalyst optimized shows the higher HI conversion in wide temperature range compared with the previously reported catalysts.

HI decomposition on Pt–Rh/TiO_2/M563 catalyst was measured under coexistence of H_2O as shown in Fig. 14.11. HI conversion was decreased by coexistence of water

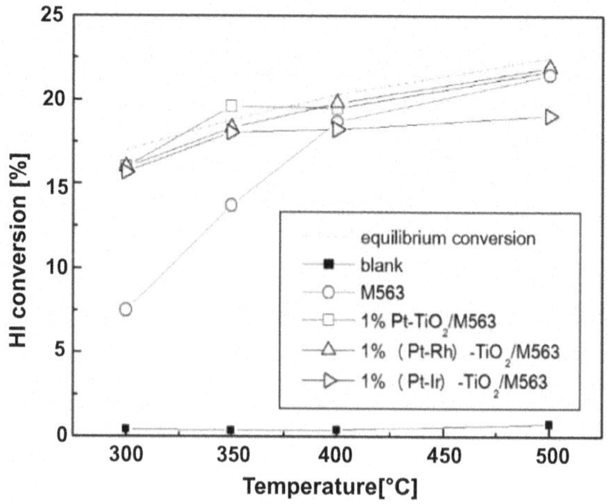

Fig. 14.9 Temperature dependence of HI conversion on Pt–TiO_2/M563 added with 0.1 wt% Rh or Ir and equilibrium one [38]

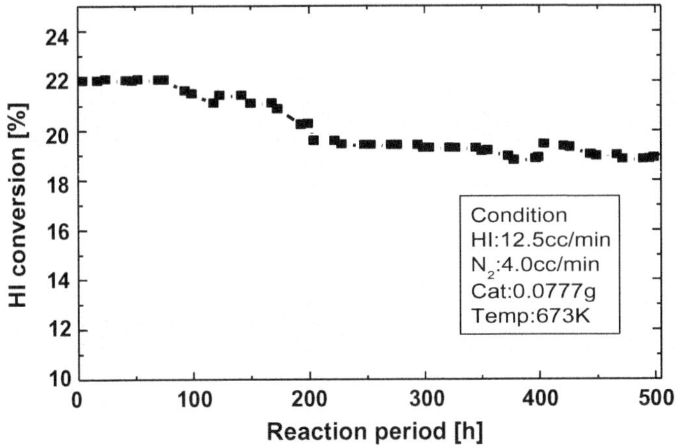

Fig. 14.10 HI conversion at 400 °C as a function of reaction period up to 500 h

at all temperatures and so the coexistence of H_2O shows the negative effects on HI decomposition activity, which is similar with the previous study, i.e., HI conversion at 500 °C was decreased from 22 to 19%. Although the reaction condition was different, Wang et al. reported that HI conversion of 17.5% was achieved on Pt-Ir/C catalyst under wet conditions [39]. Therefore, HI conversion on Pt–Rh/TiO_2/M563 shows high activity even under wet conditions, although H_2O partial pressure in HI decomposition reactor is low in the overall SI process by separation after Bunsen reaction. In summary, it is found that Pt–Rh/TiO_2/M563 is highly active and stable to HI decomposition. HI conversion on this catalyst is almost the same with that of equilibrium conversion in the temperature range from 300 to 500 °C and stably sustained more than 500 h.

14.4 Hydrogen Iodine Decomposition Reaction Assisted with Membrane

A porous alumina tube (o.d. 10 mm, i.d. 7 mm, and length 95 mm), supplied by Noritake Co., Ltd., were used as support. The effective length of membrane was 30 mm at the center of each support. Lengths of 32.5 mm at both ends of tube were glazed and calcined at 1150 °C for 1 h except for the permeating portions of 30 mm. The support tube had a three-layer structure: an α-alumina support layer (pore size of 0.82 mm and porosity of 47%), the second α-alumina layer (pore size of 0.16 mm and porosity of approximately 34%), and a γ-alumina layer. HTMOS vapor and O_2 gas were fed, respectively through the outside and inside of the tube. HTMOS was deposited by CVD method on γ-alumina-coated α-alumina tube at a selected deposition temperature of 450 °C, as determined in our previous study [12].

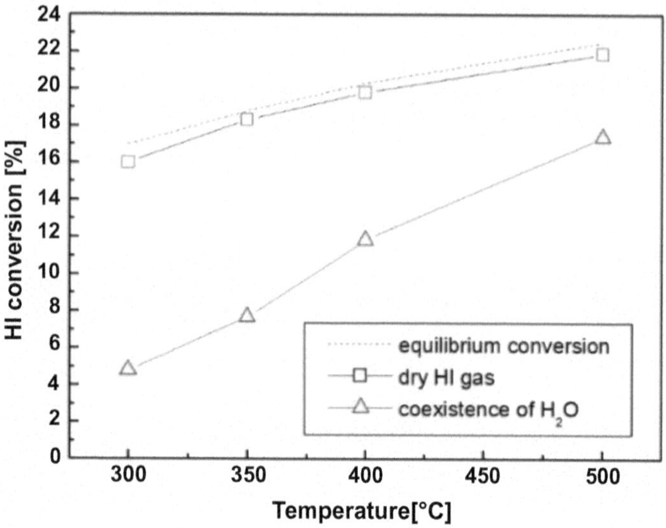

Fig. 14.11 Temperature dependence of HI conversion of Pt–Rh–TiO_2/ M563 with and without 25%vol of H_2O co-feed

Figure 14.12 shows the temperature dependence of the permeation characteristics of the support tube and HTMOS-derived membrane. In consideration of operability in the experiment, SF_6 which has approximately the same kinetic diameter as HI was used instead of HI. For the support tube, the permeance was high for all three gases and slightly decreased with an increase in temperature. The selectivity varied between 2.90 and 3.00 at temperatures in the range of 100–450 °C for H_2/N_2 and between 4.40 and 6.00 at 100–250 °C for H_2/SF_6, which were slightly lower than the theoretical values for Knudsen diffusion of $H_2/N_2 = 3.50$ and $H_2/SF_6 = 6.60$. It was assumed that the gas permeation mechanism for the support tube was Knudsen diffusion for all three gases (H_2, N_2, and SF_6). For the membrane, the permeance was lower and the permeation behavior depended on the kinetic diameter of the gas molecules (H_2: 0.28 nm, N_2: 0.36 nm, and SF_6: 0.55 nm). The H_2 permeance increased with temperature. The highest H_2 permeance of 8.2×10^{-7} mol Pa^{-1} m^{-2} s^{-1}, which was recorded at 450 °C. The H_2/N_2 and H_2/SF_6 selectivity also increased with temperature and reached maximum values of approximately 581 (H_2/N_2) and 1680 (H_2/SF_6), respectively, at 250 °C [13].

The selectivity result suggests the presence of many small pores and a few large pores on the membrane; H_2 can permeate through small pores, whereas N_2 and SF_6 cannot, and all three gases can permeate through pores. Accordingly, two types of gas diffusion may occur concurrently in the membrane: activated diffusion through micropores and Knudsen diffusion through large pores [40]. It seems that H_2 permeation through the membrane was activated, on the basis of its activation energy of 9.1 kJ mol^{-1} at 30–450 °C. The N_2 permeance was smaller because its kinetic diameter is larger, preventing it from permeating most of the available small pores. It

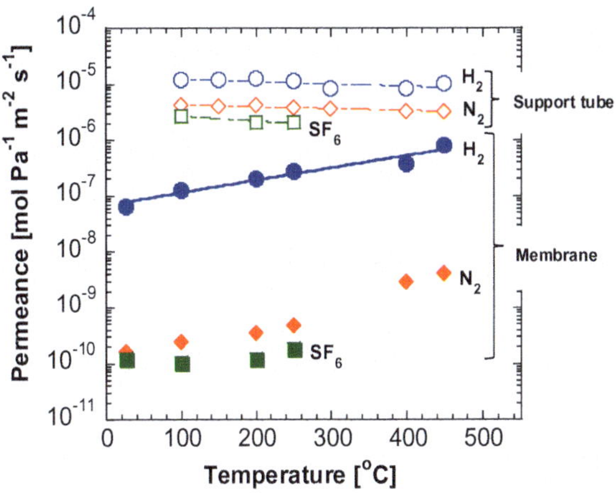

Fig. 14.12 Temperature dependence of gas permeance of H_2, N_2, and SF_6 through support tube and membrane at 30–450 °C [13]

seems that Knudsen diffusion through a few large pores partially contributed to the N_2 permeance; however, in comparison to the overall permeance, the Knudsen diffusion is insignificant at higher temperatures. For SF_6, its kinetic diameter is also larger. Hence, the existence of a few large pores offered for Knudsen diffusion contributed to the SF_6 permeance, which was therefore almost unchanged with temperature. These permeation results indicate that the silica layer in the membrane was successfully formed on the support tube.

Figure 14.13 shows a schematic diagram of the membrane reactor for HI decomposition developed for this study. The silica membrane was set in a membrane reactor made of stainless steel (SUS316) with an inner diameter of 20 mm. An active carbon catalyst was loaded outside the membrane, as shown in Fig. 14.13b. HI decomposition proceeded in countercurrent flow, where the N_2 sweep gas and HI gas feed were introduced from opposite sides of the membrane. The flow rate of HI gas was 9.7 mL min^{-1}. The permeate side of the membrane was swept by N_2 at 100 mL min^{-1}. The total pressures on both sides (permeate side and retentate side) of the membrane were the same as atmospheric pressure. When no sweep gas was applied to the permeate side of the membrane reactor, the equilibrium conversion of HI decomposition (conversion without extraction) was measured. While, the conversion with extraction was measured when N_2 was fed into the permeate side as sweep gas. The HI decomposition was carried out in the reactor continuously for 480 min at 400 °C. The H_2 concentrations on both sides were measured by a gas chromatograph (GC). The retentate (HI gas) and product (I_2) were caught in water traps before being introduced into the GC. To further investigate the effects of the HI flow rate on the conversion rate, the flow rate of HI gas was varied between 2.6 and 9.7 mL min^{-1}.

Figure 14.14 shows the conversion of HI decomposition. When no sweep gas is supplied to the permeate side of the membrane reactor, the conversion indicated by the symbol of the white circle were about 0.22 because the decomposition reaction is not accelerated, the conversion shows a reasonable agreement with the calculated equilibrium conversion in HI decomposition (0.20 at 400 °C). When the sweep gas

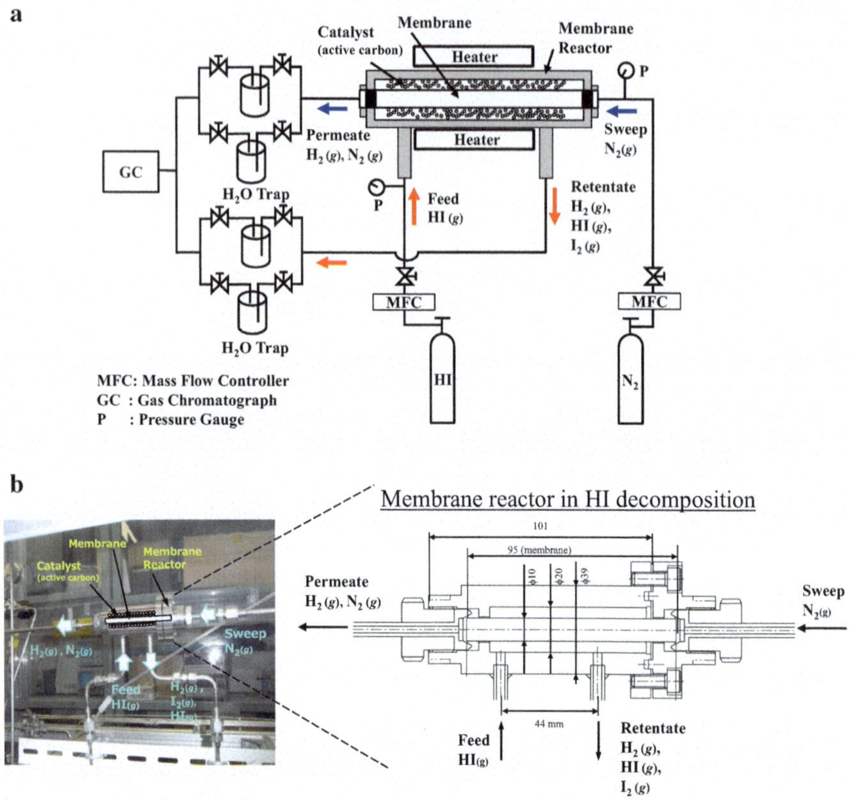

Fig. 14.13 **a** Schematic diagram of the experimental setup used for HI decomposition in countercurrent flow. **b** Membrane reactor used for HI decomposition in the laboratory [13]

was supplied to the permeate side, the conversion indicated by the symbol of the white circle were almost stable at about 0.48, which was larger than the equilibrium conversion (0.22). The conversion increased as the HI flow rate decreased, and the maximum conversion of about 0.60 was achieved at the HI flow rate of 2.6 mL min^{-1}. From these results, it was experimentally proved that the HI decomposition was accelerated by extraction of H_2 through the membrane [13].

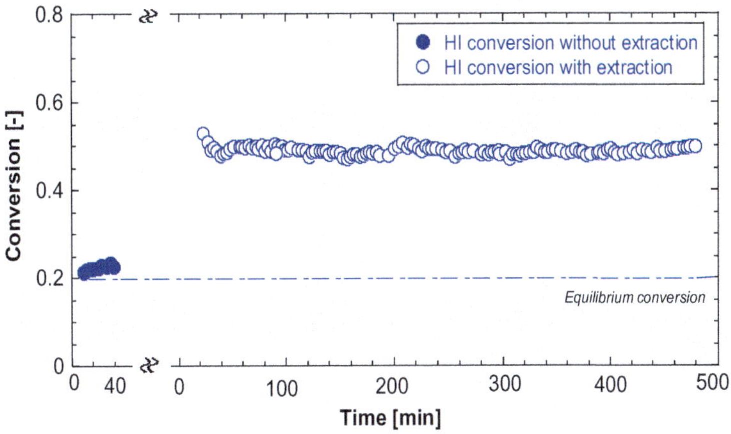

Fig. 14.14 Conversion rate of HI decomposition at 400 °C [13]

References

1. Kasahara S (2003) Effects of process parameters of the IS process on total thermal efficiency to produce hydrogen from water. J Chem Eng Jpn 36:887–899
2. Uemiya S (1999) State-of-the-art of supported metal membranes for gas separation. Sep Purif Methods 28:51–85
3. Williams GA, Ferguson JB (1922) The diffusion of hydrogen and helium through silica glass and other glasses. J Am Chem Soc 44:2160–2167
4. Kitao S, Asaeda M (1990) Separation of organic acid/water mixtures by thin porous silica membrane. J Chem Eng Jpn 23:367–370
5. Gavalas GR (1989) Deposition of H_2-permselective SiO_2 films. Chem Eng Sci 44:1829–1835
6. Sea BK (1996) Formation of hydrogen permselective silica membrane for elevated temperature hydrogen recovery from a mixture containing steam. Gas Sep Purif 10:187–195
7. Lee D (2004) Synthesis, characterization, and gas permeation properties of a hydrogen permeable silica membrane supported on porous alumina. J Membr Sci 231:117–126
8. Nakao S (2000) Preparation of microporous membranes by TEOS/O_3 CVD in the opposing reactants geometry. Micropor Mesopor Mater 37:145–152
9. O'keefe DR, Norman JH, Williamson DG (1980) Catalysis research in thermochemical water-splitting processes. Catal Rev: Sci Eng 22(3):325–369
10. Nomura M, Kasahara S, Nakao S (2004) Silica membrane reactor for the thermochemical iodine-sulfur process to produce hydrogen. Ind Eng Chem Res 43:5874–5879
11. Hwang GJ, Onuki K (2001) Simulation study on the catalytic decomposition of hydrogen iodide in a membrane reactor with a silica membrane for the thermochemical water splitting IS process. J Membr Sci 194:207–215
12. Myagmarjav O (2017) Preparation of an H_2-permselective silica membrane for the separation of H2 from the hydrogen iodide decomposition reaction in the iodine-sulfur process. Int J Hydrogen Energy 42:6012–6023
13. Myagmarjav O (2017) Hydrogen production tests by hydrogen iodide decomposition membrane reactor equipped with silica-based ceramics membrane. Int J Hydrogen Energy 42:29091–29100
14. Hwang G (2000) Separation of hydrogen from a H_2–H_2O–HI gaseous mixture using a silica membrane. AIChE J 46:92–98

15. Nomura M (2004) Silica membrane reactor for the thermochemical Iodine-Sulfur process to produce hydrogen. Ind Eng Chem Res 43:5874–5879
16. Nomura M (2005) Preparation of a stable silica membrane by a counter diffusion chemical vapor deposition method. J Membr Sci 251:151–158
17. Matsuyama E (2014) High temperature propylene/propane separation through silica hybrid membranes. Sep Purif Tech 128:25–30
18. Ikeda A (2015) High hydrogen permeance silica membranes prepared by a chemical vapor deposition method. J Membr Sep Tech 4:66–73
19. Nomura M (2007) Pore size control of a molecular sieve silica membrane prepared by a counter diffusion CVD method. J Chem Eng Jpn 40:1235–1241
20. Lee HR (2010) Evaluation and fabrication of pore-size-tuned silica membranes with tetraethoxydimethyl disiloxane for gas separation. AIChE J 57:2755–2765
21. Kim JM (2008) Decomposition of hydrogen iodide on Pt/C-based catalysts for hydrogen production. Int J Hydrogen Energy 33:4974–4980
22. Tyagi D, Varma S, Bharadwaj SR (2015) Pt/graphite catalyst for hydrogen generation by HI decomposition reaction in S–I thermochemical cycle. Int J Energy Res 39:2008–2018
23. Tyagi D (2012) Development of Pt-Carbon catalysts using MCM-41 template for HI decomposition reaction in S–I thermochemical cycle. Int J Hydrogen Energy 37:3602–3611
24. Wang L (2013) Comparisons of Pt catalysts supported on active carbon, carbon molecular sieve, carbon nanotubes and graphite for HI decomposition at different temperature. Int J Hydrogen Energy 38:109–116
25. Bo Z (2014) Metal-free and Pt-decorated graphene-based catalysts for hydrogen production in a sulfur−iodine thermochemical cycle. Ind Eng Chem Res 53:11920–11928
26. Tyagi D, Varma S, Bharadwaj SR (2015) Pt/zirconia catalyst for hydrogen generation from HI decomposition reaction of S–I cycle. Int J Energy Res 39:484–493
27. Chen Y (2010) Platinum–ceria–zirconia catalysts for hydrogen production in sulfur-iodine cycle. Int J Hydrogen Energy 35:445–451
28. Zhan Y (2008) Influence of the oxidative/reductive treatments on Pt/CeO_2 catalyst for hydrogen iodide decomposition in sulfur–iodine cycle. Int J Hydrogen Energy 33:2211–2217
29. Favuzza P (2009) Decomposition of hydrogen iodide in the S–I thermochemical cycle over Ni catalyst systems. Int J Hydrogen Energy 34:4049–4056
30. Hu S (2016) Activity and stability of monometallic and bimetallic catalysts for high-temperature catalytic HI decomposition in the iodine-sulfur hydrogen production cycle. Int J Hydrogen Energy 41:773–783
31. Wang Z (2010) Decomposition of hydrogen iodide over Pt–Ir/C bimetallic catalyst. Int J Hydrogen Energy 35:8862–8867
32. Li D (2013) Effects of the composition on the active carbon supported Pd–Pt bimetallic catalysts for HI decomposition in the Iodine-sulfur cycle. Int J Hydrogen Energy 38:6586–6592
33. Wang L (2014) Effects of the second metals on the active carbon supported Pt catalysts for HI decomposition in the iodine-sulfur cycle. Int J Hydrogen Energy 39:14161–14165
34. Singhania A (2018) Hydrogen-iodide decomposition over Pd–CeO_2 nanocatalyst for hydrogen production in sulfur iodine thermochemical cycle. Int J Hydrogen Energy 43:3886–3891
35. Zhang Y (2008) Effect of preparation method on platinum–ceria catalysts for hydrogen iodide decomposition in sulfur–iodine cycle. Int J Hydrogen Energy 33:602–607
36. Singhania A (2018) Catalytic decomposition of hydrogen-iodide over nanocrystalline ceria promoted by transition metal oxides for hydrogen production in sulfur–iodine thermo-chemical cycle. Catal Lett 148:1416–1422
37. Punkrawee W, Yamanaka A, Matsuda J, Mitoma Y, Nishiyama N, Ishihara T (2018) Pt-$Ce_{0.9}Cu_{0.1}O_2$/activated carbon as highly active and stable HI decomposition catalyst. Int J Energy Res 42(3):1088–1097
38. Punkrawee W, Yamanaka A, Matsuda J, Mitoma Y, Nishiyama N, Ishihara T (2018) Pt-Rh/TiO_2/activated carbon as highly active and stable HI decomposition catalyst for hydrogen production in sulfur-iodine (SI) process. Int J Energy Res 42(7):2494–2506

39. Wang Z, Wang L, Chen S, Zhang P, Xu J, Chen J (2010) Decomposition of hydrogen iodide over Pt–Ir/C bimetallic catalyst. Int J Hydrogen Energy 35:8862–8867
40. Okubo T, Inoue H (1989) Single gas permeation through porous glass modified with tetraethoxysilane. AIChE J 35:845–848

Chapter 15
Corrosion-Resistant Material for IS Process

Ikuo Ioka, Jin Iwatsuki, Makoto Inomata, and Yoshiro Kuriki

Abstract The hybrid material that exhibited both corrosion resistance and ductility was created using silicon powder plasma spraying and laser treatment for IS process. The round-bar specimens created using the hybrid technique exhibited excellent corrosion resistance in 47 and 95 mass% boiling H_2SO_4. No detachment of the plasma-sprayed and laser-treated layer was observed on the base metal and the welded part. The container was successfully obtained using the hybrid technique.

Keywords Acid corrosion resistance · IS process · Silicon powder · Plasma spraying · Laser treatment

15.1 Introduction

The IS process is composed of three processes: the Bunsen reaction, thermal decomposition of HI, and thermal decomposition of H_2SO_4. Sequentially, carrying out these three processes, as a net balance of the process, enables the decomposition of water into H_2 and O_2. H_2SO_4 from the Bunsen reaction is purified and concentrated and is then finally decomposed into O_2 and SO_2. Severe thermal and highly oxidizing environment for the decomposition of H_2SO_4 presents significant engineering challenges. Considering that the IS process is conducted under extremely severe environments of iodic and sulfuric acids at a wide temperature range, which is from room temperature to 650 °C, material corrosion must be considered as an important issue to achieve a successful IS process. Therefore, screening of candidate industrial materials from the viewpoint of corrosion resistance against the environments for H_2SO_4 decomposition has been conducted. Evaluation of materials for the H_2SO_4 decomposition using immersion tests has revealed that high silicon alloys, silicon carbides, and silicon nitrides exhibit excellent corrosion resistance in 98 mass% boiling H_2SO_4 [1]. These

I. Ioka (✉) · J. Iwatsuki
Japan Atomic Energy Agency, Tokai, Oarai, Ibaraki 311-1393, Japan
e-mail: ioka.ikuo@jaea.go.jp

M. Inomata · Y. Kuriki
JGC Corporation, Minatomirai, Nishi-ku, Yokohama 220-6001, Japan

K. Aika and H. Kobayashi (eds.), *CO_2 Free Ammonia as an Energy Carrier*,
https://doi.org/10.1007/978-981-19-4767-4_15

materials are brittle; thus, using them independently as structural materials is difficult. Therefore, a hybrid material that exhibits corrosion resistance and ductility was created via plasma spraying and laser treatment. In addition, a container with a welded part, chamfer, and curved surface was experimentally created via plasma spraying and laser treatment to confirm the production characteristics of the component of the IS process using the hybrid technique.

This chapter describes the outline of the manufacturing process of the hybrid material, corrosion properties of the hybrid specimens, and production of the container using the hybrid technique.

15.2 Production Technique of the Hybrid Material

The required properties of the protective layer in the hybrid material are high bonding strength between the protective layer and the substrate, Si composition of 20 mass% or more [1], and the ability to resist penetration of crack to the substrate.

Hastelloy C276 superalloy (57Ni-16Cr-16Mo-6Fe-4W, mass%) as a substrate was utilized in this study owing to its corrosion resistance, which is known in a wide range of aggressive media. The high content of molybdenum imparts resistance to localized corrosion, such as pitting. The low carbon minimizes carbide precipitation during welding to maintain resistance against intergranular attack in the heat-affected zones of the welded joints. In chemical processing, alloy is used for components, such as heat exchangers, reaction vessels, evaporators, and transfer piping.

The hybrid material was created via silicon powder plasma spraying and laser treatment [2]. The manufacturing method of the hybrid specimen is presented in Fig. 15.1. Before the plasma spraying and laser treatment, the surface of the substrate was sandblasted using alumina grit, with a size ranging from 125 to 250 μm. Argon was utilized as a carrier gas. Si powder with a size of 10 μm was deposited onto the surface via air plasma spraying. The Si layer had a thickness of ~100 μm and physically adhered to the substrate well. Numerous pores were observed in the Si layer. The deposited Si layer was treated using a pulsed Yttrium–Aluminum–Garnet (YAG) laser (ϕ200 μm) to improve the performance of the plasma-sprayed layer, such as its adhesion to the substrate, the reduction of pores in the layer, and the modification of the chemical composition of the plasma-sprayed layer. The hybrid material was created by repeating the abovementioned process twice.

Figure 15.2 presents the cross-sectional observation of the hybrid specimen in the vicinity of the plasma-sprayed and laser-treated layer. Numerous cracks were observed in the surface layers but no Si particles. It seems that these cracks were caused by the solidification shrinkage of the laser-treated layer. It is believed that if the cracks do not penetrate the plasma-sprayed and laser-treated layer, the layer is considered effective in terms of specimen corrosion resistance. All elements were uniformly distributed in each layer. The chemical compositions were 53 mass% (Si), 24 mass% (Ni), 7 mass% (Cr), 8 mass% (Mo), 3 mass% (Fe), 4 mass% (W) for the upper surface layer and 27 mass% (Si), 34 mass% (Ni), 16 mass% (Cr), 14 mass%

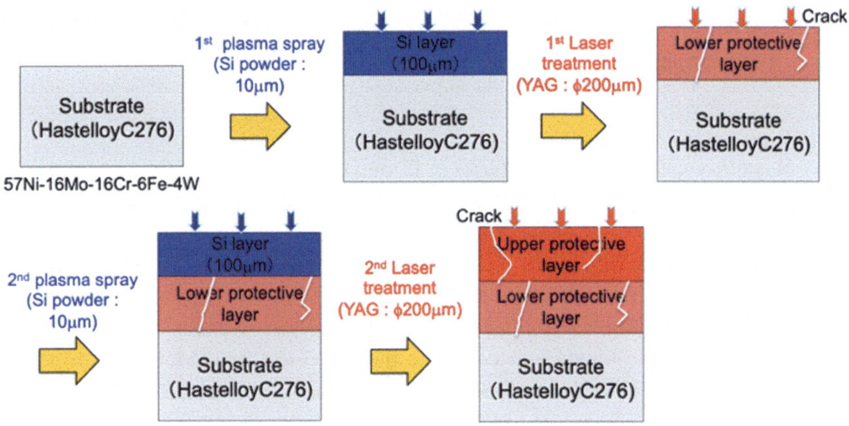

Fig. 15.1 Manufacturing method of the hybrid specimen

(Mo), 6 mass% (Fe), and 3 mass% (W) for the lower surface layer. The Si contents of the upper surface layer are higher than those of the lower surface layer. However, both layers contain sufficient Si to prevent corrosion of boiling H_2SO_4.

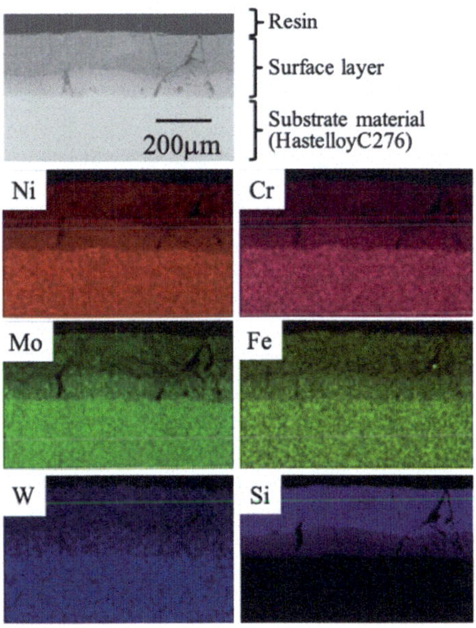

Fig. 15.2 Cross section of the hybrid specimen in vicinity of the surface layers

15.3 Corrosion Properties of Specimens Using the Hybrid Technique

1. Flat specimen

   Hastelloy C276 superalloy (ϕ70 × 6^t mm) was utilized as the substrate for the flat specimen. The hybrid processing was given to the one side of the flat specimen. Cracks were observed in the upper and lower protective layers that have Si contents of 49 and 19 mass%, respectively.

   Figure 15.3 presents a schematic diagram of the corrosion test equipment for the flat specimen. The weights before and after the corrosion test were measured using an electronic balance to calculate the corrosion rate. Moreover, to thoroughly examine the surface of the hybrid specimen, the surface roughness was measured using a 3D measuring macroscope.

   Corrosion tests were conducted in 47 mass% boiling H_2SO_4 at 124 °C for 24 h, and the results are presented in Fig. 15.4. The corrosion rate that was calculated from the weight loss was ~0.5 mm/y. The hybrid specimen's corrosion rate was found to be quite lower than that of the substrate. Moreover, a slight change in the appearance of the specimen was observed after corrosion. The average surface roughness (Ra) of both the corrosion and non-corrosion areas was 3 μm. The surface roughness after the corrosion test hardly changed. The surface of the hybrid specimen exhibited sufficient corrosion resistance in 47 mass% boiling H_2SO_4, although several cracks were found in the protective layer from the cross-sectional observation [3].

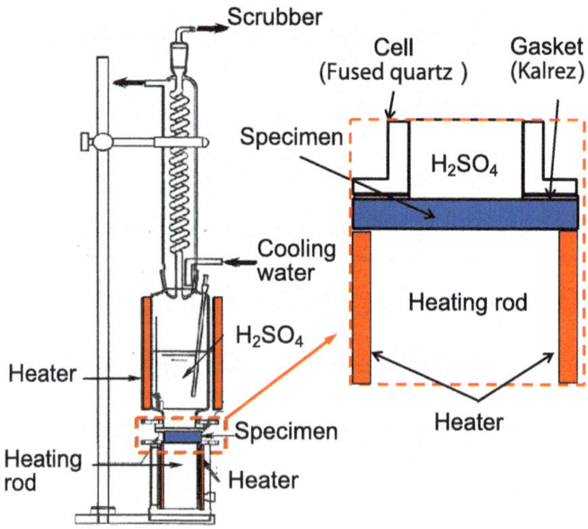

Fig. 15.3 Schematic diagram of the corrosion test equipment the flat specimen

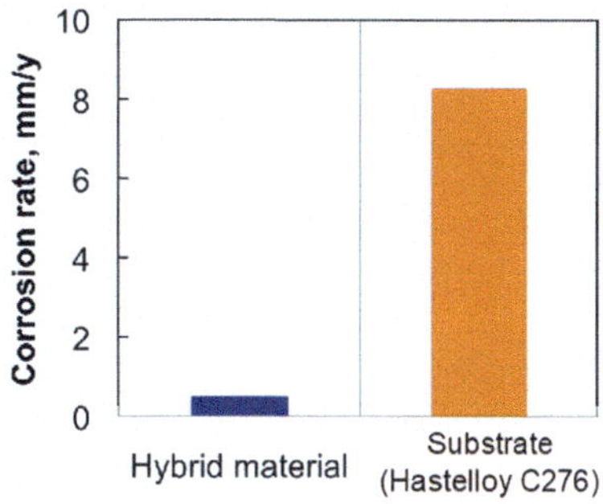

Fig. 15.4 Corrosion rate of flat specimen in 47 mass% boiling H_2SO_4

2. Round-bar specimen

The round-bar specimen with curvature was created via plasma spraying and laser treatment to confirm the applicability of the hybrid material as a structural material. It had a dimension of ϕ16 mm × 70^L mm. Hastelloy C276 superalloy was used as the substrate.

Corrosion tests were conducted in 95 and 47 mass% boiling H_2SO_4 at 300 and 124 °C for 300 h [4]. The H_2SO_4 concentration in the thermal decomposition process of H_2SO_4 ranged from ~50 to 95 mass%. Figure 15.5 presents a schematic diagram of the corrosion test equipment for the round-bar specimen. The round-bar specimen was set in the center of the corrosion cell using a special holder made of quartz. About 50 mm of the round-bar specimen was filled with boiling H_2SO_4. The solution volume to specimen area ratio was ~23 cc/cm^2. The tip of the specimen, the area that was not plasma-sprayed and laser-treated, did not directly touch the H_2SO_4 solution. There was almost no Hastelloy C276 corrosion by H_2SO_4 gas. The corrosion cell that was made of quartz, with a capacity of 500 mL, was used. It was heated using a mantle heater to maintain the boiling condition. Evaporated H_2SO_4 was liquefied in a reflex condenser to minimize evaporation loss. There was almost no change in the H_2SO_4 concentration before and after the corrosion test. The weights of the round-bar specimens before and after the corrosion test were measured using an electronic balance to calculate the corrosion rate. The cross-sectional observation of the round-bar specimen after the corrosion test was examined via scanning electron microscopy–energy-dispersive X-ray spectroscopy (SEM–EDX). The elements dissolved in the H_2SO_4 solution after the corrosion test were also analyzed via inductively coupled plasma atomic emission spectrometry (ICP-AES).

Figure 15.6 presents the corrosion rate in 95 and 47 mass% boiling H_2SO_4 as a function of immersion time. The corrosion rate in 47 mass% boiling H_2SO_4 for 24 h was ~0.5 mm/y. The corrosion rate was almost the same as that in 47 mass% boiling H_2SO_4 for 300 h. However, the corrosion rate in 95 mass% boiling H_2SO_4 decreased with the increase in immersion time. The corrosion rate in 95 mass% boiling H_2SO_4 for 300 h was ~0.01 mm/y. This decrease in the corrosion rate indicates that the hybrid material had a good protective layer

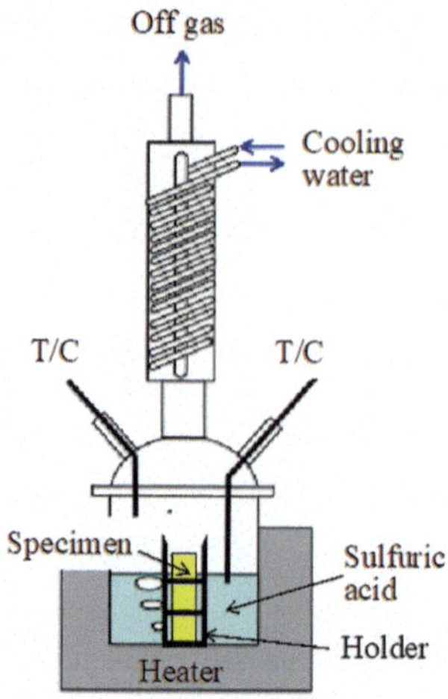

Fig. 15.5 Schematic diagram of the corrosion test equipment for round-bar specimen

[5]. The hybrid material created using the hybrid technique exhibited excellent corrosion resistance against 95 mass% boiling H_2SO_4.

The cross-sectional observation of the round-bar specimens after the corrosion test was analyzed via SEM/EDX to investigate the difference in the corrosion rate by the H_2SO_4 concentration. Figure 15.7 presents the images of the backscattered electron (BSE), O, Si, Cr, Fe, Ni, and Mo in the surface layer vicinity. Tens of microns of gap between the surface layer and the substrate material were observed after the corrosion test of 47 mass% boiling H_2SO_4, as presented in Fig. 15.7a. The gap was believed to be caused by the dissolution of the substrate due to

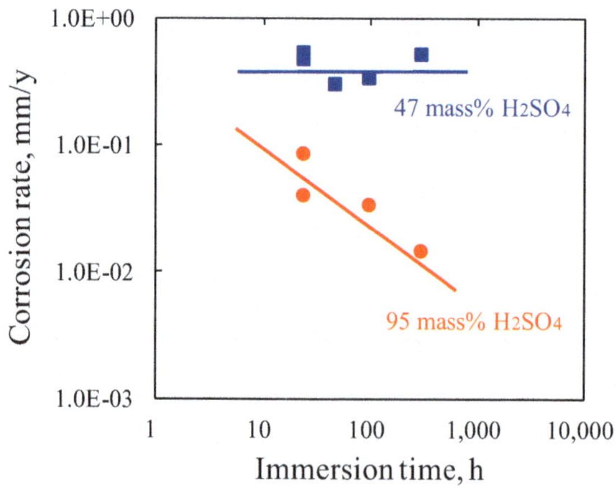

Fig. 15.6 Relationship between corrosion rate and immersion time of 95 and 47 mass% boiling H_2SO_4

the H_2SO_4 that reached the substrate through the cracks in the surface layer. As observed in Fig. 15.6, the high corrosion rate seemed to have been caused by such substrate dissolution. However, the gap has not formed between the surface layer and substrate after the corrosion test of 95 mass% H_2SO_4. SiO_2 is believed to have formed on the entire surface, since the surface was uniformly covered with O_2, as presented in Fig. 15.7b. Thus, the cracks in the surface layer were sealed by the formation of SiO_2, and it is thought that superior corrosion resistance was achieved [6]. The O_2 distribution on the surface after the corrosion test of 47 mass% boiling H_2SO_4 was scarce, as presented in Fig. 15.7a. It means that the sealing of cracks in the surface layer by the formation of SiO_2 was incomplete. Therefore, there is a possibility of the improvement of corrosion resistance in 47 mass% boiling H_2SO_4 by pre-filming.

3. Pre-filming by heat treatment

   To improve the corrosion resistance of the specimen in 47 mass% boiling H_2SO_4, the specimen was oxidized in the air for pre-filming. The pre-oxidation tests were conducted at 700, 800, 900, and 1000 °C for 24 h in the air to select the condition of the pre-filming.

The result of the cross-sectional observation is presented in Table 15.1. The pre-filming condition was selected at 800 °C for 24 h in the air from Table 15.1.

The round-bar specimen pre-oxidized at 800 °C for 24 h in the air was prepared for the corrosion test of 47 mass% boiling H_2SO_4. Figure 15.8 shows the present data, in addition to the previous results. The pre-oxidized round-bar specimen exhibited superior corrosion resistance in the condition of 47 mass% boiling H_2SO_4. It is thought that the superior corrosion resistance of the pre-oxidized round-bar specimen was achieved by the formation of SiO_2. Thus, it was confirmed that pre-oxidation was effective in improving the corrosion resistance of the specimen using the hybrid technique [7].

15.4 Production of the Container Using the Hybrid Technique

To confirm the production characteristics of the container using the hybrid technique, the container was experimentally made. It has a welded part, chamfer, curved surface, R processing, and sealing surface. R processing was performed between the inner curved surface and the inner bottom of the container. These parts are important for the fabrication of the components of the IS process plant, such as the reaction vessel, pipe, and flange. The container consists of an upper part and a lower part, as presented in Fig. 15.9 [8]. It has an inside diameter of about 150 mm, height of 120 mm, and thickness of 6 mm. The container is made of Hastelloy C276.

Plasma spraying and laser treatment were performed in the inside of the upper and lower parts, except for the surrounding parts of the welded area. The upper and lower parts were welded by tungsten inert gas. Finally, the entire welded part was

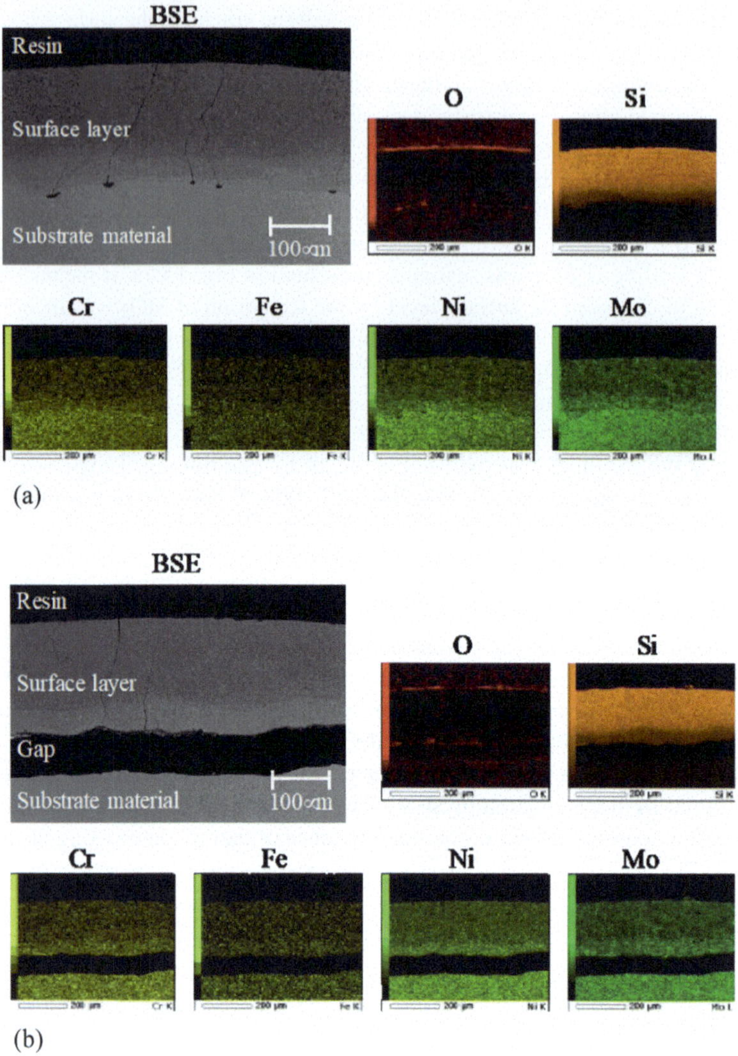

Fig. 15.7 Area analysis of the cross-sectional observation of round-bar specimens after the corrosion test for 300 h. **a** 47 mass% H_2SO_4. **b** 95 mass% H_2SO_4

Table 15.1 Results of the pre-oxidation test

Oxidation condition	Atmosphere	Layer of the surface
700 °C for 24 h	Air	No change
800 °C for 24 h	Air	SiO_2
900 °C for 24 h	Air	SiO_2 and a little Cr_2O_3
1000 °C for 24 h	Air	SiO_2 and Cr_2O_3, a part detachment

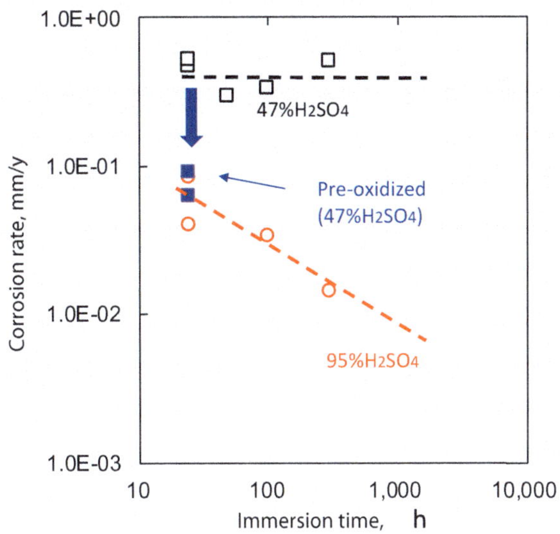

Fig. 15.8 Results of corrosion test for the pre-oxidized specimen

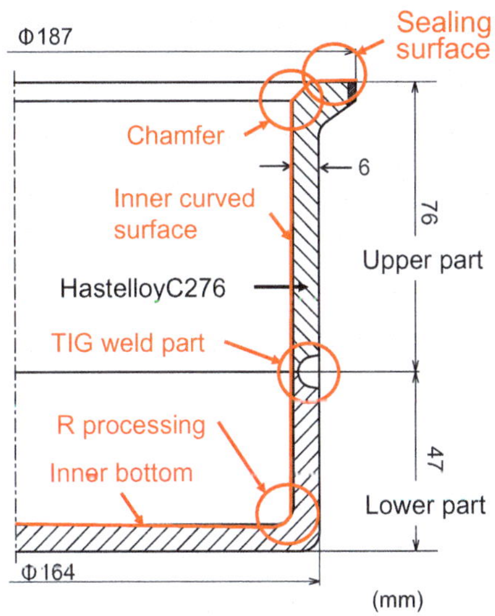

Fig. 15.9 Configuration of the container made of Hastelloy C276. Those indicated by red circles and red line are required to fabricate the components of the IS process plant [8]

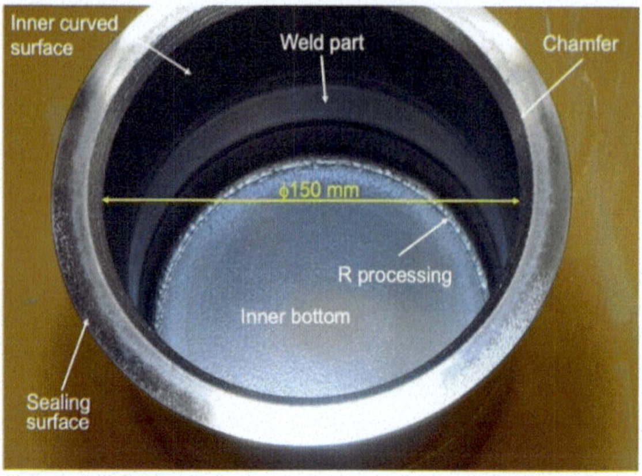

Fig. 15.10 Appearance of inner part of container [8]

covered by plasma spraying and laser treatment. The inside of the container was observed under a digital microscope to check the surface morphology.

Figure 15.10 presents an image of the inner part of the container using the hybrid technique [8]. The center of the bottom in a metal color looks stronger than the circumference. The surface of the bottom was homogeneous, according to visual observation. The difference in the center of the bottom is potentially attributed to the wavy irregularity of the surface and the undulation of the light. The inner surface condition of the container was also observed under a digital microscope. The container manufactured using plasma spraying and laser treatment did not exhibit distortion and detachment of the plasma-sprayed and laser-treated layer at the welded part, chamfer, curved surface, R processing, and sealing surface.

Figure 15.11 presents the detailed appearance of the container with the plasma-sprayed and laser-treated inner layer at each part, such as sealing surface, chamfer, R processing, inner curved surface, and inner bottom. Detachment of the plasma-sprayed and laser-treated layer on the inner part was not observed, although there were some differences in the surface morphology [8]. Only the sealing surface was mechanically polished using emery paper Grade 1200 after plasma spraying and laser treatment to achieve good contact with the O-ring.

The hybrid material that exhibited both corrosion resistance and ductility was created using silicon powder plasma spraying and laser treatment. The round-bar specimens created using the hybrid technique exhibited excellent corrosion resistance in 47 and 95 mass% boiling H_2SO_4. The hybrid container with a welded part, sealing surface, chamfer, R processing, inner curved surface, and inner bottom was experimentally developed by plasma spraying and laser treatment. No detachment of the plasma-sprayed and laser-treated layer was observed on the base metal and the welded part. Therefore, we succeeded in the production of the container using the hybrid technique. In future, a boiling H_2SO_4 test using the container will be conducted to evaluate the corrosion characteristic of the container.

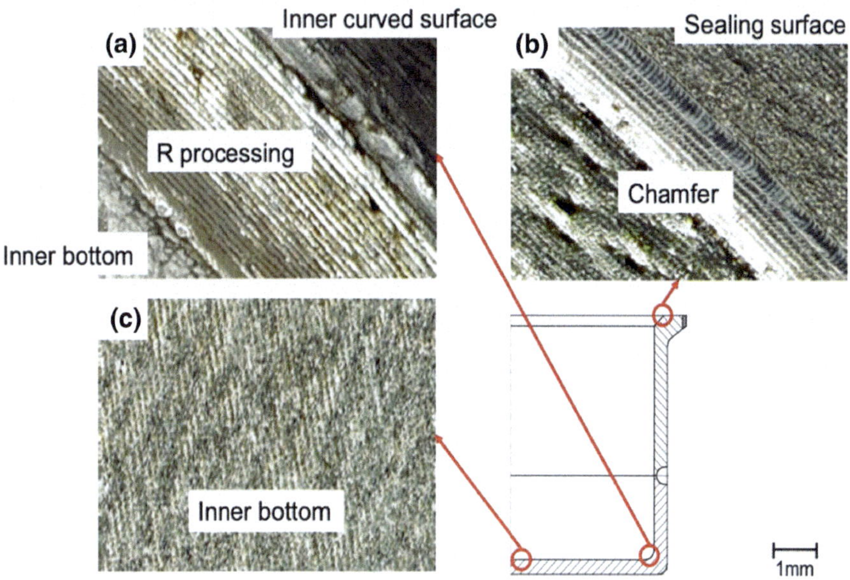

Fig. 15.11 Each part of hybrid container with plasma-sprayed and laser-treated inner layer. **a** R processing and inner curved surface. **b** Sealing surface and chamfer. **c** Inner bottom [8]

References

1. Ioka I et al (1997) Corrosion resistance of Fe–Si alloys in boiling sulfuric acid. J Soc Mater Sci Jpn 46:1041–1045
2. Japanese Patent Application No.2017-128771
3. Ioka I et al (2015) Development of hybrid material using plasma spraying and laser treatment for thermochemical water-splitting iodine-sulfur process. In: 6th world hydrogen technology convention, 11–14 Oct 2015, Sydney, pp 65–66
4. Ioka I et al (2016) Corrosion resistance of plasma spraying and laser treated material with curvature for thermochemical water-splitting iodine-sulfur process. In: 5th Asian conference on innovative energy and environment chemical engineering (ASCON-IEEChE 2016), Poster064
5. Jones DA, Greene ND (1966) Eectrochemical measurement of low corrosion rates. Corrosion 22:198–205
6. Ioka I et al (1999) The characterization of passive films on Fe–Si alloy in boiling sulfuric acid. J Mater Sci Lett 18:1497–1499
7. Ioka I et al (2017) Effect of pre-oxidation on corrosion resistance of plasma sprayed and laser treated material for thermochemical water-splitting process. In: 7th world hydrogen technology convention, 9–12 July 2017, Prague, ESN-HP-P023
8. Ioka I et al (2020) Study of container using hybrid technique for sulfuric acid decomposition of thermochemical water-splitting iodine-sulfur process. Mech Eng J 7:19–00377

Part IV
CO_2 Free Ammonia Synthesis

Chapter 16
Green Ammonia Production System

Yasushi Fujimura

Abstract Feature of green ammonia production is summarized. Recent worldwide trials of green ammonia production are reviewed. The demonstration plant of green ammonia production system built in AIST Fukushima is shown. Some of the running data including lord-changing operation are disclosed.

Keywords Green ammonia · Demonstration plant · Load-changing operation

16.1 Introduction

Ammonia is being promoted as a clean fuel because it does not produce CO_2 during combustion. "CO_2-free Ammonia," which suppresses CO_2 emission during ammonia production, is classified as "Blue Ammonia," which originates from fossil fuels, and "Green Ammonia," which is produced from renewable energy sources. CO_2-free ammonia is predicted to become a substance for energy storage and utilization of variable renewable energy sources, such as photovoltaic (PV) and wind power, in Europe and the United States. Furthermore, because ammonia manufacturing emits around 1% of CO_2 globally, fertilizer companies expect to introduce CO_2-free ammonia to decarbonize fertilizer production. Alternatively, in Japan, where energy resources are few, CO_2-free ammonia is one of the crucial measures for decarbonizing the energy industry, particularly in the power generation sector. Therefore, the Japanese government plans to produce CO_2-free ammonia overseas on a large scale and import them as a low-carbon fuel. Imported CO_2-free ammonia will be used in power generation for ammonia fueled gas turbines, co-firing with coal in coal-fired power plants, ammonia fuel cells, and fuels for marine engines and industrial furnaces. Figure 16.1 shows an overview of the CO_2-free ammonia supply chain assumed by Japan.

Y. Fujimura (✉)
JGC Corporation, Minatomirai, Nishi-ku, Yokohama 220-6001, Japan
e-mail: yfujimura@n-u.co.jp

Present Address:
Nikki-Universal Co.,Ltd., Yonnomiya, Hiratsuka, Kanagawa 254-0014, Japan

K. Aika and H. Kobayashi (eds.), *CO_2 Free Ammonia as an Energy Carrier*,
https://doi.org/10.1007/978-981-19-4767-4_16

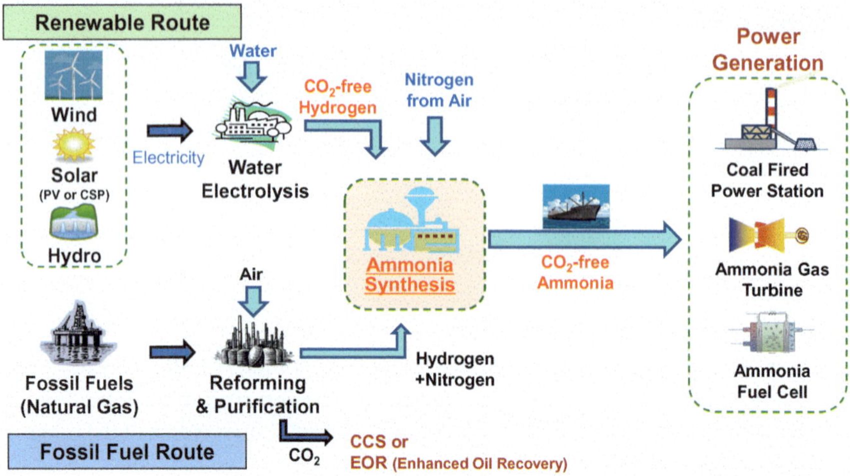

Fig. 16.1 Supply chain of CO_2-free ammonia

16.2 Challenges in Green Ammonia Production

Figure 16.2 depicts the configuration of a green ammonia production process using renewable energy. Green hydrogen is produced by water electrolyzer with renewable energy sources such as solar and wind power. Green hydrogen is compressed and combined with nitrogen, which separated from the air, and fed into the ammonia synthesis process to make green ammonia.

Compared to the production of blue ammonia from conventional fossil resources, the green ammonia manufacturing method differs in the following ways.

1. Price of renewable hydrogen

   To produce green ammonia at a reasonable price, low-cost renewable energy should be used. Additionally, lowering the cost of the water electrolyzer is critical in lowering the cost of hydrogen for ammonia synthesis.
2. Hydrogen production volatility requires a countermeasure

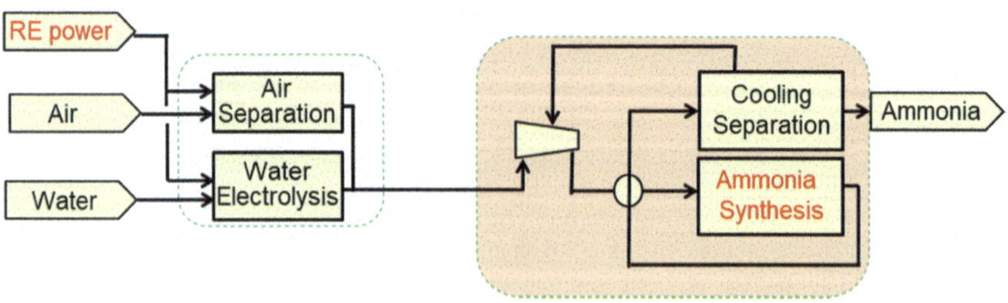

Fig. 16.2 Block flow diagram of green ammonia plant

The volatility of hydrogen production in a water electrolyzer is due to the fluctuation in the power generated by variable renewable energy. The existing ammonia synthesis process does not consider variable load operation. Therefore, buffering variation in hydrogen production or variable load operation of the ammonia synthesis process will be necessary for countermeasures.

3. Hydrogen pressure in the water electrolyzer

   The hydrogen produced in the water electrolyzer has a supply pressure of <1 MPa. The energy consumption for hydrogen compression increases since supply pressure is lower than the reaction pressure of the existing ammonia synthesis method.

4. Scale of green ammonia plant

   The availability of renewable energy determines the size of the green ammonia plant. For example, producing 2000 tons of green ammonia per day requires around 2 GW renewable energy with a capacity factor of 40% to produce hydrogen via water electrolysis.

   In general, the larger scale will be better by the scale factor in terms of the capital expense of the green ammonia plant. However, the small-scale plant will be first introduced by limiting renewable energy and the availability of water electrolyzer. To promote green ammonia production, plant expenses must be reduced, and energy efficiency must be improved.

16.3 Status of Green Ammonia Production Development

Hydropower was used to produce ammonia in the early stages of industrial ammonia manufacturing. However, the use of hydropower reduced as natural gas-based ammonia manufacturing became popular. However, water electrolysis-based ammonia production can be decarbonized by employing variable renewable energy.

1. Ammonia production using hydropower (Norway)

   In Notodden, Norway, Norsk Hydro began producing ammonia using hydrogen produced using a water electrolyzer powered by hydropower in 1927. From 1953 to 1988, the Glomfjord ammonia plant, Norway's largest plant, was in operation.

2. University of Minnesota (USA)

   In 2013, the University of Minnesota began to experiment with green ammonia manufacturing using wind energy. Their goal was to produce ammonia using local wind energy. Ammonia is injected directly into the field as fertilizer in the Midwest of the United States, which is a center of grain production. Ammonia pipelines have been installed from the Gulf Coast to the Midwest for the ease of transportation.

   Wind energy is used to power the water electrolyzer, which produces hydrogen, and an iron-based catalyst was used for ammonia synthesis. The ammonia plant had a capacity of 3.3 kg/h.

Several investigations on ammonia production have been conducted at the University of Minnesota, including ammonia synthesis, ammonia separation technologies employing inorganic adsorbents, and demonstrations of tractors powered by ammonia.

3. AIST FREA (Japan)

   From 2014 to 2018, the project team led by JGC developed novel ammonia synthesis catalysts and demonstrated the green ammonia production process as part of the Cross-ministerial Strategic Innovation Promotion Program (SIP) "Energy Career" Project, a Japanese government-funded R&D initiative.

   A demonstration plant was built by the National Institute of Advanced Industrial Science and Technology (AIST) and JGC in the Fukushima Renewable Energy Institute (FREA) in Koriyama, Fukushima Pref. Japan, and commissioned in March 2018. The project is discussed in detail in the subsequent section.

4. Siemens (UK)

   At Rutherford Appleton Laboratory in the United Kingdom, Siemens, the University of Oxford, STFC, and Cardiff University built a green ammonia plant to demonstrate on-site energy storage.

   This demonstration used a wind turbine to generate renewable power, a water electrolyzer to produce hydrogen, and the Haber-Bosch process to convert hydrogen to ammonia. The plant started operating from June 2018.

16.4 Demonstration Test in AIST FREA

1. Catalyst development

   The project team concentrated on ammonia synthesis at lower temperatures and pressures than the conventional Haber–Bosch process for acceptable green ammonia production.

   Catalyst development the project was conducted by AIST, National Institute of Technology, Numazu College, and JGC Catalysts and Chemicals Ltd. (JGC C&C). The details of the catalyst development by AIST are introduced in Chap. 17.

   The team selected Ru/CeO_2 from several ruthenium catalysts based on activity and suitability for large-scale production.

   JGC C&C established the manufacturing method on an industrial scale.

2. Demonstration plant in AIST FREA

   The ammonia production capacity of the plant was 20 kg/day using nitrogen and hydrogen gas as feedstock. The schematic process flow diagram is shown in Fig. 16.3. The method was nearly identical to the commercial green ammonia production process, except the heat management and reactor design. Hydrogen and nitrogen as the feedstock are supplied from cylinders. To preserve the ammonia synthesis catalyst, these feedstock gases were combined and processed to eliminate the trace amount of water and oxygen. The pretreated gas was compressed to the reaction pressure up to 8 MPaG and supplied to the ammonia synthesis reactor. Ammonia synthesis reactor has a dual-wall structure with a single fixed catalyst bed in the inner tube, and hot oil was flowed in the outer shell to heat feedstock gas and remove reaction heat from the catalyst bed. The volume of the reactor was ~3 L. The effluent from the reactor was cooled, and the final product recovered as liquefied ammonia. The unreacted residual gas is returned to the reactor through a recycling compressor. The schematic flow diagram and photograph of the demonstration plant is shown in Figs. 16.3 and 16.4.

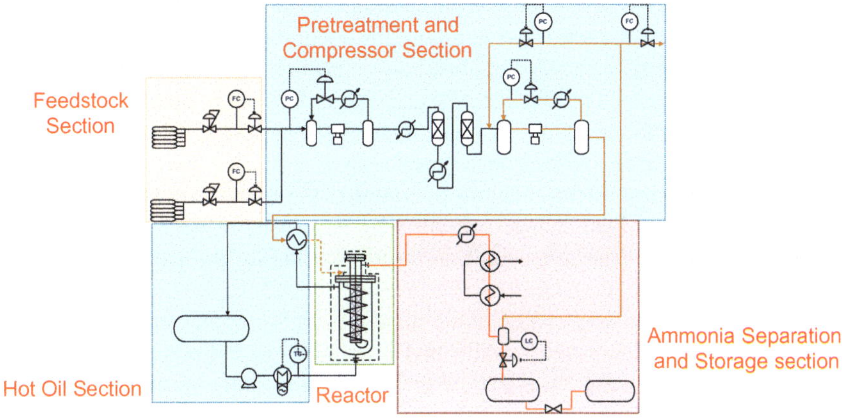

Fig. 16.3 Schematic process flow diagram of the demonstration plant in AIST FREA

Fig. 16.4 Photograph of the demonstration plant in AIST FREA (courtesy of AIST)

3. Catalyst performance evaluation

   In 2018, AIST and JGC jointly evaluated the catalyst performance of the developed Ru/CeO_2 catalyst. In the tests, 9 runs of the test were performed using operation time of 650 h, and catalyst performance data were acquired for ~130 reaction conditions. Although the startup, shut-down, and reaction conditions were frequently changed during these tests, the catalytic performance was stable.

4. Load-changing operation

   In the green ammonia production, hydrogen is produced using a water electrolyzer that is powered by renewable energy. In this instance, it is unavoidable that the amount of hydrogen produced fluctuates with the change in power generation owing to weather conditions. Adopting a battery and hydrogen tank to absorb the variation of hydrogen production is a viable option. However, to run the ammonia synthesis process at a constant load, a considerable large battery and hydrogen tank are needed to absorb long-term changes. This will result in a significant increase in production costs. To minimize battery and hydrogen tank capacity, the ammonia synthesis process should be conducted by changing load every few days to follow hydrogen production.

   To demonstrate the load changing operation of the ammonia synthesis process, we conducted the load change operation by increasing or decreasing the flow rate of hydrogen. The test conditions were as follows: the catalyst bed inlet temperature of 400 °C and the reaction pressure of 8.0 MPaG. The hydrogen flow rate was changed at 40%–100% of the standard conditions, the ratio of hydrogen and nitrogen gas flow rate was controlled as 3. Figure 16.5 shows the supply flow rate of the raw material N_2 and H_2, reaction pressure, ammonia concentration at the reactor outlet, temperature of the catalyst layer inlet and the peak position, and the amount of produced ammonia. No significant fluctuation of temperature and pressure was observed after changing the flow rate and the process can be operated stably. However, ammonia production showed a slight fluctuation when the condition changed, but it became stable in a short time. From this test, we confirmed that it was possible to cope with the load change operation of the ammonia synthesis process using the developed ammonia synthesis catalyst and process control method.

5. Demonstration of CO_2-free ammonia value chain

   We accomplished the green ammonia synthesis demonstration using green hydrogen produced via water electrolysis, although the previous tests were conducted using hydrogen cylinder gas. The green hydrogen was produced by photovoltaic power system and the water electrolyzer installed in the AIST FREA. The green hydrogen was compressed in a cylinder curdle and transported to the ammonia synthesis demonstration plant because these plants were not directly connected via pipes. In this test, we confirm that green hydrogen could be stably converted to green ammonia. Furthermore, liquefied ammonia produced in this test was used as a fuel for a 50 kW ammonia fueled gas turbine, which was also developed by the "Ammonia direct combustion" team under the SIP "Energy Career" program, and successfully generated 47 kW of green power.

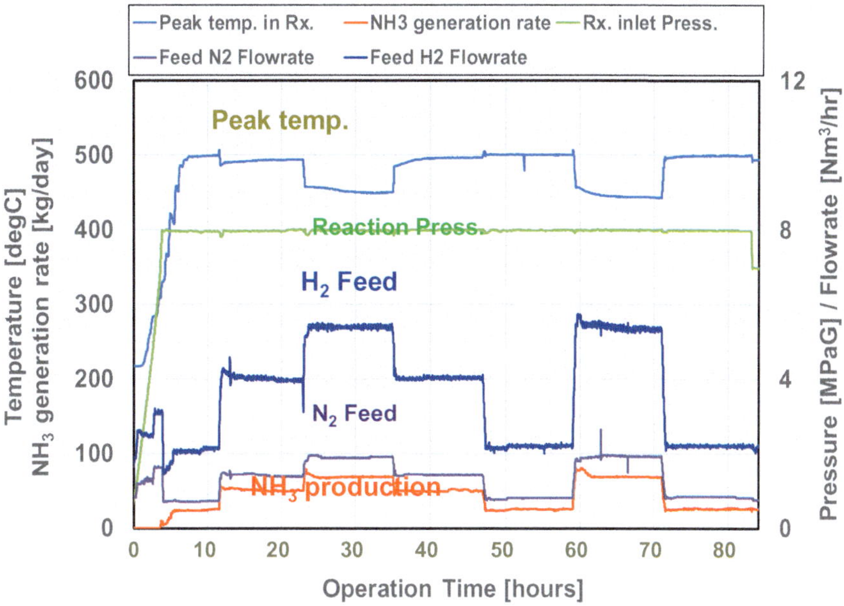

Fig. 16.5 Load-changing operation in the demonstration plant

Figure 16.6 depicts schematic diagram of the demonstration test of CO_2-Free Ammonia Value Chain.

Such a series of photovoltaic power generation, water electrolyzer for green hydrogen production, ammonia synthesis under low temperature and pressure, and green power generation by ammonia fueled gas turbines are a scheme of the CO_2-free ammonia value chain as we targeted. We succeeded in the world's first demonstrating a CO_2-free ammonia value chain.

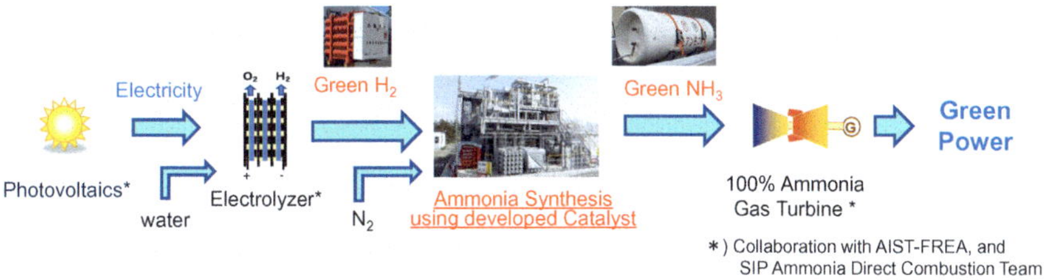

Fig. 16.6 Demonstration of CO_2-free ammonia value chain

Chapter 17
Comparison of Several Ammonia Catalysts Worked Under Industrial Conditions

Keisuke Kobayashi, Rahat Javaid, Yuichi Manaka, Tetsuya Nanba, Masayasu Nishi, Takehisa Mochizuki, Shih-Yuan Chen, and Hideyuki Takagi

Abstract Development of Ru catalysts for ammonia synthesis was investigated, assumed that hydrogen was supplied from the water electrolysis by using variable electricity from photovoltaic and wind turbine. Cerium oxide and carbon-based materials were used as the support material of ruthenium. Catalytic activity was measured under various reaction conditions, in which pressure, temperature, space velocity, and H_2/N_2 ratio were changed, supposing plant operation conditions. Optimization of catalyst composition and preparation procedure was carried out to obtain the catalysts for the demonstration test described in Chap. 16.

Keywords Catalyst · Ruthenium · Cerium oxide · Carbon · Variable hydrogen supply · Green ammonia

17.1 Introduction

Ammonia is an essential raw material for the production of fertilizers and many other nitrogen-containing chemicals. Currently, the annual global production of ammonia exceeds 150 million tons [1]. The main industrial means of producing ammonia is the Haber–Bosch process, which converts atmospheric nitrogen to ammonia by reaction with hydrogen at high temperature and pressure in the presence of a metal catalyst. The hydrogen used in the Haber–Bosch process is usually produced via the conversion of fossil fuels such as natural gas. These conversion processes are highly energy efficient; however, they also produce large amounts of CO_2, which is a major greenhouse gas. Thus, a process for the production of "green" ammonia is needed that does not utilize fossil fuels and does not result in the production of greenhouse gases [2]. It is expected that the hydrogen for such green processes will be produced by water electrolysis and catalytic water splitting powered by renewable energy.

K. Kobayashi · R. Javaid · Y. Manaka · T. Nanba · M. Nishi · T. Mochizuki · S.-Y. Chen · H. Takagi (✉)
National Institute of Advanced Industrial Science and Technology (AIST), Onogawa, Tsukuba, Ibaraki 305-8569, Japan
e-mail: hide-takagi@aist.go.jp

K. Aika and H. Kobayashi (eds.), *CO_2 Free Ammonia as an Energy Carrier*,
https://doi.org/10.1007/978-981-19-4767-4_17

However, some issues must be addressed before a viable industrial green ammonia synthesis process can be realized.

The first issue is that current sources of renewable energy have a variable supply. For example, photovoltaics does not function at night, and wind turbines do not function when there is no wind. This variability results in a hydrogen supply that is characterized by both wide and frequent fluctuations and intermittent interruptions. Such a variable hydrogen supply is not suitable for industrial ammonia production because ammonia plants must be operated continuously to remain profitable. One way to address this issue is to use very large hydrogen storage tanks from which additional hydrogen can be drawn when needed. However, these tanks are expensive and, depending on their size, may not fully prevent production shutdowns.

The second issue is that the conventional Haber–Bosch process is conducted under high pressure and temperature (e.g., 15 MPa, 500 °C) using hydrogen from the reforming of fossil fuels that is also obtained at high pressure and temperature (e.g., 8 MPa, 300 °C). This is an issue because hydrogen produced using renewable energy is obtained at ambient pressure and therefore must be brought up to pressure before it can be used, which is a costly process. Producing ammonia at ambient pressure is an option; however, the plant size increases as the amount of ammonia produced increases, which limits the scaling of production.

One way to overcome these issues is expected to be the use of specifically designed catalysts. For the production of green ammonia, a prospective catalyst must be able to perform under various hydrogen supply conditions and have a rapid response to re-start after plant shutdown. In plants using hydrogen storage tanks, the reactor operation should not be constant at the appropriate condition for the catalyst activity, but a variable operation along with input hydrogen situation is effective for minimizing buffer facilities. Our group has shown that ruthenium (Ru) catalysts have a rapid response to re-start and that they retain their catalytic activity over a wide range of reaction conditions (i.e., temperature, pressure, space velocity, and H_2/N_2 ratio) [3]. A prospective catalyst must also have high activity at low temperature and pressure to reduce energy consumption and costs. Again, Ru catalysts also show potential in this regard because they show high activity at relatively low temperatures [4–7]. Many novel support materials have also been developed that further improve the low-temperature activity of Ru catalysts. Although, the activity of these catalysts has generally been studied only at ambient pressure, the ammonia synthesis activity of Ru/Cs/MgO reportedly decreases with increasing pressure, whereas that of Ru/CeO_2 increases monotonically with increasing pressure up to 5 MPa, indicating that different support materials result in catalysts with different pressure dependences [8]. With respect to the latter catalyst, the appearance of maximum activity at a distinct pressure can be explained by inhibition of strong hydrogen adsorption [9].

Taking this information together, the ideal catalyst for the production of green ammonia should show high activity at low temperature and pressure (e.g., <10 MPa and 400 °C) and increasing activity with increasing pressure in the low-pressure region (0–10 MPa). The current literature suggests that Ru catalysts using various support materials show great potential in this regard. One final issue that should also be considered is that the hydrogen produced by water electrolysis usually contains water vapor and a small amount of oxygen as contaminants. Therefore, the ideal catalyst should have high tolerance against poisoning by these contaminants.

In this chapter, we discuss the development of Ru/CeO_2 and Ru/C catalysts for the production of green ammonia. First, we consider the activity, morphology, and reducibility of Ru/CeO_2 catalysts under various reaction and hydrogen supply conditions. Optimization of the CeO_2 support is also considered. Ru catalysts supported on novel carbon-based materials together with alkaline and alkaline earth metals have also shown potential for use in the production of green ammonia [10, 11]. Some of these catalysts have also shown high ammonia synthesis activity under high-pressure conditions [12, 13]. We therefore also consider the activity, morphology, and durability of these catalysts.

17.2 Ammonia Synthesis Over Ru/CeO_2 Catalysts

To find the optimal Ru/CeO_2 catalyst for the production of green ammonia, the effects of the catalyst and support material precursors on ammonia production activity under ambient and high-pressure conditions were investigated.

17.2.1 Catalytic Activity of Ru/CeO_2 at Ambient Pressure

The catalytic activities of 1 wt% Ru/CeO_2 catalysts, prepared using an impregnation method with various catalyst and support precursors, were examined at ambient pressure. The Ru precursors used were $Ru_3(CO)_{12}$, $Ru(acac)_3$, $Ru(NO)(NO_3)_3$, $RuCl_3$, and Na_2RuO_4. The CeO_2 precursors used were CeO_2(A) (Daiichi Kigenso,; surface area, 167 m^2/g), CeO_2(B) (Daiichi Kigenso; 133 m^2/g), and CeO_2(C) ($Ce(NO_3)_3$ calcined at 900 °C for 4 h, 10 m^2/g). After Ru loading, the catalysts were calcined in a flow of 10% H_2/N_2 at 300 °C for 1 h. The activity tests were performed in a fixed-bed flow reactor under ambient-pressure conditions. For the ambient-pressure condition, 0.2 g of catalyst was placed in a quartz reactor. The feed gas was 75% H_2 and 25% N_2 with a total flow rate of 80 mL min^{-1}. Concentration of product gases was analyzed using a Fourier transform infrared spectrometer equipped with a multi-reflection gas cell.

In this experiment, which was conducted at 325 °C, the NH_3 concentration in the effluent gas was determined as an index of low-temperature catalytic activity, and the maximum NH_3 concentration in the profile of temperature dependence of catalytic activity was determined as an index of maximum capacity (Fig. 17.1). Among the 15 catalysts examined, the highest NH_3 concentration was obtained with the combination of $Ru(acac)_3$ and CeO_2(B) (Fig. 17.1a). When CeO_2(A) was used as the support material, combination with $Ru(NO)(NO_3)_3$ produced the highest NH_3 concentration. When CeO_2(C) was used, combination with $Ru_3(CO)_{12}$ produced the

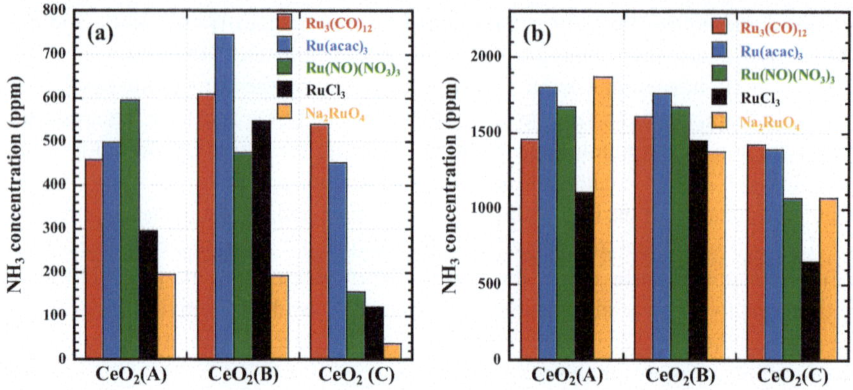

Fig. 17.1 NH_3 synthesis activity of various 1 wt% Ru/CeO_2 catalysts prepared using the indicated catalyst and support precursors. **a** NH_3 concentration in the effluent gas. **b** Maximum NH_3 concentration. Catalyst weight: 0.2 g, flow rate: 80 mL/min ($H_2/N_2 = 3$), ambient pressure, 325 °C

highest NH_3 concentration. Overall, CeO_2(B) tended to produce catalysts with higher catalytic activity than that of catalysts produced using the other support materials, suggesting that a very high surface area is not necessary for obtaining satisfactory low-temperature catalytic activity. No trend was observed with respect to the Ru precursor used.

The maximum NH_3 concentration was obtained with the combination of Na_2RuO_4 and CeO_2(A) (Fig. 17.1b). However, this combination had poor low-temperature activity, suggesting that the Ru/CeO_2 catalyst produced using Na_2RuO_4 had a high catalytic activity but narrow active temperature window. Interestingly, although Cl is a strong inhibitor of catalytic activity [14], the combination of $RuCl_3$ and CeO_2(B) did not result in a catalyst with poor low-temperature and maximum activities. Overall, in contrast to the findings for low-temperature activity, the CeO_2 supports with a high and medium surface area showed similar activities when comparing catalysts made using the same Ru precursor. Again, no trend was observed with respect to the Ru precursor used.

Together, these findings suggest that appropriate selection of Ru and CeO_2 precursors is an important part of Ru/CeO_2 catalyst development. Among the catalysts examined, although the combination of $Ru(acac)_3$ and CeO_2(B) provided a catalyst with high low-temperature and maximum activities, $Ru(acac)_3$ needs an organic solvent for its preparation. Therefore, the combination of $Ru(NO)(NO_3)_3$, which dissolves in water, is perhaps ideal for real-world applications.

17.2.2 Catalytic Activity of Ru/CeO$_2$ Under Pressurized Conditions

Next, the catalytic activities of 1 wt% Ru/CeO$_2$ catalysts prepared from Ru(NO)(NO$_3$)$_3$–CeO$_2$(A) and Ru(NO)(NO$_3$)$_3$–CeO$_2$(B) were examined under high-pressure conditions. For activity measurement, 1 mL of catalyst was placed in a tubular reactor with an inner diameter of 12 mm. Reaction condition variables were temperature (300–450 °C), pressure (0–10 MPaG), H_2/N_2 ratio (1–4), and space velocity (3000–12,000 h^{-1}). Figure 17.2 shows the pressure dependence of NH_3 synthesis activity at 400 °C, 12,000 h^{-1}, and a H_2/N_2 ratio of 3. Both catalysts showed a marked increase in activity as the pressure was increased from 0 to 3 MPaG. As the pressure was further increased, the activity of Ru/CeO$_2$(A) peaked at 5 MPaG and then began to decrease, whereas the activity of Ru/CeO$_2$(B) continued to gradually increase. These findings suggest that a support material with a high surface area results in a catalyst with high activity at low pressure. They also suggest that, for plant operation, pressures below 5 MPaG are optimal for Ru/CeO$_2$(A) and below 3 MPaG are optimal for Ru/CeO$_2$(B). However, it should be noted that optimization of the preparation conditions and Ru loading can result in a catalyst with an activity that continues to increase as the pressure is increased up to 10 MPaG [15].

Maximal NH_3 formation or a plateau in NH_3 formation at a certain pressure or NH_3 formation rate is likely due to inhibition of activity by hydrogen adsorption. The reaction order with respect to H_2 for ammonia synthesis over 1 wt% Ru/CeO$_2$(A) was determined. At ambient pressure, the reaction orders at 325 and 400 °C were 0.16 and 0.96, respectively. Both values were positive, and the value at 400 °C was

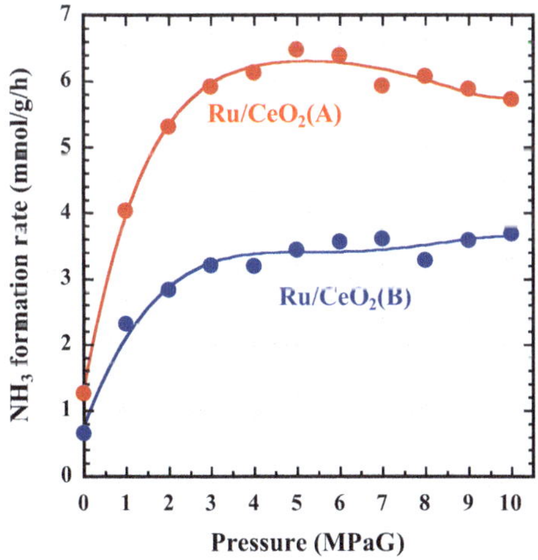

Fig. 17.2 Pressure dependence of NH_3 synthesis activity of 1 wt% Ru/CeO$_2$ catalysts produced using different support material precursors. Temperature: 400 °C, space velocity: 12,000 h^{-1}, and $H_2/N_2 = 3$

close to unity, suggesting that there is little hydrogen inhibition at this pressure and these temperatures. However, at 5 MPaG, the reaction orders at 325 and 400 °C were −0.71 and −0.37, respectively. These negative values suggest that hydrogen inhibition occurs at high-pressure conditions. Thus, these results suggest that hydrogen inhibition was increased under the pressurized conditions.

Figure 17.3 shows the effect of H_2/N_2 ratio on the NH_3 synthesis activity of 1 wt% Ru/CeO_2(A) and 1 wt% Ru/CeO_2(B) at 400 °C and 12,000 h^{-1}. For both catalysts, NH_3 synthesis activity increased with decreasing H_2/N_2 ratio at every pressure examined. At 0 MPaG, almost same activities were observed among the H_2/N_2 ratios. Comparing 1 wt% Ru/CeO_2(A) and 1 wt% Ru/CeO_2(B), the former had higher NH_3 synthesis activity at all conditions even though the activity trends of the two catalysts were similar. As the pressure was increased, the difference in NH_3 synthesis activity between a H_2/N_2 ratio of 1 and 4 increased. In real-world applications, a H_2/N_2 ratio of 3 is preferable because the amount of gas needing to be recycled is minimized. However, at 8 MPaG, the NH_3 formation rate at H_2/N_2 = 1 was 2.4 times that at H_2/N_2 = 3. Thus, operation at a low H_2/N_2 ratio will likely result in an acceptable NH_3 production rate.

Taking these findings together, the catalytic activity of Ru/CeO_2 is dramatically influenced by precursor selection. Furthermore, to obtain sufficient NH_3 synthesis activity at low temperature and pressure, and confirmation of activity over a wide range of reaction conditions is essential. Kinetic analyses are also important for estimating the catalytic behavior of Ru/CeO_2 under pressurized conditions, and characterization of the physicochemical properties of the Ru/CeO_2 catalysts can be useful for identifying how a catalyst can be improved. In the following sections, we discuss physicochemical characterization of Ru/CeO_2 catalysts.

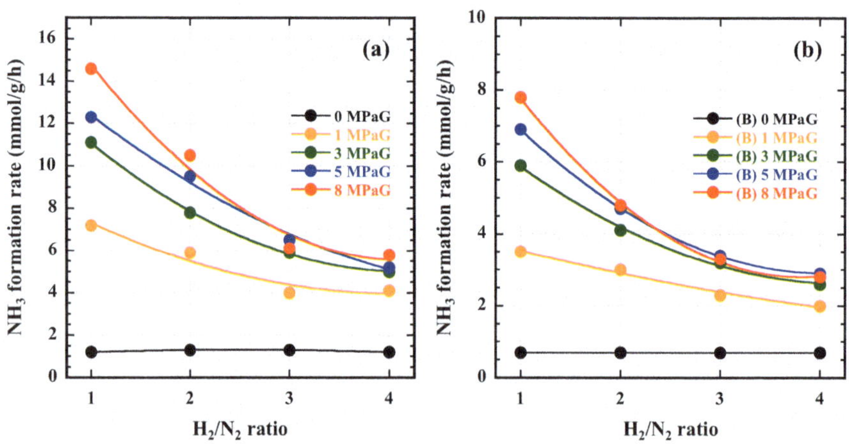

Fig. 17.3 Dependence of the NH_3 synthesis activity of 1 wt% Ru/CeO_2 catalysts on H_2/N_2 ratio. Temperature: 400 °C, space velocity: 12,000 h^{-1}

17.2.3 Characterization of Ru/CeO_2 Catalysts

A wide variety of analytical methods can be used to characterize Ru/CeO_2 catalysts. Here, we discuss recent findings obtained using two commonly used methods, both of which provide data that can be correlated with ammonia synthesis activity; these methods are electron microscopy to observe Ru morphology and temperature-programmed reduction (TPR) to observe electron donation from the support material to Ru.

17.2.3.1 Surface Morphology of Ru/CeO_2 Catalysts

Surface morphology plays a pivotal role in determining a catalyst's efficiency and activity [16–18]. Various techniques can be used to evaluate the surface morphology of catalysts. Focused ion beam transmission electron microscopy (FIB-TEM) is an advanced technology. TEM analysis provides comprehensive information about chemical composition and structure by using energy-dispersive X-ray analysis and high-angle annular dark-field imaging [19]. FIB-TEM along with energy-dispersive X-ray mapping was performed to evaluate the morphological changes occurring on the surface and in the core of Ru/CeO_2 catalysts during the ammonia synthesis reaction.

Activity tests using 3 wt% Ru/CeO_2 catalyst were performed under various temperature, pressure, H_2/N_2 ratio, and space velocity conditions. During activity testing, the durability of the catalyst was confirmed periodically under the following standard conditions: 425 °C, 2.5 MPaG, H_2/N_2 ratio = 4, and space velocity = 12,000 h^{-1}. Catalytic activity was measured under standard condition periodically, and then the activity was compared with that at the beginning of catalyst use. The ratio of catalytic activities at each measurement to the initial measurement was defined as specific activity. Total time on stream was 170 h, and specific activity increased gradually [20]. To understand the reason for this increase in activity, FIB-TEM and EDS analyses of cross-sections of fresh (as-prepared) catalyst and used catalyst after 170 h of reaction were performed (Fig. 17.4). In fresh catalyst, agglomeration of Ru particles both on the surface and in the core of the catalyst was observed, and the distribution of Ru particles throughout the catalyst (surface and core) was uniform. In contrast, used catalyst showed a markedly different Ru particle size and distribution between the surface and core of the catalyst; in the core, Ru aggregates were observed, whereas on the surface Ru was highly dispersed. Furthermore, in the used catalyst, the amount of Ru particles on the surface was higher than that in the fresh catalyst. These findings indicate that as the time on stream increases, the amount of highly dispersed Ru particles on the surface also increases, leading to an increase in the number of active sites on the catalyst surface. This likely explains the observed increase in specific activity with increasing time on stream.

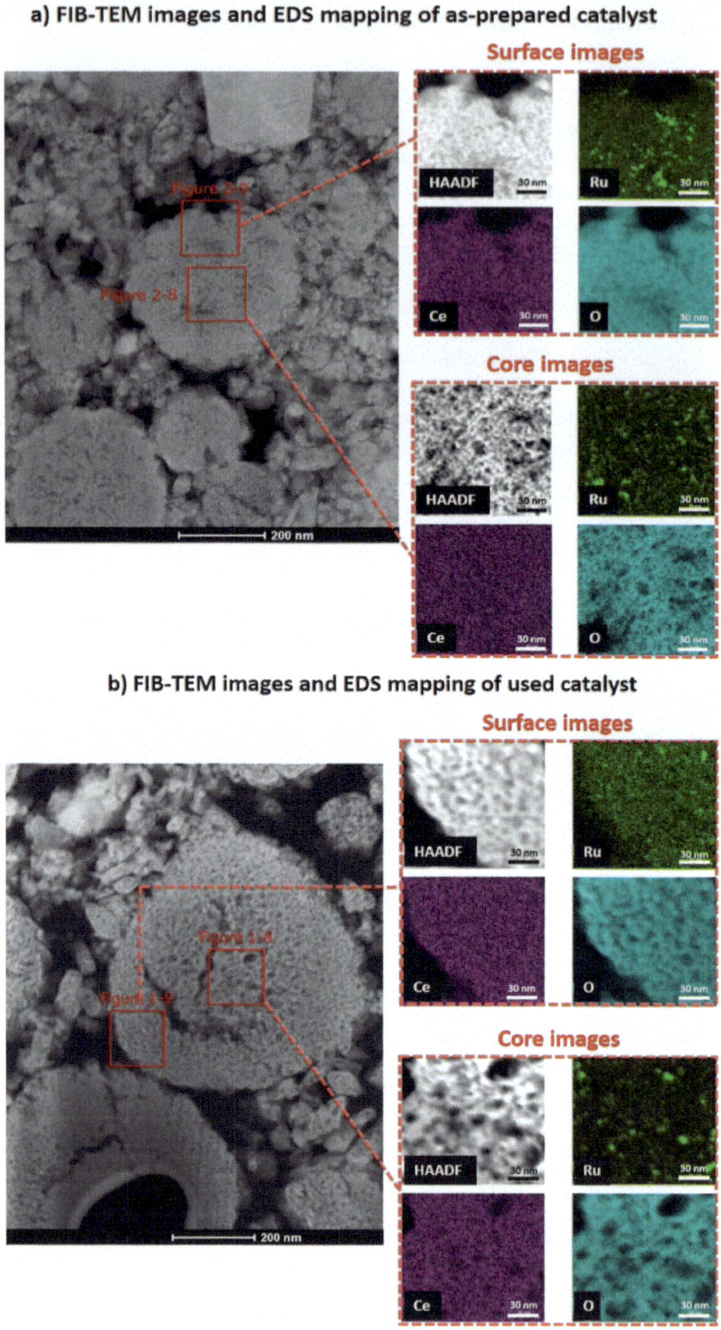

Fig. 17.4 Cross-sections of **a** fresh (as-prepared) and **b** used (time on stream: 170 h) 3 wt% Ru/CeO_2 catalyst. Reproduced with permission from [20]

17.2.3.2 *Reducibility of Ru/CeO_2*

Many analytical methods can be used to study the correlations between the ammonia synthesis activity of a catalyst and its structure and electronic state. A commonly used analytical method is temperature-programmed reduction by H_2 (H_2-TPR) [21]. Here, we discuss a H_2-TPR analysis of Ru catalysts, which revealed a correlation between the catalyst precursor and the ammonia synthesis activity of the resultant catalyst at low temperature (approx. 350 °C).

It is well known that the catalytic activity of metal catalysts depends greatly on the precursors used for their preparation. Several reasons exist for this, including contaminations, the structure of the metal derived from the precursor, and differences in the interactions between the metal precursor and the support. In Sects. 17.2.1 and 17.2.2, we have shown that this variability in catalytic activity holds true for Ru/CeO_2 catalysts. When various Ru/CeO_2 catalysts were examined using H_2-TPR, a characteristic low-temperature reduction peak was observed at a very low temperature (i.e., 50 °C). Plotting the temperature at which the reduction peaks occurred revealed a linear relationship between peak temperature and ammonia synthesis activity (Fig. 17.5) such that the lower the peak temperature, the higher the ammonia synthesis activity. The exact reason for this low-temperature reduction peak is unknown; however, it is assumed to be due to interaction between Ru and the CeO_2 support. The oxygen in CeO_2, which is enough to be reduced at low temperatures, is probably also reduced during the actual catalytic reaction (or in the Ru reduction before the reaction), which enhances electron donation from support material to Ru. The reason for the different peak temperatures for the different Ru catalysts is thought to be that each precursor affects the CeO_2 support differently during loading, i.e., the precursors alter to different extents the interaction of CeO_2 with oxygen around Ru, making it more susceptible to reduction.

17.3 Ru Catalysts Supported on Novel Carbon Materials

Novel carbon materials, such as activated carbon (AC), carbon nanotubes, and graphene, have attracted attention as potentially useful catalyst support materials because of their tunable physicochemical properties (e.g., surface area, porosity, crystallinity, electrochemical property, and framework structure) [22–24]. Pioneering work by Ozaki and Aika and others has revealed that AC-supported Ru catalysts are superior to conventional Fe-based catalysts in catalyzing the Haber–Bosch process, particularly under mild reaction conditions [13, 25–29]. However, industrial application of AC-supported Ru catalysts has been limited by the short lifetime of these materials, mainly due to the structural instability of AC under reducing atmospheres. The amorphous carbon framework of AC is easily decomposed by methanation of the framework via hydrogen spillover [30]. The framework also undergoes thermal decomposition, which is an issue for industrial application because high-temperature treatment is necessary to improve the stability of carbon materials. Also, thermally

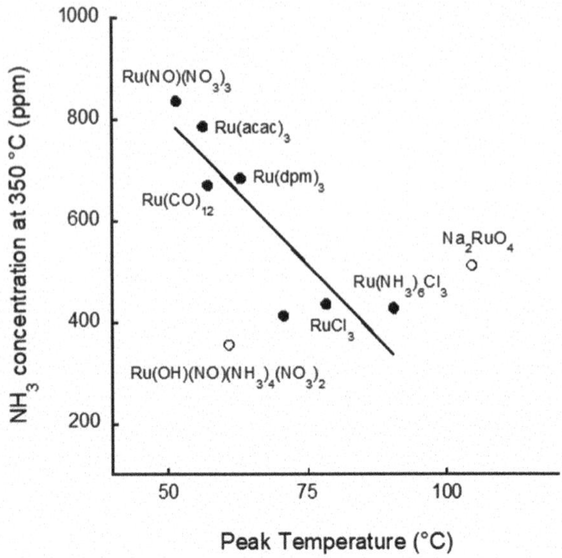

Fig. 17.5 Relationship between NH_3 synthesis activity and the temperature at which the low-temperature peak formed during H_2-temperature-programmed reduction analysis of Ru catalysts produced using the indicated catalyst precursors

treated carbon materials, particularly thermally treated AC, have low surface area and porosity, which limits their catalytic applications.

To address these issues, a novel mesoporous carbon material (MPC; CNovel P(3)010, Toyo Tanso) that is resistant to methanation and has a high surface area (930 m^2/g) and porosity (total pore volume: 2.29 mL/g, average pore size, 5.4 nm) was synthesized via a hard-templating method combined with high-temperature (1800 °C) annealing. In a typical preparation, Cs-Ru/MPC catalyst was produced by impregnating the MPC support with Cs and Ru by stepwise impregnation. The Ru loading was optimized at 10 wt% based on the weight of MPC. Cs loadings of 1.2–22 wt% were studied, which corresponded to a Cs/Ru molar ratio range of 0.1–2.5. From the view point of industrial application, $Ru(NO)(NO_3)_3$ and Cs_2CO_3 were chosen as the Ru and Cs precursors, respectively, because they are water soluble. For the preparation of a Ru reference catalyst, AC with a microporous carbon framework was used as the support material, and the AC support was impregnated with the Ru and Cs precursors using the same stepwise impregnation method. The AC was treated at 500 °C in H_2 for 3 h to clean the surface before impregnation. The surface area, total pore volume, and average pore size of the AC were 1260 m^2/g, 0.62 mL/g, and 0.8 nm, respectively. Table 17.1 summarizes the structural properties of the prepared Cs-Ru catalysts.

Table 17.1 Structural properties of the prepared Cs-Ru catalysts and reference materials [31]

Catalyst	S_{BET} (m^2/g)	V_{Total} (mL/g)	V_{Micro} (mL/g)[a]	V_{Meso} (mL/g)[b]	Average pore size (nm)	Ru size (nm)		TOF[e] (h^{-1})
						HRTEM[c]	CO Chem.[d]	
2.5Cs-Ru/MPC	430	1.33	0.17	1.16	5.4	2.1 ± 0.4	2.0 (45.8%)	224 (400 °C)
0.5Cs-Ru/MPC	750	1.84	0.27	1.57	5.4	1.9 ± 0.5	2.0 (45.0%)	114 (460 °C)
0.1Cs-Ru/MPC	840	1.91	0.35	1.56	5.4	1.9 ± 0.5	2.1 (44.2%)	63 (520 °C)
2.5Cs-Ru/AC	580	0.30	0.24	0.06	0.8	1.6 ± 0.4	6.2 (15.0%)	67 (430 °C)

[a] Microporous pore volume (V_{Micro}) was calculated using the Dubinin–Astakhov plot and α_S-plot method
[b] Mesoporous pore volume (V_{Meso}) was calculated as $V_{Total} - V_{Micro}$, where V_{Total} is the total pore volume
[c] Ru size determined by high-resolution transmission electron microscopy (HRTEM)
[d] Ru size determined by CO chemisorption. Percentages in parentheses are Ru dispersions calculated by CO chemisorption
[e] Turnover frequency. Temperatures in parentheses are the reaction temperatures at the maximum ammonia synthesis rate

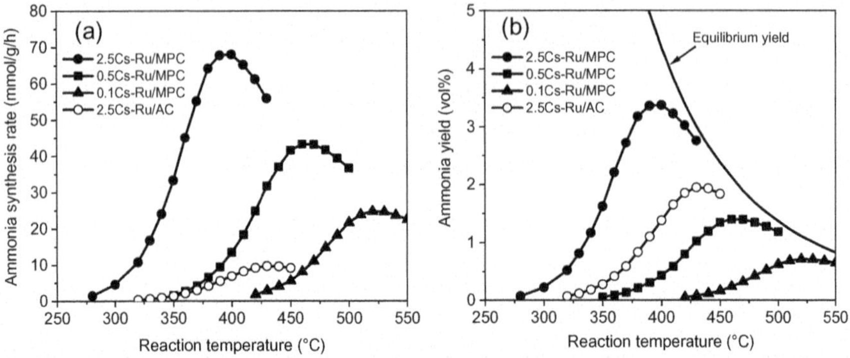

Fig. 17.6 Temperature dependence of **a** ammonia synthesis rate and **b** yield over *x*Cs-Ru/MPC catalysts and 2.5Cs-Ru/AC catalyst at 0.99 MPa, $H_2/N_2 = 3$, and gas hourly space velocity = 9000 h^{-1}. Reproduced from Ref. [31] with permission from the Royal Society of Chemistry

17.3.1 Ammonia Synthesis Activity of Ru/carbon

Figure 17.6 shows the influence of Cs/Ru molar ratio on the catalytic performance of the prepared Cs-Ru/MPC catalysts. The prepared catalysts were reduced at 450 °C under a H_2 flow for 2 h prior to use for ammonia synthesis. The reaction pressure, H_2/N_2 molar ratio, and gas hourly space velocity (GHSV) were set at 0.99 MPa, 3, and 9000 h^{-1}, respectively. The highest ammonia synthesis rate (68 mmol/g/h, corresponding to a NH_3 concentration of 3.4 vol%) was obtained with 2.5Cs-Ru/MPC (Ru loading: 10 wt%, Cs/Ru molar ratio: 2.5) at 400 °C. Lower ammonia synthesis rates and increases in reaction temperature were observed for 0.5Cs-Ru/MPC and 0.1Cs-Ru/MPC. Turnover frequency decreased in the order 2.5Cs-Ru/MPC > 0.5Cs-Ru/MPC > 0.1Cs-Ru/MPC (Table 17.1). These findings suggest that ammonia synthesis under mild conditions over Cs-Ru/MPC catalysts can be facilitated by optimization of the Cs loading.

17.3.2 Characterization of Ru/carbon

High-resolution transmission electron microscopy (HRTEM) images of 2.5Cs-Ru/MPC and 2.5Cs-Ru/AC are shown in Fig. 17.7. Few Ru particles (diameter approx. 2 nm) were observed on the 2.5Cs-Ru/MPC surface, whereas many more Ru particles were observed on the 2.5Cs-Ru/AC surface. Considering the pore sizes of the two catalysts, the images suggest that the Ru particles were loaded inside the MPC mesopores. Conversely, the HRTEM image of 2.5Cs-Ru/AC catalyst looks different from that of 2.5Cs-Ru/MPC, and it shows a high concentration of Ru particles (~2 nm) on surface. Because the micropores of AC are smaller than 1 nm, the 2 nm Ru particles are too big to enter the microporous, and they must stay on the AC

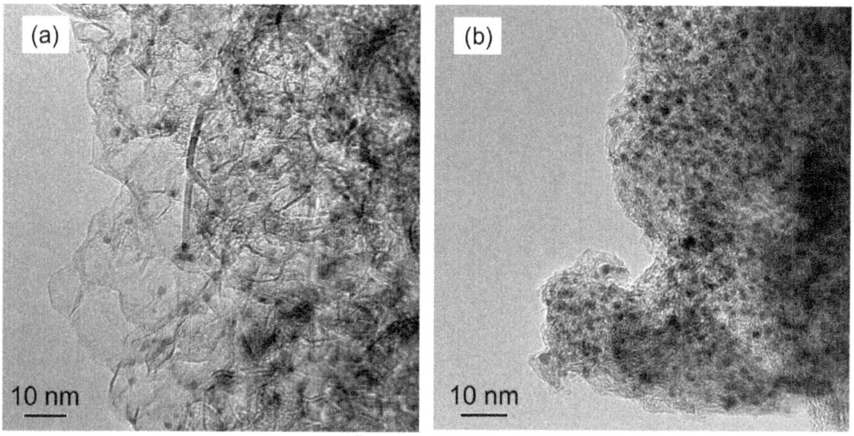

Fig. 17.7 High-resolution transmission electron microscopy images of 2.5Cs-Ru/MPC (**a**) and 2.5Cs-Ru/AC (**b**). Reproduced from Ref. [31] with permission from the Royal Society of Chemistry

external surfaces or nearby the pore mouths. Next, CO chemisorption was used to determine the size of the Ru particles in the Cs-Ru catalysts after reduction at 450 °C for 2 h. The results showed that for the Cs-Ru/MPC catalysts, Ru particle size was unaffected by the reduction treatment or Cs loading. In contrast, the Ru particle size in the Cs-Ru/AC catalyst was increased to 6.2 nm. This suggests that Ru particles on the surface of Cs-Ru/AC catalyst are unstable and that they aggregate under reducing conditions.

17.3.3 Durability Test of Ru/carbon

The changes in ammonia synthesis rate and ammonia yield during long-term ammonia synthesis (time on stream: 0–50 h) under intermittent operation conditions over 2.5Cs-Ru/MPC catalyst are shown in Fig. 17.8. The reaction temperature and GHSV ranges were 340–400 °C and 9000–18,000 h^{-1}, respectively. A total of eight 3-h fluctuation cycles (four each for temperature and GHSV) were used. When the temperature was varied, the ammonia synthesis rate remained in the range of 23–68 mmol/g/h (yield: 1.1–3.4 vol%). When the GHSV was varied, the ammonia synthesis rate remained in the range of 67.5–100 mmol/g/h (corresponding to 3.4–2.5 vol%). At 50-h time on stream, a slight decrease (about 1.5%) in catalytic performance was observed under SV-change ammonia synthesis conditions of 400 °C and 9000 h^{-1}. These findings indicate that 2.5Cs-Ru/MPC catalyst can stably catalyze ammonia synthesis under SV-change conditions. This stability is attributed to the Ru nanoparticles being confined inside the mesoporous carbon framework, as discussed in the previous section.

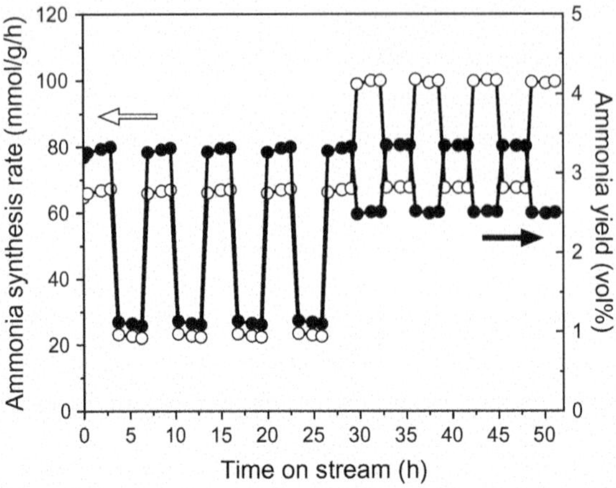

Fig. 17.8 Long-term ammonia synthesis performance over 2.5Cs-Ru/MPC under intermittent operation conditions. Reproduced from Ref. [31] with permission from the Royal Society of Chemistry

17.4 Summary

In this chapter, we have discussed the development of Ru catalysts for use in the synthesis of green ammonia using hydrogen derived from renewable energy. The production of green ammonia is characterized by an unpredictable hydrogen supply and frequent plant re-starts. In addition, renewable hydrogen is produced at ambient temperature and pressure, so measures are needed to increase the temperature and pressure before the hydrogen can be used for the production of ammonia. Ru catalysts are expected to provide the means of overcoming these issues. Here, for Ru/CeO_2 catalysts, we have shown that precursor selection has a marked effect on catalytic activity under both ambient-pressure and high-pressure conditions. We have also shown that the reducibility of CeO_2 affects the low-temperature activity of these catalysts, and that the Ru dispersion changes during the ammonia synthesis reaction. Furthermore, a kinetic analysis based on the Langmuir–Hinshelwood mechanism suggests that hydrogen inhibition and adsorption of intermediate NHx species on the catalyst surface influence the rate of ammonia formation. For Ru/C catalysts, MPC as the support material and Cs modification were used to lower the active temperature range. Cs-Ru/MPC catalysts showed good durability in a test of ammonia synthesis activity under intermittent reaction conditions over eight cycles and a time on stream of 50 h. Based on these results, optimized Ru catalysts were examined via demonstration tests using bench-scale ammonia synthesis plants; the results of these demonstration tests are discussed in other chapters.

References

1. American Institute of Chemical Engineers (2016) Introduction to ammonia production. https://www.aiche.org/resources/publications/cep/2016/september/introduction-ammonia-production. Accessed 25th Sep 2021
2. Wang Q, Guo J, Chen P (2019) Recent progress towards mild-condition ammonia synthesis. J Energy Chem 36:25–36
3. Nanba T, Nagata Y, Kobayashi K, Javaid R, Atsumi R, Nishi M, Mochizuki T, Manaka Y, Kojima H, Tsujimura T, Matsumoto H, Fujimoto T, Suzuki K, Ouchi T, Kameda S, Hoshino Y, Fujimoto S, Kai M, Fujimura Y (2021) Explorative study of a Ru/CeO_2 catalyst for NH_3 synthesis from renewable hydrogen and demonstration of NH_3 synthesis under a range of reaction conditions. J Jpn Petrol Inst 64:1–9
4. Izumi Y, Hoshikawa M, Aika K (1994) Adsorbed hydrogen effect on the adsorption and reactivity of N_2 molecules on Ru/MgO and Ru–Cs^+/MgO: hydrogen dipole effect enhanced by doped Cs^+. Bull Chem Soc Jpn 67:3191–3200
5. Kitano M, Inoue Y, Yamazaki Y, Hayashi F, Kanbara S, Matsuishi S, Yokoyama T, Kim SW, Hara M, Hosono H (2012) Ammonia synthesis using a stable electride as an electron donor and reversible hydrogen store. Nat Chem 4:934–940
6. Abe H, Niwa Y, Kitano M, Inoue Y, Sasase M, Nakao T, Tada T, Yokoyama T, Hara M, Hosono H (2017) Anchoring bond between Ru and N atoms of Ru/Ca_2NH catalyst: crucial for the high ammonia synthesis activity. J Phys Chem C 121:20900–20904
7. Sato K, Miyahara S, Ogura Y, Tsujimaru K, Wada Y, Toriyama T, Yamamoto T, Matsumura S, Nagaoka K (2020) Surface dynamics for creating highly active Ru sites for ammonia synthesis: accumulation of a low-crystalline, oxygen-deficient nanofraction. ACS Sustain Chem Eng 8:2726–2734
8. Aika K, Niwa Y (1999) Basic concepts and properties of new generation ammonia synthesis catalysts for industrial use. Stud Surf Sci Catal 121:327–332
9. Kadowaki Y, Aika K (1996) Promoter effect of Sm_2O_3 on Ru/Al_2O_3 in ammonia synthesis. J Catal 161:178–185
10. Kowalczyk Z, Krukowski M, Raróg-Pilecka W, Szmigiel D, Zielinski J (2003) Carbon-based ruthenium catalyst for ammonia synthesis: role of the barium and caesium promoters and carbon support. Appl Catal A 248:67–73
11. Aika K, Kawahara T, Murata S, Onishi T (1990) Promoter effect of alkali metal oxides and alkali earth metal oxides on active carbon-supported ruthenium catalyst for ammonia synthesis. Bull Chem Soc Jpn 63:1221–1225
12. Lin B, Guo Y, Cao C, Ni J, Lin J, Jiang L (2018) Effect of Ru loading and of Ru precursor in Ru/C catalysts for ammonia synthesis. Catal Today 316:230–236
13. Rossetti I, Mangiarini F, Forni L (2007) Promoters state and catalyst activation during ammonia synthesis over Ru/C. Appl Catal A 323:219–225
14. Lin B, Wang R, Lin J, Ni J, Wei K (2011) Effect of chlorine on the chemisorptive properties and ammonia synthesis activity of alumina-supported Ru catalysts. Catal Lett 141:1557–1568
15. Nanba T, Javaid R, Matsumoto H (2019) Ammonia synthesis by using hydrogen produced from renewable energy. Shokubai 61:66–71
16. Liu P, Niu R, Li W, Wang S, Li J (2019) Morphology effect of ceria on the ammonia synthesis activity of Ru/CeO_2 catalysts. Catal Lett 149:1007–1016
17. Lin B, Liu Y, Heng L, Wang X, Ni J, Lin J, Jiang L (2018) Morphology effect of ceria on the catalytic performances of Ru/CeO_2 catalysts for ammonia synthesis. Ind Eng Chem Res 57:9127–9135
18. Ma Z, Zhao S, Pei X, Xiong X, Hu B (2017) New insights into the support morphology-dependent ammonia synthesis activity of Ru/CeO_2 catalysts. Catal Sci Technol 7:191–199
19. Wirth R (2009) Focused Ion Beam (FIB) combined with SEM and TEM: Advanced analytical tools for studies of chemical composition, microstructure and crystal structure in geomaterials on a nanometre scale. Chem Geo 261:217–229

20. Javaid R, Nanba T (2021) Effect of reaction conditions and surface characteristics of Ru/CeO_2 on catalytic performance for ammonia synthesis as a clean fuel. Int J Hydrogen Energy 46:18107–18115
21. Manaka Y, Nagata Y, Kobayashi K, Kobayashi D, Nanba T (2020) The effect of a ruthenium precursor on the low-temperature ammonia synthesis activity over Ru/CeO_2. Dalton Trans 49:17143–17146
22. Aika K, Hori H, Ozaki A (1972) Activation of nitrogen by alkali metal promoted transition metal I. Ammonia synthesis over ruthenium promoted by alkali metal. J Catal 27:424–431
23. Guo S, Pan X, Gao H, Yang Z, Zhao J, Bao X (2010) Probing the electronic effect of carbon nanotubes in catalysis: NH_3 synthesis with Ru nanoparticles. Chem Eur J 16:5379–5384
24. Zhao J, Zhou J, Yuan M, You Z (2017) Controllable synthesis of Ru nanocrystallites on graphene substrate as a catalyst for ammonia synthesis. Catal Lett 147:1363–1370
25. Ozaki A, Aika K, Hori H (1971) A new catalyst system for ammonia synthesis. Bull Chem Soc Jpn 44:3216
26. Truszkiewicz E, Raróg-Pilecka W, Schmidt-Szałowski K, Jodzis S, Wilczkowska E, Łomot D, Kaszkur Z, Karpiński Z, Kowalczyk Z (2009) Barium-promoted Ru/carbon catalyst for ammonia synthesis: state of the system when operating. J Catal 286:181–190
27. Lin B, Qi Y, Guo Y, Lin J, Ni J (2015) Effect of potassium precursors on the thermal stability of K-promoted Ru/carbon catalysts for ammonia synthesis. Catal Sci Technol 5:2829–2838
28. Fernández C, Sassoye C, Debecker DP, Sanchez C, Ruiz P (2014) Effect of the size and distribution of supported Ru nanoparticles on their activity in ammonia synthesis under mild reaction conditions. Appl Catal A 474:194–202
29. Hansen TW, Hansen PL, Dahl S, Jacobsen CJH (2002) Support effect and active sites on promoted ruthenium catalysts for ammonia synthesis. Catal Lett 84:7–12
30. Tsyrul'nikov PJ, Iost KN, Shitova NB, Temerev VL (2016) Methanation of the carbon supports of ruthenium ammonia synthesis catalysts: a review. Catal Ind 8:341–347
31. Nishi M, Chen SY, Takagi H (2020) X-ray absorption spectroscopy of Ba- and Cs-promoted Ru/mesoporous carbon catalysts for long-term ammonia synthesis under intermittent operation conditions. Sustain Energy Fuels 4:832–842

Chapter 18
Kinetic Analysis of Ammonia Production on Ru Catalyst Under High Pressure Conditions

Rahat Javaid, Tetsuya Nanba, and Hideyuki Matsumoto

Abstract To promote deployment of ammonia production system using hydrogen generated by renewable energy, it is necessary for operation of catalytic reactor to vary process parameters, such as temperature, pressure, space velocity and ratio for flow rate of hydrogen to nitrogen, according to the situation of hydrogen supply and forecast of renewable energy supply. Therefore, development of kinetic model that can be applied to various operating conditions is important for design of the ammonia production system. In this chapter, kinetic analysis of ammonia synthesis on ruthenium-based catalysts under high pressure conditions for simulating the reactor situation under varying hydrogen supply is discussed. Various kinetic analysis on iron-based catalysts is considered, and then the reaction rate equation for ruthenium-based catalysts is estimated. In analysis of kinetics of ammonia synthesis over Ru/Cs/MgO catalyst, ammonia production on this catalyst is well explained by the Langmuir–Hinshelwood model. In kinetic analysis for Ru/CeO_2, an optimized kinetic model on the catalyst is showed to be different from that on the above-mentioned Ru/Cs/MgO catalyst. And it could be estimated that Ru/CeO_2 performs favorable activity as an ammonia synthesis catalyst under high pressure conditions. Additionally, the intermediate surface species (NH_x) on Ru/CeO_2 is suggested to be contribute to the total rate of ammonia synthesis under high pressure conditions.

Keywords Ammonia synthesis · Kinetic analysis · Ruthenium-based catalyst · Langmuir–Hinshelwood model · Renewable energy

R. Javaid · T. Nanba · H. Matsumoto (✉)
Department of Chemical Science Engineering, Tokyo Institute of Technology, Ohokayama, Meguro-ku, Tokyo 152-8550, Japan
e-mail: matsumoto.h.ae@m.titech.ac.jp

R. Javaid · T. Nanba
National Institute of Advanced Industrial Science and Technology (AIST), Onogawa, Tsukuba, Ibaraki 311-1393, Japan

K. Aika and H. Kobayashi (eds.), *CO_2 Free Ammonia as an Energy Carrier*,
https://doi.org/10.1007/978-981-19-4767-4_18

18.1 Introduction

To reduce emission of the greenhouse gases, especially for CO_2, the source of hydrogen supply is expected to replace from the conventional fossil fuels to production of hydrogen derived from renewable energy. For ammonia synthesis process, so-called green ammonia synthesis will be required. As described in Chap. 17, the green ammonia synthesis process needs should be developed by considering the different points from the traditional process to obtain high efficiency of ammonia synthesis. Especially for the operating condition of the catalytic reactor, reaction parameters, such as temperature, pressure, space velocity and H_2/N_2 ratio, should be varied according to the situation of hydrogen supply and forecast of renewable energy supply. Moreover, dynamic change in reaction parameters is expected to provide the minimization of the buffer size for the hydrogen tank and/or the buttery.

On the other hands, the transition of operating conditions makes difficult to simulate the temperature distribution of catalysts bed and to predict ammonia production, especially when the power-law model for reaction rate equation [1] that is developed by using data for catalytic activity tests. Also, the linearized model for reaction rate equation [2] could be utilized for design of the steady-state process but is difficult to be applied to simulation in wide range of operating conditions.

Therefore, development of kinetic model that can be applied to various operating conditions is important. Especially, developed model for reaction rate equation should contain the term for rate of reverse reaction, to analyze the behavior of ammonia synthesis under conditions of high concentration of ammonia inside the reactor. Moreover, kinetic analysis for the variable condition contribute to optimization of the catalytic reactor.

In this chapter, kinetic analysis of ammonia synthesis on Ru catalysts under high pressure conditions for simulating the reactor situation under varying hydrogen supply is discussed. Various kinetic analysis on Fe catalysts is considered [3–6], and then the reaction rate equation for Ru catalysts is estimated.

18.2 Kinetics of Ammonia Synthesis Over Ru/Cs/MgO Catalyst

To develop Ru/CeO_2 catalysts for the practical synthesis of ammonia from renewable hydrogen, detailed analysis of reaction kinetics is necessary for designing a reactor. Here, a kinetic analysis based on proposed elementary reaction steps was performed to obtain the reaction rate equation and to allow discussion of the reaction mechanisms. As discussed in Chap. 17, the measured value of the reaction order with respect to hydrogen varies depending on the pressure. In other words, the surface-adsorbed species are likely to strongly affect the rate of NH_3 formation. Therefore, the influence of intermediate species (e.g., NH_x) on the catalyst surface on the rate of ammonia production was investigated by using a reaction rate equation model based on the Langmuir–Hinshelwood (LH) mechanism.

First, we examined the behavior of ammonia production by using a modified LH model (with strong H adsorption in addition to nitrogen atom) derived by considering the following five elementary steps [1]:

(i) $N_2 + 2^* \leftrightarrow 2N(a)$
(ii) $H_2 + 2^* \leftrightarrow 2H(a)$
(iii) $N(a) + H(a) \leftrightarrow NH(a)+^*$
(iv) $NH(a) + H(a) \leftrightarrow NH_2(a)+^*$
(v) $NH_2(a) + H(a) \leftrightarrow NH_3 + 2^*$

where * indicates empty sites of the catalyst surface. The rate-determining step for this model was assumed to be dissociation of nitrogen (elementary reaction step (i)), because the energy of the nitrogen triple bond is 947 kJ/mol, and its dissociation requires a very large energy. The rate equation for ammonia production is expressed as

$$r = k_1 p_{\mathrm{N}} \theta_v^2 - k_2 \theta_{\mathrm{N}}^2 \tag{18.1}$$

where p is the partial pressure and θ is the concentration of adsorbed chemical species on the surface of the catalyst, which is called "coverage". The subscripts N and v indicate nitrogen and empty sites, respectively, and k_1 and k_2 represent the rate constants for the forward and backward reactions, respectively.

Because hydrogen inhibition is often observed for Ru catalysts, it was assumed that both adsorbed hydrogen and adsorbed nitrogen retarded dinitrogen activation. When the site balance on the surface of the catalyst is expressed by Eq. 18.2, Eq. 18.3 can be derived from Eq. 18.1 under the assumption that the other elementary reaction steps (ii–v) are fast enough to be at equilibrium.

$$\theta_v + \theta_{\mathrm{H}} + \theta_{\mathrm{N}} = 1 \tag{18.2}$$

$$r = \frac{k_1 \left[p_{\mathrm{N}} - p_{\mathrm{A}}^2 / \left(K_p^2 p_{\mathrm{H}}^3 \right) \right]}{\left[1 + (K_1 p_{\mathrm{H}})^{0.5} + \left(K_1^{-1.5} K_2^{-1} \right) p_{\mathrm{A}} p_{\mathrm{H}}^{-1.5} \right]^2} \tag{18.3}$$

where the subscripts H and A indicate hydrogen and ammonia, respectively; K_{p} represents the reaction equilibrium constant at a given pressure; and K_1 and K_2 are equilibrium constants that are expressed by Eqs. 18.4 and 18.5.

$$K_1 = \frac{\theta_{\mathrm{H}}^2}{p_{\mathrm{H}} \theta_v^2} \tag{18.4}$$

$$K_2 = \frac{p_{\mathrm{A}} \theta_v^4}{\theta_{\mathrm{N}} \theta_{\mathrm{H}}^3} \tag{18.5}$$

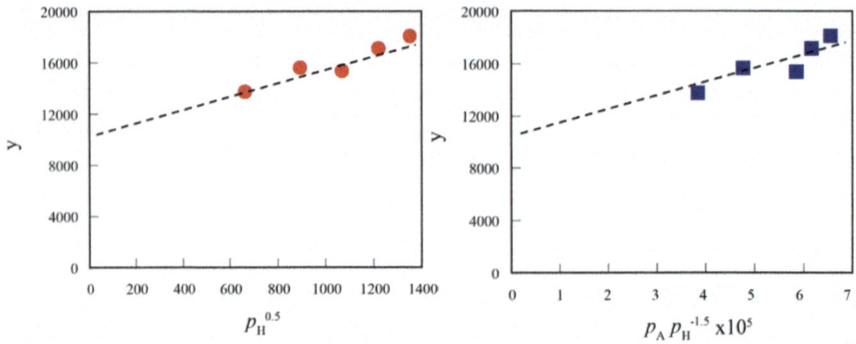

Fig. 18.1 Langmuir–Hinshelwood plots (Eq. 18.6) for the ammonia synthesis activity of Ru/Cs/MgO catalyst

For graphical analysis of the experimental data, Eq. 18.3 can be rewritten as

$$y = \left[\frac{p_N - p_A^2/(K_p^2 p_H^3)}{r}\right]^{1/2} = \frac{1}{\sqrt{k_1}} + \sqrt{\frac{K_1}{k_1}}p_H^{0.5} + \left(\frac{1}{\sqrt{k_1}}\right)\left(\frac{1}{K_1^{1.5}K_2}\right)p_A p_H^{-1.5} \tag{18.6}$$

This rate equation for the experimental data for ammonia synthesis was applied to the kinetic data for a Ru/Cs/MgO catalyst, which has been widely used as the reference Ru catalyst in many literatures. Figure 18.1 shows LH plots of rate expression (left-hand side of Eq. 18.6) versus ${p_H}^{0.5}$ or ${p_A p_H}^{-1.5}$ (right-hand side of Eq. 18.6) for a case when the operating temperature was 400 °C and the H_2/N_2 ratio was 3. The calculated value of the left-hand side of Eq. 18.6 increased with increasing pressure when plotted against ${p_H}^{0.5}$ and ${p_A p_H}^{-1.5}$. The linearity of the plot and the positive Y-intercept indicate that the ammonia production on this catalyst is well explained by the modified LH model described by Eq. 18.3.

18.3 Kinetics of Ammonia Synthesis Over Ru/CeO_2 Catalysts

Next, the modified LH model was used for kinetic analysis of ammonia synthesis over a Ru/CeO_2 catalyst (Fig. 18.2). The plot of reaction rate versus ${p_A p_H}^{-1.5}$ did not show linearity, and an inflection point was observed at 4 MPaG. Because the apparent activation energy was seen to change around 4 MPaG [7], it was inferred that the inflection point indicated kinetics not considered by the LH model above discussed. Thus, the decrease in ${p_A p_H}^{-1.5}$ at pressures over 4 MPaG suggested that reconsideration of the model was needed to account for retardation of the ammonia synthesis reaction by adsorption of hydrogen and ammonia.

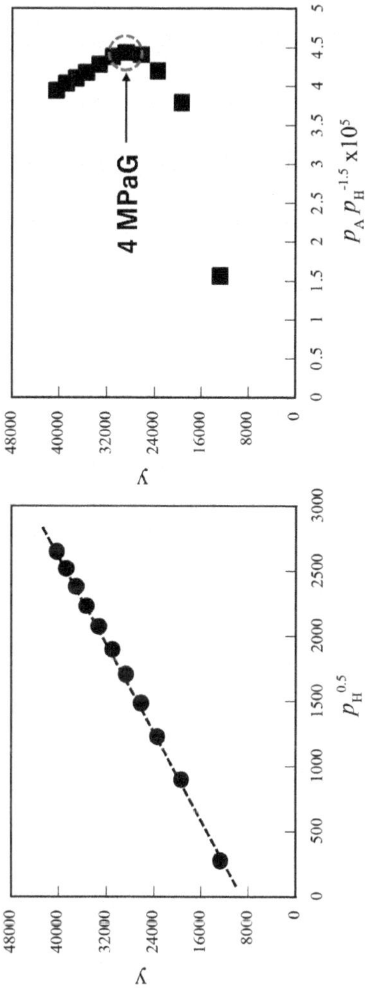

Fig. 18.2 Langmuir–Hinshelwood plot (Eq 18.6) for the ammonia synthesis activity of Ru/CeO_2 catalyst at 400 °C

Thus, the pressure dependence of the kinetic parameters was investigated by using the power-law reaction rate equation model expressed by Eq. 18.7.

$$r = k_0 \exp\left(-\frac{E}{RT}\right) p_N^l p_H^m p_A^n \tag{18.7}$$

Considering that the amount of catalyst in the reactor tube in the catalytic activity test was very small, the frequency factor k_0 was estimated using the continuous stirred tank reactor model (Eq. 18.8):

$$\frac{W}{F_0} = \left(\frac{X_f}{n}\right) \sum_{i=1}^{n} \left[\frac{1}{r}\right] \tag{18.8}$$

where W and F_0 represent the catalyst weight and feed rate of hydrogen, respectively. X_f indicates the conversion of hydrogen at the outlet of the reactor tube, which was determined from the catalytic activity tests. Table 18.1 shows the values of W and F_0 and the parameter values for Eq. 18.7 that were used. In Eq. 18.8, n in represents the number of continuous stirred tank reactors.

Figure 18.3 shows plots of estimated values for k_0 when n is 8 by referring to a previous study [8]. The apparent k_0 increased dramatically above 4 MPaG. The apparent k_0 can be considered to reflect the number of active sites on the surface of the catalyst, but it is unclear whether the observed increase in k_0 indicated an increase in active sites. Further characterization under high pressure conditions would clarify this point.

Upon reconsideration of the application of the above-mentioned LH model, it was assumed that the increase in the number of active sites was due to the contribution of intermediate NHx species on the surface of the catalyst during ammonia formation. Assuming that adsorption of NH_2 influences the coverage balance equation (Eqs. 18.9 and 18.10) can be derived from Eq. 18.1 under the same assumption as that made in the derivation of Eq. 18.3.

$$\theta_v + \theta_H + \theta_{NH_2} = 1 \tag{18.9}$$

$$r = \frac{k_1\left[p_N - p_A^2/\left(K_p^2 p_H^3\right)\right]}{\left[1 + (K_1 p_H)^{0.5} + K_a' p_A p_H^{-0.5}\right]^2} \tag{18.10}$$

Figure 18.4 shows an example of the data obtained by applying Eq. 18.11, which was derived from Eq. 18.10, to experimental data obtained at a reaction temperature

Table 18.1 Operating conditions and kinetic parameters for Eq. 18.7

W (kg)	F_0 (mol/min)	E (kJ/mol)	l (–)	m (–)	n (–)
3.66×10^{-3}	3.66×10^{-7}	81.0	0.37	0.30	−2.1

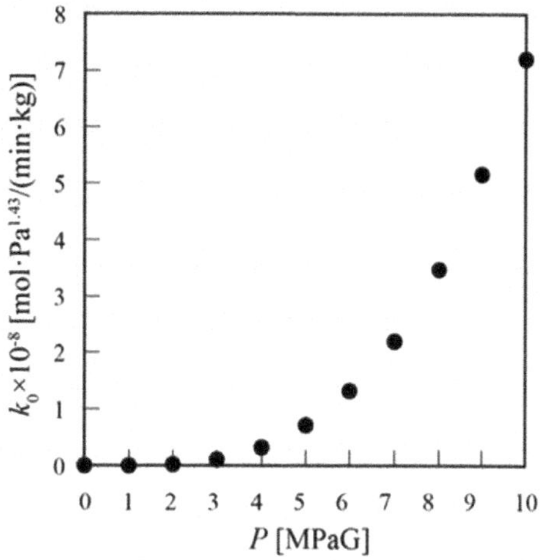

Fig. 18.3 Relationship between pressure and frequency factor for NH_3 synthesis over Ru/CeO_2 at 400 °C

of 400 °C.

$$y = \left[\frac{p_N - p_A^2/(K_p^2 p_H^3)}{r}\right]^{1/2} = \frac{1}{\sqrt{k_1}} + \sqrt{\frac{K_1}{k_1}} p_H^{0.5} + \left(\frac{K_a'}{\sqrt{k_1}}\right) p_A p_H^{-0.5} \quad (18.11)$$

The linearity with respect to $p_A p_H{}^{-0.5}$ (Fig. 18.4) suggests that the model considering the coverage of the intermediate NH_x species is applicable for interpretation of the behavior of ammonia production over Ru/CeO_2 catalysts.

From the above kinetic analysis, Ru/CeO_2 showed favorable activity as an ammonia synthesis catalyst under high pressure conditions. However, it was also revealed that more detailed analyses of the behavior of intermediate NHx species on the catalyst surface are needed. Future macro-kinetic analyses using various coverage

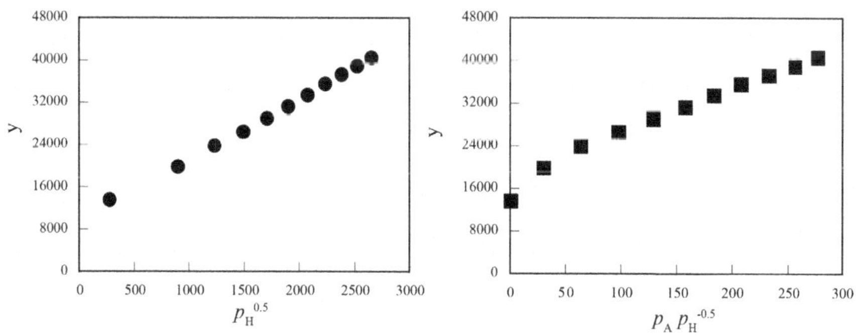

Fig. 18.4 Langmuir–Hinshelwood plot (Eq. 18.11) for the ammonia synthesis activity of Ru/CeO_2 catalyst at 400 °C

balance equations will further our understanding of the mechanisms of ammonia formation and provide kinetic parameters for the design of optimized reactors.

18.4 Summary

As a summary of kinetic analysis of ammonia synthesis on Ru catalysts, an optimized kinetic model on Ru catalysts is suggested to be different from that on Fe catalysts. Additionally, the intermediate surface species (NH_x) on Ru/CeO_2 is suggested to be contribute to the total rate of ammonia synthesis under high pressure conditions.

References

1. Aika K (2017) Role of alkali promoter in ammonia synthesis over ruthenium catalysts-Effect on reaction mechanism. Catal Today 286:14–20
2. Reese M, Marquart C, Malmali M, Wagner K, Buchanan E, McCormick A, Cusslerv EL (2016) Performance of a small-scale Haber process. Ind Eng Chem Res 55:3742–3750
3. Aika K, Ozaki A (1969) Kinetics and isotope effect of ammonia synthesis over an unpromoted iron catalyst. J Catal 13:232–237
4. Kubota H, Shindo M (1959) The rate expression of high pressure ammonia synthesis in which industrial iron catalyst is employed. Kagakukogaku 23:242–249
5. Araújo A, Skogestad S (2008) Control structure design for the ammonia synthesis process. Comput Chem Eng 32:2920–2932
6. Murase A, Roberts HL, Converse AO (1970) Optimal thermal design of an autothermal ammonia synthesis reactor. Ind Eng Chem Proc Des Dev 9:503–513
7. Nanba T, Javaid R, Matsumoto H (2019) Ammonia synthesis catalyst for storage of renewable hydrogen (in Japanese). Shokubai 61:66–71
8. Siporin SE, Davis RJ (2004) Use of kinetic models to explore the role of base promoters on Ru/MgO ammonia synthesis catalysts. J Catal 225:359–368

Chapter 19
Importance of Separation Unit in Green Ammonia Synthesis Systems

Ken-ichi Aika

Abstract The difference between the "blue ammonia" process and the "green ammonia" process mostly arises from the hydrogen production unit: methane reforming (blue) vs water hydrolysis (green). However, the synthesis loop from hydrogen and nitrogen might also be changed drastically in the green ammonia process. One of the challenges is the use of an ammonia absorber for the ammonia separator. Advances in addressing these challenges are reviewed in this chapter.

Keywords Ammonia synthesis · Separation · Absorber · Alkali earth metal halide · Recycle loop model

19.1 Introduction; Features of Green Ammonia Processes

In the latter part of Chap. 1, "Evaluation of blue and green ammonia from an energy science view" was discussed [1]. The comparisons are summarized as follows. Methane-based ammonia synthesis (i.e., blue ammonia) has been under development for many years and optimized year by year. The energy loss is said to be about 11 GJ/t-NH_3 or below. Air is used as both the oxidant and N_2 source. The steam by-product is used for compressors, etc. CO_2 from the methane reforming reactor is easily separated with an amine solution and can be used as urea source or enhanced oil recovery source for blue ammonia credit. Since the carbon footprint of blue ammonia might not be so low, it might face a carbon restriction policy, such as a tax, at some point.

The world expects green ammonia as a next technology for green fuel, replacing blue ammonia. The process is much simpler than the methane process. However, the capex of electrolysis might be high due to its layer structure with two dimensions and to low operation hours coming from an unstable renewable energy supply that may require some sort of battery to absorb the power wave. The strongest advantage of green ammonia is the cheap renewable power available in some areas of the world.

K. Aika (✉)
Professor Emeritus, Tokyo Institute of Technology, 2-12-1 Ohokayama, Meguro-ku, Tokyo 152-8550, Japan
e-mail: kenaika@c01.itscom.net

K. Aika and H. Kobayashi (eds.), *CO_2 Free Ammonia as an Energy Carrier*,
https://doi.org/10.1007/978-981-19-4767-4_19

Sunlight density differs over time depending on the geographical condition on the earth. Cheap renewable power at such districts may overcome the high capex of the green hydrogen and ammonia production units.

To supply steady H_2 to the green ammonia synthesis process, extra hydrogen tanks may be required to store the unstable H_2 produced. Despite problems due to the unstable power supply, there are several advantages for a green ammonia system, for example, (1) an easy way of supplying high-pressure H_2 by pressurizing water in the electrolysis unit, (2) the use of electric motors for utility works (cheap electric power), (3) no CO_2 absorber unit, and (4) less stringent purging of inert gases (Ar & methane) from the recycle loop. The synthesis loop itself must be so simple that it may help to reduce the whole process capex. The green ammonia synthesis loop operating at present is generally composed of a synthesis reactor and a condensation unit, as shown in Fig. 1.7 in Chap. 1 [1]. The purpose of this chapter is to review any challenges concerning the synthesis loop with a new separation unit.

19.2 Models of Ammonia Separation

$$1/2\,N_2 + 3/2\,H_2 \rightarrow NH_3 \quad \Delta H = -46.1\,\text{kJ} \tag{19.1}$$

As is well known, ammonia synthesis from hydrogen and nitrogen is a volume-decreasing and exothermic reaction (Eq. 19.1). Thus, the equilibrium concentration of ammonia is low at low pressure and high temperature. Under total pressure of 100 bar, the equilibrium concentrations are 25% at 700 K (427 °C) and 30% at 650 K (377 °C). As a standard model of green ammonia production, the reactor is assumed to be operated at around 400 °C and under 100 bars. Separation may be done either by liquification or absorption. We suppose that ammonia with partial pressure of 18.6 bar is generated from a reactor operated under 100 bar and at 400 °C. We further suppose two cases, (1) most of the outcome ammonia is quenched at −20 °C in a condenser but with some (2 bar) left as gas and recycled, and (2) most of the outcome ammonia is absorbed at some temperature (near the reaction temperature) in a separator but with some (2 bar) left as gas and recycled. The latter situation is shown in Fig. 19.1 as a model for discussion.

In this case, synthesis gas ($N_2 + 3H_2$) is fed into the reactor at the rate of 20 mol/h with produced NH_3 at a rate of 10 mol/h. Thus, introduced syngas meets with unreacted syngas (48 mol/h) and unrecovered ammonia stream (1 mol/h) making a total flow of 69 mol/h going into the synthesis reactor. Inside the reactor, 20 mol/h of syngas turns into 10 mol/h of ammonia resulting in an outlet stream of 48 mol/h of syngas and 11 mol/h of NH_3. If the total pressure is controlled at 100 bar, the partial pressures of N_2, H_2, and NH_3 are 20.3, 61, and 18.6 bar, respectively. The conversion is calculated as 29.4%.

The rate ratio of recycle gas (48 mol/h) against feed gas (20 mol/h) is 2.4 in this case with a conversion of 29.4%. The ratio is high if the conversion is low. In the case of liquification separation, the same recycle ratio is assumed, but the feed gas

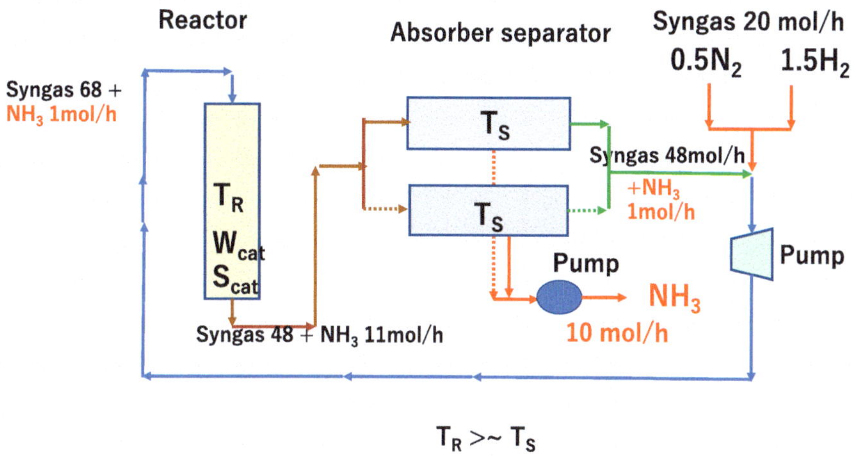

Fig. 19.1 A bench-scale model of a green ammonia synthesis system (4.1 kg/day production case) as an example. T_R: reactor temperature, T_S: separator temperature, W_{cat}: catalyst weight, S_{cat}: catalyst surface area

must pass through 400 °C and −20 °C units. This will cause much energy loss. If absorption separation can be done at a temperature near to that of the convertor, energy loss will be small. The absorption separator must serve to reduce the operation cost.

In this respect, we surveyed the past work of ammonia separation, seeking methods for practical use. Representatives of each method in the literature are summarized in Table 19.1.

The present author once proposed a small NH_3 synthesis unit for supplying NH_3 to a mobile de-NOx unit as shown in Fig. 19.2 [2]. 176 g of 5% Ru-Cs^+/MgO catalyst produces ammonia of 176 mmol/min at 350 °C under 10 bar. H_2SO_4 treated AC was selected as the NH_3 separator, 1 g of which adsorbs 5.2 mmol of NH_3 under 300 Torr (40 kPa) at 25 °C and desorbs almost completely, leaving 0.7 mmol at 200 °C, i.e., **4.5 mmol NH_3/g** is separated out as shown in Fig. 19.3. Since the reactor produces 10.56 mol (176 mmol/min × 60 min) of NH_3 in an hour, 2350 g (10.56/0.0045) of adsorber is necessary for an hour separation operation. The weight ratio of two separators against the catalyst is 26.7 due to the poor ammonia adsorption ability, which may be too high for a practical apparatus. Of course, if a shorter separation cycle time is applied, the size requirement might be reduced. H_2SO_4 or HNO_3 treatment increases the amount of NH_3 adsorption on AC, where the amount of surface organic oxygen is increased [3].

Zeolites with small pore sizes have the potential to adsorb small molecules like ammonia selectively. An example is shown in a patent [4], where six columns of Baylith SG-232 (pore size of 3A and 1.6 mm grain diameter) are used for NH_3 (3.6%) separation from N_2 (49.9%) and H_2 (46.5%) stream (4.7 NL/h) under 2.55 bar and ambient temperature. Three sets of pressure swing units were used, which successfully separated the gas mixture to 97% NH_3 stream (0.17 NL/h, 1 bar) and 99.8%

Table 19.1 Attempted ammonia separation methods other than condensation

Principle	Media (material)	Temp. °C	Separation method	Features	Literature
Adsorption	H_2SO_4/AC	0–200 °C	Pressure or temp. swing	Small adsorption amount; large separation unit	[2, 3]
Adsorption	Zeolite	Room temp.	Pressure swing	Small adsorption amount; short swing time	[4–6]
Gas absorption into solid	Solid sulfamide, $(NH_2)_2SO_2$	0 °C	Pressure swing	Deliquescence state of mixture; NH_3/sulfamide is not high: 2.48 (1 bar), 0.84 (0.4 bar)	[9]
Gas absorption into liquid	$(NH_4)_X H_{(2-X)}PO_4$ (X = 1.75)	70 °C	Temp. swing	Solution heated to 180 °C (distill. column) releasing NH_3 leaving $(NH_4)_X H_{(2-X)}PO_4$ (X = 1.25) sol	[10]
Gas absorption into ionic liquid	$[Bmim]_2[Co(NCS)_4]$/silica gel: 49/51 by wt.	Room temp.	Pressure swing	10 wt% NH_3 is absorbed into the composite in 15 min and released completely in 15 min	[11]
Film separation	Silica film	around 400 °C	Permeation with pressure difference	Separate H_2 from N_2-NH_3 by 1st film and NH_3 from N_2 by 2nd film at synthesis temp. NH_3 permeation is low	[12, 13]
Film separation	Zeolite film Si/Al: 2–10 N/Al: 0.01–4	50–500 °C	Permeation with pressure difference	One-step separation of NH_3 from N_2-H_2 NH_3 permeation is high Needs practical tests	[14, 15]

(continued)

Table 19.1 (continued)

Principle	Media (material)	Temp. °C	Separation method	Features	Literature
Gas absorption into solid	Solid alkali earth halides	Room temp. to 300 °C	Pressure or temp. swing	Molecular complex formation. MgCl(OH) separates 5.5 times more NH_3 than Na-Y-zeolite	[16, 17]

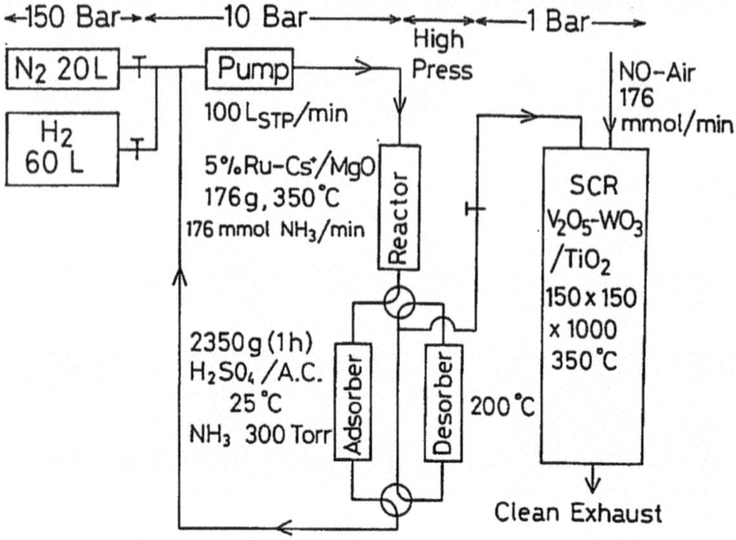

Fig. 19.2 A proposed small-scale ammonia production unit with ammonia adsorption separator [2]. Reproduced with permission from [2]

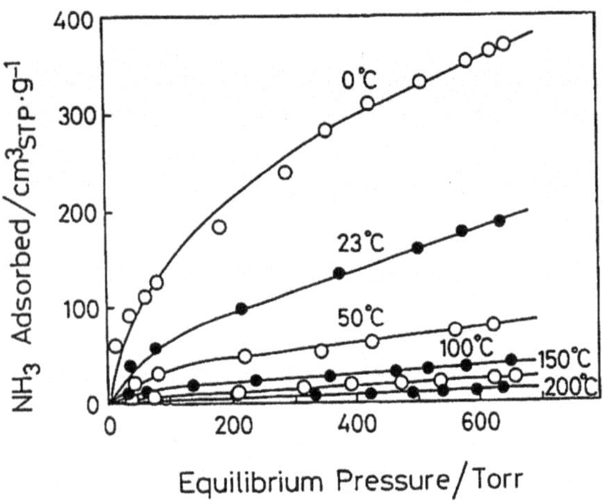

Fig. 19.3 Ammonia adsorption isotherm on H_2SO_4 treated active carbon [2]. Reproduced with permission from [2]

synthesis gas (4.61 NL/h, 2.5 bar). Two columns in a unit work in turn for adsorption and desorption of NH_3 with a cycle of 40 s. The separation is remarkable, but the amount of ammonia absorption seems to be so small that the absorbent is saturated in a short time. This might task the pumps with high load; however, the inventors claim that the energy loss is still much less than the case of any film separation methods. A

Fig. 19.4 A bench plant of green ammonia production at Institute for Environmental Science, Aomori prefecture, Japan, installed in 1994

reversible amount of NH_3 adsorption on Cu-Y-zeolite under 40 kPa has been reported as **6.49 mmol NH_3/g** [5, 6], which is a similar level to the case of H_2SO_4 treated AC [2, 3]. Though the adsorption amount might be low, the Baylith SG-232 absorber was installed in a bench plant of ammonia production (using electric power and water: a model of green ammonia production) at Institute for Environmental Science, Aomori prefecture in 1994 (Fig. 19.4). For this system, the present author supplied a synthesis catalyst, Ru-BaO/Active-carbon, which had been studied extensively [7, 8].

Gaseous NH_3 is selectively absorbed into solid sulfamide, $(NH_2)_2SO_2$, giving a deliquescence state. This phenomenon can be applied for NH_3 separation or storage [9]. NH_3 vapor pressures (bar) of the saturated absorbed state are shown as a function of temperature: 0.395 (−5 °C), 0.46 (0 °C), 0.53 (5 °C), 0.62 (10 °C), 0.68 (15 °C), 0.79 (20 °C). The adsorbed amount of NH_3 on sulfamide/alumina (weight ratio of 0.5/1) was measured. The molar ratio of adsorbed NH_3 vs sulfate at 10 °C was 1.65 under 1 bar and 0.26 under 0.4 bar of NH_3. The amount of adsorbed NH_3 on sulfamide/alumina under 0.4 bar is just **0.9 mmol/g** (260/96/3) at 10 °C. This number is much lower than that (5.2 mmol/g) for H_2SO_4/AC at 23 °C (Fig. 19.3), which may not be practical for an NH_3 separator in the synthesis system.

Ammonium dihydrogen phosphate, $(NH_4)H_2PO_4$, (acid in water) reacts with NH_3 (base) giving $(NH_4)_2HPO_4$, which releases NH_3 when heated. An example is reported in the literature [10]. A 35% $(NH_4)_XH_{(2-X)}PO_4$ ($X = 1.25$) solution reacted with 5.3% ammonia gas giving $(NH_4)_XH_{(2-X)}PO_4$ ($X = 1.75$) at 70 °C. The solution was heated up to 180 °C (using a distillation column) and released ammonia leaving solution with $(NH_4)_XH_{(2-X)}PO_4$ ($X = 1.25$). This means that **0.5 mol of NH_3 was separated using 1 mol of $(NH_4)_XH_{(2-X)}PO_4$**. The process appears to be simple, but the water solution might be troublesome in the NH_3 synthesis system. Water must be separated in the recycle loop before the catalyst column. The separation efficiency is not high either.

Ionic liquids are known to absorb NH_3; however, mass transfer is low due to their high viscosity. Silica has been used as a support for ionic liquids [11]. The $[Bmim]_2[Co(NCS)_4]$/silica gel (49/51% wt.) composites adsorbed **5.9 mmol NH_3/(g-adsorbent)** in 15 min (Bmim: 1-*n*-butyl-3-methyl-imidzolium). This value is higher than those for the composite elements used independently (synergy). The high value was attributed to hydrogen bonds between cobalt coordinated anions and NH_3, as well as to mesoporous structures. Absorbed NH_3 was found to be almost completely released in 15 min, and NH_3 absorption capacity loss was not detected during five adsorption and regeneration cycles. Though the adsorption amount is much the same as that of H_2SO_4/AC [2, 3], this material might be explored further for the pressure swing separation.

Zeolites have uniform pore sizes, which can select molecules by size. A thin film of silica colloid gel has been reported [12, 13] to separate H_2 from N_2-NH_3 by the first film and NH_3 from N_2 by the second film [12]. The separation factors are several hundreds, which is enough for NH_3 separation. The separation can be done at 400 °C (synthesis temperature), which makes total efficiency high. The challenge is the size of the separation unit or permeation rate of zeolite films. The reported value of ammonia permeation is around 3×10^{-8} (m^3/m^2-s-kPa) or **0.05 (mmol/cm^2-h-bar)**. The most active synthesis catalysts recorded an activity of 10–50 mmol/h/g-cat (340 °C, 10 bar). If we suppose a permeation tube with a diameter of 1 cm, it has an outer area of 3.14 cm^2/cm-length, which gives NH_3 permeation flow of 0.16 mmol/h per cm-length under 1 bar of concentration difference. If the corresponding volume (1 cm tube) holds 1 g of synthesis catalyst, the synthesis rate is around 50 mmol/h. In order to equip both reactor and separator with similar sizes, 100 times higher permeation rate is necessary.

One-step separation of NH_3 from N_2-H_2 is possible using zeolite films [14, 15]. At 80 °C, a high ammonia permeance (2.1×10^{-7} mol/m^2-s-Pa: **7.6 mmol/cm^2-h-bar**), and a selectivity of about 10 were found with a tubular zeolite membrane (MFI zeolite in multi-channel fiber configurations). A silica tubular membrane also gave a high ammonia permeance (7.6×10^{-7} mol/m^2-s-Pa: **27.4 mmol/cm^2-h-bar**) with a selectivity (NH_3/H_2) of about 7 [14]. Permeance values of both materials exceed 100 times of the above case [12], which may be practical for use although the selectivity is not high enough. Another group [15] has found that NH_3 permeation is promoted when the NH_3 pressure is high. They supposed that adsorbed NH_3 blocks H_2 permeation into the pores, while NH_3 travels through a hopping mechanism by way of adsorbed NH_3. They tried to create such acid sites (low Si/Al ratio: 2–10 and lower alkali metal/Al: 0.01–0.07) and modified the sites to be selective to NH_3 (N center, N/Al: 0.01–4). Their NH_4-RHO zeolite gives permeance of 1.3 and 2.8×10^{-8} mol/m^2-s-Pa (**4.7 and 10.1 mmol/cm^2-h-bar**) at 250 °C and 325 °C, respectively. The films worked at high temperatures with selectivities (NH_3/H_2) of 132 and 69 at 250 °C and 325 °C, respectively. These values might be attractive for researchers to use this film separator in industrial units (after many practice tests). Ammonia can be adsorbed on high surface area materials developed recently such as metal–organic frameworks (MOF) or MCM-41. These data are not collected under separation conditions for ammonia synthesis. Such research is expected in the future.

19.3 Absorption Separation by Alkali Earth Metal Halides

Alkali earth metal halides are known to form complexes with several molecules of NH_3, which might make them good candidates for NH_3 separation. Though there are precise thermodynamic data [16, 17], no practical data have been reported, especially for separation of ammonia. Alkali earth metal halides are known to be used in refrigeration systems [16]; however, the application pressure is higher (1 MPa) and the temperature is lower than that of those aimed for bench plant ammonia synthesis units. Practical data means data available for designing an absorber separator under practical ammonia synthesis conditions. For example, $CaCl_2$ reacts with one, two, four, and eight NH_3 molecules making $CaCl_2(NH_3)_n$ (n: 1, 2, 4, 8) depending on time and NH_3 pressure. Even though $CaCl_2(NH_3)_8$ is thermodynamically possible, it takes more than 24 h for its formation at room temperature. Liu and Aika have tried to obtain a reversible amount of absorption under ammonia synthesis conditions, which may help in designing absorption materials for practical use. They defined "quasi equilibrium pressure" as a pressure that does not change more than 0.2 Torr (27 Pa) in 40 s in the absorption measurement [17]. Some "quasi" absorption isotherms are shown as examples in Fig. 19.5.

Liu and Aika studied five samples: $MgCl_2 \bullet 6H_2O$, $CaCl_2 \bullet 2H_2O$, $CaBr_2 \bullet 2H_2O$, $SrCl_2 \bullet 6H_2O$, and $SrCl_2 \bullet 6H_2O$, most of which were heated at 523 or 723 K and turned into pure chloride or bromide [18]. One exception was $MgCl_2 \bullet 6H_2O$, which turned into MgCl(OH) at 523 K or $MgCl_{0.32}OH_{0.32}O_{0.68}$ at 723 K. Part of $MgCl_2$ reacted with H_2O ($MgCl_2 + H_2O \rightarrow MgO + 2HCl$). The absorption was saturated at 5 kPa and 67 kPa for $MgCl_2$ (723 K) and $CaCl_2$ (723 K), respectively, in the first adsorption as is shown in Fig. 19.5. If the samples form molecular complexes with two ammonia molecules, the saturation pressures would be 8×10^{-5} and 2×10^{-2} kPa for $MgCl_2$ and $CaCl_2$, respectively. For the first absorption, extra pressure (chemical potential) is necessary, and short time evacuation is not enough to desorb core ammonia. The second absorption is low but easily desorbed (reversible absorption). The following formula may be applicable for both $CaCl_2$ and $CaBr_2$, expressing irreversible (Eq. 19.2) and reversible (Eq. 19.3) processes.

$$CaCl_2 + 2NH_3 \rightarrow CaCl_2 \cdot 2NH_3 \tag{19.2}$$

$$CaCl_2 \cdot 2NH_3 + nNH_3 \leftrightarrow CaCl_2 \cdot (2 + n)NH_3 \tag{19.3}$$

The amounts of reversible absorption at 40 kPa under 298 K are 18.3 ($MgCl_2$), 19.4 ($CaBr_2$), and **22.8 mmol/g** ($SrBr_2$) [17]. Those values are 4 or 5 times higher than zeolite or active carbons discussed above. This study shows that practical data useful

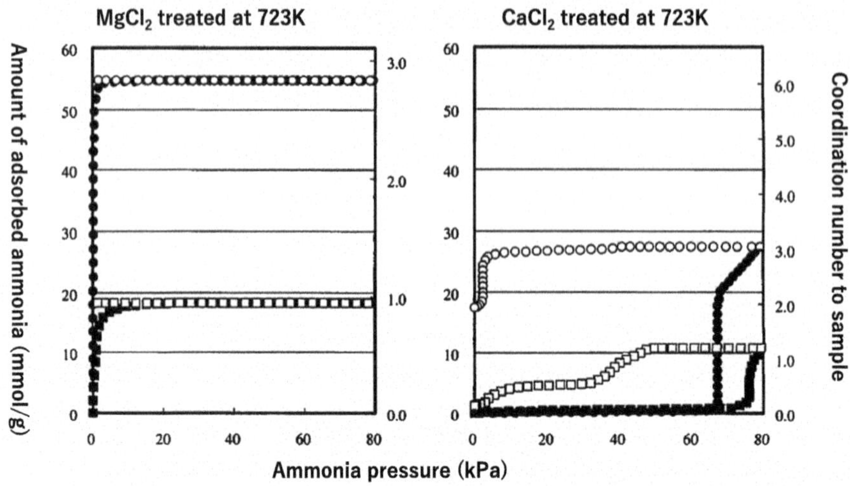

Fig. 19.5 "Quasi" absorption (closed symbols) and desorption (open symbols) isotherms of ammonia by $MgCl_2$ (left) and $CaCl_2$ (right) at 298 K [17]. It takes about 15 min to take one point of quasi adsorption and desorption. Circles show the 1st measurement and squares show the 2nd measurement after 1 h of evacuation at the same temperature. Both samples were treated at 723 K before absorption measurements. Ammonia absorption is expressed either as mmol/g (left vertical axis) and coordination number (right vertical axis) for both samples. Reproduced with permission from [17]

for designing ammonia separation are distinct from thermodynamic data because the real processes require extra chemical potential depending on temperature, pressure, and material morphology such as surface area. The first absorption starts fast at low pressure (29 kPa) for the $CaCl_2$ sample treated at 523 K, while it starts late at high pressure (67 kPa) for the $CaCl_2$ sample treated at 723 K. The rate is partly controlled by the surface area: the former (2.77 m^2/g) is higher than the latter (0.776 m^2/g). Dynamic or rate data are thus important for designing a practical process. Cussler's group studied these rate processes extensively [18]. They suggested the rate of absorption into $CaCl_2$ is controlled by the first order of ammonia pressure and by higher orders after an initial period. Desorption seems to be controlled by diffusion. Their sample (54.2 mg $CaCl_2$ + 2 g glass beads) had a higher surface area (9.4 m^2/g). The morphological change (reduction of surface area by repeated use) was also studied [18].

19.4 Material Design

Here, we show that thermodynamic data are useful to determine trends for material design. Absorption equilibrium pressures of several common ammine complexes are calculated by thermodynamic data and are shown as a function of temperature in

Fig. 19.6 [19, 20]. Figure 19.1 shows a model of a separator system, where 18.6 bar of NH_3 enters into the separator and 2 bar of NH_3 is left. If our absorber material has an NH_3 equilibrium pressure of 1 bar (101 kPa) at some temperature, 18.6 bar of NH_3 would be easily absorbed until 2 bar in an absorption column at that temperature. The desorption column is supposed to have fictitious NH_3 pressure of 1 bar at the same temperature, and thus NH_3 could be evacuated and compressed to 10 bar NH_3 liquid tanks. The temperature crossing at 1 bar level for each material can be determined from Fig. 19.6 as 40 °C for $CaCl_2$ (4 > 2), 90 °C for $CaBr_2$ (6 > 2), 140 °C for $MgCl_2$ (6 > 2), 170 °C for $CaCl_2$ (2 > 1), 220 °C for $CaBr_2$ (2 > 1) and $CaCl_2$ (1 > 0), and 280 °C for $MgCl_2$ (2 > 1) and $CaBr_2$ (1 > 0). The numbers in parenthesis show the change of ammonia coordination number. Those that work at high temperature near to that of the ammonia synthesis reactor are recommended for high heat efficiency, while those that work at low temperature are recommended with respect to high absorption capacity, which may save power.

In order to construct a new material, two halides were mixed and "quasi" absorption isotherms were studied [21–23]. When two halides having identical cations (such as $CaCl_2$-$CaBr_2$) are mixed, the two anions seem to be mixed up. Figure 19.7 shows isotherms of ammonia in a $CaCl_2$-$CaBr_2$ mixture shown as $CaCl_{2-n}Br_n$. The isotherm of the mixture is located in between those of $CaCl_2$ and $CaBr_2$. The starting pressure of the first absorption gradually moves to a lower value with increase of n (Fig. 19.7 left). It was also found that the amount of the second (reversible) absorption was high for the mixed material (Fig. 19.7 right). If we operate pressure swing separation between 60 and 10 kPa at 298 K, **27.1 mmol/g** of ammonia is separated. The authors succeeded in finding a new material for the specific PSA condition [22]. This kind of application study may be necessary to utilize ammine complexes in a practical unit where PSA conditions may be different. However, when two halides having identical anions (such as $CaCl_2$-$SrCl_2$) are mixed, the two cations do not seem

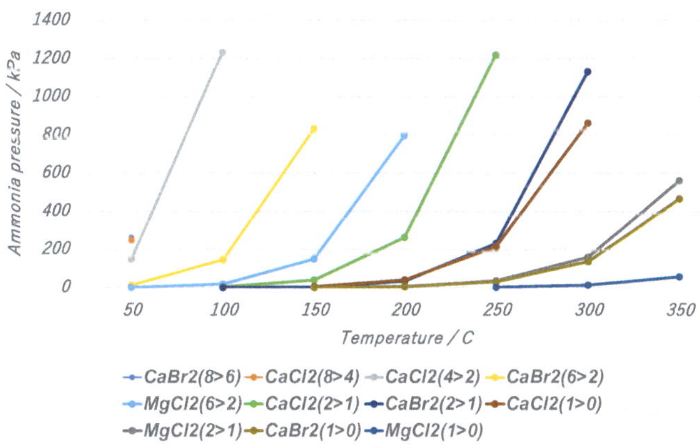

Fig. 19.6 Equilibrium ammonia pressure (kPa) between ammine complexes having different coordination numbers (shown in parenthesis) as a function of temperature

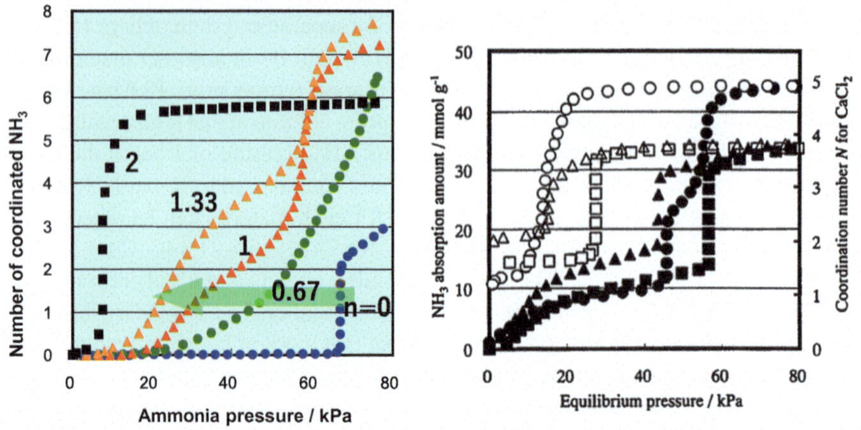

Fig. 19.7 Quasi ammonia absorption isotherm in $CaCl_{2-n}Br_n$ (treated at 523 K) as a function of ammonia pressure at 298 K. Left: 1st absorption (absorption amount is expressed by coordination number). Right: 2nd absorption (absorption amount is expressed by mmol/g*) for mixtures (closed circles for $n = 0.67$, squares for $n = 1$, triangles for $n = 1.33$) and desorption (open circles for $n = 0.67$, squares for $n = 1$, triangles for $n = 1.33$). *For obtaining the real coordination number, factors of 1.27, 1.4, and 1.53 are necessary to multiply by N on the right axis [21–23]. Reproduced with permission from [21]

to mix with each other. The "quasi" absorption isotherm was just the overlap of those of the original compounds before mixing [23].

Jensen et al. tried to design mixed metal halide ammines for practical ammonia storage, considering rate, temperature, and lifetime. They used computational techniques such as density functional theory and genetic algorithms. They proposed a novel mixed hexa ammine complex, $Ti_2CuMgCl_4Br_4$, as an example [24]. This technology may be useful to propose new materials which have not been experimentally studied. A practical examination is expected.

19.5 Kinetic Model of Synthesis Loop Including Absorption Unit (Reaction Engineering)

So far, we have discussed the effectiveness of absorption separation. For practical use, a large amount of kinetic data under the conditions of practical ammonia synthesis must be studied. Cussler's group recently published their data for such purposes [25–27]. They gathered absorption kinetic data and attempted to incorporate the data into an overall integration protocol to evaluate the process design [26, 27]. Here, we try to explain the idea of Cussler et al. using the simple model shown in Fig. 19.8, which has the same flow diagram as Fig. 19.1 but has variable letters instead of numerical numbers. The key feature of their idea is to use the first-order kinetic expressions

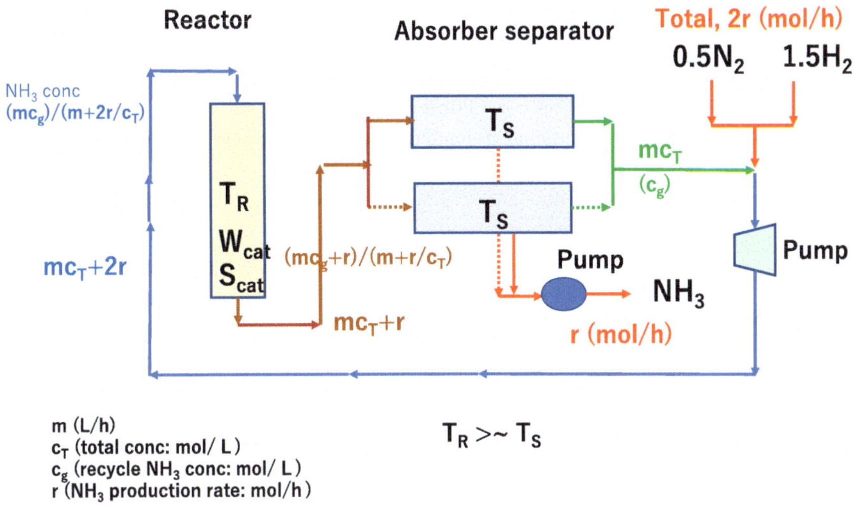

Fig. 19.8 A simplified model of an ammonia synthesis apparatus equipped with reactor and absorption separator. T_R: reactor temperature, T_S: separator temperature, W_{cat}: catalyst weight, S_{cat}: catalyst surface area

for both reactor and absorber systems. The symbol letters in Fig. 19.8 (also listed in Table 19.2) are the same as in the literature [27].

For the supposed synthesis reactor, a doubly promoted iron catalyst was used. The rate of ammonia synthesis is expressed as Eq. 19.4 [28].

$$r = kM_{\rm cat}\left[p_{\rm n} - p_{\rm a}^2/\left(K_{\rm eq}p_{\rm h}^3\right)\right]/\left[1 + p_{\rm a}/\left(K_{\rm ad}p_{\rm h}^{3/2}\right)\right]^2 \tag{19.4}$$

where $p_{\rm n}$, $p_{\rm h}$, $p_{\rm a}$, and $K_{\rm ad}$ are pressures of nitrogen, hydrogen, ammonia, and the equilibrium constant of nitrogen adsorption, respectively (Eq. 19.8 in ref. [28]). The other notation in Eqs. 19.4–19.7 and in Fig. 19.8 is listed in Table 19.2. Under low NH_3 conversion, the reaction rate is simplified as follows. The driving force of synthesis is expressed as the difference between equilibrium concentration (c^*) and outlet concentration of NH_3 (c).

$$r = k_{\rm cat}M_{\rm cat}\left(c^* - c\right) \tag{19.5}$$

We tried to visualize the kinetic relation in Fig. 19.9, where the water level means chemical potential expressed as NH_3 concentration. If the reactor outlet gate ($k_{\rm cat}$ $M_{\rm cat}$) is wide open and the difference between fictitious equilibrium NH_3 concentration at the inlet and the real one at the outlet (c^* − c) is high, the rate is high.

From the studies from Cussler's laboratory, the rate of NH_3 absorption is also presented in a simplified form. Here, the driving force of absorption is expressed as

Table 19.2 Symbols used for kinetic model (see Figs. 19.8 and 19.9, and Eqs. 19.4–19.7)

Symbol	Unit	Meaning	Note
r	mol/h	Rate of reaction (NH_3 production)	
c^*	mol/L	NH_3 concentration at equilibrium with synthesis gas (1/2 N_2 + 3/2 H_2)	Fictitious NH_3 concentration
c	mol/L	NH_3 concentration leaving the reactor	
M_{cat}	kg	Mass of catalyst	
k_{cat}		Reaction rate constant (Eq. 19.5)	$2kc^*/(RTK_{eg}c_{H2}^3)$
k		Reaction rate constant of ammonia synthesis	Equation 19.4
K_{eg}		Equilibrium constant of ammonia synthesis	
k_{abs}		Absorption rate constant	
V_{abs}	L	Volume of absorbent	
a	m^2/L	Area per volume of absorbent	
m	L/h	Volumetric flow of recycle stream	
c_T	mol/L	Total concentration in recycle stream	
c_g	mol/L	NH_3 concentration in recycle stream	NH_3 conc. at absorber surface
c_g^+	mol/L	NH_3 gas concentration equilibrium with ammonia complex at absorber core	Fictitious concentration

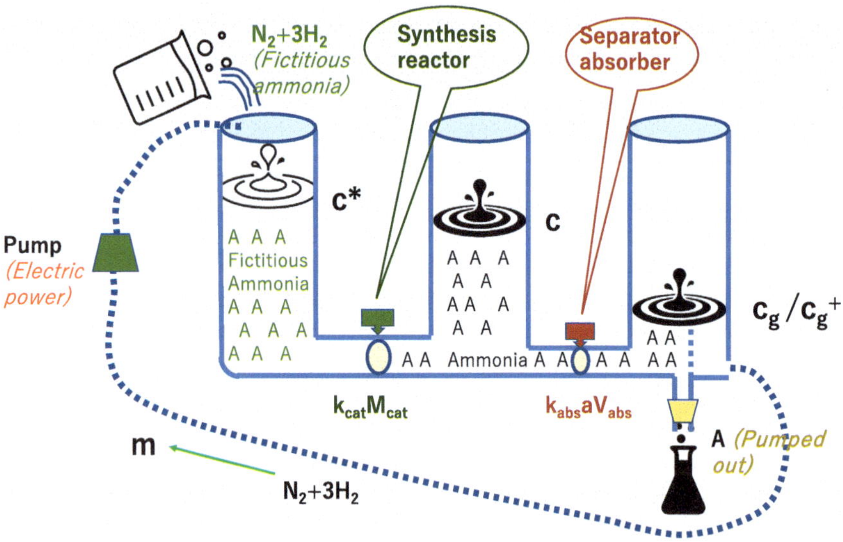

Fig. 19.9 Visualized kinetic model of ammonia production system

the difference between NH_3 concentration at the reactor outlet (absorber inlet) and that at the absorber core (c-c_g^+) (diffusion as the driving force). The gate in the flow ($k_{abs}\ a\ V_{abs}$) also controls the rate.

$$r = k_{abs} a\ V_{abs}\left(c - c_g^+\right) \tag{19.6}$$

As is shown in Fig. 19.8, after NH_3 is separated with a rate of r, the rest of the gas (mc_T) meets with syngas feed ($2r$), which maintains the level of c^* in Fig. 19.9. The material balance gives the following expression.

$$c = \left(c_g m + r\right)/[m + (r/c_T)] \tag{19.7}$$

From Eqs. 19.5–19.7, the following rate expressions are obtained.

$$r = 0.5\beta[(1 + 4\gamma/\beta^2)^{0.5} - 1] \tag{19.8}$$

$$\beta = m\ k_{cat} M_{cat}\left[1/(k_{cat} M_{cat}) + 1/(k_{abs} a\ V_{abs}) + \left(c_T - c_g\right)/m\right] \tag{19.9}$$

$$\gamma = m\ k_{cat} M_{cat}\left(c^* - c_g^+\right) \tag{19.10}$$

Now, two useful expressions for the process design are obtained. The first one comes under the condition of ($\gamma \ll \beta^2$), which seems to be practical for this system design.

$$r = \gamma/\beta = \left(c^* - c_g^+\right)/\left[1/(k_{cat} M_{cat}) + 1/(k_{abs} a\ V_{abs}) + \left(c_T - c_g\right)/m\right] \tag{19.11}$$

This expression resembles the electric flow: current (r) is proportional to voltage ($c^* - c_g^+$) and inversely related to the resistance (denominator). Here, three resistances are connected in series: reaction inefficiency [$1/(k_{cat}\ M_{cat})$], absorber inefficiency [$1/(k_{abs}\ a\ V_{abs})$] and recycle flow rate inefficiency [$(c_T - c_g)/m$]. Such resistances are visualized as gates or compressors in Fig. 19.9. When one resistance is much higher than the others, it controls the rate. When the three resistances are similar, the production proceeds effectively. This means that the design of reactor, separator, and compressor closely relate to each other. Malmali et al. have shown that $1/r$ is related to a first order of $1/m$ (reciprocal expression of Eq. 19.11) using their experimental data ($MgCl_2$ or $CaCl_2$ at 180 °C under 55 bar). Recycle flow rate and absorption rate have been identified as the major resistances in these conditions [26].

For the second case ($\gamma \gg \beta^2$; rather rare), the reaction rate is expressed as follows. We skip the comments here.

$$r = \gamma^{0.5} = \left[m\ k_{cat} M_{cat}\left(c^* - c_g^+\right)\right]^{0.5} \tag{19.12}$$

19.6 Single-Pass Reactor: In Situ Synthesis Separation Unit

If the absorbent can work at the same temperature as that of the reactor, in situ absorption separation is possible as is shown in Fig. 19.10. In a stirred tank reactor (left), ammonia is produced on the catalyst particles diffuses into absorber particles. Saturated ammonia is pumped out from another tank reactor. Ojha et al. have shown that synthesis gas reacts under 27 bars at 400 °C in a semi-batch reactor containing conventional Fe catalyst with $NiCl_2$ (or $MgCl_2$) absorbent. The reaction rate was faster than that observed without absorbent. Produced ammonia was proven to be absorbed into the absorber [27]. Smith and Torrente-Murciano have demonstrated a recycle-less process using a model reactor shown on the right in Fig. 19.10. They used Ru-Cs^+/CeO_2 as a synthesis catalyst and $MnCl_2$/SiO_2 as an absorbent. They collected data from the synthesis column (left side), but the desorption was tested separately using just the absorbent [29]. The synthesis reaction was carried out at 330–370 °C and under 20 bar. The rate was calculated from the flow rate change between inlet and outlet. The conversion exceeded the equilibrium value due to in situ absorption. However, their absorber decayed, likely due to a small amount of water released from the synthesis catalyst. These are just preliminary reports, though more attempts are expected in the future.

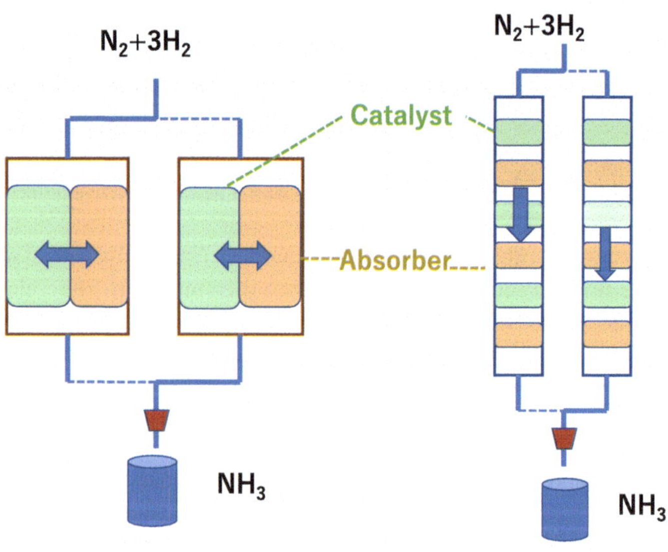

Fig. 19.10 Single-pass reactor models: a stirred tank reactor model (left) and a column-type reactor model (right)

19.7 Concluding Remarks

Absorption separation technologies are reviewed as a practical aspect for green ammonia production. Since renewable energy production is restricted in size, green ammonia units might be much smaller than conventional ammonia units for natural gas. To depress the capex of the small unit, new technology is necessary. One of those technologies must be absorption separation. Many new attempts have been made and there are several reviews that cover these issues from different viewpoints [30–32]. The author thanks a former co-worker, Professor Junichi Ryu (C. Y. Liu) of Chiba University, for instructive discussions.

References

1. Aika K (2022) Brief review of the Japanese energy carrier program and an energy science view of fuel ammonia, ch. 1 this book
2. Aika K, Kakegawa T (1991) On-site ammonia synthesis in De-NO_x process. Catal Today 10:73–80
3. Liu CY, Aika K (2003) Effect of surface oxidation of active carbon on ammonia adsorption. Bull Chem Soc Jpn 76:1463–1468
4. Sakata Y, Funada I, Numaguchi T (1994) A pressure swing apparatus for ammonia separation and an ammonia separation method. JPA 1994–296819
5. Liu CY, Aika K (2002) Modification of active carbon and zeolite as ammonia separation materials for a new de-NO_x process with ammonia on-site synthesis. Res Chem Intermed 28(5):409–417
6. Liu CY, Aika K (2003) Ammonia adsorption on ion exchanged Y-zeolites as ammonia storage material. J Jap Petroleum Institute 46(5):301–307
7. Zeng HS, Inazu K, Aika K (2002) The working state of the barium promoter in ammonia synthesis over an active-carbon-supported ruthenium catalyst using barium nitrate as the promoter precursor. J Catal 211:33–41
8. Truszkiewicz E, Rarog-Pilecka W, Schmidt-Szałowski K, Jodzis S, Wilczkowska E, Łomot D, Kaszkur Z, Karpinski Z, Kowalczyk Z (2009) Barium-promoted Ru/carbon catalyst for ammonia synthesis: state of the system when operating. J Catal 265:181–190
9. Uchida S, Tsuruga N (1973) Method of ammonia storage or separation. JPA 1973-22359
10. Mochida T (2002) Ammonia separation method, JPA 2002-012420-000001
11. Zeng S, Wang J, Li P, Dong H, Wang H, Zhang X, Zhang X (2019) Efficient adsorption of ammonia by incorporation of metal ionic liquids into silica gels as mesoporous composites. Chem Eng J 370:81–88
12. Tsuru T, Yamamoto A, Yoshioka T (2008) Ammonia separation method, ammonia synthesis method and gas separation film, JPA 2008-247654-000001
13. Kanezashi M, Yamamoto A, Yoshioka T, Tsuru T (2010) Characteristics of ammonia permeation through porous silica membranes. AIChE J 56(5):1204–1212
14. Camus O, Perera S, Crittenden B, van Delft YC, Meyer DF, Pex PPAC, Kumakiri I, Miachon S, Dalmon J-A, Tennison S, Chanaud P, Groensmit E, GrowHow K (2006) Ceramic membranes for ammonia recovery. AIChE J 52:2055–2065
15. Tanaka M, Matsuo T, Aoshima T (2018) Ammonia separation method and zeolite, WO 2018-230737
16. Wongswan W, Kumar S, Neveu P, Meunier F (2001) A review of chemical heat pump technology and applications. Appl Therm Eng 21:1489–1519

17. Liu CY, Aika K (2004) Ammonia adsorption on alkaline earth halides as ammonia separation and storage procedure. Bull Chem Soc Jpn 77:123–131
18. Smith C, Malmali M, Liu C-Y, McCormick AV, Cussler EL (2018) Rate of ammonia absorption and release in calcium chloride. ACS Sustain Chem Eng 6:11827–11835
19. Neveu P, Castaing J (1993) Solid-gas chemical heat pumps: field of application and performance of the internal heat of reaction recovery process. Heat Recovery syst CHP 13(3):233–251
20. International Critical Table of Numerical Data, Physics, Chemistry and Technology, McGraw Hill, New York (1929) 7:224–313
21. Liu CY, Aika K (2002) Absorption and desorption behavior of ammonia with alkali earth halide and mixed halide. Chem Lett 8:798–799
22. Liu CY, Aika K (2004) Effect of the Cl/Br molar ratio of a $CaCl_2$-$CaBr_2$ mixture used as an ammonia storage material. Ind Eng Chem Res 43(22):6994–7000
23. Liu CY, Aika K (2004) Ammonia absorption into alkaline earth metal halide mixtures as an ammonia storage material. Ind Eng Chem Res 43(23):7484–7491
24. Jensen PB, Lysgaard S, Quaadec UJ, Tejs Vegge T (2014) Designing mixed metal halide ammines for ammonia storage using density functional theory and genetic algorithms. Phys Chem Chem Phys 16:19732–19740
25. Reese M, Marquart C, Malmali M, Wagner K, Buchanan E, McCormick A, Cussler EL (2016) Performance of a small-scale Haber process. Ind Eng Chem Res 55:3742–3750
26. Malmali M, Reese M, McCormick AV, Cussler EL (2018) Converting wind energy to ammonia at lower pressure. ACS Sustain Chem Eng 6(1):827–834
27. Ojha DK, Kale MJ, McCormick AV, Reese M, Malmali M, Dauenhauer P, Cussler EL (2019) Integrated ammonia synthesis and separation. ACS Sustain Chem Eng 7:18785–18792
28. Aika K, Ozaki A (1969) Kinetics and isotope effect of ammonia synthesis over an unpromoted iron catalyst. J Catal 13:232–237
29. Smith C, Torrente-Murciano L (2021) Exceeding single-pass equilibrium with integrated absorption separation for ammonia synthesis using renewable energy—redefining the Haber-Bosch Loop. Adv Energy Mater 2003845:1–12
30. Zhang T, Miyaoka H, Miyaoka H, Ichikawa T, Kojima Y (2018) Review on ammonia absorption materials: metal hydrides, halides, and borohydrides. ACS Appl Energy Mater 1(2):232–242
31. Rouwenhorst KHR, Van der Ham AGJ, Lefferts L (2021) Beyond Haber-Bosch: the renaissance of the Claude Process. Intrn J Hydrogen Energy 46:21566–21579
32. Smith S, Hill AK, Torrente-Murciano L (2020) Current and future role of Haber-Bosch ammonia in a carbon-free energy landscape. Energy Environ Sci 13:331–344

Chapter 20
Rare Earth Oxide–Supported Ru Catalysts for Ammonia Synthesis

Katsutoshi Nagaoka and Katsutoshi Sato

Abstract A catalyst that can produce ammonia under mild reaction conditions (below 400 °C and 10 MPa) is needed to achieve green ammonia synthesis using renewable energy. From an industrial point of view, the catalyst should be easily prepared and can be handled in air. Ru shows unparalleled activity compared to Fe, Co, and Ni under mild reaction conditions, and Ru catalysts supported on oxides are suitable for this purpose because of their high stability and ease of handling in air. As a component of oxide supports, rare earth elements with redox properties have attracted attention because of their strong electron-donating ability and high-surface area. In this chapter, recent progress in the study of Ru catalysts supported on rare earth oxides (pure or composite oxide) is summarized with respect to their catalytic performance and physicochemical properties. In particular, the activity of Ru catalysts has been found to be high-under mild reaction conditions, and hydrogen poisoning, a typical drawback of Ru catalysts, is satisfactorily suppressed.

Keywords Green ammonia · Rare earth oxide · Ru catalyst · High activity · Strong metal–support interaction · Hydrogen poisoning

20.1 Introduction

Since the ammonia manufacturing process was developed by Haber and Bosh in Germany in the 1910s, chemical fertilizer has been produced, and crop production has increased dramatically. Currently, global production of ammonia is about 190 million tons per year, and more than 80% is used to produce fertilizers [1]. About 40% of the 7.8 billion people all over the world receive the benefits of artificially synthesized ammonia.

In addition, ammonia has been drawing attention as a potential energy carrier in the form of hydrogen produced from renewable energy because of its high-energy

K. Nagaoka (✉) · K. Sato
Department of Chemical Systems Engineering, Graduate School of Engineering, Nagoya University, Chikusa-ku, Nagoya 464-8603, Aichi, Japan
e-mail: nagaoka.katsutoshi@material.nagoya-u.ac.jp

K. Aika and H. Kobayashi (eds.), *CO_2 Free Ammonia as an Energy Carrier*,
https://doi.org/10.1007/978-981-19-4767-4_20

density and hydrogen storage capacity, ease of liquefaction at ambient temperature, and absence of CO_2 emissions when hydrogen is produced by its decomposition or ammonia is combusted in turbines and internal combustion engines [2, 3]. Furthermore, infrastructure for storage and transportation is well established.

Ammonia is manufactured primarily by the Haber–Bosh process using doubly promoted Fe-based catalysts at high temperature (>400 °C) and pressure (>20 MPa) [4]. Hydrogen used in the process is produced by steam reforming of coal or natural gas, and thus, a large amount of CO_2 (1.9 ton ton_{NH3}^{-1}) is released [5]. In order to meet the goals of the Paris Agreement on Climate Change (UNFCCC, The Paris Agreement), which established a global framework for efforts to avert climate change and curb CO_2 emissions by limiting the increase of the global average temperature [6], development of a more efficient, renewable energy-based, a so-called "green" ammonia synthesis process is needed. Furthermore, to synthesize ammonia by such a process using low-pressure, renewable hydrogen produced by water electrolysis or photocatalytic water splitting, a new catalyst that can produce ammonia under mild reaction conditions (below 400 °C and 10 MPa) must be developed. For practical reasons, the catalyst should also be easily prepared and stable in air.

In the 1970s, Ru-based catalysts were discovered by Aika and Ozaki [7, 8]. The catalysts showed much higher catalytic activity than other metals such as Fe, Co, and Ni under mild reaction conditions. After that discovery, many efforts were made to develop effective Ru catalysts. There are still two major conditions that need to be satisfied before Ru catalysts can be widely used in practical applications. The first is that the Ru catalyst must not be susceptible to hydrogen poisoning [9, 10]. This condition must be satisfied because hydrogen atoms strongly adsorb onto the Ru surface and inhibit the dissociation of N_2, which is the rate-determining step for ammonia synthesis. The second condition is that the Ru must not be expensive. This condition can be satisfied by developing highly active Ru catalysts.

Many classes of supports for Ru-based catalysts, including oxides, carbons, electrides, hydrides, nitrides, amides, and oxyhydrides, have been reported in the literature. We selected oxides because they are very stable and easy to handle in air. During the Japan Science and Technology Agency Core Research for Evolutionary Science and Technology project "Creation of Innovative Core Technology for Manufacture and Use of Energy Carriers from Renewable Energy", we worked to develop Ru-based catalysts for ammonia synthesis under mild reaction conditions. The use of redox-active rare earth elements proved to be essential to development of the catalyst. This contribution focuses on our efforts to investigate Ru catalysts supported by rare earth oxides.

20.2 Pure Rare Earth Oxide Support

Aika et al. have found that one of the rare earth oxides, Sm_2O_3, is an effective promoter of Ru/Al_2O_3 [10]. They have also reported that using rare earth oxides as the support is more effective than using an Ru/MgO promoter [11]. Furthermore, high-temperature reduction (500–600 °C) of Ru/La_2O_3, Ru/CeO_2, and Ru/Sm_2O_3 enhances the initial ammonia synthesis rate of the catalysts. The initial ammonia synthesis rate for Ru/CeO_2 was maximized after reduction at 500 °C and sustained for 14 h, although the ammonia synthesis rate of Ru/La_2O_3 and Ru/Sm_2O_3 decreased drastically with time. The high activity of the Ru/CeO_2 catalyst was attributed to the formation of reduced CeO_2 (CeO_{2-x}) during the pre-reduction. Electrons are thus donated to Ru and the antibonding π orbital of $N \equiv N$. The result is weakening of the triple bond and promotion of the ammonia synthesis reaction [12]:

$$Ce^{4+}(O^{2-})_{2-x}(e^-)_{2x} + Ru \leftrightarrow Ce^{4+}(O^{2-})_{2-x} + Ru(e^-)_{2x} \quad (20.1)$$

The reduced CeO_2 has a coordinated, unsaturated Ce^{3+} and partially covers the surface of Ru particles via a strong metal-support interaction (SMSI) to lower the surface energy [13]. In addition, the increase with pressure from 0.1 to 2 MPa of the ammonia synthesis rate of Ru/CeO_2 at 310 °C prevents hydrogen poisoning on the Ru surface. In contrast, the ammonia synthesis rate of Ru-Cs/MgO gradually decreases with increasing pressure in the same pressure range. The enhanced electron-donating ability induced by reduction of Ce^{4+} to Ce^{3+} and the strong resistance to hydrogen poisoning of the Ru/CeO_2 catalyst has motivated many researchers to further investigate use of rare earth oxide–supported Ru catalysts, such as Ru/CeO_2 [14–16] and Ru/Pr_2O_3 [17, 18], which contain rare earth elements with redox ability.

Influence of the morphology of the CeO_2 support on the activity of the Ru/CeO_2 catalyst was investigated. Loading the Ru catalyst onto CeO_2-nanorod, CeO_2-nanocube, or CeO_2-nanoparticle supports and revealed that the catalytic activity was highest when the catalyst was loaded onto nanorod supports with an exposed (111) facet because of the high-Ce^{3+}/Ce^{4+} ratio and oxygen vacancy in the catalyst [14]. It has also been reported that a high-specific surface area, basic density, and the SMSI of the CeO_2 are important factors that enhance the activity of the Ru/CeO_2 [15].

20.2.1 Pr_2O_3

Praseodymium is a rare earth element that is located to the right of Ce in the periodic table, and its +3 and +4 oxidation states give it redox ability. Praseodymium has been used to enhance the oxygen-storage and oxygen-release capacity of the so-called three-way catalysts used to purify gases emitted by automobiles [19–21]. Furthermore, reduction of Pr_6O_{11} to Pr_2O_3 occurs at much lower temperature than reduction of CeO_2 to Ce_2O_3 [22]. We have reported that the rate of ammonia synthesis is higher over Ru/Pr_2O_3 than over Ba–Ru/activated carbon (Ba–Ru/AC), which has been used in industrial processes, and over Cs–Ru/MgO, which is one of the most active Ru catalysts (entries 1, 2, and 3 in Table 20.1) [17, 23]. Furthermore, the much higher ammonia synthesis rate catalyzed by Ru/Pr_2O_3 than by Ru/CeO_2 or Ru/MgO (Fig. 20.1 and entries 4, 5, and 6 in Table 20.1) has been ascribed to a higher turnover frequency (TOF) over Ru/Pr_2O_3. Characteristics of Ru/Pr_2O_3 include the strong basicity of Pr_2O_3 and formation of low-crystalline nanolayers of Ru atoms. After pre-reduction at 400 °C, all praseodymium in the catalysts is reduced to Pr^{3+}, which is more electron-rich than Pr^{4+}, and thus, the Ru/Pr_2O_3 catalyst shows strong basicity. However, the high-angle annular dark-field scanning transmission electron microscopy (HAADF-STEM) image in Fig. 20.2 shows that most of the Ru atoms are loaded as nanolayers, not nanoparticles. The absence of lattice fringes in most parts of the Ru layer indicates that the crystallinity of the Ru layer is low. The thickness of the Ru layer ranged from 0.5 to 3 nm. We thus concluded that the surface of Pr_2O_3 is covered mainly with Ru nanolayers with low crystallinity. In this structure, the Ru atoms are not aligned correctly and may form step or terrace sites similar to a B_5-type site, which is a highly active site for ammonia synthesis. Many Ru atoms can interact with the N_2 molecule in the transition state and thereby promote the activation of N_2. Formation of the low-crystalline Ru nanolayer is due to the ease with which Pr^{4+} is reduced to Pr^{3+}; during impregnation, $Ru_3(CO)_{12}$ reacts with the O^{2-} in Pr_6O_{11}, and CO_2 is released. The fact that the peak absorbance in the Fourier transform infrared (FTIR) spectrum for the stretching vibration mode of N_2 adsorbed onto Ru/Pr_2O_3 occurred at a lower frequency than the corresponding peak absorbances of N_2 adsorbed onto Ru/MgO and Ru/CeO_2 revealed that these unique electronic and structural characteristics synergistically promote cleavage of the $N \equiv N$ bond and accelerate ammonia synthesis (Fig. 20.3). The Ru/Pr_2O_3 maintained its high activity at 390 °C and 1.0 MPa for 50 h.

20.2.2 Comparison of Pure Rare Earth Oxides

Inspired by the excellent ability of Ru/Pr_2O_3 to catalyze ammonia synthesis, we investigated the ammonia synthesis activity of different rare earth oxide–supported Ru catalysts at 400 °C and 1.0 MPa [24]. The Ru catalysts that included light rare earth support (e.g., La_2O_3, CeO_2, Pr_2O_3, Nd_2O_3, and Sm_2O_3) catalyzed higher ammonia synthesis rates than Ru catalysts supported on heavy rare earth elements or MgO (Fig. 20.4). For Ru catalysts supported on heavy rare earth oxides, the small specific surface areas of the supports induced aggregation of the Ru particles that reduced the number of active metal sites, and consequently, ammonia synthesis rates. Light rare earth oxide–supported Ru catalysts had a higher basicity than Ru/MgO, and a positive correlation was observed between the turnover frequency and basicity of the catalysts, which is an indicator of electron-donating ability. The ammonia synthesis rates of the Ru/La_2O_3, Ru/CeO_2, and Ru/Pr_2O_3 catalysts increased with pressure from 0.1 to 1.0 MPa at 400 °C, whereas the rate for Ru/MgO did not increase within the same pressure range because of hydrogen poisoning. To elucidate the kinetics of the reactions over these catalysts, we estimated the reaction orders of N_2, H_2, and NH_3 by assuming the rate expression (20.2) [25].

$$\mathrm{r} = k P_{\mathrm{N_2}}^{n} P_{\mathrm{H_2}}^{h} P_{\mathrm{NH_3}}^{a} \tag{20.2}$$

Table 20.1 Comparison of catalytic activity between various Ru catalysts

Catalyst	Entry	Ru loading	Reduction temperature	Pressure	Reaction temperature	GHSV	NH_3 synthesis rate	Ref
		(wt%)	(°C)	(MPa)	(°C)	(mL h^{-1} g^{-1})	(mmol h^{-1} g^{-1})	
Ru/Pr_2O_3	1	5	500	1	400	18,000	19	17
Ba-Ru/AC	2	9.1	400	1	400	18,000	8.3	23
Cs-Ru/MgO	3	6	400	1	400	18,000	12.1	17
Ru/Pr_2O_3	4	5	400	0.9	390	18,000	15.2	17
Ru/CeO_2	5	5	400	0.9	390	18,000	7.4	17
Ru/MgO	6	5	400	0.9	390	18,000	1.5	17
$Ru/La_{0.5}Ce_{0.5}O_{1.75}$_500red	7	5	500	1	350	72,000	22	31
Ru/La_2O_3_500red	8	5	500	1	350	72,000	11	31
Ru/CeO_2_500red	9	5	500	1	350	72,000	13	31
$Ru/La_{0.5}Ce_{0.5}O_{1.75}$_450red	10	5	450	1	350	72,000	14	31
$Ru/La_{0.5}Ce_{0.5}O_{1.75}$_650red	11	5	650	1	350	72,000	31	31
Cs-Ru/MgO	12	5	500	1	350	72,000	4	31
$Ru/La_{0.5}Ce_{0.5}O_{1.75}$_650red	13	5	650	3	350	72,000	44	31
$Ru/La_{0.5}Pr_{0.5}O_{1.75}$_650red	14	5	650	1	400	72,000	60	33
$Ru/La_{0.5}Ce_{0.5}O_{1.75}$_650red	15	5	650	1	400	72,000	64	33
$Ru/Ba/LaCeO_x$_500red	16	5	500	1	350	72,000	19	38
$Ru/Ba/LaCeO_x$_700red	17	5	700	1	350	72,000	52	38

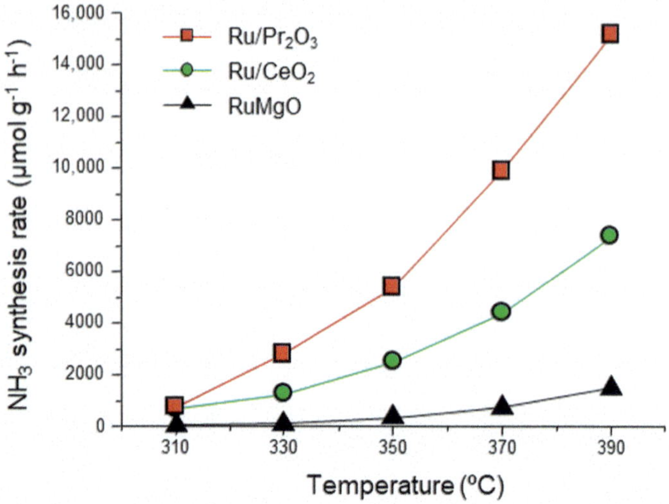

Fig. 20.1 Rate of NH_3 synthesis over supported Ru catalysts. Reaction conditions: catalyst, 0.2 g; reactant gas, $H_2/N_2 = 3$ with a flow rate of 60 mL min^{-1}; pressure, 0.9 MPa. Reproduced from Ref. [17] with permission from the Royal Society of Chemistry

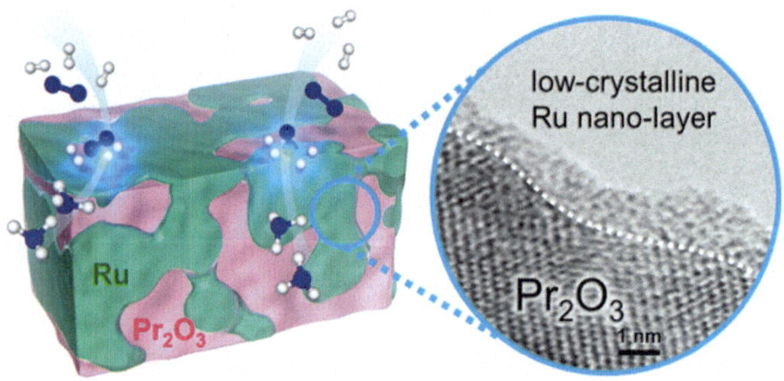

Fig. 20.2 Illustration and HAADF-STEM image of the structure of Ru/Pr_2O_3. Reproduced from Ref. [17] with permission from the Royal Society of Chemistry

The fact that the reaction orders for N_2 dissociation over Ru catalysts supported on rare earth oxides and MgO were almost unity at 400 °C and 0.1 MPa suggested that dissociation of N_2 molecules was the rate-determining step. In contrast, the fact that the reaction order for H_2 dissociation over Ru catalysts supported on rare earth oxides was nearly zero (i.e., −0.03, 0, 0.02, and −0.61 for Ru/La_2O_3, Ru/CeO_2, Ru/Pr_2O_3, and Ru/MgO, respectively) indicated that hydrogen poisoning was effectively retarded. The fact that hydrogen poisoning was sufficiently suppressed for all Ru catalysts supported on rare earth oxides indicated that rare earth elements kinetically suppress hydrogen poisoning. We are currently investigating the details

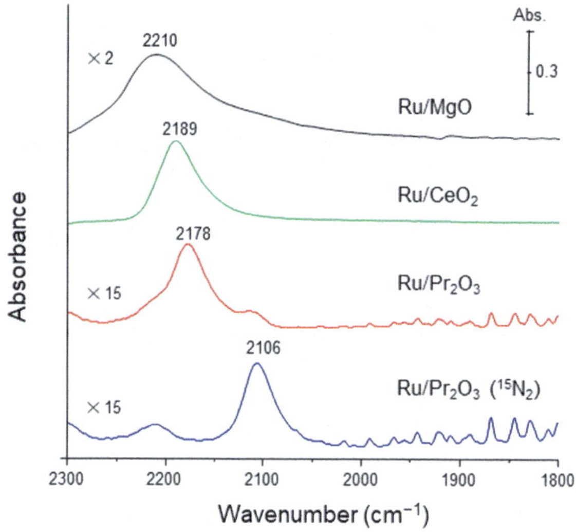

Fig. 20.3 Difference infrared spectra of N_2 molecules (before and after N_2 adsorption) on supported Ru catalysts. Spectra were collected under 6 kPa of N_2 at 25 °C. Reproduced from Ref. [17] with permission from the Royal Society of Chemistry

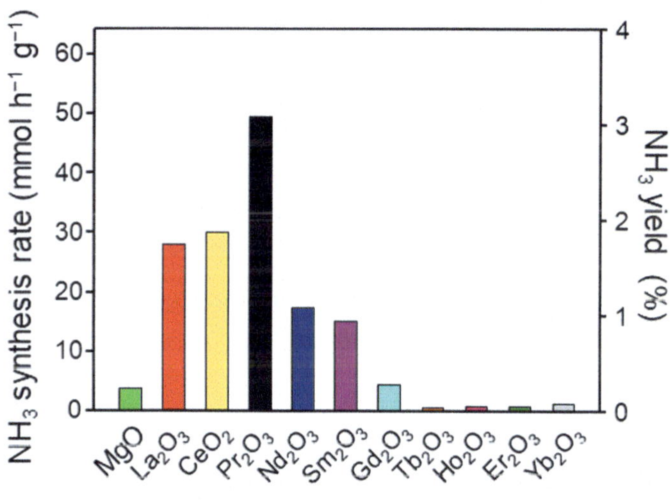

Fig. 20.4 Catalytic performance of 5 wt% Ru/rare earth oxide catalysts at 400 °C and 1.0 MPa. Original figure has been modified from [24]. Copyright (2018) Elsevier

of this finding. Further study has clarified that the reaction order of H_2 decreases with increasing reaction pressure and decreasing reaction temperature because of the strong adsorption of hydrogen atoms on Ru [18]. For example, the fact that the reaction orders of H_2 dissociation catalyzed by Ru/CeO_2 and Ru/Pr_2O_3 were –

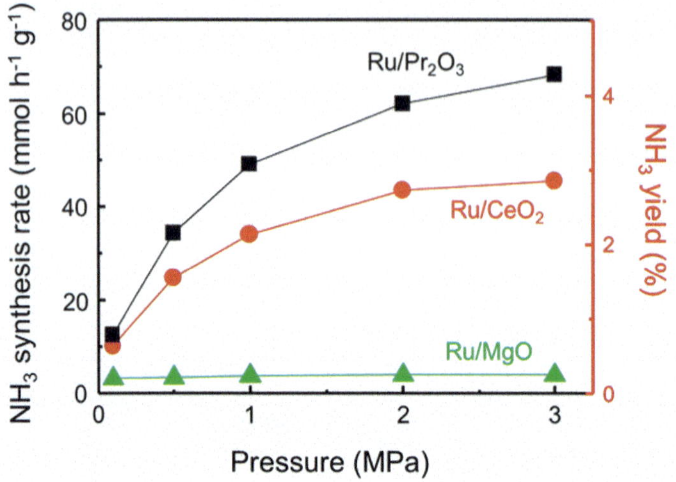

Fig. 20.5 Influence of total pressure on NH_3 synthesis activity over 5 wt% Ru/MgO, 5 wt% Ru/CeO_2, and 5 wt% Ru/Pr_2O_3 at 400 °C with a gas hourly space velocity of 72 L g^{-1} h^{-1}. Copyright (2019) Elsevier [18]

0.11 and 0.02, respectively, at 400 °C and 1.0 MPa indicated that under high pressure, hydrogen poisoning was retarded more in the reaction catalyzed by Ru/Pr_2O_3 than Ru/CeO_2. In fact, the rates of ammonia synthesis catalyzed by Ru/Pr_2O_3 and Ru/CeO_2 increased with increasing pressure from 0.1 to 3.0 MPa (Fig. 20.5). The rate was higher when the reaction was catalyzed by Ru/Pr_2O_3 than the other catalysts at all pressures and finally reached 64 mmol h^{-1} g^{-1} at 3.0 MPa. In contrast, the rate catalyzed by Ru/CeO_2 was almost the same at 2.0 and 3.0 MPa because of hydrogen poisoning at the high-partial pressure of H_2.

20.3 Composite Rare Earth Oxides

Aika et al. have found that the ammonia synthesis activity of CeO_2 increases as the reduction temperature increases from 400 to 500 °C because of an increase in the degree of reduction from Ce^{4+} to Ce^{3+}, but the activity decreases when the reduction temperature is further increased to 600 °C because of catalyst sintering [11, 12]. To increase the specific surface area of the rare earth oxides and concentration of Ce^{3+}, various researchers have used composite oxide supports containing rare earth elements, such as CeO_2-MgO [26], La_2O_3-CeO_2 [27], Sm_2O_3-CeO_2 [28], CeO_2-ZrO_2 [29], and BaO-CeO_2 [30]. However, the ammonia synthesis rates obtained with these catalysts have been too low for practical use.

20.3.1 $La_{0.5}Ce_{0.5}O_{1.75}$

In contrast to these studies in which catalyst pre-reduction has generally been carried out at temperatures below 500 °C to suppress sintering of the catalyst, we have reported that use of a composite oxide support, $La_{0.5}Ce_{0.5}O_{1.75}$, and reduction of the Ru/$La_{0.5}Ce_{0.5}O_{1.75}$ catalyst at unusually high temperature enhanced ammonia synthesis activity dramatically [31]. We found that the rate of ammonia synthesis was higher over Ru/$La_{0.5}Ce_{0.5}O_{1.75}$_500red ("500red" indicates that the catalyst had been reduced at 500 °C prior to activity tests) than Ru/La_2O_3_500red and Ru/CeO_2_500red (entries 7–9 in Table 20.1). In the case of Ru/La_2O_3, most of the support was hydrated, and the specific surface area was very low (18 m^2 g^{-1}). Ru/CeO_2 also tended to be sintered, and its specific surface area was low (24 m^2 g^{-1}). In contrast, formation of the composite oxide interfered with hydration and crystal growth of the oxidic support, and the specific surface area was high (47 m^2 g^{-1}), i.e., stability of the supports was enhanced by formation of a $La_{0.5}Ce_{0.5}O_{1.75}$ solid solution. The cause of the higher activity of Ru/$La_{0.5}Ce_{0.5}O_{1.75}$ than the other catalysts was the formation of fine Ru particles with a mean diameter less than 2 nm over the thermostable $La_{0.5}Ce_{0.5}O_{1.75}$ support. The ammonia synthesis rate over Ru/$La_{0.5}Ce_{0.5}O_{1.75}$ at 350 °C and 1.0 MPa increased 2.2-fold when the catalyst was reduced at 650 °C versus 450 °C and was much higher than the rate over Cs–Ru/MgO. (Fig. 20.6, entries 10–12 in Table 20.1). We have found that the Ru/$La_{0.5}Ce_{0.5}O_{1.75}$_650red catalyst is free from H_2 poisoning, and the rate of NH_3 synthesis over this catalyst increases gradually with an increase in reaction pressure up to 3.0 MPa (entries 11–13 in Table 20.1). Because the elemental state and structure of catalysts can be altered by exposure to air, the analysis was performed in the absence of air by moving the sample from an inert gas environment into the transmission electron microscope column using a special holder with a gas cell. In the HAADF-STEM image, we observed fine Ru particles partially covered by the support (Fig. 20.7). In addition, we observed the electron energy loss spectra of La^{3+}, Ce^{4+}, and Ce^{3+} (although Ce usually exists as Ce^{4+}) and the energy-dispersive X-ray (EDX) spectra of Ru in the center of the Ru particles. These results indicated that the Ru particles were partially covered by the reduced support and were consistent with the SMSI phenomenon. We compared the ability of Ru supported on $La_{0.5}Ce_{0.5}O_{1.75}$ after reduction at 500 and 650 °C to activate the dissociation of N_2 based on the FTIR spectrum of the adsorbed N_2. In both spectra, there was a peak at 2164 cm^{-1} and a broad peak between 1700 and 1950 cm^{-1} (Fig. 20.8). When the support was reduced at high temperature, the wavenumber of the broad peak decreased. The higher-wavenumber peaks were then assigned to the N_2 molecules on the Ru atoms that were not in contact with the support (indirect interaction in Fig. 20.7e). In contrast, we attributed the broad peak at low wavenumber to N_2 adsorbed on the Ru atoms in contact with the reduced support formed by the SMSI (direct interaction in Fig. 20.7e). The increase of the reduction of Ce^{4+} to Ce^{3+} from 23 to 43% by carrying out the reduction at higher temperature greatly enhanced the electron transfer from the Ru-mediated carrier to the antibonding π-orbitals of the N_2 molecule and further weakened the $N \equiv N$

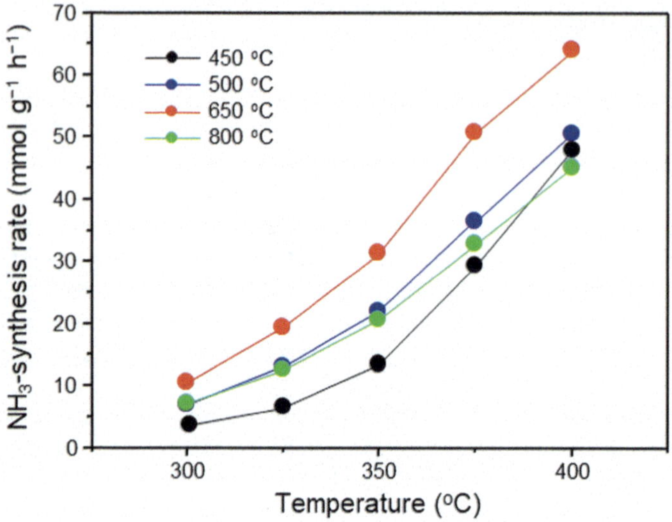

Fig. 20.6 Temperature dependence of the ammonia synthesis rate and ammonia yield at 1.0 MPa over Ru/$La_{0.5}Ce_{0.5}O_{1.75}$ after reduction at 450, 500, 650, or 800 °C. Reaction conditions: catalyst, 100 mg; reactant gas, 3:1 H_2:N_2 at a flow rate of 120 N mL min^{-1}. Reproduced from Ref. [31] with permission from the Royal Society of Chemistry

triple bond. Our results showed that the use of composite rare earth oxide supports containing reducible rare earth elements, and the reduction of their Ru catalysts at high temperatures is an efficacious strategy to obtain highly active Ru catalysts for ammonia synthesis.

We also studied the influence of the temperature of calcination of the $La_{0.5}Ce_{0.5}O_{1.75}$ support and the temperatures of reduction of the Ru/$La_{0.5}Ce_{0.5}O_{1.75}$ catalyst on the physicochemical properties of the catalyst and ammonia synthesis activity [32]. The results showed that the balance between the calcination temperature of the support and the reduction temperature of the catalyst was important for controlling the state of the catalyst (Fig. 20.9) and increasing the ammonia synthesis activity under mild reaction conditions (Fig. 20.10). Reduction at high temperature (650 °C) is preferred because the increase of the Ce^{3+}/Ce^{4+} ratio with increasing reduction temperature increases the TOF of the catalyst. However, the primary particles of $La_{0.5}Ce_{0.5}O_{1.75}$ tend to be enlarged at high temperatures, and the Ru agglomerates when the calcination temperature of the support is low (e.g., 600 °C, denoted as Ru/$La_{0.5}Ce_{0.5}O_{1.75}$_600°Ccal). To prevent enlargement of the support, the primary support particles should be sintered during calcination at a temperature higher than the temperature used for reduction of the catalyst. (Ru/$La_{0.5}Ce_{0.5}O_{1.75}$_700°Ccal). In this case, there are many fine Ru nanoparticles in the SMSI state on the reduced support, and the rate of ammonia synthesis over the catalyst is high. However, if the sintering temperature of the carrier is too high, the sintering is too advanced, and the dispersion of Ru becomes poor. The result is a very low rate of ammonia synthesis

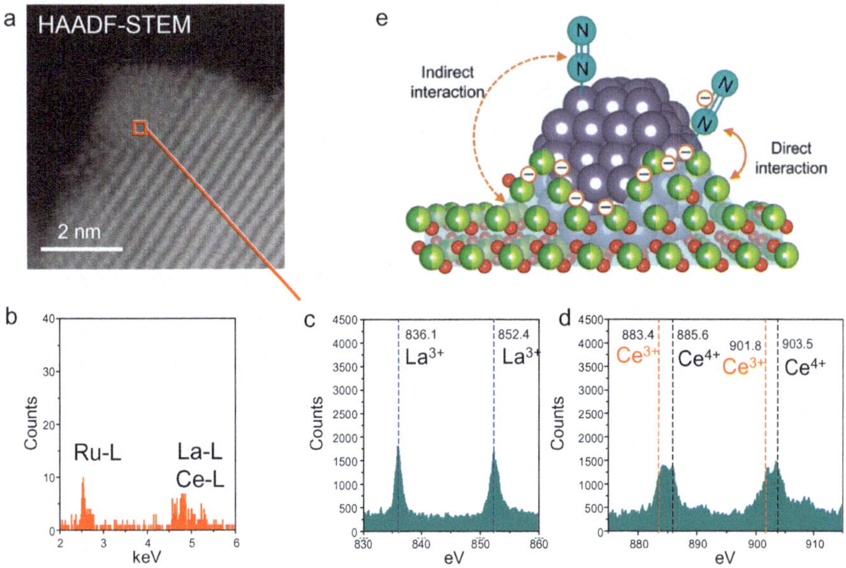

Fig. 20.7 High-magnification HAADF-STEM images, EDX spectra, and electron energy loss (EEL) spectra of the Ru/$La_{0.5}Ce_{0.5}O_{1.75}$_650red catalyst without exposure to air. **a** HAADF-STEM image. **b** EDX spectra of areas indicated by the red square in (**a**). **c** and **d** EEL spectra of La $M_{4,5}$ **c** and Ce $M_{4,5}$ **d** edges for areas indicated by the red square in (**a**). **e** Possible reaction mechanism of N_2 activation over Ru/$La_{0.5}Ce_{0.5}O_{1.75}$ reduced at 650 °C. Reproduced from Ref. [31] with permission from the Royal Society of Chemistry

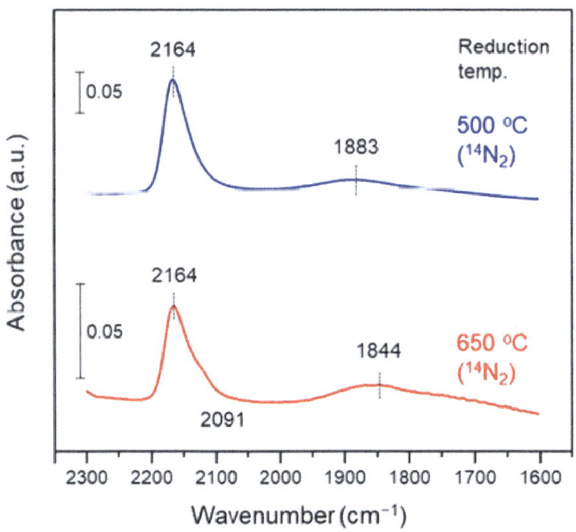

Fig. 20.8 FTIR spectra of N_2. Difference infrared spectra of N_2 before and after adsorption onto Ru/$La_{0.5}Ce_{0.5}O_{1.75}$_500red and Ru/$La_{0.5}Ce_{0.5}O_{1.75}$_650red. The spectra were measured under 6 kPa of N_2 at 25 °C. Reproduced from Ref. [31] with permission from the Royal Society of Chemistry

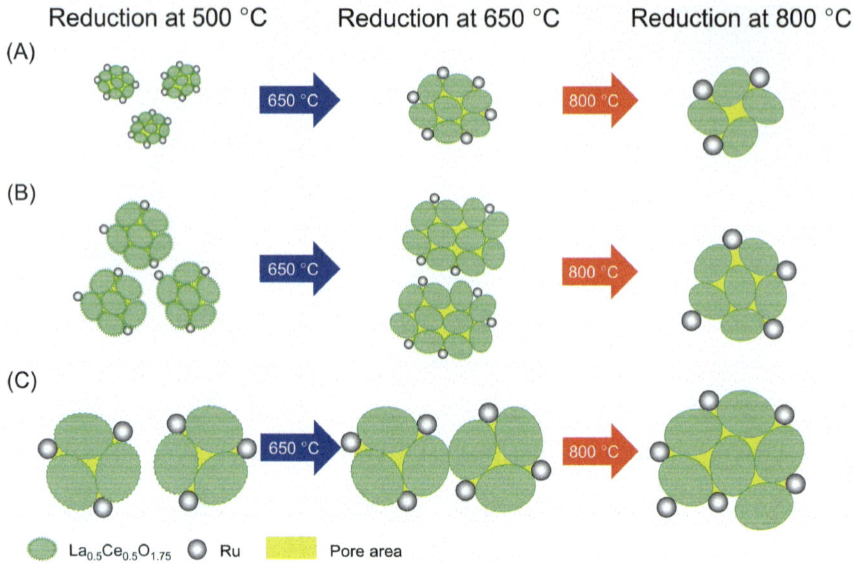

Fig. 20.9 Schematic showing the changes in the aggregation state of Ru particles and the primary support particles in the support during reduction of A) Ru/$La_{0.5}Ce_{0.5}O_{1.75}$_600°Ccal, B) _800°Ccal, and C) _900°Ccal Ref. [32] Copyright (2020) Wiley

(Ru/$La_{0.5}Ce_{0.5}O_{1.75}$_900°Ccal). The calcination of the support and reduction of the catalyst at temperatures higher than those used for the preparation of conventional Ru catalysts provide a new approach to produce highly dispersed Ru catalysts that exhibit SMSI effects and high-catalytic activity.

20.3.2 $La_{0.5}Pr_{0.5}O_{1.5}$

We then used Ru/$La_{0.5}Pr_{0.5}O_{1.5}$ to investigate our strategy of using composite oxide supports of rare earth elements, one of which could be reduced in the oxide state, and reducing the Ru catalyst at high temperatures [33]. The rate of ammonia synthesis over the Ru/$La_{0.5}Pr_{0.5}O_{1.5}$ catalyst increased with increase of the reduction temperature from 450 to 650 °C (Fig. 20.11). However, sintering of the catalyst caused the ammonia synthesis rate to decrease with further increase of the reduction temperature from 650 to 700 °C. The rates of ammonia synthesis at 400 °C and 1.0 MPa were comparable over Ru/$La_{0.5}Pr_{0.5}O_{1.5}$_650red and Ru/$La_{0.5}Ce_{0.5}O_{1.5}$_650red (entries 14 and 15 in Table 20.1). However, the activation mechanism associated with the increase of reduction temperature was completely different for these two catalysts. For the Ru/$La_{0.5}Pr_{0.5}O_{1.5}$ catalyst, the completion of the reduction of Pr^{4+} to Pr^{3+} below 500 °C indicated that the Pr^{4+} in the catalyst was completely reduced in both the Ru/$La_{0.5}Pr_{0.5}O_{1.75}$_500red and Ru/$La_{0.5}Pr_{0.5}O_{1.75}$_650red catalysts. However, part

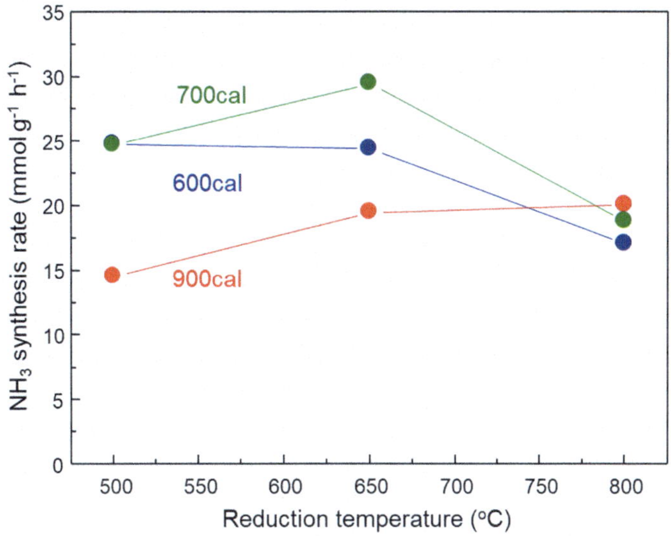

Fig. 20.10 Relationship between reduction temperature and ammonia synthesis rate over $Ru/La_{0.5}Ce_{0.5}O_{1.75}$_600–900 cal. Reaction conditions: catalyst, 100 mg; reactant gas, 3:1 H_2:N_2 at a flow rate of 120 mL min^{-1}; pressure, 1.0 MPa; temperature, 350 °C Ref. [32] Copyright (2020) Wiley

of the $La_{0.5}Pr_{0.5}O_{1.75}$ support changed from a cubic structure to a trigonal structure during reduction from 500 to 650 °C (Fig. 20.12). A Rietveld refinement calculation revealed that the percentages of trigonal $La_{0.5}Pr_{0.5}O_{1.5}$ structures increased from 38 to 77% with the increase of the reduction temperature. To understand the surface properties of $Ru/La_{0.5}Pr_{0.5}O_{1.75}$ reduced at different temperatures, we carried out CO_2 temperature-programmed desorption. We observed desorption of CO as well as CO_2 (Fig. 20.13). In the case of $Ru/La_{0.5}Pr_{0.5}O_{1.5}$_500red, we observed broad peaks of desorbed CO_2 with peak maxima at 417 °C and 778 °C. In addition, a small peak of desorbed CO with a peak maximum at 477 °C was also visible. Raising the reduction temperature from 500 to 650 °C resulted in a decrease of CO_2 desorption at around 400–800 °C and a gradual increase in CO desorption. Desorption of CO indicated that reduction of CO_2 and oxidation of the reduced carrier occurred (e.g., $0.25CO_2 + La_{0.5}Pr_{0.5}O_{1.5} \rightarrow 0.25CO + La_{0.5}Pr_{0.5}O_{1.75}$). The adsorbed CO_2 was reduced by the electrons in the reduced support, and the CO was subsequently desorbed. The oxygen defect was concurrently supplemented by the oxygen released from CO_2. It is generally accepted that the amount of CO_2 desorption reflects the number of basic sites. The desorption temperature is related to the strength of the basic sites, but not all basic sites are quantified by CO_2 desorption in this catalyst. The intensity of the peak of desorbed CO increased with the increase of the reduction temperature. This result indicates that the high-temperature reduction provided the support with electron-donating ability, and therefore, the TOF of the $Ru/La_{0.5}Pr_{0.5}O_{1.75}$ increased as the reduction temperature increased from 500 to 650

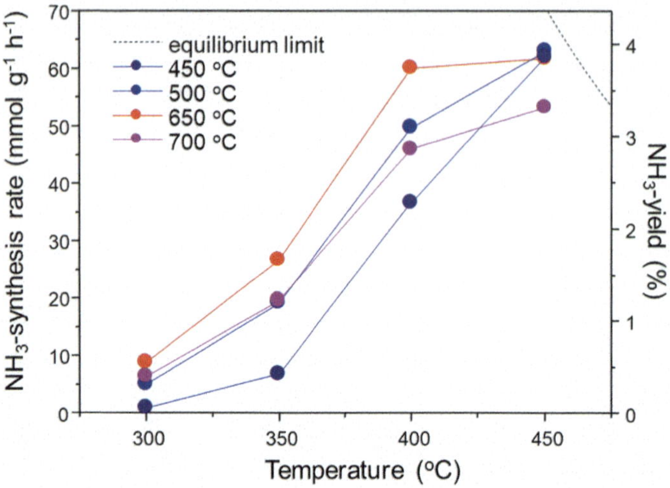

Fig. 20.11 Temperature dependence of ammonia synthesis rate and NH_3 yield at 1.0 MPa over Ru/$La_{0.5}Pr_{0.5}O_{1.75}$ after reduction at 450, 500, 650, or 700 °C. Reaction conditions: catalyst, 100 mg; reactant gas, 3:1 H_2:N_2 at a flow rate of 120 mL min^{-1} Ref. [33] Copyright (2020) American Chemical Society

and 700 °C. When Ru/$La_{0.5}Pr_{0.5}O_{1.75}$ was reduced at 650 °C, the structure of the $La_{0.5}Pr_{0.5}O_{1.5}$ support changed from cubic to trigonal. This structural change would lead to the accumulation of oxygen vacancies on the support surface and improve the electron-donating capacity of the carrier. FTIR measurements after adsorption of N_2 onto the catalyst demonstrated that electrons were efficiently donated from the trigonal pyramid of $La_{0.5}Pr_{0.5}O_{1.5}$ to the N_2 molecules adsorbed on the Ru atoms.

20.3.3 Ba/LaCeOx

Barium (Ba) is well known to be an effective promoter for the catalysis of ammonia synthesis [34–37]. Aika et al. [34, 35] have suggested that Ba species act as electronic promoters. In contrast, Muhler et al. [36] and Jacobsen et al. [37] have suggested that Ba works as a structural promoter.

We added BaO to Ru/$La_{0.5}Ce_{0.5}O_{1.75}$ (denoted as Ru/Ba/LaCeO_x) and investigated the catalytic behavior of the catalyst [38]. The rate of ammonia synthesis over the Ru/Ba/LaCeO_x catalyst at 350 °C and 1.0 MPa increased 2.7-fold with the increase of the reduction temperature from 500 to 700 °C (entries 16 and 17 in Table 20.1). The rate of ammonia synthesis was much higher over Ru/Ba/LaCeO_x_700red than Ru/$La_{0.5}Ce_{0.5}O_{1.75}$_600red and Cs^+/Ru/MgO_500red from 300 to 400 °C at 1.0 MPa (Fig. 20.14a). The Ru/$La_{0.5}Ce_{0.5}O_{1.75}$_600red catalyst is one of the most active oxide-supported Ru catalysts to date. Furthermore, the increase of the rate of ammonia synthesis over Ru/Ba/LaCeO_x_700red with an increase of the reaction pressure from

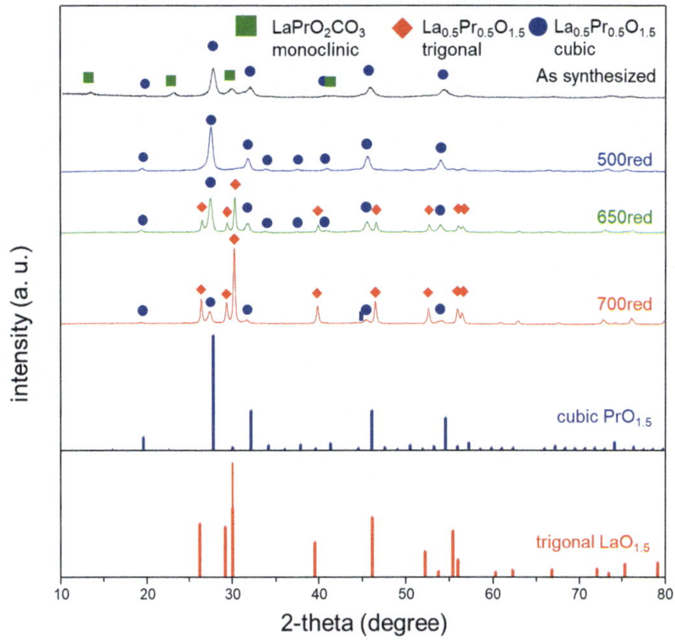

Fig. 20.12 XRD patterns measured for as-synthesized Ru/$La_{0.5}Pr_{0.5}O_{1.75}$ and without exposure to air after treatment of the catalyst in a flow of H_2 at 500, 650, or 700 °C for 1 h. Cubic Pr_2O_3 and trigonal La_2O_3 are Ref. [33] Copyright (2020) American Chemical Society

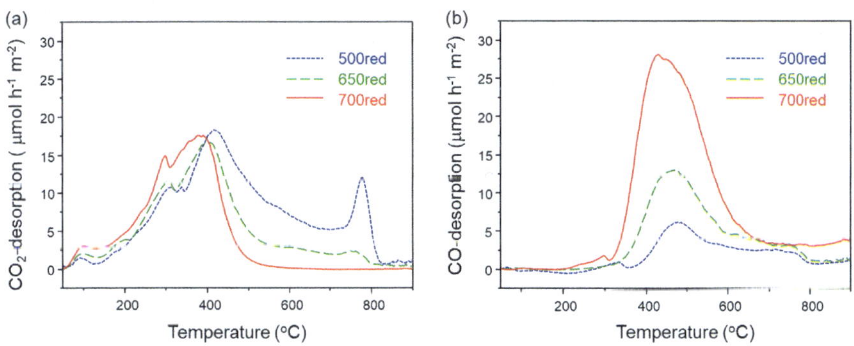

Fig. 20.13 CO_2 temperature-programmed desorption profiles measured in a flow of He after reduction of the Ru/$La_{0.5}Pr_{0.5}O_{1.75}$ at 500, 650, or 700 °C for 1 h followed by CO_2 adsorption at 50 °C. **a** CO_2 desorption, **b** CO desorption Ref. [33] Copyright (2020) American Chemical Society

0.1 to 1.0 MPa at 350 °C (Fig. 20.14b) indicated that the catalyst was free from hydrogen poisoning, even after doping of the Ba^{2+}. In contrast, hydrogen poisoning prevented the rate of ammonia synthesis over Cs^+/Ru/MgO_500red from increasing, even if the reaction pressure was increased at the same temperature. In fact, the reaction orders of H_2 synthesis over Ru/Ba/LaCeO$_x$_700red and Cs^+/Ru/MgO_500red were 0.06 and −0.76, respectively. These results revealed that rare earth elements are essential to prevent hydrogen poisoning over supported Ru catalysts. We performed STEM-EDX measurements of Ru/Ba/LaCeO$_x$_700red without exposure to air. We observed that low-crystalline fractions accumulated over the Ru nanoparticles (Fig. 20.15a). This result contrasted with that of Ru/LaCeO$_x$_650red, wherein Ru was only partially covered by the partially reduced support on the surface of the Ru nanoparticles. According to the EDX elemental map, the accumulated fractions of Ru/Ba/LaCeO$_x$_700red contained the elements Ba, Ce, and La. Electron energy loss spectra indicated that Ce^{3+} was enriched near the surface of the catalysts. These results revealed that the Ru nanoparticles were encapsulated in a composite oxide containing Ba^{2+}, La^{3+}, Ce^{3+}, and oxygen defect sites. We hypothesized that this specific structure was obtained because of the dynamics of heat-resistant rare earth oxides containing redox-active atoms and alkaline earth metal compounds when they were reduced at unusually high temperatures. During the high-temperature reduction, we used FTIR to confirm that carbonate, which reduces the electron-donating capacity of the support, was hydrogenated and removed from the catalyst (Fig. 20.16). This nanofraction showed strong electron-donating ability because of the strong basicity of the included cations, the removal of carbonate, and formation of oxygen defect sites, which eliminated O^{2-} and its electron-withdrawing effect from the interface between the Ru and the nanofraction (Fig. 20.15b–d). The number of available Ru sites on the surface of the Ru/Ba/LaCeO$_x$_700red was therefore reduced to one-third that of Ru/La$_{0.5}$Ce$_{0.45}$O$_{1.75}$_650red. In contrast, the TOF of the Ru/Ba/LaCeO$_x$_700red was enhanced five-fold versus that of Ru/La$_{0.5}$Ce$_{0.45}$O$_{1.75}$_650red. The addition of BaO significantly weakened the triple bond of the adsorbed N_2 and shifted the wavenumber of the peak attributed to the N_2 adsorbed onto Ru from 1847 cm^{-1} in the case of Ru/La$_{0.5}$Ce$_{0.45}$O$_{1.75}$_650red and to 1676 cm^{-1} in the case of Ru/Ba/LaCeO$_x$_700red. (Fig. 20.17). Note that the small peak at 2128 cm^{-1} was not attributed to N_2 because a peak shift was not observed with isotopic labeling experiments using $^{15}N_2$. In Ru/Ba/LaCeO$_x$_700red, all available Ru sites are in contact with low-crystalline nanofractions, and electrons are effectively donated to N_2 molecules via Ru. Our results clearly demonstrated that BaO functions as a chemical promoter of the Ru catalyst.

20.4 Conclusions

Use of rare earth oxide elements as components of catalyst supports and control of the structure of the metal-support boundary is an effective way to obtain Ru catalysts with high activity and strong resistance to hydrogen poisoning. Such control can be achieved by adjusting the type of precursor, the order of addition of each component, and the method of preparation, such as the calcination and reduction temperatures.

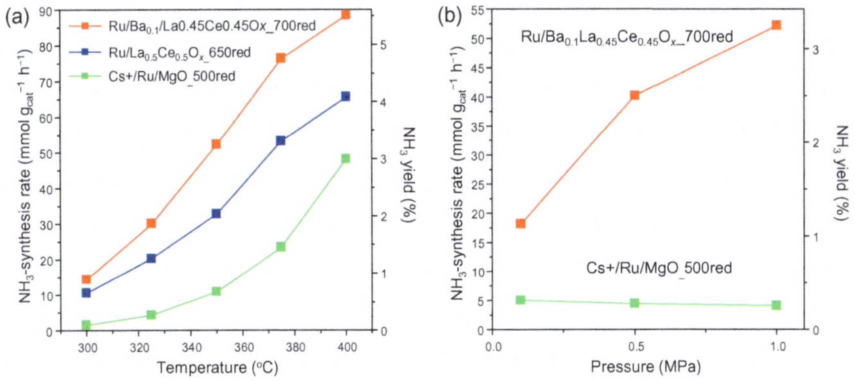

Fig. 20.14 Evaluation of catalyst activities for ammonia synthesis. **a** Influence of reduction temperature on ammonia synthesis activity at 350 °C and 1.0 MPa. **b** Pressure dependence of the ammonia synthesis rates and ammonia yield. (xxxred means that the catalyst was pre-reduced at xxx °C) Ref. [38] Copyright (2020) American Chemical Society

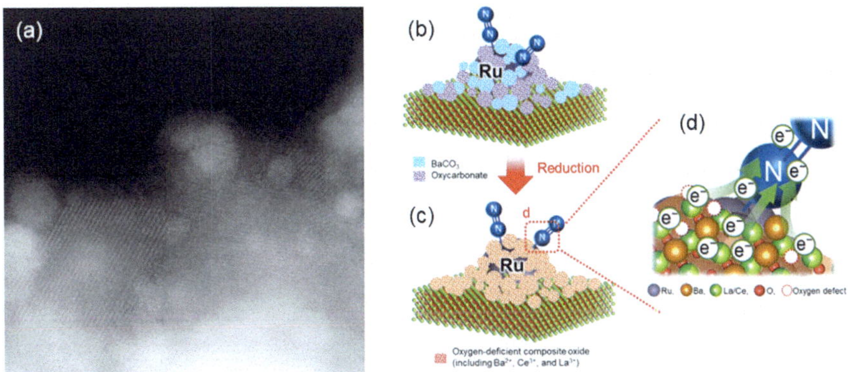

Fig. 20.15 HAADF-STEM image and schematic illustration of Ru/Ba/LaCeO$_x$. **a** HAADF-STEM image of Ru/Ba/LaCeO$_x$_700red catalyst without exposure to air. **b–d** Schematic illustration of the surface morphology of supported Ru catalysts. **b** Ru/Ba/LaCeO$_x$_500red, **c** Ru/Ba/LaCeO$_x$_700red, and **d** a possible mechanism of N_2 activation at a surface-available Ru atom in contact with the nanofraction in Ru/Ba/LaCeOx_700red Ref. [38] Copyright (2020) American Chemical Society

In future, it is expected that the performance of Ru catalysts will be dramatically improved to meet the requirements for green ammonia production.

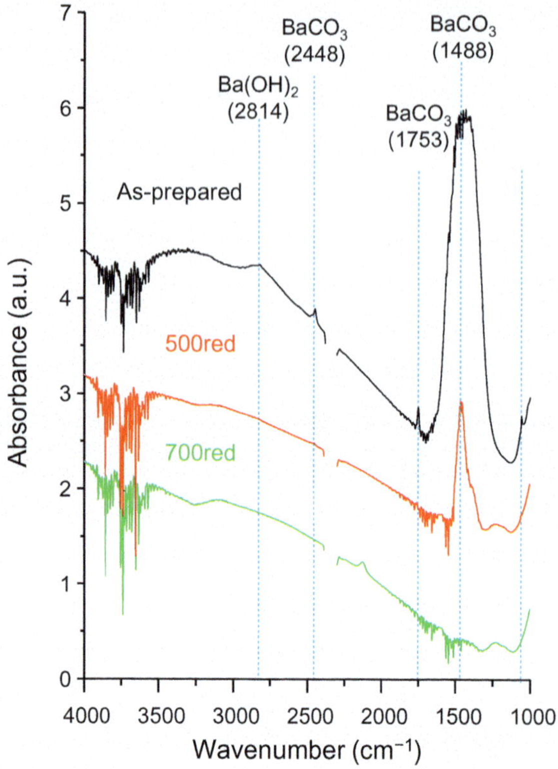

Fig. 20.16 IR spectra for as-prepared Ru/Ba/LaCeOx and after reduction at 500 and 700 °C. [38] Copyright (2020) American Chemical Society

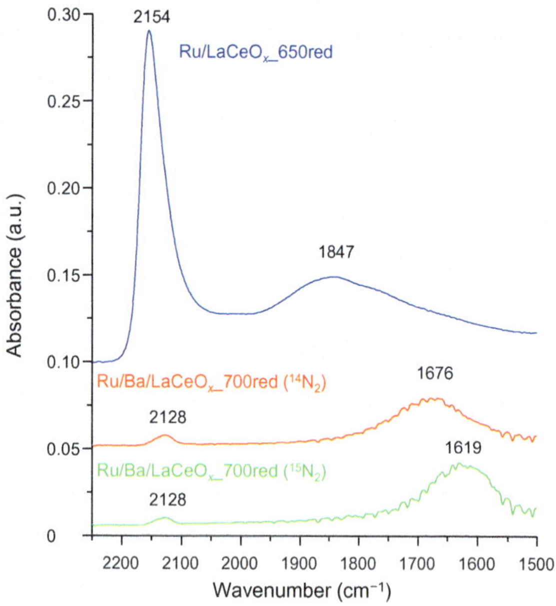

Fig. 20.17 FTIR spectra of N_2. Difference infrared spectra of $^{14}N_2$ or $^{15}N_2$ before and after adsorption onto Ru/Ba/LaCeO$_x$_700red and Ru/LaCeO$_x$_650red Ref. [38] Copyright (2020) American Chemical Society

Acknowledgements This research was partially supported by the Core Research for Evolutionary Science and Technology program of the Japan Science and Technology Agency [grant no. JPMJCR1341], a Japan Society for the Promotion of Science grant-in-aid for scientific research (KAKENHI) 20H02522 JP, and the Toyota Mobility Foundation. K.S. thanks the Program for Elements Strategy Initiative for Catalysts and Batteries commissioned by the Japanese Ministry of Education, Culture, Sports, Science, and Technology.

References

1. World Fertilizer Trends and Outlook to 2022, Food & Agriculture Organization, 2020
2. Aika K (2016) Expecting LCA data of "energy carrier": CO_2 free hydrogen and ammonia as examples. J Life Cycle Assess Jpn 12:161–169
3. Smith C, Hill AK, Torrente-Murciano L (2020) Current and future role of Haber-Bosch ammonia in a carbon-free energy landscape. Energy Environ Sci 13:331–344
4. Mittasch A, Frankenburg W (1950) Early studies of multicomponent catalysts. Adv Catal 2:81–104
5. Foster SL, Bakovic SIP, Duda RD, Maheshwari S, Milton RD, Minteer SD, Janik MJ, Renner JN, Greenlee LF (2018) Catalysts for nitrogen reduction to ammonia. Nat Catal 1:490–500
6. The Paris Agreement, in, United Nations Framework Convention on Climate Change
7. Ozaki A, Aika K, Hori H (1971) A new catalyst system for ammonia synthesis. Bull Chem Soc Jpn 44:3216–3216
8. Aika K, Hori H, Ozaki A (1972) Activation of nitrogen by alkali metal promoted transition metal I. Ammonia synthesis over ruthenium promoted by alkali metal. J Catal 27:424–431
9. Aika K, Kumasaka M, Oma T, Kato O, Matsuda H, Watanabe N, Yamazaki K, Ozaki A, Onishi T (1986) Support and promoter effect of ruthenium catalyst. Appl Catal 28:57–68
10. Aika K, Kadowaki Y (1996) Promoter effect of Sm_2O_3 on Ru/Al_2O_3 in ammonia synthesis. J Catal 161:178–185
11. Niwa Y, Aika K (1996) Ruthenium catalyst supported on CeO_2 for ammonia synthesis. Chem Lett 25:3–4
12. Niwa Y, Aika K (1996) The effect of lanthanide oxides as a support for ruthenium catalysts in ammonia synthesis. J Catal 162:138–142
13. Tauster SJ (1987) Strong metal-support interactions. Acc Chem Res 20:389–394
14. Lin B, Liu Y, Heng L, Wang X, Ni J, Lin J, Jiang L (2018) Morphology effect of ceria on the catalytic performances of Ru/CeO_2 catalysts for ammonia synthesis. Ind Eng Chem Res 57:9127–9135
15. Li W, Liu PC, Niu RY, Li JP, Wang S (2020) Influence of CeO_2 supports prepared with different precipitants over Ru/CeO_2 catalysts for ammonia synthesis. Solid State Sci 99:105983
16. Ma Z, Zhao S, Pei X, Xiong X, Hu B (2017) New insights into the support morphology-dependent ammonia synthesis activity of Ru/CeO_2 catalysts. Catal Sci Technol 7:191–199
17. Sato K, Imamura K, Kawano Y, Miyahara SI, Yamamoto T, Matsumura S, Nagaoka K (2017) A low-crystalline ruthenium nano-layer supported on praseodymium oxide as an active catalyst for ammonia synthesis. Chem Sci 8:674–679
18. Imamura K, Miyahara S, Kawano Y, Sato K, Nakasaka Y, Nagaoka K (2019) Kinetics of ammonia synthesis over Ru/Pr_2O_3. J Taiwan Inst Chem Eng 105:50–56
19. Luo MF, Yan ZL, Jin LY (2006) Structure and redox properties of $Ce_xPr_{1-x}O_{2-\delta}$ mixed oxides and their catalytic activities for CO, CH_3OH and CH_4 combustion. J Mol Catal A: Chem 260:157–162
20. Song Z, Liu W, Nishiguchi H, Takami A, Nagaoka K, Takita Y (2007) The Pr promotion effect on oxygen storage capacity of Ce–Pr oxides studied using a TAP reactor. Appl Catal A 329:86–92

21. Zhao B, Wang Q, Li G, Zhou R (2013) Effect of rare earth (La, Nd, Pr, Sm and Y) on the performance of Pd/$Ce_{0.67}Zr_{0.33}MO_{2-\delta}$ three-way catalysts. J Environ Chem Eng 1:534–543
22. Nagaoka K, Sato K, Takita Y (2012) Effect of the nature of rhodium catalyst supports on initiation of H_2 production during n-butane oxidative reforming at room temperature. J Catal 287:86–92
23. Kitano M, Inoue Y, Yamazaki Y, Hayashi F, Kanbara S, Matsuishi S, Yokoyama T, KimSW HM, Hosono H (2012) Ammonia synthesis using a stable electride as an electron donor and reversible hydrogen store. Nat Chem 4:934–940
24. Miyahara S, Sato K, Kawano Y, Imamura K, Ogura Y, Tsujimaru K, Nagaoka K (2021) Ammonia synthesis over lanthanoid oxide–supported ruthenium catalysts. Catal Today 376:36–40
25. Aika A, Kubota J, Kadowaki Y, Niwa Y, Izumi Y (1997) Molecular sensing techniques for the characterization and design of new ammonia catalysts. Appl Surf Sci 121–122:488–491
26. Saito M, Itoh M, Iwamoto J, Li CY, Machida K (2006) Synergistic effect of MgO and CeO_2 as a Support for ruthenium catalysts in ammonia synthesis. Catal Lett 106:107–110
27. Luo X, Wang R, Ni J, Lin J, Lin B, Xu X, Wei K (2009) Effect of La_2O_3 on Ru/CeO_2-La_2O_3 catalyst for ammonia synthesis. Catal Lett 133:382–387
28. Zhang L, Lin J, Ni J, Wang R, Wei K (2011) Highly efficient Ru/Sm_2O_3-CeO_2 catalyst for ammonia synthesis. Catal Commun 15:23–26
29. Ma Z, Xiong X, Song C, Hu B, Zhang W (2016) Electronic metal–support interactions enhance the ammonia synthesis activity over ruthenium supported on Zr-modified CeO_2 catalysts. RSC Adv 6:51106–51110
30. Yang XL, Zhang WQ, Xia CG, Xiong XM, Mu XY, B, (2010) Low temperature ruthenium catalyst for ammonia synthesis supported on $BaCeO_3$ nanocrystals. Catal Commun 11:867–870
31. Ogura Y, Sato K, Miyahara SI, Kawano Y, Toriyama T, Yamamoto T, Matsumura S, Hosokawa S, Nagaoka K (2018) Efficient ammonia synthesis over a Ru/$La_{0.5}Ce_{0.5}O_{1.75}$ catalyst pre-reduced at high temperature. Chem Sci 9:2230–2237
32. Ogura Y, Asai T, Sato K, Miyahara S-i, Toriyama T, Yamamoto T, Matsumura S, Nagaoka K (2020) Effect of calcination and reduction temperatures on the catalytic activity of Ru/$La_{0.5}Ce_{0.5}O_{1.75}$ for ammonia synthesis under mild conditions. Energy Technol 8:2000264
33. Ogura Y, Tsujimaru K, Sato K, Miyahara S, Toriyama T, Yamamoto T, Matsumura S, Nagaoka K (2018) Ru/$La_{0.5}Pr_{0.5}O_{1.75}$ Catalyst for low-temperature ammonia synthesis. ACS Sustainable Chem Eng 6:17258–17266
34. Aika K, Kawahara T, Murata S, Onishi T (1990) Promoter effect of alkali metal oxides and alkali earth metal oxides on active carbon-supported ruthenium catalyst for ammonia synthesis. Bull Chem Soc Jpn 63:1221–1225
35. Zeng H, Inazu K, Aika K (2002) The working state of the barium promoter in ammonia synthesis over an active-carbon-supported ruthenium catalyst using barium nitrate as the promoter precursor. J Catal 211:33–41
36. Bielawa H, Hinrichsen O, Birkner A, Muhler M (2001) The ammonia-synthesis catalyst of the next generation: barium-promoted oxide-supported ruthenium. Angew Chem Int Ed 40:1061–1063
37. Hansen TW, Hansen PL, Dahl S, Jacobsen CJH (2002) Support effect and active sites on promoted ruthenium catalysts for ammonia synthesis. Catal Lett 84:7–12
38. Sato K, Miyahara SI, Ogura Y, Tsujimaru K, Wada Y, Toriyama T, Yamamoto T, Matsumura S, Nagaoka K (2020) Surface dynamics for creating highly active ru sites for ammonia synthesis: accumulation of a low-crystalline, oxygen-deficient nanofraction. ACS Sustainable Chem Eng 8:2726–2734

Chapter 21
Electride Catalysts for Ammonia Synthesis

Hideo Hosono

Abstract This chapter describes the motivation and progress in electride-based catalysts for ammonia synthesis at mild conditions. C12A7 electride (C12A7:e$^-$) has unique properties of having low work function and chemical/thermal stability. We utilize this unique nature to an active support for activation of N_2 adsorbed on the surface of metal nanoparticles loaded. The activation energy and TOF of Ru: C12A7:e$^-$ are half and higher an order of magnitude, respectively, and robustness to hydrogen poisoning is improved, compared with the previous Ru-loaded catalysts. This chapter outlines of Ru/C12A7:e$^-$ first and subsequently describes various types of electride materials including 2-dimensional(2D) and surface electrides and their catalysis applications.

Keywords Electride · Ammonia synthesis · Model catalyst · Surface electride

21.1 Introduction

Most heterogeneous catalysts are composed of active metal nanoparticles and a support that immobilizes them. An inert solid having a large specific surface area and thermal and chemical stability is used as the support, in general. alumina, silica, and carbon are typical examples. Introduced here are catalysts that use an electride which is a compound having electrons as anions, as a support. They exhibit unique catalytic activity arising from electron-active support with loaded metals. This chapter describes the motivation, characteristics, and mechanism of electride catalysts for NH_3 synthesis at mild conditions.

H. Hosono (✉)
Tokyo Institute of Technology, Nagatsuta-cho, Midori-ku, Yokohama 226-8503, Japan
e-mail: hosono@mces.titech.ac.jp

K. Aika and H. Kobayashi (eds.), *CO_2 Free Ammonia as an Energy Carrier*,
https://doi.org/10.1007/978-981-19-4767-4_21

21.2 C12A7 Electride

Electride is a general term for solid materials in which electrons act as anions, and was first synthesized by James Dye in 1983 as an organic material [1]. When an alkali metal is dissolved in liquid ammonia, which is a polar solvent, ionized electrons are coordinated by NH_3 molecules to generate "solvated electrons". James Dye was fascinated by the beautiful color of "solvated electron" and began research to crystallize it [2]. This approach was successful by dissolving metallic cesium in a crown ether solution and cooling it. Crown ethers strongly encapsulate alkaline ions to form large-sized cation complex ions. By filling with a large size cation, a space of ~4A in diameter was created at the interstitial positions and ionized electrons are entrapped to be stabilized as anions. This electride attracted interest of materials researchers as a material of a new concept and was described in college inorganic chemistry textbooks, but the physical property research was hardly progressed due to its extremely unstable chemical and thermal nature. Therefore, the realization of a stable electride at an ambient condition had been the most important challenge in this field.

The present authors focused on the crystal structure of $12CaO{\cdot}7Al_2O_3$ (C12A7), in which positively charged sub-nanometer-sized cages are connected by sharing the thin wall of mono-layer oxide. The oxygen ion (O^{2-}), which is loosely bounded by 6 Ca^{2+} ions constituting the cage wall, exists as the counter anion to keep electro-neutrality, and it can be replaced with an electron by a reductive treatment without breaking the cages [3]. The resulting material, C12A7:e, is the first stable electride material at ambient condition. This success opened the material research on electride to examine the electronic structure and physical properties. We found several unique properties for C12A7:e^-. First, unlike insulating nature of C12A7:O^{2-}, it conducts electricity well. When the electron concentration is ~1×10^{21} cm^{-3} or less, it is semiconducting but when it exceeds this value, it exhibits metallic conduction (resistance increases with temperature) [4]. When such a metallic conducting C12A7:e^- is cooled down to low temperatures, it transforms into a superconductor [5]. What is even more interesting for chemical application is that the work function is as low as 2.4 eV. This value is close to that of metal potassium which has extremely high reactivity and low melting temperature. Materials with a small work function are generally less stable chemically and thermally. The exceptionally low work function and thermal/chemical stability of C12A7:e^- may be understood from its unique crystal and electronic structure. As expected from such a low work function, C12A7:e^- exhibits excellent field electron-emission property [6]. Figure 21.1 shows the crystal structure and band structure of C12A7:e^- [7, 8]. In addition to the conventional conduction band which is similar to the case of CaO crystals, there is another conduction band called "cage conduction band". This band originates from three-dimensionally connected cages that allow electrons to pass through monolayer oxide walls through tunneling. The electrons introduced by replacing O^{2-} in the cage partially occupy this band, thereby they show high electrical conductivity and have a low work function defined as the energy difference between Fermi level and

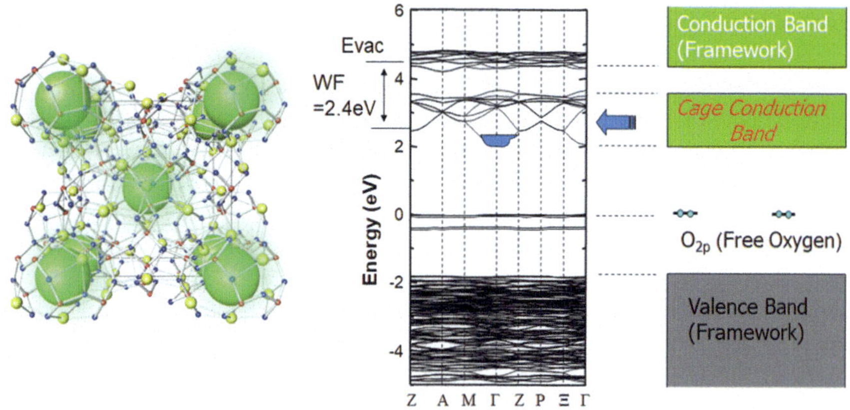

Fig. 21.1 Crystal and band structure of C12A7:e$^-$. Large circle indicates an anionic electron replacing oxygen ions. When O^{2-} as the counter anion is extracted by chemical treatment, electrons are injected into the cages. Since each cage is connected by sharing monoxide layer, injected electrons can pass through the thin cage wall by tunneling, giving rise to electronic conduction. *WF* work function; E_{vac} vacuum level

vacuum level. In conventional metals, the outermost electrons dominate most of the properties, but in C12A7:e$^-$, the electronic properties are determined by the anion electrons existing in the crystallographic cages, but the thermal and chemical properties are mostly controlled by the cage skeleton structure. Therefore, it maintains a high melting point (~1250 °C) of C12A7:O^{2-}, and neither oxygen molecule nor water molecule can enter from atmosphere at around room temperature because the opening mouse of the cage is small compared to O_2 and H_2O. As a result, thermal and chemical stabilities are preserved irrespective of a low work function comparable to alkali metals. In other words, both low work function and thermal/chemical stability are achieved by sharing the functions between the electrons in the crystallographic cages and the skeleton of crystal structure [9].

21.3 C12A7 Electride Catalyst

21.3.1 Why Catalyst Application

The most unique physical properties of C12A7:e$^-$ composed only of earth abundant elements are to have low work function, thermal and chemical stability. In order to utilize the intrinsic properties, we examined the possibility of several applications such as field electron emission sources [10]. One of them is the activation of inert molecules represented by carbon dioxide and di-nitrogen. Di-nitrogen activation is the key process as catalysts for ammonia synthesis. As for NH_3 synthesis, Haber-Bosch method (HB method) is regarded as a completed technology. We did not think it was a theme to tackle anymore, but when I read a book on HB process in 2010 and noted a description that the catalyst is Swedish iron ore, and the key is that it contains potassium and aluminum. This description suggests an essential role of electron donation in the catalytic process. To dissociate the strong N ≡ N bond, it should be possible to increase the electron donating ability and put an electron in the antibonding orbital. Alkali metals are the strongest electron-donating materials, but they are difficult to use as catalysts because they are highly reactive to form stable nitrides and the vapor pressure is high at the reaction temperatures. On the other hand, C12A7:e^- has a work function similar to that of alkali metals, but is thermally and chemically stable, so I thought catalyst for NH_3 synthesis would be a suitable application for C12A7:e. In addition, several years before that, the surface atomic structure of C12A7:e^- was studied in detail using a low-temperature scanning tunneling microscope (STM) on clean surfaces prepared from a single crystal under an ultra-high vacuum. Unfortunately, C12A7 has no distinct cleavage surface. When it is mechanically ruptured, the cages are broken and electrons are escaped from the cages to give insulating surfaces. However, if an appropriate heat treatment is performed at a temperature, it would be reconstructed to recover the cage structure. Fortunately, we could find such a condition that high conductivity and low work function are restored [11]. Therefore, we judged that the preparation was ready, and started the study on C12A7:e as an ammonia synthesis catalyst.

21.3.2 Catalytic Activity

Since our objective is to examine C12A7:e as the catalyst for NH_3 synthesis at low temperature and low pressure, ruthenium whose effectiveness had been reported by Ozaki and Aika [12, 13] was chosen first as the loaded metal catalyst. Ruthenium is located almost at the peak of the volcano plot described in ref [14]. A Ru-bearing compound was thermally decomposed and supported as nanoparticles on C12A7:e particles, and reduced in a hydrogen stream to form metal nanoparticles (Fig. 21.2). The catalytic properties are shown in Fig. 21.3 [15]. Compared with the cases of conventional supports such as MgO reported so far, the apparent activation energy was reduced by less than half, and the TOF value representing the catalytic activity per active site was increased by an order of magnitude. It is well known that ruthenium catalysts are prone to hydrogen poisoning in which the active sites are covered with hydrogen adatom and the reaction rate is saturated even if the partial pressure of hydrogen is increased. Industrially, the produced ammonia is required to be collected as a liquid. It is thus necessary to react under a pressure of 10 atm or higher. At that time, hydrogen poisoning of ruthenium catalyst became an obstacle. Interestingly, Ru/C12A7:e^- does not exhibit distinct H-poisoning as shown in Fig. 21.4. Effectiveness of C12A7:e^- for NH_3 synthesis was confirmed for Co as well [16].

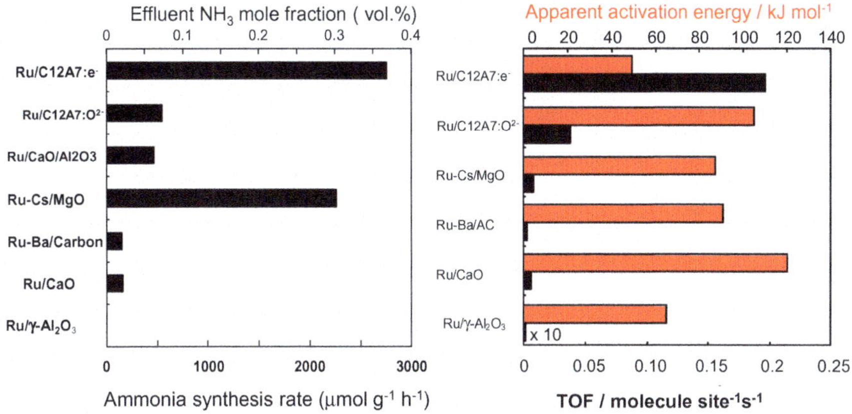

Fig 21.2 Activity of various Ru-loaded catalysts for ammonia synthesis. Reaction conditions; 400 °C, 0.1 MPa, 1 wt% Ru-loading, velocity = 60 mlmin^{-1}. Reproduced with permission from [15]

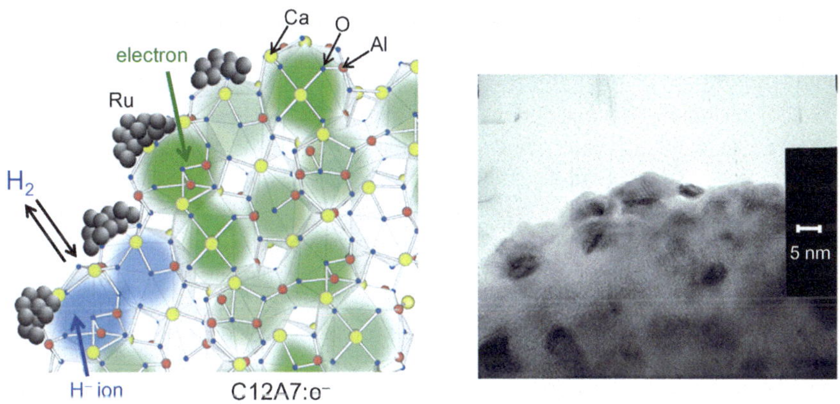

Fig. 21.3 TEM photo and schematic drawing od Ru-loaded C12A7:e^{-}. Reproduced with permission from [15]

21.3.3 Reaction Mechanism

Ruthenium-supported C12A7:e (Ru/C12A7:e) has an order of magnitude higher TOF and activity than MgO supporting Ru and Cs (Ru/Cs-MgO), which was the most active Ru-supporting catalyst at the stage of this research (~2011). The activation energy is halved, and hydrogen poisoning is less likely to occur. To clarify the cause of these results, the catalytic activity was examined using the isotope exchange reaction of $^{14}N_2$ and $^{15}N_2$ ($^{14}N_2 + {}^{15}N_2 = 2{}^{14}N^{15}N$) [17]. In this reaction, the formation of N_2 molecules by the association of N adatoms adsorbed on the Ru surface is

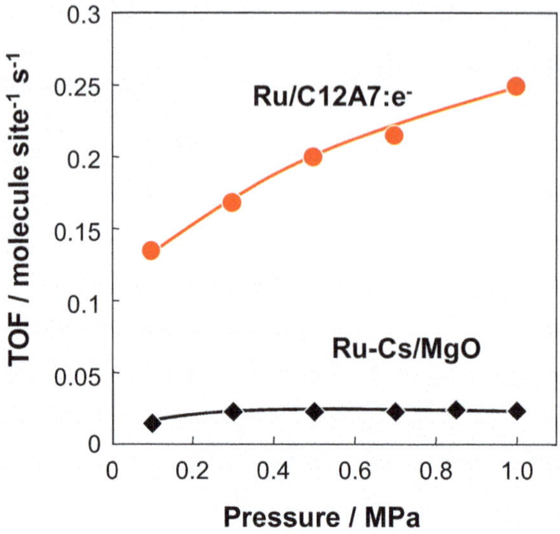

Fig. 21.4 Catalytic activity as a function of reaction pressure. Temp. = 360 °C, $H_2/N_2 = 3$. Reproduced with permission from [15]

the rate-determining process, and the activation energy obtained was 58 kJ.mol^{-1} for C12A7:e$^-$, which is distinctly lower than that (~135 kJmol^{-1}) for Ru/C12A7:O^{2-} (electron-free) or Ru-loaded Ru-Cs/MgO. Using these results, we estimated the activation energy of the dissociation of a certain N_2 on Ru-loaded C12A7:e$^-$ does not exceed 29 kJmol^{-1} in ammonia synthesis by taking it consideration the N-adsorption energy on Ru/C12A7:e is exothermic. Remembering the apparent activation energy of NH_3 synthesis is ~50 kJmol^{-1} for Ru / C12A7:e$^-$, we may conclude that the dissociation of N_2 is not rate-determining.

Over conventional Ru-supported catalysts such as Ru-Cs / MgO and Ru/C12A7:O^{2-}, the activation energy of the NH_3 synthesis reaction is 20–40 kJmol^{-1} smaller than that of nitrogen isotope exchange, indicating the dissociation of N_2 is rate-determining over these catalysts. When Ru is supported on C12A7:O^{2-}, which does not contain electrons in the cage, the activation energy is similar to that for Ru-Cs/MgO. Figure 21.5 shows the energy profile for Ru/C12A7:e$^-$ and conventional Ru-catalysts for NH_3 synthesis [17].

This result clarified that the high electron donating property originating from the presence of low work function electrons remarkably promotes the dissociation of N_2 molecules adsorbed on the surface of Ru particles. This idea was substantiated by DFT calculations on a simplified model [18]. It is inferred that the reason why it is less susceptible to hydrogen poisoning, which is another advantage, is considered as follows: The enthalpy of formation of C12A7: H$^-$ from C12A7:e$^-$ and H_2 is given $\Delta H = -434$ kJmol^{-1}[19]. Thus, C12A7:e$^-$ should be converted to C12A7:H$^-$ under H_2-Ar atmosphere at 633 K, and the high electron donating property should disappear. Actually, as shown in Fig. 21.6, the hydrogenation reaction

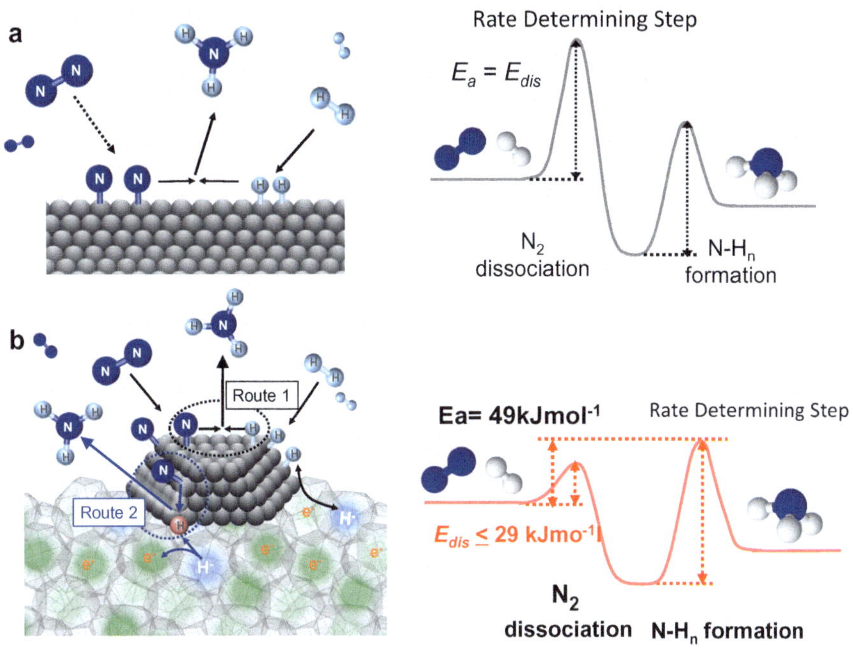

Fig. 21.5 Schematic energy profile for Ru-loaded C12A7:e$^-$ catalyst (**b**) and conventional Ru-catalysts (**a**) for NH_3 synthesis. Reproduced with permission from [17]

of Ru-unloaded C12A7:e$^-$ proceeds linearly with the heating duration. However, hydrogenation hardly proceeds for Ru/C12A7:e$^-$ and the content of H incorporated is easily saturated.

This result indicates hydrogenation occurs only near the surface of Ru-supported C12A7:e$^-$, and H^0 adatom generated by H_2 dissociation on Ru reacts with anion electrons in the cage on the surface, and H^- is formed in the cage [17]. When the N adatom generated by the dissociation of N_2 on Ru comes close to this H^-, it leaves an electron in the cage and forms an NH bond. That is, since the hydrogen adatom rapidly reacts with the electron having a low work function. When N adatom comes to the peripheral of Ru nanoparticle, it escapes from the cage as H^o with much smaller size to react with the N to form N–H bonding leaving an electron in the cage. As a consequence, covering the active site on the Ru nanoparticles with hydrogen can be suppressed. The reactions described above are expressed in the following:

H_2 (on Ru) = $2H^o$ (on Ru)
H^0 (on Ru) + e^- (in surface cage) = H^- (in surface cage)
N_2 (on Ru) = 2 N (on Ru)
N (on Ru) + H^-(in surface cage) = N–H (on Ru) + e^-(in surface cage).

We examined the activity as a function of electron concentrations (*Ne*) in C12A7:e and found the catalytic activity changes drastically with the metal–insulator (MI)

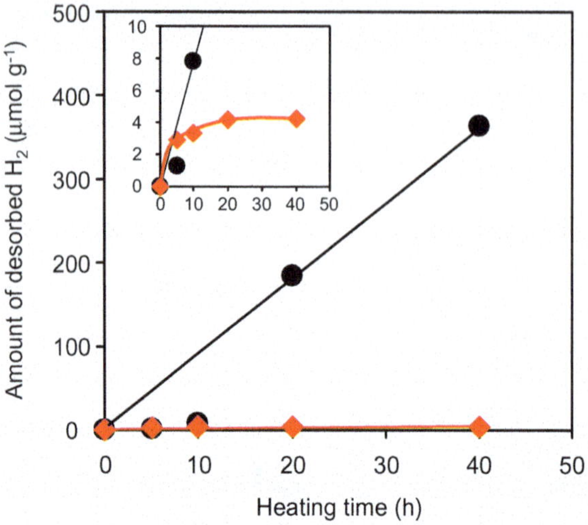

Fig. 21.6 Effect of Ru-loading on formation of C12A7:H$^-$ by reaction of C12A7:e$^-$ with H_2 at 360 °C in 75%H_2-25%Ar. Desorbed amount expresses the incorporated H during the heating duration. Reproduced with permission from [17]

transition [20]. That is, the activity for Ru:C12A7:e with $Ne < Nc$, where Ne is the critical electron concentration for metal–insulator (MI) transition, is similar to the conventional Ru-catalysts, while the activity is jumped for $N > Nc$. No catalytic support materials exhibit MI transition so far to our knowledge. So, we think the obtained result will be useful for catalyst design concept based on a model proposed recently [21].

21.3.4 Two-Dimensional Electride Ca_2N Catalyst

Based on the findings obtained with Ru/C12A7:e$^-$, the system was extended to a 2D-electride in which anionic electrons occupy a two-dimensional nanospace sandwiched by cationic slabs. 2D-electride materials may be regarded as a bulk structure of a two-dimensional electron gas (2DEG) with extraordinarily large electron mobility. [22–24] Here, the catalyst based on Ca_2N is introduced as a typical example. Ca_2N has a composition of $(Ca^{2+})_2(N^{3-})(e^-)$, and anionic electrons are sandwiched between slabs called $[Ca_2N]^+$. The interlayer distance is about 4 Å. The work function of this material is 2.6 eV in the plane perpendicular to the anion electron, which is similar to C12A7:e$^-$. When Ru was supported and the catalytic activity was examined, the activity was high as shown in Fig. 21.7, hydrogen poisoning was less likely to occur, and the activation energy was about the same [25]. When X-ray diffraction was taken after the reaction, it was found that the anion electron was replaced with H$^-$ to convert to Ca_2NH. The work function of Ca_2NH itself is 2.8 eV, but in the

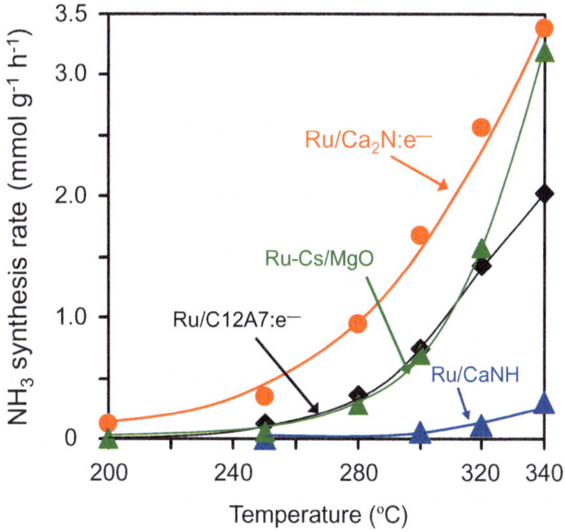

Fig. 21.7 Activity of various Ru-loaded catalysts as a function of temperature. Pressure:0.1 MPa, Ru:2 wt%, and catalyst:0.1 g. Reproduced with permission from [25]

composition where anion electrons and H^- coexist, it drops to 2.3 eV, so the reactions between the anion electrons and hydrogen adatom are described in the previous section occurs in this system as well (Fig. 21.8).

Recently, this reaction mechanism was substantiated by microkinetic analysis using DFT calculation [26]. Furthermore, we considered based on this result that the same catalytic activity could be obtained by supporting Ru on the hydride, and investigated Ru/CaH_2[25]. As expected, almost the same activity was obtained. The work function of CaH_2 itself is 4.0 eV, but when Ru is loaded, a bond is formed between Ru and H^-, and electrons at the site of H^- vacancy are occupied in the antibonding state, so that desorption of hydrogen atoms is enhanced from the lattice leaving anionic electron at the H^- vacancy site. The work function of the electrons drops significantly to 2.3 eV, so it can be understood that the activity is similar to that of electrides such as C12A7:e^- and Ca_2N. According to this model, it is suggested that hydrides are effective support materials, but proton compounds should show inferior activity. In fact, for Ru-loaded CaNH, which can be described as (Ca^{2+}) (N^{3-}) (H^+), the activity was low as shown in Fig. 21.7.

21.4 Expansion of Electride Catalyst

It is sufficient that the hydrogen-related anion species are replaced with electrons in the region in the vicinity of the metal when Ru is supported. If the work function of the anion electron is low in that category and the related band crosses the Fermi

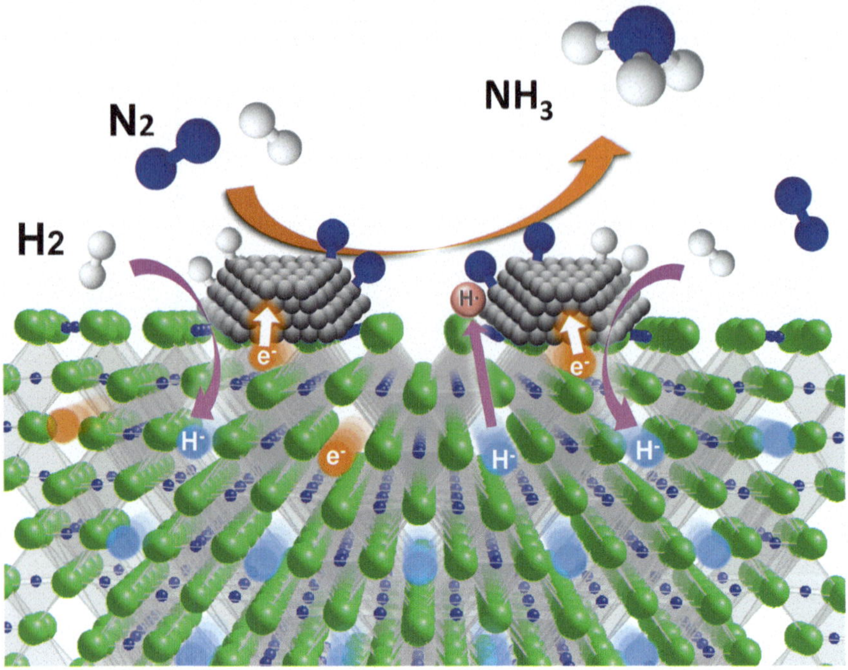

Fig. 21.8 Mechanism for Ru/CaNH$_{1-x}$e$_x$ catalyst. Reproduced with permission from [25]

level, it is considered to be effective as an ammonia catalyst. That is, the search was performed on the assumption that the band structure of C12A7:e$^-$ should exist near the surface. This idea led to the concept of surface electride [27]. Monovalent anions-metal cations have less binding and lattice cohesive energy, in general. Thus, we focused metal hydrides and amines as the candidate of surface electride. As a result, we found Ru-Ba/Ca $(NH_2)_2$ catalyst [28], which exhibits the highest activity to date. Also, the parent material of electride was expanded from an ionic insulating material with an opened bandgap to an intermetallic compound in which electron density is populated at the crystallographic void sites, and a new catalyst was searched for. One of such intermetallic electrides was LaCoSi, which showed the highest activity and the lowest activation energy among the Co-supported catalysts reported so far [29].

The progress in electride materials and electride catalysts for NH_3 synthesis is summarized in Fig. 21.9. The parent material system of electride has been extended from insulating ionic oxide to surface and intermetallic. Recently, a design concept.

of Ru-free catalysts was born from the surface electride. Catalytic activity of transition metal for NH_3 synthesis is primarily determined by the "volcano plot" in which Ru is located around the peak position when the activation of N_2 occurs on the surface of the transition metal loaded. We reported Ni-based catalysts having activity comparable to conventional Ru-based catalysts utilizing surface electrides of rare-earth nitrides such as LaN and CeN [30, 31]. High concentrations of nitrogen

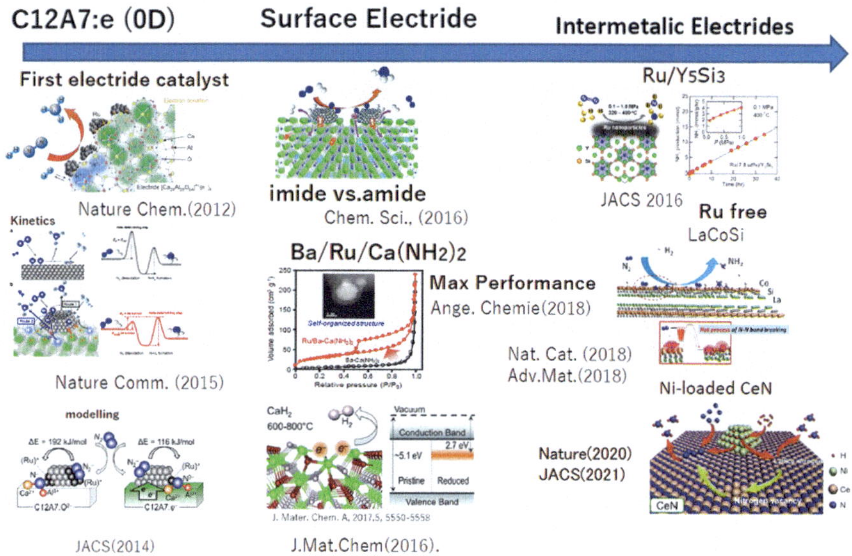

Fig. 21.9 Progress in electride materials and their catalysts for NH_3 synthesis

vacancies at the surface are created on the surface by deposition of Ni particles. Since electrons trapped at the site of the N-vacancy have low work function, N_2 molecule is activated by this N-vacancy, not on loaded Ni. The role of loaded Ni is not needed for activation of N_2 but is enough to activate H_2.

Role of cesium is well known as an effective electronic promoter for NH_3 synthesis but the action mechanism remains unclear. In many cases, cesium is added in the form the nitrate, carbonate or hydroxide, but these compounds cannot be reduced to metal cesium with low work function (2.1 eV) under NH_3 synthesis conditions. The formation of stable suboxides is plausible in the reaction process. Recently, we reported Cs_3O, which may be described as $(Cs^+)_3(O^{2-})(e^-)$, is an electride having lower work function (1.7 eV) than metal Cs [21].

21.5 Next Development

Electride is excellent not only in the ammonia synthesis catalyst described in this chapter, but also in the ammonia decomposition reaction [32, 33], the selective hydrogenation reaction of nitro groups and carbonyl groups [34], and the coupling reaction of forming carbon–carbon bonds [35]. It selectively adsorbs CO_2 at room temperature and decomposes it into CO + O [36], but unfortunately it did not function well as a catalyst. Recently, research on electride has become active worldwide, and by combining genetic algorithm and DFT calculation, the existence of a stable new substance is predicted, and it is actually synthesized to examine physical properties [37]. In addition, two-dimensional electrides are of interest to condensed matter physics because they tend to become "topological substances" when the anionic electron band is mediated the band inversion between the valence band maximum and conduction band minimum [38]. We would like to find an important chemical reaction that makes full use of the characteristics of electride. A comprehensive review article focused on electride materials and their chemical applications was published in 2021 [39]. This review is recommended when the readers have interested in this class of materials science and its chemical applications.

Acknowledgements The author thanks Professor Masaaki Kitano of Materials Research Center for Element Strategy, Tokyo Institute of Technology for his continuous collaboration. Professor Michikazu Hara of Tokyo Institute of Technology is acknowledged for his kind acceptance of my proposal for joint research of C12A7:e^- catalyst. We would also like to thank the collaborators involved in this research. This work was supported by a grant from the JST ACCEL Program and JSPS Kakenhi Grants-in-Aid (No. 17H06153). Finally, the author appreciates professor emeritus Ken-ichi Aika for providing me an opportunity to write this account.

References

1. Ellaboudy A, Dye JL, Smith PB (1983) Cesium 18-crown-6 compounds. A crystalline ceside and a crystalline electride. J Am Chem Soc 105(21):6490–6491
2. Dye JL (2009) Electrides: early examples of quantum confinement. Acc Chem Res 42(10):1564–1572
3. Matsuishi S, Toda Y, Miyakawa M, Hayashi K, Kamiya T, Hirano M, Tanaka I, Hosono H (2003) High-Density electron anions in a nanoporous single crystal: $[Ca_{24}Al_{28}O_{64}]^{4+}(4e^-)$. Science 301(5633):626–629
4. Kim SW, Matsuishi S, Nomura T, Kubota Y, Takata M, Hayashi K, Kamiya T, Hirano M, Hosono H (2007) Metallic state in a lime$^-$ alumina compound with nanoporous structure. Nano Lett 7(5):1138–1143
5. Miyakawa M, Kim SW, Hirano M, Kohama Y, Kawaji H, Atake T, Hosono H (2007) Superconductivity in an inorganic electride $12CaO{\cdot}7Al_2O_3$: e. J Am Chem Soc 129(23):7270–7271
6. Toda Y, Yanagi H, Ikenaga E, Kim JJ, Kobata M, Ueda S, Kamiya T, Hirano M, Hosono H (2007) Work function of a room-temperature, stable electride $[Ca_{24}Al_{28}O_{64}]^{4+}(e^-)_4$. Adv Mater 19(21):3564–3569
7. Sushko PV, Shluger AL, Hayashi K, Hirano M, Hosono H (2003) Electron localization and a confined electron gas in nanoporous inorganic electrides. Phys Rev Lett 91(12):126401
8. Toda Y, Matsuishi S, Hayashi K, Ueda K, Kamiya T, Hirano M, Hosono H (2004) Field emission of electron anions clathrated in subnanometer-sized cages in $[Ca_{24}Al_{28}O_{64}]^{4+}(4e^-)$. Adv Mater 16(8):685–689
9. Kim SW, Hosono H (2012) Synthesis and properties of $12CaO{\cdot}7Al_2O_3$ electride: review of single crystal and thin film growth. Phil Mag 92(19–21):2596–2628
10. Toda Y, Kim SW, Hayashi K, Hirano M, Kamiya T, Hosono H, Yasuda H (2005) Intense thermal field electron emission from room-temperature stable electride. Appl Phys Lett 87(25):254103
11. Toda Y, Kubota Y, Hirano M, Hirayama H, Hosono H (2011) Surface of room-temperature-stable electride $[Ca_{24}Al_{28}O_{64}]^{4+}(e^-)_4$: preparation and its characterization by atomic-resolution scanning tunneling microscopy. ACS Nano 5(3):1907–1914
12. Ozaki A, Aika KI, Hori H (1971) A new catalyst system for ammonia synthesis. Bull Chem Soc Jpn 44(11):3216–3216

13. Ozaki A (1981) Development of alkali-promoted ruthenium as a novel catalyst for ammonia synthesis. Acc Chem Res 14(1):16–21
14. Jacobsen CJ, Dahl S, Clausen BS, Bahn S, Logadottir A, Nørskov JK (2001) Catalyst design by interpolation in the periodic table: bimetallic ammonia synthesis catalysts. J Am Chem Soc 123(34):8404–8405
15. Kitano M, Inoue Y, Yamazaki Y, Hayashi F, Kanbara S, Matsuishi S, Hara M, Hosono H (2012) Ammonia synthesis using a stable electride as an electron donor and reversible hydrogen store. Nat Chem 4(11):934–940
16. Inoue Y, Kitano M, Tokunari M, Taniguchi T, Ooya K, Abe H, Niwa Y, Sasase M, Hara M, Hosono H (2019) Direct activation of cobalt catalyst by 12CaO $7Al_2O_3$ electride for ammonia synthesis. ACS Catal 9(3):1670–1679
17. Kitano M, Kanbara S, Inoue Y, Kuganathan N, Sushko PV, Yokoyama T, Hara M, Hosono H (2015) Electride support boosts nitrogen dissociation over ruthenium catalyst and shifts the bottleneck in ammonia synthesis. Nat Commun 6(1):1–9
18. Kuganathan N, Hosono H, Shluger AL, Sushko PV (2014) Enhanced N_2 dissociation on Ru-loaded inorganic electride. J Am Chem Soc 136(6):2216–2219
19. Trofymluk O, Toda Y, Hosono H, Navrotsky A (2005) Energetics of formation and oxidation of microporous calcium aluminates: a new class of electrides and ionic conductors. Chem Mater 17(22):5574–5579
20. Kanbara S, Kitano M, Inoue Y, Yokoyama T, Hara M, Hosono H (2015) Mechanism switching of ammonia synthesis over Ru-loaded electride catalyst at metal–insulator transition. J Am Chem Soc 137(45):14517–14524
21. Hosono H (2022) Electron transfer from support/promotor to metal catalyst: requirements for effective support. Catal Lett 152(2):307–314
22. Lee K, Kim SW, Toda Y, Matsuishi S, Hosono H (2013) Dicalcium nitride as a two-dimensional electride with an anionic electron layer. Nature 494(7437):336–340
23. Inoshita T, Jeong S, Hamada N, Hosono H (2014) Exploration for two-dimensional electrides via database screening and ab initio calculation. Phys Rev X 4(3):031023
24. Zhang X, Xiao Z, Lei H, Toda Y, Matsuishi S, Kamiya T, Hosono H (2014) Two-dimensional transition-metal electride Y_2C. Chem Mater 26(22):6638–6643
25. Kitano M, Inoue Y, Ishikawa H, Yamagata K, Nakao T, Tada T, Matsuishi S, Yokoyama T, Hara M, Hosono H (2016) Essential role of hydride ion in ruthenium-based ammonia synthesis catalysts. Chem Sci 7(7):4036–4043
26. Nakao T, Tada T, Hosono H (2019) First-principles and microkinetic study on the mechanism for ammonia synthesis using Ru-loaded hydride catalyst. J Phys Chem C 124(3):2070–2078
27. Ong PV, Johnson LE, Hosono H, Sushko PV (2017) Structure and stability of CaH_2 surfaces: on the possibility of electron-rich surfaces in metal hydrides for catalysis. J Mater Chem A 5(11):5550–5558
28. Kitano M, Inoue Y, Sasase M, Kishida K, Kobayashi Y, Nishiyama K, Hara M, Hosono H (2018) Self-organized ruthenium–barium core–shell nanoparticles on a mesoporous calcium amide matrix for efficient low-temperature ammonia synthesis. Angew Chem Int Ed 57:2648–2652
29. Gong Y, Wu J, Kitano M, Wang J, Ye TN, Li J, Kobayashi Y, Kishida K, Abe H, Niwa Y, Yang H, Hosono H (2018) Ternary intermetallic LaCoSi as a catalyst for N_2 activation. Nature Catal 1(3):178–185
30. Ye TN, Park SW, Lu Y, Li J, Sasase M, Kitano M, Tada T, Hosono H (2020) Vacancy-enabled N_2 activation for ammonia synthesis on an Ni-loaded catalyst. Nature 583(7816):391–395
31. Ye TN, Park SW, Lu Y, Li J, Sasase M, Kitano M, Hosono H (2020) Contribution of nitrogen vacancies to ammonia synthesis over metal nitride catalysts. J Am Chem Soc 142(33):14374–14383
32. Hayashi F, Toda Y, Kanie Y, Kitano M, Inoue Y, Yokoyama T, Hara M, Hosono H (2013) Ammonia decomposition by ruthenium nanoparticles loaded on inorganic electride C12A7:e^-. Chem Sci 4(8):3124–3130
33. Ogasawara K, Nakao T, Kishida K, Ye TN, Lu Y, Abe H, Miwa Y, Kitano M, Hosono H (2021) Ammonia decomposition over CaNH-supported Ni catalysts via an NH^{2-}-vacancy-mediated Mars–van Krevelen mechanism. ACS Catal 11(17):11005–11015

34. Ye TN, Lu Y, Li J, Nakao T, Yang H, Tada T, Kitano M, Hosono H (2017) Copper-based intermetallic electride catalyst for chemoselective hydrogenation reactions. J Am Chem Soc 139(47):17089–17097
35. Ye TN, Lu Y, Xiao Z, Li J, Nakao T, Abe H, Miwa Y, Kitano M, Tada T, Hosono H (2019) Palladium-bearing intermetallic electride as an efficient and stable catalyst for Suzuki cross-coupling reactions. Nat Commun 10(1):1–10
36. Toda Y, Hirayama H, Kuganathan N, Torrisi A, Sushko PV, Hosono H (2013) Activation and splitting of carbon dioxide on the surface of an inorganic electride material. Nat Commun 4(1):1–8
37. Wang J, Hanzawa K, Hiramatsu H, Kim J, Umezawa N, Iwanaka K, Tada T, Hosono H (2017) Exploration of stable strontium phosphide-based electrides: theoretical structure prediction and experimental validation. J Am Chem Soc 139(44):15668–15680
38. Hirayama M, Matsuishi S, Hosono H, Murakami S (2018) Electrides as a new platform of topological materials. Phys Rev X 8(3):031067
39. Hosono H, Kitano M (2021) Advances in materials and applications of inorganic electrides. Chem Rev 121(5):3121–3185

Chapter 22
Ammonia Synthesis from Nitrogen and Water Using an Electrochemical Hydrogen-Membrane Reactor, Ru Catalyst, and Phosphate Electrolytes

Jun Kubota

Abstract The development of a direct synthesis of ammonia (NH_3) from nitrogen (N_2) and water (H_2O) using electricity is expected to promote the development of new synthetic processes without any need for carbon resources. Although the direct electrochemical reduction of N_2 to form ammonia in an aqueous medium has been extensively investigated, its performance is still far from that required on an industrial scale. Our research group has designed a new electrochemical system for ammonia synthesis from N_2 and H_2O using electricity, which involves a combination of a phosphate-based water electrolyzer, Ru catalyst, and H_2-permeable membrane at 200–250 °C. In this chapter, the significance of the direct synthesis of ammonia from N_2 and H_2O using electricity and not from N_2 and H_2 is described and the potential of electrochemical hydrogen-membrane reactors is discussed.

Keywords Electrochemical ammonia synthesis · Renewable electricity · Ru catalysts

22.1 Introduction

As mentioned in the other chapters, there is no doubting the importance of NH_3 as a chemical fuel in the future energy society. Even though NH_3 is a carbon-free fuel, it is currently synthesized from H_2, which is mainly obtained from carbon-based fossil resources, such as natural gas and coal [1, 2]. The latest industrial NH_3 synthesis process based on the Haber–Bosch process consumes only 28 GJ/ton-NH_3 for the production of ammonia (NH_3) [1, 2]. Considering that 176 kg of H_2 is contained in 1 ton of NH_3 and 176 kg of H_2 has 25 GJ as the heat of combustion is 286 kJ/mol, the modern industrial NH_3 synthesis process has little room for energy saving. NH_3 is an energy medium with a heat of combustion of 383 kJ/mol, which is equivalent to 22.5 GJ/ton-NH_3, therefore this energy is essentially required for NH_3

J. Kubota (✉)
Department of Chemical Engineering, Fukuoka University, Nanakuma, Jonan-ku, Fukuoka 814-0180, Japan
e-mail: jkubota@fukuoka-u.ac.jp

K. Aika and H. Kobayashi (eds.), *CO_2 Free Ammonia as an Energy Carrier*,
https://doi.org/10.1007/978-981-19-4767-4_22

synthesis. From another perspective, the energy of NH_3 is slightly lower than that of H_2. Therefore, if NH_3 is synthesized from a hydrogen source that has a lower energy than H_2, it can be an energy-saving process. The direct synthesis of NH_3 from N_2 and H_2O, but not from N_2 and H_2, has the potential to reduce the energy consumption of NH_3. A detailed explanation of the benefits of thermodynamics is provided in the next section. The NH_3 synthesis process based on the Haber–Bosch process is sometimes referred to as a waste of energy, but this perception is incorrect. The NH_3 synthesis reaction from N_2 and H_2 is a well-known exothermic reaction. Therefore, no energy is required even if the temperature is high or low. Most of the energy for NH_3 synthesis is consumed during the preparation of H_2, thereby energy saving and carbon-free methods are expected to be developed for the preparation of H_2.

Solar and wind energy are expected to be core energies in the future carbon–neutral society. However, the biggest drawbacks of these renewable energy sources are the daily, weekly, and monthly fluctuations, and uneven global distribution. Electricity based on these renewable energies can be converted to chemical fuels, such as NH_3, for storage and transportation because of the high energy density of chemical fuels. Traditionally, some NH_3 has been synthesized from H_2, which originates from water electrolysis coupled with hydropower plants. These NH_3 plants are located in India, Egypt, Zimbabwe, Peru, Iceland, and Canada [1]. The development of a carbon-free NH_3 synthesis method based on renewable energy is the goal of humankind.

In this chapter, a system for the direct synthesis of NH_3 from N_2 and H_2O using electricity is introduced. In the near future, NH_3 should be synthesized from electricity obtained from renewable energy sources, but not from carbon-based H_2.

22.2 Thermodynamics of NH_3 Synthesis from H_2O

The combination of water electrolysis and the Haber–Bosch NH_3 synthesis process is a conventional and practical method to synthesize NH_3 from N_2 and H_2O using electricity. However, the first weakness of this system is that the large-scale Haber–Bosch NH_3 process is not suitable for fluctuating operation. Water electrolysis can somehow follow the fluctuation because low-temperature electrochemical systems can change their output, as shown in batteries and fuel cells. However, heat management in the Haber–Bosch NH_3 process conducted at high temperature and pressure, which includes an exothermic reactor, gas/liquid separator for the reacted and unreacted gases, and heat exchangers in the recycling system, is designed for steady-state operation, so that equipment with a sufficient amount of H_2 reservoir tank is required for the fluctuating flow of H_2 from water electrolysis. However, H_2 is hard to store on a large scale because it requires cryogenic temperatures for liquefaction. There are many examples of NH_3 synthesis based on hydropower electricity, as mentioned in the previous section. Because hydropower plants do not fluctuate, the combination of water electrolysis and the Haber–Bosch NH_3 synthesis process is relatively suitable. However, the output of solar and wind power plants fluctuates depending on the weather. If all chemical processes are performed in a single electrochemical cell, it has the advantage for fluctuating operation.

Figure 22.1 shows the thermodynamic parameters for the combination of water electrolysis and the Haber–Bosch NH_3 process and the direct electrochemical synthesis of NH_3 [3]. All of the parameters were considered under standard conditions. Water electrolysis is an endothermic reaction ($\Delta_r H^\circ = 286$ kJ/mol-H_2), which is a thermally balanced potential of 1.48 V with 2 electron reactions. As an electrochemical reaction, the energy of the electrons is $\Delta_r G^\circ = 237$ kJ/mol-H_2, which is an equilibrium potential of 1.23 V. A difference of 49 kJ/mol-H_2 can be supplied as heat. Water electrolysis is an endothermic reaction, even if the electrochemical energy is excluded. On the other hand, the Haber–Bosch NH_3 process is a well-known exothermic reaction ($\Delta_r H^\circ = -30$ kJ/mol-H_2). A considerable amount of heat is released from the reactor at 400–500 °C. The low-temperature water electrolysis of alkaline electrolytes or proton-conductive polymers is the most conventional method of water electrolysis and the operational temperature is typically <100 °C. Recently, solid oxide electrolysis cells have been extensively developed and operated at ~1000 °C. The temperature of water electrolysis does not match that used in NH_3 synthesis. This exothermic heat is typically utilized somewhere in large-scale chemical plants. However, considering only water electrolysis and the Haber–Bosch NH_3 process, the exothermic heat of the Haber–Bosch NH_3 process cannot be utilized. Summarizing the combination of water electrolysis and the Haber–Bosch NH_3 process, 1.23 V of electricity and 49 kJ/mol-H_2 of heat are required at the temperature used for water electrolysis. As a result, 30 kJ/mol-H_2 of heat at the temperature of the Haber–Bosch NH_3 process is wasted.

Considering the direct conversion reaction from N_2 and H_2O to NH_3 in a single and heat-balanced cell at a constant temperature, the thermodynamic parameters are different from those of the combination of water electrolysis and Haber–Bosch NH_3 processes. Based on the chemical reaction shown in Fig. 22.1, the reaction

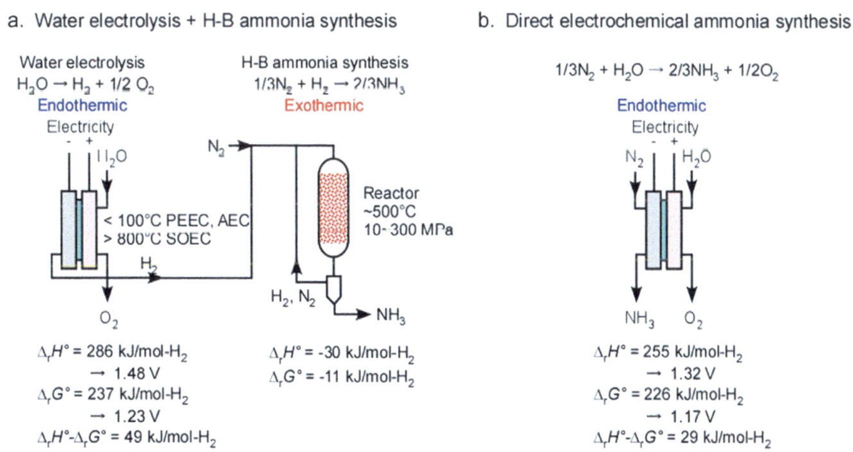

Fig. 22.1 Thermodynamic parameters of the combination of water electrolysis and the Haber-Bosch NH_3 process and the direct electrochemical synthesis of NH_3

is endothermic [$\Delta_r H^\circ = 286$ kJ/mol-H_2 (1.32 V)] and the electrochemical part is $\Delta_r G^\circ = 226$ kJ/mol-H_2 (1.17 V). The difference is 29 kJ/mol-H_2. All parameters are slightly smaller than those for water electrolysis and the direct electrochemical conversion has a theoretical advantage over separating water electrolysis and the Haber–Bosch NH_3 process. NH_3 is a less energetic material than H_2, so that it can be synthesized with lower energy than H_2 when obtained from H_2O.

Thus, the direct electrochemical synthesis of NH_3 from N_2 and H_2O has attracted the attention of researchers [4–6]. Another electrochemical advantage is that the limit of the chemical equilibrium between $N_2 + 3H_2$ and $2NH_3$ can be avoided if an appropriate function is designed in the cell. Electrochemical systems can be separated using anodes, electrolytes, and cathodes, so that the fractions of the product species can exceed the chemical equilibrium limits. In water electrolysis, H_2O can be split into H_2 and O_2 using an electrochemical system because they are separated between the cathode and anode.

22.3 Electrochemical NH_3 Synthesis

As discussed in the previous section, the electrochemical synthesis of NH_3 from N_2 and H_2O has the potential to be an alternative method of NH_3 synthesis with fluctuating electricity obtained from renewable energy sources [4–6]. Electrochemical NH_3 synthesis can be categorized into several groups by the electrolytes. These categories are presented in Table 22.1. In the first group, NH_3 is synthesized in aqueous and non-aqueous electrolytes at ambient temperatures. Ion-conducting polymers are sometimes used as the electrolyte. N_2 reduction at ambient temperatures is regarded as artificial nitrogen fixation, which is one of the dream reactions in chemistry. A large number of attempts have been reported [4–6], but there is no consensus among scientists [7, 8]. In any case, the efficiency and current density of electrochemical NH_3 synthesis at ambient temperatures at present are far from that required on an industrial scale.

Table 22.1 Various electrochemical synthesis of NH_3

Type	Example of electrolyte	Temperature/°C
Aqueous solution	$HClO_4$, $LiClO_4$, KOH	~25
Ion-conducting polymer	Fluorine-type (Nafion™)	60–100
Phosphate	CsH_2PO_4, $CsH_5(PO_4)_2$	200–300
Molten salt (Li^+ containing)	LiCl-KCl LiCl-KCl-CsCl	~450
Solid oxides (H^+ conductive)	$BaCe_{1-x}Y_xO_3$ (BCY) $BaCe_{1-x}Z_xO_3$ (BCZ)	500–700
Solid oxides (O^{2-} conductive)	$Y{:}ZrO_2$ (YSZ) $Gd{:}CeO_2$ (GDC)	600–1000

Electrochemical NH_3 synthesis using Li ions in molten salts or similar non-aqueous electrolytes has been demonstrated [6, 9]. N_2 is readily reduced to N^{3-} when it is attached to metallic Li, and N^{3-} can react with H_2O to form NH_3. Although the utilization of Li ions is a reliable method, the redox of Li requires significant electrochemical potential. Therefore, it is not easy to operate with a cell voltage similar to that required for water electrolysis.

The application of solid oxide electrolytes at 500–1000 °C toward the electrochemical synthesis of NH_3 is another possibility [4–6]. Typically, H^+ conductive oxides can operate at a lower temperature than O^{2-} conductive oxides. The temperature required is higher than the Haber–Bosch NH_3 synthesis and the chemical equilibrium between $N_2 + 3H_2$ and $2NH_3$ is reversed at such temperatures, resulting in the formation of N_2 and H_2. In electrochemical cells, it is difficult to pressurize the cells up to 100–200 atm at such high temperature, which is in contrast to the Haber–Bosch NH_3 synthesis.

The subject of this chapter is a phosphate electrolyte. H^+ conductive phosphates are effective in the temperature range of 200–300 °C and the electrochemical synthesis of NH_3 using phosphate electrolytes has great potential for practical application [10–14]. There is an advantage in the chemical equilibrium between $N_2 + 3H_2$ and $2NH_3$ at 200–300 °C. The activation of N_2 molecules via a mechanism similar to that of the Haber–Bosch NH_3 synthesis is also possible in this temperature range. The properties of the electrochemical cell equipped with a phosphate electrolyte are explained in the next section.

22.4 Electrochemical Hydrogen-Membrane Reactor

Our research group has designed an electrochemical cell equipped with a Ru catalyst, Pd-Ag membrane cathode, phosphate electrolyte, and Pt anode. A schematic illustration of the electrochemical cell is presented in Fig. 22.2 [12–14]. The method used for the preparation of the cell is described in detail in the literature [12–14]. Oxide-supported Ru catalysts with promoters comprised of alkali metal compounds are effective NH_3 synthesis catalysts, especially on a laboratory-scale [15, 16]. 0.5–0.8 mm granules were used, the primary particles of the oxide support were 20–40 nm in diameter, and the size of the Ru particles was 2–4 nm. Pd-Ag membranes are known to be hydrogen-permeable membranes. The addition of Ag strengthens the Pd membrane. A composite of CsH_2PO_4 and SiP_2O_7 was used as the phosphate electrolyte. CsH_2PO_4 is a proton-conducting electrolyte. A few phase transitions of the crystal and melting with dehydration are observed for CsH_2PO_4 at 200–300 °C, and the addition of stable SiP_2O_7 is desirable. New phosphate phases with a high H^+ conducting nature are formed at the interfaces between CsH_2PO_4 and SiP_2O_7 at 200–300 °C.

This electrochemical reactor uses a hydrogen-permeable membrane. Due to its H^+ conducting nature, CsH_2PO_4 is an acid that can easily absorb NH_3. The hydrogen-permeable membrane separates the electrolyte from the NH_3 synthesis catalyst and

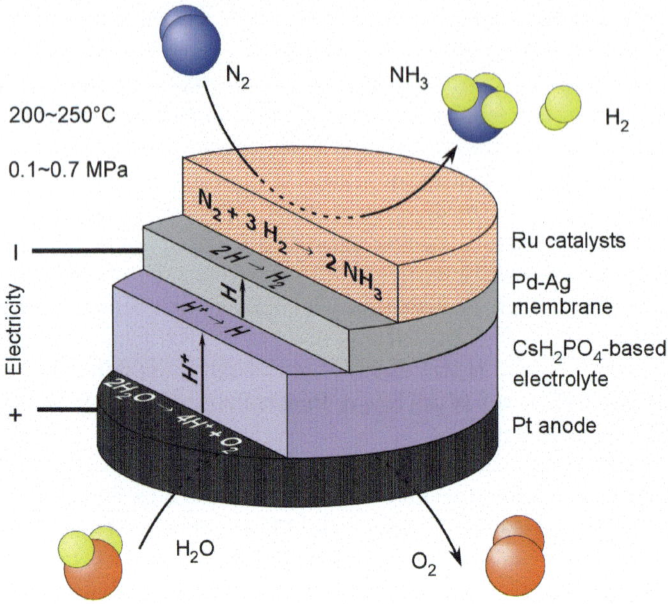

Fig. 22.2 A schematic illustration of the electrochemical reactor equipped with a hydrogen-permeable membrane. Reproduced from Ref. [14] with permission from the Royal Society of Chemistry

completely blocks the absorption of NH_3 by the electrolyte. The second reason for using a hydrogen-permeable membrane is to avoid contamination of NH_3 with H_2O. To split water, H_2O vapor is supplied to the electrolyte, but the affinity between H_2O and NH_3 is significant. Therefore, NH_3 is obtained with a considerable amount of H_2O impurity. The presence of a hydrogen-permeable membrane prevents NH_3 contamination by H_2O. Because deliquescent alkali metal compounds are used in Ru catalysts, the presence of H_2O vapor in the chamber for NH_3 synthesis may also strongly suppress the activity of the Ru catalyst. Furthermore, CsH_2PO_4 has the properties of melting with dehydration and deliquescence, and CsH_2PO_4 electrolytes are better isolated from the Ru catalyst. The Ru catalyst plays a key role in the activation of N_2, followed by NH_3 synthesis. Therefore, it is better to be isolated from the electrolyte. Practically, the electrochemical cell without a hydrogen-permeable membrane has been examined several times and it is difficult to obtain NH_3 formation with sufficient reproducibility for a long period of time. The use of a hydrogen-permeable membrane enhances the NH_3 synthesis performance.

The presence of a hydrogen-permeable membrane indicates that there is no electrochemical potential for the Ru catalyst. N_2 only reacts with H_2, which penetrates the bulk of the hydrogen-permeable membrane. N_2 is not electrochemically reduced, but is catalytically hydrogenated. Therefore, NH_3 is not electrochemically synthesized but is catalytically synthesized. However, the thermodynamic advantage of the direct synthesis of NH_3 from N_2 and H_2O discussed in Sect. 22.2 is unchanged. The present system is a combination of water electrolysis and NH_3 synthesis. However, the endothermic and exothermic heats of the respective elemental steps are thermally equilibrated at a specific temperature. The advantage of thermodynamics is equivalent to the pure electrochemical synthesis of NH_3, but not to the combination of two separate processes: Water electrolysis and NH_3 synthesis. In the future, if the H atoms that penetrate the hydrogen-permeable membrane react with N_2 before desorption as H_2, a new concept for NH_3 synthesis beyond the equilibrium limit is expected to be obtained.

The experimental setup is shown in Fig. 22.3a. The cell size of the electrodes and electrolyte was 20 mm in diameter and the outer diameter of the cell vessel was 50 mm. The Ru catalyst, electrodes, membrane, and electrolyte were mounted in the cathode and anode vessels constructed from stainless steel and titanium, respectively. Although the cathode and anode are illustrated at the upper and lower sides, respectively, as shown in Fig. 22.2, the practical position in the photograph was opposite; the cathode and anode vessels were positioned on the upper and lower sides. The bottle behind the cell is the evaporation chamber for H_2O. H_2O was introduced into the line as a liquid and the humidity was stabilized after passing through this bottle. The cell and evaporation chamber were positioned inside an electric oven at 100 °C to avoid any condensation of H_2O. The photograph in Fig. 22.3b was obtained under electrolysis at 10 mA (3.2 mA cm^{-2}) and 250 °C with a cell voltage of 1.32 V. The exhaust gas from the cathode vessel was introduced into a phenolphthalein solution and the color of the solution changed from clear to purple–red, indicating that alkaline NH_3 gas was produced. Usually, the amount of NH_3 formed is measured from the change in the electroconductivity of 1 mmol/L H_2SO_4, which continuously absorbs the exhaust gas from the cathode vessel.

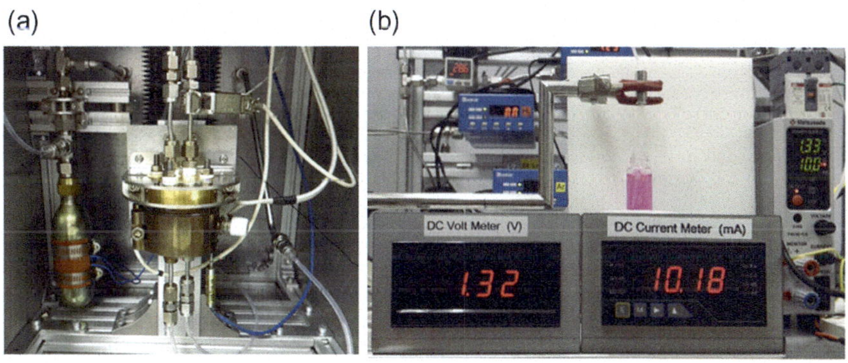

Fig. 22.3 Photographs of the exterior of the electrochemical hydrogen membrane reactor (**a**) and bubbling of the exhaust gas from the cathode side into a phenolphthalein solution (**b**). Reproduced from Ref. [14] with permission from the Royal Society of Chemistry

22.5 Properties of the Electrochemical Hydrogen-Membrane Reactor

An example of the time course of the amount of NH_3 produced by the electrochemical hydrogen membrane reactor is shown in Fig. 22.4. A 2 wt%-Ru/Cs^+/MgO catalyst was used in this experiment. For the anode, Ar at 10 $cm_{STP}{}^3$ min^{-1} using a mass flow controller with the addition of H_2O at 0.016 $cm_{liq}{}^3 min^{-1}$ was supplied. The amount of NH_3 monotonically changes with time, indicating that NH_3 was synthesized at a steady-state. Subsequently, H_2 production was monitored using gas chromatography. The respective values obtained for the formation rates, current efficiencies, and attainments of equilibrium for this experiment are summarized in Table 22.2. The current efficiency, also known as the Faradaic efficiency, for NH_3 and H_2 production can be derived as follows:

$$FE_{NH_3} = \frac{3 \times N \times F}{Q} \times 100(\%)$$

$$FE_{H_2} = \frac{3 \times N \times F}{Q} \times 100(\%)$$

where N, F, and Q are the amounts of NH_3 and H_2 generated (mol), Faraday constant (96,500 C mol^{-1}), and electric charge through the cell (C), respectively.

The sum of the current efficiencies for NH_3 and H_2 under each condition studied was ~100%, indicating that the entire electric current was used for the production of NH_3 and H_2. The sum of the current efficiencies was far from 100%, especially

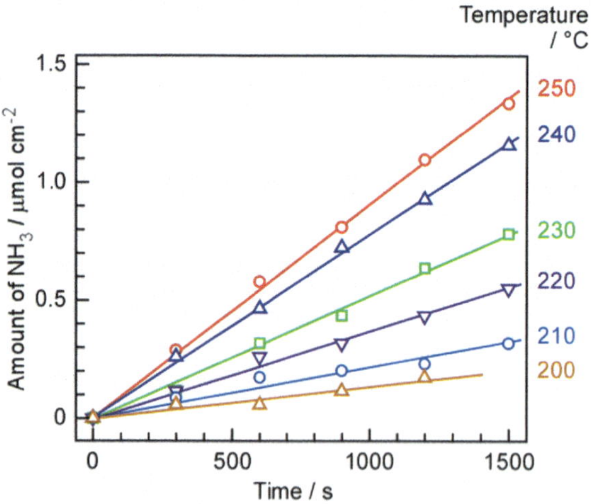

Fig. 22.4 Time course of NH_3 formation at various temperatures and 10 mA cm^{-2} using N_2 as the cathode gas at 3 $cm_{STP}{}^3$ min^{-1}. Reproduced from Ref. [13] with permission from the Royal Society of Chemistry

Table 22.2 NH_3 formation rates and current efficiencies at various current densities, N_2 flow rates and temperatures

Current	N_2 Flow[a]	Temp	r_{NH3}[b]	FE_{NH3}[c]	AE[d]	FE_{H2}[e]
mA cm^{-2}	cm^3 min^{-1}	°C	nmol s^{-1} cm^{-2}	%	%	%
10	3.0	200	0.14	0.39	2.2	85
		210	0.22	0.62	4.3	95
		220	0.37	1.1	9.4	100
		230	0.52	1.5	16	101
		240	0.78	2.3	29	102
		250	0.90	2.6	41	103
3.2	0.5	250	0.17	1.5	24	65
	1.0		0.28	2.5	40	82
	2.0		0.27	2.5	38	95
	5.0		0.18	1.6	25	103
	7.0		0.15	1.4	21	100

Reproduced from Ref. [13] with permission from the Royal Society of Chemistry
[a]Flow rates are in standard temperature and pressure, $cm^3{}_{STP}$ min^{-1}
[b]Rate of NH_3 formation.
[c]Current efficiency for NH_3 formation.
[d]Degree of attainment of thermodynamic equilibrium
[e]Current efficiency for H_2 formation.

at low temperatures and low N_2 flow rates. At a low temperature of 200 °C, it was proposed that there was a problem with the sealing of the cell. At low temperatures, the sealing materials, as well as the electrolyte, harden and leakage of gas may occur. The amount of H_2 was estimated using gas chromatography from the unreacted N_2 as an internal standard, so the accuracy of the N_2 flow rate at a low flow rate was not sufficient.

The temperature dependence of the electrochemical synthesis of NH_3 at another current densities is shown in Fig. 22.5a as an Arrhenius plot. In this experiment, 5 wt% Ru was loaded onto the oxide supports and several catalysts were examined. The equilibrium limit is indicated by the dashed line. The degree of attainment of thermodynamic equilibrium was estimated to be 77% at 250 °C and the Arrhenius plots indicate that the rate of NH_3 formation was strongly limited by the equilibrium.

Figure 22.5b shows the dependence of the rate of NH_3 formation on temperature when a mixture of $N_2 + 3H_2$ was flowed into the cathode chamber without any current flow. Only catalytic formation from $N_2 + 3H_2$ gas occurred in this case. It should be noted that the apparent activation energy for the catalytic reaction from $N_2 + 3H_2$ was remarkably higher than that for electrochemical NH_3 synthesis, as shown in Fig. 22.5a. Practically, the apparent activation energies for $Ru/Cs^+/CeO_2$, $Ru/Cs^+/MgO$, and Ru/CeO_2 were 36, 31, and 36 kJ mol^{-1}, respectively for the electrochemical synthesis of NH_3, but 129, 131, and 125 kJ mol^{-1}, respectively for the catalytic synthesis of NH_3 from $N_2 + 3H_2$ [14]. An apparent activation energy in the

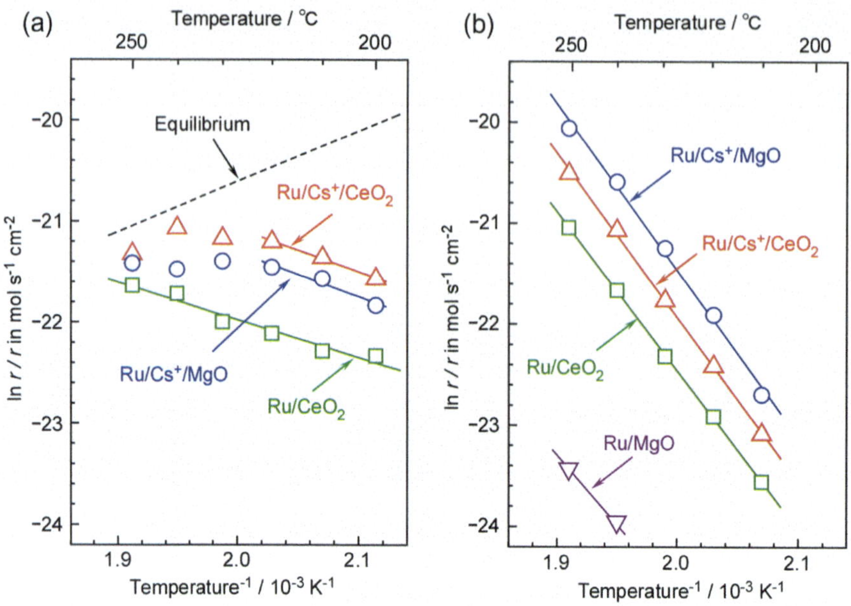

Fig. 22.5 Arrhenius plots of the rate of NH_3 formation for the electrochemical synthesis at a current density of 3.2 mA cm^{-2} under ambient pressure using 5 wt%-Ru/Cs^+/CeO_2 (Cs/Ru = 1 mol), 5 wt%-Ru/Cs^+/MgO (Cs/Ru = 1 mol), and 5 wt%-Ru/CeO_2, respectively. The electrochemical hydrogen-membrane reactor was operated using N_2 as the cathode gas (1 $cm^3{}_{STP}$ min^{-1}) and Ar as the anode gas (10 $cm^3{}_{STP}$ min^{-1}) in the presence of H_2O (4 μL-liquid min^{-1}). (B) Arrhenius plots of the rate of NH_3 formation from a mixed gas comprised of $N_2 + 3H_2$ at ambient pressure using 5 wt%-Ru/Cs^+/CeO_2 (Cs/Ru = 1 mol), 5 wt.%-Ru/Cs^+/MgO (Cs/Ru = 1 mol), 5 wt%-Ru/CeO_2, and 5 wt%-Ru/MgO, respectively. The flow of N_2 and H_2 was supplied at 7 and 21 $cm^3{}_{STP}$ min^{-1}, respectively. Reproduced from Ref. [14] with permission from the Royal Society of Chemistry

range of 100–150 kJ mol^{-1} has been previously reported for Ru catalysts [15, 16] and the electrochemical synthesis of NH_3 has a somewhat small apparent activation energy.

The dependence of the rate of NH_3 formation on the N_2 flow rate is shown in Fig. 22.6a. Independent from the kind of catalyst used, ~ 1 $cm^3{}_{STP}$ min^{-1} of N_2 led to the maximum formation of NH_3. More or less N_2 flow suppressed NH_3 formation. The conversion of a current density of 3.2 mA cm^{-2} with the H_2 flow rate and a cross-sectional area of 3.14 cm^2 was equivalent to 0.071 $cm^3{}_{STP}$ min^{-1} of H_2. The stoichiometric ratio of NH_3 synthesis is, of course, $H_2/N_2 = 3$, but the experimentally optimized point was $H_2/N_2 = 0.071$, which was 42 times smaller than the stoichiometric value of 3. Generally, Ru catalysts used for NH_3 synthesis are known to have a negative reaction order for the pressure of H_2 in terms of their kinetics [15, 16], which has been referred to as hydrogen poisoning. This indicates that the rate of NH_3 synthesis increases upon decreasing the H_2 pressure. In the electrochemical synthesis of NH_3 reported in this work, the reaction temperature is much lower than that used for the Haber–Bosch NH_3 process. Therefore, the optimized N_2 flow rate was much

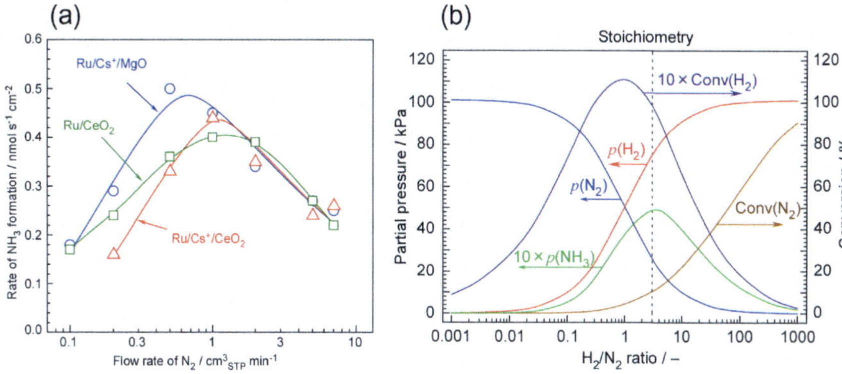

Fig. 22.6 A Dependence of the rate of NH_3 formation on the N_2 flow rate observed for the electrochemical synthesis at a current density of 3.2 mA cm^{-2} and ambient pressure using 5 wt%-$Ru/Cs^+/CeO_2$ (Cs/Ru = 1 mol), 5 wt%-$Ru/Cs^+/MgO$ (Cs/Ru = 1 mol), and 5 wt%-Ru/CeO_2. The electrochemical hydrogen-membrane reactor was operated at 250 °C using N_2 as the cathode gas at the flow rates stated and Ar as the anode gas (10 $cm^3{}_{STP}$ min^{-1}) in the presence of H_2O (4 μL-liquid per min). (B) Partial pressures of the various species at chemical equilibrium during the synthesis of NH_3 as a function of the initial H_2/N_2 ratio at a total pressure of 101 kPa and 250 °C. Partial pressures of N_2 [$p(N_2)$], H_2 [$p(H_2)$], and NH_3 [$p(NH_3)$] at equilibrium and conversion of N_2 [$Conv(N_2)$] and H_2 [$Conv(H_2)$] are plotted. Note that both $p(NH_3)$ and $Conv(H_2)$ are multiplied by 10 in these plots. Reproduced from Refs. [13, 14] with permission from the Royal Society of Chemistry

lower than that estimated from the stoichiometric ratio. The small activation energy for electrochemical synthesis also originates from the small hydrogen pressure of $H_2/N_2 = 0.071$. In other words, the rate of NH_3 formation is limited by the hydrogen poisoning nature of Ru catalysts.

The equilibrium between $N_2 + H_2$ and NH_3 with the H_2/N_2 ratio was calculated, as shown in Fig. 22.6b. The maximum NH_3 yield was obtained at $H_2/N_2 = 3$. At $H_2/N_2 = 0.071$, the equilibrium limit for NH_3 formation was suppressed by ca. 1/10. Therefore, the use of catalysts with a reaction rate close to $H_2/N_2 = 3$ is desirable. The development of catalysts with resistance to hydrogen poisoning is considered a strategy for further development.

The rate of NH_3 synthesis in the present electrochemical cell was found to be limited by the chemical equilibrium, which is attributed to the H_2/N_2 ratio. To improve the formation rate and current efficiency of NH_3 synthesis, increasing the operating pressure is a promising method. Figure 22.7a shows the pressure dependence of the rate of NH_3 formation. Upon increasing the pressure, the rate of NH_3 formation deviates from the equilibrium limit, indicating that the activity of the Ru catalyst was not sufficient to reach equilibrium. The degree of attainment of thermodynamic equilibrium was 61–77% at total pressure of 0.1 MPa, but it decreased to 28–65% at 0.5 MPa [14]. $Ru/Cs^+/MgO$ was found to have the highest reaction rate, so that the electrochemical NH_3 synthesis using $Ru/Cs^+/MgO$ was further examined by increasing the current density at 0.7 MPa. Finally, 12.4 nmol s^{-1} cm^{-2} of NH_3 was synthesized with

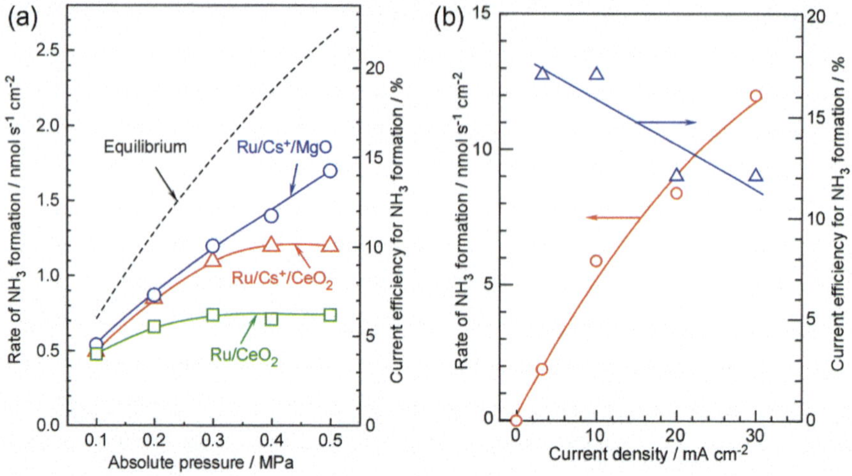

Fig. 22.7 Rate of NH_3 formation with elevated pressure at a current density of 3.2 mA cm^{-2} using 5 wt%-Ru/Cs^+/CeO_2 (Cs/Ru = 1 mol), 5 wt%-Ru/Cs^+/MgO (Cs/Ru = 1 mol), and 5 wt%-Ru/CeO_2. The current efficiency for NH_3 formation is shown in the vertical scale on the right-hand side. The electrochemical hydrogen-membrane reactor was operated at 250 °C using N_2 as the cathode (1 $cm^3{}_{STP}$ min^{-1}) and Ar as the anode gas (10 $cm^3{}_{STP}$ min^{-1}) in the presence of H_2O at 4 μL-liquid per min. (B) Rate of NH_3 formation upon increasing the current density at 0.7 MPa using 5 wt%-Ru/Cs^+/MgO (Cs/Ru = 1 mol). The electrochemical hydrogen-membrane reactor was operated at 250 °C using N_2 as the cathode gas, while maintaining the H_2/N_2 ratio at 0.071 and Ar as the anode gas (10 $cm^3{}_{STP}$ min^{-1}) in the presence of H_2O at 4 μL-liquid per min. Reproduced from Ref. [14] with permission from the Royal Society of Chemistry

a current efficiency of 12% at 30 mA cm^{-2}. The attainment of equilibrium was 41%, indicating a shortage of catalyst activity under strong conditions.

22.6 Concluding Remarks

Using an electrochemical hydrogen-membrane reactor, the synthesis of NH_3 from N_2 and H_2O utilizing electric energy is now realistic. However, various challenges remain for the development of this electrochemical system. In the present electrochemical system, the chemical equilibrium mainly controls the rate of formation and current efficiency of NH_3 synthesis because NH_3 is formed from N_2 and H_2 on the surfaces of the same catalyst. For example, methanation of CO_2, in which the reactants shift almost to the product with the chemical equilibrium, this type of cell can convert electricity to the target chemical fuel (*FE* >90%) with a negligible amount of by-product formation [17]. However, in the synthesis of NH_3, the restrictions of chemical equilibrium are strict. Even at low temperature (200 °C) and 1 MPa, the equilibrium indicates the formation of NH_3 and H_2 at 60 and 40%, respectively [14]. Further pressurizing or decreasing the temperature is required to build a practical electrochemical cell for NH_3 synthesis. It is not easy to establish this cell as a one-pass reactor. Another possibility for the realization of this cell is that the cell is equipped with an NH_3 separation system to recover the unreacted gas, which is recycled to the cell. Practically, the Haber–Bosch NH_3 process will not be possible without recycling the unreacted gas. However, the presence of the NH_3 separation component complicates the electrochemical NH_3 synthesis system and makes it difficult to accommodate fluctuations in the electrical energy.

According to the United States Department of Energy (DOE), the targets of electrochemical NH_3 synthesis are energy efficiency > 60%, Faradaic efficiency > 90%, and current density >300 mA cm^{-2} [18]. The current density of 30 mA cm^{-2} must be improved tenfold.

Although various challenges remain for the practical use of this method, it can be expected to be established as a technology for producing chemical fuel from electricity based on renewable energy sources.

Acknowledgements This work was supported as a project under the "Creation of Innovative Core Technology for Manufacture and Use of Energy Carriers from Renewable Energy" CREST program of the Japan Science and Technology Agency (JST).

References

1. Rafiqul I, Weber C, Lehmann B, Voss A (2005) Energy efficiency improvements in ammonia production-perspectives and uncertainties. Energy 30:2487–2504
2. Valera-Medina A, Amer-Hatem F, Azad AK, Delouse IC, de Joannon M, Fernandes RX, Glarborg P, Hashemi H, He X, Mashruk S, McGowan J, Mounaim-Rouselle C, Ortiz-Prado A, Ortiz-Valera A, Rossetti I, Shu B, Yehia M, Xiao H, Costa M (2021) Review on ammonia as a potential fuel: from synthesis to economics. Energy Fuels 35:6964–7029
3. Ed. J. R. Rumble (2018), CRC Handbook of chemistry and physics, 99th edn. CRC Press, Boca Raton
4. Giddey S, Badwal SPS, Kulkarni A (2013) Review of electrochemical ammonia production technologies and materials. Int J Hydrogen Energy 38:14576–14594
5. Martín AJ, Shinagawa T, Pérez-Ramírez J (2018) Electrocatalytic reduction of nitrogen: from haber-bosch to ammonia artificial leaf. Chem 5:1–21
6. Qing G, Ghazfar R, Jackowski ST, Habibzadeh F, Ashtiani MM, Chen CP, Smith MR III, Hamann TW (2020) Recent advances and challenges of electrocatalytic N2 reduction to ammonia. Chem Rev 120:5437–5516
7. Greenlee LF, Renner JN, Foster SL (2018) The use of controls for consistent and accurate measurements of electrocatalytic ammonia synthesis from dinitrogen. ACS Catal 8:7820–7827
8. Andersen SZ, Čolić V, Yang S, Schwalbe JA, Nielander AC, McEnaney JM, Enemark-Rasmussen K, Baker JG, Singh AR, Rohr BA, Statt MJ, Blair SJ, Mezzavilla S, Kibsgaard J, Vesborg PCK, Cargnello M, Bent SF, Jaramillo TF, Stephens IEL, Nørskov JK, Chorkendorff I (2019) A rigorous electrochemical ammonia synthesis protocol with quantitative isotope measurements. Nature 570:504–508
9. Ito Y, Nishikiori T, Tsujimura H (2016) Advances toward industrialization of novel molten salt electrochemical processes. Faraday Discuss 190:307–326
10. Kishira S, Qing G, Suzu S, Kikuchi R, Takagaki A, Oyama ST (2017) Ammonia synthesis at intermediate temperaturesin solid-state electrochemical cells using cesiumhydrogen phosphate based electrolytes and noblemetal catalysts. Int J Hydrogen Ener 42:26843–26854

11. Yuan Y, Tada S, Kikuchi R (2021) Ammonia synthesis using Fe/BZY–RuO_2 catalysts and a cesium dihydrogen phosphate based electrolyte at intermediate temperatures. Mater. Adv. 2:793–803
12. Imamura K, Matsuyama M, Kubota J (2017) Ammonia Synthesis from Nitrogen and Water by Electricity Using an Electrochemical Cell with Ru Catalyst, Hydrogen-Permeable Pd-Ag Membrane, and Phosphate-Based Electrolyte. Chem. Select 2:11100–11103
13. Imamura K, Kubota J (2018) Electrochemical membrane cell for NH_3 synthesis from N_2 and H_2O by electrolysis at 200 to 250°C using a Ru catalyst, hydrogen-permeable Pd membrane and phosphate-based electrolyte. Sustain. Ener. Fuels 2:1278–1286
14. Imamura K, Kubota J (2019) Ammonia synthesis from nitrogen and water at intermediate temperatures and elevated pressures by using an electrochemical hydrogen-membrane reactor with supported Ru catalysts and phosphate electrolytes. Sustain. Ener. Fuels 3:1406–1417
15. Aika K (2017) Role of alkali promoter in ammonia synthesis over rutheniumcatalysts—effect on reaction mechanism. Catal. Toady 286:14–20
16. Aika K, Takano T, Murata S (1992) Preparation and characterization of chlorine-free ruthenium catalysts and the promoter effect in ammonia synthesis: 3. A magnesia-supported ruthenium catalyst. J Catal 136:126–140
17. Kubota J, Okumura T (2021) Methane synthesis from CO_2 and H_2O with electricity using H-permeable membrane electrochemical cells with Ru catalyst and phosphate electrolyte. Sustain Ener Fuels 5:935–940
18. Fernandez CA, Hatzell MC (2020) Editors' choice—economic considerations for low-temperature electrochemical ammonia production: achieving haber-bosch parity. J Electrochem Soc 167:143504

Part V
Ammonia Source Hydrogen Station

Chapter 23
Production Technology of Hydrogen Fuel for Fuel Cell Vehicles from Ammonia

Yoshitsugu Kojima

Abstract We have developed component technologies to produce 1Nm^3/h-scale high-purity hydrogen from ammonia with high efficiency (NH_3 <0.02 ppm, N_2 <1 ppm, H_2 >9.98%, H_2 purification efficiency 90%, H_2 conversion efficiency 80%) for fuel cell vehicles. The high-purity hydrogen can be produced by the combination of a microchannel NH_3 cracker including ruthenium supported on MgO with a heat supply unit, NH_3 remover including zeolite and four tower type H_2 purifier combined new process with off-gas supply unit. 10 Nm^3/h-scale H_2 purifier with off-gas supply unit has been also developed.

Keywords Ammonia · Catalyst · Zeolite · Fuel cell · Hydrogen · 0.1 ppm

23.1 Background

On December 12, 2015, the Paris Agreement was adopted to reduce the emission of greenhouse gases such as carbon dioxide (CO_2) to virtually zero in the latter half of this century at the 21st Conference of the Parties to the United Nations Framework Convention on Climate Change (COP21). The Paris Agreement has been set the goal to keep the global average temperature rise below 2 °C compared to before the Industrial Revolution and to pursue efforts for keeping it at 1.5 °C [1].

Since the Industrial Revolution, about 2 trillion tons of CO_2 have already been released and the temperature has risen by about 1 °C [2, 3]. The remaining temperature rise is up to 1 °C. This is 1 trillion tons when converted to CO_2 emissions. Japan aimed to reduce greenhouse gas emissions by 26% compared to FY2013 by 2030 and 80% by 2050 [2, 3]. Recently, Japan is aiming toward decarbonized society in 2050.

Under this background, ammonia contains large amounts of hydrogen about 18 wt% and is a widely traded chemical [4]. Blue ammonia is defined as the version of ammonia whose production involves carbon capture and storage [5, 6]. Saudi

Y. Kojima (✉)
Natural Science Center for Basic Research and Development, Hiroshima University, 1-3-1, Kagamiyama, Higashi-Hiroshima, Hiroshima 739-8530, Japan
e-mail: kojimay@hiroshima-u.ac.jp

K. Aika and H. Kobayashi (eds.), *CO_2 Free Ammonia as an Energy Carrier*,
https://doi.org/10.1007/978-981-19-4767-4_23

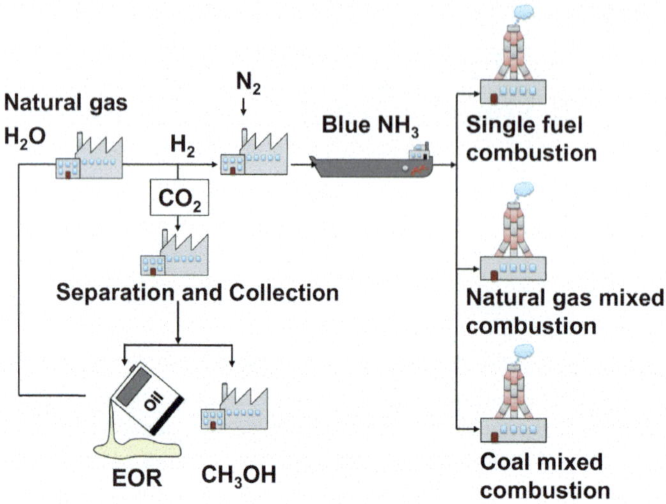

Fig. 23.1 Blue ammonia supply chain flow diagram in demonstration test (August 2020-October 2020)

Aramco and the Institute of Energy Economics, Japan (IEEJ), in partnership with Saudi Basic Industries Corporation (SABIC) have demonstrated the production and shipment of blue ammonia from Saudi Arabia to Japan with support from the Japan Ministry of Economy, Trade and Industry. Forty tons of blue ammonia have already been dispatched to Japan for use in zero-carbon power generation (Fig. 23.1). In future, large amounts of ammonia will be used for combustion as an energy carrier. After ammonia is used for power generation, it also can be used as a hydrogen carrier for a polymer electrolyte membrane (PEM) fuel cell vehicle (FCV). In this review, component technologies to develop ammonia cracking and high-purity hydrogen supply systems will be introduced.

23.2 Hydrogen Fuel Cost for Fuel Cell Vehicles (FCVs)

The higher heating value (HHV) of ammonia is 22.5 MJ/kg and 13.6 MJ/L (1 MPa, density 0.603 g/cm^3 at 298 K). HHV includes the latent heat of condensation when the water vapor in the combustion gas is condensed. The higher heating value (HHV) of hydrogen is 142 MJ/kg. The price of 70 MPa hydrogen fuel for FCV is set to 1000–1100 yen/kg. The energy price will be 7.4 yen/MJ (27 yen/kWh), and the price of hydrogen per 1 kWh will be about twice compared with that of gasoline.

The efficiency of a fuel cell vehicle (FCV) is about twice compared with that of a gasoline vehicle (ICE), and the mileage per unit price is about the same. Therefore, the price of hydrogen is equivalent to the price of gasoline. If the HHV of gasoline is 34.6 MJ/L and the selling price is 143 yen/L (actual sales price on August 27, 2018),

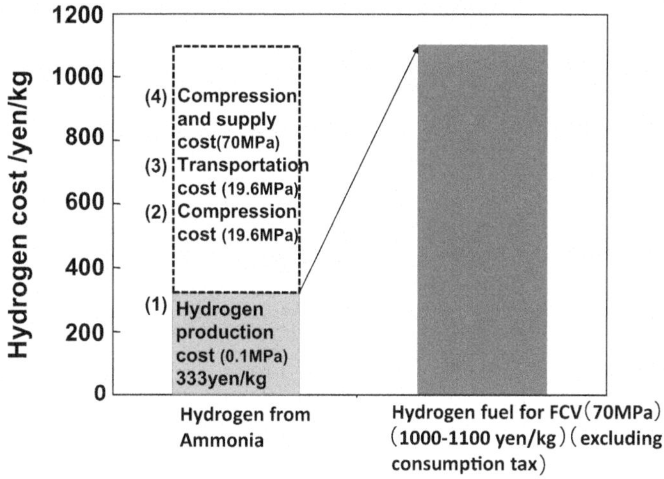

Fig. 23.2 Hydrogen cost from ammonia and 70 MPa hydrogen fuel cost for FCV [7]

the gasoline price per unit energy is 4.1 yen/MJ (15 yen/kWh) (1 MJ = 0.278 kWh). Energy company Shell is considering ammonia as a renewable energy carrier [8]. The cost of converting the renewable energy to ammonia, transporting it, and reconverting it to hydrogen (ammonia-derived hydrogen cost, H_p) was estimated. The cost was 3 dollars/kg hydrogen (333 yen/kg, 1 dollar: 111 yen, August 27, 2018), which was equivalent to liquid hydrogen and was reported to be slightly lower than that of organic hydride [8].

Figure 23.2 shows a comparison of the hydrogen cost in green ammonia produced by a renewable energy and a hydrogen fuel price for FCV. The hydrogen cost in ammonia is less than 1/3 of hydrogen fuel price. After the hydrogen derived from ammonia is compressed, it can be transported to the hydrogen station, decompressed and recompressed to 70 MPa at the hydrogen station. It is possible to be supplied to the FCV at a cost of about 1100 yen/kgH_2.

23.3 Hydrogen Fuel Cell Strategy Roadmap

On December 15, 2014, the world's first FCV was sold for about 7 million yen from Toyota Motor Corporation, and the new FCV has started to be sold on December 9, 2020. On March 10, 2016, Honda (Honda Motor Co., Ltd.) also marketed FCV. The cumulative domestic sales of FCVs in 2019 were 3695. According to the hydrogen fuel cell strategy roadmap revised in March 2016, it aims to spread about 40,000 units by 2020, about 200,000 units by 2025, and about 800,000 units by 2030. Figure 23.3 shows the relationship between the number of FCVs N_f and the year x.

The following equation is established from Fig. 23.3.

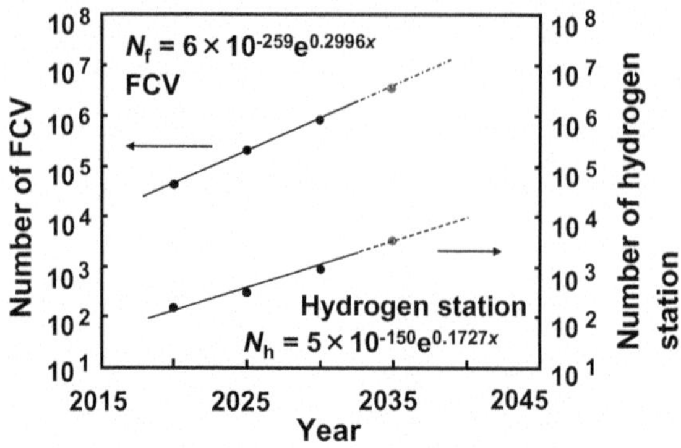

Fig. 23.3 Number of FCV and number of hydrogen station (hydrogen fuel cell strategy roadmap, revised in March 2016) [7]

$$N_f = 6 \times 10^{-259}\exp(0.2996x) \quad (R^2 = 0.9982) \tag{23.1}$$

Here, R^2 is a coefficient of determination. From Eq. (23.1), it can be estimated that the number of FCV in 2035 will be 3.6 million. On the other hand, 162 hydrogen stations were installed by the end of 2020. The target number of hydrogen station will be 320 and 900 by 2025 and 2030, respectively. Figure 23.3 also shows the relation between the number of hydrogen stations N_h and the year x. The following equation was obtained.

$$N_h = 5 \times 10^{-150}\exp(0.1727x) \quad (R^2 = 0.9872) \tag{23.2}$$

It is predicted that there will be about 2100 stations by 2035. Electric forklifts require a spare battery storage area to replace and use the batteries because the charging time is long, but fuel cell (FC) forklifts do not need it. Approximately 9000 units (end of FY2015) are in widespread use in USA, and domestic sales began in November 2016. The FC bus was also sold in February 2017 and delivered to the Tokyo Metropolitan Bureau of Transportation. The driving demonstration of FC heavy-duty trucks is scheduled to start around spring of 2022. FCVs, FC forklifts, FC buses, and FC trucks use a polymer electrolyte membrane fuel cell (PEMFC) composed of perfluorosulfonic acid-based proton exchange membranes.

23.4 Component Technologies for High-Purity Hydrogen Production from Ammonia

Thermodynamically, about 1000 ppm of unreacted ammonia is remained in the mixed gas of hydrogen and nitrogen produced by ammonia cracking at 0.1 MPa. Power decline after 1 h has been reported for PEMFC when the ammonia concentration in the fuel gas is below 13 ppm [9]. It has been reported that hydrogen contained 1 ppm of ammonia deteriorates a PEMFC for about a week [10]. The effect of the trace impurities in hydrogen on the power generation performance of a PEM fuel cell was evaluated at the Japan Automobile Research Institute (JARI). Based on the results obtained, a quality standard for hydrogen for FCV was proposed. In December 2012, the international standard for hydrogen fuel for FCV (ISO14687-2) came into effect. Table 23.1 shows hydrogen fuel specifications for fuel cell vehicles (FCV) (ISO14687-2) [11]. When using ammonia decomposition gas as a hydrogen fuel for fuel cell vehicles (FCVs), hydrogen purity is 99.97% or more. It is necessary to reduce the ammonia concentration below 0.1 ppm and the nitrogen concentration including Ar below 100 ppm.

Figure 23.4 shows a conceptual picture of ammonia decomposition and high-purity hydrogen supply system. The ammonia hydrogen station team composed of Hiroshima University, National Institute of Advanced Industrial Science and Technology (AIST), Toyota Industries Corp., Taiyo Nippon Sanso Co., Ltd., and Showa Denko Co., Ltd., developed (1) a durable ammonia decomposition catalyst and (2) an ammonia storage material that can reduce the residual ammonia concentration below 0.1 ppm, and established (3) a purification technology that can achieve a hydrogen purity above 99.97% and nitrogen concentration below 100 ppm in the research project "Ammonia Hydrogen Station Basic Technology" of SIP (Strategic Innovation promotion Program) Energy Carrier. Using these technologies, an ammonia

Table 23.1 Hydrogen fuel specifications for fuel cell vehicles (international standard: ISO 14687-2) analysis results of industrial ammonia (general grade) and high-purity ammonia [12–15]

Species	ISO 14687-2	Industrial ammonia		High purity ammonia
Purity of H_2 /%	99.97*	0.09 ppm	48.9 ppm	<0.5 ppm
Total hydrocarbons(C1) /ppm	2**	5.91	46.0	<0.5(CH_4)
Water(H_2O)/ppm	5**	89	19	<1.0
Oxygen(O_2)/ppm	5**	0.29 (O_2 + Ar)	40.3 (O_2 + Ar)	<0.5
N_2, Ar/ppm	100**	3.06(N_2)	64.4(N_2)	<2.0(N_2)
He/ppm	300**	–	–	–
Carbon dioxide(CO_2) /ppm	2**	0.03	0.05	–
Total sulphur compounds/ppm	0.004**	–	–	–
Formaldehyde/ppm	0.01**	–	–	–
Formic acid/ppm	0.2**	–	–	–
Ammonia/%	0.1**	>99.9%	>99.995%	>99.999
Total halogenated compounds/ppm	0.05**	–	–	–

* Minimum concentration, ** Maximum concentration

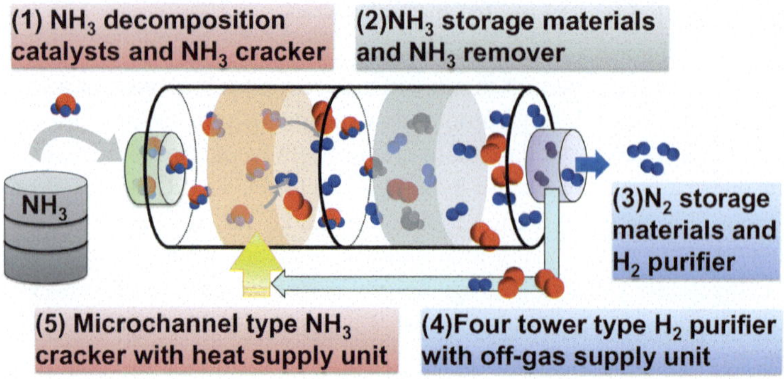

Fig. 23.4 Conceptual picture of ammonia decomposition and high-purity hydrogen supply system [15]

cracker, an ammonia remover, and a hydrogen purifier with 1 Nm^3/h scale were developed (Fig. 23.4). The component technologies to improve the efficiency such as an ammonia cracker with heat supply unit and a hydrogen purifier with off-gas supply unit were also developed as shown in Fig. 23.4.

23.4.1 *Ammonia Decomposition Technology (Ammonia Decomposition Catalysts and 1 Nm³/h Scale Ammonia Cracker)*

The decomposition of ammonia can be represented by Eq. (23.3).

$$NH_3 \rightarrow (1/2)N_2 + (3/2)H_2 \tag{23.3}$$

The mixed gas concentration of nitrogen, hydrogen, and ammonia at the chemical equilibrium state from 323 to 923 K in the ammonia decomposed gas was calculated using Outotec's calculation software HSC Chemistry 6.0. The hydrogen concentration increased with temperature at 0.1 MPa. The residual ammonia concentration at 773 K, 823 K, and 923 K are 1300 ppm, 800 ppm, and 340 ppm, respectively. A high-performance catalyst is required to promote the decomposition of ammonia at a practical rate, and various catalysts have been synthesized and evaluated (see Table 23.2).

It can be seen that the ammonia conversions of the ruthenium-based catalyst at 673–873 K are higher than that of the iron, cobalt, and nickel-based catalysts. Ammonia conversion at 673 K shows about 50% using Ru-CNT-KNO_3 catalyst, and the conversion improves up to 80% at 673 K using Ru-KNO_3/MgO-CNTs. When iron, cobalt, and nickel-based catalysts are used, the conversion at 673 K is below 10%. Although the conversion increases with temperature, the temperature above

Table 23.2 Ammonia conversion rate using various catalysts

Catalysts	NH_3 conversion/%			References
	673 K	773 K	873 K	
Ru/MgO (coprecipitation method) (impregnation method) Ru: 3wt%, SV: 15000 h^{-1}	60.2 24.2	99.8 66.8	99.9 98	A. Takahashi and T. Fujitani, JP2016-159209
$Ru/Cs_2O/Pr_6O_{11}$	91	99.8	> 99.8	K. Nagaoka et al., Int. J. Hydrogen Energy, 39, 20,731–20,735 (2014)
Ru/LSZ-DP 30,000 mLg^{-1} h^{-1} 4000 mLg^{-1} h^{-1}	18 54	75 99.8	> 99.8 > 99.8	B. Lorenzut et al., Chem. Cat. Chem., 2, 1096 -1106 (2010)
$Ru/CNT\text{-}KNO_3$ $Ru\text{-}KNO_3/MgO\text{-}CNTs$	50 80	– –	– –	S.F. Yin et al., Appl. Catal. A: Gen., 277, 1–9 (2004)
$Nano\text{-}Ru/SiO_2$ $Nano\text{-}Ni/SiO_2$ $Nano\text{-}Fe/SiO_2$ $Nano\text{-}Co/SiO_2$	34 8.5 0.5 1.0	90 35 2.4 15	> 99.8 82 83 56	L.H. Yao et al., Catalysis Today, 158, 401–408 (2010)
Ni/MWCNTs Co/MWCNTS	7 7	58 61	- -	H. Zhang et al., Int. J. Hydrogen Energy, 39, 277–287 (2014), H. Zhang et al., Int. J. Hydrogen Energy, 39, 17,573–17,582 (2014)
$Nano\text{-}Fe/mesoSiO_2$	7	27	86	Y. Li et al., Catal. Commun., 11, 368–372 (2010)
Ni/Al_2O_3 Ni/La_2O_3	5 6	34 33	97 90	H. Muroyama et al., Appl. Catal. A: Gen., 443–444,119–124 (2012)
$Fe_2O_3/CMK\text{-}5$ NiO/Al_2O_3	9* 6*	26 17	96 80	A. Lu et al., J. Am. Chem. Soc., 132, 14,152–14,162 (2010)

*623 K, LSZ: Lanthanum-stabilized zirconia, CMK-5: Ordered mesoporous carbon

873 K is required to achieve a conversion above 80%. A high-performance ruthenium based catalyst (Ru/MgO) that decomposes to a residual ammonia concentration of approximately 1000 ppm at 773 K was prepared by the coprecipitation method in the National Institute of Advanced Industrial Science and Technology (AIST). The ammonia to hydrogen conversion efficiency by the catalyst at 773 K was similar to the theoretical value (theoretical conversion value at 773 K: 99.74%, theoretical conversion value at 823 K: 99.84%). This indicates that residual ammonia concentration in the decomposed NH_3 is approximately 1000 ppm at 773–823 K (theoretical ammonia concentration: 1300 ppm at 773 K, 800 ppm at 823 K, as shown in Fig. 23.5), although the conventional ruthenium-based catalyst had a residual ammonia concentration of about 70,000 ppm (space velocity: 15,000 h^{-1}). No deterioration was observed in the 500 h durability test of the Ru/MgO catalyst prepared by the coprecipitation method [16, 17].

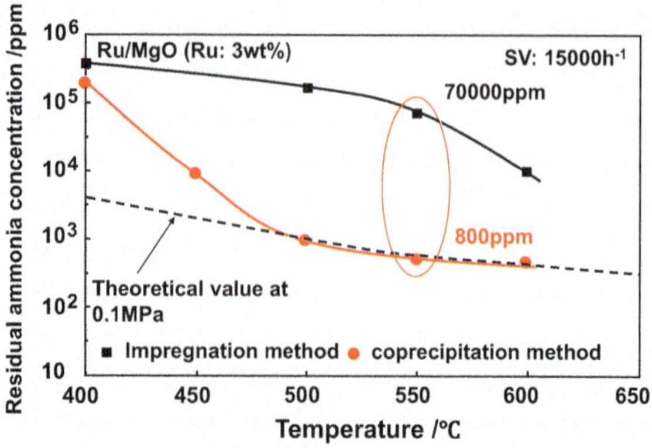

Fig. 23.5 Residual ammonia concentration using Ru/MgO as a function of reaction temperature [12, 15, 18]

Industrial ammonia is lower in cost than high-purity ammonia, and it will be used as the raw material for the hydrogen carrier. However, industrial ammonia contains water (see Table 23.1) [13, 19]. An accelerated test of the decrease in activity of the ammonia decomposition catalyst in the presence of water was conducted, and it was suggested that the decrease in the catalyst activity due to the presence of water was small (water content: below 500 ppm) [19]. An ammonia cracker using this catalyst with shell and tube has been developed (Fig. 23.6). Ammonia gas was supplied to the ammonia cracker at a flow rate of 1 Nm^3/h, and the decomposed gas (hydrogen 75%, nitrogen 25%) having an ammonia concentration below 1000 ppm was obtained at a flow rate of 2 Nm^3/h at 823 K.

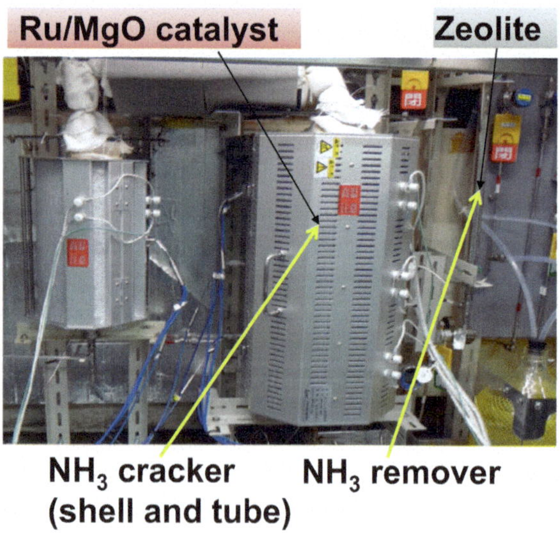

Fig. 23.6 Overview of ammonia cracker and ammonia remover [19]

23.4.2 Ammonia Removal Technology (Ammonia Storage Material, 1Nm^3/h Scale Ammonia Remover)

Ammonia pressure-composition isotherms of various zeolites were measured [20]. The equilibrium plateau pressure was around 8 Pa for Li-X (X-type zeolite whose exchangeable cation is lithium, pore size 0.8–1.0 nm) and Ca-X (X-type zeolite whose exchangeable cation is calcium, pore size 0.8–1.0 nm). The amount of adsorbed ammonia (pressure: 0.1 MPa) changed from 11 wt% to 16 wt% depending on the kinds of exchangeable cation (Na-A: 11.0 wt%, Ca-A: 11.1 wt%, Na-X: 11.9 wt%, Ca-X: 14.5 wt%, Li-X: 15.5 wt%) [20]. The adsorption amount of X-type zeolite was larger than that of A-type zeolite. The pore diameter of A-type zeolite is 0.3–0.5 nm, and that of X-type zeolite is 0.8–1.0 nm [21]. It is considered that the X-type zeolite with a large pore volume has a larger ammonia adsorption capacity. In addition, the amount of adsorbed ammonia increased in the order of Na-X < Li-X among the X-type zeolites with same valence number of cation. Electronegativity of metal cation increased in this order. It is well known that the metal cation and nitrogen of ammonia form coordinate bond. We can presume that the smaller the electronegativity difference, the larger the energy of coordination bond. However, even with Li-X, which has the largest ammonia adsorption capacity, the ammonia plateau pressure was about 8 Pa (ammonia concentration: 80 ppm) by the static (equilibrium) method (ammonia adsorption capacity at 8 Pa, Na-A: 0.44 wt%, Ca- A: 1.21 wt%, Na-X: 0.83 wt%, Ca-X: 3.38 wt%, Li-X: 4.50 wt%). This indicates that ammonia removal concentration is 80 ppm. Therefore, we examined the reduction of ammonia removal concentration by a dynamic method. A simulated ammonia decomposition gas consisting of 75% hydrogen, 25% nitrogen, and 1000 ppm ammonia was passed through a zeolite (Li-X) packed column at a constant flow rate (dynamic adsorption method, non-equilibrium and unsteady adsorption method). The ammonia concentration after passing through the column was measured by an ammonia analyzer using the wavelength scanning cavity ring-down spectroscopy (WS-CRDS method) [22]. It was found that the residual ammonia concentration was below 0.02 ppm, which sufficiently satisfied the FCV hydrogen fuel specifications (Fig. 23.7).

The ammonia adsorption amount calculated from the breakthrough time was 5.7 wt% (34 g/L). Even if Ca-X or Na-X was used, it had similar ammonia removing ability, but the ammonia adsorption capacity calculated from the breakthrough time was decreased.

The dynamic ammonia adsorption amount changed in the order of Na-X < Li-X having the same valence number [20]. The measurement interval of the ammonia analyzer was as short as 2 s, and it became possible to incorporate it in the ammonia removal system in the practical use. We have developed a 1Nm^3/h scale ammonia remover using Li-X as an ammonia storage material (Figs. 23.6 and 23.8) [19, 20].

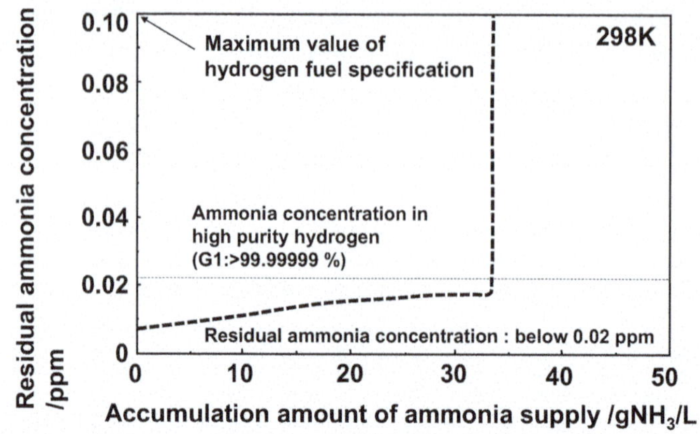

Fig. 23.7 Relation between residual ammonia concentration and accumulated amount of ammonia supply (zeolite 0.5 g, flow rate: 500 cm^3/min) [12, 15, 20, 22]

Fig. 23.8 Overview of ammonia removal system [20]

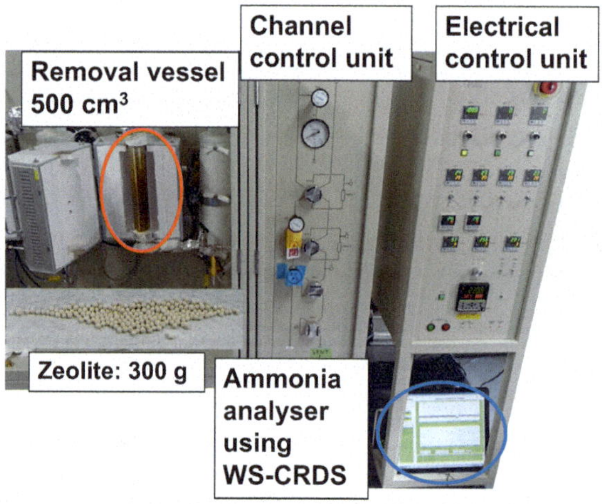

23.4.3 Hydrogen Purification Technology (Nitrogen Adsorbing Material, 1Nm³/h Scale Hydrogen Purifier

A simulated ammonia decomposition gas which contains 25% nitrogen and 75% hydrogen was used. Various zeolites (Ca-A, Ca-X, Na-X, Li-X) were investigated to adsorb ammonia. The adsorption capacity of X-type zeolite Ca-X was larger than that of A-type zeolite Ca-A. It is suggested that the amount of nitrogen adsorbed in the X-type zeolite with a large pore volume has larger capacity.

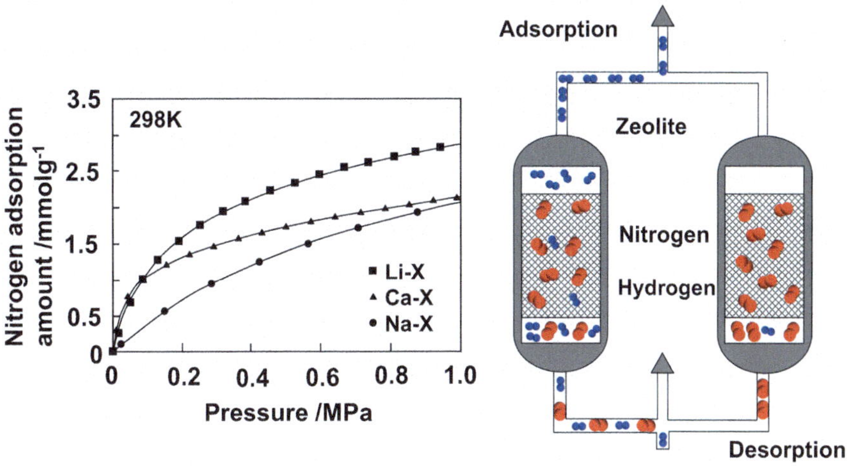

Fig. 23.9 Nitrogen adsorption isotherms for X-type zeolites and conceptual picture of hydrogen purifier (two-column VPSA method) [23]

Figure 23.9 shows the nitrogen adsorption isotherms of X-type zeolites (298 K) [23]. Nitrogen is adsorbed on the surface of zeolite. But, hydrogen is passed through the zeolite without adsorption.

Among the X-type zeolites, the amount of adsorbed nitrogen increased with the order Na-X < Li-X < Ca-X in the low pressure region [23]. In the ammine complex salt, it has been reported that the equilibrium vapor pressure during storage of ammonia decreases as the electronegativity of the cation increases [24]. In the case of zeolite, it is indicated that the larger the electronegativity, the more the nitrogen adsorption amount increases. However, in the high pressure region, the adsorption amount of Li-X increased compared with Ca-X. This is probably because the ionic radius of Li was smaller than that of Ca and the pore volume increased. Therefore, we used Li-X as the nitrogen adsorbent for hydrogen purifier as well as the ammonia remover.

A mixed gas of hydrogen 75% and nitrogen 25% was fed to a hydrogen purification experimental apparatus having two-column vacuum pressure swing adsorption method (two-column VPSA method). The nitrogen concentration passed through the purifier was below 1 ppm, and other impurities also satisfied FCV hydrogen fuels specification. Hydrogen purification efficiency of this system is 70% and produces high purity of hydrogen at a rate of 1.05 Nm^3/h. Then, we have developed a 1Nm^3/h scale two-column hydrogen purifier using Li-X as a nitrogen adsorbent (Fig. 23.10).

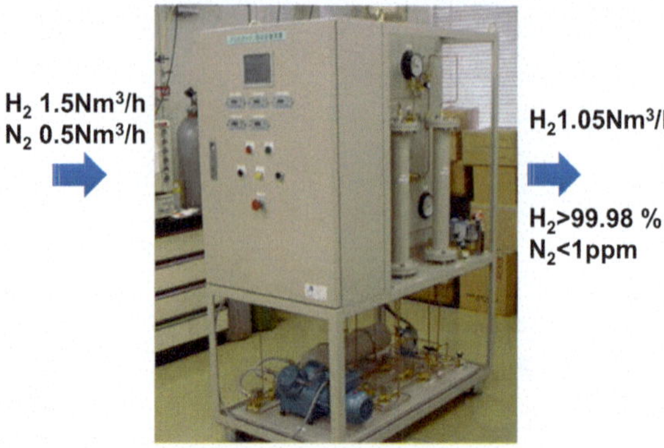

Fig. 23.10 Overview of two-column hydrogen purifier

23.4.4 Hydrogen Purity and Hydrogen Conversion Efficiency

The purity of hydrogen, the ammonia concentration, the nitrogen concentration in the fuel obtained by the NH_3 cracker, the NH_3 remover, and the H_2 purifier were above 99.98%, below 0.02 ppm, and below 1 ppm, respectively.

The energy balance of the ammonia decomposition / high-purity hydrogen supply system was examined. The hydrogen conversion efficiency (conversion efficiency from ammonia to hydrogen) E_c of this system can be expressed by Eq. (23.4).

$$E_c = \frac{G E_{H_2}}{E_{NH_3} + E_a + E_e} \tag{23.4}$$

Here, E_{H2} is the heat of combustion of hydrogen obtained by ammonia decomposition (429 kJ / 1.5 mol H_2, 429 kJ / molNH_3), and G is 70% when using the two-column hydrogen purifier (hydrogen purification efficiency: 0.7), E_{NH3} is the combustion heat of ammonia (383 kJ/ molNH_3), E_a is the amount of additional heat for ammonia cracking, and E_e is the amount of electricity supplied externally. By ignoring the amount of electricity used to operate the apparatus E_e,

$$E_c = \frac{G E_{H_2}}{E_{NH_3} + E_a} \tag{23.5}$$

According to the simulation, 129 kJ/molNH_3 is required as the additional heat E_a in order to decompose ammonia [25]. When the additional heat is supplied with fossil fuel, the hydrogen conversion efficiency E_c is obtained by substituting the values of E_a (129 kJ/ molNH_3), E_{H2} (429 kJ / 1.5 molH_2), G (0.7), and E_{NH3} (383 kJ/ molNH_3) into Eq. (23.5) was 59%.

23.5 Highly Efficient Hydrogen Production Technology from Ammonia

23.5.1 Hydrogen Purifier with Off-Gas Supply Unit

Hydrogen purification efficiencies of hydrogen purifiers using VPSA methods are shown in Fig. 23.11. The hydrogen purification efficiency (high-purity hydrogen amount divided by hydrogen amount in ammonia) increased with the number of tower as shown in Fig. 23.11. Furthermore, a new process (reducing the amount of off-gas by precisely controlling the opening and closing of the switching valve of the hydrogen purification experimental device) is combined with the hydrogen purifier (1 Nm^3/h scale) using the 4 tower VPSA method. As a result, hydrogen purification efficiency is improved up to 90% (Fig. 23.11). High-purity hydrogen can be obtained with the 4 tower type hydrogen purifier including the new process. The hydrogen purification efficiency is 90% and it also produces off-gas, which is a mixed gas of 100% nitrogen (0.5 Nm^3/h) and 10% hydrogen (0.15N m^3/h). We have developed an off-gas supply unit to utilize this as a heat source for ammonia decomposition.

Figure 23.12 shows overview of a 4 tower type purifier combined new process with off-gas supply unit [23]. 10Nm^3/h scale 4 tower type hydrogen purifier combined new process with off-gas supply unit has been also developed [23]. Figure 23.13 shows the energy balance of a high-efficiency ammonia decomposition and high purity hydrogen supply system. The hydrogen conversion efficiency E_c of this system can be expressed by Eq. (23.6).

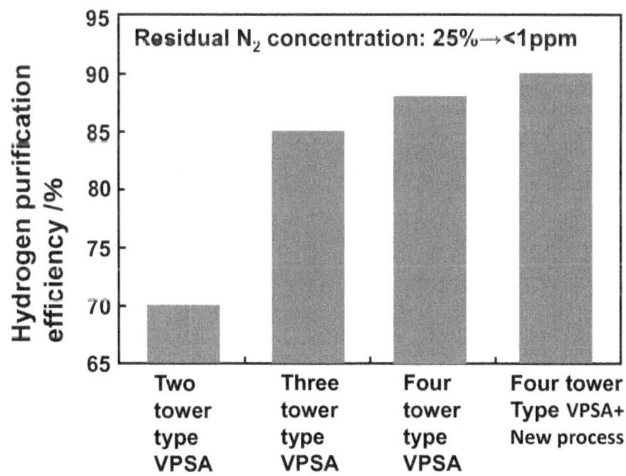

Fig. 23.11 Hydrogen purification efficiency of hydrogen purifiers using VPSA methods (vacuum pressure swing adsorption methods, 1 Nm^3/h scale) [12, 23]

Fig. 23.12 Overview of a 4 tower type hydrogen purifier combined new process with off-gas supply unit with 1 Nm^3/h scale [23]

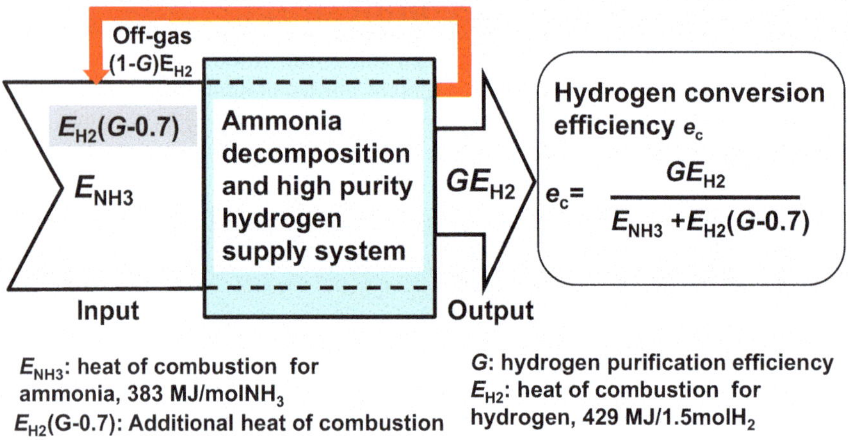

Fig. 23.13 Energy balance of high-efficiency ammonia decomposition and high-purity hydrogen supply system [12]

$$E_c = \frac{GE_{H2}}{E_{NH3} + (G - 0.7)E_{H2}} \tag{23.6}$$

Here, E_{H2} is the heat of combustion of hydrogen (429 kJ/1.5molH_2) obtained by ammonia decomposition, and G is 0.9 (90%) when a four tower type hydrogen purifier with new process is used (Fig. 23.12). E_{NH3} is the heat of combustion of ammonia (383 kJ/molNH_3). When the hydrogen purification efficiency is 90% (10% of hydrogen and 100% of nitrogen), additional ammonia is also partially supplied as a heat source. In that case, the amount of heat required is 86 kJ/mol of ammonia. This amount of heat is equal to the combustion heat of 0.22 mol of ammonia. From this, 1.35 mol of high purity hydrogen is produced using 1.22 mol of ammonia. The hydrogen conversion efficiency is calculated to be 82% from Eq. (23.6). This value is about 1.4 times higher than that of the 2 tower type hydrogen purifier (59%) without using off-gas.

23.5.2 Microchannel Type Ammonia Cracker with Heat Supply Unit

A microchannel refers to a narrow flow path processed by microfabrication technology. The size of the flow path is below several millimeters. Heat transfer performance was improved by the microchannel, so that the heat exchange is possible with high efficiency. It has higher heat transfer performance than the shell & tube type ammonia cracker (see Fig. 23.6) and is expected to be compact (A4 size SV = 10,000 h^{-1}, 1.3Nm/ h scale microchannel type). The ammonia cracker with a heat supply unit was designed and manufactured. Figures 23.14 and 23.15 shows the prototyped microchannel ammonia cracker with a heat supply unit [25]. Simulated 10% off-gas (H_2:0.15 Nm^3/h) and additional ammonia were mixed and catalytically burned to be used for ammonia decomposition, and the conversion efficiency at 823 K was 99.6–99.7% (residual ammonia concentration 1500–2000 ppm). The experimental value of the hydrogen conversion efficiency was 80%, and it was possible to decompose ammonia with high efficiency even in a microchannel with a small scale. The performance did not change after 200 h.

The power generation efficiency E_g of polymer electrolyte membrane FCV was reported to be about 60% [26]. Energy efficiency E_f is calculated by the following equation.

$$E_f = E_c \times E_g \tag{23.7}$$

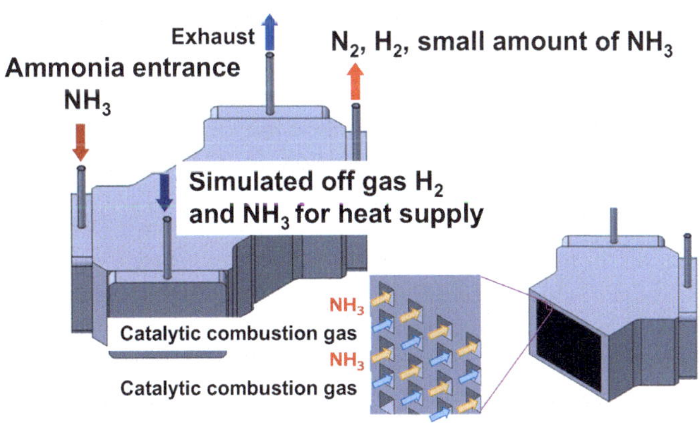

Fig. 23.14 Schematic picture of the prototyped microchannel ammonia cracker with a heat supply unit [25]

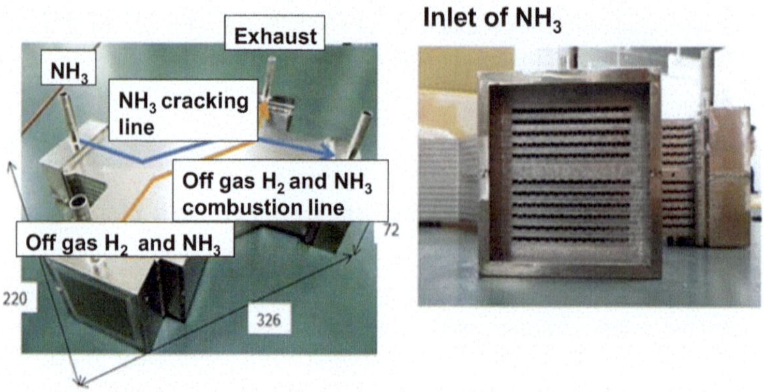

Fig. 23.15 Overview of the prototyped microchannel ammonia cracker with a heat supply unit [15, 25]

From this, it is suggested that when high-purity hydrogen is used in a polymer electrolyte membrane fuel cell (PEMFC), the energy efficiency will be about 50% by assuming that the compression energy of H_2 is small.

23.6 Hydrogen Fuel Cost for FCV

Hydrogen stations, like gas stations, will be installed mainly in densely populated residential area. However, the equipment cost is unknown for measures against the toxicity (odor) of ammonia. Therefore, we evaluated the hydrogen fuel cost at the off-site ammonia hydrogen station. Figure 23.16 shows a conceptual picture of the off-site type ammonia hydrogen station. The hydrogen fuel cost at the hydrogen station consists of (1) hydrogen production cost from ammonia derived from renewable energy, (2) compression cost, (3) transportation cost, and (4) 70 MPa hydrogen compression and dispensing cost.

We assumed that the hydrogen production costs composed of ammonia synthesis from renewable energy, ammonia transportation from the production area to Japan, and ammonia to high-purity hydrogen conversion, are 333 yen/kg [8]. The hydrogen conversion capacity at the plant is assumed to be 12 tons/day (500 kg/h, 5600 Nm^3/h) [27], and the hydrogen produced in the plant is compressed from 0.1 MPa to 19.6 MPa for transportation. In this case, the compression cost H_c is 66.3 yen/kgH_2 [7, 27]. All of the hydrogen will be transported to the off-site hydrogen station and then supplied to FCV. The average hydrogen transportation distance is 50 km of one way. Land transportation from the hydrogen supply base to the off-site hydrogen station will be carried out. For example, 20-ton trailer can transport 245 kg (2740 Nm^3) of hydrogen.

If the hydrogen transportation cost is H_t and the annual hydrogen filling amount per hydrogen station is W, the relationship between them can be approximately

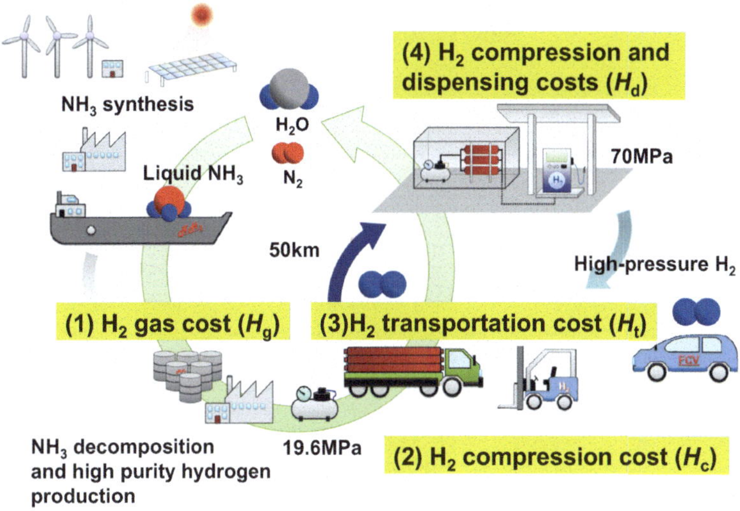

Fig. 23.16 Conceptual picture of off-site ammonia hydrogen station [7, 15]

expressed by the following straight line [27].

$$W \leq 30 \quad H_t = -56.4\ W + 1930 \tag{23.8}$$

$$W \geq 30 \quad H_t = -0.326\ W + 249 \tag{23.9}$$

Assuming that the hydrogen filling amount in the FCV is 100 kg/year (fuel consumption: 100 km/kgH_2, mileage: 10,000 km/year), the annual required hydrogen amount is 4000 tons in 2020, 20,000 tons in 2025, 80,000 tons in 2030, and 360,000 tons in 2035. According to hydrogen fuel cell strategy roadmap, the number of FCV is 40,000 and 160 hydrogen stations will be introduced with annual hydrogen filling amount per hydrogen station (25 tons) in 2020. The number of FCV is 200,000 and 320 hydrogen stations will be introduced with annual hydrogen filling amount per hydrogen station (62.5 tons) in 2025. The number of FCV is 800,000 and 900 hydrogen stations will be introduced with annual hydrogen filling amount per hydrogen station (88.9 tons) in 2030. The number of FCV is 3.6 million and 2100 hydrogen stations will be introduced with annual hydrogen filling amount per hydrogen station (171 tons) in 2035. The hydrogen transportation cost was calculated by substituting the hydrogen amount per one hydrogen station into the Eqs. (23.8) and (23.9). The results are shown in Table 23.3. The higher the annual hydrogen filling amount, the lower the transportation cost.

If the compression supply cost of 70 MPa is H_s and the annual hydrogen filling amount per hydrogen station is W, the relationship between them can be approximately expressed by the following equation [27].

Table 23.3 High-pressure hydrogen fuel supply cost for FCV at hydrogen stations [7]

Cost	2020 FCV 40,000 units 160 H_2 stations	2025 FCV 200,000 units 320 H_2 stations	2030 FCV 800,000 units 900 H_2 stations	2035 FCV 3,600,000 units 2100 H_2 stations
Annual filling amount of H_2 divided by number of H_2 station	25 ton/year	62.5 ton/year	88.9 ton/year	171 ton/year
(1) H_2 gas cost (H_g)	333 yen/kg	333 yen/kg	333 yen/kg	333 yen/kg
(2) Compression cost at 19.6 MPa (H_c)	66.3 yen/kg	66.3 yen/kg	66.3 yen/kg	66.3 yen/kg
(3) Transportation cost (H_t)	520 yen/kg	229 yen/kg	220 yen/kg	193 yen/kg
(4) 70 M Pa compression and dispensing cost (H_d)	2936 yen/kg	1178 yen/kg	829 yen/kg	432 yen/kg
H_2 fuel cost/kg (70 MPa) (H_f)	3855 yen	1806 yen	1448yen	1024 yen

$$H_d = 72{,}700\ W^{-0.997} \quad (R^2 = 0.9999) \tag{23.10}$$

The annual hydrogen filling amount per hydrogen station was substituted into Eq. (23.10) to calculate 70 MPa compression and dispensing cost. As shown in Table 23.3, the higher the annual hydrogen filling amount, the lower the compression and dispensing cost (H_d). Table 23.3 also shows hydrogen gas cost at 0.1 MPa (H_g) which is obtained by adding up the hydrogen production, ammonia synthesis, transportation and ammonia to hydrogen conversion costs from renewable energy, the hydrogen compression cost at 0.1 MPa (H_c) and the hydrogen transportation cost (H_t). The annual hydrogen filling amount per number of hydrogen station increases as the number of FCVs and hydrogen stations increases. As a result, transportation costs (H_t) and the dispensing costs (H_d) will decrease as the amount of hydrogen supply increases, and the hydrogen fuel cost (H_f) will be below 1100 yen/kg after 2035. The compression and dispensing cost (H_d) at 70 MPa is higher than that of hydrogen gas (H_g), compression (H_c), and transportation (H_t). This is probably because the construction cost of the hydrogen station is more than 400 million yen (435 million yen), which is expensive [27].

23.7 Conclusions

(1) We have developed component technology for ammonia decomposition and high-purity hydrogen production that fully satisfies international standard (this work: H_2 purity >99.98%, NH_3 <0.02 ppm, N_2 <1 ppm, international standard: H_2 purity ≥99.97%, NH_3 <0.1 ppm, N_2 <100 ppm).
(2) We have also developed component technology for producing hydrogen fuel from ammonia with high efficiency and have achieved a hydrogen purification efficiency of 90%. The hydrogen conversion efficiency was calculated to be about 80%, and it was confirmed with a microchannel cracker.
(3) When high-purity hydrogen was produced from ammonia and used in a fuel cell vehicle, the energy efficiency was considered to be about 50%.

Acknowledgements This work was partially supported by Council for Science, Technology and Innovation (CSTI), Cross-ministerial Strategic Innovation Promotion Program (SIP), "energy carrier" (funding agency: JST). The author is greatly indebted to Professor Takayuki Ichikawa and Associate Professor Hiroki Miyaoka of Hiroshima University, Dr. Tadahiro Fujitani of National Institute of Advanced Industrial Science and Technology, Mr. Hideto Kubo of Toyoda Industries, Mr. Takayoshi Adachi of Taiyo Nissan, Mr. Tsunekichi Kuriyama of Showa Denko for their help and discussion. And, this work was carried out as a study in Research Center for Nitrogen Recycling Energy Carrier of Hiroshima University.

References

1. Summary of Paris Agreement. http://www.env.go.jp/earth/ondanka/cop21_paris/paris_conv-a.pdf
2. Ministry of the Environment (2017) For overview of global warming. http://www.chugoku.meti.go.jp/event/enetai/pdf/170529_5.pdf
3. Aizawa H (2019) Toward the realization of a hydrogen area and society utilizing renewable energy. J Hydrogen Energy Syst Soc Jpn 44:1–2
4. Kojima Y (2019) Hydrogen storage materials for hydrogen and energy carriers. Int J Hydrogen Energy 44:18179–18192
5. World's first blue ammonia shipment opens new route to a sustainable future, Dhahran, 27 Sept 2020. https://www.aramco.com/en/news-media/news/2020/first-blue-ammonia-shipment
6. Crolius SH (2019) Technology advances for blue hydrogen and blue ammonia. https://www.ammoniaenergy.org/articles/technology-advances-for-blue-hydrogen-and-blue-ammonia/
7. Kojima Y (2019) Development trend of ammonia decomposition and high purity hydrogen supply system. In: Hydrogen production, transportation, storage, technology and material development, pp 186–194, Apr 26 (2019) NTS
8. Gupta N (2018) Ammonia as a renewable energy carrier: opportunities & challenges. In: Expectations and prospects of ammonia fuel derived from renewable energy grand renewable energy 2018 international conference and exhibition, June 21 2018, Pacifico Yokohama
9. Uribe FA, Zawodzinski TA, Gottesfeld S (2002) Effect of ammonia potential fuel impurity on proton exchange membrane fuel cell performance. J Electrochem Soc 149:A293–A296
10. Halseid R, Vie PJS, Tunold R (2006) Effect of ammonia on the performance of polymer electrolyte membrane fuel cells. J Power Sources 154:343–350

11. Takagi Y (2008) ISO standard for hydrogen energy technology-activities of hydrogen fuel specification standardization WG12 for fuel cell vehicles. J Hydrogen Energy Syst Soc Jpn 33:17–22. https://www.iso.org/standard/55083.html
12. Kojima Y (2020) Production technologies of hydrogen fuel for fuel cell vehicles from ammonia. Ceram Jpn 55:302–306
13. Abe T, Uemori K, Kurita E, Yoneda T (2004) Development of industrial ammoni are fining technology. Tech Rep Taiyo Nissan 23:86–87
14. High-purity ammonia product information, Showa Denko. https://www.sdk.co.jp/products/45/76/1114.html
15. Kojima Y, Hiroshima University, Theme *Final Report*, SIP (Cross-minisreal Strategic Innovation Promotion Program) Subject "Energy Carrier" R&D Theme "Ammonia Hydrogen Station Basic Technology" July 1, 2014–Mar 31, 2017. https://www.jst.go.jp/sip/dl/k04/end/team4-1.pdf
16. Fujitani T, Takahashi A (2016) Ammonia decomposition catalyst, method for producing the catalysr, and method for decomposing ammonia using the catalyst. JP 2016-159209
17. Fujitani T, National Institute of Advanced Industrial Science Technology, Final Report, SIP (Cross-minisreal Strategic Innovation Promotion Program) Subject "Energy Carrier" R&D Theme "Ammonia Hydrogen Station Basic Technology" Research Subject "Ammonia Decomposition Technology", July 1, 2014 to Mar 31, 2017, https://www.jst.go.jp/sip/dl/k04/end/team4-3.pdf
18. Kojima Y (2016) Production of hydrojen fuel for fuel cell vehicles using ammonia as a raw material. J Hydrogen Energy Syst Soc Jpn 41:290–291
19. Kuriyama T, Showa Denko, Final Report, SIP (Cross-minisreal Strategic Innovation Promotion Program) Subject "Energy Carrier" R&D Theme "Ammonia Hydrogen Station Basic Technology" Research Subject " Ammonia utilization technology for hydrogen station", Apr 1, 2015 to Mar 31, 2018. https://www.jst.go.jp/sip/dl/k04/end/team4-7.pdf
20. Kojima Y, Hiroshima University, Final Report, SIP (Cross-minisreal Strategic Innovation Promotion Program) Subject "Energy Carrier" R&D Theme "Ammonia Hydrogen Station Basic Technology" Research Subject "Ammonia removal technology development", July 1, 2014 to Mar 31, 2018, https://www.jst.go.jp/sip/dl/k04/end/team4-4.pdf
21. Auerbach SM, Carrado KA, Dutta PK (2003) Handbook of zeolite science and technology. In: Sircar S, Myers AL (eds) Gas separation by zeolite, vol 22. CRC Press
22. Miyaoka H, Ichikawa T, Kojima Y (2018) Highly purified hydrogen production from ammonia for PEM fuel cell. Int J Hydrogen Energy 43:14486–14492
23. Adachi T, Taiyo Nissan, Final Report, SIP (Cross-minisreal Strategic Innovation Promotion Program) Subject "Energy Carrier" R&D Theme "Ammonia Hydrogen Station Basic Technology" Research Subject "Hydrogen refining method", Oct 1, 2014 to Sept 30, 2018, https://www.jst.go.jp/sip/dl/k04/end/team4-6.pdf
24. Aoki T, Ichikawa T, Miyaoka H, Kojima Y (2014) Thermodynamics on ammonia absorption of metal halides and borohydrides. J Phys Chem C 118:18412–18416
25. Kubo H, Toyota Industries Coorporation, Final Report, SIP (Cross-minisreal Strategic Innovation Promotion Program) Subject "Energy Carrier" R&D Theme "Ammonia Hydrogen Station Basic Technology" Research Subject "Combustion heating type ammonia cracker", July 1, 2014–Mar 31, 2018, https://www.jst.go.jp/sip/dl/k04/end/team4-5.pdf
26. Kobuki S, Initiatives for sustainable mobility and development of fuel cell vehicles, FCEXPO2011, Keynote Lecture
27. Incorporated Administrative Agency, New Energy and Industrial Technology Development Organization, technology development such as hydrogen production/transportation/storage system, feasibility study research and development related to technology scenarios such as next-generation technology development/feasibility study, feasibility study according to hydrogen carrier, Outsourcing contractors Institute of Applied Energy, Iwatani Corporation, Kawasaki Heavy Industries, Ltd., Kansai Electric Power Co., Inc., Shimizu Corporation, Mitsubishi Heavy Industries, Ltd., Chiyoda Corporation, 2008 Research Report, Mar 2009

Chapter 24
Ruthenium Catalyst for Ammonia Decomposition

Tadahiro Fujitani and Isao Nakamura

Abstract Ammonia is of interest as a hydrogen transport carrier because of the high hydrogen storage capacity and easy liquefaction, and its decomposition to obtain hydrogen is a key reaction in energy and environmental industries. This review is focused on the ammonia decomposition on Ru, which is regarded as the most active metal in catalyzing its reaction. Previous studies for both the ammonia decomposition properties over Ru single crystals and the effects of catalyst preparation, support, and promoter on the NH_3 decomposition over Ru powder catalysts are summarized.

Keywords Ru catalyst · Ru single crystal · Ammonia · Decomposition · Hydrogen

24.1 Introduction

Recently, with the increasing social concern about exhaustion of fossil fuels and pollution problems, the utilization of hydrogen (H_2) has attracted much attention. H_2 is recognized as one of the most promising energy resources because it is a clean fuel with a high energy capacity, and can be used in a wide range of applications such as high-efficiency power generation systems by fuel cells, automobile transport, and distributed energy generation. However, the transport and storage of H_2 is a major obstacle to its practical application as an energy resource. Although the transport and storage by liquid H_2 are possible, there is a major economic problem for the use of liquid H_2 because liquefaction and boiloff losses consume large amounts of energy [1]. Thus, various alternative H_2 carriers such as methanol, hydrazine, ammonia (NH_3) hydrides, and formic acid have been proposed to acquire pure H_2 [2–4]. Among these materials, the NH_3 has received attention as an excellent candidate for the H_2 carrier because of the high hydrogen storage capacity (about 18 wt%), the high energy density (3000 Wh/kg), and easy liquefaction [2]. Furthermore, from an economical standpoint, it has been reported that, compared to methanol-reforming, NH_3 decomposition is a more economical process for H_2 generation [5]. Therefore,

T. Fujitani (✉) · I. Nakamura
Interdisciplinary Research Center for Catalytic Chemistry, National Institute of Advanced Industrial Science and Technology (AIST), Onogawa, Tsukuba, Ibaraki 305-8569, Japan
e-mail: t-fujitani@aist.go.jp

K. Aika and H. Kobayashi (eds.), *CO_2 Free Ammonia as an Energy Carrier*,
https://doi.org/10.1007/978-981-19-4767-4_24

the NH_3 has the potential to play an important role as an H_2 carrier in the realization of the future H_2 economy.

The NH_3 is a very stable molecule, and the thermal NH_3 decomposition is difficult to take place because of its extremely high activation energy (E_a) (ca. 1172 kJ/mol). Thus, when NH_3 is used as the H_2 carrier, a catalyst for producing H_2 with high efficiency from the decomposition reaction of NH_3 is required. So far, various catalysts have been developed to catalyze NH_3 decomposition, and the direct catalytic decomposition of NH_3 over supported Ru nanoparticles has been found to be the most efficient one for this process [6–9].

In this chapter, we will review the studies for the fundamental reaction properties of the NH_3 decomposition using Ru single crystals, as well as the reports concerning the effects of catalyst preparation method, support, and promoter addition on the NH_3 decomposition activity using Ru powder catalysts.

24.2 Ru Single Crystal

Fundamental reaction properties such as the adsorption, desorption, and dehydrogenation of NH_3 using Ru single crystals, especially Ru(0001) crystal plane which is the most stable surface structure, have been widely reported from both experiments and calculations. In addition, the influences of the Ru step site and co-adsorbed species on these reaction properties have been investigated. In this section, the previous reports concerning the following three topics are summarized: (i) adsorption and desorption of NH_3, (ii) NH_3 dehydrogenation and formation of H_2 and nitrogen (N_2), and (iii) effect of coadsorbed species on NH_3 reactivity.

24.2.1 Adsorption and Desorption of NH_3

The adsorption and desorption properties of NH_3 on Ru(0001) have been examined in detail using various surface analysis methods such as high-resolution electron energy loss spectroscopy (HREELS), infrared reflection absorption spectroscopy (IRAS), low energy electron diffraction (LEED), thermal desorption spectroscopy (TDS), electron stimulated desorption ion angular distribution (ESDIAD), and scanning tunneling microscopy (STM) [10–14] as well as theoretical calculations [15–18].

Benndorf and Madey [10] investigated the adsorption state and orientation of NH_3 on Ru(0001) at 80 K. They detected four different binding states (α_1, α_2, β, and γ) with TDS, depending on the NH_3 coverage (Fig. 24.1). At low coverages ($\theta_{NH_3} \leqq 0.15$), NH_3 desorbs from the α_1 state with a TDS peak maximum at 310 K, which is randomly oriented or freely rotating NH_3 monomers, bounded via the N atoms to the surface with the H atoms pointing away from the surface. At higher NH_3 coverages ($\theta_{NH_3} \geqq 0.15$), a second desorption peak (α_2) develops at 180 K, accompanied by a (2 $\times$ 2) LEED structure. The NH_3 saturation coverage for the first layer was estimated to be approximately 0.25. With further increase of NH_3 coverage ($\theta_{NH_3} > 0.25$), a sharp desorption peak (β state) appears at 140 K, and is interpreted as due to NH_3 species desorbing from a second adsorption layer. Finally, a desorption peak due to multilayer adsorption (γ state) is observed at 115 K. Parmeter et al. [11] also demonstrated that multilayer, second layer, and monolayer NH_3 states desorb near 115, 130 and between approximately 150 and 350 K, respectively. Furthermore, they characterized these three states vibrationally by means of HREELS. Monolayer NH_3 on Ru(0001) exhibited EEL spectra with $\delta_s(NH_3)$ (1070 cm^{-1}), $\delta_a(NH_3)$ (1580 cm^{-1}), $\nu_s(NH_3)$ (3260 cm^{-1}), and $\nu_a(NH_3)$ (3370 cm^{-1}). These frequencies agreed with the harmonic frequencies calculated from the global minimum structure of NH_3 on the Ru model cluster [15].

On the other hand, the IRA spectra of NH_3 chemisorbed on Ru(0001) at 80 K were characterized by the presence of a sharp peak in the 1050–1150 cm^{-1} region ($\delta_s(NH_3)$) and the absence of any appreciable signal in the 3200–3400 cm^{-1} region ($\nu(NH_3)$) [12], which was due to that only the $\delta_s(NH_3)$ mode generates the large dipole moment perpendicular to the surface that is necessary to produce an appreciable cross-section in infrared spectroscopy for the adsorption geometry of NH_3 on Ru(0001) [10]. In addition, the IR intensity of the $\delta_s(NH_3)$ mode is shown to be a maximum when the α_1 state is saturated ($\theta_{NH_3} = 0.15$), and then the population of

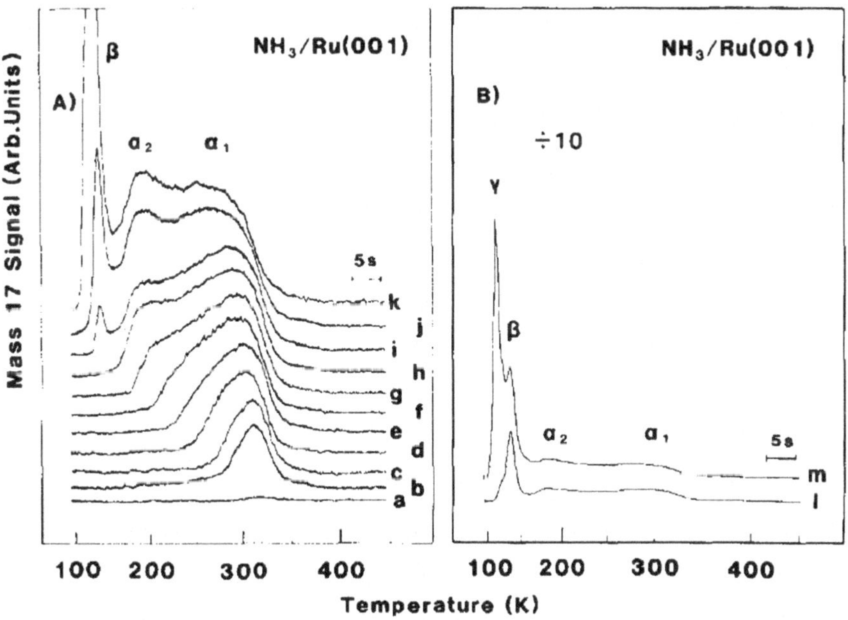

Fig. 24.1 NH_3 thermal desorption spectra for NH_3 on Ru(0001): A) TDS for low NH_3 coverages (α_1, α_2 and β state); B) TDS for high NH_3 coverages including the γ state. The NH_3 coverage corresponding to each spectrum is: **a** 0.00; **b** 0.04; **c** 0.05; **d** 0.07; **e** 0.09; **f** 0.12; **g** 0.15; **h** 0.18; **i** 0.2; **j** 0.28; **k** 0.35; **l** 0.52; **m** 0.76. Published with permission from Ref. [10]

the α_2 state results in a decrease of the δ signal. Results of ESDIAD [10] indicate that in the α_1 state the NH_3 molecules are bonded with their symmetry axes perpendicular to the surface. In contrast, when the α_2 state is present, the symmetry axes of the NH_3 species are to some extent inclined toward the surface. Thus, this reorientation reduces the perpendicular component of the dipole moment associated with the $\delta_s(NH_3)$ mode. Similar to results of Refs. [11, 12], Widdra et al. [13] observed a drastic redshift of 100 cm^{-1} with increasing NH_3 coverage for $\delta_s(NH_3)$ mode in NH_3 chemisorbed on Ru(0001) by HREELS. They concluded that the origin of this shift can be understood as a Stark shift due to the electric fields produced by the static NH_3 dipoles within the adlayer, based on density functional theory (DFT) model calculations. As a consequence of the electric field, the NH_3 molecule opens the H–N–H angle and a softening of the $\delta_s(NH_3)$ mode occurs.

The adsorption sites of NH_3 on Ru(0001) were investigated using STM by Maier et al. [14]. They confirmed that NH_3 adsorbs on-top sites with the N−H bond directions projecting in the direction of the neighboring Ru atoms, which was consistent with adsorption sites for NH_3 predicted by DFT calculations [16, 17]. This is the first direct experimental determination of the adsorption site of NH_3 on Ru, which has been the subject of disagreement [10, 11, 18].

24.2.2 NH_3 Dehydrogenation and Formation of H_2 and N_2

The dehydrogenation of NH_3 to NH_2, NH, and N adsorbed species on the Ru(0001) and Ru step surfaces as well as the formation of H_2 and N_2 by the associative desorption of the atomic H and N on Ru has been investigated using HREELS, IRAS, X-ray photoelectron spectroscopy (XPS), TDS and DFT.

Dietrich et al. [19, 20] studied the (N + NH)/Ru(0001) surface, where the atomic N and NH species ($\theta_N = 0.47$) were prepared by NH_3 decomposition at sample temperatures decreasing from 500 to 350 K during NH_3 exposure. HREEL measurements showed that the peaks for atomic N and NH species were observed at 72 and 410 meV, which were due to the ν(Ru–N) and ν(N–H) vibrations, respectively. From the TDS for (N + NH)/Ru(0001), the N_2 desorption peaks were detected at 500 and 680 K, which were attributed to the NH decomposition and associative desorption of atomic N, respectively. At 500 K, the desorption of H_2 is also observed in parallel with the N_2 desorption. Carabineiro et al. [21] have also examined the decomposition of NH_3 on the Ru(0001) surface in the temperature range from 150 up to 800 K. TDS results indicated that most of the NH_3 is dissociatively adsorbed between 150 and 300 K, with formation of H_2 around 390 K and N_2 between 600 and 800 K.

The effect of step sites on NH_3 decomposition over Ru surface has been studied. Egawa et al. [22] reported that during the adsorption at 120 K, NH_3 was molecularly adsorbed both on Ru(0001) flat surface and on Ru(1,1,10) stepped surface which possesses one step per five terrace widths, where molecularly adsorbed NH_3 was confirmed from the N 1 s core level peak observed at 400.0 eV by means of XPS. On subsequent heating, NH_3 desorbed from the Ru(0001) flat surface, although the NH_2(a) species was formed on the Ru(1,1,10) stepped surface at 300 K, which displayed the N 1 s peak at 398.1 eV. Under NH_3 flow conditions, no surface species were observed on the Ru(0001) surface at 300 K, although atomic N was formed by heating up to 360 K, which exhibited the N 1 s level at 397.4 eV. In contrast, atomic N besides the NH_3(a) complex was formed on the Ru(1,1,10) stepped surface even at 300 K. Furthermore, they demonstrated that the initial formation rate of N_2 on the stepped Ru(1,1,10) surface is an order of magnitude faster than that on the flat Ru(0001) surface under the steady-state of NH_3 decomposition at 2.7×10^{-5} Pa and 520 K [23]. These results clearly indicate that the scission of the N–H bond proceeds preferably on a step site of Ru surface.

In addition, the NH_3 decomposition properties have been investigated even on other Ru step surfaces. Dietrich et al. [24, 25] reported that the open Ru($11\bar{2}1$) surface is much more reactive for the thermal decomposition of NH_3 was than the Ru(0001) surface. They demonstrated that NH_3 decomposes completely during annealing of the Ru($11\bar{2}1$) surface for small NH_3 coverages up to 0.25. At higher coverages, molecular desorption is also observed. The dehydrogenation gives rise to sharp peaks at α(420 K), β(360 K), γ(320 K), and δ(220 K) in the H_2 TD spectra, and atomic N associatively desorbed at 610–670 K (Fig. 24.2). Furthermore, they identified the adsorbed species of NH_3, NH_2, NH, and N on the annealed NH_3-adsorbed Ru($11\bar{2}1$) surface using HREELS. Above results, they have clarified four steps of decomposition [25]: (i) at 220 K: $NH_{3,ad} + 2H_{ad} \rightarrow NH_3 + H_2$, (ii) at 320 K: $2NH_{3,ad} \rightarrow 2NH_{2,ad} + H_2$, (iii) at 360 K: $2NH_{2,ad} \rightarrow 2NH_{ad} + H_2$, and (iv) at 420 K: $2NH_{ad} \rightarrow 2N_{ad} + H_2$.

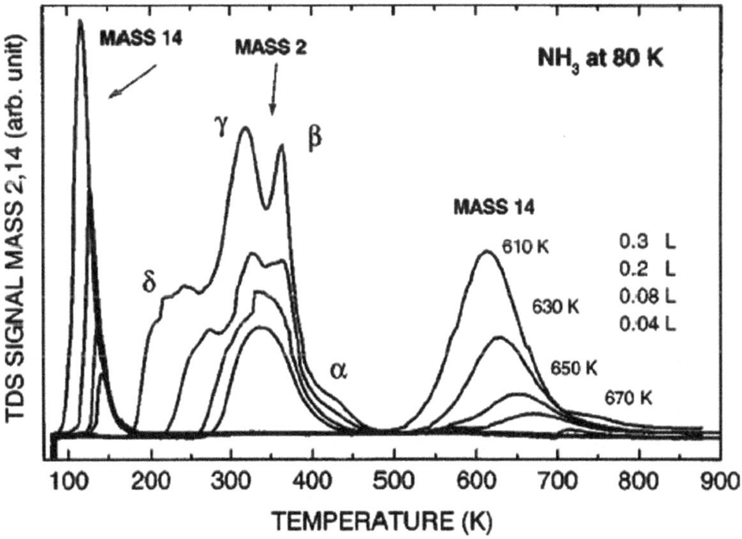

Fig. 24.2 TD spectra for a sequence of NH_3 exposures on Ru($11\bar{2}1$) at 80 K. Published with permission from Ref. [25]

On the other hand, the adsorption and dehydrogenation of NH_3 on Ru(11$\bar{2}$0) were examined by Wang et al. [26]. Following NH_3 exposure at 85 K, annealing to 140 K leads to desorption of the multilayer and second layer leaving the NH_3 monolayer which is characterized by two adsorption states: α_1-NH_3 at a threefold hollow site and a more weakly bonded α_2-NH_3 at an on-top site. As the temperature is increased, the α_2-NH_3 desorbs molecularly, while partial dehydrogenation occurs for the α_1-NH_3. This process proceeds at threefold hollow sites in successive steps finally leading to atomic nitrogen, which agreed with the results of the theoretical calculation [27]. Furthermore, from the HREELS measurements, they succeeded in isolating successively NH_2, NH, and N on the surface, which was consistent with the result of the Ru(11$\bar{2}$1) surface [24, 25]. However, the associative desorption temperature of atomic hydrogen produced by NH_x dehydrogenation was lower than those for Ru(11$\bar{2}$1), indicating that the Ru(11$\bar{2}$0) surface is more active for the NH_3 decomposition than the Ru(11$\bar{2}$1) surface.

24.2.3 *Effect of Coadsorbed Species on* NH_3 *Reactivity*

The effects of coadsorption of O [28–30], CO [31, 32], N [29], Na [28], B [33], and S [30] on the adsorption, desorption, and decomposition properties of NH_3 on the Ru(0001) surface have been investigated. Benndorf and Madey [28] examined the interaction of NH_3 with adsorbed O (electronegative) and Na (electropositive) on Ru(0001), and found that the coadsorption of NH_3 with O leads to an increase in the NH_3 binding energy; Na causes a decrease. Sun et al. [29] showed that the binding energy of NH_3 on Ru increased on the N-modified surface relative to the clean Ru(0001) surface, and an even greater increase in the binding energy to the O-modified surface. On the other hand, the coadsorption of CO and NH_3 on Ru(0001) and the dissociation of NH_3 in the mixed adsorbed layer were investigated by Sasaki et al. [31, 32]. From TDS results, the amount of dissociated NH_3 decreased with increasing CO precoverage up to $\theta_{CO} = 0.25$, whereas a further increase of CO coverage promoted the NH_3 dissociation. The HREEL spectra indicated that the molecular axis of C_{3V} NH_3 is tilted in the coadsorbed layers, the tilting being most pronounced in the (2×2) phase with a high CO partial coverage. This observation suggests that the tilting of NH_3 due to the interaction with CO facilitates electron donation from Ru 4d to LUMO of NH_3, leading to the N–H bond dissociation.

Rodriguez et al. [33] studied the bonding and chemistry of NH_3 on clean and B-covered Ru(0001) surfaces. They demonstrated that the B adatoms enhance the binding energies of NH_3 on Ru(0001). This is a consequence of attractive interactions between the negatively charged B adatoms and the positively charged admolecules. Furthermore, the decomposition of NH_3 on O- and S-modified Ru(0001) surfaces was investigated by Huang et al. [30]. On an O-covered Ru(0001) surface after NH_3 dose, two thermal desorption peaks of N_2 were observed. In addition to the broad feature centered around 750 K, a sharp peak was observed at 640 K, increasing in intensity with oxygen coverage, which is due to the repulsive interaction between the adsorbed nitrogen and oxygen atoms. They also showed that the NH_3 decomposition is significantly inhibited by S adatoms, although the presence of O adatoms on Ru(0001) only slightly affected the amount of NH_3 decomposed. This difference was explained to be attributed to the larger physical size of the S-atom.

24.3 Ru Powder Catalyst

A number of studies for the NH_3 decomposition have been carried out using Ru powder catalysts supported on various metal oxides, carbon materials, and so on. In addition, the effect of addition of alkali metal and alkaline earth metal to the supported Ru catalysts has also been examined. In this section, we have focused on the following three topics, which have been studied to improve the NH_3 decomposition activity on Ru, and summarized the previous reports: (i) catalyst preparation method, (ii) effect of support, and (iii) effect of promoter.

24.3.1 Catalyst Preparation Method

The effects of Ru precursor [34–36], support preparation method [37–41], and Ru supporting method [35, 42–47] on NH_3 decomposition activity have been investigated.

Yin et al. [34] examined the effect of Ru precursors (Ru acetylacetonate and $RuCl_3$) on NH_3 decomposition over the Ru/ZrO_2-KOH catalysts. They found that the Ru catalyst performs better when the Cl-free Ru precursor is adopted, which was due to direct as well as indirect effects of Cl on the electronic state of Ru. It has been reported that the Cs-Ru/C catalyst prepared using $Ru(NO)(NO_3)_3$ precursor showed higher H_2 formation rate for NH_3 decomposition compared with $RuCl_3$ precursor [35]. The catalyst prepared from $RuCl_3$ may have residual Cl^- on the catalyst, which reacts with NH_3 and produces NH_4Cl precipitation on the catalyst surface, resulting in the decrease of the NH_3 decomposition activity. Armenise et al. [36] prepared the Ru/Al_2O_3/monoliths catalyst employing three different Ru precursors, which are $RuCl_3$, $Ru(NO)(NO_3)_3$, and Ru acetylacetonate, by an equilibrium adsorption method. Among these Ru catalysts, the highest conversion and lowest E_a for NH_3 decomposition correspond to that prepared using $Ru(NO)(NO_3)_3$ as precursor. This catalyst exhibited the largest mean particle size (3.5 nm) and the highest ratio (69.1%) of $Ru^0/Ru_{(total)}$. The highest activity of Ru prepared from $(NO)(NO_3)_3$ was explained to be due to the higher proportion of B_5- type sites on the catalyst surface.

The carbon nanotubes (CNT) and the carbon nanofibers (CNF) in the NH_3 decomposition on the Ru catalyst are often used as an effective support. The NH_3 decomposition activity of Ru particles has been known to be significantly improved when supported on CNT and CNF doped with nitrogen [37–40]. The cause for the enhancement of the NH_3 decomposition activity has been explained to be the improvement of the basicity and the electronic density for the CNT support [37], the increase of the dispersion of Ru particles [38], the stabilization of small and uniform Ru metal nanoparticles [39], and the electronic modification for the Ru active sites by the increase of the conductivity and basicity of the CNT support [40]. Furthermore, it has been reported that the NH_3 decomposition activity of the Ru/La_2O_3 catalyst differs depending on the preparation method of the La_2O_3 support [41]. Among four kinds of La_2O_3 synthesized via the citric acid complex, KOH-K_2CO_3 coprecipitation, pyrolysis, and urea method, the Ru/La_2O_3 catalyst prepared using the citric acid complex showed the highest NH_3 conversion. It has been concluded that the formation of pure La_2O_3 phase is likely responsible for the high catalytic activity.

The supporting method of Ru onto MgO, which is known as one of the effective metal oxide support, has been studied. Ju et al. [42] examined the catalytic performance of NH_3 decomposition on Ru/MgO prepared by the deposition–precipitation (DP) and incipient wetness impregnation (IWI) methods. They demonstrated that the Ru/MgO-DP catalyst is highly active in comparison with Ru/MgO-IWI, which was attributed to unique mesoporous structure and abundant exposed Ru sites. We also investigated the influence of catalyst preparation method on NH_3 decomposition over Ru/MgO [43]. We revealed that the Ru/MgO catalysts prepared from DP and coprecipitation methods exhibited higher catalytic activity than that prepared from IWI method, and concluded that the formation of MgO mesopores of about 3–5 nm and highly dispersed Ru species improved the catalytic performance. In addition, a high NH_3 decomposition activity has been obtained by supporting highly dispersed Ru nanoparticles strongly interacted with the MgO support using polyol reduction [44] and modified colloidal deposition [45] methods. Furthermore, various supporting methods of Ru onto ZrO_2 [46], carbon powder [35], and graphene nanocomposite [47] have been evaluated, and suitable catalyst preparation methods of Ru for NH_3 decomposition reaction have been proposed.

24.3.2 *Effect of Support*

The examination of effective support for Ru catalysts has been performed by comparing the NH_3 decomposition activity using Ru catalysts supported on various metal oxides, carbon materials, mesoporous materials, and zeolites.

At first, for metal oxides support, it has been reported that the turnover frequency (TOF) for NH_3 decomposition was an order of magnitude higher on the Ru/Al_2O_3 catalyst, compared to the carbon-supported Ru catalyst, while E_a did not vary much (88–100 kJ/mol), where both NH_2–H bond cleavage and recombinative nitrogen desorption are slow kinetic steps, and nitrogen atoms are the most abundant reaction intermediate [48]. On the other hand, Choudhary et al. [49] demonstrated that the NH_3 decomposition activity per metal site for Ru/SiO_2 was greater than that of Ru/Al_2O_3, even though the dispersion of Ru is greater on the Al_2O_3 support. The E_a values, unlike the catalytic activity, were essentially the same for both Ru/Al_2O_3 and Ru/SiO_2 (79.5 and 81.6 kJ/mol), and the E_a value for Ru/Al_2O_3 was in good agreement with a kinetic study on an identical catalyst (88 kJ/mol) [48]. Concerning SiO_2 supports, it has been known to exhibit a high activity and high stability against the aggregation of Ru metal during the reaction by core-shelling (Ru@SiO_2) [50, 51]. The TOF values for H_2 formation of Ru@SiO_2 in the 400–600 °C range are shown to be essentially higher than those of supported Ru catalysts such as Ru/Al_2O_3 and K-Ru/MCM-41 [51]. La_2O_3 [41], CeO_2 [52, 53], MgO [42–45, 54], and Cr_2O_3 [55] have been also reported as effective metal oxide supports. Huang et al. [41] compared the NH_3 decomposition activities of Ru catalysts supported on various metal oxide supports. They demonstrated that the activities were in the order of Ru/La_2O_3 > Ru/Er_2O_3 > Ru/Al_2O_3 > Ru/SiO_2 > Ru/TiO_2. This result resonates with previous work that the activity of Ru particles correlates with the alkaline strength of the support [6]. Hu et al. [52] found that Ru supported on CeO_2 nanorods exhibited activity over eight times higher than Ru/CNT. This enhanced activity is believed to be related to a strong metal-support interaction promoting the formation of small (~3 nm) Ru particles and the electronic modification of the Ru active sites by CeO_2.

On the other hand, Ru catalysts using various kinds of carbon materials such as carbon black (CB), activated carbon (AC), graphitic carbon (GC), CNT, CNF, and graphene as supports have been also widely studied [6, 8, 37, 47, 56–62]. Among them, the CNT is shown to be the effective support compared to the conventional oxides. Yin et al. [58, 59] investigated the effect of support (CNT, AC, Al_2O_3, MgO, ZrO_2, TiO_2) on the catalysis of NH_3 decomposition, and found that the Ru catalyst using CNT as support shows the highest NH_3 conversion. In addition, they demonstrated that the dispersion of Ru on CNT is the highest among the examined support materials, and the particle size of Ru is the smallest (2–5 nm). They concluded that the excellent catalytic activity of Ru/CNT was ascribed to the high dispersion of Ru particles, the high graphitization, and the high purity of the CNT.

Several carbon-based materials have been reported to exhibit higher catalytic performance than CNT as a support for Ru catalysts. The NH_3 decomposition activity of Ru catalysts supported on different kinds of carbons was examined by Li et al. [60]. They showed that the GC is an excellent carbon support for Ru catalyst, and the activity follows this sequence: Ru/GC > Ru/CNT > Ru/synthesized CB > Ru/commercial CB > Ru/mesoporous carbon $\approx$ Ru/AC. It has been concluded that the graphitic structure of the carbon support is more important for the catalytic activity compared to the pore structures, and surface area of carbon is not a crucial factor. Xuezhi et al. [61] reported that the Ru nanoparticles on the CNF support are more active for NH_3 decomposition than those on CNT. They have explained that the peculiar surface edge structure on the CNF improves the electronic property of Ru. In addition, they found that the presence of oxygen groups on the CNF clearly enhances the catalytic activity over the Ru/CNFs, which was due to the promotion of the associative desorption of N atoms by the electron-withdrawing properties of the oxygen groups on the CNF. Yin et al. [62] demonstrated that the Ru catalyst supported by the MgO–CNT nanocomposite exhibits higher dispersion and higher catalytic activity for NH_3 decomposition as compared to that of Ru supported by MgO or CNT. The high activity is ascribed to a combined result of the high graphitization and high surface area of CNTs, basicity of MgO, and interaction between

CNT and MgO. In addition to metal oxides and carbon materials, mesoporous materials (MCM-41) [63] and Y-form zeolite [64] have also been reported to be effective supports in Ru catalysts for NH_3 decomposition.

24.3.3 Effect of Promoter

It is well known that the NH_3 decomposition activity is greatly improved by adding alkali metals such as Li [65, 66], Na [52, 67, 68], K [34, 41, 57–59, 63, 69, 70], and Cs [35, 44, 47, 71, 72] to the supported Ru catalyst.

Bajus et al. [65] reported that the modification of the Ru/Al_2O_3 catalyst with LiOH led to an increase in NH_3 decomposition activity by a factor of nearly 6.5 times compared to the unmodified catalyst, which was caused by the formation of basic aluminates by support transformation with the LiOH modifier. García-García et al. [67] showed that the addition of Na to the Ru/AC catalyst promoted the NH_3 decomposition activity. This enhanced catalytic activity is explained to be related with the formation of B5 surface sites, which consist of an arrangement of three Ru atoms in one layer and two further Ru atoms in an internal layer. The B5 surface sites are presented for average Ru particles in the order 3–5 nm, as well as electronic modifications of the Ru nanoparticles. In addition, Hu et al. [52] reported that the addition of Na to the Ru catalyst as a promoter leads to the formation of smaller Ru particle sizes in addition to the modification of the electronic environment of Ru, enhancing the NH_3 decomposition activity at low temperatures. This promotional effect of Na is particularly enhanced when using CNT as support due to their high conductivity, allowing distant electronic modification of the Ru active sites.

Yin et al. [34, 58, 59] demonstrated that the modification of CNT and ZrO_2 with KOH results in a significant enhancement in the catalytic activity of supported Ru catalyst. The investigation on the relationship between support basicity and catalytic activity disclosed that a support material of strong basicity is essential for high catalytic performance. Furthermore, from the N_2-TPD studies of supported Ru catalysts (Fig. 24.3), the desorption was shown to be promoted over catalysts of strong basicity, suggesting that N_2 desorption is the rate-determining step in NH_3 decomposition. They have concluded that it is possible to develop a highly efficient Ru catalyst for NH_3 decomposition by using electron-conductive materials of strong basicity as support. Hill and Torrente-Murciano [71, 72] examined the influence of Cs addition to the Ru/CNT catalyst on NH_3 decomposition. They revealed that the low-temperature activity is significantly improved as the Cs loading increases, reducing the E_a from 96.7 kJ/mol in the absence of Cs to 59.3 kJ/mol with a Cs/Ru molar ratio of 3. The catalytic enhancement is shown to be due to the electronic modification of Ru by the electron-donating Cs promoter located on the Ru surface and in close proximity on the CNT surface.

In addition to alkali metals, alkaline earth metals are also reported to promote the activity for NH_3 decomposition, and comparison with the promotional effect of alkali metals has been carried out. The effect of Ba and Cs addition has been investigated

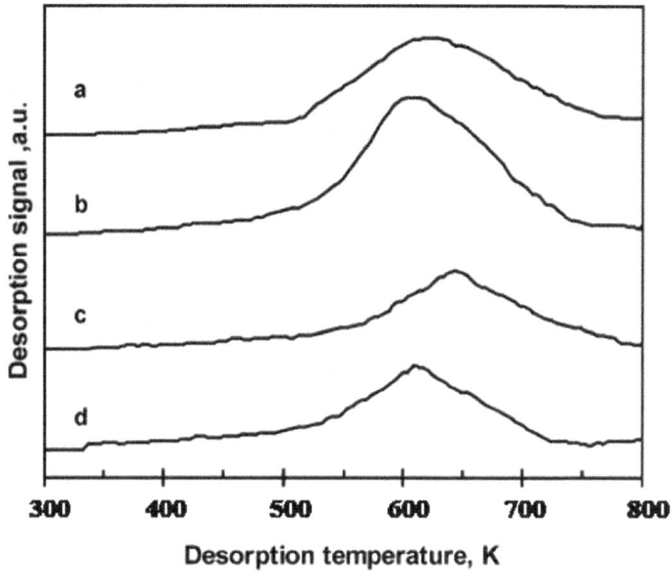

Fig. 24.3 N_2-TPD curves over Ru catalysts (**a** Ru/CNTs; **b** Ru-K/CNT; **c** Ru/ZrO_2-BD; **d** Ru-K/ZrO_2-BD). Published with permission from Ref. [58]

using Ru/C [73, 74] and Ru/$MgAl_2O_4$ [75] catalysts, and it has been shown that Cs is more active promoter than Ba in both catalysts. The N_2-TPD studies revealed that the amount of N_2 desorbed from Cs-Ru/C was much larger and the peak position was shifted to lower temperature when compared to Ba–Ru/C [73]. On the other hand, an improvement in NH_3 decomposition activity and thermal stability for the Ru/C catalyst has been observed by adding both Ba and Cs [56].

The modification of Ru/CNT with rare earth, alkali, and alkaline earth compounds was systematically studied by Wang et al. [76]. They showed that the modification does not change the size and morphology of the particles of metallic Ru, but affects the electronic property of Ru. For the catalysts modified by metal nitrates, activities were found in the order of K–Ru > Na–Ru > Li–Ru > Ce–Ru > Ba–Ru > La–Ru > Ca–Ru > Ru, signifying that within the same groups (K, Na, and Li; Ba and Ca), the higher the electronegativity of the promoter, the lower is the NH_3 conversion. They also demonstrated that addition of the electron-withdrawing groups, F^{-1}, Cl^{-1}, Br^{-1}, SO_4^{2-}, and PO_4^{3-} are inhibitors of Ru catalysts. From the results of N_2-TPD investigation, they concluded that the promotional effect of a modifier is a combined result of: (i) enhancing combinative desorption of nitrogen atoms and (ii) decreasing of the apparent E_a of the decomposition reaction. Li et al. [77] also examined the additional effect of alkali or earth alkali metals on NH_3 decomposition over Ru/C and demonstrated that K is the most effective modifier for the Ru/C catalyst, which was consistent with above report [76]. On the other hand, Nagaoka et al. [78] prepared Ru/Pr_6O_{11} catalysts doped with various kinds of basic oxides such as alkali metal oxides, alkaline earth metal oxides, and rare earth oxides, and found that Ru/Pr_6O_{11}

doped with Cs_2O, which is the strongest base, showed the highest NH_3 conversion among the catalysts, which was ascribed to the effect of electron donation to Ru.

24.4 Conclusions

In this review, we summarized the previous reports for the NH_3 decomposition on Ru catalyst. So far, various studies such as the influence of the support and the addition effect of the promoter have been conducted on Ru, which is the most active metal in the NH_3 decomposition reaction. As a result, it has been found that the activity and stability of the Ru catalyst are remarkably enhanced by optimization of the Ru supporting method, improvement of the CNT support, addition of alkali metals, and so on. However, for the utilization of H_2 to be transported by NH_3 and produced on-demand, further improvements of both economics and performance of currently developed Ru-based catalysts would be required in future.

Acknowledgements This work was partially supported by the Cross-ministerial Strategic Innovation Promotion Program (SIP) of the Cabinet Office, Government of Japan.

References

1. Sherif SA, Barbir F, Veziroglu TN (2005) Towards a hydrogen economy. Electr J 18:62–76
2. Lan R, Irvine JTS, Tao S (2012) Ammonia and related chemicals as potential indirect hydrogen storage materials. Int J Hydrogen Energy 37:1482–1494
3. Orimo S, Nakamori Y, Eliseo JR, Züttel A, Jensen CM (2007) Complex hydrides for hydrogen storage. Chem Rev 107:4111–4132
4. Yu WY, Mullen GM, Flaherty DW, Mullins CB (2014) Selective hydrogen production from formic acid decomposition on Pd–Au bimetallic surfaces. J Am Chem Soc 136:11070–11078
5. Metkemeijer R, Achard P (1994) Comparison of ammonia and methanol applied indirectly in a hydrogen fuel cell. Int J Hydrogen Energy 19:535–542
6. Yin SF, Xu BQ, Zhou XP, Au CT (2004) A mini-review on ammonia decomposition catalysts for on-site generation of hydrogen for fuel cell applications. Appl Catal A 277:1–9
7. Schüth F, Palkovits R, Schlögl R, Su DS (2012) Ammonia as a possible element in an energy infrastructure: catalysts for ammonia decomposition. Energy Environ Sci 5:6278–6289
8. Duan X, Zhou X, Chen D (2013) Structural manipulation of the catalysts for ammonia decomposition. Catalysis 25:118–140
9. Lamb KE, Dolan MD, Kennedy DF (2019) Ammonia for hydrogen storage; a review of catalytic ammonia decomposition and hydrogen separation and purification. Int J Hydrogen Energy 44:3580–3593
10. Benndorf C, Madey TE (1983) Adsorption and orientation of NH_3 on Ru(001). Surf Sci 135:164–183
11. Parmeter JE, Wang Y, Mullins CB, Weinberg WH (1988) Electron energy loss spectroscopy of ammonia on Ru(001). J Chem Phys 88:5225–5236
12. Rodriguez JA, Kuhn WK, Truong CM, Goodman DW (1992) A FT-IRAS study of ammonia adsorbed on Ru(0001). Surf Sci 271:333–339

13. Widdra W, Moritz T, Kostov KL, König P, Staufer M, Birkenheuer U (1999) Identification of a vibrational Stark shift within an adsorbate layer: NH_3 on Ru(001). Surf Sci 430:L558–L564
14. Maier S, Stass I, Cerda JI, Salmeron M (2012) Bonding of ammonia and its dehydrogenated fragments on Ru(0001). J Phys Chem C 116:25395–25400
15. Tautermann CS, Wellenzohn B, Clary DC (2006) The thermodesorption mechanism of ammonia from Ru(0001). Surf Sci 600:1054–1059
16. Herron JA, Tonelli S, Mavrikakis M (2013) Atomic and molecular adsorption on Ru(0001). Surf Sci 614:64–74
17. Hu X, Yang M, Xie D, Guo H (2018) Vibrational enhancement in the dynamics of ammonia dissociative chemisorption on Ru(0001). J Chem Phys 149:044703
18. Neyman KM, Staufer M, Nasluzov VA, Rösch N (1997) Structure and vibrations of adsorption complexes NH_3/Ru(001): density functional model cluster studies. J Mol Catal A 119:245–251
19. Dietrich H, Jacobi K, Ertl G (1996) Coverage, lateral order, and vibrations of atomic nitrogen on Ru(0001). J Chem Phys 105:8944–8950
20. Dietrich H, Jacobi K, Ertl G (1997) Vibrational analysis of the NH + N coadsorbate on Ru(0001). Surf Sci 377–379:308–312
21. Carabineiro SAC, Matveev AV, Gorodetskii VV, Nieuwenhuys BE (2004) Selective oxidation of ammonia over Ru(0001). Surf Sci 555:83–93
22. Egawa C, Naito S, Tamaru K (1984) Adsorption and decomposition of NO and NH_3 on Ru(001) and Ru(1,1,10) surfaces. Surf Sci 138:279–291
23. Egawa C, Nishida T, Naito S, Tamaru K (1984) Ammonia decomposition on (1 1 10) and (001) surfaces of Ruthenium. J Chem Soc Faraday Trans 1(80):1595–1604
24. Dietrich H, Jacobi K, Ertl G (1996) Decomposition of NH_3 on Ru($11\bar{2}1$). Surf Sci 352–354:138–141
25. Jacobi K, Wang Y, Fan CY, Dietrich H (2001) Adsorption and thermal dehydrogenation of ammonia on Ru($11\bar{2}1$). J Chem Phys 115:4306–4313
26. Wang Y, Jacobi K (2004) Adsorption and reaction of ammonia on the Ru($11\bar{2}0$) surface. J Phys Chem B 108:14726–14732
27. Liu S, Hao C, Li S, Wang Z (2009) Adsorption characteristics of atomic nitrogen on ruthenium surfaces. Appl Surf Sci 255:4232–4238
28. Benndorf C, Madey TE (1983) Interaction of NH_3 with adsorbed oxygen and sodium on Ru(001): Evidence for both local and long-range interactions. Chem Phys Lett 101:59–64
29. Sun YK, Wang YQ, Mullins CB, Weinberg WH (1991) Interaction of ammonia with chemically modified Ru(001) surfaces. Langmuir 7:1689–1694
30. Huang HH, Seet CS, Xu GQ (1994) Decomposition of ammonia on O- and S-covered Ru(001) surfaces. Surf Sci 317:353–360
31. Sasaki T, Aruga T, Kuroda H, Iwasawa Y (1989) Coadsorption of NH_3 and CO on Ru(001): the ordering in mixed layers and the effect of intermolecular interactions on NH_3 dissociation. Surf Sci Lett 224:L969–L978
32. Sasaki T, Aruga T, Kuroda H, Iwasawa Y (1990) Interaction between CO and NH_3 coadsorbed on Ru(001): its effects on the ordering in mixed adlayers and the ammonia dissociation. Surf Sci 240:223–244
33. Rodriguez JA, Truong CM, Goodman DW (1992) Synthesis of boron nitride ultrathin films: The bonding and chemistry of ammonia and hydrazine on Ru(0001) and B/Ru(0001) surfaces. J Vac Sci Technol A 10:955–959
34. Yin SF, Xu BQ, Wang SJ, Au CT (2006) Nanosized Ru on high-surface-area superbasic ZrO_2-KOH for efficient generation of hydrogen via ammonia decomposition. Appl Catal A 301:202–210
35. Huang DC, Jiang CH, Liu FJ, Cheng YC, Chen YC, Hsueh KL (2013) Preparation of Ru–Cs catalyst and its application on hydrogen production by ammonia decomposition. Int J Hydrogen Energy 38:3233–3240
36. Armenise S, Cazaña F, Monzón A, García-Bordejé E (2018) In situ generation of CO_x-free H_2 by catalytic ammonia decomposition over Ru-Al-monoliths. Fuel 233:851–859

37. García-García FR, Álvarez-Rodríguez J, Rodríguez-Ramos I, Guerrero-Ruiz A (2010) The use of carbon nanotubes with and without nitrogen doping as support for ruthenium catalysts in the ammonia decomposition reaction. Carbon 48:267–276
38. Chen J, Zhu ZH, Wang S, Ma Q, Rudolph V, Lu GQ (2010) Effects of nitrogen doping on the structure of carbon nanotubes (CNTs) and activity of Ru/CNTs in ammonia decomposition. Chem Eng J 156:404–410
39. Armenise S, Roldán L, Marco Y, Monzón A, García-Bordejé E (2012) Elucidation of catalyst support effect for NH_3 decomposition using Ru nanoparticles on nitrogen-functionalized carbon nanofiber monoliths. J Phys Chem C 116:26385–26395
40. Bell TE, Zhan G, Wu K, Zeng HC, Torrente-Murciano L (2017) Modification of ammonia decomposition activity of ruthenium nanoparticles by N-doping of CNT supports. Top Catal 60:1251–1259
41. Huang C, Yu Y, Yang J, Yan Y, Wang D, Hu F, Wang X, Zhang R, Feng G (2019) Ru/La_2O_3 catalyst for ammonia decomposition to hydrogen. Appl Surf Sci 476:928–936
42. Ju X, Liu L, Yu P, Guo J, Zhang X, He T, Wu G (2017) Mesoporous Ru/MgO prepared by a deposition-precipitation method as highly active catalyst for producing CO_x-free hydrogen from ammonia decomposition. Appl Catal B 211:167–175
43. Fujitani T, Nakamura I, Hashiguchi Y, Kanazawa S, Takahashi A (2020) Effect of catalyst preparation method on ammonia decomposition reaction over Ru/MgO catalyst. Bull Chem Soc Japan 93:1186–1192
44. Zhang J, Xu H, Ge Q, Li W (2006) Highly efficient Ru/MgO catalysts for NH_3 decomposition: Synthesis, characterization and promoter effect. Catal Commun 7:148–152
45. Hu XC, Wang WW, Si R, Ma C, Jia CJ (2019) Hydrogen production via catalytic decomposition of NH_3 using promoted MgO-supported ruthenium catalysts. Sci China Chem 62:1625–1633
46. Furusawa T, Shirasu M, Sugiyama K, Sato T, Itoh N, Suzuki N (2016) Preparation of Ru/ZrO_2 catalysts by $NaBH_4$ reduction and their catalytic activity for NH_3 decomposition to produce H_2. Ind Eng Chem Res 55:12742–12749
47. Li G, Nagasawa H, Kanezashi M, Yoshioka T, Tsuru T (2014) Graphene nanosheets supporting Ru nanoparticles with controlled nanoarchitectures form a high-performance catalyst for CO_x-free hydrogen production from ammonia. J Mater Chem A 2:9185–9192
48. Bradford MCJ, Fanning PE, Vannice MA (1997) Kinetics of NH_3 decomposition over well dispersed Ru. J Catal 172:479–484
49. Choudhary TV, Sivadinarayana C, Goodman DW (2001) Catalytic ammonia decomposition: CO_x-free hydrogen production for fuel cell applications. Catal Lett 72:197–201
50. Yao LH, Li YX, Zhao J, Ji WJ, Au CT (2010) Core–shell structured nanoparticles (M@SiO_2, Al_2O_3, MgO; M= Fe Co, Ni, Ru) and their application in COx-free H_2 production via NH_3 decomposition. Catal Today 158:401–408
51. Yao L, Shi T, Li Y, Zhao J, Ji W, Au CT (2011) Core–shell structured nickel and ruthenium nanoparticles: very active and stable catalysts for the generation of CO_x-free hydrogen via ammonia decomposition. Catal Today 164:112–118
52. Hu Z, Mahin J, Datta S, Bell TE, Torrente-Murciano L (2019) Ru-based catalysts for H_2 production from ammonia: effect of 1D support. Top Catal 62:1169–1177
53. Hu XC, Fu XP, Wang WW, Wang X, Wu K, Si R, Ma C, Jia CJ, Yan CH (2020) Ceria-supported ruthenium clusters transforming from isolated single atoms for hydrogen production via decomposition of ammonia. Appl Catal B 268:118424
54. Takahashi A, Fujitani T (2016) Kinetic analysis of decomposition of ammonia over nickel and ruthenium catalysts. J Chem Eng Japan 49:22–28
55. Li L, Wang Y, Xu ZP, Zhu Z (2013) Catalytic ammonia decomposition for CO-free hydrogen generation over Ru/Cr_2O_3 catalysts. Appl Catal A 467:246–252
56. Iost KN, Borisov VA, Temerev VL, Smirnova NS, Surovikin YV, Trenikhin MV, Arbuzov AB, Gulyaeva TI, Shlyapin DA, Tsyrulnikov PG, Vedyagin AA (2019) Effect of the carbon support graphitization on the activity and thermal stability of Ru-Ba-Cs/C ammonia decomposition catalysts. React Kinet Mech Catal 127:85–102

57. Yin SF, Xu BQ, Ng CF, Au CT (2004) Nano Ru/CNTs: a highly active and stable catalyst for the generation of CO_x-free hydrogen in ammonia decomposition. Appl Catal B 48:237–241
58. Yin SF, Zhang QH, Xu BQ, Zhu WX, Ng CF, Au CT (2004) Investigation on the catalysis of CO_x-free hydrogen generation from ammonia. J Catal 224:384–396
59. Yin SF, Xu BQ, Zhu WX, Ng CF, Zhou XP, Au CT (2004) Carbon nanotubes-supported Ru catalyst for the generation of CO_x-free hydrogen from ammonia. Catal Today 93–95:27–38
60. Li L, Zhu ZH, Yan ZF, Lu GQ, Rintoul L (2007) Catalytic ammonia decomposition over Ru/carbon catalysts: the importance of the structure of carbon support. Appl Catal A 320:166–172
61. Xuezhi D, Jinghong Z, Gang Q, Ping L, Xinggui Z, De C (2010) Carbon nanofiber-supported Ru catalysts for hydrogen evolution by ammonia decomposition. Chin J Catal 31:979–986
62. Yin SF, Xu BQ, Wang SJ, Ng CF, Au CT (2004) Magnesia–carbon nanotubes (MgO–CNTs) nanocomposite: novel support of Ru catalyst for the generation of CO_x-free hydrogen from ammonia. Catal Lett 96:113–116
63. Li XK, Ji WJ, Zhao J, Wang SJ, Au CT (2005) Ammonia decomposition over Ru and Ni catalysts supported on fumed SiO_2, MCM-41, and SBA-15. J Catal 236:181–189
64. Hashimoto K, Toukai N (2000) Decomposition of ammonia over a catalyst consisting of ruthenium metal and cerium oxides supported on Y-form zeolite. J Mol Catal A 161:171–178
65. Bajus S, Agel F, Kusche M, Bhriain NN, Wasserscheid P (2016) Alkali hydroxide-modified $Ru/\gamma\text{-}Al_2O_3$ catalysts for ammonia decomposition. Appl Catal A 510:189–195
66. Lamb K, Hla SS, Dolan M (2019) Ammonia decomposition kinetics over LiOH-promoted, $\alpha\text{-}Al_2O_3$-supported Ru catalyst. Int J Hydrogen Energy 44:3726–3736
67. García-García FR, Guerrero-Ruiz A, Rodríguez-Ramos I (2009) Role of B5-type sites in Ru catalysts used for the NH_3 decomposition reaction. Top Catal 52:758–764
68. García-García FR, Guerrero-Ruiz A, Rodríguez-Ramos I, Goguet A, Shekhtman SO, Hardacre C (2011) TAP studies of ammonia decomposition over Ru and Ir catalysts. Phys Chem Chem Phys 13:12892–12899
69. Raróg W, Kowalczyk Z, Sentek J, Składanowski D, Szmigiel D, Zieliñski J (2001) Decomposition of ammonia over potassium promoted ruthenium catalyst supported on carbon. Appl Catal A 208:213–216
70. Pyrz W, Vijay R, Binz J, Lauterbach J, Buttrey DJ (2008) Characterization of K-promoted Ru catalysts for ammonia decomposition discovered using high-throughput experimentation. Top Catal 50:180–191
71. Hill AK, Torrente-Murciano L (2014) In-situ H_2 production via low temperature decomposition of ammonia: Insights into the role of cesium as a promoter. Int J Hydrogen Energy 39:7646–7654
72. Hill AK, Torrente-Murciano L (2015) Low temperature H_2 production from ammonia using ruthenium-based catalysts: synergetic effect of promoter and support. Appl Catal B 172–173:129–135
73. Raróg-Pilecka W, Szmigiel D, Kowalczyk Z, Jodzis S, Zielinski J (2003) Ammonia decomposition over the carbon-based ruthenium catalyst promoted with barium or cesium. J Catal 218:465–469
74. Sørensen RZ, Klerke A, Quaade U, Jensen S, Hansen O, Christensen CH (2006) Promoted Ru on high-surface area graphite for efficient miniaturized production of hydrogen from ammonia. Catal Lett 112:77–81
75. Szmigiel D, Raróg-Pilecka W, Miśkiewicz E, Kaszkur Z, Kowalczyk Z (2004) Ammonia decomposition over the ruthenium catalysts deposited on magnesium–aluminum spinel. Appl Catal A 264:59–63
76. Wang SJ, Yin SF, Li L, Xu BQ, Ng CF, Au CT (2004) Investigation on modification of Ru/CNTs catalyst for the generation of CO_x-free hydrogen from ammonia. Appl Catal B 52:287–299
77. Li L, Zhu ZH, Lu GQ, Yan ZF, Qiao SZ (2007) Catalytic ammonia decomposition over CMK-3 supported Ru catalysts: effects of surface treatments of supports. Carbon 45:11–20
78. Nagaoka K, Eboshi T, Abe N, Miyahara S, Honda K, Sato K (2014) Influence of basic dopants on the activity of Ru/Pr_6O_{11} for hydrogen production by ammonia decomposition. Int J Hydrogen Energy 39:20731–20735

Chapter 25
Production of High-Purity Hydrogen Through Ammonia Decomposition and Gas Removal

Takayoshi Adachi

Abstract Ammonia gas has potential as an energy carrier because it is easy to liquefy and transport and has a very high hydrogen density. Therefore, it could be applied to the development of technology to produce fuel hydrogen for fuel cell vehicles by thermally decomposing the ammonia for hydrogen extraction. International standards for hydrogen for fuel cell vehicles require high-purity hydrogen. To achieve this, ammonia decomposition gas (75% H_2 + 25% N_2) was generated using a pressure swing adsorption method, which reduced nitrogen concentration to 10 ppm or less and methane concentration to 0.1 ppm or less. A highly efficient refining technology also was developed that increased the hydrogen recovery rate to 90% or more [1].

Keywords Ammonia decomposition · Hydrogen purification · Pressure swing · Adsorption · Fuel cell vehicles

25.1 Background

In Japan, fuel cell vehicles (FCVs) have been on the market since 2014, and hydrogen stations are being installed to supply the vehicles. Two types of hydrogen stations exist: an off-side type that transports hydrogen produced at the factory to fill the FCV and an on-site type that produces the hydrogen by steam reforming from city gas at the station to fill the FCV [2]. In the on-site type, a PSA hydrogen purification device is installed, and the hydrogen is loaded into the FCV after refining to FCV standards.

The use of ammonia as a hydrogen carrier in Japan was examined by the Energy Carriers of the Cabinet Office Strategic Innovation Creation Program (SIP) Phase 1 (2014–2018). The SIP promotes the development of technology to extract hydrogen from ammonia and use it for FCVs and fuel cell forklifts (FCFL) along with the direct combustion technology of ammonia and ammonia fuel cells. The technology for reforming ammonia into high-purity hydrogen was established through this project.

T. Adachi (✉)
Gas Separation Development Department, Yamanashi Technology Solution Center, R & D Unit, TAIYO NIPPON SANSO Corporation, 3054-3, Shimokurosawa, Takane-cho, Hokuto-shi, Yamanashi 408-0015, Japan
e-mail: adachit.qgl@tn-sanso.co.jp

K. Aika and H. Kobayashi (eds.), *CO_2 Free Ammonia as an Energy Carrier*,
https://doi.org/10.1007/978-981-19-4767-4_25

Table 25.1 Fuel cell vehicle (FCV) hydrogen specifications

	FCV Hydrogen specifications ISO14687-2 (ppm)
Total hydrocarbon	<2
H_2O	<5 (DP<−66 °C)
O_2	<5
He	<300
$Ar + N_2$	<100
CO_2	<2
CO	<0.2
Sulfur	<0.004
HCHO	<0.01
HCOOH	<0.2
NH_3	<0.1
Halogen	<0.05

25.2 Hydrogen for Fuel Cell Vehicles: Standard of Hydrogen

The fuel cell used in FCVs and FCFLs is a polymer electrolyte fuel cell (PEFC). Therefore, electricity is easy to start and stop and can be generated at a lower temperature than that used by other types of fuel cells. However, these fuel cells are easily affected by impurities in the hydrogen [3]. Therefore, the hydrogen used for FCVs must meet strict purity standards. Table 25.1 shows the international standards for hydrogen used for FCVs.

Hydrogen for FCVs must maintain a level of purity that is not affected by impurities even after 5000 h of operation; the limit for NH_3 is 0.1 ppm, along with limits of 0.2 ppm for CO and 0.004 ppm for H_2S. The inert gases Ar and N_2 do not affect PEFC adversely but are limited to 100 ppm due to problems with the FCV system and errors in measurement at hydrogen stations [4].

25.3 Ammonia Decomposition Gas

The ammonia decomposition gas is called AX gas and is used for bright annealing treatment of stainless steel and other purposes. The AX gas is synthesized by thermally decomposing ammonia with a nickel catalyst, which requires a high temperature of about 700 °C and an electric furnace.

However, to use ammonia as an energy (hydrogen) carrier, the energy efficiency of decomposition must be increased, and the ammonia must be decomposed at low temperature and with high efficiency. For this reason, a ruthenium catalyst with a

Table 25.2 Analysis of ammonia cracked gas and FCV hydrogen

	Cracked gas from ammonia (ppm)	FCV hydrogen specifications ISO14687-2 (ppm)
Total hydrocarbon	**6**	<2
H_2O	N.A	<5 (DP <−66 °C)
O_2	<1	<5
He	<100	<300
$Ar + N_2$	**25%**	<100
CO_2	<1	<2
CO	<0.2	<0.2
Sulfur	<0.004	<0.004
HCHO	<0.01	<0.01
HCOOH	<0.02	<0.2
NH_3	**870**	<0.1
Halogen	<0.02	<0.05

decomposition capability close to chemical equilibrium at about 500 °C has been developed for SIP [5].

For purifying the ammonia decomposition gas, ammonia was decomposed with a newly developed ruthenium catalyst, and the impurity components in the decomposition gas were analyzed. Results are shown in Table 25.2, with the FCV hydrogen standard in the right column for comparison. Components that do not meet the FCV hydrogen fuel standard are shown **in bold**. Nitrogen concentration was **25%**, and ammonia concentration was **870** ppm, which is nearly the chemical equilibrium value under the reaction conditions. In addition, **6** ppm of methane was found, likely because this ammonia was synthesized by the Haber process and the methane may be derived from the raw material.

The FCV hydrogen standard is strict in that many impurities must be reduced to very low levels. However, the hydrogen derived from ammonia decomposition gas cannot meet the standard for nitrogen, ammonia, and methane.

25.4 Development of Hydrogen Purification Method

As mentioned above, hydrogen from ammonia decomposition requires removal of nitrogen, ammonia, and methane to reach low concentrations. Zeolite is used as a nitrogen remover in the oxygen PSA device that produces oxygen from air, and oxygen can be produced separately by adsorbing nitrogen in the air. Since zeolite adsorbs very little hydrogen, the hydrogen can be purified by PSA using zeolite as in oxygen PSA. In addition, zeolite generally has stronger adsorptive power for methane compared to nitrogen and can remove methane at the same time as nitrogen.

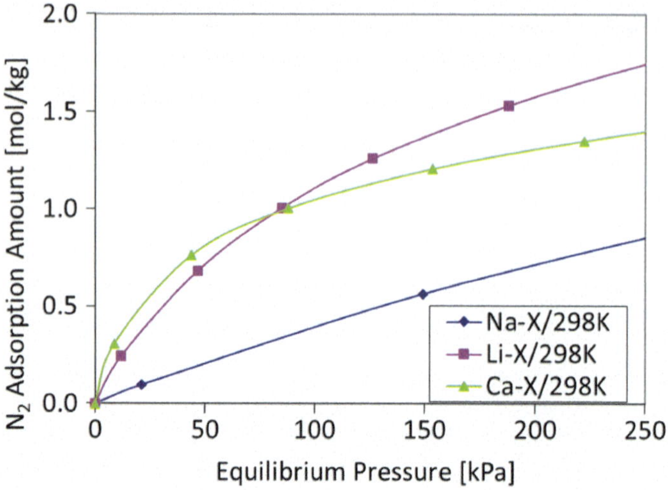

Fig. 25.1 Adsorption isotherms of nitrogen for an X-type zeolite (298 K)

Recently, X-type zeolites that have been ion-exchanged with lithium or calcium, which have a particularly large amount of nitrogen adsorbed, have been developed. These are used for oxygen PSA and also can be applied to hydrogen PSA [6].

Figure 25.1 shows adsorption isotherms of nitrogen of general sodium X-type zeolite and X-type zeolite ion-exchanged with lithium or calcium.

Adsorption isotherms show the amount of gas adsorbed as a function of the pressure of the adsorbed gas. A large amount of nitrogen adsorption is observed for every zeolite. Among them the Li-X type zeolite is suitable for the hydrogen PSA device when the adsorption partial pressure is 100 kPa or greater.

Another component that needs to be removed from the hydrogen is ammonia, which is strongly adsorbed to the zeolite. A SIP study showed that an X-type zeolite, ion-exchanged with lithium or calcium and with the nitrogen removed, was optimal [7]. However, since ammonia is strongly adsorbed to the zeolite, heat is needed to desorb any adsorbed ammonia. Therefore, the ammonia cannot be removed by PSA at the same time as nitrogen and must be removed using a thermal swing adsorption (TSA) method. For the ammonia decomposition gas, then, first the ammonia is removed with a TSA refiner, followed by removal of the nitrogen and methane using a PSA refiner.

During gas purification by PSA, exhaust gas is generated, and hydrogen is discharged. Since the discharge of hydrogen causes a loss of energy, a system was devised that utilizes the heat generated by reacting this exhaust gas with air over a catalyst to decompose the ammonia. The schematic flow from the decomposition of ammonia to the purification is shown in Fig. 25.2.

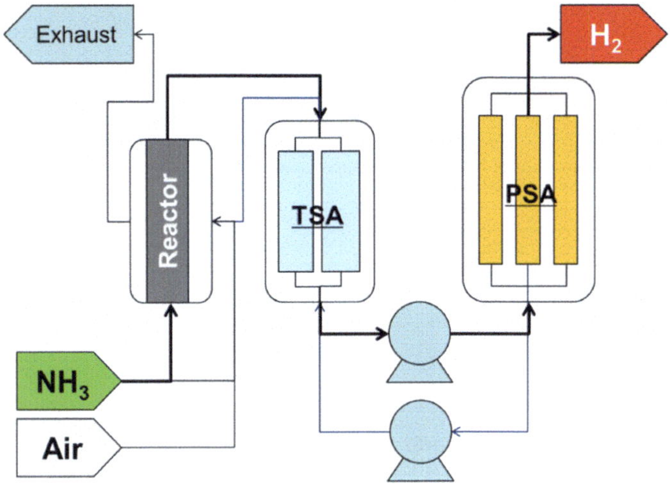

Fig. 25.2 Ammonia decomposition and purification system

25.5 Development of PSA System

25.5.1 Removal of Nitrogen and Methane Using PSA

A simple two-tower vacuum regeneration PSA (also called VPSA) was used to perform a removal test on hydrogen containing 25% nitrogen and 1000 ppm using a Li-X type zeolite as an adsorbent. Results indicated that the nitrogen concentration could be reduced to 100 ppm or less and the methane concentration to 1 ppm or less with a hydrogen recovery rate of 80% or more (Fig. 25.3), which confirmed that methane concentration can be lowered to 1 ppm or less under the PSA operating conditions that also reduce nitrogen gas concentration to 100 ppm or less.

In addition, since the heat of decomposition reaction of ammonia can be supplied by burning 30% of the hydrogen decomposed from ammonia, the performance target of the PSA device was set to a gas recovery efficiency of 70% or more.

25.5.2 Development of a Multi-column PSA System

An experimental multi-column PSA hydrogen purification device (Fig. 25.4) with a raw material gas scale of 2 Nm^3/h was manufactured, and the PSA process was examined.

Figure 25.5 shows the results of the process using the multi-tower PSA device. Results confirmed that an increase in the number of adsorption towers improved hydrogen recovery rate and hydrogen purification capacity. Furthermore,

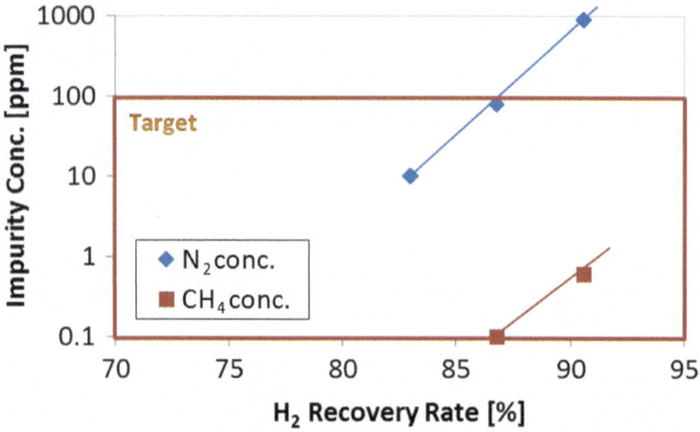

Fig. 25.3 Results of the two-column PSA system test

Fig. 25.4 Experimental multi-tower PSA hydrogen purification device

the hydrogen recovery rate could be improved to 90% using a four-tower PSA process in which the pressure equalization process was provided in two stages (4 tower PSA 2Eq).

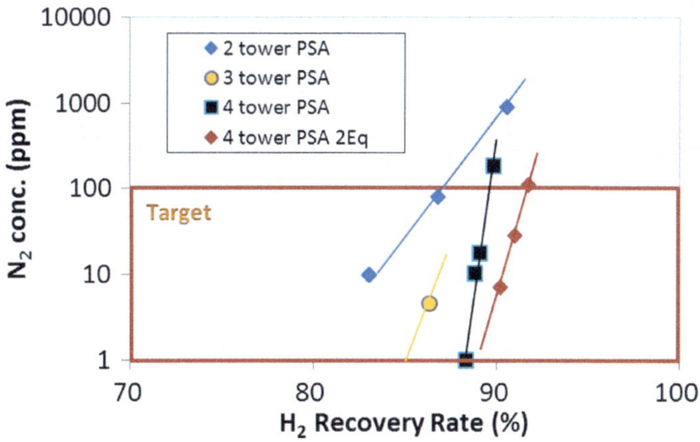

Fig. 25.5 Results of multi-column PSA hydrogen purification test

25.6 PSA Hydrogen Purification Pilot Test

25.6.1 Development of a PSA Hydrogen Purification Pilot System

Then, the PSA hydrogen purification device (Fig. 25.6) capable of purifying 20 Nm^3/h of ammonia decomposition gas as a raw material was manufactured, and a pilot test was conducted.

Fig. 25.6 PSA hydrogen purification pilot device

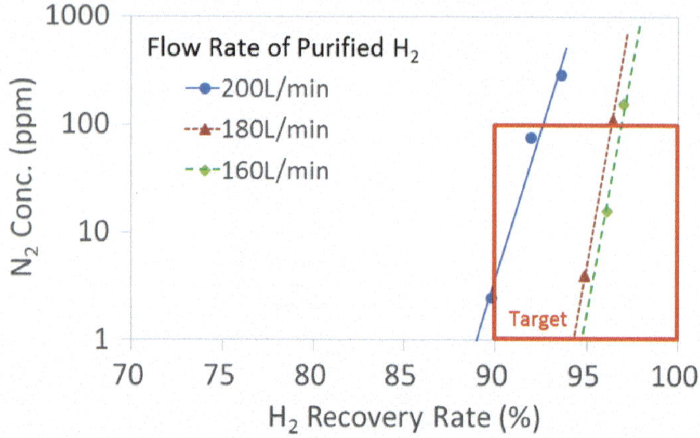

Fig. 25.7 Results of PSA hydrogen purification pilot test

A hydrogen purification test was conducted using a simulated ammonia decomposition gas (75% hydrogen + 25% nitrogen) in a PSA process that adopted four-tower two-stage pressure equalization. Results are shown in Fig. 25.7.

To completely decompose ammonia, 30% of the generated hydrogen was burned. However, further studies showed that the total thermal efficiency could be improved by increasing the recovery rate of purified hydrogen and supplementing the insufficient amount of heat with ammonia. Therefore, the target value for hydrogen recovery rate was improved to 90%.

Generally, as the PSA device is scaled up and the adsorption tower becomes larger, the influence in heat due to adsorption/desorption increases, and the separation performance tends to decrease. For this reason, a decrease in separation ability was observed in this pilot test. However, by suppressing the flow rate of the raw material gas, a hydrogen recovery rate of 90% or more could be achieved. Results also demonstrated that the hydrogen recovery rate could be improved up to 95% by further reducing the flow rate of the raw material gas.

25.6.2 Development of PSA Exhaust Gas Supply System

In this system, PSA exhaust gas is used for the heat of reaction of ammonia decomposition, so catalytic combustion must be done with air. However, the flow rate and hydrogen concentration of the PSA exhaust gas fluctuate widely, so that the reaction temperature fluctuates, and ammonia does not decompose adequately.

Therefore, the PSA device was redesigned as shown in Fig. 25.8. The PSA exhaust gas is collected in a tank, boosted, and then introduced into a catalytic combustion device. Through this design, the hydrogen concentration is stabilized by recycling a part of the pressurized gas to the tank.

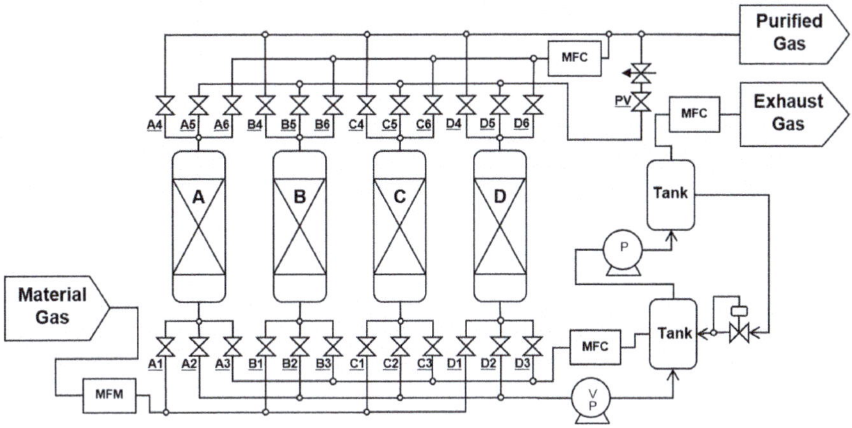

Fig. 25.8 Flow diagram of PSA hydrogen purification pilot device

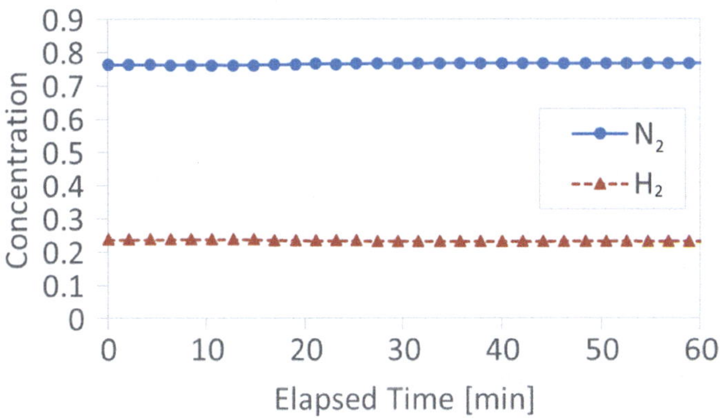

Fig. 25.9 Change in exhaust gas concentration of pilot test device

This system allowed the stable introduction of PSA exhaust gas into the catalyst combustion system at a constant hydrogen concentration. Figure 25.9 shows the changes in the concentration of hydrogen and nitrogen in the PSA exhaust gas. Results confirmed that very little or no fluctuation in hydrogen concentration [8].

25.7 Summary

Technology was developed to produce hydrogen for fuel cell vehicles (FCVs) from ammonia decomposition gas. The process was able to generate high-purity hydrogen with a hydrogen recovery rate of at least 90%. In addition, burning exhaust gas from pressure swing adsorption with a catalyst and using the heat for the ammonia decomposition reaction allows the generation of hydrogen with high energy efficiency.

In Japan, the technology for using ammonia as an energy carrier is being researched at SIP, and attempts to import renewable energy from overseas using ammonia are being considered.

This technology shows promise for realizing renewable energy from ammonia, which could allow operation of FCVs with true carbon-free hydrogen.

Acknowledgements This research was developed by the "Ammonia Hydrogen Station Basic Technology" of the Cabinet Office's Strategic Innovation Creation Program (SIP) energy carrier and is the result of research commissioned by the Japan Science and Technology Agency.

Participating institutions: Hiroshima University, National Institute of Advanced Industrial Science and Technology, Toyota Industries, Showa Denko, Taiyo Nippon Sanso.

References

1. Adachi T (2018) Hydrogen purification for FCV from ammonia cracking gas. Taiyo Nippon Sanso Technical Report 37:25–27
2. Hydrogen Utilization Technology Collection (2003) NTS CO., LTD., p5
3. Hydrogen Energy Association (2014) Hydrogen dictionary. Asakura Shoten CO., LTD, p111
4. Tomioka H (2012) International standardization for hydrogen fuel cell vehicles. Hydrogen Energy Syst 37(1):25–28
5. Fufitani T, Takahashi A (2016) Ammonia decomposition catalyst and method for producing the catalyst and decomposition method of ammonia using the catalyst. JP 2016–159209
6. Ono Y, Yashima T (2000) Zeolite science and engineering. Kodansha Sci 107–110
7. Kojima Y, Aoki T, Fujitani T, Adachi T, Kubo H, (2017) Ammonia removal material, ammonia removal method and hydrogen gas production method for fuel cell vehicles. JP 2017–104778
8. Adachi T (2018) Hydrogen gas production apparatus and hydrogen gas production method. JP 2018–177615

Chapter 26
Development of Vanadium Alloy Membrane for Hydrogen Purification

Chikashi Nishimura

Abstract As described in previous chapters, high performance catalysts for decomposing ammonia to generate hydrogen are being developed vigorously. The generated hydrogen should be purified thoroughly before the use in fuel cells, especially in polymer electrolyte fuel cells (PEFC), which is most widely commercialized, and where the platinum (Pt) catalysts prone to be damaged by impurity gases are used. A prominent way to produce high purity hydrogen gas is based on hydrogen-selective membranes used to separate hydrogen from decomposed gas mixtures. Metallic membranes, which operate via a solution/diffusion mechanism, play a key role in the massive production of high purity hydrogen. At present, commonly and commercially used membranes for hydrogen separation are palladium (Pd) alloys. However, the excessive cost of palladium alloys limits its practical application on a large scale. Vanadium (V) has been proposed as promising alternatives to Pd alloys in view of its high hydrogen permeabilities. Here in this chapter, development situations of V membranes for hydrogen separation and purification are described. Mainly, outcomes of a JST-CREST project, "Development of Innovative Hydrogen Production Technology from Energy Carriers Based on Vanadium Alloy Membranes" are briefly introduced.

Keywords Vanadium · Permeation · Diffusion · Membrane · Fuel cells

26.1 Advantage of Dense Metallic Membrane

Metallic membranes for hydrogen separation and purification can selectively permeate hydrogen without nano-sized pores, via a mechanism of hydrogen solution/diffusion in metal lattice. Figure 26.1 shows the results of analysis of the gas permeated from a gas mixture of hydrogen (H_2) and helium (*He*), through a vanadium-nickel (V-Ni) alloy as a separation membrane, which was previously measured by the author using a quadrupole mass spectrometer (Q-Mass). There

C. Nishimura (✉)
National Institute for Materials Science (NIMS), Sengen, Tsukuba, Ibaraki 305-0047, Japan
e-mail: nishimura.chikashi@emeritus.nims.go.jp

K. Aika and H. Kobayashi (eds.), *CO_2 Free Ammonia as an Energy Carrier*,
https://doi.org/10.1007/978-981-19-4767-4_26

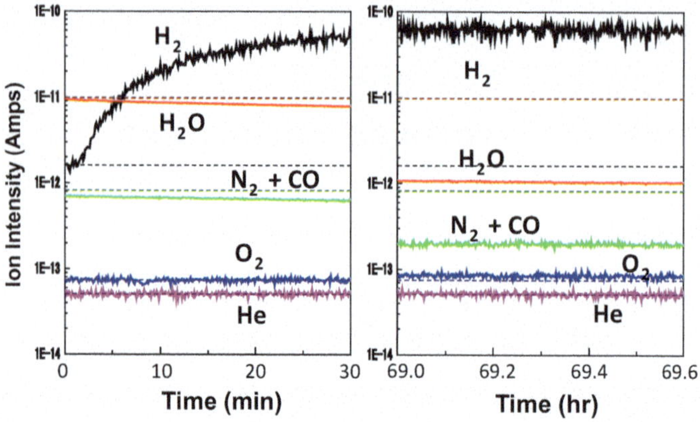

Fig. 26.1 Analysis of permeated gas through V-15Ni alloy membrane at 573 K, using gas mixture of H_2 + 10% He, at 200 kPa

observed a slight time-lag for hydrogen to permeate the membrane, and then the pressure gradually increased. The partial pressure of hydrogen in the permeated gas rose by about two orders of magnitude 30 min after the start of hydrogen permeation run. On the other hand, *He* whose molecular diameter is smaller than that of H_2 does not change at all as in the case of other background gases, which clearly means no *He* permeation is observed through the membrane. The right figure shows the ion intensities after 70 h after the start of the experiment. No change in signals is observed for H_2 and *He*: hydrogen permeates the membrane constantly, but He does not even after a prolonged period. This is a unique and excellent feature of metallic membrane, which cannot be performed by porous membranes.

Here in this chapter, recent development situations of V membranes for hydrogen separation and purification are described. Mainly, results of JST-CREST project, "Development of Innovative Hydrogen Production Technology from Energy Carriers Based on Vanadium Alloy Membranes," lead by the author, are briefly introduced.

26.2 Pd Membranes and V Membranes

26.2.1 Historic Works on Benchmark Material, Pd Alloys

The benchmark alloy materials for V membranes are long-known Pd membranes. Graham discovered that Pd absorbed a large amount of hydrogen in 1866 [1], more than 150 years ago. In 1916, Snelling obtained the first patent for hydrogen purification using Pd membranes [2]. Since then, a substantial number of works has been done on development of hydrogen separation membranes based on Pd and application of them to membrane reactors. History and applications of Pd alloy membranes up to about 2005 are piled up and overviewed in [3]. Pd is a precious metal and the main drawback of Pd is an extremely excessive cost. Pd is mainly used for chemical synthesis catalysts, automotive catalysts, and medical dental materials. Pd alloy membranes are being commercially used for hydrogen separation and purification, but they are used in limited applications such as semiconductor manufacturing and optical fiber manufacturing processes.

26.2.2 *Hydrogen Permeability of V and Pd*

Arrhenius plots of diffusion coefficients, solution constants and permeabilities for V and Pd are shown in Fig. 26.2. Those for iron (Fe) is also shown for comparison. As described in the upper part of the figure, the hydrogen permeability (Φ) is defined by the product of the diffusion coefficient (D) and solution constant (K) of hydrogen. When expressed in a logarithmic scale, the product can be converted into a sum form, and therefore, the result of adding the value of the diffusion coefficient (left) and solution constant (center) on the graph is represented to the right as the hydrogen permeability.

V and Fe have the body-centered cubic (BCC) lattice structure that is suited for fast diffusion of hydrogen. On the other hand, Pd is a face-centered cubic (FCC) metal having the most densely packed lattice structure, and hence the interstitial hydrogen diffusion is not as fast as V and Fe. Hydrogen solution behavior in these three metals is substantially different, one another. V absorbs hydrogen strongly exothermically, and hydrogen content becomes remarkably higher at lower temperatures under a same hydrogen pressure. Pd absorbs some amount of hydrogen, exceptionally among noble metals. But the heat of hydrogen solution (enthalpy change) is small, and the

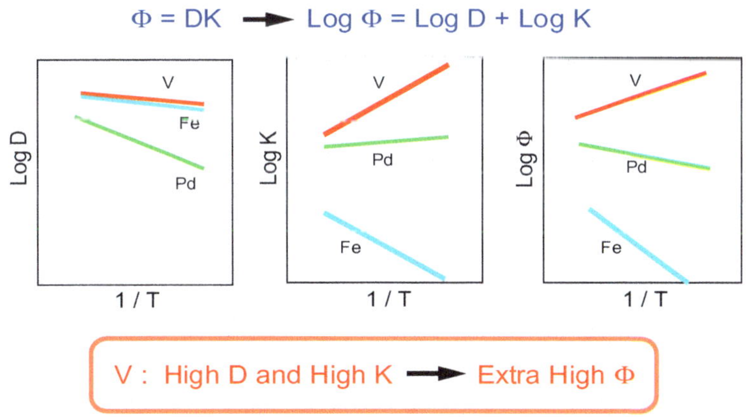

Fig. 26.2 Diffusion coefficient, solution constant and permeability of hydrogen in vanadium (V), palladium (Pd) and Fe (iron)

hydrogen content is almost independent of temperature. Fe absorbs little amount of hydrogen: the absorption of hydrogen is a large endothermic reaction, and hydrogen solution content becomes extremely low at lower temperatures.

As a result, V shows an extremely high hydrogen permeability, and has an interesting feature where the hydrogen permeability increases as temperature decreases. Pd shows a fair hydrogen permeability, but the intrinsic hydrogen permeation performance is considerably lower than that of V. Fe shows an extremely low hydrogen absorption content, and the hydrogen permeability is quite low despite the fast hydrogen diffusion.

In an actual hydrogen permeation process, too much hydrogen solution or too high hydrogen content in a metallic lattice of V results in fracture of membranes. So, in many cases, alloying of one or two metallic elements to some extent is required to reduce the hydrogen solution moderately, as shown in the following.

26.3 Previous Studies on V Alloys

Research and development of V alloys for hydrogen separation membranes have been continuously conducted in the former National Research Institute for Metals and the current National Institute for Materials Science, from around 1990 [4–7]. It has been shown that V-Ni alloys can be easily fabricated to cold-rolled sheets at elevated temperatures, do not crack even under several atmospheres of hydrogen pressure in the temperature range of 623 K and 463 K, and maintain high hydrogen permeation properties stable [5]. Many studies have been conducted on vanadium alloys for hydrogen separation all over the world and are summarized in the reviews [8–10]. There was an early attempt to apply V alloys for membrane reactors [11].

26.4 JST-CREST Project: Development of Innovative Hydrogen Production Technology from Energy Carriers Based on Vanadium Alloy Membranes

26.4.1 Synopsis of the Project

The research area "Creation of Innovative Core Technology for Manufacture and Use of Energy Carriers from Renewable Energy" had been established under JST-CREST program, Japan, between 2013 and 2021 [12]. Under this research area, the project "Development of Innovative Hydrogen Production Technology from Energy Carriers Based on Vanadium Alloy Membranes" had been conducted from 2014 to 2021 [13]. The target of this project was to develop an innovative hydrogen separation device which is optimized for energy carriers including ammonia and methylcyclohexane by using vanadium alloy membranes to generate highly pure hydrogen at a high rate, 3 Nm^3/hr.

26.4.2 *Survey of Alloying Elements to V*

Suzuki and his co-workers succeeded to describe hydrogen permeation of metal membranes based on chemical potential through a careful analysis and detailed consideration of pressure-composition-isotherms (PCT) on hydrogen absorption and hydrogen permeation flux of niobium (Nb), Pd alloys and V alloys [14]. By those findings, it became possible to describe the hydrogen permeation performance of metallic membranes, properly in a wide range of hydrogen pressures applied on a membrane. Suzuki and co-workers have searched alloying elements to V, and proposed V-10Fe alloy for hydrogen separation membrane to be operated at around 623 K [15]. Combining the consistent description with a criterion for ductile/brittle transition of 5a-group metals on hydrogen absorption [16] made them succeed to propose the suitable alloy composition. Figure 26.3 shows a schematic diagram of PCT curves for V, V-5%M alloy and V-10%M alloy at an operation temperature of membrane separation device. As described in 26.2.2, vanadium absorbs pretty much hydrogen. When a hydrogen pressure of gas permeation operation is applied to pure vanadium membrane, the hydrogen content well exceeds the ductile limit, the vertical dashed line in Fig. 26.3, and goes into the brittle region, resulting in the fracture of vanadium membrane. As alloying content of M increases, hydrogen content of V-M alloy decreases. V-10%M alloy absorbs a hydrogen lower than the ductile/brittle transition content at an operation hydrogen pressure. Thus, the alloy membrane does not fracture at the pressure. It is reported that V-10Fe achieved a five times higher hydrogen flux than Pd-23Ag alloy, without fracture for over one thousand hours [15]. There appears a new and comprehensive review on consistent analysis of hydrogen permeability on metallic membranes by Suzuki and Yukawa [17].

26.4.3 *Effects of Ammonia on Hydrogen Permeation of V Alloy Membrane*

Figure 26.4 shows a result of hydrogen permeation experiments using a gas mixture of 74% H_2 + 25% N_2 + 1% NH_3, which is a simulated gas of ammonia decomposition. Hydrogen permeation was conducted at 623 K, using a 0.1 mm thick disk sample of V-10Fe alloy coated with thin Pd-25Ag layer (about 150 nm) on both sides of the disk. The Pd–Ag alloy layers were for catalysts for decomposition of hydrogen molecules and for barriers of vanadium oxidation. V-10Fe alloy membrane shows hydrogen permeation flux five times as high as commercial Pd–Ag alloy membrane

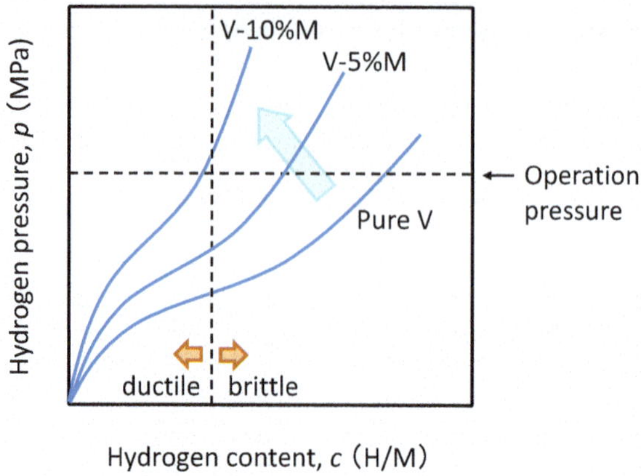

Fig. 26.3 A schematic diagram of Pressure-composition-isotherms (PCT) for V, V-5%M and V-10%M alloys

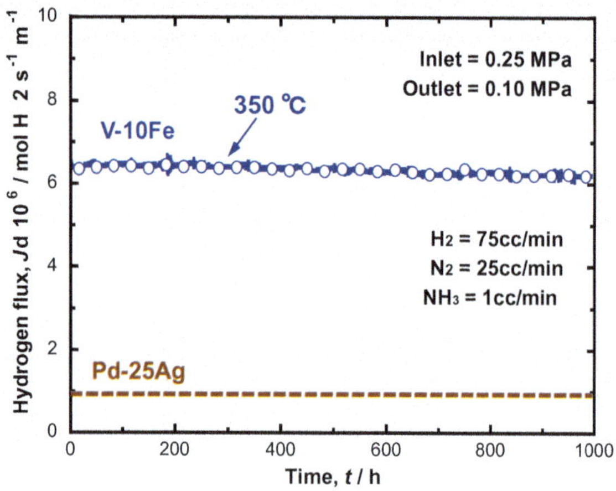

Fig. 26.4 Hydrogen permeation performance of V-10Fe alloy using a gas mixture of 74% hydrogen + 25% nitrogen + 1% ammonia

for over one thousand hours, like a result obtained with pure hydrogen gas (Fig. 26.3). No detrimental effects of ammonia are observed on gas permeation run for over one thousand hours, which is a promising and important experimental data for applications of V alloy membranes. Similar results, no detrimental effects of ammonia, have also been observed in V-Ni alloy [18].

26.4.4 Hydrogen Separation Device Equipped with Flat V-10Fe Membranes

As the project progressed, membrane devices were fabricated using V-10Fe alloy designed to achieve high flow rates. We designed a device capable of separating and purifying hydrogen from ammonia decomposition gas at a high flow rate, by enlarging a surface area, preventing hydrogen-induced fracture, and minimizing the stagnation of gas molecules other than hydrogen, such as nitrogen and ammonia, on the feed side surface during hydrogen permeation using mixed gas [19].

Figure 26.5 shows the appearance of a hydrogen separation device incorporating four flat membranes of V-10Fe alloy with an effective diameter of 90 mm. Figure 26.6 shows hydrogen permeation performance of this membrane device using a gas mixture of 74.5% hydrogen and 25.5% nitrogen at 0.54 MPa; partial pressure of hydrogen is 0.4 MPa.

A maximum hydrogen permeation flow rate of 9.3 L/min. (=558 L/hr.) was achieved by increasing the feed gas flow rate up to 15.5 L/min., with a separation efficiency over 80%. This flow rate exceeds the hydrogen flow rate necessary to operate a common household co-generation system based on fuel cell, ENE-FARM.

At the end of the JST-CREST project period, a hydrogen separation device incorporating six membranes of V-10Fe alloy, three-dimensionally formed and saucer-shaped, was fabricated. These saucer-shaped membranes were so designed to withstand the gas pressure, and to relax the stress caused by lattice expansion due to hydrogen absorption. FEM calculations were extensively used for designing the shape and for determining the plastic work conditions. Figure 26.7 shows the appearance of the device with six membranes, three-dimensionally formed. The device achieved a hydrogen flow rate of 0.74 Nm^3/hr. using a gas mixture of hydrogen and nitrogen under 0.54 MPa. The device showed a fair repeat durability over fifty times. Using four sets of this device, the target value of the project, 3 Nm^3/hr. could be achieved.

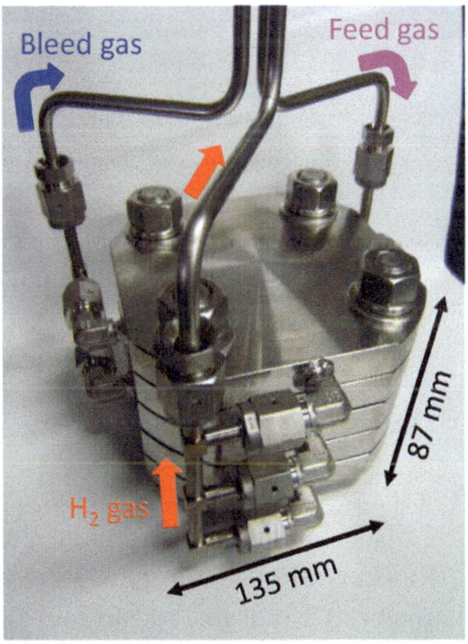

Fig. 26.5 Hydrogen separation device equipped with four flat membranes of V-10Fe

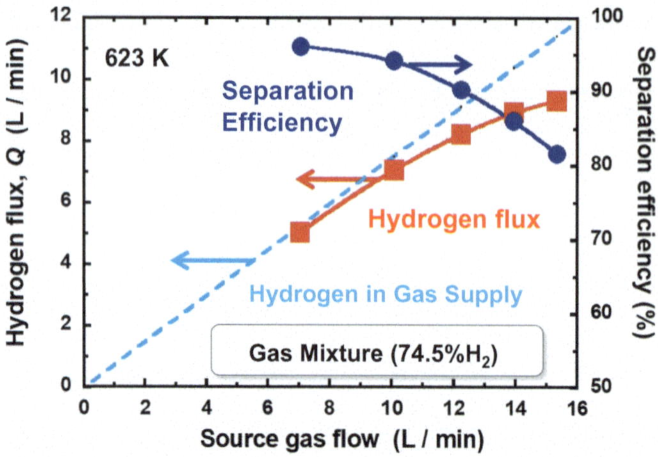

Fig. 26.6 Hydrogen permeation performance of a separation device equipped with quadruple-layered flat membranes of V-10Fe

Fig. 26.7 Hydrogen separation device equipped with six membranes of V-10Fe, three-dimensionally formed and saucer shaped

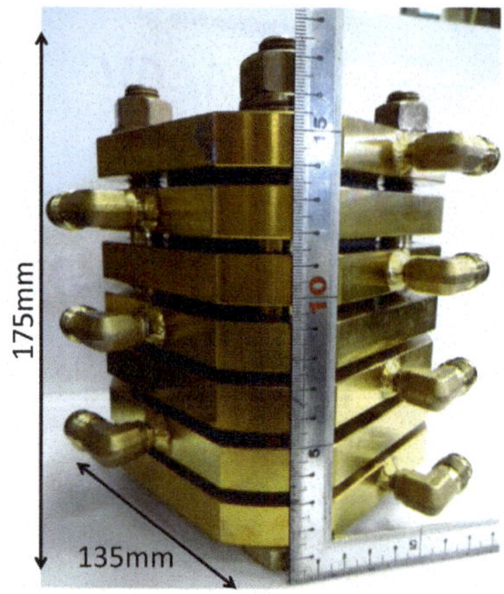

26.5 Competing Works in the World

A research group lead by Dolan at CSIRO, Australia is the most active group in this field, delivering clean hydrogen fuel from ammonia using metal membranes [20]. They fabricated a membrane module with nineteen tubes of V alloy, 10 mm diameter, 0.2 mm thick, 500 mm long. They appeared to use a V-Al alloy for the tubes according to a report [21]. As of August 2018, they successfully produced high purity hydrogen from ammonia, refueled commercial Vehicles Toyota MIRAI and Hyundai NEXO.

26.6 Concluding Remarks

Starting from initial permeation experiments using coin-sized disk samples to explore alloying elements [2], we came to the stage to fabricate hydrogen separation device capable of generating high-purity hydrogen from an ammonia-decomposition simulated gas, at a flow rate exceeding 550 L/hr. This is the achievement of our research team consisting of academia, national research institute and industry. Toward practical application and commercialization,

both fundamental and engineering efforts are required to continue. One important challenge is a high temperature durability and longevity of membranes and devices. There is an attempt to clarify the degradation phenomena as a first step to overcome this issue [22].

References

1. Graham T (1866) Absorption and separation of gases by colloid septa. Philos Trans R Soc 156:426–431
2. Snelling WO (1916) Apparatus for separating gases. U.S. Patent 1 174 631, Mar 7
3. Sammuells AF, Mundschau (2006) Nonporous Inorganic Membranes. Wiley 77 p
4. Nishimura C, Komaki M, Amano M (1991) Hydrogen permeation characteristics of vanadium-nickel alloys. Mater Trans JIM 32:501–507
5. Nishimura C, Komaki M, Hwang S, Amano M (2002) V-Ni alloy membranes for hydrogen purification. J Alloy Comp 330–332:902–908
6. Ozaki T, Zhang Y, Komaki M, Nishimura C (2003) Preparation of palladium-coated V and V-15Ni membranes for hydrogen purification by electroless plating technique. Int J Hydr Energy 28:297–302
7. Zhang Y, Ozaki T, Komaki M, Nishimura C (2003) Hydrogen permeation of Pd-Ag coated V-15Ni composite membrane: effects of overlayer composition. J Membr Sci 224:81–91
8. Dolan MD (2010) Non-Pd BCC alloy membranes for industrial hydrogen separation. J Membr Sci 362:12–28
9. Ockwig NW, Nenoff TM (2007) Membranes for hydrogen separation. Chem Rev 107:4078–4110
10. Sammuells AF, Mundschau (2006) Nonporous inorganic membranes. Wiley 107 p
11. Matsuka M, Higashi M, Ishihara T (2013) Hydrogen production from methane using vanadium-based catalytic membrane reactors. Int J Hydr Energy 38:6673–6680
12. Energy Carrier CREST jst.go.jp (2021) Creation of innovative core technology for manufacture and use of energy carriers from renewable energy. https://www.jst.go.jp/kisoken/crest/research_area/ongoing/bunyah25-1.html Accessed 28 Jan 2022
13. Nishimura C (2021) Development of innovative hydrogen production technology from energy carriers based on vanadium alloy membranes. in Energy Carrier CREST jst.go.jp https://www.jst.go.jp/kisoken/crest/project/40/14532003.html Accessed 28 Jan 2022
14. Suzuki A, Yukawa H, Murata Y (2017) Consistent description of hydrogen permeation through metal membrane based on hydrogen chemical potential and its application for alloy design. J Mater Res 32:227–238
15. Suzuki A, Yukawa H, Nambu T, Matsumoto Y, Murata Y (2016) Quantitative evaluation of hydrogen solubility and diffusivity in V-Fe alloys toward the design of hydrogen permeable membrane for low operating temperature. Mater Trans 57:1823–1831

16. Matsumoto Y, Yukawa H, Nambu T (2016) Evaluation of ductile-to-brittle transition for tantalum in hydrogen environment by small punch test. In: Proceedings of the 4th international conferences on small sample test techniques (SSTT2016) pp 65–69
17. Suzuki A, Yukawa H (2020) A review for consistent analysis of hydrogen permeability through dense metallic membranes. Membranes 10:120–140
18. Tsai JT, Kimura H, Nishimura C, Wang SF (2016) Hydrogen permeation characteristics of V-15Ni membrane from gas mixtures of Hydrogen, Nitrogen and Ammonia. In: Proceeding of the 9th Pacific Rim international conference on advanced materials and processing (PRICM9) pp 576–579
19. Nishimura C (2019) Development of innovative hydrogen production technology from energy carriers based on vanadium alloy membranes. Impact 2019:29–31. Available on line https://doi.org/10.21820/23987073.2019.5.29. Accessed 28 Jan 2022
20. Dolan MD (2022) Available online 1715-AIChE-Dolan-v2.pdf (ammoniaenergy.org). Accessed 28 Jan 2022
21. Dolan MD, Viano D, Hla SS, Liang D, Kellam ME, Morpeth LD, Fogliaresi L, Langley M, Song G, Meysel P, Weiss F (2015) Alloy membrane reactor for pre-combustion CO_2 capture, report prepared for ANCLE R&D. Available online Alloy CMR for H_2 production—ANLEC R&D (anlecrd.com.au). Accessed 28 Jan 2022
22. Yukawa H, Nambu T, Matsumoto Y (2020) Thermal degradation behavior of hydrogen permeability of Pd-coated V-alloy membrane for hydrogen separation and purification. Mater Sci Forum 1016:1710–1714

Part VI
Ammonia Fuel Cell

Chapter 27
Ammonia Decomposition Catalysts for Fuel Cell Application

Hiroki Muroyama

Abstract Ammonia has attracted interest as a hydrogen source for fuel cells. Hydrogen production from ammonia is mainly carried out using decomposition reaction. Integration of ammonia decomposition reactor and fuel cell can enhance total energy conversion efficiency. Permissible concentrations of unreacted ammonia in hydrogen-rich fuel gas are relatively high for solid oxide fuel cells and anion exchange membrane fuel cells. Thus, Ni metal can be candidates of ammonia decomposition catalyst although it is inferior to Ru metal in activity. It was suggested that the particle size and surface structure of Ni particles are important factors determining the catalytic activity. Moreover, support materials and additives strongly affected the catalyst performance. Rare-earth components promoted ammonia decomposition as support materials and additives of Ni catalysts due to the alleviation of hydrogen inhibition. Alkaline-earth metal elements also contributed to high activity as the catalyst components.

Keywords Ammonia · Decomposition · Nickel catalyst · Support · Additive · Rare-earth elements · Alkaline-earth metal elements

27.1 Ammonia Fuel for Fuel Cells

Ammonia has attracted attention as one of the promising fuels for fuel cells. Internal and external decompositions of ammonia can be considered as fuel supply methods. In the external decomposition system, hydrogen is produced from ammonia via the following reaction in the reactor and then supplied to the anode compartment of fuel cells.

$$NH_3 \rightarrow 1/2N_2 + 3/2H_2 \quad \Delta H^\circ = +46 \text{ kJ mol}^{-1} \tag{27.1}$$

H. Muroyama (✉)
Department of Energy and Hydrocarbon Chemistry, Graduate School of Engineering, Kyoto University, Nishikyo-ku, Kyoto 615-8510, Japan
e-mail: muroyama.hiroki.5c@kyoto-u.ac.jp

K. Aika and H. Kobayashi (eds.), *CO_2 Free Ammonia as an Energy Carrier*,
https://doi.org/10.1007/978-981-19-4767-4_27

According to thermodynamic calculation, the equilibrium ammonia conversion achieves above 99% at 400 °C and 1 atm [1]. Many studies have demonstrated that various metals, alloys, nitrides, and carbides are active for ammonia decomposition. Among them, Ru metal is widely known as the most active species. It has been already reported that the supported Ru-based catalysts reached almost the equilibrium conversion at 400 °C [2].

Ammonia decomposition is endothermic, while operation of fuel cells generates heat as well as electricity. Thus, integration of ammonia decomposition reactor and fuel cell can enhance total energy conversion efficiency. In conventional polymer electrolyte fuel cells (PEFCs) employing an acidic electrolyte, even a trace amount of ammonia in the fuel gas causes serious deterioration of cell performance. It was suggested that the replacement of H^+ by NH_4^+ lowered the proton conductivity of electrolyte and anode catalyst layer [3]. The catalytic activity for oxygen reduction reaction was also degraded in the presence of NH_4^+ [4]. Considering that it is impossible to completely decompose ammonia even using highly active catalysts, removal of unreacted ammonia from the produced hydrogen-rich gas is essential. On the other hand, in anion exchange membrane fuel cells (AEMFCs), much higher concentration of ammonia in fuel gas could be permissible because the electrolyte membrane is a basic polymer [5]. The performance of cell fed with 7% NH_3-containing H_2 fuel was comparable to that with 100% H_2 fuel. Moreover, ammonia in the supplied fuel gas poses little problem for solid oxide fuel cells (SOFCs). Ammonia can be decomposed over the conventional Ni-based anodes in SOFCs due to their high operating temperature above 700 °C. Thus, power generations of SOFCs have been successfully demonstrated even with a supply of pure ammonia fuel to the anode [6]. The integrated system of SOFCs and ammonia decomposition reactor is more effective in terms of energy conversion efficiency due to the close operating temperatures of these processes compared with low-temperature fuel cells.

Ru catalysts with high activity are desirable to be used in ammonia decomposition reactors combined with PEFCs because of the necessity for the removal of unreacted ammonia from hydrogen-rich fuel gas and their low operating temperatures. In contrast, since ammonia-containing hydrogen gases are available as fuels in AEMFCs and SOFCs, lower-active catalysts can be candidates for the reactor of the integrated systems. Parts of base metals are active for ammonia decomposition although they are inferior to Ru metal in activity. Ni metal has attracted interest owing to the highest activity among base metals. Moreover, its low cost is attractive for the industrial application.

In this chapter, recent researches on Ni catalysts for ammonia decomposition are introduced.

27.2 Ammonia Decomposition Over Nickel Catalysts

In ammonia decomposition reaction, ammonia adsorbed on the active sites undergoes sequential hydrogen dissociation, and then, the resultant atoms are recombinatively desorbed as hydrogen and nitrogen molecules. It has been proposed that the nitrogen desorption, or the first or second hydrogen dissociation is a kinetically slow process [1, 7]. Ganley et al. investigated the relationship between turnover frequency (TOF) of ammonia decomposition and activation energy of N–H bond dissociation [8]. The reaction rate was decreased among Ru, Rh, and Pt catalysts as the activation energy of N–H bond dissociation increased. In contrast, base metal catalysts such as Ni, Co, and Fe showed the opposite tendency. Therefore, it was concluded that the nitrogen desorption process is the rate-determining step, not the N–H bond dissociation process for base metal catalysts. In addition, the DFT calculation for the reaction mechanism over Fe(110), Co(111), and Ni(111) also indicated that the energy barrier for the desorption of nitrogen atoms as a molecule from the metal surface is high, and that this process is the rate-determining step [9].

Zhang et al. studied the relationship between Ni particle size and TOF [10]. Ni catalysts with average particle sizes less than 2.9 nm exhibited high activity. The maximal TOF value was obtained at 2.3 nm. For Ru catalysts, a specific atomic arrangement of B5 site induced high activity for ammonia decomposition [11]. The existence probability of this B5 site was high in Ni particles with 1.8–2.5 nm [12]. Therefore, it is suggested that the relationship between average particle size of Ni and TOF was ascribable to the formation of B5 sites.

The structural sensitivity of the ammonia decomposition over Ni catalysts has been investigated by the DFT calculation [13]. The reaction mechanism at the Ni(211) step site and the Ni(111) terrace site was examined. NH_x ($x = 0$–3) and H species adsorbed more strongly at the Ni(211) site than at the Ni(111) site. The barrier of recombinative desorption of nitrogen atoms at the Ni(211) site was higher by 1.10 eV than that at the Ni(111) site, indicating that the nitrogen desorption could not readily proceed at the Ni(211) site. However, the Ni(211) site promoted the dehydrogenation of NH_3 to NH_2. Thus, the ratio of step sites to terrace sites, which is affected by the particle size, will be one of the important factors determining the ammonia decomposition activity.

27.3 Influence of Support Material on Ammonia Decomposition for Nickel Catalysts

The properties of support material strongly affect the activity of the supported metal catalyst, and various support materials have been studied for Ni catalysts. Carbon-based materials such as carbon nanotube have been reported to be effective for supported Ru catalysts [14–16]. In contrast, there are not many studies on Ni catalysts supported on carbon-based material [17–19], probably because the positive effect of carbon-based materials on the performance was not remarkable.

Ni catalysts supported on metal oxides such as SiO_2 and Al_2O_3 have been investigated extensively. Mesoporous materials of silica (MCM–41, SBA–15) contributed to high dispersion of Ni particles, resulting in high ammonia conversion [20]. Zeolites were also effective as support materials [21]. Ni/ZSM–5 showed high activity and stability because strong support–metal interaction enabled Ni nanoparticles to be loaded in a highly dispersed state.

We evaluated the ammonia decomposition activity of Ni catalysts supported on various metal oxides [22, 23]. Figure 27.1 shows the relationship between the specific surface area of support and the ammonia conversion at 550 °C. The conversion was enhanced with increasing the surface area of supports, indicating a correlation between the catalytic activity and the dispersion of Ni particles. On the other hand, the rare-earth oxide-supported Ni catalysts exhibited high conversion in spite of low surface area of support. This clarified that these oxides have a positive effect on the catalyst performance. Other group has reported that Ni catalysts supported on MgO and rare-earth oxides showed high activity [24]. Accordingly, basic oxides are effective as support materials for supported Ni catalysts.

We focused on Ni catalysts supported on rare-earth oxide and investigated the factors of their high activity in more detail [23]. Several rare-earth oxides, Y_2O_3, La_2O_3, CeO_2, Sm_2O_3, and Gd_2O_3, were selected as the support materials. Table 27.1 shows the specific surface area of the support measured by N_2 adsorption and Ni surface area estimated by the CO pulse method. Compared with Al_2O_3 support, the specific surface area of the rare-earth oxide was very small. Generally, high surface area of support facilitates the dispersion of the metal catalyst. However, a part of rare-earth oxide-supported catalysts attained high Ni surface area. In particular, the

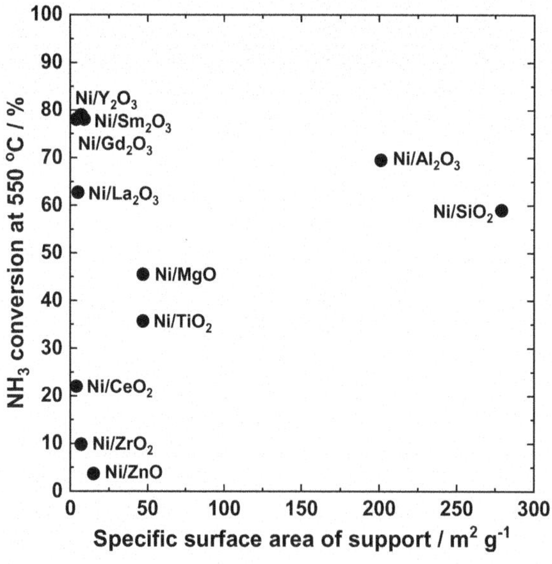

Fig. 27.1 Ammonia conversion at 550 °C as a function of specific surface area of supports for 10 wt% Ni catalysts supported on various oxides. Reaction conditions; 100% NH_3, Space velocity (S. V.): 6000 l kg^{-1} h^{-1} [22, 23]

Table 27.1 Specific surface area of support, Ni surface area, and reaction orders according to Eq. (27.2) for ammonia decomposition

Catalyst (10 wt% Ni)	BET surface area of support/$m^2\ g^{-1}$	Ni surface area/$m^2\ g^{-1}$	Reaction order[a]	
			α	γ
Ni/Y_2O_3	7	1.62	0.77	−0.48
Ni/La_2O_3	5	0.42	1.00	−0.45
Ni/CeO_2	4	0.18	0.74	−0.86
Ni/Sm_2O_3	9	0.74	0.98	−0.66
Ni/Gd_2O_3	4	1.06	0.85	−0.46
Ni/Al_2O_3	201	1.88	0.37	−0.88

Reproduced from Ref. [23], Copyright (2016) with permission from Wiley–VCH
[a] Reaction temperature: 450 °C, reaction gas: NH_3–H_2–Ar (NH_3: 10–30%, H_2: 10–30%, and Ar balance), S. V.: 10,000 l $kg^{-1}\ h^{-1}$

value of Ni/Y_2O_3 was comparable to that of Ni/Al_2O_3, suggesting that Y_2O_3 is an effective support for the dispersion of Ni particles.

Activity test of ammonia decomposition was conducted using these Ni catalysts. As shown in Fig. 27.2, ammonia decomposition was initiated at 350 °C over all the catalysts. Higher ammonia conversion was achieved for Ni/Y_2O_3, Ni/Sm_2O_3, and Ni/Gd_2O_3 than for Ni/Al_2O_3. Considering the Ni/Al_2O_3 catalyst had the highest Ni surface area, the rare-earth oxide supports of these catalysts positively affected the activity. Among the rare-earth oxide-supported catalysts, higher surface area of Ni metal resulted in superior performance, and the Ni/Y_2O_3 catalyst was the most active. Therefore, small particles over the Y_2O_3 support would be responsible for high catalytic activity.

Kinetic analysis was performed for ammonia decomposition over the Ni catalysts to clarify the effect of rare-earth oxide supports. The reaction rate (r) was defined using the partial pressures of ammonia (P_{NH_3}) and hydrogen (P_{H_2}) as follows.

$$r = k\, P_{NH_3}^{\alpha}\, P_{H_2}^{\gamma} \tag{27.2}$$

where k is the reaction rate constant, and α and γ are reaction orders. It was reported that the reaction rate does not depend on the partial pressure of nitrogen [1]. The ammonia decomposition test was conducted at 450 °C by supplying mixtures of NH_3–H_2–Ar. The reaction orders are also summarized in Table 27.1. The order of ammonia partial pressure was positive for all the catalysts. In contrast, a negative dependence of reaction rate on the hydrogen partial pressure was confirmed, indicating that hydrogen inhibited ammonia decomposition. Since ammonia and hydrogen are competitively adsorbed on the active sites in ammonia decomposition, the covering of the sites by hydrogen prevents the reaction from proceeding [1]. The absolute value of γ was much smaller for Ni/Y_2O_3, Ni/La_2O_3, and Ni/Gd_2O_3 than for Ni/CeO_2 and Ni/Al_2O_3. Thus, hydrogen had a less effect on ammonia decomposition over these catalysts. Previous study on the Ru surface found that the presence of Cs species significantly

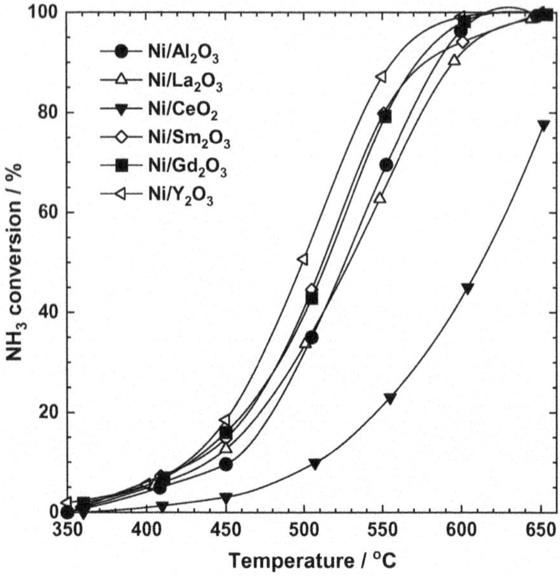

Fig. 27.2 Ammonia conversion for ammonia decomposition over 10 wt% Ni/Y_2O_3, Ni/La_2O_3, Ni/CeO_2, Ni/Sm_2O_3, Ni/Gd_2O_3, and Ni/Al_2O_3. Reaction conditions; 100% NH_3, S. V.: 6000 l kg^{-1} h^{-1}. Reproduced from Ref. [23], Copyright (2016) with permission from Wiley–VCH

decreased the sticking coefficient of hydrogen [25]. Thus, the basic property of rare-earth oxide supports will be one of the factors for this phenomenon. Consequently, the high activity of some rare-earth oxide-supported catalysts was ascribable to the suppression of hydrogen inhibition.

According to the kinetic analyses, the support materials influenced the hydrogen adsorption on the catalysts. Then, the hydrogen desorption over the surface of Ni/Al_2O_3, Ni/Y_2O_3, and respective supports during ammonia decomposition was examined by ammonia temperature programmed surface reaction (NH_3–TPSR). After heat treatment under the reducing condition, the sample was exposed to NH_3 and then heated in Ar. The desorption behavior of hydrogen was recorded by a mass spectrometer during the heating, as shown in Fig. 27.3. No desorbed nitrogen was detected for both supports, indicating that ammonia was not decomposed under this condition. Hydrogen was formed above ca. 300 °C, which was mainly derived from the hydrogen atoms strongly adsorbed on the surface during the pretreatment under the reducing atmosphere. The Ni catalysts exhibited different behavior from the corresponding support materials for both systems. Hydrogen was newly desorbed from Ni/Al_2O_3 at 90–290 °C and 320–500 °C, whereas new hydrogen desorption was confirmed only at 90–290 °C for Ni/Y_2O_3. In these processes, hydrogen atoms generated by ammonia dissociation were desorbed as hydrogen molecules from both Ni catalysts. Considering the desorption process at high temperatures for Ni/Al_2O_3, dehydrogenation of NH_x (x = 1–3) and subsequent hydrogen desorption will be

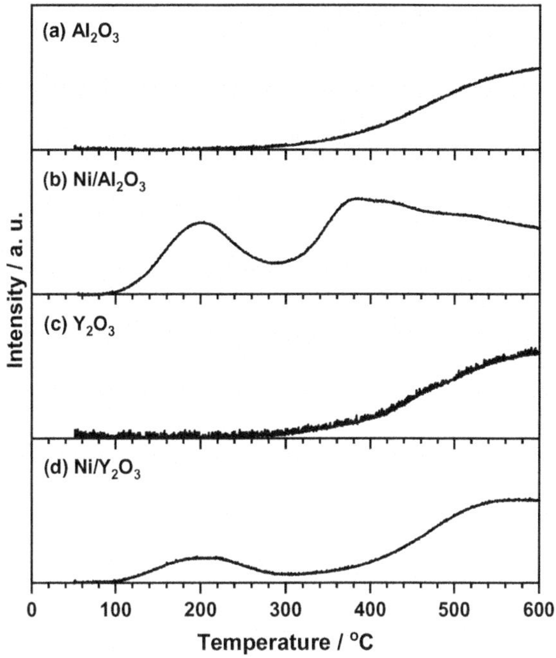

Fig. 27.3 Desorption profiles of hydrogen (m/z = 2) for **a** Al_2O_3, **b** 40 wt% Ni/Al_2O_3, **c** Y_2O_3, and **d** 40 wt% Ni/Y_2O_3 obtained by NH_3–TPSR in Ar flow at a heating rate of 5 °C min^{-1} Reproduced from Ref. [23], Copyright (2016) with permission from Wiley–VCH

facilitated more readily over the Ni particles supported on Y_2O_3. Consequently, the Y_2O_3 support alleviated hydrogen inhibition, leading to the high catalytic activity of Ni/Y_2O_3.

The basic components contributed to the activity of supported Ni catalysts for ammonia decomposition. Then, we employed perovskite-type oxides including alkaline metal, alkaline-earth metal, and rare-earth elements as support materials for the catalysts [26]. The perovskite-type oxide is generally expressed as a chemical formula of ABO_3, in which two types of A and B sites are occupied by cations. Figure 27.4 shows the ammonia conversion at 550 °C for the prepared Ni catalysts. The Ni/*RE*AlO_3 (*RE* = La, Sm, Gd), Ni/*AE*TiO_3, and Ni/*AE*ZrO_3 (*AE* = Sr, Ba) catalysts exhibited relatively high activity. For Ni/*AE*TiO_3 and Ni/*AE*ZrO_3 (*AE* = Ca, Sr, Ba), the catalytic activity was strongly dependent on the A-site element of the perovskite-type oxides, and the Sr and Ba elements were more effective than the Ca one in each system. The Ni/*AE*ZrO_3 catalysts exhibited better performance than Ni/*AE*TiO_3 in the case of the same A-site cation. The NH_3–TPSR measurements revealed a clear difference in the desorption behavior of nitrogen molecules over Ni/*AE*ZrO_3 (*AE* = Ca, Sr, Ba). Compared with $Ni/CaZrO_3$, desorbed nitrogen was detected at lower temperatures for $Ni/SrZrO_3$ and $Ni/BaZrO_3$. As a result, the

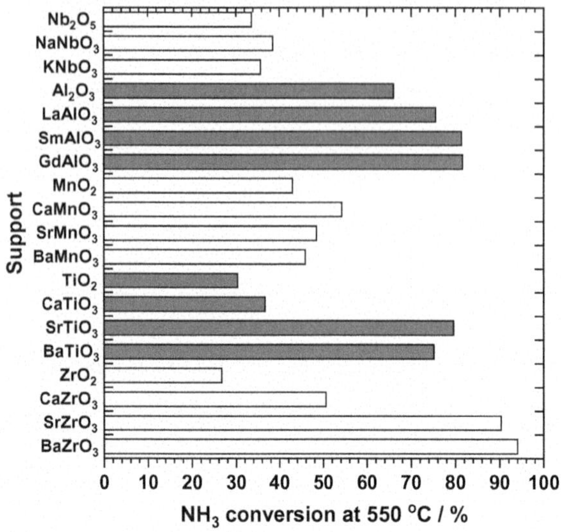

Fig. 27.4 Ammonia conversion for ammonia decomposition at 550 °C for 40 wt% Ni catalysts supported on various oxides. Reaction conditions; 100% NH_3, S. V.: 6000 l kg^{-1} h^{-1}. Reproduced from Ref. [26] with permission from the Royal Society of Chemistry

promotion of nitrogen desorption over these catalysts should contribute to their high activity for ammonia decomposition.

The researches introduced in this part so far investigated supported Ni catalysts fabricated by an impregnation method and so on. On the other hand, some researchers have developed Ni catalysts prepared from composite oxides containing Ni species. In this case, Ni metal is precipitated from the parent phase by a reduction treatment. Among them, layered double hydroxides (LDHs) containing Ni element were examined for ammonia decomposition catalysts [27–29]. This composite oxide is generally composed of divalent and trivalent metal ions (Mg^{2+}, Ca^{2+}, Ni^{2+}, Al^{3+}, Fe^{3+}, etc.). The catalytic activity depended on the amount of precipitated Ni, the size and morphology of the precipitated Ni particles, and the basic properties of the catalyst.

27.4 Effect of Additive Component on Ammonia Decomposition for Nickel Catalysts

Effects of additive components on the activity of Ru catalysts have been examined, and the addition of basic components such as alkaline metal, alkaline-earth metal, and rare-earth elements is effective for the improvement of their activity [1]. This is attributable to the promotion of recombinative desorption of nitrogen atoms by the electron donation from the basic component to Ru metal. Therefore, Ni-based catalyst system has also been studied for the additive effect of basic components.

No improvement in activity by the addition of K component to Ni/SiO_2 catalyst was confirmed [20]. However, there has not been many reports on the modification of Ni catalysts by alkaline metal elements. On the other hand, it has been reported that the activity of Ni catalysts was improved by the addition of alkaline-earth metal and rare-earth elements. The addition of La and Ce components enhanced the catalytic activity of Ni/Al_2O_3 [10, 30, 31]. These additive components would improve the dispersion of Ni particles, resulting in an increase in active sites. In addition, the promotion of recombinative desorption of nitrogen atoms by the electron donation from the additive component to Ni metal was regarded as one of the factors for the improved activity.

We also studied the additive effect of various rare-earth components on catalytic activity and physical properties of Ni/Al_2O_3 [32]. Figure 27.5 shows the ammonia conversion for the prepared catalyst at 550 °C. The addition of rare-earth components remarkably enhanced the ammonia conversion. In contrast, no significant difference in the surface area of metallic Ni was observed depending on the absence or presence of the additive. According to the previous reports, the addition of rare-earth components enlarged the Ni surface area [30, 33]. In these cases, the Ni particle size of unmodified Ni/Al_2O_3 catalysts was ca. 5 nm. In our study, the particle size was ca. 30 nm. Thus, the expansion of Ni surface area by rare-earth elements would be affected by the Ni particle size of the unmodified catalyst.

The desorption behavior of chemical species was evaluated by NH_3–TPSR measurement for the rare-earth-modified Ni/Al_2O_3 catalysts. The desorption profile of hydrogen is shown in Fig. 27.6. Hydrogen was desorbed from the unmodified Ni/Al_2O_3 at 100–200 °C and above 300 °C. On the other hand, the desorption was confirmed only at 100–200 °C for the modified catalysts. These results suggested that the rare-earth component decreased the amount of hydrogen species strongly interacting with Ni. The addition of the rare-earth component should promote the hydrogen dissociation from the adsorbed ammonia and the subsequent desorption of hydrogen molecules, and suppress the hydrogen poisoning on Ni catalysts, as in the case of using the rare-earth oxide as a support. The suppression of hydrogen adsorption at the interface of Ni–Al_2O_3 by the rare-earth components should be one of the possible reasons for the promotion of hydrogen desorption. As mentioned above, two hydrogen desorption processes were observed at 100–200 °C and above 300 °C for the unmodified Ni/Al_2O_3. Considering that hydrogen was desorbed at ca. 300 °C from Al_2O_3 support (Fig. 27.3), the former and latter processes would be derived from the hydrogen adsorbed onto the Ni surface and onto the Ni–Al_2O_3 interface, respectively. Thus, the rare-earth components located around Ni particles could inhibit the diffusion of hydrogen atoms from Ni surface to the interface, leading to the significant decrease in the fraction of hydrogen desorbed at high temperatures. Consequently, the enhancement of catalytic activity by the rare-earth component was attributable to the alleviation of hydrogen inhibition for ammonia decomposition.

In addition to Ni/Al_2O_3 system, we investigated the additive effect of alkaline-earth metal component on the catalytic activity of Ni/Y_2O_3 [34]. Figure 27.7 shows

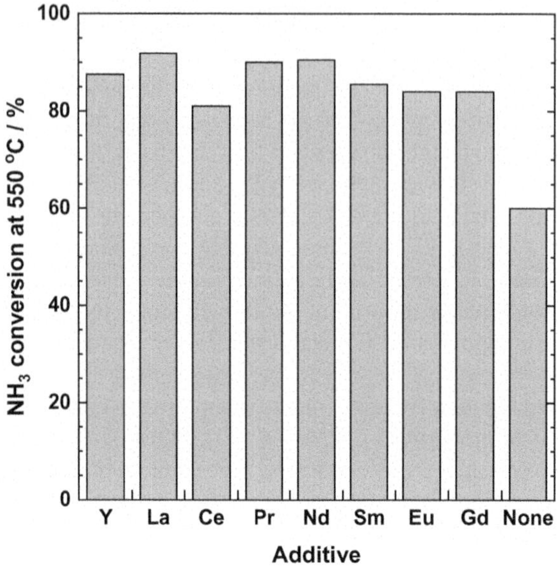

Fig. 27.5 Ammonia conversion at 550 °C for ammonia decomposition over 12 wt% MO_x–9 wt% Ni/Al_2O_3 (M = Y, La, Ce, Pr, Nd, Sm, Eu, Gd) and 10 wt% Ni/Al_2O_3 catalysts. Reaction conditions; 100% NH_3, S. V.: 6,000 l kg^{-1} h^{-1} Reproduced from Ref. [32], Copyright (2015) with permission from Elsevier

the ammonia conversion at 500 °C as a function of the loading amount of alkaline-earth metal oxide. The activity of the Mg and Ca-modified catalysts was lower than that of the unmodified catalyst at all the loading amounts tested. On the other hand, the ammonia conversion was significantly improved by adding a small amount of Sr and Ba components. Therefore, it was found that the additive effect of these components could be distinguished into two types: Mg, Ca components and Sr, Ba components.

Crystal phases in each catalyst were examined after preparation and after subsequent reduction treatment. In all catalysts, in addition to the Y_2O_3 phase, NiO and Ni phases were confirmed after preparation and after reduction treatment, respectively. In contrast, the crystal phase derived from the additive component was different depending on the catalyst. The Mg and Ca components were present as MgO and CaO in as-prepared and as-reduced catalysts for the respective systems. In the Sr- and Ba-modified catalysts as prepared, composite oxides including the additive component and Ni were formed. These composite oxides were decomposed by the reduction treatment, implying the Sr and Ba components would be located in the vicinity of Ni particles in the reduced catalysts.

As mentioned above, the basic component in the catalysts contributes to the promotion of recombinative desorption of nitrogen atoms over active sites. Then, the basic property of the alkaline-earth metal-modified Ni/Y_2O_3 catalysts was evaluated by the desorption behavior of CO_2. As a result, desorption peak of CO_2 was confirmed in the higher temperature range over all the modified catalysts than over

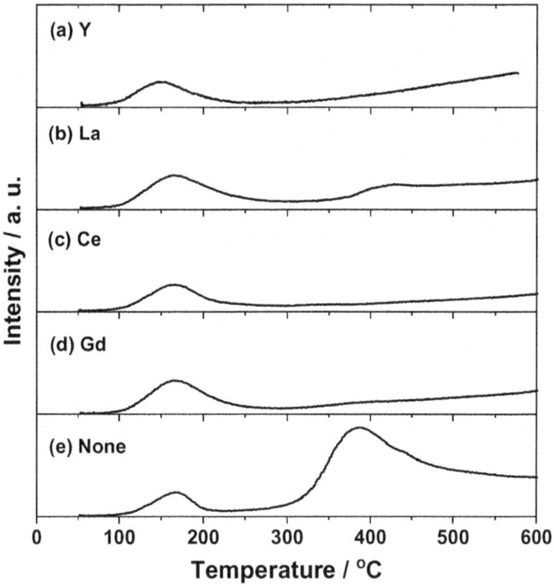

Fig. 27.6 Desorption profiles of hydrogen ($m/z = 2$) for **a** 12 wt% Y_2O_3–, **b** 12 wt% La_2O_3–, **c** 12 wt% CeO_2–, **d** 12 wt% Gd_2O_3–9 wt% Ni/Al_2O_3, and **e** 10 wt% Ni/Al_2O_3 obtained by NH_3–TPSR in He flow at a heating rate of 10 °C min^{-1} Reproduced from Ref. [32], Copyright (2015) with permission from Elsevier

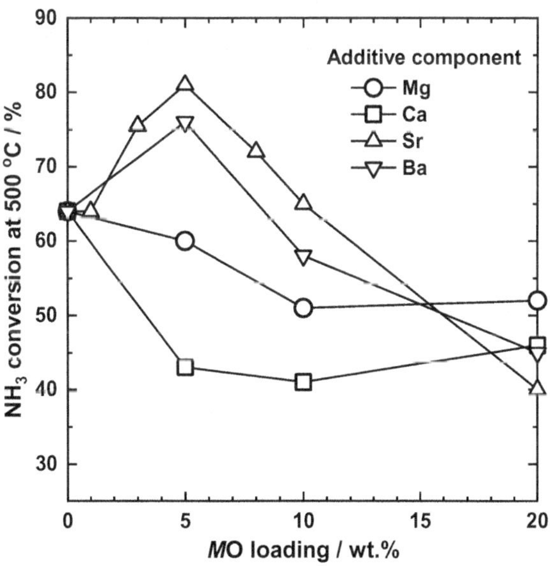

Fig. 27.7 NH_3 conversion for ammonia decomposition at 500 °C as a function of *M*O loading over 0–20 wt% *M*O–40 wt% Ni/Y_2O_3 (M = Mg, Ca, Sr, Ba). Reaction conditions; 100% NH_3, S. V.: 6,000 l kg^{-1} h^{-1} Reproduced from Ref. [34] with permission from the Royal Society of Chemistry

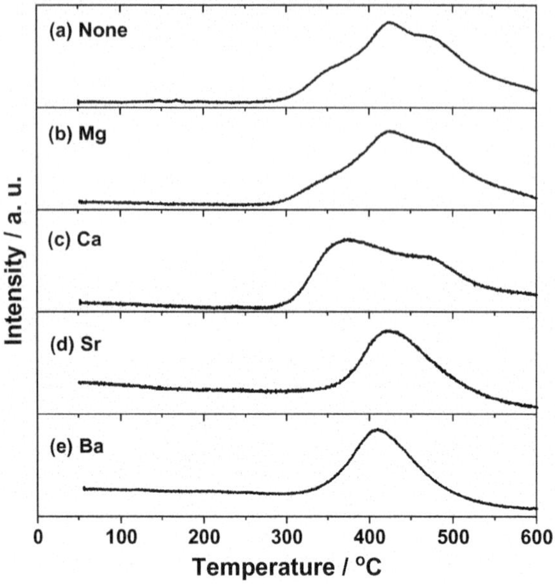

Fig. 27.8 N_2 desorption profiles for **a** 40 wt% Ni/Y_2O_3, **b** 5 wt% MgO–, **c** 5 wt% CaO–, **d** 5 wt% SrO–, and **e** 5 wt% BaO–40 wt% Ni/Y_2O_3 obtained by NH_3–TPSR measurement in Ar flow under a heating rate of 5 °C min^{-1}. Reproduced from Ref. [34] with permission from the Royal Society of Chemistry

the unmodified catalyst, indicating the increase in the basic strength of catalyst by the modification. In particular, this tendency was remarkable for the Sr- and Ba-modified catalysts. Moreover, the desorption behavior of nitrogen molecules from the catalysts was examined by NH_3–TPSR measurement (Fig. 27.8). The nitrogen desorption continuously occurred over the Mg- and Ca-modified catalysts from 300 °C to more than 600 °C. These behaviors were similar to that of the unmodified catalyst, suggesting that these additive components had little effects on nitrogen desorption. In contrast, the desorption peak was observed at 340–560 °C for the Sr- and Ba-modified catalysts. Compared with the unmodified catalyst, the desorption was terminated at lower temperature although its onset temperature was higher. Accordingly, the basic strength of Sr and Ba components decreased the nitrogen species strongly adsorbed on Ni, resulting in the enhancement of catalytic activity of Ni/Y_2O_3.

References

1. Yin SF, Xu BQ, Zhou XP, Au CT (2004) A mini-review on ammonia decomposition catalysts for on-site generation of hydrogen for fuel cell applications. Appl Catal A Gen 277:1–9
2. Nagaoka K, Eboshi T, Abe N, Miyahara S, Honda K, Sato K (2014) Influence of basic dopants on the activity of Ru/Pr_6O_{11} for hydrogen production by ammonia decomposition. Int J Hydrogen

Energy 39:20731–20735
3. Uribe FA, Gottesfeld S, Zawodzinski TA (2002) Effect of ammonia as potential fuel impurity on proton exchange membrane fuel cell performance. J Electrochem Soc 149:A293
4. Halseid R, Vie PJS, Tunold R (2006) Effect of ammonia on the performance of polymer electrolyte membrane fuel cells. J Power Sources 154:343–350
5. Katayama Y, Yamauchi K, Hayashi K, Okanishi T, Muroyama H, Matsui T, Kikkawa Y, Negishi T, Watanabe S, Isomura T, Eguchi K (2017) Anion-exchange membrane fuel cells with improved CO_2 tolerance: impact of chemically induced bicarbonate ion consumption. ACS Appl Mater Interfaces 9:28650–28658
6. Yang J, Akagi T, Okanishi T, Muroyama H, Matsui T, Eguchi K (2015) Catalytic influence of oxide component in Ni-based cermet anodes for ammonia-fueled solid oxide fuel cells. Fuel Cells 15:390–397
7. Schüth F, Palkovits R, Schlögl R, Su DS (2012) Ammonia as a possible element in an energy infrastructure: catalysts for ammonia decomposition. Energy Environ Sci 5:6278–6289
8. Ganley JC, Thomas FS, Seebauer EG, Masel RI (2004) A priori catalytic activity correlations: the difficult case of hydrogen production from ammonia. Catal Letters 96:117–122
9. Duan X, Ji J, Qian G, Fan C, Zhu Y, Zhou X, Chen D, Yuan W (2012) Ammonia decomposition on Fe(110), Co(111) and Ni(111) surfaces: a density functional theory study. J Mol Catal A Chem 357:81–86
10. Zhang J, Xu H, Li W (2005) Kinetic study of NH_3 decomposition over Ni nanoparticles: the role of La promoter, structure sensitivity and compensation effect. Appl Catal A Gen 296:257–267
11. García-García FR, Guerrero-Ruiz A, Rodríguez-Ramos I (2009) Role of B5-type sites in Ru catalysts used for the NH_3 decomposition reaction. Top Catal 52:758–764
12. van Hardeveld R, van Montfoort A (1966) The influence of crystallite size on the adsorption of molecular nitrogen on nickel, palladium and platinum. Surf Sci 4:396–430
13. Duan X, Qian G, Liu Y, Ji J, Zhou X, Chen D, Yuan W (2013) Structure sensitivity of ammonia decomposition over Ni catalysts: a computational and experimental study. Fuel Process Technol 108:112–117
14. Raróg-Pilecka W, Szmigiel D, Kowalczyk Z, Jodzis S, Zielinski J (2003) Ammonia decomposition over the carbon-based ruthenium catalyst promoted with barium or cesium. J Catal 218:465–469
15. Yin SF, Xu BQ, Ng CF, Au CT (2004) Nano Ru/CNTs: a highly active and stable catalyst for the generation of CO_x-free hydrogen in ammonia decomposition. Appl Catal B Environ 48:237–241
16. Li L, Zhu ZH, Yan ZF, Lu GQ, Rintoul L (2007) Catalytic ammonia decomposition over Ru/carbon catalysts: the importance of the structure of carbon support. Appl Catal A Gen 320:166–172
17. Zhang H, Alhamed YA, Kojima Y, Al-Zahrani AA, Miyaoka H, Petrov LA (2014) Structure and catalytic properties of Ni/MWCNTs and Ni/AC catalysts for hydrogen production via ammonia decomposition. Int J Hydrogen Energy 39:277–287
18. Ji J, Duan X, Qian G, Zhou X, Chen D, Yuan W (2013) In situ production of Ni catalysts at the tips of carbon nanofibers and application in catalytic ammonia decomposition. Ind Eng Chem Res 52:1854–1858
19. Meng T, Xu QQ, Li YT, Chang JL, Ren TZ, Yuan ZY (2015) Nickle nanoparticles highly dispersed on reduced graphene oxide for ammonia decomposition to hydrogen. J Ind Eng Chem 32:373–379
20. Li XK, Ji WJ, Zhao J, Wang SJ, Au CT (2005) Ammonia decomposition over Ru and Ni catalysts supported on fumed SiO_2, MCM-41, and SBA-15. J Catal 236:181–189
21. Hu ZP, Weng CC, Chen C, Yuan ZY (2018) Catalytic decomposition of ammonia to CO_x-free hydrogen over Ni/ZSM-5 catalysts: a comparative study of the preparation methods. Appl Catal A Gen 562:49–57
22. Muroyama H, Saburi C, Matsui T, Eguchi K (2012) Ammonia decomposition over Ni/La_2O_3 catalyst for on-site generation of hydrogen. Appl Catal A Gen 443–444:119–124

23. Okura K, Okanishi T, Muroyama H, Matsui T, Eguchi K (2016) Ammonia decomposition over nickel catalysts supported on rare-earth oxides for the on-site generation of hydrogen. ChemCatChem 8:2988–2995
24. Nakamura I, Fujitani T (2016) Role of metal oxide supports in NH_3 decomposition over Ni catalysts. Appl Catal A Gen 524:45–49
25. Shi H, Geng P, Jacobi K (1994) Hydrogen chemisorption on cesium-predosed Ru(001) surfaces. Surf Sci 315:1–8
26. Okura K, Miyazaki K, Muroyama H, Matsui T, Eguchi K (2018) Ammonia decomposition over Ni catalysts supported on perovskite-type oxides for the on-site generation of hydrogen. RSC Adv 8:32102–32110
27. Zhao J, Deng L, Zheng W, Xu S, Yu Q, Qiu X (2020) Nickel-induced structure transformation in hydrocalumite for enhanced ammonia decomposition. Int J Hydrogen Energy 45:12244–12255
28. Sato K, Abe N, Kawagoe T, Miyahara S, Honda K, Nagaoka K (2017) Supported Ni catalysts prepared from hydrotalcite-like compounds for the production of hydrogen by ammonia decomposition. Int J Hydrogen Energy 42:6610–6617
29. Su Q, Gu L, Yao Y, Zhao J, Ji W, Ding W, Au CT (2017) Layered double hydroxides derived $Ni_x(Mg_yAl_zO_n)$ catalysts: enhanced ammonia decomposition by hydrogen spillover effect. Appl Catal B Environ 201:451–460
30. Zhang J, Xu H, Jin X, Ge Q, Li W (2005) Characterizations and activities of the nano-sized Ni/Al_2O_3 and $Ni/La–Al_2O_3$ catalysts for NH_3 decomposition. Appl Catal A Gen 290:87–96
31. Zheng W, Zhang J, Ge Q, Xu H, Li W (2008) Effects of CeO_2 addition on Ni/Al_2O_3 catalysts for the reaction of ammonia decomposition to hydrogen. Appl Catal B Environ 80:98–105
32. Okura K, Okanishi T, Muroyama H, Matsui T, Eguchi K (2015) Promotion effect of rare-earth elements on the catalytic decomposition of ammonia over Ni/Al_2O_3 catalyst. Appl Catal A Gen 505:77–85
33. Liu H, Wang H, Shen J, Sun Y, Liu Z (2008) Promotion effect of cerium and lanthanum oxides on Ni/SBA-15 catalyst for ammonia decomposition. Catal Today 131:444–449
34. Okura K, Okanishi T, Muroyama H, Matsui T, Eguchi K (2016) Additive effect of alkaline earth metals on ammonia decomposition reaction over Ni/Y_2O_3 catalysts. RSC Adv 6:85142–85148

Chapter 28
Component Materials and Reactions for Ammonia-Fueled Solid Oxide Fuel Cells (SOFCs)

Koichi Eguchi

Abstract Ammonia is suitable fuel for solid oxide fuel cells (SOFCs) with respect to high generation efficiency and carbon free characteristics. Excellent fuel flexibility of SOFCs enables to supply ammonia to the fuel electrode in direct internal cracking mode of operation. Combination of ammonia cracking and SOFCs gives rise to highly efficient generation systems. Ammonia is thermally cracked on Ni in the fuel electrode, and then, electrochemical oxidation of hydrogen subsequently proceeds. Conversion of ammonia cracking approaches to 100% with a sufficient contact time with Ni-cermet layer at the normal operation temperatures of SOFCs. Several fuel electrode materials composed of Ni-oxide cermet have been proposed for high ammonia cracking activities. Proton-conducting ceramic fuel cells are also operated with direct supply of ammonia. Basicity of proton conductor in Ni-cermet leads to high activity for ammonia cracking.

Keywords Ammonia · Solid oxide fuel cell · SOFC · Fuel electrode · Proton conducting ceramic fuel cell · PCFC

28.1 Background of Ammonia-Fueled Solid Oxide Fuel Cells

Fuel cells are known as energy conversion devices with high generation efficiency. Among the fuel cell systems, solid oxide fuel cells (SOFCs) achieve highest generation efficiency and excellent fuel flexibility. SOFC systems have been commercially produced either as household application or business use. On the other hand, ammonia has been considered as a clean fuel and energy carrier which contributes to realize carbon neutral energy system and connection between renewable energy and energy consumption [1]. If ammonia supply chains are established, the combination of ammonia fuel and SOFC systems is regarded as most environmentally benign and efficient generation systems. Fuel flexibility of SOFC is superior to other

K. Eguchi (✉)
Graduate School of Engineering, Kyoto University, Nishikyo-ku, Kyoto 615-8510, Japan
e-mail: eguchi.koichi.y76@kyoto-u.jp

K. Aika and H. Kobayashi (eds.), *CO_2 Free Ammonia as an Energy Carrier*,
https://doi.org/10.1007/978-981-19-4767-4_28

fuel cell systems because of the high operation temperatures. It is known that SOFC systems can achieve high generation efficiency exceeding 50% LHV even with small-scale unit [2]. It is expected that the generation efficiency will be further enhanced with fuel recycling or multi-stage stack arrangement. In the natural gas (NG)-fueled SOFCs, direct supply of NG fuel to anode is known as internal reforming operation. The operation temperature of SOFCs at 700–900 °C is very suitable to combine internal catalytic ammonia cracking like internal reforming of NG. The component materials for NG-fueled SOFC have been well established. Thus, the development of ammonia-fueled SOFCs is started by use of cells and the set of materials for NG-fueled SOFC.

28.2 Basic Properties of Ammonia-Fueled SOFCs

Ammonia-fueled cells are tested, and thermal integration with cracking has been investigated. The three possible combinations of SOFC unit and ammonia cracker are shown in Fig. 28.1. For fuel cells operative at lower temperatures than those of cracking reaction, the external cracking is the sole possible combination. Autothermal cracking reactors are also installed externally to which a mixture of ammonia and air is supplied. Rapid start-up, e.g., for a few minutes can be possible when the autothermal reactor is operated in the exothermic heating condition. After autothermal catalytic cracker is heated up at sufficient temperature for ammonia decomposition, high-temperature hydrogen is supplied to the SOFC stack for start-up.

Figure 28.2 summarizes operation temperatures of several fuel cell systems and temperature of catalytic reactors for hydrogen production [3]. Catalytic hydrogen production processes have been known as steam reforming of hydrocarbons and oxygenated compounds, dehydrogenation of organic hydrides, and cracking of ammonia. These hydrogen production processes are endothermic reactions. On the other hand, the reactions of fuel cells are electrochemical oxidation of fuels with

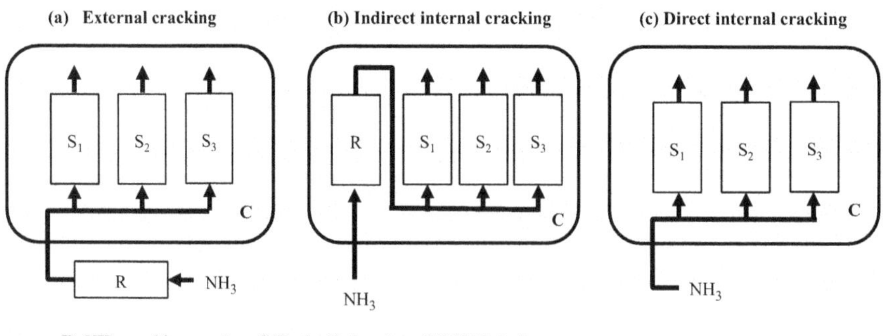

Fig. 28.1 Three arrangements of NH_3 cracker and SOFC stacks

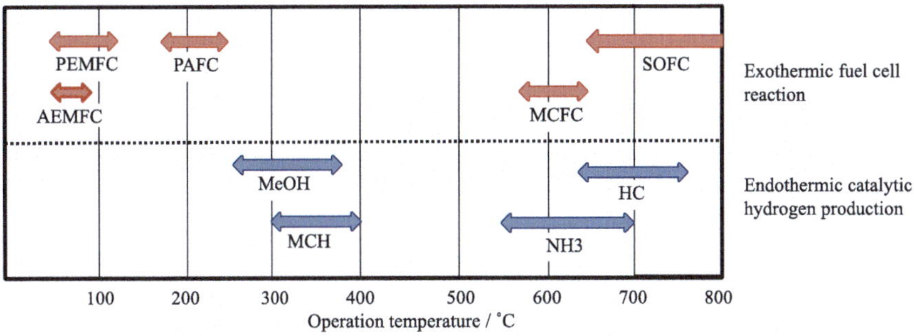

Fig. 28.2 Operating temperature ranges of fuel cells and catalytic hydrogen production processes. PEMFC: proton exchange membrane fuel cell. MeOH: methanol steam reforming. AEMFC: anion exchange membrane fuel cell. MCH: methyl cyclohexane dehydrogenation. PAFC: phosphoric acid fuel cell. NH_3: ammonia cracking. MCFC: molten carbonate fuel cell. HC: hydrocarbon steam reforming. SOFC: solid oxide fuel cell

exothermic heat. The operation temperature of SOFCs is the highest among the fuel cell systems and the temperature range 700–900 °C is similar to that of the catalytic hydrocarbon reforming or the ammonia cracking. This means that catalytic reactor for steam reforming of hydrocarbon can be coupled and the catalytic reactor can be installed in the hot module of SOFC. This is known as the indirect internal reforming operation for hydrocarbon supplied SOFC. Similarly internal cracking operation of ammonia is advantageous because exothermic heat from the fuel cell generation can be utilized for the endothermic ammonia decomposition (Fig. 28.2).

In addition to the indirect internal decomposition operation, direct supply of ammonia to SOFCs is possible because of the high operation temperatures. The most popular fuel electrode of SOFC is Ni-yttria-stabilized zirconia (YSZ) cermet. The Ni-based cermet possesses sufficient activity for ammonia decomposition at the operation temperature of SOFCs. Ammonia is thermodynamically unstable at this temperature and pressure. Two possibilities of reactions on the fuel electrode can be considered for direct ammonia-fueled SOFC.

(a) Ammonia is directly oxidized electrochemically with oxide ion,

$$2NH_3 + 3O^{2-} \rightarrow N_2 + 3H_2O + 6e^- \quad \text{This reaction is not operative.} \tag{28.1}$$

Or (b) ammonia is cracked thermally on an Ni-cermet catalyst followed by electrochemical oxidation.

$$2NH_3 \rightarrow N_2 + 3H_2 \quad \text{catalytic ammonia cracking} \tag{28.2}$$

$$3H_2 + 3O^{2-} \rightarrow 3H_2O + 6e^- \quad \text{electrochemical oxidation of } H_2 \tag{28.3}$$

In either case, the cathodic reaction is the reduction of gaseous oxygen.

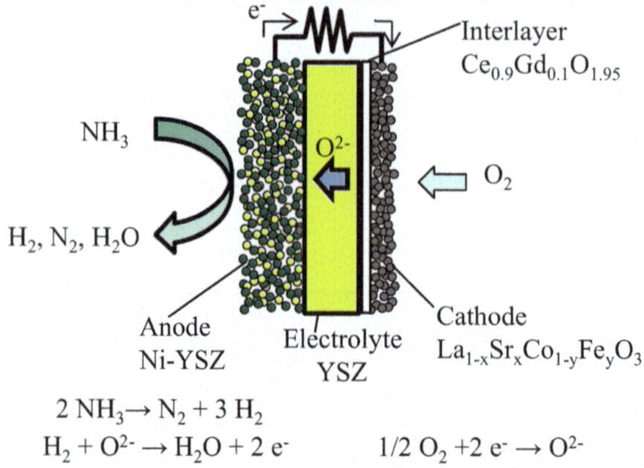

Fig. 28.3 Standard component materials and reactions of direct ammonia-fueled SOFC

$$1/2O_2 + 2e^- \rightarrow O^{2-} \tag{28.4}$$

Considering from thermodynamic instability of ammonia at the operation temperatures of SOFCs, the case (b) is operative (Fig. 28.3).

The open circuit voltage of direct ammonia SOFC was equal to the theoretical electro motive force (EMF) of H_2-O_2 fuel cell (Fig. 28.4). The open circuit voltage of humidified NH_3 is expected from the H_2/H_2O ratio assuming the complete decomposition of ammonia at high temperature such as at 750 °C. This means that ammonia has been completely decomposed to H_2 and N_2 on the Ni-YSZ layer in the vicinity of the reaction site (triple phase boundary). At lower temperatures, e.g., at 600 °C, the open circuit voltage (OCV) is lower than the expected value. Ammonia decomposes incompletely and does not directly participate in the electrochemical reaction. OCV in this case is determined by the formed H_2/H_2O, while undecomposed ammonia behaves as inert coexistence. Based on this fact, one can estimate the hydrogen partial pressure and then conversion of ammonia on the surface of Ni-YSZ from the OCV value at lower temperatures or insufficient residence time as schematically shown in Fig. 28.4. The theoretical OCV is determined by the partial pressure of H_2O (P_{H2O}) supplied and that of H_2 (P'_{H2}) formed by the complete decomposition. Conversion of ammonia cracking deviates from equilibrium at low temperatures or with insufficient residence time with Ni-based cermet. Partial pressure of H_2 (P''_{H2}) can be measured from EMF. From the ratio of P''_{H2}/P'_{H2} and total flow rate of respective case, one can estimate the conversion of ammonia cracking [4]. Such a behavior can be also observed for the supply of humidified CH_4. The OCV of CH_4/H_2O is determined by formed H_2, CO, CO_2 and H_2O content. The reactivity on the Ni-YSZ cermet is higher for ammonia cracking than methane steam reforming. It is possible to supply dry ammonia directly to the anode. This situation is quite different from the supply of methane or natural gas where the surplus amount of water vapor is added to the supplied gas to avoid carbon deposition. In the case of dry ammonia

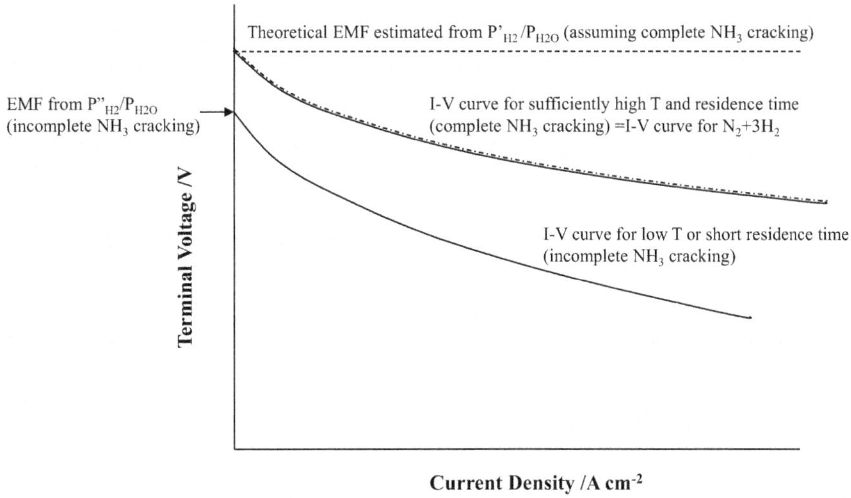

Fig. 28.4 Outline of I-V curves for direct NH_3 supply with complete and incomplete cracking

supply, however, one cannot estimate the theoretical OCV because the steam partial pressure is unknown.

It is well known and actually utilized that the internal reforming of SOFCs gains the enthalpy of the fuel combustion when its endothermic heat is supplied to the system from the environment [5]. The gain of energy for internal reforming operation of methane-fueled SOFC is thermodynamically calculated to be 23%. Figure 28.5 also compares the enthalpy for combustion of ammonia and that of hydrogen formed by decomposition at 500 K and 1000 K, i.e., the typical operation temperatures for phosphoric acid fuel cells and SOFCs. Upon cracking of ammonia, formed hydrogen possesses higher enthalpy of combustion than ammonia combustion at two temperatures. The overall combustion energy is gained by 17% after decomposition. This internal cracking operation is possible with the system setup of Fig. 28.1b, c. Figure 28.5 also indicates maximum efficiency of heat engines for ammonia combustion and electrochemical cell fueled with formed hydrogen. Carnot efficiency is defined by $1\text{-}T_L/T_H$, whereas the output of electrochemical energy is given by ΔG, since remaining thermal energy $T\Delta S$ only leads to thermal energy. It is clear that temperature dependences of Carnot efficiency and Gibbs efficiency, $\Delta G/\Delta H$, are opposite. The ideal efficiency of heat engines increases with temperature, whereas the ideal efficiency of fuel cells decreases with temperature.

It is advantageous that the fuel flow path is completely separated from the oxidant flow path in SOFC like other fuel cell systems. This means that in the SOFC with practical size and sufficiently high fuel utilization, the supplied ammonia is contacted with Ni-cermet anode with sufficient residence time around 700 °C. Ammonia is decomposed into hydrogen and nitrogen nearly equilibrium concentration with thermal catalytic cracking after passing through the cell. The equilibrium ammonia concentration at 750 °C is *ca.* 60 ppm, and oxidizing species of NO_x is scarcely included in

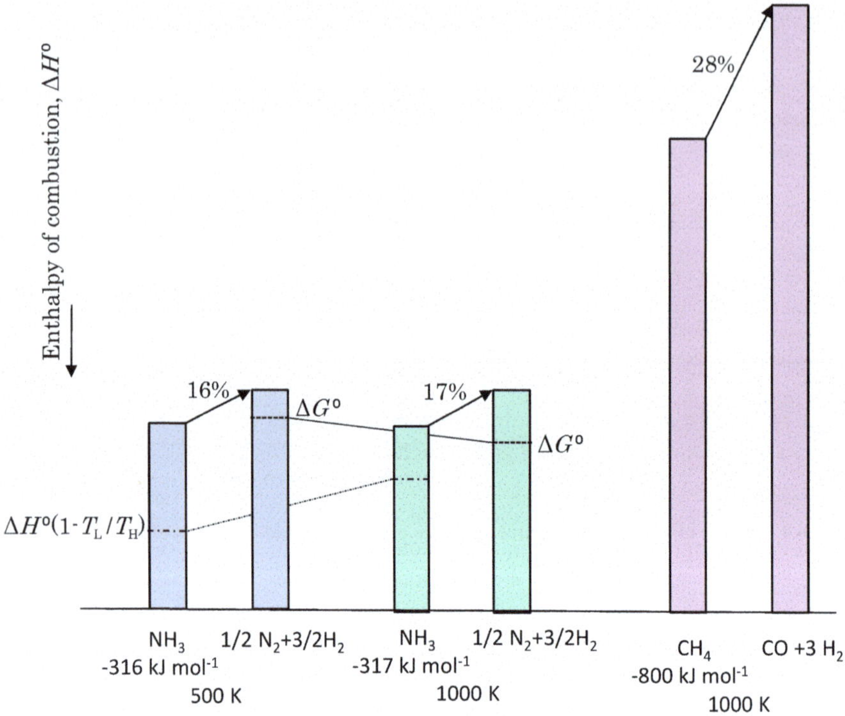

Fig. 28.5 Enthalpy of combustion for methane and ammonia

the exhaust from fuel gas. This situation is contrasting with the flame combustion of ammonia, where fuel NOx formation is expected depending on the condition. Small amount of ammonia included in the exhaust from SOFC can be further eliminated by catalytic oxidation after mixing with air before release to the atmosphere.

28.3 Component Material for Ammonia-Fueled SOFCs

The standard materials for hydrogen or NG-fueled SOFC are the following (Fig. 28.3),

$$\text{Fuel, Ni} - \text{YSZ}|\text{YSZ}|\text{GDC}|(\text{La, Sr})(\text{Fe, Co})\text{O}_3\text{, air} \tag{28.5}$$

YSZ: Yttria-stabilized zirconia, GDC: Gadolinium-doped ceria.

Anode supported cell configuration is popular for SOFC in which Ni-YSZ cermet serves as anode for electrochemical oxidation of hydrogen, catalyst for fuel decomposition, i.e., methane reforming or ammonia cracking, and support material for the cell. The YSZ electrolyte layer with the typical thickness of 10–50 μm is deposited on the thick Ni-YSZ fuel electrode with the thickness of ca. 1 mm. The Ni-YSZ cermet is often composed of two layers with different microstructures, i.e., thin active Ni-YSZ layer with fine grains of components in contact with the electrolyte and thick structural layer with coarse grains exposed to the fuel gas flow. The typical operation temperature of this standard cell is 700–750 °C. Thin GDC layer is inserted between YSZ electrolyte and porous air electrode for promotion of cathodic reaction and to avoid solid-state reaction between the electrolyte and cathode. The cathode material is the perovskite type oxide, ABO_3, consisting of $La_{1-x}Sr_xCo_{1-y}Fe_yO_3$ and also denoted as $(La,Sr)(Co,Fe)O_3$. The selection of the cathode perovskite materials depends on the operation temperature irrespective of the use of ammonia fuel. $La_{1-x}Sr_xCoO_3$-based cathode material has been used as the more active component, but its high thermal expansion coefficient limits the use of this material at low temperatures, though at higher operation temperatures, the solid-state reaction and thermal expansion mismatch become the problem. Thus, $La_{1-x}Sr_xMnO_3$ is more suitable material for the operation at relatively high temperatures of 800–900 °C. Alternative electrolyte materials, such as scandia-stabilized zirconia or proton-conducting perovskite, enable operation at lower temperatures than the YSZ electrolyte.

Anode supported cells are most popular for the commercial SOFCs fueled with NG reformate. The thick Ni-YSZ layer serves as the mechanical support for thin electrolyte layer. The anode thick layer is also preferable as direct fuel supply, including direct ammonia-fueled cell, because of the sufficient contact time with Ni catalysts. The combination of the components for the standard cell can be used for the direct supply of ammonia. The performance of anode supported cell with the supply of ammonia has been compared with the supply of H_2+N_2 mixture as shown in Fig. 28.6. In every case, almost similar performance was attained by the direct supply of ammonia to that of the H_2+N_2 mixture. At the typical operation temperature of SOFC, ammonia is readily decomposed into N_2 and H_2. Thus, the performance of small button type cell is the same as in the case of hydrogen fuel. The open circuit voltage is close to the expected value from the Nernst equation because of the sufficient contact time to decompose ammonia. Water vapor was added to estimate the oxygen potential from formed H_2/H_2O. The power density in this case was 350 mW cm^{-2} which was slightly lower than the case of N_2+H_2 mixture but acceptable level for the practical application.

For larger cells, on the other hand, directly supplied ammonia starts to decompose with a contact on the fuel electrode, and then, rapid progress of cracking cools down the inlet part of the large cell. This situation leads to a decrease of temperature in the inlet zone of ammonia-fueled cell. Some metallic parts in this zone are attacked by corrosive effect of ammonia in the specific temperature region. It is not observable for the test of small sized cells, but practical cell size will be 10 cm by 10 cm or larger cell area. Direct supply of ammonia to this larger cell gives rise to sudden cracking of ammonia at the inlet zone of the fuel electrode of the cells. This situation results in cooling effect because of the endothermic cracking reaction [6]. Significant drop of temperature at the ammonia inlet was sometimes observed which also results in slight lowering of the cell performance because of the temperature gradient. It is

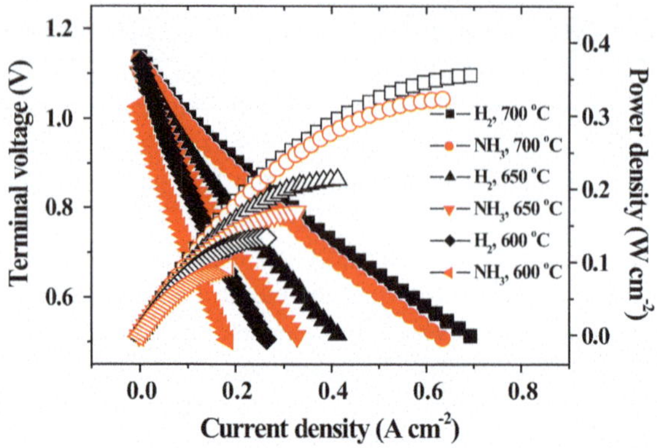

Fig. 28.6 Performance of anode-supported cell: Ni/YSZ|YSZ|GDC|LSCF fueled with **a** 66.7% NH_3-1.4% H_2O-31.9% N_2 and **b** 60.0% H_2-1.0% H_2O-39.0% N_2 gases; cathode gas: O_2

recommended that pre-cracker is installed in the hot module of the cell chamber to avoid this temperature gradient in the cell.

Many metallic parts including gas supply and exhaust, flow controller, and manifold are exposed to concentrated ammonia during start-up, shut-down, and operation. Metallic components for SOFC are sometimes corroded with ammonia. Therefore, piping and tubing in contact with ammonia at various concentrations and temperatures should be selected to avoid corrosion with ammonia as well as flow controller and supply tubing. Aluminum-coated metallic parts cannot be used because of the chemical instability against ammonia. Nitride can be formed at specific temperatures on ferritic steels. Ni-cermet anode is deteriorated at around 500 °C on exposure to NH_3 [7]. At lower temperatures, nitride reaction is too slow to affect the nitridation. At higher temperatures, nitride is thermally unstable; therefore, the metallic Ni is stable at normal operation temperatures. Most of cell components for NG-fueled cell can be used for ammonia-fueled cell. A care should be taken to the possibility that chemical reaction between glass sealant and ammonia causes deterioration and leakage.

28.4 Fuel Electrode Materials for Ammonia Decomposition

Modification of Ni-based cermet materials

Various materials active for ammonia cracking have been proposed as fuel electrodes for direct ammonia SOFCs. The anode materials developed for direct ammonia-fueled SOFCs with oxide ion conductor electrolytes are summarized in Table 28.1. Basically, Ni-based cermet generally possesses sufficient activities at general operation temperatures of 700–900 °C.

Table 28.1 Anode materials for direct ammonia-fueled SOFCs with oxide ion conductor electrolytes

Anode material	Electrolyte	Reference
Ni-YSZ	YSZ	Standard materials
Ni-SDC	SDC	[8, 9]
Ni-SDC	ScCeSZ	[10]
Reduced LSTNC-infiltrated SDC	SDC	[11]
Ni-SSZ	SSZ	[12]
Ba-Ni-YSZ	YSZ	[13]
Nitrided Ni-Cr-SDC	LSGM	[14]
Nitrided Ni-Mo-SDC	LSGM	[15, 16]

YSZ: $(Y_2O_3)_{0.08}(ZrO_2)_{0.98}$
SDC: $Ce_{0.8}Sm_{0.2}O_{1.9}$
ScCeSZ: $(ScO_{1.5})_{0.18}(CeO_2)_{0.01}(ZrO_2)_{0.81}$
LSTNC: $La_{0.52}Sr_{0.28}Ti_{0.94}Ni_{0.03}Co_{0.03}O_{3-\delta}$
LSGM: $La_{0.9}Sr_{0.1}Ga_{0.8}Mg_{0.2}O_{2.85}$
SSZ: $Sc_{0.1}Zr_{0.9}O_{1.95}$

Several anode materials proposed for direct ammonia supply are summarized in Table 28.1. Alternative anode materials are necessary when the cell is operated at low temperatures. Ruthenium-based catalysts have been known as most active catalysts for ammonia decomposition. The metal content in the cermet, however, should be larger than 40 vol% to ensure the electric conduction path with metal percolation. This means costly Ru metal cannot be used as the main metallic component. The activity of nickel, being lower than Ru, is relatively high among base metals. Infiltration of Ru or other active components is effective in enhancing the high activity. The solution of Ru salt was infiltrated in the pores of Ni-YSZ cermet scaffold dried and calcined. The performance of Ru-infiltrated Ni-YSZ was higher than the Ni-YSZ.

As was the supported catalyst, the activities of Ni-based cermets were significantly affected by the oxide components. Ni-YSZ cermet is the most standard electrode material. Although the additive effect on promotion was not observed for YSZ. This material can be used without further modification.

It is reported that the Ni-SDC is more active for the ammonia decomposition [8, 9]. Ni-Mo is active Mo component in the working state and is converted to nitride. It is known that the nitrided Mo is efficient additive to promote the decomposition reaction [15].

28.5 Ammonia-Fueled Proton Conducting Ceramic Fuel Cells

Perovskite type oxides such as $BaZrO_3$ and $BaCeO_3$ have been known as ceramic proton conductors. They are attracting as electrolytes of solid oxide fuel cells because of their higher ionic conductivities at low temperatures than YSZ. Use of proton conducting oxides for fuel cells leads to efficient fuel utilization, since water vapor is produced in the cathode side. This means that fuel is not diluted with formed water even with high fuel utilization. Similar high efficiency due to this advantage is also expected for the ammonia-fueled proton conducting ceramic fuel cells (PCFCs). The components and reaction of the typical ammonia-fueled PCFC are schematically shown in Fig. 28.7 [17].

The typical operation temperature of PCFCs is 550–650 °C which is slightly lower than that of YSZ-based SOFCs due to high conductivity and low activation energy of conduction. Ammonia is readily cracked at these temperatures on Ni-based cermets partially because of higher cracking activity than Ni-YSZ [17, 18]. Many researchers have already confirmed the performance of ammonia fed PCFC. The Ni-based cermets with proton conducting ceramics have been used for the anode. Ni-ceria was also employed as a fuel electrode. These electrodes possess sufficient activity for decomposition of ammonia.

A unique direct ammonia-fueled PCFC developed by Aoki et al. [19] consisted of electrolyte layer with Pd-based hydrogen separation membrane on the anode face of $BaZr_{0.1}Ce_{0.7}Y_{0.2}O_{3-\delta}$, which is called as a hydrogen membrane fuel cell. Ammonia is directly supplied to the dense Pd membrane anode. Hydrogen formed by decomposition of ammonia on the Pd surface is dissolved in the bulk Pd and transferred to the proton conductor in the form of atomic hydrogen through Pd membrane, and then, the electrochemical oxidation of atomic hydrogen to proton proceeded at the interface of Pd membrane and electrolyte. This system supplies dry

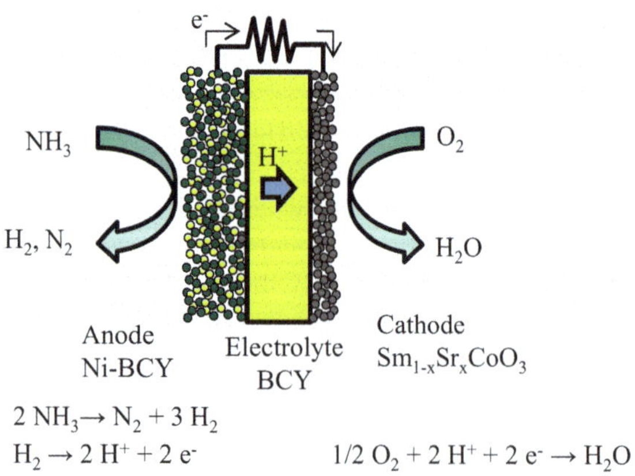

Fig. 28.7 Standard component materials and reactions of direct ammonia-fueled PCFC

hydrogen to the electrolyte, and the cell is operative at relatively low temperatures. Operation at 550 °C was possible because of the use of the hydrogen separation membrane and a very thin film of the proton conductor.

The standard PCFC is also used for direct ammonia supply. The electrolyte materials are perovskite type oxides based on $BaZrO_3$, $BaCeO_3$, and $Ba(Zr,Ce)O_3$ containing B site additives, i.e., $Ba(Ce,Gd)O_3$, $Ba(Ce,Nd)O_3$, $Ba(Ce,Y)O_3$, $Ba(Zr,Ce,Y)O_3$ [20–22]. The best composition of the proton conductive electrolyte has been under investigation. Therefore, variety of the candidate electrolyte materials have been investigated. The anode materials for ammonia-fueled PCFC generally composed of proton conductor ceramics and Ni-cermets, but the proton conductor is similarly different depending on the electrolyte materials.

The catalytic activities of cermet candidate materials for ammonia decomposition are compared in Table 28.2. Ni/alumina characterized by high surface area is listed in the table for comparison with cermet materials. Although Ni/YSZ is the most popular cermet material for the YSZ electrolyte, the catalytic activity of Ni/YSZ for ammonia decomposition was the lowest among the four cermets. On the other hand, $Ni/BaZrO_3$ and $Ni/BaCeO_3$ are expected as the candidate cermet electrode for ammonia-fueled PCFC. These cermet materials are more active than other cermet materials. The basic property of Ba-based perovskite is effectively promoting the ammonia decomposition reaction. Higher activity for ammonia cracking is requested for the anode of PCFC because of the lower operation temperatures, but some Ni/proton conductor cermet fulfill this requirement.

The I-V and I-P characteristics of the following cell have been tested (Fig. 28.8) [17].

$$NH_3, Ni^{-}BaCe_{0.75}Y_{0.25}O_{3-\delta}|BaCe_{0.9}Y_{0.1}O_{3-\delta}|Sm_{0.5}Sr_{0.5}CoO_3, O_2 \qquad (28.6)$$

The open circuit voltage was slightly lower than the theoretical one because of the hole-proton mixed conduction in an air atmosphere, but the current and power obtained were the sufficiently acceptable level. The use of ceramic proton conductor has been attracting great attention for highly efficient natural gas-fueled cells. Proton conductors and PCFCs have been regarded as materials for highly efficient fuel cells. The cell and component materials should be further developed before commercialization, e.g., transference number of electrolyte, active cathode materials, and mechanical strength of the cells. Reformate fuel from hydrocarbon has been also considered

Table 28.2 Catalytic activities for ammonia cracking on Ni (40 wt%) cermet as candidate anode materials and Ni/Al_2O_3 [17, 18]

Catalyst	Ni/YSZ	Ni/CeO_2	Ni/BZY10	Ni/BCY10	Ni/Al_2O_3
Temperature at 80% NH_3 Conversion (°C)	600	650	530	540	570
Surface area (m^2/g)	4.0	4.0	2.6	1.8	80

BCY10: $BaCe_{0.9}Y_{0.1}O_3$, BZY10: $BaZr_{0.9}Y_{0.1}O_3$

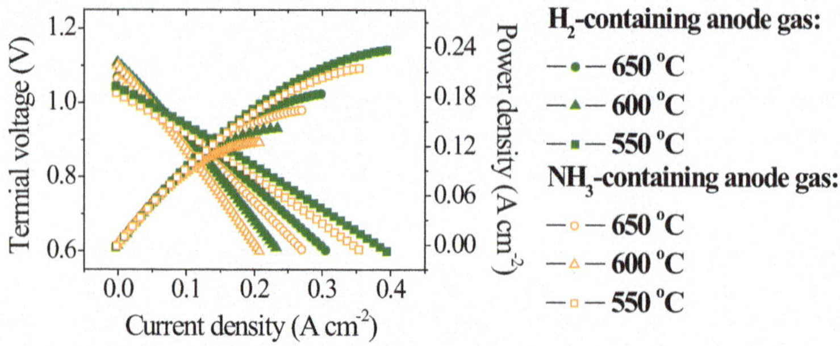

Fig. 28.8 Cell performance of Ni-$BaCe_{0.75}Y_{0.25}O_{3-\delta}$ | $BaCe_{0.9}Y_{0.1}O_{3-\delta}$ | $Sm_{0.5}Sr_{0.5}CoO_3$. Cathode gas: pure O_2; anode gas: 43% NH_3-1.4% H_2O-Ar or equivalent decomposed gas; Thickness: Anode 1.5 mm^t, Electrolyte 60 μm^t, Cathode 30 μm^t. Reproduced from Sustainable Energy & Fuels, 4, 5241 (2020) with permission from Royal Society of Chemistry

as the fuel for PCFC. However, the proton conductor materials containing Ba are potentially weak in the CO_2-containing atmosphere due to the formation of carbonate. In this respect, ammonia is suitable carbon free fuel for PCFC.

28.6 Summary

The use of ammonia as a fuel for solid oxide fuel cells is attracting from their high generation efficiency and clean energy conversion. The temperature zone of SOFC operation is suitable for ammonia cracking reactors. When ammonia supply chain is established, the SOFC systems provide electricity with high efficiencies even from small units to large generation facilities. The use of green or blue ammonia is regarded as carbon neutral energy system. The combination of the components for NG-fueled SOFC can be utilized without further modification, whereas some parts, e.g., fuel supply system and sealants, should be tolerant to corrosion with ammonia. Materials for SOFC have been developed including PCFC application. This investigation is also applicable to the ammonia-fueled SOFC.

References

1. Jeerh G, Zhang M, Tao S (2021) Recent progress in ammonia fuel cells and their potential applications. J Mater Chem A 9:727–752
2. Okanishi T, Okura K, Srifa A, Muroyama H, Matsui T, Kishimoto M, Saito M, Iwai H, Yoshida H, Saito M, Koide T, Iwai H, Suzuki S, Takahashi Y, Horiuchi T, Yamasaki H, Matsumoto S, Yumoto S, Kubo H, Kawahara J, Okabe A, Kikkawa Y, Isomura T, Eguchi K (2017) Comparative study of ammonia-fueled solid oxide fuel cell systems. Fuel Cells 3:383–390
3. Eguchi K (2016) Concept of energy carrier, candidate materials, and reactions. In: Koyama M, Fukushima Y, Nakagawa T (eds) Kato Y. Energy Technology Roadmap of Japan, Springer, pp

167–181

4. Molouk AFS, Okanishi T, MuroyamaH MT, Eguchi K (2015) Electrochemical and catalytic behaviors of Ni–YSZ anode for the direct utilization of ammonia fuel in solid oxide fuel cells. J Electrochem Soc 162:F1268-1274
5. Eguchi K (2003) Internal Reforming. In: Vielstich W, Lamm A, Gasteiger HA (eds) Handbook of fuel cells. Wiley, Fundamentals Technology and Applications, pp 1057–1069
6. Cinti G, Desider U, Penchini D, Discepoli G (2014) Experimental analysis of SOFC fuelled by ammonia. Fuel Cells 14(2):221–230
7. Yang J, Molouk AFS, Okanishi T, Muroyama H, Matsui T, Eguchi K (2015) A stability study of Ni/Yttria-stabilized zirconia anode for direct ammonia solid oxide fuel cells. ACS Appl Mater Interfaces 7:28701–28707
8. Ma Q, Peng RR, Tian L, Meng G (2006) Direct utilization of ammonia in intermediate-temperature solid oxide fuel cells. Electrochem Commun 8:1791–1795
9. Meng G, Jiang C, Ma J, Ma Q, Liu X (2007) Comparative study on the performance of a SDC-based SOFC fueled by ammonia and hydrogen. J Power Sources 173:189–193
10. Itagaki Y, Cui J, Ito N, Aono H, Yahiro H (2018) Effect of Ni-loading on Sm-doped CeO_2 anode for ammonia-fueled solid oxide fuel cell. J Ceram Soc Jpn 126:870–876
11. Song Y, Li H, Xu M, Yang G, Wang W, Ran R, Zhou W, Shao Z (2020) Infiltrated NiCo alloy nanoparticle decorated perovskite oxide: a highly active, stable, and antisintering anode for direct-ammonia solid oxide fuel cells. Small 16:2001859
12. Liu L, Sun K, Wu X, Li X, Zhang M, Zhang N, Zhou X (2012) Improved performance of ammonia-fueled solid oxide fuel cell with SSZ thin film electrolyte and Ni-SSZ anode functional layer. Int J Hydrogen Energy 37:10857–10865
13. Wang Y, Yang J, Wang J, Guan W, Chi B, Jia L, Chen J, Muroyama H, Matsui T, Eguchi K (2020) Low-temperature ammonia decomposition catalysts for direct ammonia solid oxide fuel cells. J Electrochem Soc 167(6):064501
14. Hashinokuchi M, Zhang M, Doi T, Inaba M (2018) Enhancement of anode activity and stability by Cr addition at Ni/Sm-doped CeO_2 cermet anodes in NH_3-fueled solid oxide fuel cells. Solid State Ionics 319:180–185
15. Hashinokuchi M, Yokochi R, Akimoto W, Doi T, Inaba M, Kugai J (2016) Enhancement of anode activity at Ni/Sm-doped CeO_2 cermet anodes by Mo addition in NH_3-fueled solid oxide fuel cells. Solid State Ionics 285:222–226
16. Akimoto W, Fujimoto T, Saito M, Inaba M, Yoshida H, Inagaki T (2014) Ni–Fe/Sm-doped CeO_2 anode for ammonia-fueled solid oxide fuel cells. Solid State Ionics 256:1–4
17. Miyazaki K, Muroyama H, Matsui T, Eguchi K (2020) Impact of the ammonia decomposition reaction over an anode on direct ammonia-fueled protonic ceramic fuel cells. Sustain Energy Fuels 4:5238–5246
18. Miyazaki K, Okanishi T, Muroyama H, Matsui T, Eguchi K (2017) Development of Ni–Ba(Zr, Y)$_O$3 cermet anodes for direct ammonia-fueled solid oxide fuel cells. J Power Sources 365:148–154
19. Aoki Y, Yamaguchi T, Kobayashi S, Kowalski D, Zhu C, Habazaki H (2018) High-efficiency direct ammonia fuel cells based on $BaZr_{0.1}Ce_{0.7}Y_{0.2}O_{3-\delta}$/Pd oxide-metal junctions. Glob Challenges 2:1700088
20. Lin Y, Ran R, Guo Y, Zhou W, Cai R, Wang J, Shao Z (2010) Proton-conducting fuel cells operating on hydrogen, ammonia and hydrazine at intermediate temperatures. Int J Hydrogen Energy 35:2637–2642
21. Zhang L, Yang W (2008) Direct ammonia solid oxide fuel cell based on thin proton-conducting electrolyte. J Power Sources 179:92–95
22. Xie K, Ma Q, Lin B, Jiang Y, Gao J, Liu X, Meng G (2007) An ammonia fuelled SOFC with a $BaCe_{0.9}Nd_{0.1}O_{3-\delta}$ thin electrolyte prepared with a suspension spray. J Power Sources 170:38–41

Chapter 29
Ammonia-Fueled SOFC Stack

Masashi Kishimoto

Abstract Solid oxide fuel cell (SOFC) is a type of fuel cell that can convert the chemical energy of fuels into electrical energy in a highly efficient manner. Since the electrolyte materials used in SOFCs have oxygen ion conductivity, SOFCs can utilize not only hydrogen but also other hydrogen-containing fuels, including ammonia. In addition, nickel-based catalysts, which are typically used for the anode material, have catalytic activity for ammonia decomposition reaction, therefore, ammonia can be directly supplied to the cell and hydrogen is produced within the cell. Since the ammonia decomposition reaction is endothermic, the heat released from the electrochemical reaction in the cell can be effectively used, which can improve the overall efficiency of the system. Performance of the ammonia-fueled SOFC cell/stacks are found to be almost comparable to those with hydrogen fuel, and their durability over 1000 h operation is also confirmed. However, nitriding of metallic components in the stacks is not negligible and may affect the stability of the stack performance in a much longer time scale.

Keywords SOFC stack · Direct supply · External decomposition · Conversion efficiency · Durability

29.1 Introduction

In this chapter, the utilization of ammonia for power generation by SOFC stacks is discussed. Although the most common fuel for fuel cells is hydrogen, SOFCs can also utilize hydrocarbon and ammonia fuels because the nickel-based catalyst used in the anode also acts as a reforming/decomposition catalyst. Therefore, hydrocarbon and ammonia fuels can be directly supplied to the anode side of SOFCs and generate hydrogen. Both steam reforming reaction of hydrocarbon fuels and decomposition reaction of ammonia are endothermic, these reactions can effectively utilize the excess heat generated from the electrochemical reactions in the SOFC stacks. This

M. Kishimoto (✉)
Department of Mechanical Engineering and Science, Graduate School of Engineering, Kyoto University, Kyoto 615-8540, Japan
e-mail: kishimoto.masashi.3m@kyoto-u.ac.jp

K. Aika and H. Kobayashi (eds.), *CO_2 Free Ammonia as an Energy Carrier*,
https://doi.org/10.1007/978-981-19-4767-4_29

can improve the overall efficiency of the SOFC systems. However, several issues were reported when hydrocarbon fuels are directly supplied to SOFCs. First, since the reaction rate of steam reforming reaction is high on the nickel-based anode, a steep temperature gradient tends to form on the anode, which may cause mechanical failure of the cell [1, 2]. Second, solid carbon is liable to form through the cracking reaction of hydrocarbon fuels, and it may cover up the electrochemical reaction sites and also block the gas diffusion pathways in the porous anode [3, 4]. To avoid the formation of solid carbon deposition, an excessive amount of steam needs to be added to the fuel with the steam-to-carbon ratio typically over two, however, this may complicate the system configuration and decrease the conversion efficiency of the entire system. Therefore, commercialized SOFC systems have not yet employed the direct internal reforming operation with hydrocarbon fuels. Alternatively, they use an external reformer to generate hydrogen outside the SOFC stack and hence avoid such adverse impacts on the SOFC performance.

In comparison to hydrocarbon fuels, ammonia fuel is considered to have several advantages. First, since the endothermic heat of the ammonia decomposition reaction is lower than that of the steam reforming reaction of hydrocarbons, the temperature gradient formed on the anode is expected to be limited [5]. Second, since the ammonia decomposition reaction does not require a steam supply and only yields gaseous hydrogen and nitrogen, the system configuration can be simplified. Last, but not least, ammonia is a carbon-free fuel, therefore, no additional device for carbon capture and storage is required.

In this chapter, investigations of practical-scale ammonia-fueled SOFC systems with a power output of up to 1 kW are introduced.

29.2 Comparative Study of Ammonia-Fueled SOFC Stack

29.2.1 Stack and System Configuration

In this section, the performance of practical-size ammonia-fueled SOFC cell/stacks reported in the literature is introduced. Table 29.1 summarizes the specification of the cell/stacks used in the experiments in Refs. [6–12]. Since the main focus of these studies was the direct utilization of ammonia fuel in the SOFC stacks, anode-supported cells with conventional nickel-based anode catalysts were employed to provide sufficient reaction sites for the ammonia decomposition reaction. Conventional oxygen-ion conductive zirconia-based materials were used for the electrolyte, whereas the state-of-the-art cathode materials, such as LSCF and LSC were used to pursue higher cell/stack performance. The operation temperature was in the range from 700 to 850 °C, and maintained using an electric furnace; in other words, none of the cell/stacks in the literature were thermally self-sustaining. Note that pressurized operation up to 5 atm was performed by Hung et al. [10].

Table 29.1 Specifications of ammonia-fueled SOFC cell/stacks in the literature

	Cinti (2016)	Barelli (2020)	Stoeckl (2019)	Hagen (2019)	Hung (2020)	Okanishi (2017)	Kishimoto (2020)
Number of cells	4	6	1	1	1	10	30
Cell size	10 × 10 cm^2	Not provided	10 × 10 cm^2	5.3 × 5.3 cm^2	5 × 5 cm^2	ϕ 12 cm	ϕ 12 cm
Cell active area	80 cm^2	80 cm^2	80 cm^2	16 cm^2	4 × 4 cm^2	ϕ11 cm	ϕ 11 cm
Anode material	Ni/YSZ	Ni/YSZ	Ni/YSZ & Ni/ZrO_2	Ni/YSZ	Ni/YSZ	Ni/YSZ	Ni/YSZ
Electrolyte material	YSZ	YSZ	YSZ	YSZ	YSZ	ZrO_2-based	ZrO_2-based
Cathode material	LSM/YSZ & LSM	LSCF–GDC	LSCF	LSM–YSZ	LSC–GDC	Perovskite	Perovskite
Provider	Jülich	SOLIDPower	CeramTec	Not provided	Not provided	Noritake	Noritake
Temperature	700–800 °C	750 °C	700–800 °C	750–850 °C	750–800 °C	770 °C	750 °C

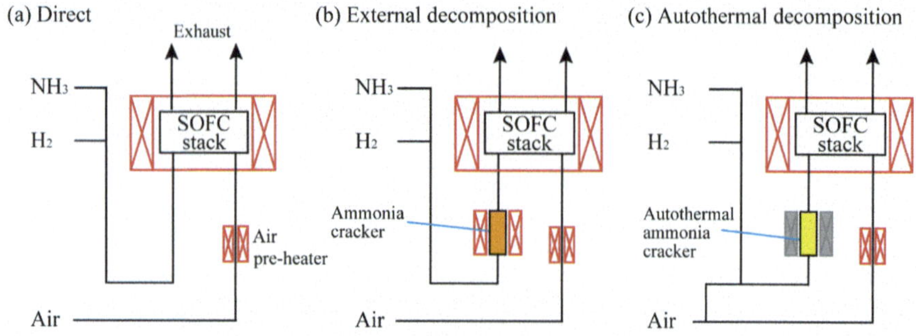

Fig. 29.1 Schematic flow diagram of ammonia-fueled SOFC systems

To utilize ammonia in SOFC systems, several operation modes can be employed, and their schematic diagrams are shown in Fig. 29.1, namely, the direct ammonia mode, the external decomposition mode, and the autothermal decomposition mode. Although fuel gas is ideally pure ammonia, an auxiliary hydrogen supply may be needed during the reduction of anode material at the beginning of the operation and also during a hibernation mode of the systems. Note that the diagrams are intended for laboratory-scale investigation. In more realistic systems for commercialization, the heaters for the external ammonia cracker and the air pre-heating should be replaced by heat recovery components, such as heat exchangers, to fully utilize the heat in the high-temperature exhaust gas from the SOFC stack. Also, the SOFC stack should be thermally insulated so that the stack can be thermally self-sustaining; the exothermic heat from the electrochemical reaction covers the endothermic heat of the ammonia decomposition reaction and also heat loss from the stack. Therefore, thermal management is of significant importance to achieve a sufficiently high energy conversion efficiency of the entire system.

29.2.2 Direct Internal Decomposition

Figure 29.1a shows the configuration of a direct-ammonia-fueled SOFC system, where ammonia is directly supplied to the SOFC stack and decomposed on the nickel-based catalyst in the anode. This configuration is among the simplest and can most effectively utilize the heat generated from the electrochemical reaction in the SOFC stack. However, thermal management of the stack becomes inevitable to mitigate the excessive temperature gradient formed on the anode.

In Refs. [6–12], the performance of SOFC stacks fueled with direct ammonia was reported to be almost comparable to that fueled with hydrogen–nitrogen mixture gas, whose composition corresponds to the fully decomposed ammonia fuel. Also, the open-circuit voltage (OCV) of the cell with the direct ammonia was almost the same as that with the hydrogen–nitrogen mixture fuel. This indicates that the utilization of ammonia proceeds in two steps in SOFCs; first ammonia is decomposed into hydrogen and nitrogen, subsequently, the produced hydrogen is electrochemically consumed in the electrochemical reaction. If ammonia was assumed to be directly oxidized in the anode, the OCV should have been much higher.

The exhaust gas from the SOFC stack was analyzed. Barelli et al. [7] reported that ammonia was almost fully decomposed in the stack and the remaining ammonia was on the ppm order, whereas Stoeckl et al. [8] reported that a maximum of 50% ammonia was detected at 700 °C. This difference can be attributed to several factors, such as the anode thickness, supplied gas flow rate, fuel channel geometry, etc. Also, it should be noted that the metallic components used for the piping and interconnectors also have catalytic activity for the ammonia decomposition reaction.

Owing to the ammonia decomposition reaction inside the stacks, the stack temperature was reported to decrease: 3 °C in Ref. [6], 20 °C in Ref. [8], and 15 °C in Ref. [12]. This variation in the temperature drop may be due to the stack configuration and its thermal boundary condition. The decrease in cell temperature is likely to lower the electrochemical activity of the electrodes and the ionic conductivity of the electrolyte phase. In fact, a noticeable increase in the ohmic resistance was observed by Stoeckl et al. [8] by electrochemical impedance spectroscopy.

To develop practical SOFC systems that can thermally self-sustain without the help of an electric furnace, thermal management is important to balance the various heat loss with the exothermic electrochemical reaction. Generally, since the stacks generate excess heat during operation, a large amount of air is supplied to the cathode side, which consumes a non-negligible amount of electric energy in the air supply pump. Since the ammonia decomposition reaction can effectively utilize the heat generated from the electrochemical reaction, the internal decomposition operation can reduce the air flow rate and hence increase the total efficiency of the system. Cinti et al. [6] developed a mathematical model of an ammonia-fueled SOFC stack and reported that the air flow rate can be reduced by 30% compared with the hydrogen-fueled SOFC systems.

The largest electrical power output as of 2021 obtained from the ammonia-fueled SOFC stack was reported by Kishimoto et al. [12] as 1048 W (= 0.368 W/cm^2) at the current load of 50 A from a stack consisting of 30 planar cells as shown in Fig. 29.2. The conversion efficiency was 52.3%, which is comparable to the commercial SOFC systems fueled with reformed methane. Further improvement in the conversion efficiency can be achieved by optimizing the operation condition.

The pressurized operation was performed by Hung et al. [10]. They found that pressurization decreases the concentration resistance in the anode and that ammonia can be decomposed into hydrogen and nitrogen even under pressurized conditions, which are thermodynamically unfavorable for the ammonia decomposition reaction.

29.2.3 External Decomposition

Figure 29.1b shows the configuration of a combined system of an external ammonia cracker and an SOFC stack. In this configuration, ammonia is decomposed in the

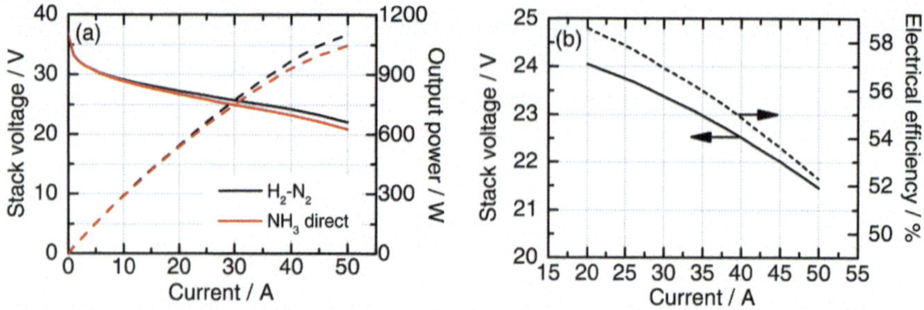

Fig. 29.2 Performance of the SOFC stack fueled with direct ammonia fuel, **a** *I*–*V* and *I*–*P* characteristics of the stack with direct ammonia fuel (F_{NH_3} = 8.19 L min^{-1}, F_{Air} = 50 L min^{-1}) and hydrogen–nitrogen mixture fuel (F_{H_2} = 12.3 L min^{-1}, F_{N_2} = 4.10 L min^{-1}, F_{Air} = 50 L min^{-1}), **b** dependency of the stack voltage and electrical efficiency on the current load at a constant fuel utilization of 80%. Reprinted from [12]

ammonia cracker before being supplied to the SOFC stack. In this configuration, designated catalysts are used in the ammonia cracker and its operating temperature can be independently determined to achieve the highest ammonia decomposition rate. On the other hand, the external ammonia cracker requires heat to sustain the endothermic decomposition reaction, therefore, proper thermal management of the entire system is indispensable to maintain the temperature of the cracker within a suitable range by effectively recycling the heat in the exhaust gas from the SOFC stack.

Various metal catalysts have been investigated for the ammonia decomposition reaction, and Yin et al. [13] found that ruthenium-based catalysts have the highest catalytic activity for ammonia decomposition and that use of catalyst supports and promoters can further enhance the catalyst activity. However, ruthenium is categorized as a rare transition metal and its use in SOFC systems will raise an issue of the cost competitiveness of the systems. Therefore, the development of alternative catalysts using abundant materials is required. Also, the long-term stability of the catalysts needs to be confirmed for the stable operation of the SOFC systems. Okanishi et al. [11] developed a nickel-based catalyst using Y_2O_3 as a catalyst support and SrO as an additive. It was confirmed that the ammonia decomposition rate was almost 100% at temperatures over 600 °C at a space velocity (SV) of 2400 h^{-1}. The stability of the catalyst was also investigated for over 1000 h at 515 °C and 565 °C and no degradation was detected. After that, Kishimoto et al. [12] developed a BaO/Ni/Sm_2O_3/MgO catalyst to pursue the further enhancement of the nickel-based catalyst. The ammonia decomposition rate was found to increase at lower temperatures and almost 100% decomposition was achieved at around 550 °C. Long-term stability was also confirmed at 700 °C.

The performance of the SOFC stack accompanied with an external ammonia cracker has not been widely investigated on a practical scale. The only available data was reported by Kishimoto et al. [12] using a 30-cell stack with the aforementioned nickel-based catalyst. As anticipated, the performance of the SOFC stack fueled with

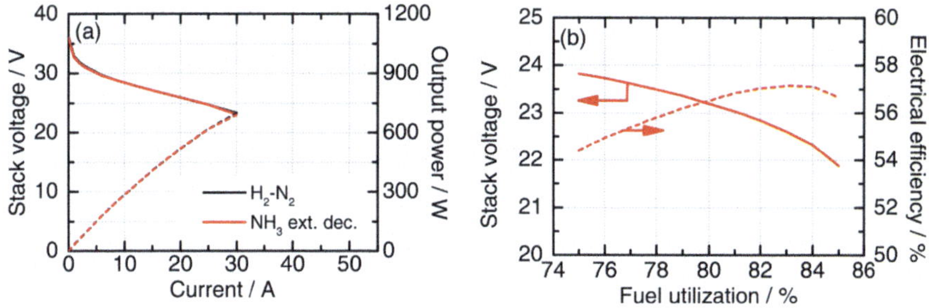

Fig. 29.3 Performance of the SOFC stack fueled with externally decomposed ammonia. **a** *I–V* and *I–P* characteristics of the stack with externally decomposed ammonia ($F_{NH_3} = 5.22$ L min^{-1}, $F_{Air} = 30$ L min^{-1}) and with hydrogen–nitrogen mixture fuel, **b** dependency of the stack voltage and the electrical efficiency on the fuel utilization at the current load of 30 A. Reprinted from [12]

externally decomposed ammonia shows identical performance with that fueled with hydrogen–nitrogen mixture fuel (H_2:N_2 = 3:1). The electrical conversion efficiency of 57.1% was achieved at a current load of 30 A, fuel utilization of 83%, and power output of 678 W (= 0.238 W/cm^2) as shown in Fig. 29.3.

29.2.4 Autothermal Decomposition

Figure 29.1c shows the configuration of a combined system of an autothermal ammonia cracker and an SOFC stack. Similar to the external ammonia cracker, the supplied ammonia is decomposed in the autothermal ammonia cracker, and hydrogen–nitrogen mixture fuel is supplied to the SOFC stack. The difference is that a small amount of air or oxygen is added to the supplied ammonia and the corresponding amount of ammonia is combusted to produce the heat required for decomposing the remaining ammonia. In this way, heat recovery from the exhaust gas is not mandatory. Therefore, the autothermal ammonia cracker can prove its merits especially during the start-up of an SOFC system when exhaust heat from the stack is not obtained. The autothermal ammonia cracker can heat not only the cracker itself but also the SOFC stack to its operating temperature. However, since a part of ammonia is consumed by combustion and heat recycling is not performed, the energy conversion efficiency of the system becomes lower than that of the other configurations. Therefore, the utilization of the autothermal ammonia cracker should be limited only to the start-up of the system.

The performance of an SOFC stack fueled with autothermal ammonia cracker was reported by Kishimoto et al. [12]. They used a Co-Ce-Zr composite oxide as a catalyst and supplied a mixture of ammonia and air with an air-to-fuel ratio (AFR) of 0.45–075. They confirmed that the autothermal ammonia cracker was able to thermally self-sustain without using a heater. The OCV values of the 30-cell stack obtained in the experiment were 31.3, 30.9, and 30.6 V for the AFR of 0.45, 0.60, and 0.75, respectively, which are in good agreement with the theoretical values of 31.3, 30.8, and 30.5 V, respectively, which indicates that the oxygen was fully used for the catalytic ammonia combustion and that the remaining ammonia was fully decomposed into hydrogen and nitrogen in the reactor. Although the OCV values were lower than that from the direct ammonia fuel, the terminal voltage and output power during the operation were almost identical among the four cases investigated. This indicates that the combustion products do not have an adverse influence on the stack performance.

29.3 Long-Term Durability

The stability of the performance of the ammonia-fueled SOFC stacks is of importance for their practical applications. Although the long-term stability of the cell/stacks has rarely been investigated in the literature, Hagen et al. [9] and Kishimoto et al. [12] performed a durability test with ammonia fuel for over 1000 h. Hagen et al. operated a single planar cell with direct ammonia fuel for 1500 h, and found that the degradation rate was about 2–4%/1000 h in the terminal voltage. Since a similar degradation rate was also found in the cell fueled with hydrogen–nitrogen gas mixture, they concluded that there was no additional degradation mechanism unique to the ammonia fuel. Kishimoto et al. [12] performed a 1000 h durability test of a 30-cell stack fueled with direct ammonia and with externally decomposed ammonia. They also reported that there was no noticeable degradation in the cell performance in both fuel-supplying modes. The output power of the stack was maintained at 690 W with an electrical efficiency of 55.9%.

Considering the performance and durability of the ammonia-fueled SOFC stacks reported in the literature, ammonia seems to be a promising fuel for SOFC operation. However, when metallic components of the stack, including the anode catalyst, piping, and interconnectors, are exposed to ammonia fuel at high temperatures, these materials are liable to be nitrided, which can alter their surface morphology and even deform their structure. Yang et al. [14] exposed a Ni-YSZ anode material in an ammonia atmosphere at 600–700 °C and analyzed its microstructure by scanning electron microscope (SEM) and energy-dispersive X-ray spectroscopy (EDX). They found that the nickel surface was significantly roughened and many small pores with a diameter less than 100 nm were formed in the particles. Also, the existence of nitrogen was confirmed on the nickel surface. These are considered to affect the electrochemical activity and electric conductivity of the anode material. In fact, an increase in the ohmic resistance and activation resistance was confirmed on the button-cell scale. However, quantitative information that correlates the performance with the change in the anode morphology is still limited. Further investigation is required to clarify the influence of nickel nitriding on cell performance.

In the stack scale, nitriding of metallic interconnectors was also reported by Kishimoto et al. [12]. After the 1000 h durability test, the stack was disassembled and the interconnectors (SUS430) were analyzed using an electron probe microanalyzer

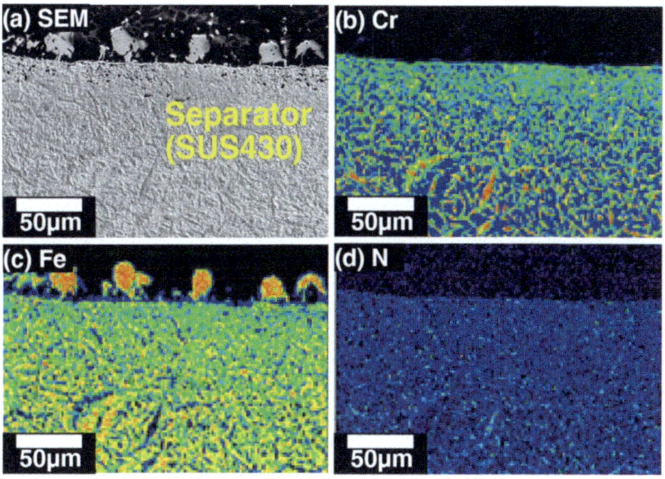

Fig. 29.4 Cross-sectional images of the separator on the anode side after 1000 h durability test with direct ammonia fuel, **a** scanning electron microscopy image, and the distribution of **b** Cr, **c** Fe, and **d** N. Reprint from [12]

(EPMA). The obtained images indicated that the surface of the interconnector on the anode side was significantly affected by the ammonia fuel. Figure 29.4 shows the cross-section of the separator including its surface on the anode side. Fe-rich particles with a diameter of around 30 μm were found on the surface of the interconnector, and also nitrogen element was detected inside the interconnector. These indicate that the interconnector materials are also prone to react with ammonia and nitriding occurs. Although this morphology change did not influence the stack performance during the durability test for 1000 h, it can alter the fuel flow on the anode side, and eventually degrade the stack performance in a much longer time scale. Although it is too early to draw a decisive conclusion on the effect of nitriding on the cell/stack performance, countermeasures may need to be taken to realize the stable operation of the cell/stacks. Development of new interconnector materials and barrier coating techniques are considered as possible approaches.

References

1. Iwai H, Yamamoto Y, Saito M, Yoshida H (2011) Numerical simulation of intermediate-temperature direct-internal-reforming planar solid oxide fuel cell. Energy 36:2225–2234
2. Sugihara S, Iwai H (2020) Experimental investigation of temperature distribution of planar solid oxide fuel cell: Effects of gas flow, power generation, and direct internal reforming. Int J Hydrogen Energy 45:25227–25239
3. Koh J-H, Yoo Y-S, Park J-W, Lim HC (2002) Carbon deposition and cell performance of Ni-YSZ anode support SOFC with methane fuel. Solid State Ionics 149:157–166
4. Stoeckl B, Subotić V, Reichholf D, Schroettner H, Hochenauer C (2017) Extensive analysis

of large planar SOFC: operation with humidified methane and carbon monoxide to examine carbon deposition based degradation. Electrochim Acta 256:325–336
5. Ni M (2011) Thermo-electrochemical modeling of ammonia-fueled solid oxide fuel cells considering ammonia thermal decomposition in the anode. Int J Hydrogen Energy 36:3153–3166
6. Cinti G, Discepoli G, Sisani E, Desideri U (2016) SOFC operating with ammonia: Stack test and system analysis. Int J Hydrogen Energy 41:13583–13590
7. Barelli L, Bidini G, Cinti G (2020) Operation of a solid oxide fuel cell based power system with ammonia as a fuel: experimental test and system design. Energies 13:6173
8. Stoeckl B, Subotić V, Preininger M, Schwaiger M, Evic N, Schroettner H, Hochenauer C (2019) Characterization and performance evaluation of ammonia as fuel for solid oxide fuel cells with Ni/YSZ anodes. Electrochim Acta 298:874–883
9. Hagen A, Langnickel H, Sun X (2019) Operation of solid oxide fuel cells with alternative hydrogen carriers. Int J Hydrogen Energy 44:18382–18392
10. Hung YT, Shy SS (2020) A pressurized ammonia-fed planar anode-supported solid oxide fuel cell at 1–5 atm and 750–850°c and its loaded short stability test. Int J Hydrogen Energy 45:27597–27610
11. Okanishi T, Okura K, Srifa A, Muroyama H, Matsui T, Kishimoto M, Saito M, Iwai H, Yoshida H, Saito M, Koide T, Iwai H, Suzuki S, Takahashi Y, Horiuchi T, Yamasaki H, Matsumoto S, Yumoto S, Kubo H, Kawahara J, Okabe A, Kikkawa Y, Isomura T, Eguchi K (2017) Comparative study of ammonia-fueled solid oxide fuel cell systems. Fuel Cells 17:383–390
12. Kishimoto M, Muroyama H, Suzuki S, Saito M, Koide T, Takahashi Y, Horiuchi T, Yamasaki H, Matsumoto S, Kubo H, Takahashi N, Okabe A, Ueguchi S, Jun M, Tateno A, Matsuo T, Matsui T, Iwai H, Yoshida H, Eguchi K (2020) Development of 1 kw-class ammonia-fueled solid oxide fuel cell stack. Fuel Cells 20:80–88
13. Yin SF, Xu BQ, Zhou XP, Au CT (2004) A mini-review on ammonia decomposition catalysts for on-site generation of hydrogen for fuel cell applications. Appl Catal A 277:1–9
14. Yang J, Molouk AF, Okanishi T, Muroyama H, Matsui T, Eguchi K (2015) A stability study of Ni/yttria-stabilized zirconia anode for direct ammonia solid oxide fuel cells. ACS Appl Mater Interfaces 7:28701–28707

Chapter 30
Numerical Modeling of Ammonia-Fed SOFCs

Hiroshi Iwai

Abstract For the safe and efficient operation of SOFCs, it is important to understand and properly control the internal phenomena during operation. It is, however, generally difficult to measure the internal state of a fuel cell operating at a high temperature such as SOFC. For this reason, numerical analysis plays an important role, and analytical models at various scales have been developed. Another advantage of using numerical models includes that the geometry of cells and stacks can be easily changed, and predictions can be made under extreme conditions that are difficult to achieve experimentally. This chapter outlines a numerical model for an ammonia-fueled SOFC. For ammonia-fueled SOFCs, the basic framework of the numerical model is the same as for conventional fuels, with the inclusion of the ammonia decomposition reaction.

Keywords Numerical modeling · Solid oxide fuel cells · Ammonia · Reaction rate

30.1 Introduction

In this chapter, the numerical analyses of ammonia-fueled SOFCs are discussed. When ammonia is supplied as fuel to the SOFCs, most of the supplied ammonia is first decomposed into hydrogen and nitrogen through a thermal decomposition reaction in the operating temperature range of a typical SOFC.

$$NH_3 \rightarrow \frac{1}{2}N_2 + \frac{3}{2}H_2 \tag{30.1}$$

The generated hydrogen is used for the electrochemical reaction at the fuel electrode to generate electricity.

H. Iwai (✉)
Department of Mechanical Engineering and Science, Graduate School of Engineering, Kyoto University, Nishikyo-ku, Kyoto 615-8540, Japan
e-mail: iwai.hiroshi.4x@kyoto-u.ac.jp

K. Aika and H. Kobayashi (eds.), *CO_2 Free Ammonia as an Energy Carrier*,
https://doi.org/10.1007/978-981-19-4767-4_30

$$H_2 + \frac{1}{2}O^{2-} \rightarrow H_2O + 2e \quad \text{(SOFC)} \tag{30.2}$$

$$H_2 \rightarrow 2H^+ + 2e \quad \text{(PCFC: proton-conducting ceramic fuel cell)} \tag{30.3}$$

Since the decomposition reaction of ammonia is an endothermic reaction, the use of the heat generated by the power generating electrochemical reaction and various losses in the cell would be an advantage in terms of heat utilization. One may notice that this reforming of fuels through endothermic reactions to produce hydrogen and the use of the generated hydrogen in power generation reactions is similar to the SOFC operation involving steam reforming of hydrocarbon fuels such as methane. As in the case of hydrocarbon fuels, there are mainly two types of reforming operations when using ammonia as a fuel: external reforming, which uses an external heat source to decompose ammonia, and internal reforming, which uses the heat generated in the SOFCs. The latter is further classified into indirect internal reforming, in which a reformer dedicated to ammonia decomposition is installed near the SOFC, and direct internal reforming, in which ammonia is directly supplied to the fuel electrode of the SOFC. One of the advantages of ammonia fuel over hydrocarbon fuels is that the reforming of hydrocarbon fuels requires the supply of a sufficient amount of water vapor to prevent carbon formation, while ammonia does not require water vapor because there is no risk of coking.

There have been many fundamental studies on the power generation performance of SOFCs when ammonia is supplied as fuel. Some use dedicated catalysts for ammonia decomposition (e.g. [1]) and others directly supply ammonia (e.g. [2, 3]). It has been reported that the performance of the ammonia-fed cells is close to that of cells supplied with mixed hydrogen/nitrogen fuel when the ammonia decomposition reaction progresses to some extent at the elevated temperature of SOFCs. The OCV of the ammonia-fed cell is also close to that of the hydrogen-fed cell and its temperature dependency indicates that the hydrogen kinetics, not direct oxidation of ammonia, is the main contributor to the electrochemical reaction on the anode electrode [3]. Therefore, in the numerical analysis of ammonia-fueled SOFCs, the ammonia decomposition reaction and the power generation reaction at the anode, which is the electrochemical oxidation reaction of hydrogen in SOFC, are usually modeled separately and implemented in the simulation.

For the electrochemical oxidation of hydrogen, the same model can be used as in conventional SOFC simulations using hydrogen or hydrocarbon reforming gas as fuel. In particular, under conditions where ammonia decomposition is already almost complete at the time of being fed to the SOFC by external reforming or indirect internal reforming, there is no significant change from conventional simulations, although there is a dilution effect of the fuel by generated nitrogen. Therefore, in the following sections of this chapter, unless otherwise noted, we will mainly focus on direct internal reforming SOFCs.

30.2 Reaction Model for Ammonia Decomposition on Anode

The fuel electrode of a SOFC often contains nickel, which also acts as a catalyst for the ammonia decomposition reaction. When ammonia is supplied to the SOFCs, some of the ammonia may undergo thermal decomposition in the gas phase, but most of it is decomposed into hydrogen and nitrogen using the nickel in the anode electrode as a catalyst in the temperature range of typical SOFCs. In order to incorporate this reaction into a numerical analysis, it is necessary to model the reaction rate. There are various methods available. The choice depends on the accuracy required for the calculation, the allowable computational load, and other conditions.

Assumptions that all ammonia decomposes on the surface of the fuel electrode [4, 5] or that the composition of the gas mixture is always in equilibrium [6] can be found in the literature as simple methods to describe ammonia decomposition. These assumptions are, however, not always valid, as they may require special conditions, such as extremely high reaction rates.

Kinetics of ammonia decomposition reaction on the nickel and other metal catalysts at high temperatures has been investigated by several research groups and the ammonia decomposition rate has been found to depend on gas partial pressures and catalyst temperature [7, 8]. On the basis of the experimental data for Ni/SBA-15 (SBA: Santa Barbara Amorphous) and Ni/Al_2O_3 catalysts, considering that the reaction rate mainly depends on the temperature and the partial pressure of ammonia, Ni proposed the following equation as the reaction rate equation for ammonia decomposition in a typical Ni-YSZ cermet with a Ni loading of about 50 vol% [9–11].

$$r = 4.0 \times 10^{15} \exp\left(-\frac{196{,}200}{RT}\right) p_{NH_3} \text{ mol m}^{-3}\,\text{s}^{-1} \text{ of } NH_3 \tag{30.4}$$

where R, T, p_{NH_3} denote the gas constant, temperature, and the partial pressure of ammonia, respectively.

Kishimoto et al. [12] conducted reaction experiments on ammonia decomposition using Ni-YSZ porous cermet, which is one of the typical SOFC fuel anodes and observed the microstructure of the porous cermet after the experiments by FIB-SEM. From the structural data obtained by FIB-SEM, the characteristic values of porous cermets, such as the volume fractions, tortuosity factors, surface area density, and contact area density of each phase were quantified. By combining the results of the reaction experiments with the structural data, the following equation was proposed as the reaction rate equation per unit nickel-pore contact area in Ni-YSZ cermet. Note that the inhibition effect of hydrogen is considered in this model.

$$r = 98.4 \exp\left(-\frac{1.20 \times 10^5}{RT}\right) p_{NH_3}^{0.69} \left(p_{H_2} + 750\right)^{-0.39} \text{ mol m}^{-2}\,\text{s}^{-1} \text{ of } NH_3 \tag{30.5}$$

Since the model treats the reaction and microstructure separately, it is possible to predict the ammonia decomposition reaction at various Ni-YSZ anodes by measuring or assuming the microstructure.

The reaction models presented above are the overall reaction models of ammonia decomposition. Although these reaction models are simple and computationally inexpensive, they may not be suitable for a detailed study of the reaction. Luo et al. [13] proposed an elementary reaction model of the ammonia decomposition reaction and incorporated it into the numerical analysis along with the power generation reactions. This allows for more detailed discussions, such as rate-limiting steps, which are not possible with the overall reaction models, while it is inevitably computationally expensive.

30.3 Numerical Modeling

Numerical simulation has been a powerful tool to investigate the SOFC cells because the direct measurements of physicochemical properties within the microporous electrodes are generally difficult, which is also true for ammonia-fueled SOFCs. Some early examples of numerical simulations of ammonia-fueled SOFCs include the work of Ni and his co-workers. They developed numerical models for ammonia-fed SOFC in both oxide-ion-conducting and proton-conducting electrolyte cases. With the advancement of commercial software, numerical analysis of SOFCs is now performed at various levels such as button cell, single-cell, stack, and system.

The basic structure of the numerical analysis of ammonia-fueled SOFCs is the same as that of SOFCs using conventional fuels such as hydrogen and hydrocarbons. It would be ideal if a time-dependent 3D analysis with sufficient spatial and temporal resolution could be performed, but the computational load of such a calculation would be inevitably high, and depending on the application, such detailed results may not be necessary. Therefore, calculations are often carried out using appropriate simplified models by introducing various assumptions. The following sections provide an overview at a different level.

30.3.1 1D Model

One-dimensional models are usually used to investigate the fundamental phenomena in the thickness direction of the cell or electrode to be often compared with experimental data of a small button cell. The electrodes are assumed to be homogeneous porous materials. Since the simulation focuses on a small part of the cell, it is common to assume that the temperature field is uniform in the computational domain. As a result, the governing equations for the transport phenomena are the species conservation equations for the multi-component gas system and the charge conservation equations for electrons and ions. In ammonia-fueled SOFCs, the gas at the fuel electrode side is a multi-component mixture, so a gas diffusion model that can handle multi-component gases, such as the Stefan-Maxwell model, Stefan-Maxwell-Knudsen model, dusty-gas model, or cylindrical pore interpolation model, is adopted. Note that, for simplicity, some of the conservation equations may be omitted, for example, assuming that the electron potential is uniform in the direction of the electrode thickness. Those conservation equations are connected through the electrochemical reaction model. A Butler–Volmer-type model or its simplified model is most commonly employed, which describes the correlation among the gas concentrations, electron potential, ion potential, temperature, and charge transfer current at the reaction site. As explained in the previous section, there are options for handling the ammonia decomposition reaction. It can be assumed that the reaction occurs completely at the anode surface, or a model can be adopted in which the reaction proceeds in the porous anode.

An example of a one-dimensional model framework for a Ni-YSZ anode fueled by ammonia is briefly introduced below. In this model, gas diffusion, ionic conduction, and electronic conduction in a homogeneous porous anode are considered. Local ammonia decomposition and electrochemical reactions in the electrode are taken into account. Figure 30.1 schematically shows the Ni-YSZ anode and reactions considered. The coordinates are set in the thickness direction, for example, from the anode surface to the anode-electrolyte interface.

Conservation of electrons and ions is described as follows using the electrical potential of electron, ϕ_{el}, and that of ion, ϕ_{io}, in the solid phase:

$$\nabla \cdot \left(\sigma_{\mathrm{el}}^{\mathrm{eff}} \nabla \phi_{\mathrm{el}}\right) = i_{\mathrm{ct}} \tag{30.6}$$

$$\nabla \cdot \left(\sigma_{\mathrm{io}}^{\mathrm{eff}} \nabla \phi_{\mathrm{io}}\right) = -i_{\mathrm{ct}} \tag{30.7}$$

where $\sigma_{\mathrm{el}}^{\mathrm{eff}}$, $\sigma_{\mathrm{io}}^{\mathrm{eff}}$, and i_{ct} denote the effective electronic and ionic conductivities, and the volumetric density of the charge-transfer current. Employing the dusty-gas model, the multi-component gas diffusion in the porous anode is described as follows

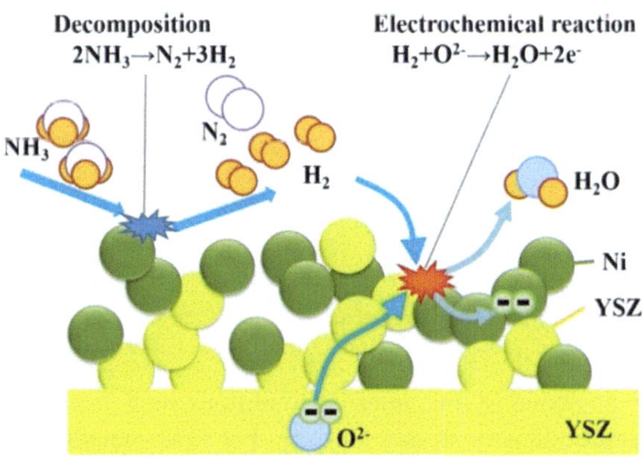

Fig. 30.1 Schematic diagram of an SOFC anode with direct ammonia supply

$$\nabla \cdot N_i = \frac{z_i i_{ct}}{F} + z_i' R_{dec} \tag{30.8}$$

$$\frac{N_i}{D_{i,K}^{eff}} + \sum_{j \neq i} \frac{X_j N_i - X_i N_j}{D_{ij}^{eff}} = -\frac{P_t}{RT} \nabla X_i - \frac{X_i}{RT} \left(1 + \frac{K P_t}{\mu D_{i,K}^{eff}}\right) \nabla P_t \tag{30.9}$$

X_i and N_i are the molar fraction and molar flux of gas species i, respectively. P_t is the total pressure. R and F are the gas constant and Faraday constant. μ and K are the mixture viscosity and permeability, while $D_{i,K}^{eff}$ and D_{ij}^{eff} are the effective Knudsen and molecular diffusion coefficients, respectively. z_i is the stoichiometric coefficient for the electrochemical oxidation of hydrogen while z_i' is the stoichiometric coefficient for the ammonia decomposition reaction. R_{dec} is the ammonia decomposition rate, which is obtained by multiplying the volume density of Ni-Pore contact surface area at the anode, $S_{Ni\text{-}Pore}$, by the reaction rate expressed in Eq. (30.5).

$$R_{dec} = 98.4 \exp\left(-\frac{1.20 \times 10^5}{RT}\right) P_{NH_3}^{0.69} \left(P_{H_2} + 750\right)^{-0.39} S_{Ni\text{-}Pore} \tag{30.10}$$

The charge-transfer current density, i_{ct}, can be modeled by the Butler–Volmer-type equation

$$i_{ct} = i_{0,TPB} l_{TPB} \left\{ \exp\left(\frac{\alpha n_e F}{RT} \eta_{act}\right) - \exp\left(-\frac{(1-\alpha) n_e F}{RT} \eta_{act}\right) \right\} \tag{30.11}$$

where $i_{0,TPB}$ and l_{TPB} are the exchange current density per unit TPB (triple-phase boundary) length and the TPB density. $S_{Ni\text{-}Pore}$ and l_{TPB} are characteristic values determined by the microstructure of the porous anode. The effective electronic and ionic conductivities, effective Knudsen and molecular diffusion coefficients, and permeability also depend on the microstructure of the anode. By numerically solving the above governing equations with appropriate boundary conditions, the distribution of the local reaction rate, gas partial pressure, potential, and various losses in the fuel electrode can be obtained.

Some examples of 1D simulations of ammonia-fed SOFCs include the work by Ni and his co-workers [4, 10, 11], Luo et al. [13], and Kalinci et al. [14] for both SOFC and PCFC fed with ammonia. Among these, Luo et al. [13] developed an elementary reaction kinetic model for NH_3-fueled SOFCs allowing for discussion of the rate-determining step in the system.

30.3.2 2D/3D Model

In an actual SOFC, internal conditions such as gas concentration, temperature, and current density are not uniform. For example, since fuel and oxygen are consumed

Fig. 30.2 Computational domain for two-dimensional simulation of an anode-supported planar SOFC

by electrochemical reactions, their concentration decreases as they flow downstream of each flow path. To capture these effects, CFD-based two- or three-dimensional numerical analysis is effective. Owing to the high computational load of 3D simulations, 2D models are relatively commonly used for SOFC simulations.

Similar to the 1D simulation in the previous section, various modeling can be done in 2D simulation. Here, an example of a two-dimensional model framework for a planar SOFC fueled by ammonia is introduced [15]. Figure 30.2 schematically shows the computational domain to be considered. It consists of seven subdomains: upper metal separator, fuel channel, porous anode, dense electrolyte, porous cathode, air channel, and lower metal separator. A co-flow configuration is adopted, where the fuel and air flow in the same direction. Mass, momentum, energy, species, and charge conservation equations are considered and those governing equations are connected by the electrochemical reactions and the ammonia decomposition reaction. The anode and cathode electrodes are assumed to be homogeneous porous materials, respectively.

The following equations are the governing equations for thermo-fluid fields. The fuel and air flows are assumed to be laminar. The gas properties depend on the local temperature and concentration of chemical species in the gas mixture. All of the variables in the equations are intrinsic volume averages except for the velocity components u and v; the total volume averages are used for these velocity components. Forchheimer-extended Darcy model is employed in the momentum equations so that the same equations are to be used both for the gas channels and the porous regions. For the energy equation, a local thermal non-equilibrium model is employed.

The continuity equation in the gas phase is expressed as follows.

$$\frac{\partial(\rho u)}{\partial x}+\frac{\partial(\rho v)}{\partial y}=S \tag{30.12}$$

where S is the source/sink term associated with the reactions in the electrodes.

The momentum conservation equation in the fluid phase is expressed as follows.

$$\left[\frac{\partial}{\partial x}\left(\frac{\rho u u}{\varepsilon}\right)+\frac{\partial}{\partial y}\left(\frac{\rho u v}{\varepsilon}\right)\right]=-\varepsilon\frac{\partial p}{\partial x}+\frac{\partial}{\partial x}\left(\mu\frac{\partial u}{\partial x}\right)+\frac{\partial}{\partial y}\left(\mu\frac{\partial u}{\partial y}\right)$$

$$
\begin{aligned}
&+\frac{1}{3}\frac{\partial}{\partial x}\left(\mu\frac{\partial u}{\partial x}\right)-\frac{2}{3}\frac{\partial}{\partial x}\left(\mu\frac{\partial v}{\partial y}\right)+\frac{\partial}{\partial y}\left(\mu\frac{\partial v}{\partial x}\right)\\
&-\varepsilon\frac{\mu}{K}u-\varepsilon\frac{\rho f}{\sqrt{K}}u\sqrt{u^2+v^2} \qquad (30.13)
\end{aligned}
$$

$$
\begin{aligned}
\left[\frac{\partial}{\partial x}\left(\frac{\rho vu}{\varepsilon}\right)+\frac{\partial}{\partial y}\left(\frac{\rho vv}{\varepsilon}\right)\right] = &-\varepsilon\frac{\partial p}{\partial y}+\frac{\partial}{\partial x}\left(\mu\frac{\partial v}{\partial x}\right)+\frac{\partial}{\partial y}\left(\mu\frac{\partial v}{\partial y}\right)\\
&+\frac{1}{3}\frac{\partial}{\partial y}\left(\mu\frac{\partial v}{\partial y}\right)-\frac{2}{3}\frac{\partial}{\partial y}\left(\mu\frac{\partial u}{\partial x}\right)+\frac{\partial}{\partial x}\left(\mu\frac{\partial u}{\partial y}\right)\\
&-\varepsilon\frac{\mu}{K}v-\varepsilon\frac{\rho f}{\sqrt{K}}v\sqrt{u^2+v^2} \qquad (30.14)
\end{aligned}
$$

where ε is the porosity of the electrodes; μ is the mixture viscosity; K is the permeability of the electrodes; f is the Forchheimer coefficient. Momentum conservation in the channels was obtained by applying $\varepsilon = 1$ and $K = \infty$.

The energy equation in the fluid phase in the porous electrodes is expressed as follows.

$$
\frac{\partial(\rho c_p u T_f)}{\partial x}+\frac{\partial(\rho c_p v T_f)}{\partial y}=\frac{\partial}{\partial x}\left(\lambda_f^{\mathrm{eff}}\frac{\partial T_f}{\partial x}\right)+\frac{\partial}{\partial y}\left(\lambda_f^{\mathrm{eff}}\frac{\partial T_f}{\partial y}\right)+h_{sf}a_{sf}(T_s-T_f) \qquad (30.15)
$$

where T_f and T_s are the temperature of the fluid phase and the solid phase, respectively; c_p and λ_f^{eff} is the heat capacity and effective thermal conductivity of the gas mixture, respectively; h_{sf} is the heat transfer coefficient at the gas–solid interface; a_{sf} is the specific surface area of the porous electrode. The energy equation in the gas phase in the channels is obtained by using the bulk property for λ_f^{eff} and omitting the last term in Eq. (30.15).

The energy equation in the solid phase of the porous electrodes is expressed as follows.

$$
0=\frac{\partial}{\partial x}\left(\lambda_s^{\mathrm{eff}}\frac{\partial T_s}{\partial x}\right)+\frac{\partial}{\partial y}\left(\lambda_s^{\mathrm{eff}}\frac{\partial T_s}{\partial y}\right)+h_{sf}a_{sf}(T_f-T_s)+Q_{\mathrm{ct}}+Q_{\mathrm{dec}}+Q_{\mathrm{loss}} \qquad (30.16)
$$

where λ_s^{eff} is the effective thermal conductivity of the solid phase; and Q_{ct}, Q_{dec}, and Q_{loss} are the source/sink term associated with the exothermic electrochemical reaction, the endothermic ammonia decomposition reaction, and the heat generation associated with various losses in the cell, respectively. Energy equations in the other bulk solid phases are obtained by using the bulk property for λ_s^{eff} and omitting the third, fourth, and fifth terms in Eq. (30.16).

Species conservation in the fluid phase in the porous electrodes is expressed as follows.

$$\frac{\partial}{\partial x}(\rho u Y_i) + \frac{\partial}{\partial y}(\rho v Y_i) = \frac{\partial}{\partial x}\left(\rho D_{i,\mathrm{mix}}^{\mathrm{eff}} \frac{\partial Y_i}{\partial x}\right) + \frac{\partial}{\partial y}\left(\rho D_{i,\mathrm{mix}}^{\mathrm{eff}} \frac{\partial Y_i}{\partial y}\right) + S_i \quad (30.17)$$

where Y_i is the mass fraction of the chemical species; $D_{i,\mathrm{mix}}^{\mathrm{eff}}$ is the diffusivity in the gas mixture; and S_i is the source/sink term associated with the electrochemical reaction and the ammonia decomposition reaction. Species conservation in the channels is obtained by applying the bulk molecular diffusivity to $D_{i,\mathrm{mix}}^{\mathrm{eff}}$.

Charge conservation equations in the cell are expressed as follows, similar to the 1D model in the previous section.

$$\frac{\partial}{\partial x}\left(\sigma_{\mathrm{el}}^{\mathrm{eff}} \frac{\partial \phi_{\mathrm{el}}}{\partial x}\right) + \frac{\partial}{\partial y}\left(\sigma_{\mathrm{el}}^{\mathrm{eff}} \frac{\partial \phi_{\mathrm{el}}}{\partial y}\right) = i_{\mathrm{ct}} \quad (30.18)$$

$$\frac{\partial}{\partial x}\left(\sigma_{\mathrm{io}}^{\mathrm{eff}} \frac{\partial \phi_{\mathrm{io}}}{\partial x}\right) + \frac{\partial}{\partial y}\left(\sigma_{\mathrm{io}}^{\mathrm{eff}} \frac{\partial \phi_{\mathrm{io}}}{\partial y}\right) = -i_{\mathrm{ct}} \quad (30.19)$$

where ϕ_{el} and ϕ_{io} are the electrical potential of electron and that of ion in the solid phase, respectively; $\sigma_{\mathrm{el}}^{\mathrm{eff}}$ and $\sigma_{\mathrm{io}}^{\mathrm{eff}}$ are the effective electron and ion conductivity, respectively; and i_{ct} is the charge-transfer current density. The electrochemical and ammonia decomposition reactions can be modeled similarly to the 1D analysis in the previous section, namely Eqs. (30.10) and (30.11). The results are reflected in the source terms in the governing equations. These equations are to be solved under appropriate boundary conditions to obtain detailed information in the operating cell.

As an example of the simulation results when ammonia is directly supplied, the distributions of mole fractions of ammonia, hydrogen, and water vapor on the anode side are shown in Fig. 30.3. Note that only the fuel flow channel and the anode are depicted in the figure. Ammonia is immediately decomposed near the cell inlet, which causes an increase of hydrogen in the upstream region. As the hydrogen is consumed by the electrochemical reaction, it starts to decrease downstream. The steam produced by the electrochemical reaction increases downstream. The gradient of the mole fraction in the cell thickness direction is large inside the anode, while it is very small in the fuel channel.

The rapid decomposition reaction of ammonia near the cell inlet, seen in Fig. 30.3a, leads to a decrease in the local cell temperature in that region. The steep decrease of ammonia concentration near the anode inlet and the formation of a 'cold spot' were also reported by Ni [9] as a result of 2D simulation. Ni et al. [6] also performed a 2D simulation to analyze the relationship between the fuel utilization and the electrochemical performance of the cells in both SOFC and PCFC, showing that PCFC had superior electrochemical performance to SOFC when fed with ammonia. The difference between them became more significant with the increase of fuel utilization and temperature owing to the dilution of fuel by the produced water in the anode side of SOFC.

By extending the above governing equations to three dimensions, a 3D analysis can be performed. However, since the computational load is high, the computational domain is often limited to a part of the cell, such as a single repeating unit of the cell

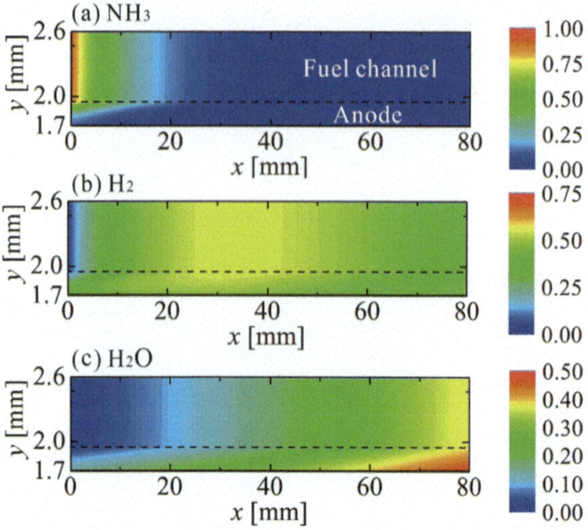

Fig. 30.3 Example of 2D simulation results: distribution of the mole fractions on the anode side

[16]. In the analysis of an entire cell or a stack containing multiple cells, it is generally difficult to ensure sufficient spatial resolution to understand the phenomena inside the thin porous electrode. Some kind of simplification is necessary so that the analysis can be performed with a realistic computational load. For example, an electrode reaction can be modeled as a surface reaction without considering the phenomena inside the electrode, or a quasi-3D analysis can be performed at the expense of spatial resolution in a specific direction [17].

30.4 Conclusions

The framework of 1D and 2D/3D analytical models for ammonia-fueled SOFC was introduced. These are just examples; more detailed or more concise models are also possible. It is important to use the most appropriate model according to the purpose of the numerical simulation. All the models introduced were steady-state models, but transient calculations are also possible by considering unsteady terms. Being a multiphysics system, it is necessary to pay attention to the characteristic times of the various phenomena occurring in the cell/stack and set the time resolution to suit the purpose.

The introduced numerical models are essentially the same as that for SOFCs using conventional fuels such as hydrogen or hydrocarbon fuels. A unique point about the case of ammonia fuel is the part of the modeling of the decomposition reaction. The dilution effects of the generated nitrogen and the endothermic effects of the decomposition reaction on the power generation performance and internal conditions of the cell/stack are often discussed. Numerical analysis has been contributing greatly to the visualization of internal phenomena such as temperature, gas concentration, and current density distributions, and the understanding of their impact on performance. On the other hand, numerical analysis of thermal stresses and associated deformations [18, 19] and degradation phenomena associated with nitridation by ammonia [20, 21] is still limited at present and is a direction that should be developed in the future.

References

1. Wojcik A, Middleton H, Damopoulos I, Van Herle J (2003) Ammonia as a fuel in solid oxide fuel cells. J Power Sources 118:342–348
2. Fournier GGM, Cumming IW, Hellgardt K (2006) High performance direct ammonia solid oxide fuel cell. J Power Sources 162:198–206
3. Ma Q, Ma J, Zhou S, Yan R, Gao J, Meng G (2007) A high-performance ammonia-fueled SOFC based on a YSZ thin-film electrolyte. J Power Sources 164:86–89
4. Ni M, Leung DYC, Leung MKH (2008) Mathematical modeling of ammonia-fed solid oxide fuel cells with different electrolytes. Int J Hydrogen Energy 33:5765–5772
5. Ni M, Leung DYC, Leung MKH (2008) Electrochemical modeling of ammonia-fed solid oxide fuel cells based on proton conducting electrolyte. J Power Sources 183:687–692
6. Ni M, Leung DYC, Leung MKH (2008) Thermodynamic analysis of ammonia fed solid oxide fuel cells: comparison between proton-conducting electrolyte and oxygen ion-conducting electrolyte. J Power Sources 183:682–686
7. Yin SF, Xu BQ, Zhou XP, Au CT (2004) A mini-review on ammonia decomposition catalysts for on-site generation of hydrogen for fuel cell applications. Appl Catal A 277:1–9
8. Molouk AFS, Yang J, Okanishi T, Muroyama H, Matsui T, Eguchi K (2016) Comparative study on ammonia oxidation over Ni-based cermet anodes for solid oxide fuel cells. J Power Sources 305:72–79
9. Ni M (2011) Thermo-electrochemical modeling of ammonia-fueled solid oxide fuel cells considering ammonia thermal decomposition in the anode. Int J Hydrogen Energy 36:3153–3166
10. Ni M, Leung DYC, Leung MKH (2008) An improved electrochemical model for the NH_3 fed proton conducting solid oxide fuel cells at intermediate temperatures. J Power Sources 185:233–240
11. Ni M, Leung MKH, Leung DYC (2009) Ammonia-fed solid oxide fuel cells for power generation—a review. Int J Energy Res 33:943–959
12. Kishimoto M, Furukawa N, Kume T, Iwai H, Yoshida H (2017) Formulation of ammonia decomposition rate in Ni-YSZ anode of solid oxide fuel cells. Int J Hydrogen Energy 42:2370–2380
13. Luo Y, Shi Y, Liao S, Chen C, Zhan Y, Au CT, Jiang L (2019) Coupling ammonia catalytic decomposition and electrochemical oxidation for solid oxide fuel cells: a model based on elementary reaction kinetics. J Power Sources 423:125–136
14. Kalinci Y, Dincer I (2018) Analysis and performance assessment of NH_3 and H_2 fed SOFC with proton-conducting electrolyte. Int J Hydrogen Energy 43:5795–5807
15. Kishimoto M, Kume T, Iwai H, Yoshida H (2017) Numerical analysis of ammonia-fueled planar solid oxide fuel cells. ECS Trans 78:2845–2853
16. Ni M (2009) Three-dimensional computational fluid dynamics modeling of a planar solid oxide fuel cell. Chem Eng Technol 32:1484–1493
17. Tan WC, Iwai H, Kishimoto M, Brus G, Szmyd JS, Yoshida H (2018) Numerical analysis on effect of aspect ratio of planar solid oxide fuel cell fueled with decomposed ammonia. J Power Sources 384:367–378

18. Schlegl H, Dawson R (2017) Finite element analysis and modelling of thermal stress in solid oxide fuel cells. Proc Inst Mech Eng Part A J Power Energy 231:654–665
19. Xie J, Hao W, Wang F (2017) Analysis of anode functional layer for minimizing thermal stress in solid oxide fuel cell. Appl Phys A 123:656
20. Yang J, Molouk AFS, Okanishi T, Muroyama H, Matsui T, Eguchi K (2015) A stability study of Ni/Yttria-stabilized Zirconia anode for direct ammonia solid oxide fuel cells. ACS Appl Mater Interfaces 7:28701–28707
21. Yokochi R, Hashinokuchi M, Doi T, Inaba M (2015) A stability study of Ni/Yttria-stabilized Zirconia anode for direct ammonia solid oxide fuel cells. ECS Trans 68:2745–2750

Chapter 31
Low-Temperature Ammonia Fuel Cells

Toshiaki Matsui

Abstract Low-temperature fuel cells are one of the promising energy conversion devices that use ammonia as a fuel. In this chapter, we focus on fuel cells that operate below the decomposition temperature of ammonia, especially at room temperature to 400 °C; mainly, an anion exchange membrane fuel cell and a fuel cell using a molten salt as an electrolyte. Since in these fuel cell systems, the direct electrochemical oxidation reaction of ammonia proceeds on the electrode, the understanding of the reaction mechanism is of importance to design high performance electrocatalysts. Pt is the most promising single metal electrode for the ammonia oxidation among various candidates. Therefore, the electrochemical properties of Pt in an alkaline media in the presence of ammonia will be described in detail.

Keywords Ammonia · Direct/indirect supply · Electrochemical oxidation · Low-temperature fuel cells

This chapter introduces fuel cells that operate below the decomposition temperature of ammonia, especially at room temperature to 400 °C. In 1960s, direct ammonia fuel cells were reported firstly [1–3]. Cairns et al. have demonstrated the power generation with feeding ammonia directly to an alkaline fuel cell employing a KOH aqueous solution as an electrolyte [1]. However, since then there has been no significant progress due to several drawbacks, such as (1) a large amount of precious metals used for electrodes and (2) carbonate precipitation caused by CO_2 contaminant in air. It is also difficult to apply existing PEFC systems for the direct use of ammonia fuel. This is because the acidic proton exchange membrane is used as an electrolyte. In this case, ammonium ions occupy proton exchange site in the membrane, resulting in the lowering of proton conductivity as well as anode catalytic activity. Furthermore, ammonia crossover to the cathode leads to the inhibition of the oxygen reduction reaction. Therefore, the ammonia concentration in the fuel must be less than 0.1 ppm.

T. Matsui (✉)
Department of Energy and Hydrocarbon Chemistry, Graduate School of Engineering, Kyoto University, Nishikyo-ku, Kyoto 615-8510, Japan
e-mail: matsui@elech.kuic.kyoto-u.ac.jp

K. Aika and H. Kobayashi (eds.), *CO_2 Free Ammonia as an Energy Carrier*,
https://doi.org/10.1007/978-981-19-4767-4_31

On the other hand, in principle, ammonia can be directly used in anion exchange membrane fuel cells, in which hydroxide ions serve as charge carriers. The anion exchange membrane is a basic polymer having a quaternary ammonium base, or a quaternary pyridinium base, etc. as an ion-exchange groups; e.g., polysulfone adding a quaternary ammonium base [4–6], and poly(hexafluoropropylene-co-tetrafluoroethylene) film adding a quaternary ammonium base obtained by graft polymerization [7–9]. This fuel cell system has advantages; (i) non-precious metals can be applied as electrodes [10] and (ii) the carbonate precipitation does not occur due to the absence of cation sources such as K^+ in the electrolyte [11, 12]. Although CO_2 dissolution into the membrane causes the formation of carbonate ion species, the self-purging ability under discharge conditions alleviates the performance degradation [13, 14]. Therefore, the anion exchange membrane is a promising electrolyte for direct ammonia fuel cells [15].

When ammonia is directly supplied to the anode of anion exchange membrane fuel cells, the following reaction occurs at each electrode;

$$\text{Anode: } 2NH_3 + 6OH^- \rightarrow N_2 + 6H_2O + 6e^- \quad (31.1)$$

$$\text{Cathode: } O_2 + 2H_2O + 4e^- \rightarrow 4OH^- \quad (31.2)$$

In this case, the theoretical electromotive force is 1.17 V at 25 °C.

The electrochemical oxidation reaction of NH_3 has been studied in an alkaline aqueous electrolyte, and several studies reported the reactivity of ammonia on various electrocatalysts such as Ni electrode [16], precious metal electrodes (Ru, Rh, Pd, Ir, Pt, Au, Ag) [17, 18], and Pt alloy electrodes [19–22]. From these reports, it can be concluded that Pt shows the highest activity among pure metals.

Figure 31.1 shows the time course of open circuit voltage (OCV) for anion exchange membrane fuel cells (AEMFCs) employing Pt/C, Ru/C, and Pt–Ru/C as anodes at 50 °C; Pt/C was used as a cathode [23]. The OCV was identical regardless of electrode material, when humidified hydrogen was supplied to the anode. However, the OCV dropped immediately after switching the hydrogen fuel to ammonia. Furthermore, the OCV differed depending on the difference in electrocatalytic activity of each electrode. The decrease in OCV is also caused by the permeation of ammonia through the membrane. In fact, N_2 and NO were detected in addition to NH_3 at the cathode when the cathode exhaust gas was monitored simultaneously with a mass spectrometer during feeding the ammonia fuel to the anode. This suggests that the ammonia oxidation reaction occurred at the cathode. Therefore, the improvement of electrocatalytic activity and the suppression of ammonia permeation will be development subjects to be settled.

Figure 31.2 shows the current–voltage characteristics of AEMFC employing Pt/C as an anode measured at 50 °C with feeding humidified hydrogen or ammonia [23]. A drastic reduction in cell performance was confirmed when an ammonia fuel was supplied, as compared to under hydrogen supply. Furthermore, the current density turned to decrease at the terminal voltage of 0.15 V or less. When the discharge

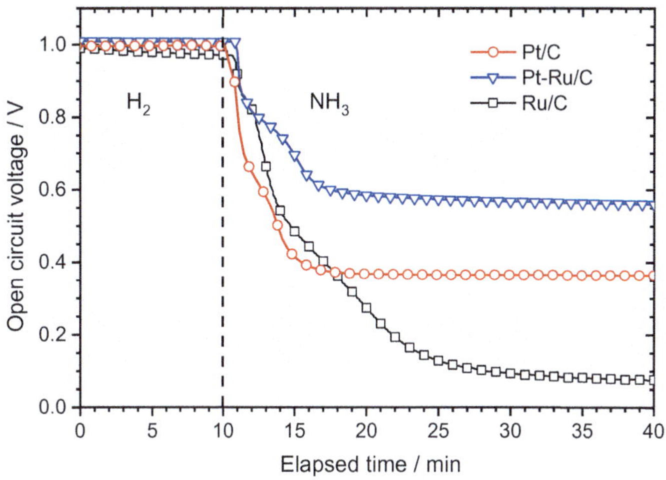

Fig. 31.1 Time course of open circuit voltage for AEMFCs employing Pt/C, Ru/C, and Pt–Ru/C as anodes. Temperature; 50 °C, Cathode; Pt/C, Anode gas; H_2 or NH_3 (humidified at 50 °C), Cathode gas; 21% O_2–79% N_2 (humidified at 50 °C). Reproduced from Ref. [23], Copyright (2012) with permission from Elsevier

test from OCV to the terminal voltage of 50 mV was repeated three times consecutively with a 2 min interval in between, the cell performance deteriorated with each measurement [23]. However, the degraded performance was recovered completely after the subsequent discharge treatment in a hydrogen fuel (0.1 A cm^{-2} for 1 min). This indicates that the performance deterioration was derived from ammonia fuel. Such a peculiar phenomenon can be understood by examining the electrode reaction mechanism.

The electrochemical ammonia oxidation over Pt electrode in alkaline aqueous solutions has been studied extensively by many researchers [17, 18, 21, 24–30]. The most widely accepted reaction mechanism for ammonia oxidation over Pt electrode was proposed by Gerischer and Mauerer as follows [24]:

$$NH_3(aq) \rightarrow NH_{3,ad} \tag{31.3}$$

$$NH_{3,ad} + OH_{ad}^{\delta-} \rightarrow NH_{2,ad} + H_2O + \delta e^- \tag{31.4}$$

$$NH_{2,ad} + OH_{ad}^{\delta-} \rightarrow NH_{ad} + H_2O + \delta e^- \tag{31.5}$$

$$NH_{x,ad} + NH_{y,ad} \rightarrow N_2H_{x+y,ad} \tag{31.6}$$

$$N_2H_{x+y,ad} + (x + y)OH_{ad}^{\delta-} \rightarrow N_2 + (x + y)H_2O + (x + y)\delta e^- \tag{31.7}$$

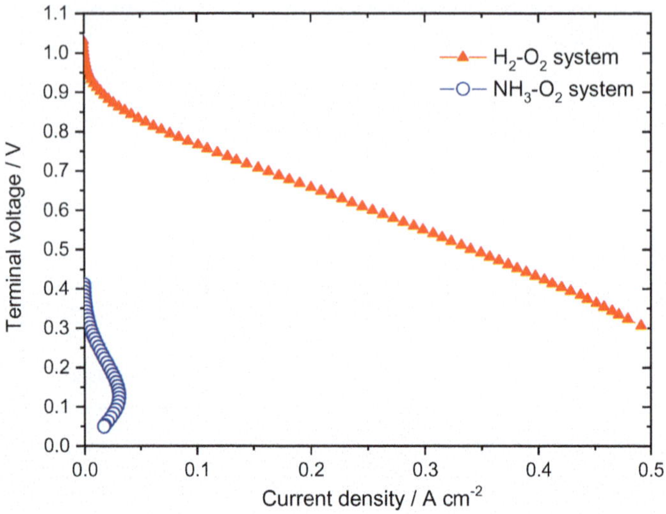

Fig. 31.2 Current–voltage characteristics of AEMFC employing Pt/C as an anode. Temperature; 50 °C, Cathode; Pt/C, Anode gas; H_2 or NH_3 (humidified at 50 °C), Cathode gas; 21% O_2–79% N_2 (humidified at 50 °C). Reproduced from Ref. [23], Copyright (2012) with permission from Elsevier

$$NH_{ad} + OH_{ad}^{\delta-} \rightarrow N_{ad} + H_2O + \delta e^- \tag{31.8}$$

where $x = 1$ or 2 and $y = 1$ or 2.

This reaction mechanism has been supported by the coulometric experiment [25], differential electrochemical mass spectroscopy [17, 26], surface enhanced Raman spectroscopy [18, 27]. In situ attenuated total reflection infrared (ATR-IR) spectroscopy is also a powerful tool to identify the adsorbates, such as reaction intermediates, on the electrode. Figure 31.3 shows the potential dependence of current density of Pt electrode and the intensity of three characteristics bands at 1269 cm^{-1} ($N_2H_{4,ad}$), 1497–1508 cm^{-1} (NO_{ad}), and 1662–1674 cm^{-1} ($NH_{3,ad}$); time resolved IR spectra of the Pt surface was acquired simultaneously with the linear sweep voltammogram in 0.1 M NH_3–1 M KOH [31]. The band assigned to adsorbed NH_3 was detected at 1662–1674 cm^{-1} in the potential range of 0.1–1.1 V. The intensity of this band decreased continuously with a rise in potential, which suggested the consumption of adsorbed ammonia through the oxidation reaction. In response to the behavior of the NH_3 band, the band at 1269 cm^{-1}, which can be assigned to the NH_2 wagging mode of N_2H_4, appeared alternatively above 0.2 V. Note that the intensity of the N_2H_4 band reached the local maximal value at ca. 0.4 V, where the ammonia oxidation current began to be observed clearly in the linear sweep voltammogram. The observed behavior supports the mechanism described above, suggesting the N_2H_4 adspecies was one of the active intermediates of the ammonia oxidation reaction [26]. According to previous studies, $NH_{2,ad}$ and NH_{ad} species are produced on the Pt electrode by the sequential ammonia dehydrogenation with the aid of OH_{ad}

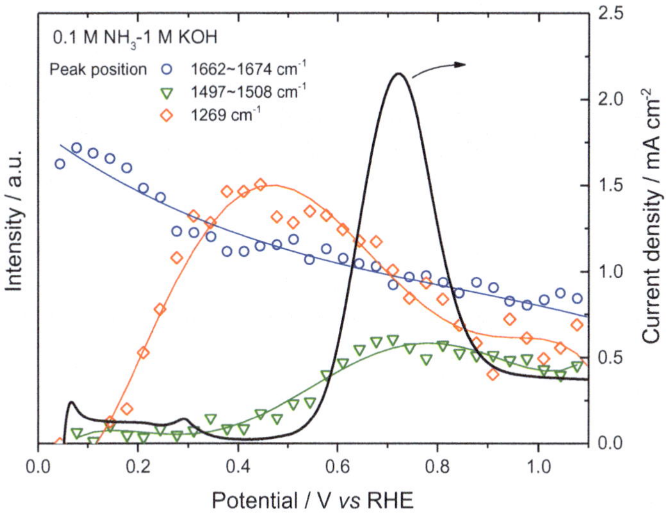

Fig. 31.3 Potential dependence of the current density of Pt electrode and the intensity of three characteristic bands at 1269 cm^{-1}, 1497–1508 cm^{-1}, and 1662–1674 cm^{-1}. Reprinted with permission from Ref. [31]. Copyright (2015) American Chemical Society

species, and subsequently combined each other to form $N_2H_{x+y,ad}$ ($x + y = 2$–4). The further dehydrogenation of $N_2H_{x+y,ad}$ leads to the formation of N_2. Unfortunately, it is known that the formation of poisonous species proceeds in parallel with the N_2 formation reaction, which is the major disadvantages of ammonia fuel. The fully dehydrogenated adsorbate of atomic nitrogen (N_{ad}) is inert for N_2 formation and serves as a poisoning species on Pt electrode; at the current state, however, the direct detection of N_{ad} has not been achieved yet. The formation of NO and NO_2 species was suggested in several studies [25, 26] in situ ATR-IR studies also indicated the formation of bridged NO above the onset potential of ammonia oxidation, ca. 0.5 V (Fig. 31.3). Such adsorbed NO species probably inhibit the electrochemical reaction due to the occupation of reaction sites at higher potential [31]. Figure 31.4 displays cyclic voltammograms of Pt/C measured in 0.1 M NH_3-1 M KOH at 25 °C [23]. The onset potential of ammonia oxidation was ca. 0.45 V as shown in Fig. 31.4a. This voltammogram was reproducible upon repeated cycling between 0.05 and 0.90 V. On the other hand, the current density of ammonia oxidation decreased with a rise in the number of measurement times, when the sweep range was changed to 0.50–0.90 V. This can be explained by the accumulation of N_{ad} and/or NO over the Pt surface at 0.56 V or higher [25], considering the discussion about the reaction mechanism. It has been reported that the reductive stripping of the accumulated adsorbed species on Pt occurs in the hydrogen adsorption region [25]; this behavior can be confirmed as the reduction current at ca. 0.40 V in Fig. 31.4a. Therefore, it can be concluded that the peculiar behavior observed in Figs. 31.2 during power generation was caused by N_{ad} and/or NO adsorption on Pt electrode.

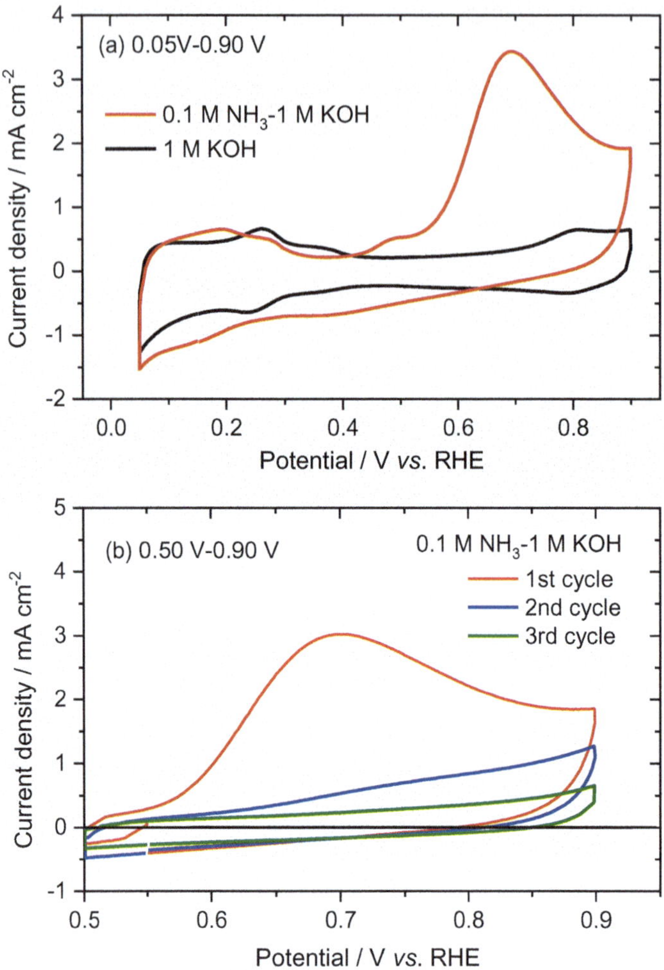

Fig. 31.4 Cyclic voltammograms of Pt/C in 0.1 M NH_3–1 M KOH at 25 °C in the scanning range of **a** 0.05–0.90 V and **b** 0.50–0.90 V. Reproduced from Ref. [23], Copyright (2012) with permission from Elsevier

From the obtained results, we can say that it is difficult to directly supply NH_3 fuel to AEMFCs and the combination with an ammonia cracker will be a promising system. Then, it is necessary to understand the allowable amount of residual ammonia in the fuel. Figure 31.5 shows the current–voltage characteristics measured with feeding humidified H_2–NH_3 gaseous mixture to the Pt/C anode at 50 °C [32]. The OCV dropped by about 80 mV upon the supply of mixed fuel due to the mixed potential. When the NH_3 concentration in the fuel was 1%, the cell performance is almost the same as hydrogen fuel. On the other hand, the power generation behavior changed when 7% NH_3 was mixed in the fuel; the performance was comparable to

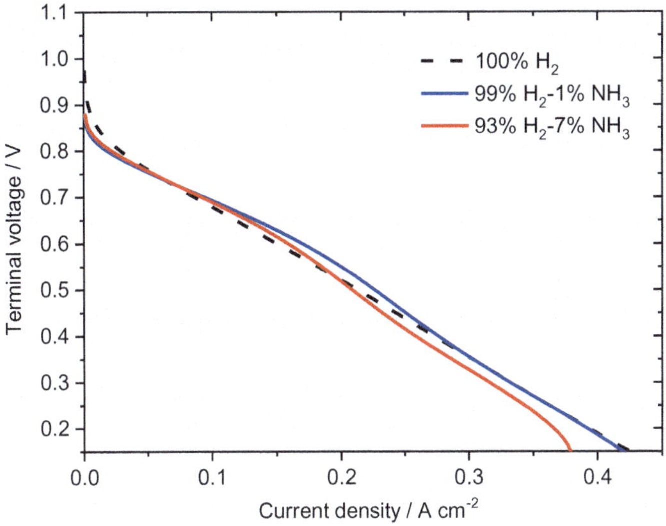

Fig. 31.5 Current–voltage characteristics of AEMFC employing Pt/C as an anode. Temperature; 50 °C, Cathode; Pt/C, Anode gas; $(100 - x)\%$ H_2–$x\%$ NH_3 ($x = 0, 1, 7$) (humidified at 50 °C), Cathode gas; 100% O_2 (humidified at 50 °C). Reproduced with permission from Ref. [32], Copyright (2017) American Chemical Society

that of hydrogen from the open circuit voltage to the terminal voltage of 0.5 V, while the deviation became large in the low terminal voltage region. This behavior will be due to the generation of N_{ad} and/or NO above-mentioned on the Pt electrode. Since the generation of these adsorbed species depends on the potential, a certain amount of ammonia contamination can be tolerated if the operating conditions are well controlled. In such a case, the performance required for the cracker is greatly relaxed.

The effect of adsorption species such as N_{ad} can be mitigated by raising the operating temperature of the fuel cell. A direct ammonia fuel cell composed of a molten NaOH–KOH electrolyte and lithiated-Ni electrodes was reported by Ganley et al. [33]. Although the cell structure was not optimized, the OCV of ca. 0.8 V and the maximum power density of 40 mW cm^{-2} were achieved at 450 °C. Then, the electrode reaction over Pt electrode upon discharge was investigated in molten NaOH–KOH electrolyte at 200 °C with feeding ammonia fuel [34]. Figure 31.6 shows the time course of the current density during polarization at −0.25 V versus Ag reference electrode and the corresponding mass signals of hydrogen and nitrogen in the exhaust gas. A rapid decrease in current density due to the formation of diffusion layer was observed within 40 s from the beginning of polarization, and then the current density became stable. Correspondingly, the nitrogen production was confirmed while no noticeable change was observed in the hydrogen signal. Thus, this result indicates that the direct oxidation of ammonia proceeded on the Pt electrode without the negative influence of adsorption species. In fact, the direct

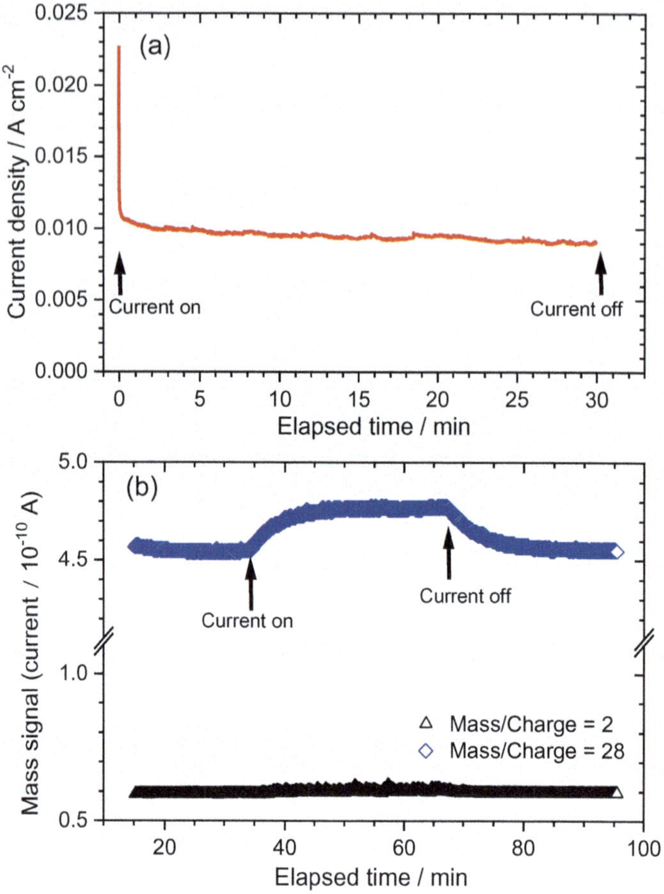

Fig. 31.6 Time course of **a** current density for Pt plate electrode polarized at –0.25 V versus Ag reference electrode in the molten NaOH–KOH electrolyte at 200 °C with feeding NH_3 and **b** corresponding mass signals of hydrogen and nitrogen (mass/charge = 2 and 28) in the exhaust gas. Reproduced from Ref. [34], Copyright (2014) with permission from Elsevier

ammonia fuel cell employing Pt electrodes and a molten NaOH–KOH electrolyte can be operated at 200 °C without the performance degradation observed in AEMFCs, especially in the low terminal voltage. However, the application of molten hydroxide is not suitable for the practical use because the vulnerability to CO_2 is inevitable. If a new electrolyte applicable in this temperature range is developed successfully in the near future, the realization of an attractive system, which can use ammonia as a fuel directly or indirectly, is expected.

References

1. Cairns EJ, Simons EL, Tevebaugh AD (1968) Ammonia–oxygen fuel cell. Nature 217:780–781
2. Simons EL, Cairns EJ, Surd DJ (1969) The performance of direct ammonia fuel cells. J Electrochem Soc 116:556–561
3. Mckee DW, Scarpellino AJ Jr, Danzig IF, Pak MS (1969) Improved electrocatalysts for ammonia fuel cell anodes. J Electrochem Soc 116:562–568
4. Vico S, Palys B, Herman CB (2003) Hydration of a polysulfone anion-exchange membrane studied by vibrational spectroscopy. Langmuir 19:3282–3287
5. Fang J, Shen PK (2006) Quaternized poly(phthalazinon ether sulfone ketone) membrane for anion exchange membrane fuel cells. J Membr Sci 285:317–322
6. Zhou J, Unulu M, Vega JA, Kohl PA (2009) Anionic polysulfone ionomers and membranes containing fluorenyl groups for anionic fuel cells. J Power Sources 190:285–292
7. Herman H, Slade RCT, Varcoe JR (2003) The radiation-grafting of vinylbenzyl chloride onto poly(hexafluoropropylene-co-tetrafluoroethylene) films with subsequent conversion to alkaline anion-exchange membranes: optimisation of the experimental conditions and characterization. J Membr Sci 218:147–163
8. Varcoe JR, Slade RCT, Yee ELH, Poynton SD, Driscoll DJ, Apperley DC (2007) Poly(ethylene-*co*-tetrafluoroethylene)-derived radiation-grafted anion-exchange membrane with properties specifically tailored for application in metal-cation-free alkaline polymer electrolyte fuel cells. Chem Mater 19:2686–2693
9. Danks TN, Slade RCT, Varcoe JR (2003) Alkaline anion-exchange radiation-grafted membranes for possible electrochemical application in fuel cells. J Mater Chem 13:712–721
10. Asazawa K, Yamada K, Tanaka H, Oka A, Taniguchi M, Kobayashi T (2007) A platinum-free zero-carbon-emission easy fuelling direct hydrazine fuel cell for vehicles. Angew Chem 119:8170–8173
11. Adams LA, Poynton SD, Tamain C, Slade RCT, Varcoe JR (2008) A carbon dioxide tolerant aqueous-electrolyte-free anion-exchange membrane alkaline fuel cell. Chemsuschem 1:79–81
12. Watanabe S, Fukuta K, Yanagi H (2010) Determination of carbonate ion in MEA during the alkaline membrane fuel cell (AMFC) operation. ECS Trans 33:1837–1845
13. Lang CM, Kim K, Kohl PA (2006) High-energy density, room-temperature carbonate fuel cell. Electrochem Solid-State Lett 9:A545–A548
14. Unlu M, Zhou J, Kohl PA (2009) Anion exchange membrane fuel cells: experimental comparison of hydroxide and carbonate conductive ions. Electrochem Solid-State Lett 12:B27–B30
15. Lan R, Tao S (2010) Direct ammonia alkaline anion-exchange membrane fuel cells. Electrochem Solid-State Lett 13:B83–B86
16. Yao K, Cheng YF (2008) Investigation of the electrocatalytic activity of nickel for ammonia oxidation. Mater Chem Phys 108:247–250
17. de Vooys ACA, Koper MTM, van Santen RA, van Veen JAR (2001) The role of adsorbates in the electrochemical oxidation of ammonia on noble and transition metal electrodes. J Electroanal Chem 506:127–137
18. de Vooys ACA, Mrozek MF, Koper MTM, van Santen RA, van Veen JAR, Weaver MJ (2001) The nature of chemisorbates formed from ammonia on gold and palladium electrodes as discerned from surface-enhanced Raman spectroscopy. Electrochem Commun 3:293–298
19. de Mishima BAL, Lescano D, Holgado TM, Mishiima HT (1998) Electrochemical oxidation of ammonia in alkaline solutions: its application to an amperometric sensor. Electrochim Acta 43:395–404
20. Endo K, Katayama Y, Miura T (2004) Pt–Ir and Pt–Cu binary alloys as the electrocatalyst for ammonia oxidation. Electrochim Acta 49:1635–1638
21. Endo K, Nakamura K, Katayama Y, Miura T (2004) Pt–Me (Me=Ir, Ru, Ni) binary alloys as an ammonia oxidation anode. Electrochim Acta 49:2503–2509
22. Vitse F, Cooper M, Botte GG (2005) On the use of ammonia electrolysis for hydrogen production. J Power Sources 142:18–26

23. Suzuki S, Muroyama H, Matsui T, Eguchi K (2012) Fundamental studies on direct ammonia fuel cell employing anion exchange membrane. J Power Sources 208:257–262
24. Gerischer H, Mauerer A (1970) Untersuchungen zur anodischen oxidation von ammoniak an platin-elektroden. J Electroanal Chem 25:421–433
25. Gootzen JFE, Wonders AH, Visscher W, van Santen RA, van Veen JAR (1998) A DEMS and cyclic voltammetry study of NH_3 oxidation on platinized platinum. Electrochim Acta 43:1851–1861
26. Vidal-Iglesias FJ, Solla-Gullón J, Feliu JM, Baltruschat H, Aldaz A (2006) DEMS study of ammonia oxidation on platinum basal planes. J Electroanal Chem 588:331–338
27. Vidal-Iglesias FJ, Solla-Gullon J, Perez A, Aldaz A (2006) Evidence by SERS of azide anion participation in ammonia electrooxidation in alkaline medium on nanostructured Pt electrodes. Electrochem Commun 8:102–106
28. Vidal-Iglesias FJ, Solla-Gullón J, Montiel V, Feliu JM, Aldaz A (2007) Screening of electrocatalysts for direct ammonia fuel cell: ammonia oxidation on PtMe (Me: Ir, Rh, Pd, Ru) and preferentially oriented Pt (100) nanoparticles. J Power Sources 171:448–456
29. Katayama Y, Okanishi T, Muroyama H, Matsui T, Eguchi K (2015) Electrochemical oxidation of ammonia over rare earth oxide modified platinum catalysts. J Phys Chem C 119:9134–9141
30. Okanishi T, Katayama Y, Muroyama H, Matsui T, Eguchi K (2015) SnO_2–modified Pt electrocatalysts for ammonia–fueled anion exchange membrane fuel cells. Electrochim Acta 173:364–369
31. Matsui T, Suzuki S, Katayama Y, Yamauchi K, Okanishi T, Muroyama H, Eguchi K (2015) In situ attenuated total reflection infrared spectroscopy on electrochemical ammonia oxidation over Pt electrode in alkaline aqueous solutions. Langmuir 31:11717–11723
32. Katayama Y, Yamauchi K, Hayashi K, Okanishi T, Muroyama H, Matsui T, Kikkawa Y, Negishi T, Watanabe S, Isomura T, Eguchi K (2017) Anion-exchange membrane fuel cells with improved CO_2 tolerance: impact of chemically induced bicarbonate ion consumption. ACS Appl Mater Interfaces 9:28650–28658
33. Ganley JC (2008) An intermediate-temperature direct ammonia fuel cell with a molten alkaline hydroxide electrolyte. J Power Sources 178:44–47
34. Yang J, Muroyama H, Matsui T, Eguchi K (2014) Development of a direct ammonia-fueled molten hydroxide fuel cell. J Power Sources 245:277–282

Part VII
Ammonia Combustion Mechanism and Application for Internal Combustion Engines

Chapter 32
Significance and Impact of Ammonia Combustion Research

Hideaki Kobayashi and Akihiro Hayakawa

Abstract The Cross-ministry Strategic Innovation Program (SIP) "Energy Carriers" was implemented for five years starting in 2014 with the aim of establishing a hydrogen value chain. The project involved significant research and development in ammonia combustion and produced a number of significant results on the use of ammonia as a carbon-free fuel, including feasibility studies on gas turbines, internal combustion engines, industrial furnaces, co-firing of pulverized coal combustion, as well as ammonia combustion in large-scale power plants. The project will be implemented in the energy and industrial sectors. This project has brought renewed interest in ammonia combustion to the international combustion community in terms of reducing greenhouse gas emissions from combustion in the energy and industrial sectors. In addition, ammonia combustion has influenced not only energy and environmental policies in Japan but also international energy organizations that influence governments and industries in many countries.

Keywords Ammonia combustion · Energy carriers · Social impact · Carbon-free fuel

32.1 Introduction

Combustion is a fundamental energy transformation technology that supports our society. However, as a result of the international recognition of the correlation between greenhouse gas emissions and global warming since the Industrial Revolution, unrealistic discussions such as "de combustion" have been made in recent years. Although combustion researchers who wish to pursue not only technology but also purely physicochemical processes have been facing a difficult time these

Originally published in Journal of the Combustion Society of Japan, Vol. 61 (2019), pp. 277–282 (in Japanese);

H. Kobayashi (✉) · A. Hayakawa
Institute of Fluid Science, Tohoku University, 2-1-1 Katahira, Aoba-ku, Sendai, Miyagi 980-8577, Japan
e-mail: hideaki.kobayashi.a3@tohoku.ac.jp

K. Aika and H. Kobayashi (eds.), *CO_2 Free Ammonia as an Energy Carrier*,
https://doi.org/10.1007/978-981-19-4767-4_32

years, fuel ammonia and its combustion have reemerged. The success of ammonia rocket engines [1] and research of ammonia gas turbines[2] in the United States in the 1960s remain in history. After that, although the process of ammonia combustion was mainly discussed in research on nitrogen oxide production, its usefulness as a hydrogen carrier did not receive special interest until around COP3 in the late 1990s, when the issue of global warming came into focus.

The argument that hydrogen is a clean energy because it does not emit CO_2 when burnt can be seen everywhere. Actually, hydrogen combustion is not particularly difficult for combustion researchers, but the difficulty of transportation and storage due to the thermophysical properties of hydrogen has led to the development of these technologies for almost a quarter of a century. However, as can be inferred from Japan's fossil fuel imports, the most important perspectives in energy issues are total quantity and cost. In the midst of such discussions, the usefulness of ammonia as a hydrogen carrier, which is easy to store and transport and can already be used with low-cost infrastructure, and at the same time does not emit CO_2 when it burns, has been recognized both domestically and internationally. These led to the Cabinet Office's Strategic Innovation Program (SIP) "Energy Carriers," which was implemented for five years from 2014.

The detailed reaction mechanism of ammonia combustion was proposed in the 1980s, and research in the field of combustion chemistry has been progressing. However, in order to maximize the use of ammonia in existing energy systems as fuel, it is necessary to reexamine combustion characteristics using the latest technology and knowledge in combustion science, including numerical analysis. The number of conference presentations and academic papers on ammonia combustion has been rapidly increasing in recent years. Life cycle assessment (LCA) from ammonia production using renewable energy or natural gas as raw material, combined with CCS and CCUS, to utilization of ammonia as a fuel is also essential, as is hydrogen. Research and development of low-fuel NO_x emission while overcoming the low flammability of ammonia have provided the international combustion community with excellent topics that directly relate to environmental and energy issues, as well as the economy, in terms of both flame chemistry and flame dynamics.

In 2016, after the start of the SIP Energy Carriers Project, the Journal of the Combustion Society of Japan published a series of articles on ammonia combustion [3]. After the completion of the project, special feature articles that give overviews of the remarkable achievements were issued. The details of the achievements will be described in detail in each part, so we will only give an overview here. In addition, I will outline some of the interesting aspects of ammonia combustion that have come to light as a result of our work as fundamental research.

32.2 Overview of the SIP Energy Carriers Project and Its Impact on Policy in Japan

This project was implemented as part of the SIP and was completed in March 2019. The objective of the project was to establish a value chain of hydrogen carriers, which is indispensable for Japan, an energy-importing country, to introduce hydrogen in large quantities in the future [4]. During the five years of the project, under the strong leadership of Program Director Shigeru Muraki, the issues to be addressed during the research period were reexamined, taking into account the strict evaluation of results and collaboration with other projects. At the beginning of the project, the number of participating institutions in the ammonia direct combustion team was seven, including two universities and one national institute, but by the end of the project, the number had increased to 18. One noteworthy aspect of the project was the participation of four electric power companies in Japan, indicating their expectations for ammonia as a key energy source. The main R&D themes were internal combustion engines such as gas turbines and reciprocating engines, industrial furnaces for heat utilization, and ammonia combustion in existing thermal power plants as shown in Fig. 32.1. For large-scale thermal power generation, the team conducted legal reviews and feasibility studies as well as demonstration tests for the introduction of large amounts of ammonia to the power plants. The activities of the ammonia direct combustion team not only focused on individual themes, but also involved a wide range of collaborations, as related organizations such as internal combustion engines, industrial furnaces, pulverized coal boilers at power plants, and power plants under operation actively exchanged information and engaged each other in research and development. The project also served as a role model for inter-company cooperation toward the future large-scale use of ammonia.

Notable achievements include a 50 kW micro gas turbine for dedicated ammonia power generation, a 2 MW natural gas turbine for power generation with an ammonia heat ratio of 20%, decomposed ammonia combustion in a large-scale GTCC, ammonia co firing in a pulverized coal power generation boiler and demonstration in actual operation, ammonia combustion in automobile engines, application to marine diesel engines, and a 300 kW micro gas turbine, 10 and 100 kW ammonia combustion industrial furnaces, steel plate degreasing furnaces, cement manufacturing furnaces, and feasibility study of existing large-scale thermal power plants, etc. The participating universities, Tohoku University, and Osaka University, have played a role in clarifying and evaluating the fundamental combustion processes of internal combustion engines and industrial furnaces, respectively, and have been active in disseminating information [5, 6].

Another important outcome of this project is its impact on policy of Japan and international organizations. In the "Basic Hydrogen Strategy," which was decided at the Ministerial Conference on Renewable Energy and Hydrogen in December 2017, specific goals were clearly stated for the first time, such as aiming to introduce and start using CO_2-free ammonia by the middle of the 2020s, starting ammonia co-firing power generation at coal power plants by around 2020, and expanding the use of

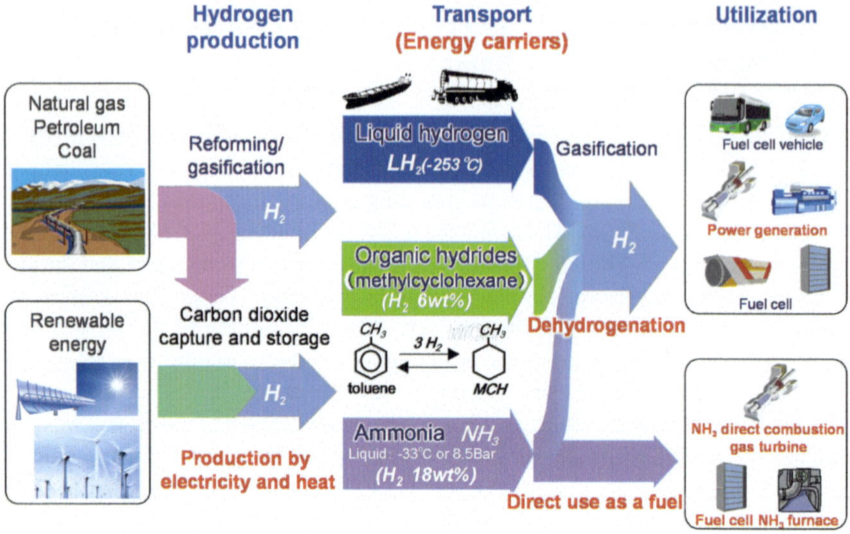

Fig. 32.1 Schematics of the hydrogen value chain aiming by the SIP Energy Carriers Project modified from [4]

ammonia in gas turbines by around 2030 [7]. In addition, the Fifth Energy Masterplan approved by the Cabinet in July 2018 includes the promotion of technology development for direct use of ammonia as a fuel [8], and the Integrated Innovation Strategy 2019 approved by the Cabinet in June 2019 sets a target of importing 3 million tons of ammonia in 2030. In the environmental energy field of the "Integrated Innovation Strategy 2019" approved by the Cabinet in June 2019, the import of 3 million tons of ammonia in 2030 was specified as a target [9].

In the report Renewable Energy for Industry published in November 2017 by the International Energy Agency (IEA), which also influences the energy policies of various countries, the SIP Energy Carriers project was introduced extensively. The report, Renewable Energy for Industry, published by the IEA in November 2017, introduces the SIP Energy Carriers Project and specifically describes the effectiveness of ammonia utilization in the industrial and energy sectors [10]. In addition, the IEA released a report, The Future of Hydrogen, in conjunction with the Energy and Environment-related Ministerial Meeting of the G20 in Japan in 2019, detailing the usefulness of ammonia as well as the results of a comparative study of the future potential, including costs comparisons with various energy carriers [11]. The use of ammonia as an energy source received interest in Europe, where the Netherlands and other countries have been active in curbing global warming, and a wide range of companies, from well-known multinationals to venture companies, are participating. In Japan, the Green Ammonia Consortium (presently, Clean Fuel Ammonia Association) was established in April 2019 to take over the SIP Energy Carriers Project, with the participation of about 80 domestic and foreign companies, trading companies, local governments, and state governments in other countries. The consortium aims to commercialize blue ammonia (produced by capturing and reusing CO_2 generated in the ammonia production process) and green ammonia (produced from renewable energy sources), which are mainly produced overseas and imported into Japan.

32.3 Fundamental Combustion Characteristics of Ammonia

32.3.1 Enhancement of Ammonia Flames

For instance, the maximum laminar burning velocity of ammonia/air flame at atmospheric pressure and room temperature condition is about 7 cm/s, which is almost 1/5 of that of methane/air flame [3, 12]. In addition, flame temperature of ammonia/air flame is almost 200 K lower [3] than that of methane/air flames, and the rich and lean flammable limits are between 0.63 and 1.4 [13]. Thus, improvement of flame intensity is needed to apply ammonia as for fuel of a practical combustor.

Examples of previous attempts for flame enhancement of ammonia flames were co-firing with hydrogen [14, 15] or methane [16, 17], and oxygen-enriched combustion [18, 19]. In the ammonia-firing gas turbine power generation facility at the Fukushima Renewable Energy Institute AIST (FREA), inlet air temperature is increased by applying a heat-regenerative cycle and thus flame enhancement was achieved [20]. In addition, employment of turbulent combustion was also the way to improve flame intensity of ammonia flames [21, 22]. Flame stabilization in a swirling flow is also effective for an improvement of flame stability. An increase in the characteristic time due to a recirculation and turbulent mixing between unburnt mixture and burnt gas containing high-temperature active radicals improves flame stability and enables stable combustion of ammonia for wider equivalence ratio and mixture inlet velocity conditions [23].

Figure 32.2 shows the relationship between the ratio of turbulent burning velocity and laminar burning velocity, S_T/S_L, and the ratio of turbulence intensity and laminar burning velocity, u'/S_L, of methane/ammonia/air premixed turbulent flames at the mixture pressure of 0.5 MPa for the equivalence ratio of 0.9 [21]. Here, E_{NH3} is the heat fraction of ammonia in the binary fuel of ammonia and methane. As shown in Fig. 32.2, the value of S_T/S_L increased with an increase in u'/S_L even for the case of co-firing of ammonia and methane. Even for the case of ammonia heat fraction of 0.2, the value of S_T/S_L increased up to around 10. Therefore, improvement of flame intensity can be achieved by employing turbulent combustion, and thus it is considered that the ammonia flame can be available in conventional internal combustion engines.

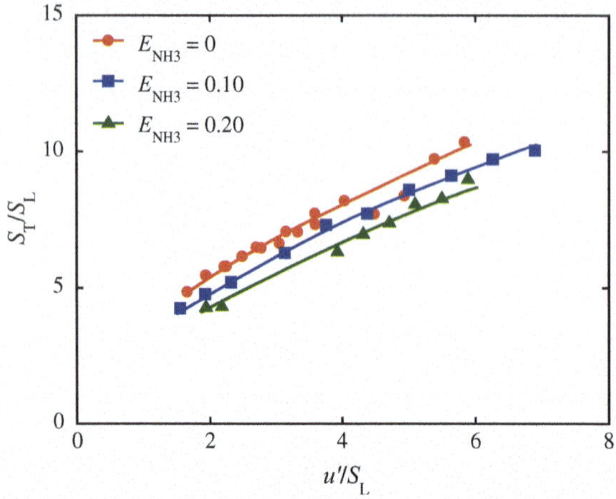

Fig. 32.2 Variation of normalized turbulent burning velocity, S_T/S_L, with normalized turbulence intensity, u'/S_L of CH_4/NH_3/air flames with permission from Elsevier [21]

32.3.2 Characteristics of Product Gas from Ammonia Flames

Since ammonia contains nitrogen atoms in its molecule, NO_x (Fuel NO_x) originates of nitrogen in the fuel may be produced when ammonia burns. In addition, ammonia is classified as deleterious substance because of irritating odor of ammonia. Emission revel of NO_x is restricted by the Air Pollution Control Law. In Japan, the emission level of NO_x from gas turbine combustors must be lower than 70 ppm under the oxygen concentration of 16% [24]. For ammonia, the emission is regulated by the Offensive Odor Control Law, and the concentration of ammonia must be less than a few ppm at the site boundary [25]. Therefore, product gas characteristics need to be well understood for the use of ammonia as fuel in a combustor.

Figure 32.3 shows the product gas characteristics of NO and unburnt NH_3 obtained from ammonia/air premixed laminar flames stabilized in stagnation flows in terms of the equivalence ratio, ϕ [26]. Here, the curves in Fig. 32.3 represent numerical results calculated by using the detailed reaction mechanism developed by Okafor et al. [16]. The maximum NO mole fraction within the examined equivalence ratio range was about 3500 ppm at the equivalence ratio of around 0.9. For lean conditions, there was no unburnt ammonia emission. The NO concentration decreased with an increase in the equivalence ratio, ϕ. However, the mole fraction of unburnt ammonia increased for rich conditions. Therefore, there is a trade-off relation between NO and unburnt ammonia emissions. A narrow equivalence ratio window for simultaneous reductions of NO and unburnt ammonia simultaneously is observed around equivalence ratio of 1.07. Numerical results calculated using the reaction mechanism developed by Okafor et al. predicted well the trend of unburnt ammonia. However, the amount of NO emission was slightly overestimated. The detailed reaction model developed by Okafor et al. well predicts of the laminar burning velocity of methane/ammonia

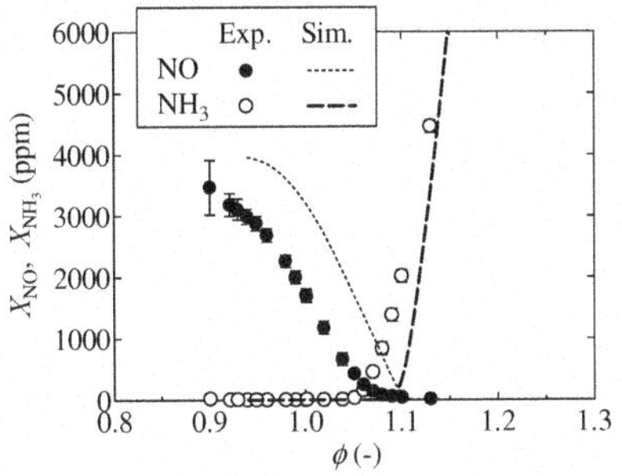

Fig. 32.3 Mole fractions of NO and unburnt NH_3 in product gas of ammonia/air mixture for various equivalence ratios, ϕ, at atmospheric pressure [26]

fuels at atmospheric pressure, and thus the detailed reaction model developed by Okafor et al. was validated in terms of equivalence ratio. Therefore, in order to improve the prediction accuracy of NO, the reactions having large sensitivity for NO but having small sensitivity for laminar burning velocity shall be updated. In the case of ammonia flame, NO reduction occurs through the reaction of $N + NO = N_2 + O$, which is known as one of the extended Zeldovich mechanisms and one of the important reactions for thermal NO for a conventional fuel. This reaction has relatively small sensitivity for laminar burning velocity but has large sensitivity for NO. Thus, further investigation for this elementary reaction is required. Although it was not shown in Fig. 32.3, the emissions of NO_2 which is harmful to humans and N_2O which has large global warming potential (GWP) of around 300 were small enough compared to NO emission for ammonia/air flames stabilized in a stagnation flow within the examined conditions in Fig. 32.3. Thus, major NO_x components from ammonia flame for this condition is NO.

32.3.3 Reaction Flow in Ammonia Flames

Understanding of the reaction process of ammonia is important to understand NO_x formation from fuel and reduction mechanisms. Figure 32.4 shows the reaction flow diagrams of ammonia/air premixed laminar flame calculated using the reaction mechanism developed by Tian et al. [27]. The major reaction paths were: from NH_3 to NH_2, NH and reaches to N_2; the path via HNO to NO and converted to N_2 directly or in directory which throughs N_2O. There were three major reaction flows in ammonia/air flames: reactions. In the case of methane/air flames, methane (CH_4) is converted to CH_3, then the reactions from CH_3 to CH_2O and the reaction from CO to CO_2 are known as the reactions generates large heat release [28]. However, in the case of ammonia/air flame, the reaction of R81 which generates NH_2 from NH_3 which reacts with OH is the major heat release reaction and plays an important role in ammonia/air flames. The NO production from ammonia/air flames is that the

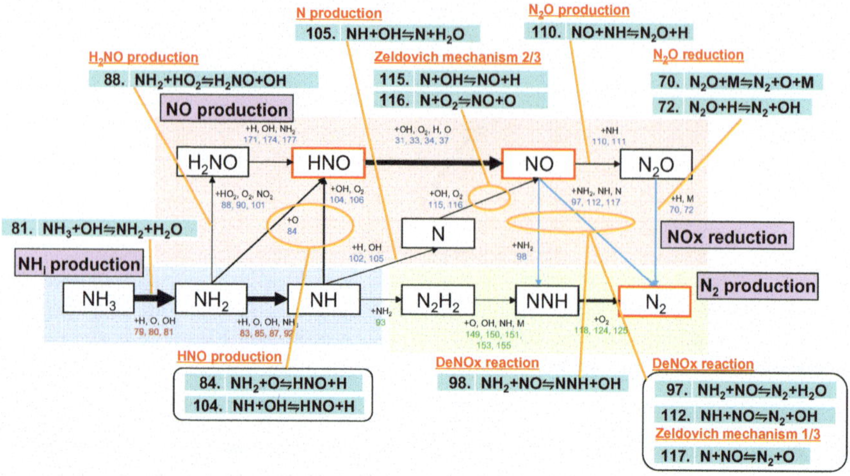

Fig. 32.4 Reaction flow diagram for $\phi = 0.8$ obtained using Tian's detailed reaction mechanism modified from [6], © Combustion Society of Japan 2019

reaction flow from NH_2 or NH to NO via HNO is considered to a major fuel NO production pathway. In the NO formation pathway, the radicals of O and OH are important. In particular, the behavior of OH is important for ammonia flames, and it is considered as one of the reasons for small laminar burning velocity of ammonia flame because of relatively small diffusivity of OH radical.

In the case of methane/air flames, thermal NO formation occurs mainly through the extended Zeldovich mechanism. However, as it was described in Sect. 32.3.2, NO is reduced through one of the reactions in the extended Zeldovich mechanism, $N + NO = N_2 + O$ (R117 in the Tian's detailed reaction mechanism), and generates N_2 in the case of ammonia flames.

NO formation pathways are different between methane/air flames and ammonia/air flames. It is well known that an employment of RQL (Rich-burn/Quick-quench/Lean-burn) combustion for the reduction of NO from conventional hydrocarbon flames [29]. The RQL combustion is the reduction procedure of NO formation through Prompt and Thermal NO formation.

The rich-lean two-stage ammonia combustion concept was proposed for the simultaneous reduction procedure for NO and unburnt ammonia from ammonia flame. Figure 32.5 shows the schematic figure of the rich-lean two-stage ammonia combustion concept in an ammonia-fueled gas turbine combustor. This combustion concept employs rich flame for primary combustion zone and lean flame for secondary combustion zone, and the combustion process seems to be similar with the RQL combustor described above. However, in the case of ammonia flame, Fuel NO is the major NO production pathway and the formation process of NO is different from hydrocarbon flames. Therefore, the reduction mechanisms for NO were different.

In the rich-lean two-stage ammonia combustion concept, slightly rich flame is generated for the primary combustion zone. This equivalence ratio is close to the

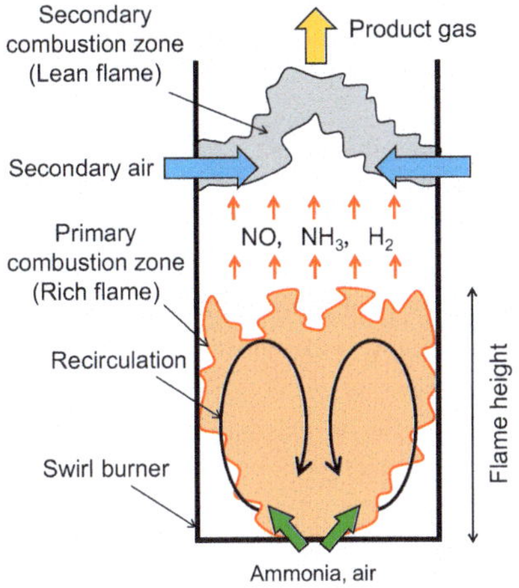

Fig. 32.5 Schematic of the concept of rich-lean two-stage ammonia combustion for simultaneous reduction of NO and unburnt NH_3 in a gas turbine combustor © Combustion Society of Japan 2019

equivalence ratio that NO and unburnt ammonia emission reaches minimum value simultaneously, as shown in Fig. 32.3. Both combustion configurations, premixed and non-premixed, are applicable. From the primary zone, NO, unburnt NH_3, and H_2 were produced. Then, the secondary air is injected to the products from the primary zones, and lean combustion occurs in the secondary zone. Unburnt ammonia and hydrogen burn completely by the secondary air injection. In this combustion concept, the control of primary zone equivalence ratio is important. If primary zone is too lean, the secondary air injection just behaves as dilution of the products from the primary zone by secondary air. On the other hand, if the primary zone is too rich, large amount of unburnt ammonia and hydrogen burns and generates NO in the secondary zone. This combustion concept is validated by numerical simulation [30], and by experiments using swirl combustor [31] and the gas turbine combustor at FREA [32, 33]. Therefore, even for the case of ammonia is used as fuel, simultaneous reductions of NO and unburnt ammonia can be achieved by an appropriate control of combustion process.

32.4 Concluding Remarks

In this chapter, the outline of SIP Energy carriers project, and the impact of ammonia combustion study to the policy and the international organizations were explained. In addition, from the standpoint of fundamental combustion study, ammonia combustion characteristics that clarified recently were explained. It is also explained that the simultaneous reductions of NO and unburnt ammonia from ammonia flame can be achieved when appropriate combustion process, i.e., the rich-lean two-stage ammonia combustion concept, is employed.

Acknowledgements The research results presented in this chapter were supported by the Cabinet Office, the Strategic Innovation Program (SIP) of the Council for Science, Technology and Innovation, the Energy Carriers, and the Priority Research Program of the Institute of Fluid Science, Tohoku University. The experimental and computational results were made possible by the cooperation of Mr. Taku Kudo, Dr. K. D. Kunkuma A. Somarathne, Dr. Ekenechukwu C. Okafor (currently with Kyushu University), and many students at the Institute of Fluid Science, Tohoku University. We would like to express our gratitude to them.

References

1. NASA. https://www.nasa.gov/centers/armstrong/news/FactSheets/FS-052-DFRC.html. Accessed 26 Aug 2019
2. Pratt DT (1967) Performance of ammonia-fired gas-turbine combustors. Technical report no. 9, DA-04-200-AMC-791(x). Berkley University of California. http://www.dtic.mil/dtic/tr/fulltext/u2/657585.pdf. Accessed 26 Aug 2018
3. Kobayashi H, Hayakawa A (2016) Carbon-free ammonia combustion. J Combust Soc Jpn 58:41–48 (in Japanese)
4. JST (2019) Cross-ministerial strategic innovation promotion program (SIP) energy carriers. https://www.jst.go.jp/sip/pdf/SIP_energycarriers2016_en.pdf. Accessed on 26 Aug 2019
5. Kobayashi H, Hayakawa A, Somarathne KDKA, Okafor EC (2019) Science and technology of ammonia combustion. Proc Combust Inst 37(1):09–133
6. Kobayashi H, Hayakawa A (2019) Ammonia direct combustion. J Hydrogen Energy Syst Soc Jpn 44:105–113
7. Ministerial Council on Renewable Energy (2017) Basic hydrogen strategy. http://www.meti.go.jp/english/press/2017/pdf/1226_003b.pdf. Accessed 26 Aug 2019
8. Agency for Natural Resources and Energy (2018) Strategic energy plan. https://www.enecho.meti.go.jp/category/others/basic_plan/pdf/180703.pdf. Accessed on 26 Aug 2019
9. Cabinet Office (2019) Integrated innovation strategy. https://www8.cao.go.jp/cstp/togo2019_honbun.pdf. Accessed 26 Aug 2019
10. Philibert C (2017) Renewable energy for industry from green energy to green materials and fuels. Insight series 2017. https://webstore.iea.org/insights-series-2017-renewable-energy-for-industry. Accessed 26 Aug 2019
11. IEA (2019) The future of hydrogen. https://webstore.iea.org/the-future-of-hydrogen. Accessed 26 Aug 2019
12. Hayakawa A, Goto T, Mimoto R, Arakawa Y, Kudo T, Kobayashi H (2015) Laminar burning velocity and Markstein length of ammonia/air premixed flames at various pressures. Fuel 159:98–106
13. Law CK (2006) Combustion physics. Cambridge University Press, p 347
14. Ichikawa A, Hayakawa A, Kitagawa Y, Somarathne KDKA, Kudo T, Kobayashi H (2015) Laminar burning velocity and Markstein length of ammonia/hydrogen/air premixed flames at elevated pressures. Int J Hydrogen Energy 40:9570–9578
15. Lee JH, Kim JH, Park KH, Kwon OC (2010) Studies on properties of laminar premixed hydrogen-added ammonia/air flames for hydrogen production. Int J Hydrogen Energy 35:1054–1064
16. Okafor EC, Naito Y, Colson S, Ichikawa A, Kudo T, Hayakawa A, Kobayashi H (2018) Experimental and numerical study of the laminar burning velocity of CH_4–NH_3-air premixed flames. Combust Flame 187:185–198

17. Okafor EC, Naito Y, Colson S, Ichikawa A, Kudo T, Hayakawa A, Kobayashi H (2019) Measurement and modelling of the laminar burning velocity of methane-ammonia-air flames at high pressures using a reduced reaction mechanism. Combust Flame 204:162–175
18. Takeishi H, Hayashi J, Kono S, Arita W, Iino K, Akamatsu F (2015) Characteristics of ammonia/N_2/O_2 laminar flame in oxygen-enriched air condition. Trans JSME (Jpn) 81:14–00423
19. Li J, Huang H, Kobayashi N, He Z, Osaka Y, Zeng T (2015) Numerical study on effect of oxygen content in combustion air on ammonia combustion. Energy 93:2053–2068
20. Kurata O, Iki N, Matsunuma T, Inoue T, Tsujimura T, Furutani H, Kobayashi H, Hayakawa A (2017) Performances and emission characteristics of NH_3-air and NH_3–CH_4-air combustion gas-turbine power generations. Proc Combust Inst 36:3351–3359
21. Ichikawa A, Naito Y, Hayakawa A, Kudo T, Kobayashi H (2019) Burning velocity and flame structure of CH_4/NH_3/air turbulent premixed flames at high pressure. Int J Hydrogen Energy 44:6991–6999
22. Ichimura R, Hadi K, Hashimoto N, Hayakawa A, Kobayashi H, Fujita O (2019) Extinction limits of an ammonia/air flame propagating in a turbulent field. Fuel 246:178–186
23. Hayakawa A, Arakawa Y, Mimoto R, Somarathne KDKA, Kudo T, Kobayashi H (2017) Experimental investigation of stabilization and emission characteristics of ammonia/air premixed flames in a swirl combustor. Int J Hydrogen Energy 42:14010–14018
24. Ministry of the Environment (2019). https://www.env.go.jp/air/osen/law/t-kise-6.html. Accessed on 5 Aug 2019 (in Japanese)
25. e-GOV. https://elaws.e-gov.go.jp/search/elawsSearch/elaws_search/lsg0500/detail?lawId=346AC0000000091. Accessed on 5 Aug 2019
26. Hayakawa A, Hirano Y, Okafor EC, Yamashita H, Kudo T, Kobayashi H (2021) Experimental and numerical study of product gas characteristics of ammonia/air premixed laminar flames stabilized in a stagnation flow. Proc Combust Inst 38:240–2417
27. Tian Z, Li Y, Zhang L, Glarborg P, Qi F (2009) An experimental and kinetic modeling study of premixed NH_3/CH_4/O_2/Ar flames at low pressure. Combust Flame 156:1413–1426
28. Colson S, Hayakawa A, Kudo T, Kobayashi H (2016) Extinction characteristics of ammonia/air counterflow premixed flames at various pressures. J Therm Sci Technol 11:16–00384
29. Lefebvre AH, Ballal DR (2010) Gas turbine combustion, 3rd edn. CRC Press, p 25
30. Somarathne KDKA, Hatakeyama S, Hayakawa A, Kobayashi H (2017) Numerical study of a low emission gas turbine like combustor for turbulent ammonia/air premixed swirl flames with a secondary air injection at high pressure. Int J Hydrogen Energy 42:27388–27399
31. Hayakawa A, Tsukamoto M, Somarathne KDKA, Kudo T, Kobayashi H (2019) Simultaneous NO and unburnt ammonia reductions of ammonia/air premixed flames using two stage combustion. Proc Nat Heat Transf Symp Jpn 56:F235 (in Japanese)
32. Kurata O, Iki N, Inoue T, Matsunuma T, Tsujimura T, Furutani H, Kawano M, Arai K, Okafor EC, Hayakawa A, Kobayashi H (2019) Development of a wide range-operable, rich-lean low-NO_x combustor for NH_3 fuel gas-turbine power generation. Proc Combust Inst 37:4587–4595
33. Okafor EC, Somarathne KDKA, Hayakawa A, Kudo T, Kurata O, Iki N, Kobayashi H (2019) Towards the development of an efficient low-NO_x ammonia combustor for a micro gas turbine. Proc Combust Inst 37:4597–4606

Chapter 33
Demonstration of Ammonia Fuel with Small Gas Turbines

Norihiko Iki

Abstract AIST carried out demonstration tests with the aim to show the potential of ammonia-fired power plant. 50 kW class turbine system firing kerosene is selected as a base model. A standard combustor is replaced by a prototype combustor which enables a bi-fuel supply of kerosene and fuel gas. The gas turbine started by burning kerosene and increased its electric power output. After achievement of stable power output, ammonia gas was started to be supplied and its flow rate increased gradually. Ammonia combustion in the prototype bi-fuel combustor was enhanced by supplying hot combustion air and by modifying the air inlets. However, the exhaust gases from the ammonia flames had high NO_x concentrations. NO_x removal equipment using selective catalytic reduction can reduce NO_x emission levels to below 10 ppm from more than 1,000 ppm (converted value of NO_x to 16% O_2) as already reported. However, downsizing of NO_x removal equipment should be achieved for practical use. Therefore, a low-NO_x combustor was developed. We modified the combustor to the rich-lean burning method and found the condition that the NO emission can be reduced to less than half, but the NO_x emission reduction was insufficient as compared with the small-scale flow test at the lab scale. Therefore, each part of the combustor was redesigned so that mixing of fuel and air improves uniformity. Low-NO_x combustor (Step 2) was designed using knowledge of prototype bi-fuel combustor for ammonia (Step 0) and low-NO_x combustor (Step 1). As a result, we found a condition that the NO emission can be reduced to 200 ppm or less as compared with before redesigning for rich-lean combustion.

Keywords Micro gas turbine · Ammonia combustion · NO_x

Originally published in Journal of the Combustion Society of Japan, Vol. 61 (2019), pp. 283–288 (in Japanese);

N. Iki (✉)
National Institute of Advanced Industrial Science and Technology, 2-2-9 Machiikedai, Koriyama, Fukushima 963-0298, Japan
e-mail: n-iki@aist.go.jp

K. Aika and H. Kobayashi (eds.), *CO_2 Free Ammonia as an Energy Carrier*,
https://doi.org/10.1007/978-981-19-4767-4_33

33.1 Introduction

Ammonia is not only used in large quantities in the chemical industries but has also attracted attention as an energy carrier in recent years. Efforts to use ammonia fuel in gas turbines are not only recent and were carried out for vehicles in the United States in the 1960s, but their combustion efficiencies were low and practical application was not realized [1, 2]. At that time, development of a catalytic combustor and NO_x measurement was also carried out. In the 2010s, when a gas turbine for ammonia-fired power generation was proposed, a jet engine-based combustor test began [3]. In Japan, Tohoku University and the National Institute of Advanced Industrial Science and Technology (AIST) demonstrated successfully the first ammonia-fueled gas turbine power generation in the Cross-ministerial Strategic Innovation Promotion Program (SIP), "Energy Carriers" by the Cabinet Office [4, 5]. In this way, small gas turbines could be tackled relatively quickly in the development of gas turbines using ammonia as fuel. In addition, Tohoku University, AIST, and Toyota Energy Solutions Inc. have been developing ammonia-fueled micro gas turbines [6–8]. In this paper, these results are introduced.

33.2 Ammonia Gas Turbine Test Facility

Basic research on gas turbine combustion has started by Tohoku University [6] and AIST has started research and development of a 50 kW class micro gas turbine mainly [7]. Based on these results, Toyota Energy Solutions Inc. has been developing 50 kW class micro gas turbines and a 300 kW class micro gas turbine [8]. In the start of this research and development at AIST, the installation of an ammonia direct combustion gas turbine has been proceeded almost simultaneously with the opening of the Fukushima Renewable Energy Institute, AIST (FREA) in April 2014. Since the test operation of the gas turbine has been carried out in parallel with the maintenance of the research facility, it is roughly divided into the next phase.

- Phase1 (year 2013–2014): Power generation tests cofiring kerosene-ammonia using a temporary ammonia supply facility.
- Phase2 (year 2014–2016): Power generation tests firing ammonia and power generation tests cofiring methane-ammonia, using permanent ammonia supply facilities.
- Phase3 (year 2016–2019): Combustor testing with combust test rigs, Development of low-NO_x combustors.

Since ammonia is harmful to the human body and liquefied gas, various laws and regulations are received, so the period and expense required for installation are greatly different depending on the installation place of the test facility. AIST decided to install a 50 kW class gas turbine outdoors, so there were few applicable regulations, saving the time and money required before the start of the tests. In addition, by supplying liquid ammonia from the ammonia cylinder and evaporating it, even a 500 kg cylinder can supply sufficient ammonia gas for ammonia-fired power generation tests of 50 kW. However, if larger amount of ammonia supply were planned for application of a 300 kW class gas turbine, installation of the facility would be delayed and costly. In addition, by installing a supply facility for methane gas, it is becoming able to test methane-ammonia cofiring. The gas turbine was remodeled based on the TPC50 micro gas turbine of Toyota Turbine System (currently Toyota Energy Solutions Inc.). This gas turbine has a built-in regenerator and uses kerosene as a fuel for starting up. The main factor that the gas turbine started up with kerosene was that there was no city gas supply at the test site. Subsequently, as results of grasping the methane gas consumption of the gas turbine in the demonstration tests, it was replaced in 2018 with a new micro gas turbine that can be started up with methane gas. This gas turbine was manufactured by Toyota Energy Solutions Inc.

33.3 Test Results

33.3.1 Ammonia Direct Combustion

As a method of using ammonia that is difficult to ignite as a fuel, it is considered that ammonia is supplied to the combustion region after pyrolysis at high temperatures. In the swirl combustor used in the gas turbine, the upstream part of the combustor becomes high temperature by the recirculation flow of the combustion gas, and the effect of helping the pyrolysis of ammonia can be expected. In the basic research of Tohoku University, it succeeded in controlling ammonia flame by swirl combustor, and the gas turbine was considered as an internal combustion engine suitable for the demonstration test of ammonia direct combustion. However, in practice, it was necessary to (1) add a gas fuel injector to the combustion, (2) add control system of gas fuel (ammonia, methane), and (3) change the combustor's shape. Prototype bi-fuel combustor for ammonia (Step 0) was prepared. At first, as a shape change, the combustion air hole was added to the combustor liner as shown in Fig. 33.1(b) for the kerosene combustion in Fig. 33.1(a). This can be expected to reduce the flow velocity of the upstream part of the combustor and increase the temperature of the combustion area so that even ammonia flame with a slow-burning velocity can be held. In addition, the diffusion combustion system is adopted so that the ammonia flame becomes easy to hold, and the combustion air is preheated with a regenerator. It is necessary that the gas turbine is started for preheating when the gas turbine is cold. Therefore, as shown in Fig. 33.2, the gas turbine was started with kerosene, warmed sufficiently, and then switched the fuel to ammonia to demonstrate ammonia combustion.

As shown in Fig. 33.3, there is an operating range of rotating speed and power output for power generation by ammonia-only firing. When the rotating speed is constant and the power output increase, the turbine inlet temperature increases, and

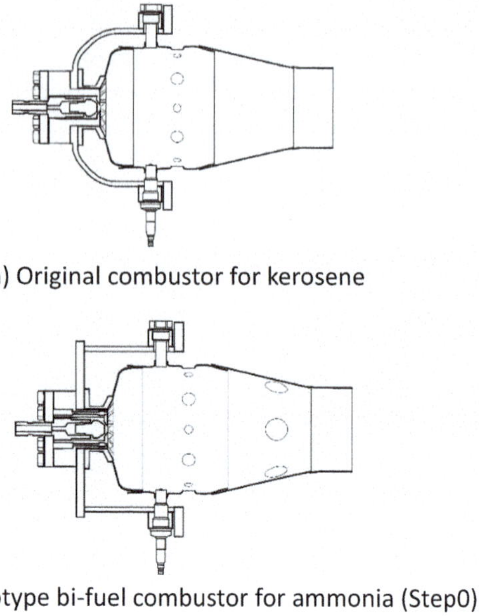

Fig. 33.1 Prototype combustors for ammonia combustion © Combustion Society of Japan 2019

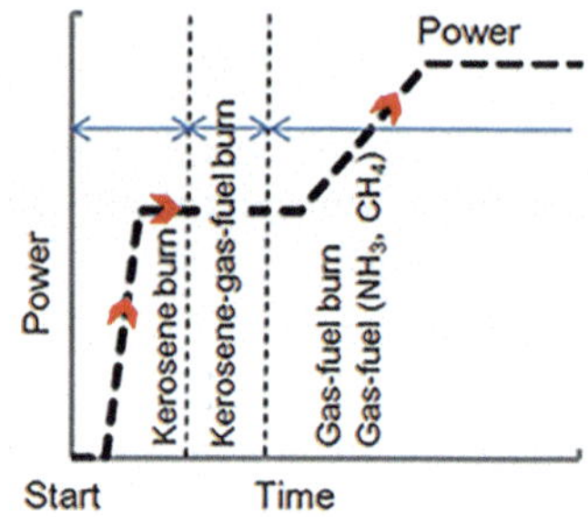

Fig. 33.2 Operation sequence © Combustion Society of Japan 2019

the combustion gas temperature at the turbine outlet exceed the allowable temperature of the regenerator, so there is a limit to the operating range. On the other hand, when the power output decreases with the constant rotating speed, the turbine inlet temperature decreases, and the combustion air temperature preheated by the regenerator decreases due to the decrease in the turbine outlet temperature, and the unburned ammonia emission increases. In this research and development, unburned ammonia can be used for NO_x treatment with NO_x removal equipment using selective catalytic reduction (SCR). Therefore, the limit of the operating range was when unburned ammonia increased to the range that can be used for NO_x treatment. And, in the condition that the rotating speed was high, and the power output was high, there was a limit by the influence of remodeling on the generator. Subsequently, by improving the electrical system, we succeeded in generating 47 kW [9].

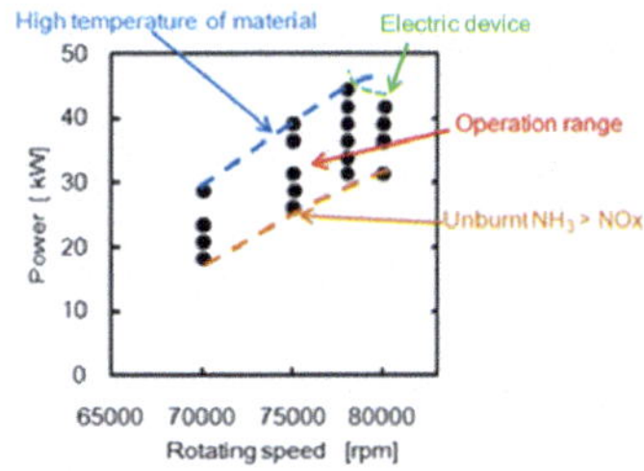

Fig. 33.3 Operation range for NH_3-air combustion © Combustion Society of Japan 2019

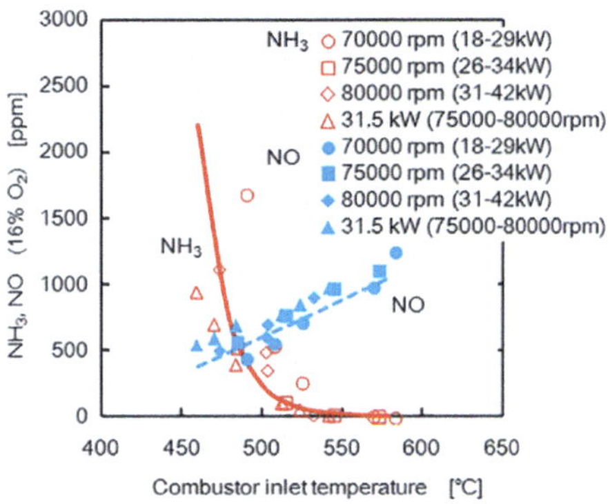

Fig. 33.4 Variations of emitted NO and NH_3 concentrations with combustor inlet temperature for the NH_3-air combustion © Combustion Society of Japan 2019

Figure 33.4 shows the trend of unburned ammonia and NO during ammonia-only firing. NO accounts for about 80–90% of NO_x. NO increases linearly with the combustion air temperature supplied to the combustor, but unburned ammonia increases rapidly when the combustion air temperature drops below 500 °C, and unburned ammonia is rarely emitted above 550 °C of the combustion air temperature. That is, preheating of combustion air is very important in ammonia burning.

In the case of methane-ammonia cofiring, an example of the emission component is shown in Fig. 33.5. When the power output increases as methane and ammonia as 1:1 at ratio of lower heating value, NO increases but the rate of NO increase slightly decreases. NO_2 decreases slightly with the increase in power output. CO decreases with the increase in power output. Unburned ammonia is generated under conditions where the power output is low, but the lower limit of the power output is lower than that during ammonia-only firing, and methane-ammonia cofiring can expand the operating range. Unburned ammonia is emitted under conditions where the power output is low, but the lower limit of the power output is lower than that during ammonia-only firing, and methane-ammonia cofiring can expand the operating range. Figure 33.6 shows the change of exhaust gas components and the combustion air temperature supplied to the combustor regarding to ammonia ratio of the lower heating value. Here is an example of unburned ammonia emission during ammonia-only firing. Along with only a slight addition of ammonia, NO and NO_2 increase rapidly compared to methane firing, but when the ammonia ratio increases, the NO increase becomes moderate and has a peak, and NO_2 has slight change and is almost constant. When the ammonia ratio increases, unburned ammonia is emitted following N_2O emission. N_2O has a problem because the global warming coefficient is high, but in fact it does not occur if the power output is high. Since NO_x can be lowered during ammonia-only firing than the peak value of NO_x during methane-ammonia firing, it is clear that the conversion ratio to NO_x decreases when the nitrogen content of fuel increases in ammonia combustion. This combustor is a combustion state with low uniformity from the structure, and in some view, it consists of flames with various as-you-go ratios and NO_x generation states. Since NO_x emissions change due to changes in these combinations, there is a high possibility that fuel NO_x emission during ammonia firing can be reduced by devising combustion.

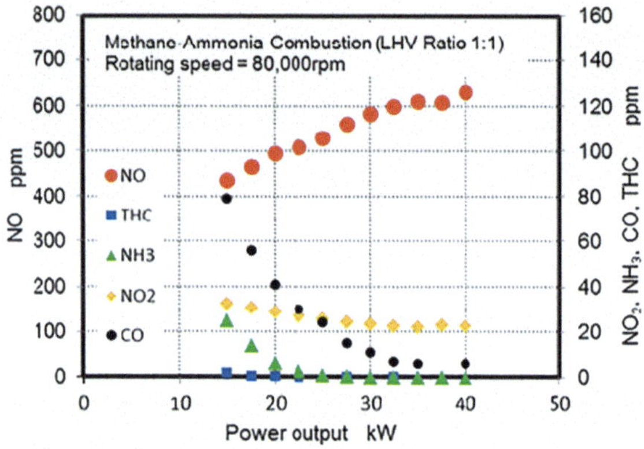

Fig. 33.5 Emission of CH_4- NH_3-air combustion © Combustion Society of Japan 2019

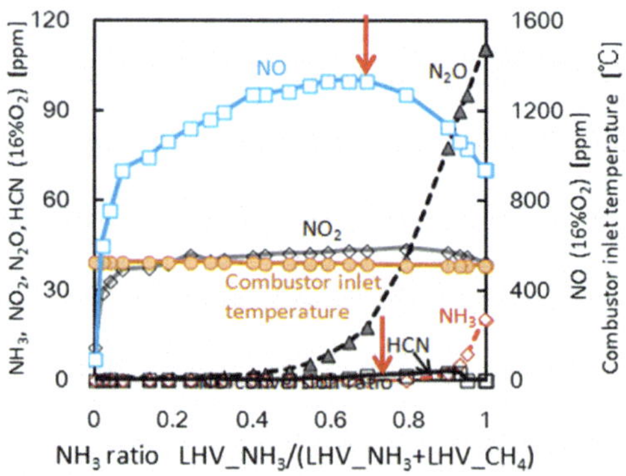

Fig. 33.6 Emission characteristics for varied NH_3 fuel ratio in the case of NH_3–CH_4-air combustion gas turbine operation keeping the electric power of 31.4 kW and the rotating speed of 75,000 rpm © Combustion Society of Japan 2019

33.3.2 Research and Development of Low-NO_x Combustors

Based on the above knowledge, the development of low-NO_x combustor was started. For this reason, we first prepared a combustor test rig so that we could pursue a combustor shape that can reduce NO_x emissions without assuming gas turbine start-up. Since basic research at Tohoku University has shown that NO_x decreases when the equivalence ratio is 1 or more, we aimed for low-NO_x combustion with the concept of rich-lean combustor as shown in Fig. 33.7(a). And we applied rich-lean combustion to the gas turbine combustors by segmenting the combustion area as shown in (b). Since the air supplied to the primary combustion zone is reduced to realize the fuel-rich region, there is a risk that the recirculation flow of the combustion gas caused by the swirling flow may be weakened. Therefore, the injection direction of the fuel was changed to the outside direction. To increase the equivalence ratio of the primary combustion zone, the changes shown in (c) were made, and their effects were confirmed. Turbine inlet temperature of a small gas turbine is low, and the total equivalence ratio is lower than that of other internal combustion engines. It was an issue to be able to form a primary combustion zone with equivalence ratio over 1, and on the other hand, to show the NO_x reduction effect even if it was affected by decreasing local equivalence ratio due to mixing of the cooling air of the combustor liner.

As the result that the combustion test was carried out using the combustion test rig, as shown in Fig. 33.8, it was found that NO emission could be greatly reduced in the fuel flow range of 250–300 kW by sealing the primary air holes and by halving of the flow path in the swirler. In Fig. 33.8, since ammonia of 250 kW or more could not be supplied due to the convenience of the test equipment, methane-ammonia mixed firing data is also added, so there are not small variations of the data of NO concentration. In addition, when secondary air holes are added to double the area, fuel flow rate at the minimum value of the NO concentration shifts to lower fuel flow rate as shown in Fig. 33.9, but the minimum value is around 400 ppm (at 16% O_2), and there is no big difference from Fig. 33.8. Furthermore, since the cooling air flowing in from the fitting part of the base and the combustor liner was not mixed into the primary combustion zone, the extremely small value of the NO concentration decreased when the sleeve was fitted. From the above, by changing the ratio of air supplied to the primary combustion zone and the secondary combustion zone, the flow range of ammonia that can obtain the NO_x reduction effect, in other words, the power output range of the gas turbine can be adjusted. If the mixing of cooling air into the primary combustion zone is prevented, the NO_x reduction effect can be increased.

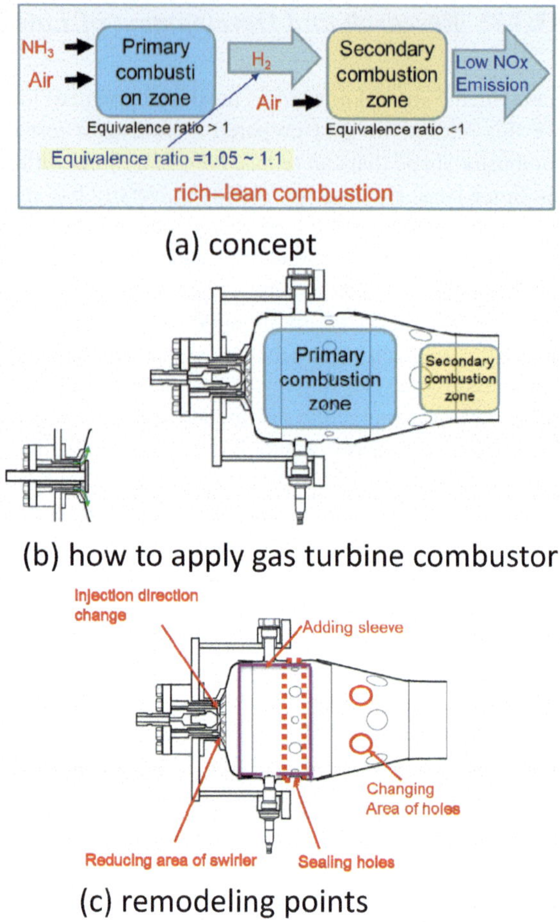

Fig. 33.7 Rich-lean combustion

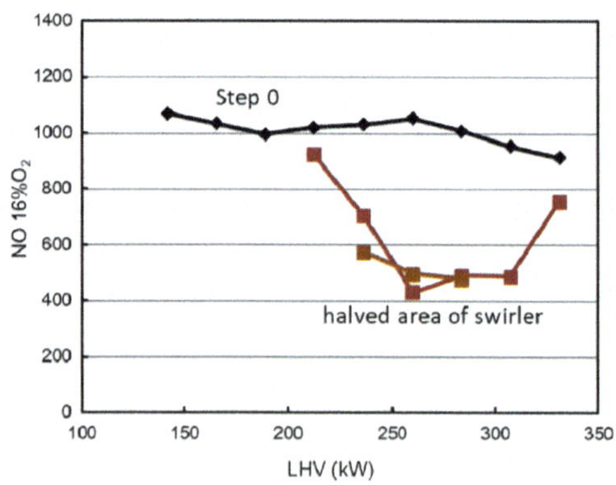

Fig. 33.8 Effect of sealed holes and area of Swirler © Combustion Society of Japan 2019

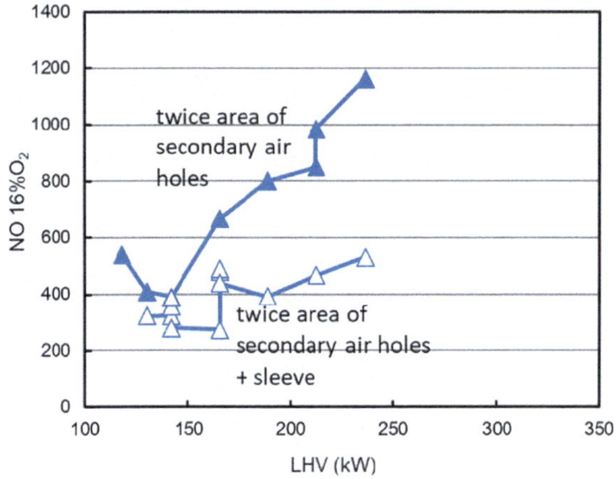

Fig. 33.9 Effect of sleeve © Combustion Society of Japan 2019

33.3.3 Demonstration Test of Low-NO_x Combustor

Regarding combustors aiming at rich-lean combustion, we designed and prototyped a low-NO_x combustor for gas turbine power generation demonstration tests, based on the knowledge that the possibility of NO_x reduction and handling of cooling air was important in micro gas turbine combustors as shown in Fig. 33.10. In some cases, the liner temperature of the low-NO_x combustor (Step 1) reached to 1000 °C. Therefore, the operating range of the gas turbine was limited, and sufficient NO_x reduction effect could not be obtained. For this result, we improved the cooling capacity of the liner and manufactured the low-NO_x combustor (Step 2) with the cooling air hole outlet extending further downstream. Figure 33.11 shows the NO emission and the liner temperature of the low-NO_x combustor (Step 2). Step 2 shows improved performance compared to Step 1 so that NO was significantly reduced to 222 ppm. However, the liner temperature has reached 1000 °C, and it is necessary to devise a design for the cooling method. Toyota Energy Solutions Inc., AIST and Tohoku University developed the low-NO_x combustor (Step 3) combustor with further improvements. Since the low-NO_x combustor can allow the less amount of SCR catalyst, a new 50 kW gas turbine package was designed and manufactured with a built-in compact NO_x removal unit using SCR. And Its NO_x emission performance was evaluated. This new gas turbine package can start up with methane and needs no kerosene fuel injector. As a result of the test, the gas turbine was able to start up safely, and the power generation of 50 kW was succeeded. At a power output of 48.6 kW, NO_x before SCR achieved 164 ppm (at 16% O_2) and was reduced to 25 ppm (at 16% O_2) after SCR. More details were reported in the final report of SIP energy carrier by Toyota Energy Solutions Inc. [8].

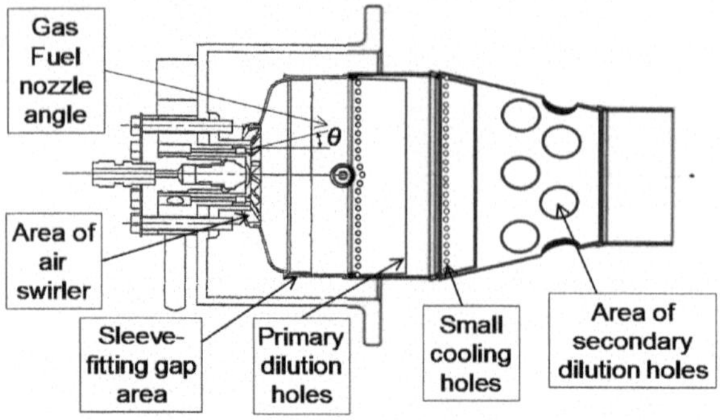

	Primary dilution holes	Area of air swirler	Sleeve fitting gap area	Primary dilution holes	Area of secondary dilution holes	Designed primary equivalence ratio	Fuel nozzle angle
Case A	Without	Less than Step 0	Without	Air emitting to secondary combustion zone	6 holes	0.7	45°
Case B					9 holes	0.9	45°
Case C					12 holes	1.1	45°
Case D					12 holes	1.1	30°
Case E					12 holes	1.1	0°
Case F					12 holes	1.1	60°

(a) Redesign points

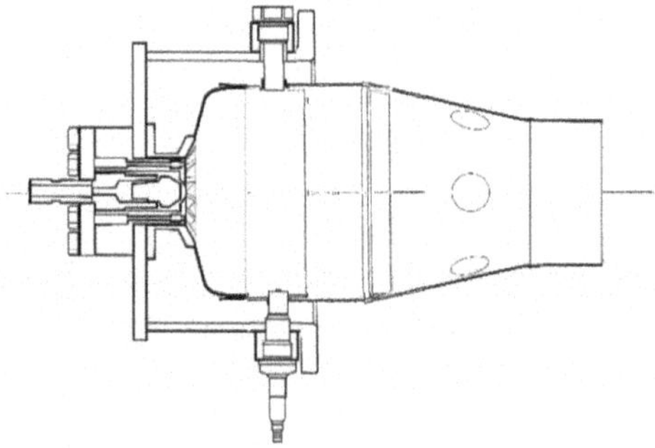

(b) Low-NOx combustor (Step1)

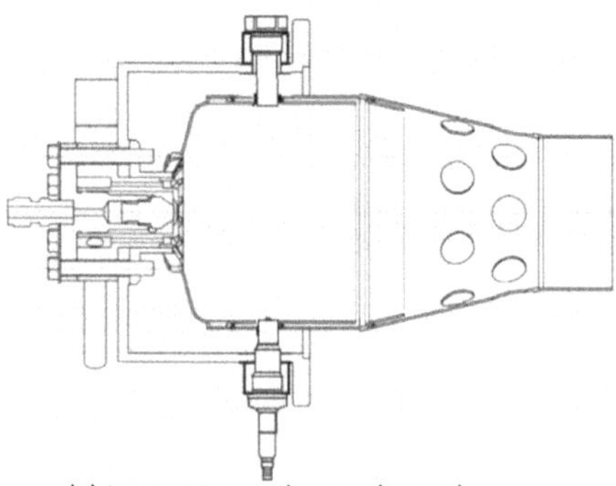

(c) Low-NOx combustor (Step2)

Fig. 33.10 Prototype combustors for ammonia combustion

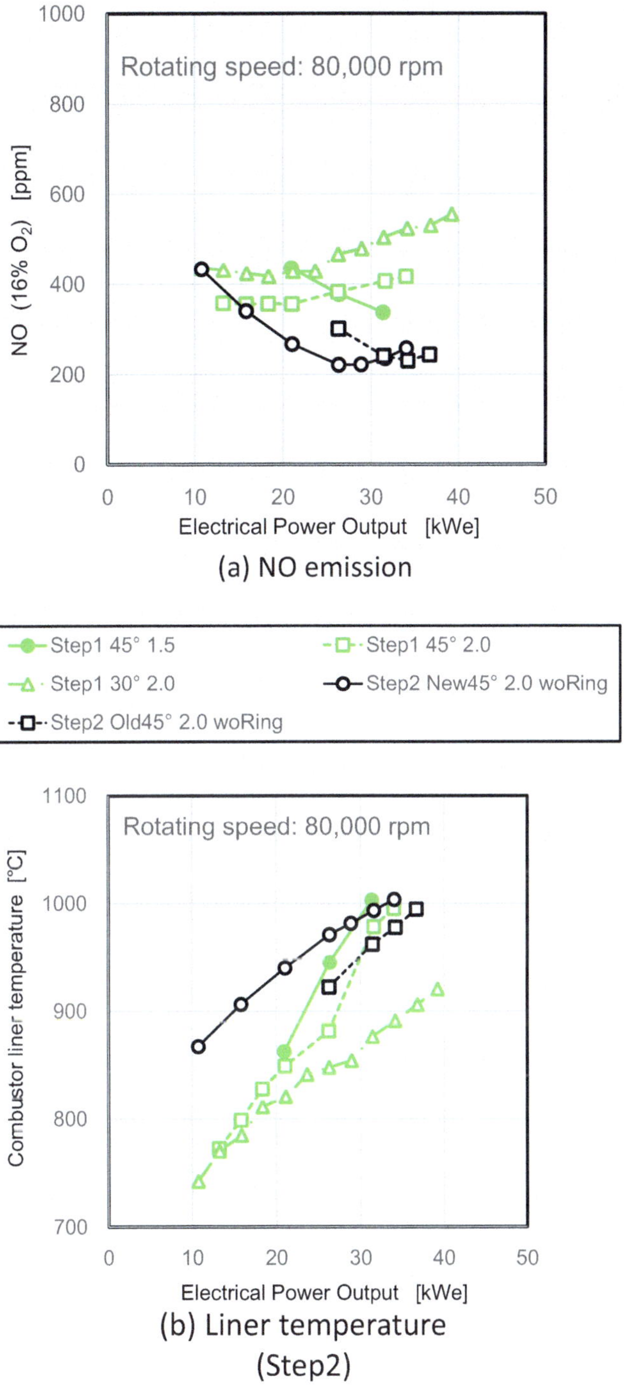

Fig. 33.11 Low NO_x combustor (Step 1, Step 2) © Combustion Society of Japan 2019

33.4 Conclusions and Outlook

The attempt to use ammonia as a gas turbine fuel is not new, but we succeeded in generating electric power by a 50 kW class gas turbine using swirl combustor, regenerator, SCR technology, etc. The amount of NO_x generated from ammonia combustion is higher than that from other fuels, and over 1000 ppm (at 16% O_2) of NO_x is emitted in some cases, but it can be handled by NO_x removal equipment using SCR. In micro gas turbines, low-NO_x combustor can be developed based on the concept of rich-lean combustion, and it can be reduced to 200 ppm (at 16% O_2) or less. However, it is expected to improve the cooling technology of the combustor and further reduce NO_x.

In addition, Toyota Energy Solutions Inc. succeeded in ammonia-fired 300kW-class micro gas turbines without regenerator [10]. By rising the temperature of the combustion air, emission of unburned ammonia can be suppressed even in the case of ammonia-only firing. Therefore, ammonia can be expected to be used as a fuel even in larger high-pressure gas turbines with higher combustion air temperatures. In addition, AIST and JGC have succeeded for the first time in the world in demonstrating an energy system that includes ammonia synthesis and power generation using hydrogen derived from renewable energy [9]. However, kerosene was still used to start up gas turbines. By development and demonstration of gas turbine start-up technology using ammonia or hydrogen, practical application of ammonia energy system can be prospected.

Acknowledgements This work was supported by the Council for Science, Technology, and Innovation (CSTI), the Cross-ministerial Strategic Innovation Promotion Program (SIP), "Energy Carriers" (Funding agency: Japan Science and Technology Agency (JST)). I would like to thank Professor Hideaki Kobayashi of Tohoku University and many other persons for their great cooperation.

References

1. Pratt DT (1967) Performance of ammonia-fired gas-turbine combustors. Technical report no. 9. DA-04-200-AMC-791(x)
2. Solar Division of International Harvester Company (1968) Development of an ammonia-burning gas turbine engine. Final technical report. DA-44-009-AMC-824(T)
3. Evans B (2013) Using local green energy and ammonia to power gas turbine generators. In: The tenth annual NH_3 fuel conference. Sacramento, CA
4. Iki N et al (2014) Micro gas turbine operation with kerosene and ammonia. In: The 11th annual NH_3 fuel conference. Des Moines, IA
5. Iki N et al (2015) Micro gas turbine firing ammonia. In: The 12th annual NH_3 fuel conference. Lemont, IL
6. SIP Energy Carriers (2019) Basic combustion characteristics and fundamental technology R&D of ammonia combustion. Ammonia direct combustion final report. https://www.jst.go.jp/sip/dl/k04/end/team6-1.pdf
7. SIP Energy Carriers (2019) Technology development of ammonia internal combustion engine. Ammonia direct combustion final report. https://www.jst.go.jp/sip/dl/k04/end/team6-7.pdf

8. SIP Energy Carriers (2019) Ammonia combustion micro gas turbine. Ammonia direct combustion final report. https://www.jst.go.jp/sip/dl/k04/end/team6-19.pdf
9. JGC (2018) News release. https://www.jgc.com/jp/news/assets/pdf/20181019.pdf
10. Toyota Energy Solutions (2019) Press-release. https://www.toyota-energy.co.jp/wp-content/uploads/2019/04/190408_press-release.pdf

Chapter 34
Development of 300 kW Ammonia Combustion Micro Gas Turbine

Keisuke Arai, Masato Kawano, Naoki Kudomi, Hiroshi Sugiura, Yoshinori Yasui, and Fumitake Ishiwaki

Abstract Recently, ammonia has been focused as one of the hydrogen carriers, and the research and demonstration of direct combustion utilization have been reported. Toyota Energy Solutions Inc. (TE) participated in a research project on ammonia combustion power generation using a 50 kW micro gas turbine (MGT) in collaboration with Tohoku University and the National Institute of Advanced Industrial Science and Technology, and obtained many results. This time, TE has begun research on a 300 kW MGT for industrial use, and has achieved 100% ammonia-only combustion as a result of evaluating various prototype combustors for ammonia combustion.

Keywords Micro gas turbine · Ammonia combustion · Diffusion combustion · Premixed combustion · NO_x · Fuel control

34.1 Introduction

It is critical for Japan to move away from the dependence on fossil fuels and reduce CO_2 emissions. In the Basic Energy Plan of Japan, hydrogen is positioned as a new source of energy, and measures to realize "Hydrogen Society" have been accelerated. On the other hand, the high costs of transportation and storage have been raised as issues hindering the penetration of hydrogen in the society. In the meantime, ammonia has been attracting attentions as a carrier of hydrogen energy as it has high hydrogen content and is easy to be liquified, and various research and evaluation projects have been underway including the study on the use of ammonia direct combustion. Ammonia has a potential for playing a key role in realizing future low-carbon society if its procurement cost is reduced as its use as energy source increases and there are more advanced technologies to produce CO_2-free ammonia.

Originally published in Journal of Gas Turbine Society of Japan, Vol. 48 (2020), pp. 234–239 (in Japanese);

K. Arai (✉) · M. Kawano · N. Kudomi · H. Sugiura · Y. Yasui · F. Ishiwaki
Toyota Energy Solutions Inc., 1 Motomachi, Toyota, Aichi 471-8573, Japan
e-mail: keisuke_arai@toyota-energy.co.jp

K. Aika and H. Kobayashi (eds.), *CO_2 Free Ammonia as an Energy Carrier*,
https://doi.org/10.1007/978-981-19-4767-4_34

Toyota Energy Solutions Inc. (TE) participated in the "Energy Carrier" program, one of the programs of the Cabinet Office's Cross-Ministerial Strategic Innovation Promotion Program (SIP), and jointly worked with Tohoku University and the National Institute of Advanced Industrial Science and Technology (AIST) as part of the Ammonia Direct Combustion team to generate electricity by combusting ammonia with a micro gas turbine (MGT) [1]. (*TE participated in the project as joint researcher from 2017 to 2018.) In the previous experiments using a 50 kW MGT installed at the Fukushima Renewable Energy Institute, AIST (FREA), we successfully achieved the co-combustion of kerosene and ammonia [*30% co-combustion rate (LHV)] in 2014 and the co-combustion of LNG and ammonia [*50% co-combustion rate (LHV)] as well as 100% ammonia-only combustion in 2015. In 2017–2018, we achieved the reduction of nitrogen oxide (200 ppm @16% O_2 or less) solely with the MGT using the combustor design proposed by Tohoku University and AIST that allowed the reduction of fuel nitrogen oxide (NO_x), subsequently completing the development of a new package embedded with down-sized NO_x removal catalysts [2, 3].

This time (in 2018), TE conducted research and development of a 300 kW MGT in anticipation of its wider use in industrial and large-scale commercial facilities [4], and successfully achieved 100% ammonia-only combustion.

The development details and the evaluation results are reported below.

34.2 Overview of the Demonstration of Ammonia Combustion Using 300 kW MGT

In demonstrating ammonia combustion using a 300 kW MGT, the numerical targets were set taking into consideration that the development period was only for a year. The target ammonia co-combustion rate was set to be 50% or above (LHV) and the amount of nitrogen oxide in the exhaust gas was set to be below the environmental standard specified by the Air Pollution Control Act (70 ppm @16% O_2).

The demonstration was conducted at Chita Thermal Power Station of Chubu Electric Power Company, Inc. (its fuel business, overseas power generation and energy infrastructure business and thermal power generation business were transferred to JERA Co., Inc. in April 2019) so that the existing ammonia supply system used for NO_x removal at the thermal power station could be used as an ammonia supply system for MGT combustion. As we were able to use the existing ammonia supply system, the installation of the MGT was regarded as a minor change to the gas consumption equipment under the High Pressure Gas Safety Act, and thus required minimum regulatory compliance work, such as changes and notifications related to the addition of relevant equipment to the control/inspection items, allowing us to start the construction in a relatively short period of time.

The appearance of the 300 kW ammonia combustion MGT package is shown in Fig. 34.1. The power generation package is an all-in-one package consisting of a micro gas turbine, a synchronous generator, a reduction gear, a fuel gas compressor, lubricant system and its cooling system, a control panel for MGT as well as a high voltage panel for grid interconnection.

For the demonstration project, an ammonia supply unit consisting of a large capacity fuel shut-off valve and a fuel control valve was added to the system though the existing co-generation package was used as it was (Fig. 34.2).

Fuels (ammonia and LNG), auxiliary power and nitrogen gas (N_2) for purging and instrumentation were supplied from the existing infrastructure within the power plant. The generated power was consumed by the load resistor without being interconnected to grid. A NO_x removal unit was installed in the downstream of the MGT for NO_x reduction. Then exhaust gas was finally released into the air via exhaust silencer as there is no heat recovery boiler for cogeneration in this system.

The specification of the 300 kW MGT is shown in Table 34.1. As the 50 kW MGT, which had been used for ammonia combustion, was a regenerative cycle unit, the combustor inlet temperature was as high as 500 °C or above. On the other hand, with this test unit being a simple cycle unit, it was necessary to combust more ammonia at a lower inlet temperature around 300 °C. It was expected to be difficult to achieve the targets as ammonia as a fuel was considered to be relatively disadvantageous due to its low calorific value of around a third of LNG, low flame temperature, and slow combustion speed. Against this backdrop, three different types of prototype combustor (STEP 1–3) were created based on the standard LNG combustor (STEP 0) for ammonia combustion while the fundamental components from the standard MGT, such as compressor and turbine, were kept as they were (Fig. 34.3).

As mentioned above, the existing ammonia supply facility (Fig. 34.4) in the power plant was used for the demonstration. The unit consists of a liquid ammonia tank (Tank capacity: 50 t) and a vaporizer (Vaporizing capacity: 600 kg/h), and has a sufficient ammonia supply capacity required to run a 300 kW MGT.

However, the ammonia supply pressure from the ammonia supply unit was around 0.2 MPa, which was low against the pressure of the MGT, and it was impossible to supply fuel directly to the MGT, and therefore, a newly built ammonia gas compressor was installed between the ammonia supply unit and the MGT for the demonstration (Fig. 34.5).

The ammonia gas compressor uses an electric motor to drive the horizontally-opposed pistons, and is capable of supplying ammonia at a flow rate of 330 kg/h or above that is required to operate a 300 kW MGT at the rated output while at the same time increasing the ammonia supply pressure up to 0.8 MPa at the MGT inlet. Although the ammonia gas compressor controls the flow rate with unloaded operation (by turning the intake valve on/off) while maintaining the motor rotation at a constant rate, the buffer tanks (inlet snubber/outlet snubber) were installed in order to control the fluctuating pressure on the inlet and the outlet of the gas. In addition to the ON/OFF control of the cooling fan and monitoring of oil and gas pressures, the ancillary control panel has functions, such as the protective stop in case of abnormality and the emergency alarm sent to the MGT control panel in order to protect the overall system.

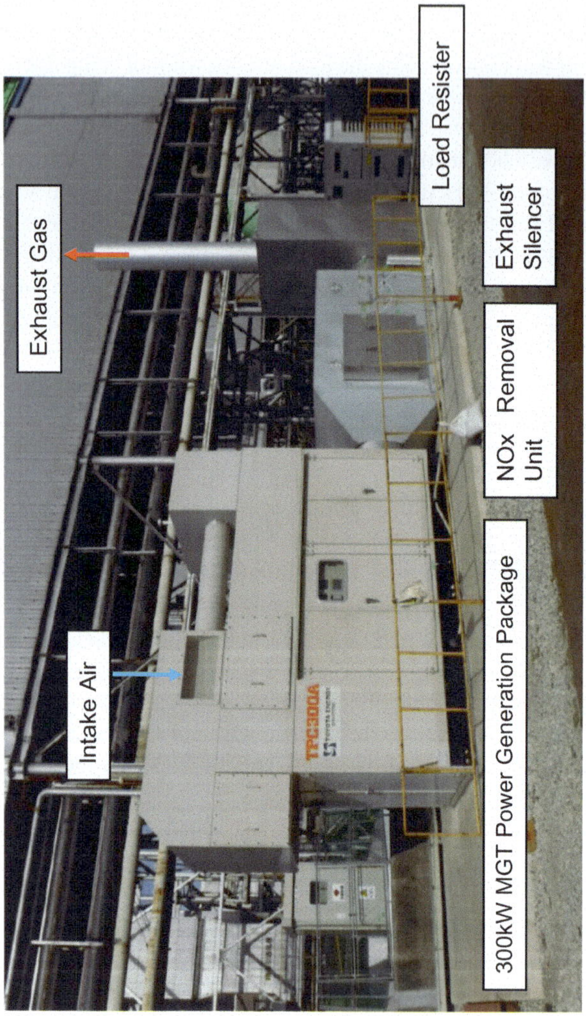

Fig. 34.1 300 kW ammonia combustion MGT package

Fig. 34.2 Ammonia supply unit in the package

Table 34.1 Specification of 300 kW MGT

Model		300 kW MGT
Generator output		295 kW (25 °C)
Power generation efficiency		17.4%
Motor	Type	Simple open cycle Single shaft gas turbine
	Rotating speed	40,000 min^{-1}
	Air flow rate	2.59 kg/s (7210 Nm^3/h)
	Pressure ratio	6.6
	Combustor inlet temperature	300 °C
	Fuel consumption (LNG)	150.2 Nm^3/h
	Fuel consumption (Converted to NH_3)	328.9 kg/h
	Exhaust gas temperature	496 °C
Generator type		Synchronous generator

A NO_x removal unit was installed in the downstream of the MGT as ammonia was used as a fuel this time and it was expected that a large amount of fuel NO_x would be released into the air during the MGT exhaust (Fig. 34.6). The NO_x removal unit is selective catalyst reduction (SCR) type, and has a capacity to reduce the amount of NO_x in the exhaust gas released from the MGT from 1000 to 30 ppm. The catalyst chosen this time would be active at around 500 °C that corresponds to the exhaust temperature from the 300 kW MGT. The total of 576 catalysts with honeycomb

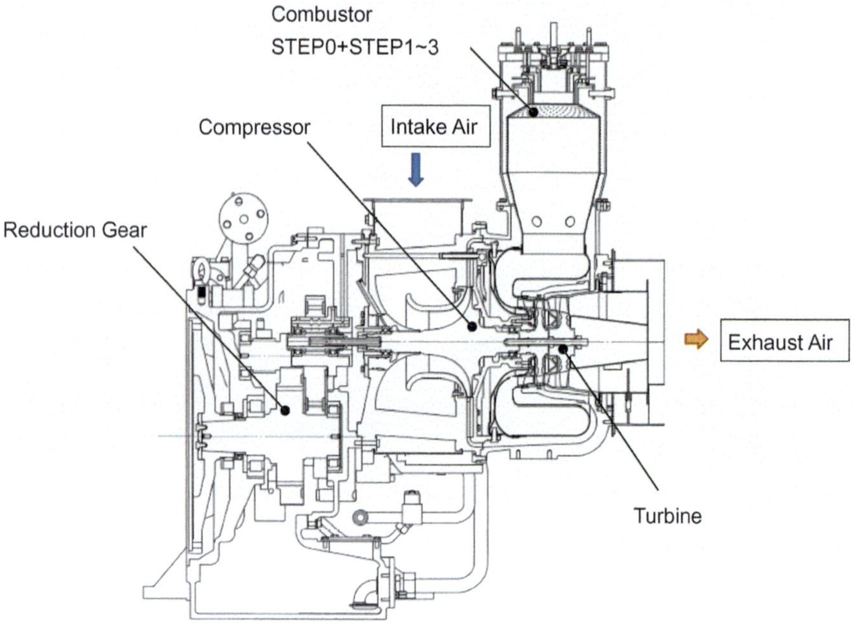

Fig. 34.3 300 kW MGT cross-section

Fig. 34.4 Ammonia supply facility @ Gas Turbine Society of Japan 2020

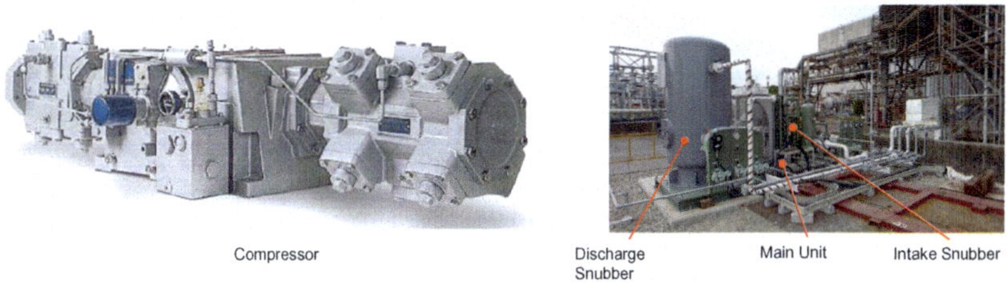

Fig. 34.5 Ammonia gas compressor

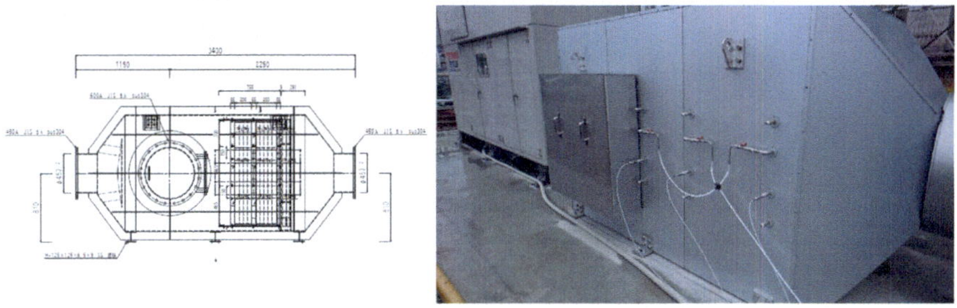

Fig. 34.6 NO_x removal unit (SCR) @ Gas Turbine Society of Japan 2020

structure (8 pieces × 8 pieces × 9 stages) were mounted in the catalyst cell with the size of 150 mm × 150 mm × 50 mm. The sampling ports were created in the upstream, the midstream (in-between the 4th and 5th stages) and the downstream for the catalyst performance evaluation. Exhaust gas temperature and exhaust gas properties were measured.

The state quantities related to the MGT performance appear on the engine control unit (ECU), and together with other data including fuel consumption data (ammonia/LNG), exhaust gas analysis (O_2, CO_2, CO, THC, NO_x and N_2O) before and after the NO_x removal unit as well as the temperature data of the combustor liner, the data is collected and stored in the measurement PC and the data logger installed in the temporary facility next to the MGT package.

34.3 Demonstration Results

The appearance, characteristics, and evaluation results of the combustors STEP 0 and STEP 1–3 are shown in Table 34.2.

Table 34.2 Prototype combustors for ammonia combustion

Combustor type	Base combustor (STEP 0)	STEP 1	STEP 2	STEP 3
Appearance	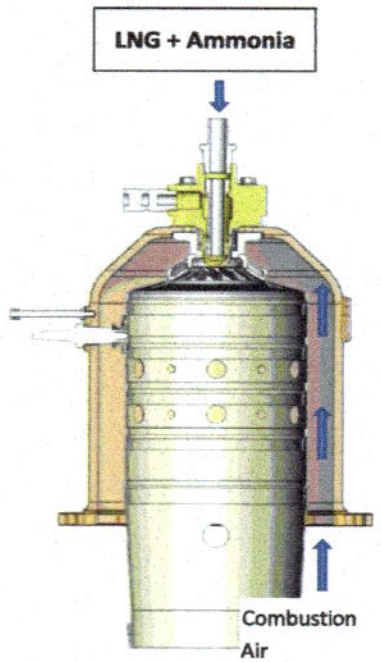	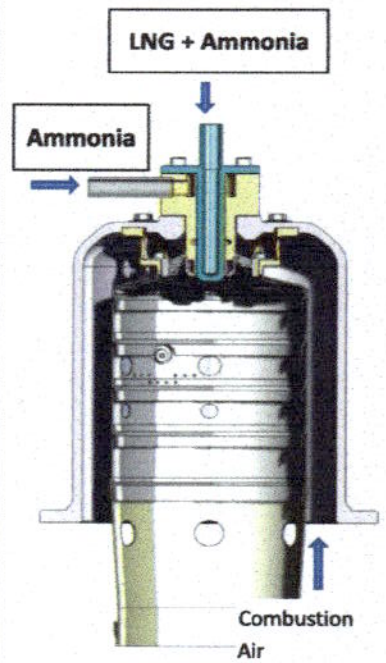	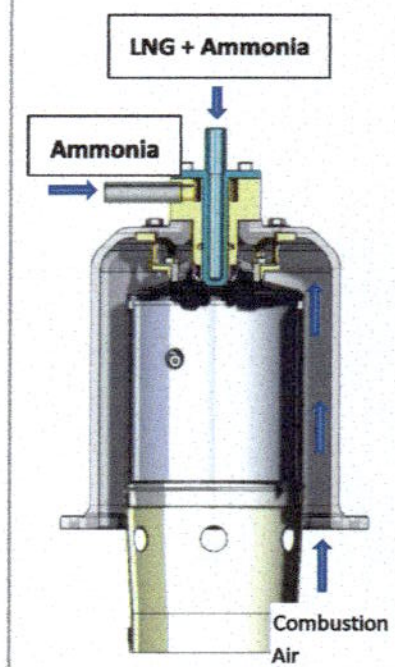	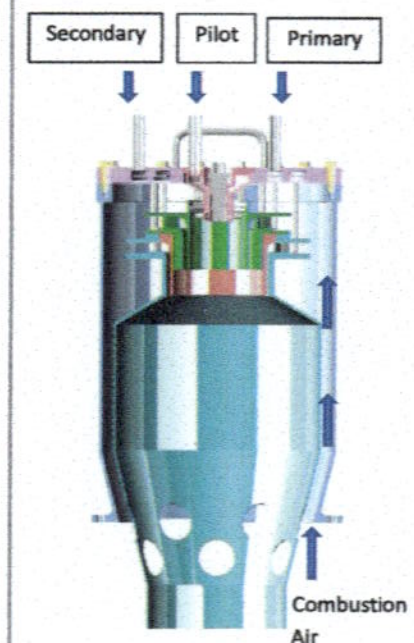
Characteristics	• Diffusion combustion (original) • 1 fuel line • (LNG-ammonia mixed injection) • Film-cooling liner	• Enrichment in combustion zone • 2 fuel lines (possible to inject LNG/ammonia separately) • Film-cooling liner	• The same as left • The same as left • Impingement cooling liner	• Premix combustion • 3 fuel lines (pilot: diffusion, primary and secondary: premix) • Non-cooled liner
Results	• LNG-ammonia co-combustion • **>Co-combustion rate: 34.2% (LHV)** • **NO_x at catalyst outlet: 15 ppm or less**	• LNG-ammonia co-combustion • **>Co-combustion rate: 50.9% (LHV)** • **NO_x at catalyst outlet: 15 ppm or less**	• LNG-ammonia co-combustion • **>Co-combustion rate: 71.4% (LHV)** • **NO_x at catalyst outlet: 15 ppm or less**	• **Achieved 100% ammonia-only combustion** • **NO_x at catalyst outlet: 15 ppm or less**

34.3.1 STEP 0 Combustor

A 300 kW diffusion combustor for cogeneration was used as STEP 0 combustor. The combustor's air distribution and liner cooling structure (film cooling type) remain the same as the original one. The ammonia line merges into LNG system before the inlet port of the fuel injection nozzle, and LNG-ammonia mixed gas is injected from the common fuel injection hole into the combustor. The combustor was operated with up to 30% ammonia co-combustion rate (LHV), and it was confirmed that ammonia supply unit was safely operated and controlled and that the NO_x at the exit of the NO_x removal unit was below the regulation value.

34.3.2 STEP 1 Combustor

STEP 1 combustor was designed based on the rich-lean, two-stage combustion studied by Tohoku University and AIST, where over-rich combustion and lean combustion of fuel are performed in a single combustor, and its fuel equivalent ratio was enriched in the combustion zone. As there are two fuel lines (fuel injection nozzles), LNG and ammonia can be injected not only in the form of mixed gas, but also separately as individual gas. The liner cooling of the combustor is film-cooling type like STEP 0. The combustor was able to achieve 50% ammonia co-combustion rate (LHV), but it was decided that it would be difficult to supply ammonia further as there was a noticeable increase of unburnt gas when the co-combustion rate went above 50%.

During the experiment, it was found that the unburnt ammonia from the outlet of the MGT (combustor) helped reduce NO_x in the NO_x removal unit, but it was decided that unburnt ammonia should be allowed up to the amount equivalent to the amount of ammonia to be injected into the inlet of the NO_x removal unit for the purpose of denitrification (based on the assumption that it would be the molar quantity equivalent to that of NO_x to be reduced), and thus the point where the NO_x value hit the lowest at the outlet of the NO_x removal unit was set as the upper limit of ammonia co-combustion rate (Fig. 34.7).

34.3.3 STEP 2 Combustor

Like STEP 1 combustor, STEP 2 combustor has the air distribution based on the rich-lean combustion technique, and the fuel lines (fuel nozzles) are based on the same specification as the one for STEP 1. The liner cooling mechanism was changed from film-cooling to impingement cooling in order to prevent the dilution of fuel gas and the over-cooling of the liner wall due to the inflow of cooling air into the combustion zone and improve the ammonia's combustion performance by reducing

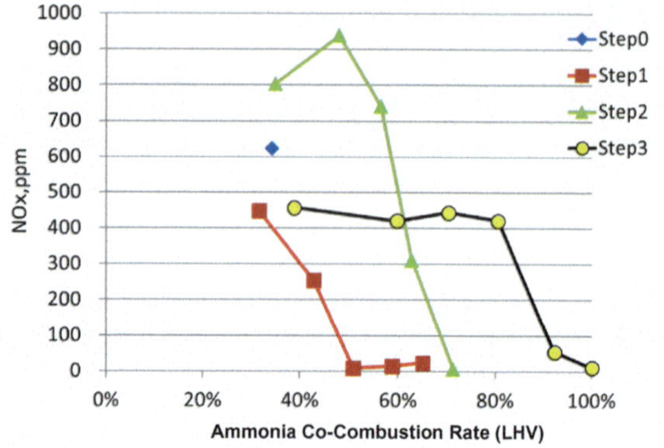

Fig. 34.7 NO_x emissions (after NO_x removal unit)

unburnt ammonia near the liner wall. The combustor was able to achieve 70% co-combustion rate (LHV).

34.3.4 *STEP 3 Combustor*

STEP 3 combustor was built based on a 300 kW premix combustor for LNG cogeneration and was modified to be used for ammonia combustion. The combustor has three fuel lines: the pilot fuel line is for diffusion combustion, and the primary and secondary fuel lines are for premix combustion that are used by the ammonia combustor. The pilot line is for LNG-ammonia mixed gas while the primary and secondary lines are for ammonia only. The equivalent ratios in the diffusion combustion zone and premix combustion zone were enriched for ammonia combustion. In addition, the residence time in the combustor was adjusted to be longer as the volume of the combustor was larger than that of the diffusion combustor. For liner cooling, non-cooled liner was used (=cooled only through convective heat transfer) so that the temperature decrease in the combustion zone and at the liner wall would be mitigated and the ammonia combustion performance would be further improved. The combustor showed a better performance than that of the diffusion combustion type, and it successfully achieved 100% ammonia-only combustion, meaning no CO_2 emission (Fig. 34.8).

34.3.5 *Operation with 100% Ammonia-Only Combustion*

Figure 34.9 shows the operation data of STEP 3 combustor when it achieved 100% ammonia-only combustion. LNG was used to start up the MGT. The loading was started after the engine speed reached the rated rotation speed, and ammonia feeding

Fig. 34.8 CO_2 emissions (after NO_x removal unit)

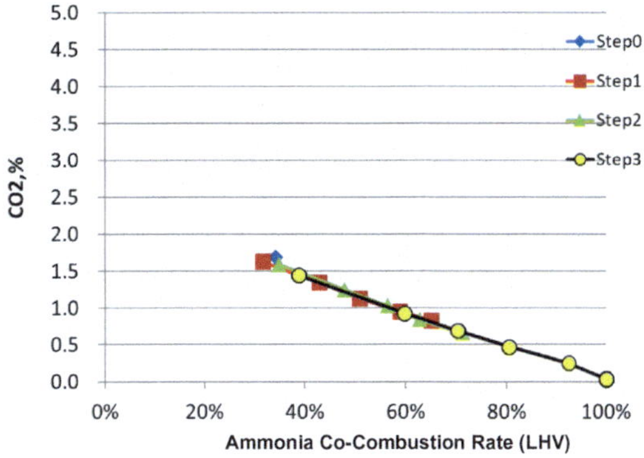

was started when the engine output reached around the rated output (295 kW). The ammonia flow rate was gradually increased with manual control while the LNG flow rate was decreased automatically with feedback control. When the LHV rate of LNG and ammonia became approximately 50–50, the target of feedback control was switched from LNG to ammonia. (Switched from LNG control to ammonia control at the dotted line in Fig. 10). After the switch, the LNG flow rate was manually reduced, and 100% ammonia-only combustion was achieved when the LNG flow rate became zero in the end.

Table 34.3 shows the engine performance and the exhaust gas properties. It was confirmed a certain level of engine performance was achieved, and that CO_2 and

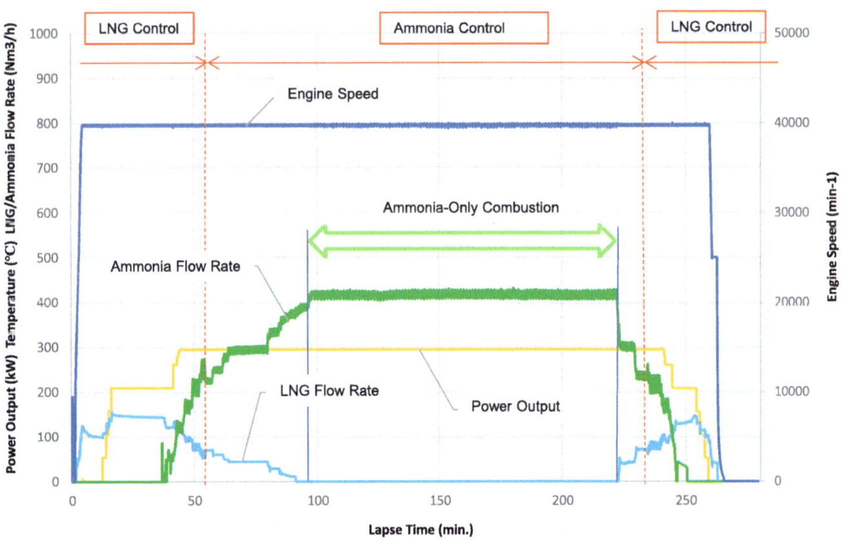

Fig. 34.9 300 kW MGT operation with 100% ammonia combustion

Table 34.3 Test result at 100% ammonia combustion

Engine performance		Measurement value
Power output	kW	294.7
Intake air temperature	°C	14.6
Turbine outlet temperature	°C	427.2
Compressor discharge pressure	kPa A	651 0.9
Compressor discharge temperature	°C	281.7
LNG flow rate	Nm^3/h	0.0
Ammonia flow rate	Nm^3/h	419.8
Exhaust gas properties (catalyst outlet)		
O_2	%	17.4
CO_2	%	0.04
NO_x	ppm	10.6
NO_x (@ 16% O_2)	ppm	14.7
CO	ppm	0

NO_x in the exhaust gas was zero (*The measured CO_2 concentration of 0.04% is equivalent to CO_2 concentration of 400 ppm in the air.) and less than 15 ppm @ 16% O_2 respectively, successfully achieving below the regulation value of 70 ppm required by the Air Pollution Control Act.

Although the previous tests found a tendency that the liner temperature went down and the combustion deteriorated as the amount of ammonia supply was increased, STEP 3 combustor successfully realized stable combustion to achieve 100% ammonia combustion by manually adjusting the fuel distribution among the three fuel lines (1 diffusion line and 2 premix lines) while minimizing the liner temperature reduction and the deterioration of combustion (Fig. 34.10).

34.4 Future Development

TE achieved 100% ammonia combustion using a 300 kW MGT. Moving forward, it plans to work on the evaluation of the durability and reliability of the equipment under long-hour operation, the expansion of ammonia's stable combustion zone, the automation of fuel distribution control as well as further advancement of ammonia combustion technologies including the deployment of low NO_x combustion technology based on the expertise obtained through 50 kW MGT. Also, it intends to optimize the overall system, such as realizing a single fuel system without the use of LNG at the start-up and using exhaust heat from MGT for ammonia supply unit (vaporizer)

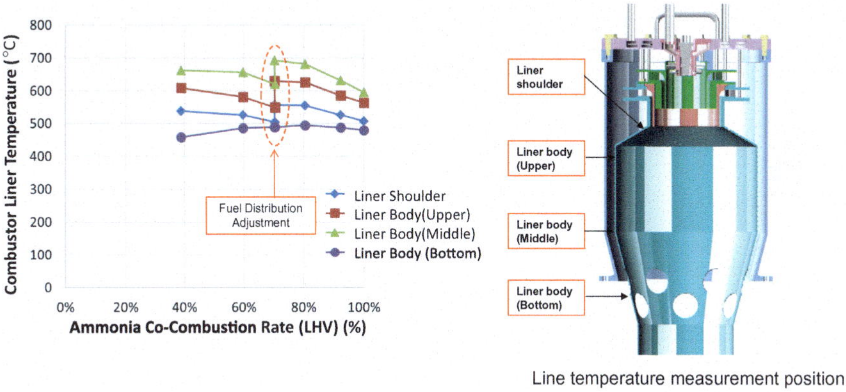

Fig. 34.10 Combustor liner wall temperature

with the aim of its commercialization and the enhancement of its merchantability in anticipation of the prevalence of CO_2-free ammonia in the future.

Acknowledgements A part of this work was supported by the Council for Science, Technology and Innovation (CSTI), Cross-ministerial Strategic Innovation Promotion Program (SIP), "Energy Carriers" [Funding agency: Japan Science and Technology Agency (JST)]. We thank Professor Hideaki Kobayashi of Tohoku University and all the relevant parties for their guidance and cooperation.

We also thank Chubu Electric Power Company, Inc. (Current JERA Co., Inc.) and its Chita Thermal Power Station for providing a test location and allowing us to use the ammonia supply unit.

References

1. Toyota Energy Solutions Inc (2019) Final report of the "Energy Carriers" program under the cross-ministerial Strategic Innovation Promotion Program (SIP) "ammonia direct combustion", "ammonia combustion micro gas turbine"
2. Iki N, Kurata O (2016) Gas turbine firing ammonia. J Combust Soc Jpn 58:215–222
3. Iki N et al (2016) Ammonia utilization for gas turbine. In: Proceedings of the 44th GTSJ annual meeting, vol C-12. Sakata, Japan, pp 259–264
4. Hasegawa T (2016) Highly efficient IGCC power generation with ammonia co-production and CO_2 capture—current status of ammonia use and challenges in the development of gas turbine. In: Proceedings of the 44th GTSJ annual meeting, vol C-11. Sakata, Japan, pp 253–257

Chapter 35
Demonstration Tests of Ammonia/Natural Gas Co-firing Power Generation with a 2-MW-Class Gas Turbine

Shintaro Ito, Masahiro Uchida, Toshiyuki Suda, and Toshiro Fujimori

Abstract To develop an ammonia/natural gas co-fired gas turbine which clears environmental regulations, power generation tests have been conducted using a 2 MW class gas turbine. The test unit consists of an ammonia supply unit, a natural gas compressor, a gas turbine, and a NO_x removal device. In the ammonia supply unit, ammonia is pressurized in liquid state by a pump and gasified with a hot water evaporator. The high-pressure ammonia gas is supplied to the gas turbine. For the gas turbine, IM270 gas turbine manufactured by IHI Corporation is used. Only the combustor is modified for ammonia/natural gas co-firing. In the NO_x removal device, a catalyst which used in a natural gas firing plant is applied. Low-pressure ammonia gas is supplied from the ammonia supply unit as a NO_x reducing agent. As a result of the tests, the successful power generation of 2 MW has been demonstrated for the first time with ammonia mixing ratio of 20% LHV. It has further been confirmed that CO_2 emissions can be reduced about 23%. Although the NO_x concentration at the outlet of the turbine is higher than that of the natural gas fired gas turbine, it has also been confirmed that the NO_x removal device can reduce it to 7 ppm or less.

Keywords Ammonia · Natural gas · Gas turbine · CO_2 · NO_x

35.1 Introduction

Ammonia generated using hydrogen derived from renewable energy is a carbon-free fuel whose transportation and storage technology has been established; it can also be burned directly. Therefore, it is expected to serve as an alternative fuel, thereby reducing the amount of CO_2 emitted from thermal power plants. Natural gas-fired gas turbines, in particular, are expected to continue to be used as a power source for

Originally published in Journal of the Combustion Society of Japan, Vol. 61 (2019), pp. 289–292 (in Japanese);

S. Ito (✉) · M. Uchida · T. Suda · T. Fujimori
IHI Corporation, 1 Shin-nakahara-cho, Isogo-ku, Yokohama 235-8501, Japan
e-mail: ito7931@ihi-g.com

K. Aika and H. Kobayashi (eds.), *CO_2 Free Ammonia as an Energy Carrier*,
https://doi.org/10.1007/978-981-19-4767-4_35

load adjustments. Then, converting a certain amount of this natural gas to ammonia is expected to contribute toward significant reductions in CO_2 emissions.

Ammonia exhibits combustion characteristics that are significantly different from those of natural gas. The laminar flame velocity of ammonia is lower than that of natural gas, which may reduce flame stability and combustion efficiency. The N atoms contained in ammonia molecules may also generate a large amount of fuel NO_x during combustion. Therefore, it is difficult to directly apply low-NO_x combustion technologies that have been developed for natural gas-fired gas turbines to ammonia combustion. A low-NO_x combustion technology that is suitable for ammonia combustion is needed.

At IHI, a low-NO_x combustor has been developed for ammonia/natural gas co-firing with the objective of establishing combustion technology that conforms with environmental regulations by co-firing ammonia at a heat quantity ratio of 20% in a 2-MW-class natural gas-fired gas turbine [1–3]. This combustor was applied to a 2-MW-class gas turbine, and power generation demonstration tests were conducted. This paper introduces an outline of the ammonia/natural gas co-firing gas turbine equipment and the power generation demonstration test results.

35.2 Test Equipment and Method

35.2.1 System Flow

Figure 35.1 shows the system flow diagram of the power generation test facility. The equipment mainly consists of four parts: the natural gas compressor, ammonia supply unit, gas turbine equipment, and de-nitration equipment. Each facility is described in the subsequent sections.

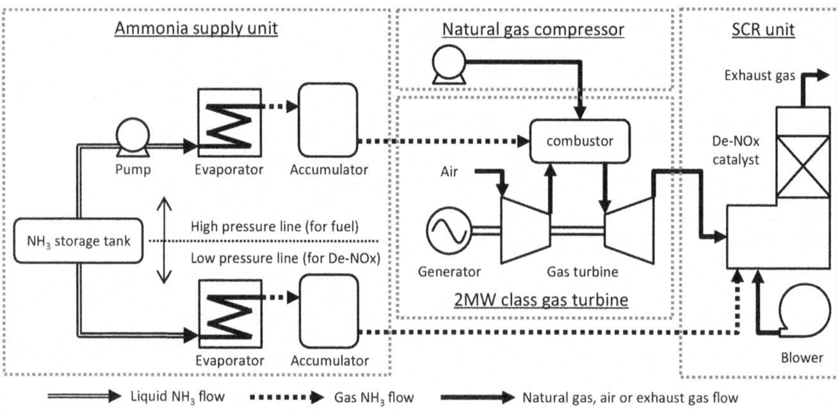

Fig. 35.1 System flow diagram. © Combustion Society of Japan 2019

Table 35.1 Capacity of ammonia supply unit.

	For fuel	For De-NO_x
Liquid NH_3 tank (ton)	4.5	
Maximum supply pressure (MPaG)	2.0	0.3
Temperature of gasified NH_3 (°C)	70	Ambient temperature
Maximum supply rate (kg/h)	400	60

35.2.2 Ammonia Supply Unit

Table 35.1 and Fig. 35.2 present the capacity and photographs of the ammonia supply unit. The unit has two ammonia supply lines that supply high-pressure ammonia gas as the fuel (supply pressure of 2 MPaG) and low-pressure ammonia gas for the de-nitration unit. Ammonia is stored in a liquid ammonia tank with a storage capacity of 4.5 t and supplied to each line as a liquid. Ammonia used as the fuel is pressurized in the liquid form by using a pump; this approach requires lower compression power than that required for pressurizing ammonia after vaporization. This pressurized liquid ammonia is then vaporized using the hot water evaporator, and subsequently, its pressure fluctuations are suppressed using the accumulator. Thereafter, it is supplied to the combustor of the gas turbine. Meanwhile, as the ammonia for the reducing agent can be supplied at the storage pressure of the liquid ammonia tank, it was supplied to the de-nitration unit after decompressing it to 0.3 MPaG with a pressure-regulating valve installed downstream of the hot water evaporator.

This test facility implements safety measures based on the Electricity Business Act in order to handle ammonia, which is a deleterious substance. The main safety measures are as follows:

(1) A gas leak detection alarm and sprinkler are installed in the ammonia supply unit. These are linked to adsorb the ammonia to water when the ammonia leaks.
(2) The secondary piping and vent piping of the safety valve are connected to a water-sealed dilution tank in order to eliminate the ammonia discharged from these piping systems.
(3) Ammonia water generated from (1) and (2) is then stored in a drainage tank and treated as industrial waste.

35.2.3 Gas Turbine Equipment

Figure 35.3 shows a photograph of the gas turbine equipment used in the test. The gas turbine used in the test was IM270 [4], which is a 2-MW-class simple-cycle

Fig. 35.2 Ammonia supply unit: **a** tank side and **b** evaporator side © Combustion Society of Japan 2019

gas turbine manufactured by IHI Corp. The heat recovery boiler, which is typically installed at the turbine outlet, was omitted in this test facility; the turbine outlet was directly connected to the de-nitration device. The standard products for natural gas combustion were directly used as the components, except for the combustor of the gas turbine. The combustor was tested after replacing it with a low-NO_x combustor developed for ammonia/natural gas co-firing [1–3].

35.2.4 De-Nitration Unit

A de-nitration unit that uses selective catalytic reduction (SCR) was installed at the turbine outlet in order to reduce the concentration of NO_x in the exhaust gas of the gas turbine and ensure that it was within the limits prescribed by environmental regulations. The catalysts used were standard materials for natural gas-fired gas turbines, and the reducing agent was ammonia gas from the ammonia supply facility. The temperature was adjusted by diluting the exhaust gas with air from a blower in

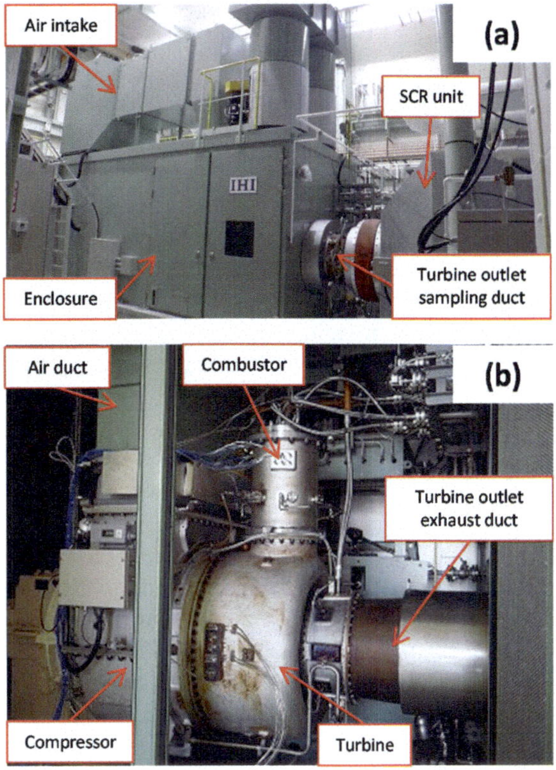

Fig. 35.3 Gas turbine: **a** appearance; **b** engine body. © Combustion Society of Japan 2019

order to ensure that the catalyst temperature remained below the active temperature. The outlet of the de-nitration device was connected to the exhaust tower, and the exhaust gas after de-nitration was released to the atmosphere through the exhaust tower.

35.2.5 Test Method

First, the entire operation, i.e., from starting the gas turbine to obtaining the rated load, was performed using natural gas alone. The power generation output was confirmed to be stable at 2 MW. Thereafter, the fuel gas ammonia was supplied while the power generation output was kept constant. The system was configured such that the flow rate of natural gas was automatically controlled according to the flow rate of the fuel gas ammonia. In this test, the ratio of the heat quantity of the supplied fuel ammonia to the total heat quantity of the supplied fuel was defined as the ammonia mixing ratio (shown below), and this varied between 0% LHV and 20% LHV. Here, the heat quantity was based on the lower heating value (LHV).

$$r_{NH_3}\,[\%\,LHV] = 100 \times \frac{\text{Heat quantity of supplied ammonia}}{\text{Total heat quantity of supplied fuel}} \tag{35.1}$$

Gas sampling probes were installed at the turbine outlet, and connected to Gas analyzers MEXA-ONE-D1 and VIA-3001, manufactured by Horiba, Ltd., where the concentrations of six components—CO, CO_2, O_2, total hydrocarbon (THC; measured value converted to CH_4), NO_x, and NH_3—were measured. Furthermore, gas was sampled at the de-nitration unit outlet with the ENDA-C9430A, manufactured by Horiba, Ltd., which analyzed the concentrations of three components—NO_x, O_2, and NH_3. Concentrations in the exhaust gas were evaluated by converting them into a dry concentration excluding water vapor and into an oxygen concentration of 16%, according to the evaluation method prescribed in the Air Pollution Control Act.

35.3 Test Results

Figure 35.4 shows the relationship between the ammonia mixing ratio and the generator output and fuel flow rate. The ammonia mixing ratio could gradually be increased from natural gas-exclusive combustion to power generation under conditions of an ammonia mixing ratio exceeding 20% LHV. Increases in the ammonia flow ratio resulted in a monotonic decrease of the natural gas flow rate, while maintaining a constant generator output. Abnormal noise, vibrations, or temperatures were not observed in the combustor or the gas turbine components in response to the change in the ammonia mixing ratio. It was confirmed that there were no issues pertaining to the power generation operations caused by ammonia co-firing.

Figure 35.5 shows the relationship between the ammonia mixing ratio and the CO_2 emissions per 1 kWh of power generation output at the turbine outlet. CO_2 emissions decreased monotonically as the ammonia mixing ratio increased. It was

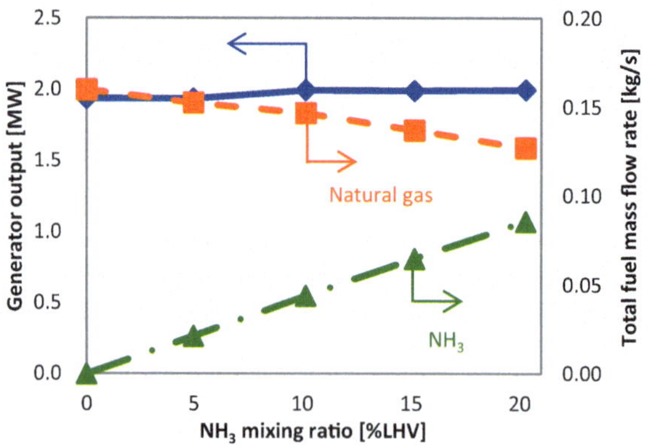

Fig. 35.4 Relationship between ammonia mixing ratio and generator output and fuel flow rate. © Combustion Society of Japan 2019

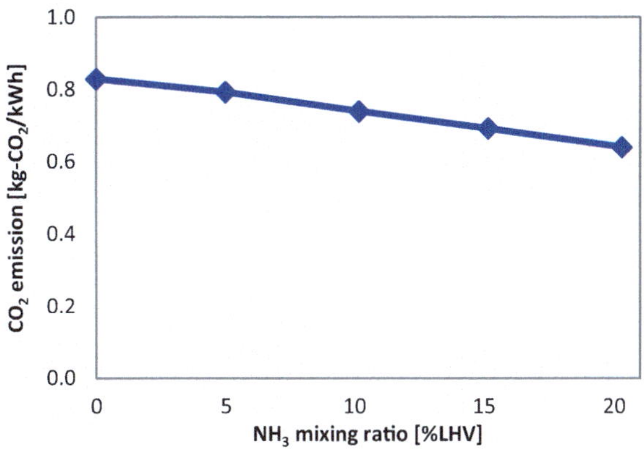

Fig. 35.5 Relationship between ammonia mixing ratio and CO_2 emissions per unit power generation output. © Combustion Society of Japan 2019

confirmed that the CO_2 emissions decreased by approximately 23% at an ammonia mixing ratio of 20% LHV. The CO_2 reduction rate was higher than the ammonia mixing ratio because the power generation efficiency of the gas turbine increases due to the ammonia co-firing.

Figure 35.6 shows the relationship between the ammonia mixing ratio and the NO_x concentrations at the turbine outlet and the de-nitration unit outlet. The NO_x concentration at the turbine outlet increased sharply after commencing ammonia co-firing, but the change was small when the ammonia mixing ratio was 5% LHV or higher. The NO_x concentration at the turbine outlet was maximum at an ammonia mixing ratio of 20% LHV, at approximately 290 ppm (converted with oxygen concentration of 16%). Meanwhile, the NO_x concentration at the outlet of the de-nitration unit remained below 7 ppm (converted with oxygen concentration of 16%), which is the environmental regulation limit for Yokohama city, despite changes in the ammonia mixing ratio. The NO_x concentration at the turbine outlet under ammonia co firing was higher than that of a general natural gas-fired gas turbine. However, it was confirmed that this could be further reduced to below the environmental regulation limits using a de-nitration device. Additionally, the unburned NH_3, CO, and THC concentrations at the turbine outlet were all 3 ppm or less; the introduced ammonia and natural gas were almost completely burned in the combustor, and it was confirmed that there were no other issues pertaining to the exhaust gas.

35.4 Conclusions

A low-NO_x combustor for ammonia/natural gas co-firing was installed in a 2-MW-class gas turbine, and power generation demonstration tests were conducted under the conditions of a power output of 2 MW and an ammonia mixing ratio of up

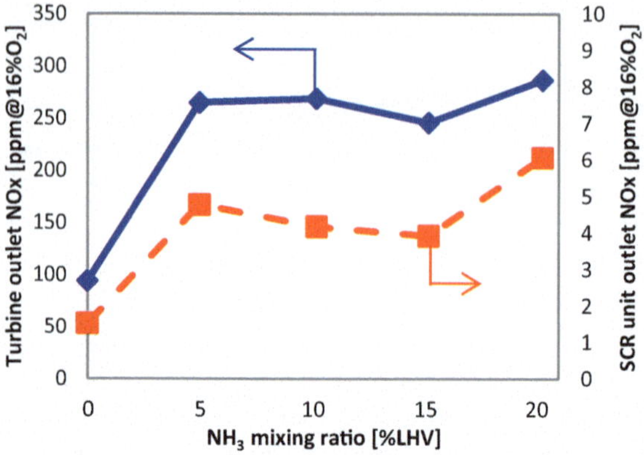

Fig. 35.6 Relationship between ammonia mixing ratio and NO_x concentrations at the turbine and de-nitration unit outlets. © Combustion Society of Japan 2019

to 20% LHV. The test results confirmed that CO_2 emissions could be reduced by ammonia co-firing and NO_x concentrations could be reduced below the limits stated in environmental regulations by using a general de-nitration device; furthermore, the injected ammonia was almost completely combusted. Meanwhile, it was clarified that the NO_x concentrations at the turbine outlet were significantly higher than those of a general natural gas-fired gas turbine. Reducing the NO_x concentrations enabled a reduction in the operation and introduction costs of the gas turbine, such as by reducing the flow rate of ammonia for de-nitration and the size of the de-nitration unit. Therefore, we intend to further develop a combustor that can reduce NO_x and contribute toward the realization of a low-carbon society through the use of ammonia.

Acknowledgements This work was supported by the Council for Science, Technology and Innovation (CSTI), Cross-Ministerial Strategic Innovation Promotion Program (SIP), "Energy Carriers" (Funding agency: JST).

References

1. Ito S, Uchida M, Kato S, Fujimori T, Kobayashi H (2019) Emission characteristics of a lean-premixed ammonia/natural-gas gas-turbine combustor and effect of secondary ammonia injection. Mech Eng J 6:19–00266
2. Ito S (2019) Japan Society of mechanical engineers. TED Newsletter 87 (in Japanese). https://www.jsme.or.jp/ted/NL87/TED-Plaza_NL87_01_itou.pdf. Accessed 11 Apr 2019
3. Ito S, Uchida M, Kato S, Fujimori T, Kobayashi H (2019) A study of low NO_x combustion method for ammonia/natural gas co-fired gas turbine using detailed chemistry. J. Combust. Soc. Jpn. 61:368–375 (in Japanese)
4. IHI Power Systems Co., Ltd (2019) Homepage. https://www.ihi.co.jp/powersystems/en/lineup/IM270/index.html. Accessed 22 July 2019

Chapter 36
Development of Ammonia Utilization Technology for Large Gas Turbines for Power Generation

Masakazu Nose, Hidefumi Araki, Noriaki Senba, Hiroyuki Furuichi, and Satoshi Tanimura

Abstract This paper describes research and future plan of ammonia cracking gas turbine system, which burns hydrogen decomposed from ammonia using exhaust heat of gas turbine. The system has two alternative configurations: a co-firing system which burns mixtures of natural gas and ammonia cracked gas, and a mono-firing system which burns ammonia cracked gas. The co-firing system has fewer development subjects, because heat necessary to crack ammonia is fed by steam branched from HRSG (Heat Recovery Steam Generator). The gross/net thermal efficiency of the mono-firing system was 102.4 and 101.7% of that of the natural gas fired GTCC, respectively, owing to the increase in the calorific value of the fuel during ammonia cracking and to the increase in the flow rate of the turbine working fluid. Among several known catalysts for cracking ammonia, a base metal catalyst and a noble metal catalyst which have high catalytic activity even under relatively low temperature conditions were selected and tested under at 5.2 MPa, which is the assumed operating condition of the system. Then, the rate equation as a function of temperature and pressure was derived from the measurement of ammonia cracking ratio. A laminar combustion burner which forms a flat plate flame was prepared and NO_x concentration was measured. Also, NO_x concentration was calculated by a one-dimensional laminar premixed flame calculation model combining the PREMIX code of CHEMKIN and the GRI 3.0 mechanism. These results indicate that the NO_x

Originally published in Journal of the Combustion Society of Japan, Vol. 61 (2019), pp. 293–298 (in Japanese);

M. Nose (✉) · S. Tanimura
Mitsubishi Power, Ltd., 2-1-1 Shinhama, Arai-cho, Takasago 676-8686, Hyogo, Japan
e-mail: masakazu.nose.dt@mhi.com

H. Araki
Mitsubishi Power, Ltd., 3-1-1 Minatomirai, Nishi-ku, Yokohama 220-8401, Japan

N. Senba
Mitsubishi Heavy Industries, Ltd., 5-717-1 Fukahori-machi, Nagasaki 851-0392, Japan

H. Furuichi
Mitsubishi Heavy Industries Engineering, Ltd., 3-1-1 Minatomirai, Nishi-ku, Yokohama 220-8401, Japan

K. Aika and H. Kobayashi (eds.), *CO_2 Free Ammonia as an Energy Carrier*,
https://doi.org/10.1007/978-981-19-4767-4_36

concentration increases in proportion to the increase in the residual ammonia concentration. They also show that most of the residual ammonia in the fuel is converted into NO_x.

Keywords Gas turbine · Ammonia · Hydrogen · Cracking · Combustion · Catalyst

36.1 Introduction

Efforts to reduce the amount of CO_2 emissions have been made around the world for global environmental conservation. For large gas turbines, we have continuously worked to reduce CO_2 emissions by improving the efficiency by increasing combustion gas temperature [1]. In addition, developments have also been made to improve the operability of gas turbines in response to increasing renewable energy deployment [2]. Furthermore, the Basic Strategy for Hydrogen [3] specifies the use of 300,000 tons of hydrogen per year by 2030, and research and development of hydrogen-fired gas turbines, which do not emit CO_2 during combustion, is underway.

In this strategy, not only the method for transporting and storing hydrogen by liquefying it, but also the utilization of energy carriers such as ammonia and organic hydride are presented. For small gas turbines, in SIP (Strategic Innovation Creation Program) of the Cabinet Office, research and development of direct ammonia combustion [4] was carried out. On the other hand, for large gas turbines, due to the restrictions shown in Table 36.1, study of a system that thermally cracks ammonia to hydrogen and burns it in a hydrogen-fired gas turbine [5] was started. This paper describes the current research status of the ammonia cracking gas turbine system and our efforts toward future social implementation of this technology.

Table 36.1 Characteristics of Ammonia combustion and consideration for large gas turbines

Characteristics of ammonia combustion	Considerations for large gas turbines
Slow combustion speed (about 1/5 of that of methane)	• The size of the combustor increases to secure the time necessary for completing the combustion • Since the large gas turbine is composed of a multi-can combustor, there is a restriction on the size expansion of the combustor
Nitrogen contained in fuel	• There is little room to allow Fuel NO_x because the combustion gas temperature of a large gas turbine has been increased to the extent permitted by Thermal NO_x • Lowering of NO_x by two-stage combustion is considered, but in the case of a large gas turbine, there are many technical problems such as upsizing and complication of the combustor

36.2 Ammonia Cracking GTCC System

The concept of the ammonia cracking GTCC (Gas Turbine Combined Cycle) system is shown in Fig. 36.1. The system uses exhaust heat from gas turbine to crack ammonia into hydrogen and nitrogen. Then cracked gas is burned in a hydrogen co-firing gas turbine combustor in practical use or in a hydrogen mono-firing combustor under development. Ammonia cracking requires about 46 kJ heat per 1 mol of raw ammonia, but the reaction results in an increase of calorific value of the product fuel by 1.14 times that of ammonia due to chemical recuperation. Therefore, there is no reduction in thermal efficiency in principle if the exhaust heat of the gas turbine is used until the same temperature as that of a conventional GTCC.

Table 36.2 shows a comparison of two systems: (a) a co-firing system which buns mixtures of natural gas and ammonia cracked gas, and (b) a mono-firing system which burns ammonia cracked gas. In the case of co-firing, the hydrogen concentration is 20 vol% at the combustor inlet, which is within the range of hydrogen concentrations that ensures stable combustion in a state of the art, premixed DLN (Dry Low NO_x) combustor. As the calorific value per volume of hydrogen is lower than that of natural gas, the calorific value of the fuel derived from ammonia is approx. 6.5%. The co-firing system has fewer development subjects because heat necessary to crack ammonia is relatively small and can be fed by steam branched from ordinary HRSG (Heat Recovery Steam Generator).

On the other hand, in the case of a mono-firing system of the cracked gas, which burns 75% hydrogen and 25% nitrogen (both by volume) using a multi-cluster burner under development, reduction of about 1.1 million tons of CO_2 emissions per year per

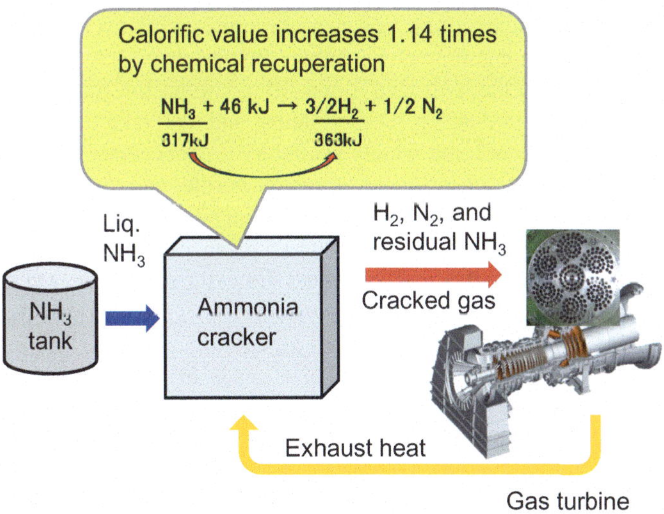

Fig. 36.1 Concept of ammonia cracking GTCC (Gas Turbine Combined cycle) © Combustion Society of Japan 2019

Table 36.2 Comparison of co-firing system and mono-firing system of the ammonia cracking gas turbine

Type of the system	(a) Co-firing system (power output: 430 MW)	(b) Mono-firing system (power output: 530 MW)
Composition of fuel at the combustor	Mixture of natural gas (almost methane) and ammonia cracked gas (H_2 = 75 vol%, N_2 = 25 vol%)	Ammonia cracked gas (H_2 = 75 vol%, N_2 = 25 vol%)
Ratio of calorific value of initial NH_3	6.5% (when H_2 = 20 vol%)	100%
Required equipment	• Small NH_3 cracker (10 t/h) • Pre-mix combustor (state of the art) 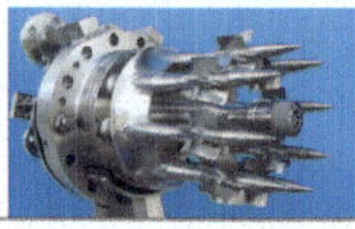	• Large NH_3 cracker (170 t/h) • Multi cluster burner (under development) 
Reduction of CO_2 emission	30,000 t/year/unit	1,100,000 t/year/unit
Utilization of H_2	10,000 t/year/unit	180,000 t/year/unit
Remarks	• Modification of existing systems (retrofit) is possible • Fewer R&D is needed compared with mono-firing system	Larger contribution for CO_2 reduction

gas turbine is expected when capacity factor is 70% [6]. The mono-firing system can provide significant CO_2 reductions; however, it requires a relatively large ammonia cracker (where the ammonia flow rate is about 170 t/h when GTCC output is about 530 MW) in addition to the hydrogen mono-firing combustor. Since it will take a certain amount of time to develop these systems, and also take time for ammonia supply infrastructures to become widely available, it is necessary to develop a co-firing system first because of lower development hurdle, and then accumulate the technology and experience necessary to realize more advanced systems.

Figure 36.2 shows schematic heat and material balances of the ammonia cracking gas GTCC (mono-firing system), and Table 36.3 shows the performance of the system compared to a natural gas-fired GTCC. The system is designed to supply superheated steam of 620 °C generated by the HRSG to an ammonia cracker. The pressure of liquid ammonia fed to the ammonia cracker is made higher than the pressure of the gas turbine combustor, and therefore no fuel gas compressor is required. The fuel supplied to the gas turbine increases in its calorific value by approx. 1.14 times during ammonia cracking, so that the calorific value of ammonia at the system inlet is reduced by approx. 93.6% compared to a natural gas-fired gas turbine. The power output of the gas turbine is approx. 1.18 times higher compared to a natural gas-fired turbine, due to the increase in the flow rate of the turbine working fluid using

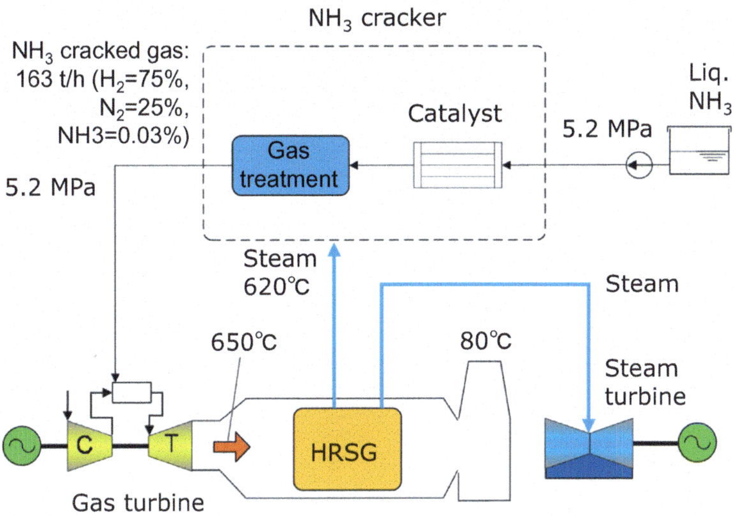

Fig. 36.2 Heat and material balances schematic of the ammonia cracking gas turbine (mono-firing system) © Combustion Society of Japan 2019

low-calorie fuel, ammonia cracked gas. On the other hand, the steam from HRSG is diverted to the ammonia cracker, reducing the output of the steam turbine by less than half. As a result, the power output of the mono-firing system as GTCC is about 96% of that of natural gas-fired GTCC.

In addition, a gas treatment system consisting of an ammonia scrubber and an ammonia stripper is to be installed downstream of an ammonia cracking reactor to remove some unreacted residual ammonia. When the heat balance of the gas treatment system was integrated, in this case, the gross/net thermal efficiency of the mono-firing system is rated at 102.4 and 101.7% of that of the natural gas fired GTCC, respectively, since the effect of the increase in the calorific value of the fuel and the increase in the flow rate of the turbine working fluid exceeded the effect of the decrease in the power output of the steam turbine and the increase in the auxiliary power of the gas treatment system.

36.3 Ammonia Cracker

The ammonia cracker, which thermally cracks ammonia, is an important equipment of this system. However, the type of ammonia cracking catalyst and operating conditions suitable for this system have not been studied, and therefore the test on some ammonia cracking catalysts was conducted to gain a better understanding of these conditions [7]. Figure 36.3 shows the results of the catalyst screening test under atmospheric conditions, indicating the volume ratio of ammonia cracked gas (N_2 +

Table 36.3 Performance prediction of the ammonia cracking GTCC (mono-firing system)

Items	Unit	NG-fired GTCC	Ammonia cracking GTCC	Remarks
Heat of fuel before NH_3 cracking (relative)	%	100	93.6	
Heat of fuel after NH_3 cracking (relative)	%	100	107.1	Heat increases 1.14 times by chemical recuperation
GT output	MW	381	448	Increased due to larger combustion gas flow
ST output	MW	176	86	Decreased due to extraction of HP steam flow
GTCC (GT + ST) output	MW	557	534	
Aux. power of NH_3 cracker	MW	–	3.4	Power of the gas treatment system (fuel-gas compressor is not required)
Gross efficiency (relative)	%	100	102.4	Higher electrical efficiency is possible owing to chemical recuperation of NH_3
Net efficiency (relative)	%	100	101.7	

H_2) at the exit of catalyst layer. Table 36.4 shows the summary of each catalyst. As a result of the screening test, the non-precious metal catalyst D and the precious metal catalyst E, which have high catalytic activity even under relatively low temperature conditions, were selected as candidates for the high-pressure ammonia cracking catalyst test at 5.2 MPa, which is the assumed operating condition of the system.

Based on the results of the high-pressure measurements, reaction rate equations as a function of temperature and pressure were established for each catalyst. Figure 36.4 shows a comparison of measured residual ammonia concentrations at the exit of ammonia cracker for the non-precious metal catalyst D and the predicted residual ammonia concentrations using the rate equation. The figure indicates that the rate equation accurately predicts the residual ammonia concentration at the exit of the ammonia cracker. This rate equation was used to determine the required specifications of the gas treatment system installed downstream of the ammonia cracker and to predict residual ammonia concentrations flowing into the gas turbine combustor.

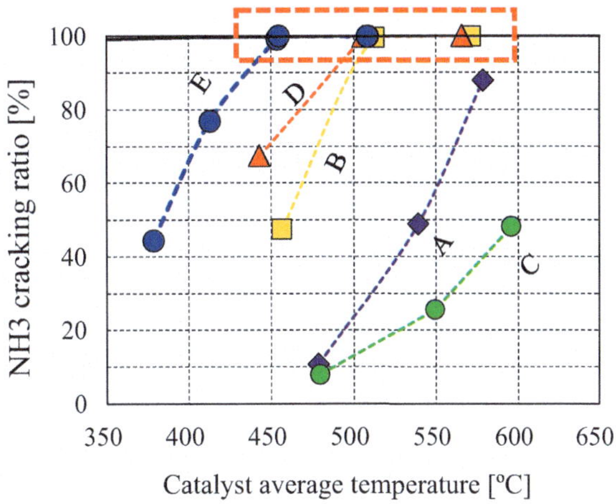

Fig. 36.3 Measurement of ammonia cracking ratio © Combustion Society of Japan 2019

Table 36.4 Comparison of some catalysts for ammonia cracking

Symbol	A	B	C	D	E
Type of catalyst	Fe-base	Ni-base	Ni-base	Non-precious metal	Precious metal
Price	Low	Medium	Medium	Medium	High
Cracking performance	Low	Medium	Low	High	High

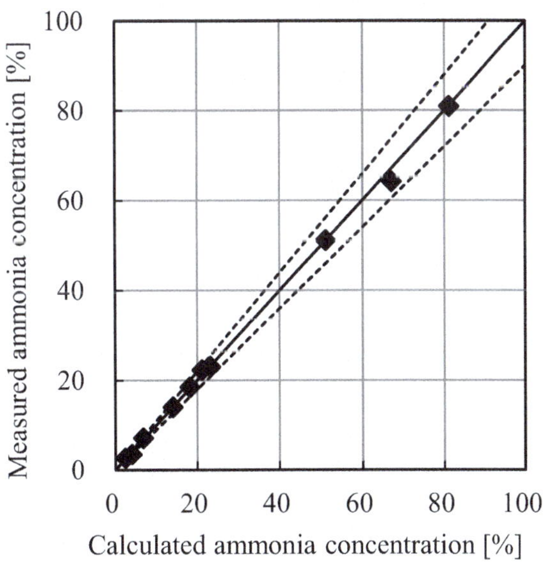

Fig. 36.4 Comparison of calculated and measured residual ammonia concentrations at the exit of ammonia cracker © Combustion Society of Japan 2019

36.4 Gas Turbine Combustors

In order to handle the increase in turbine inlet temperature (combustion temperature) and NO_x emissions resulting from higher efficiency and increased power output of gas turbines, a dry-type premixing combustor has been used in gas turbines [8]. Following the development of the 1600 °C class gas turbine, the next generation 1650 °C class gas turbine is under development. For combustors, further development and commercialization of technologies for low NO_x emissions and stable combustion, such as suppression of combustion vibrations and prevention of flashback, are underway.

Compared to the combustion of hydrocarbons such as methane, ammonia is less combustible and more prone to produce fuel-derived NO_x. Therefore, when burning ammonia in a premixed combustor of a large gas turbine using natural gas as the main fuel, the issues listed in Table 36.1 must be considered, which faces higher hurdles in addition to the existing technical issues. For this reason, in the ammonia cracking GTCC systems, hydrogen and nitrogen decomposed from ammonia will be used instead of directly burning ammonia, and both the newly developed hydrogen co-firing combustor and the hydrogen mono-firing combustor, which is under development, will be applied. Hydrogen, which accounts for 75 vol% of the ammonia cracked gas, burns faster and has lower minimum ignition energy than natural gas, which increases the risk of flashback during pre-mixing combustions. For this reason, the following measures have been implemented [9]:

Ammonia Cracked Gas Co-firing Combustor (DLN (Dry Low NO_x) Multi-nozzle Combustor for Hydrogen Co-firing)
An overview of the newly developed hydrogen co-firing combustor based on the conventional DLN combustor is shown in Fig. 36.5. The air supplied from the compressor to the combustor passes through the turning vane (swirler) to generate a swirling flow. The fuel is supplied through a small hole on the surface of the swirler and is rapidly mixed with the surrounding air by the swirling flow. Meanwhile, a flashback possibly occurs due to the flame backing up in the low velocity area at the swirling flow center (vortex core). In the new combustor, air is injected from the tip of the nozzle to increase the flow velocity in the low-flow area at the vortex core, thereby preventing the flashback. Combustion tests using one can of the new combustor under the operating pressure of the actual system shows that NO_x is within the operational range even under conditions where 30 vol% hydrogen is mixed, and the combustor can be operated without flashback or a significant increase in combustion vibration.

Ammonia Cracked Gas Mono-firing Combustor (Multi-cluster Combustor for Hydrogen Mono-firing)
As hydrogen concentrations increase, the risk of flashback in the low-velocity area at the vortex core becomes even higher in the fuel–air mixing method using a swirling flow, which is applied to the hydrogen co-firing combustor described in this paper. Therefore, we are currently developing a new mixing method capable of mixing air and hydrogen on a smaller scale without using a swirling flow. Based on the multi-cluster combustor shown in Fig. 36.6, which has a larger number of nozzles

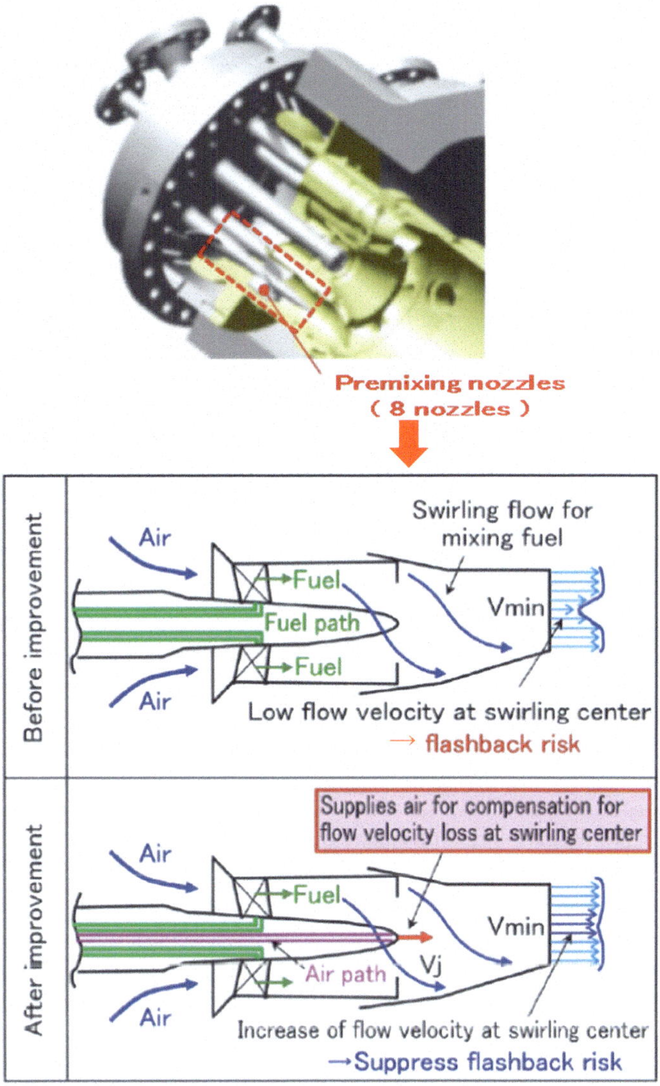

Fig. 36.5 Outline of new combustor for hydrogen co-firing © Combustion Society of Japan 2019

compared to 8 fuel supply nozzles of the hydrogen co-firing combustor, it is possible to achieve both high flashback resistance and low NO_x combustion by reducing the size of one nozzle hole, sending air and blowing hydrogen into it to mix on a small scale, and dispersing the flame.

In the combustion of ammonia cracked gas, it must be noted that a small amount of residual ammonia during ammonia cracking is converted to NO_x in the combustor. This emission is considered to be added to the NO_x originally generated in the

Fig. 36.6 Multi-cluster combustor (under development) © Combustion Society of Japan 2019

combustor, and therefore the increase in NO_x due to the amount of residual ammonia needs to be determined in order to meet the NO_x emission limits.

In this study, NO_x concentration in combustion exhaust gas was determined by a simple chemical reaction calculation (a one-dimensional laminar premixed flame calculation model combining the PREMIX code of CHEMKIN and the GRI 3.0 mechanism) and an elemental test of a plane flame on a laminar combustion burner under atmospheric pressure and temperature conditions as shown in Fig. 36.7. Figure 36.8 shows calculation results and measurement results of NO_x concentration when ammonia concentration in the fuel is changed at the equivalence ratio corresponding to the turbine inlet temperature of 1650° C with three different percentages of cracked gas in the fuel (0, 20, and 50% of hydrogen). Both chemical reaction calculations and exhaust gas measurements indicate that the NO_x concentration increases in proportion to the increase in the residual ammonia concentration.

With increasing the cracked gas ratio in the fuel, the calorific value per unit volume of the fuel decreases, and more fuel is required to achieve the same test conditions (flame temperature). The figure shows that the NO_x concentration increases in proportion to the percentage of cracked gas in the fuel, due to the increase in the ammonia concentration in the premixed gas.

In addition, Fig. 36.9 shows the conversion ratio of residual ammonia to NO_x (the percentage of residual ammonia converted to NO_x by combustion). The parameters are the combustion gas temperature (equivalent to 1400, 1500, 1600, and 1650 °C at the turbine inlet) and the ammonia concentration (0.03, 0.1, and 0.3%) under co-firing conditions with the hydrogen concentration of 50% in the fuel. In all cases, a high percentage of residual ammonia was converted into NO_x. Based on these results, combustion tests on a real combustor scale will be carried out under real gas turbine

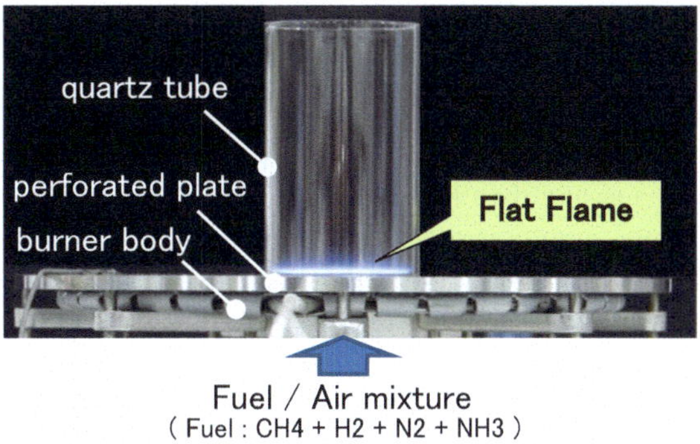

Fig. 36.7 Photo of the flat flame on the laminar flame burner © Combustion Society of Japan 2019

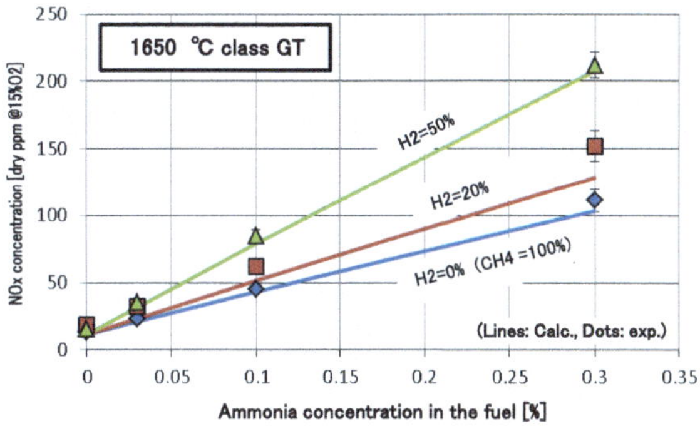

Fig. 36.8 NO_x concentrations in the flue gas when ammonia content in the fuel was varied. (Dots: exp., Lines: calc.) © Combustion Society of Japan 2019

pressure and temperature conditions to more accurately determine the relationship between residual ammonia and NO_x emissions.

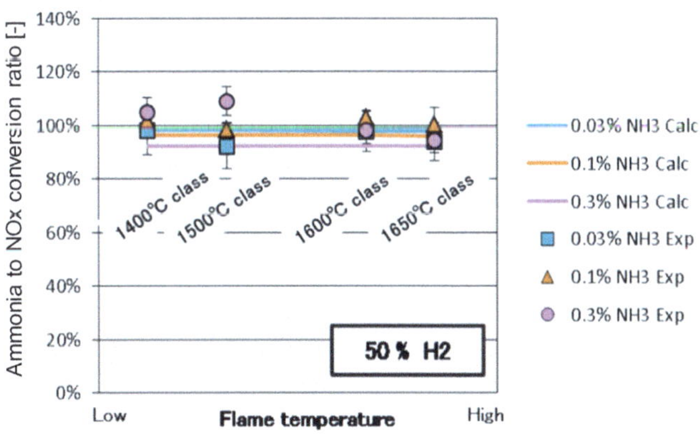

Fig. 36.9 Ammonia to NO_x conversion ratios when ammonia content in the fuel was varied. (Dots: exp., Lines: calc.) © Combustion Society of Japan 2019

36.5 Conclusion

This paper describes the current status of a study on a large gas turbine for power generation in which ammonia, an energy carrier, is cracked into hydrogen and nitrogen and then burned in a hydrogen-fired gas turbine, focusing on the GTCC system, ammonia cracker, and gas turbine combustors. As for the whole system, the potential of the power generation efficiency and the minimum required system configuration were clarified by the SIP research until 2018. For the future social implementation, we are planning to study the improved system configuration and operating conditions in consideration of its operation and economic efficiency.

As for the ammonia cracker, the basic system configuration and the catalysts to be used have been determined by the SIP research, and further studies will be carried out on the specific structure of the system, catalyst life, and toxic substances.

As for the gas turbine combustor, the effect of residual ammonia on NO_x emissions was evaluated based on chemical reaction calculations and NO_x measurements using a simple flame burner in the SIP research. For future studies, a co-firing test of ammonia cracked gas and natural gas will be conducted to confirm the combustibility of a hydrogen co-fired combustor, simulating actual operating conditions. In preparation for the coming hydrogen society, our development efforts are directed toward the practical application of hydrogen-fired gas turbines and gas turbine systems utilizing ammonia as an energy carrier for hydrogen.

Acknowledgements The work described in this paper is an outcome of the Council for Science, Technology and Innovation (CTSI), Cross-ministerial Strategic Innovation Promotion Program (SIP), "Energy Carriers" (Funding agency: JST). We would like to express our gratitude for their support.

References

1. Ishizaka K, Saitoh K, Ito E, Yuri M, Masada J (2017) Key technologies for 1700 °C class ultra high temperature gas turbine. Mitsubishi Heavy Ind Tech Rev 54(3)
2. Matsumi I, Kawamura K, Komori T (2019) MHPS's gas turbine technologies for rapid load change. J Gas Turbine Soc Jpn 47:14–20 (in Japanese)
3. The Ministerial Council on Renewable Energy (2017) The basic hydrogen strategy. http://www.meti.go.jp/press/2017/12/20171226002/20171226002-1.pdf (in Japanese)
4. Iki N et al (2016) Ammonia utilization for gas turbine. In: Proceedings of the 44th GTSJ annual meeting, vol C-12, Sakata, Japan, pp 259–264 (in Japanese)
5. SIP Energy Carriers (2019) Final report. Mitsubishi Hitachi Power Systems, Ltd., Technology development of gas turbine using ammonia (system and combustor). https://www.jst.go.jp/sip/dl/k04/end/team6-15.pdf (in Japanese)
6. Ichikawa Y, Yuasa A, Uechi H, Inoue K, Miura K, Tada K, Tanimura S (2019) Development of hydrogen-fired gas turbine combustor. J Combust Soc Jpn 61:15–23 (in Japanese)
7. SIP Energy Carriers (2019) Final report, Mitsubishi Heavy Industries Engineering, Ltd., Technology development of gas turbine using ammonia (study of ammonia cracking equipment). https://www.jst.go.jp/sip/dl/k04/end/team6-16.pdf (in Japanese)

8. Tanimura S, Saitoh K (2016) Development and current status of combustors for industrial gas turbines. J Soc Mech Eng 119:442–445 (in Japanese)
9. Nose M, Kawakami T, Araki H, Senba N, Tanimura S (2018) Hydrogen-fired gas turbine targeting realization of CO_2-free society. Mitsubishi Heavy Ind Tech Rev 55(4)

Chapter 37
Combustion in Ammonia-Hydrogen-Fueled Spark-Ignition Reciprocating Engines

Makoto Koike, Hiroshi Miyagawa, Tetsunori Suzuoki, Seiji Yamamoto, and Kazuto Ogasawara

Abstract The use of NH_3 as a fuel in reciprocating engines differs from that of other alternative fuels which have been studied so far. Firstly, reciprocating engines require an auxiliary fuel, as few operating conditions allow NH_3 to be used as a single fuel. H_2 can be an ideal combustion promotor for S.I. engines. The study using a constant volume vessel showed that the laminar flame speed of an NH_3-H_2 mixture containing 40% H_2 was comparable to that of CH_4, for a stoichiometric air-fuel mixture. Under the mixture conditions with the same laminar flame speed, the difference between the NH_3-H_2 mixture and CH_4 was found to be small in terms of MBT ignition timing as well as combustion phasing, which is important for engine performance. On the other hand, combustion efficiency and stability were better for higher H_2 ratios, and thermal efficiency was better for lower H_2 ratios. Secondly, some NO_x emission characteristics were different from conventional S.I. engines, such as an increase in NO_x when the ignition timing is delayed, due to the large contribution of fuel NO_x. The N_2O concentration in the exhaust gas of concern was about 100 ppm. For exhaust purification, the aftertreatment system combining a three-way catalyst and a theoretical air-fuel ratio mixture worked effectively.

Keywords Internal combustion engine · Premixed ammonia-hydrogen combustion · Thermal efficiency · Nitrogen oxides emissions

37.1 Introduction

An early use of ammonia (NH_3) as fuel for a spark-ignition (S.I.) engine was reported in the 1930s [1]. In 1943, an NH_3-coal-gas-fueled engine was used for route buses in Belgium because of an extreme shortage of diesel fuel [2]. Other reports on these engines using NH_3 have been released since then, indicating that NH_3 is feasible as an alternative fuel for reciprocating engines [3–6].

M. Koike · H. Miyagawa (✉) · T. Suzuoki · S. Yamamoto · K. Ogasawara
Toyota Central R&D Labs., Inc., 41-1 Yokomichi, Nagakute, Aichi 480-1192, Japan
e-mail: e1041@mosk.tytlabs.co.jp

K. Aika and H. Kobayashi (eds.), *CO_2 Free Ammonia as an Energy Carrier*,
https://doi.org/10.1007/978-981-19-4767-4_37

However, every report has indicated the difficulty of solo NH_3 combustion under various load and speed conditions because of the low flame speed of an air-NH_3 mixture. Currently, mixed combustion with hydrocarbon fuels or hydrogen (H_2) are considered a necessary condition for engine operation. Although a simple solution is probably a dual-tank fuel system with conventional hydrocarbon fuels, H_2 is more attractive from a long-term perspective because it does not produce carbon dioxide (CO_2) from combustion. Furthermore, a single-tank (mono-fuel) system is possible if H_2 is produced from NH_3 dissociation using an on-board reformer.

The present report examines the combustion characteristics of an NH_3-H_2-fueled S.I. engine, including a comparison of the differences and the similarities with hydrocarbon fuels. The most reliable S.I. engine system combines stoichiometric air-fuel mixture control and a three-way catalyst (TWC). Good robustness and high ability for reduction of exhaust gas emissions are expected for NH_3-H_2-fueled engines in a manner similar to those of current gasoline engines. Therefore, the air-fuel mixture of all the tests in this report was stoichiometric.

37.2 Laminar Burning Velocity of NH_3-H_2 Mixtures

The effect of H_2 on laminar burning velocity S_{L0} was investigated before testing engine combustion. Figure 37.1 shows a schematic view of the experimental system [7]. A cylindrical combustion vessel with a diameter of 80 mm and a height of 30 mm was used in this test. Optical quartz windows were mounted on the upper and lower ends. Electrodes entering through the cylinder wall were used to enable spark-ignition with 60 mJ within a gap of 2.2 mm at the center of the vessel. A Schlieren measurement technique combined with a high-speed camera (Vision Research, 4800 fps, 800 × 600 pixels) was applied to visualize the flame. A 500 W Xenon lamp (Ushio) was used as a light source. The pressure in the chamber was measured with a pressure transducer (AVL, GU12P).

Mixtures of NH_3-H_2-nitrogen (N_2)-air and methane (CH_4)-CO_2-N_2-air were investigated. Here, N_2 was an additional component that was distinguished from

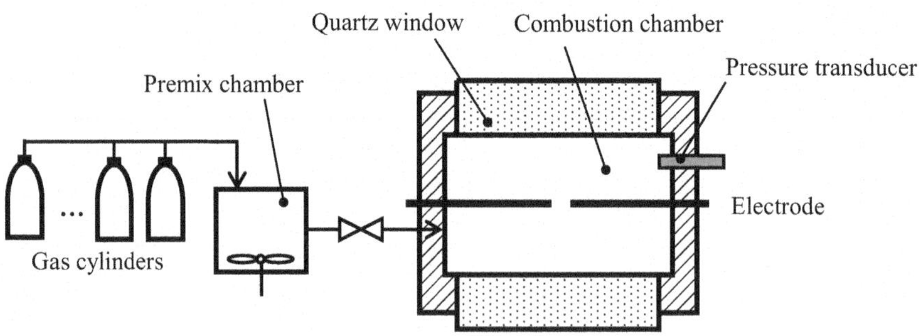

Fig. 37.1 Experimental setup of the closed combustion vessel [7]

the N_2 in the air. The ratio of N_2 to H_2 was fixed at 1:3 assuming NH_3 dissociation. In contrast, the ratio of N_2 to CO_2 was fixed at 7.55:1 in the latter mixture, assuming dilution by the CH_4-air mixture combustion gas. The dilution ratio was defined by the following equation:

$$[\text{Dilution ratio}] = [\text{Dilution gas}]/[\text{Air} + \text{Dilution gas}] \tag{37.1}$$

where the dilution gas was the total amount of CO_2 and additional N_2 and initial temperature and pressure were 298 K and 0.1, 0.4 MPa, respectively.

Figure 37.2 shows flame images of NH_3-H_2-N_2-air combustion. Flame image processing was done at less than 15 mm in radius, to avoid interactions between the flame and the wall. As the pressure was almost constant in this range, the test was considered to be conducted under constant pressure. The flame images were not perfect circles, so the equivalent spherical radius was calculated assuming it was equivalent to the radius of a circle with an area identical to that of the measured flame area. The S_{L0} was calculated with the following equations [8]:

$$\kappa = \frac{2}{r} \cdot \frac{dr}{dt} \tag{37.2}$$

$$S_L = \frac{\rho_b}{\rho_{ub}} \cdot \frac{dr}{dt} \tag{37.3}$$

$$S_L = S_{L0} - L_b \cdot \kappa \tag{37.4}$$

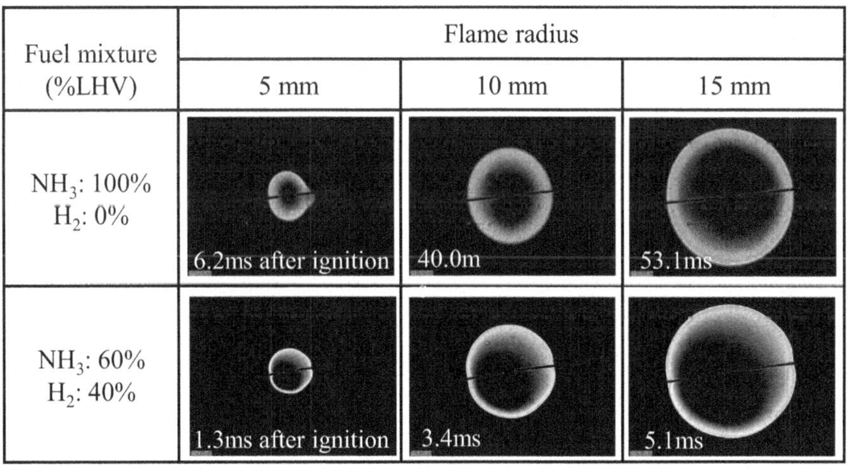

Fig. 37.2 Flame images of NH_3 and NH_3-H_2 at pressure of 0.1 MPa_abs

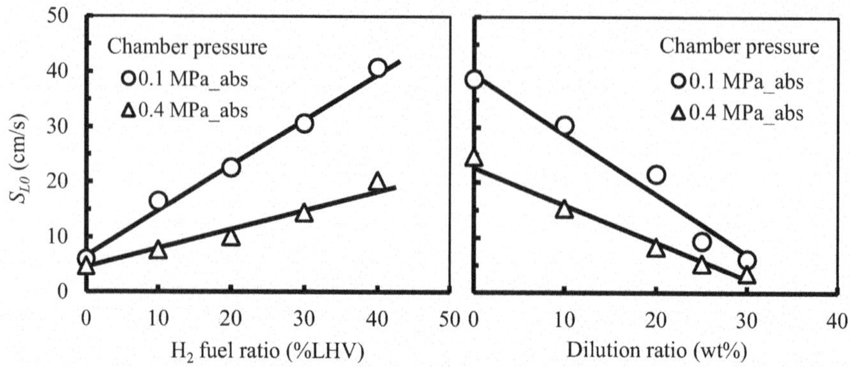

Fig. 37.3 Laminar burning velocities of the NH_3-H_2 and CH_4 flame

where κ is flame stretch ratio, r is equivalent spherical radius, t is time, S_L is flame speed, ρ_b is burned gas density, ρ_{ub} is unburned gas density, and L_b is Markstein length.

Figure 37.3 shows the test results. For the NH_3-H_2 fuel, S_{L0} increased nearly proportional to the H_2 fuel ratio [on a lower heating value (LHV) basis], while S_{L0} decreased inversely proportional to the dilution ratio (wt%) with CH_4. The S_{L0} of solo NH_3 was very similar to that of CH_4 with a dilution ratio of 30%. A H_2 fuel ratio of 40% was required for the NH_3-H_2 fuel to achieve the same S_{L0} of CH_4 without dilution.

37.3 Combustion Performance in a S.I. Engine

37.3.1 Experimental

A single-cylinder research engine used in this study had a four-valve, pent-roof cylinder head with a central spark plug. The cylinder bore and stroke were both 86 mm and the static compression ratio was set to 12:1. Fuel gases, as well as other mixing gas components, were controlled with mass flow controllers and were supplied to the intake port. Engine speed was 1200 rpm in this report.

Engine combustion was analyzed by measuring cylinder pressures and exhaust gas emissions. The indicated mean effective pressure (IMEP) and the heat release rate (HRR) were calculated from the cylinder pressure. A Fourier-transform infrared spectrometer (FT-IR) (Iwata, FAST-2000) and an engine exhaust analyzer (Horiba, MEXA7100D) were used to analyze NH_3-H_2 combustion and CH_4 combustion, respectively.

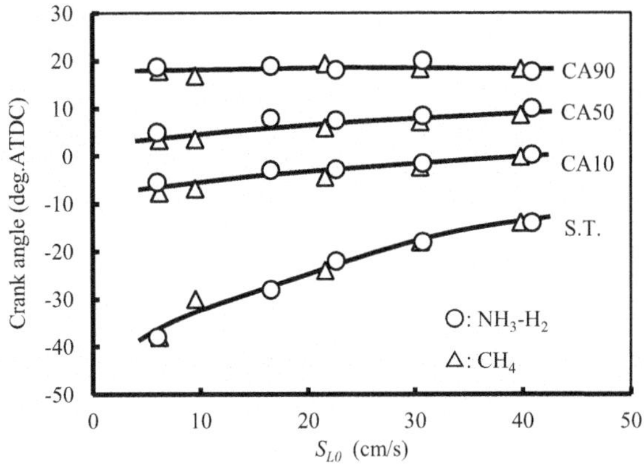

Fig. 37.4 Comparison of MBT spark timing (S.T.) and combustion phasing

37.3.2 Heat Release Rate

The heat release rate of NH_3-H_2 combustion was evaluated and compared with that of CH_4 as a critical parameter of engine combustion. The H_2 fuel ratio was changed from 0 to 40% on LHV basis, while the dilution ratio was changed from 0 to 30 wt%. The dilution gas components and their ratios were the same as described in the previous section. Total fuel energy of the engine was 740 J/st for both types of fuel. Spark timing for the maximum brake torque (MBT) was used throughout the test.

Figure 37.4 shows MBT spark timings and combustion phasing: 10%, 50%, and 90% completion of heat release, abbreviated as CA10, CA50, and CA90, respectively. To compare two different fuels with different characteristics, results are arranged by laminar burning velocity S_{LO}. Combustion phasing and MBT spark timings of both fuel types corresponded at every S_{LO}. This sorting indicated that combustion in a NH_3-H_2-fueled engine was similar to that of conventional hydrocarbon-fueled engines with exhaust gas recirculation (EGR).

The CA50 value was found between 5 and 10 degrees ATDC, which is the typical value obtained from the empirical process of determining the optimum start of combustion [9, 10]. However, compared to the duration between CA10 and CA90, that between spark timing and CA10 became large when S_{LO} became small, and so spark timing was forced to advance significantly. Using a small H_2 fuel ratio is difficult when engine speed is high.

37.3.3 Combustion Fluctuations

Figure 37.5 shows the coefficient of variation (CoV) of IMEP for three different levels of fuel supply energy. Minimal H_2 fuel ratios for combustion stability varied

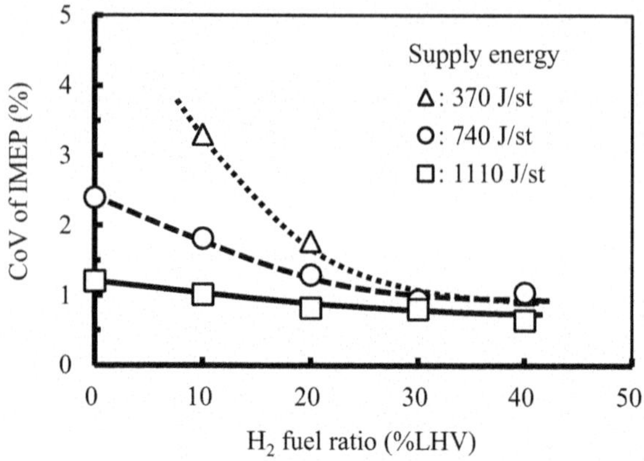

Fig. 37.5 Effect of H_2 fuel ratio on CoV of IMEP under different fuel supply energy

with the amount of fuel supply energy (i.e., load): the greater the load, the smaller the minimal H_2 fuel ratio. When the supply energy was 1110 J/st (IMEP $\approx$ 870 kPa), the engine ran stably without H_2 assistance. This load dependence probably was due to residual gas in the cylinder. As the pressure at the intake port was low at low load, a large amount of burned gas inevitably remained at low load. A reduced S_{L0} caused by a large dilution with burned gas probably caused combustion instabilities.

37.3.4 Combustion Efficiency

The combustion efficiency of the NH_3-H_2-fueled engine was calculated based on the following assumptions.

The stoichiometric combustion reaction was used:

$$
\begin{aligned}
&(0.79\mathrm{N_2} + 0.21\mathrm{O_2}) + (0.28 - b_n)\mathrm{NH_3} \\
&\quad + 0.5b_n(\mathrm{N_2} + 3\mathrm{H_2}) \rightarrow 0.93\mathrm{N_2} + 0.42\mathrm{H_2O}
\end{aligned} \tag{37.5}
$$

where b_n is the mol value of NH_3 allotted for H_2 generation ($0 \leq b_n \leq 0.28$). Assuming a portion of the air-fuel mixture remains unburned, the gas components in the exhaust gas can be described by

$$
\begin{aligned}
&\chi\{(0.79\mathrm{N_2} + 0.21\mathrm{O_2}) + (0.28 - b_n)\mathrm{NH_3} + 0.5b_n(\mathrm{N_2} + 3\mathrm{H_2})\} \\
&\quad + (1 - \chi)(0.93\mathrm{N_2} + 0.42\mathrm{H_2O})
\end{aligned} \tag{37.6}
$$

where χ is the ratio of unburned air-fuel mixture. The measured NH_3 concentration $[NH_3]_{ex}$ can be described by Eq. (37.7), while combustion efficiency η_c can be

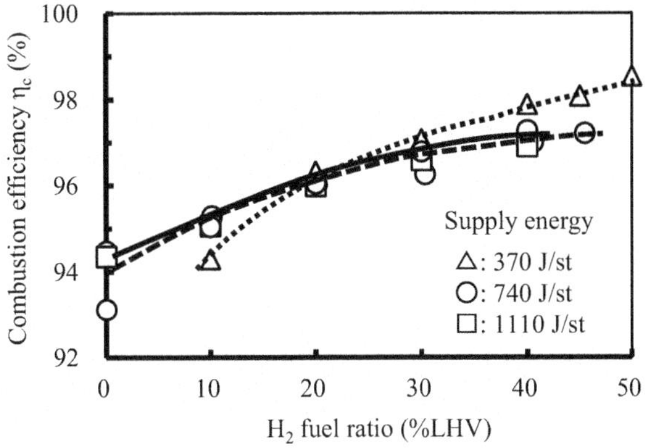

Fig. 37.6 Effect of H_2 fuel ratio on combustion efficiency under different fuel supply energy values

calculated using Eq. (37.8). This estimate assumes that the unburned ratio of NH_3 and H_2 are the same.

$$[NH_3]_{ex} = (0.28 - b_n)\chi/(1.35 + (b_n - 0.07)\chi) \quad (37.7)$$

$$\eta_c = 1 - \chi = 1 - 1.35[NH_3]_{ex}/(0.28 - b_n - (b_n - 0.07)[NH_3]_{ex}) \quad (37.8)$$

Figure 37.6 shows the effect of H_2 fuel ratio on η_c. The value of η_c decreased when the H_2 fuel ratio was reduced. As described in the subsection 37.3.2, stable combustion was possible without H_2 at high load (supply energy = 1110 J/st); however, η_c decreased to less than 95%, probably due to wall quenching. Since the Péclet number represents the ratio of heat release from the flame to the heat loss to the wall, a small S_{L0} will cause increasing wall quenching distance. However, the η_c of CH_4 combustion upon dilution was apparently better than NH_3-H_2 combustion, even if the value of S_{L0} was the same. Further research is needed to improve η_c. A H_2 fuel ratio of approximately 40% was required to achieve the same level of η_c as conventional S.I. engines with an η_c of 98% or more.

37.3.5 *Thermal Efficiency*

Figure 37.7 shows net thermal efficiency η_i for the spark timing sweep of each H_2 fuel ratio. Results of two different fuel energies are shown in the graph. The amount of fuel energy was adjusted to be constant throughout the test, even as the H_2 fuel ratio varied. The indicated mean effective pressure (IMEP) at maximum η_i was 203 kPa for a fuel energy of 370 J/st and 551 kPa for a fuel energy of 740 J/st.

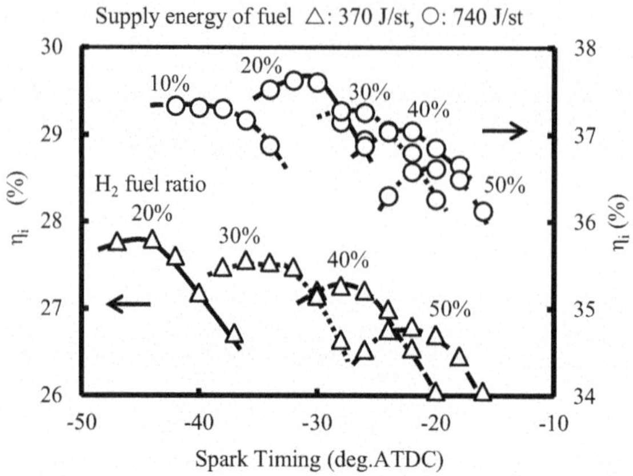

Fig. 37.7 Effect of H_2 fuel ratio on thermal efficiency η_i

It is notable that the smaller the H_2 fuel ratio, the larger the η_i, despite that a small H_2 fuel ratio has been reported to cause a low η_c value and long combustion duration. One reason for the difference in η_i was due to the changes in the number of molecules before and after combustion. The reaction $2H_2+O_2 \rightarrow 2H_2O$ occurs from a diatomic molecule to a triatomic molecule, so the number of molecules decreases, while the reaction $4NH_3 + 3O_2 \rightarrow 6H_2O + 2N_2$ includes a four-atom molecule in the reactant and a diatomic molecule in the product, so the number of molecules increases. Changes in the ratio of specific heats as well as heat losses were additional reasons for the difference in η_i, because the smaller the H_2 fuel ratio, the lower the flame temperature, despite the constant chemical composition ratio of H/N/O for all test conditions.

37.4 Exhaust Gas Emissions

Figure 37.8 shows the effect of H_2 fuel ratio and spark timing on NH_3 and nitrogen oxide (NO, NO_2, N_2O) engine emissions. Note that the NH_3 fraction in the air-fuel mixture was changed when the H_2 fuel ratio was varied. Therefore, NH_3 concentration in the exhaust gas includes the effect of the NH_3 fraction in the air-fuel mixture. When the H_2 fuel ratio was changed from 20 to 40%, the NH_3 volumetric fraction in the air-fuel mixture was reduced by 25%. Although NH_3 concentration in the exhaust gas was reduced by 60% when the H_2 fuel ratio was increased from 20 to 40%, 50% of which was due to the difference of NH_3 volumetric fractions in the air-fuel mixture. Changes in spark timing resulted in a tendency for NH_3 emissions to decrease with ignition retardation, similar to that of HC emissions from typical S.I. engines.

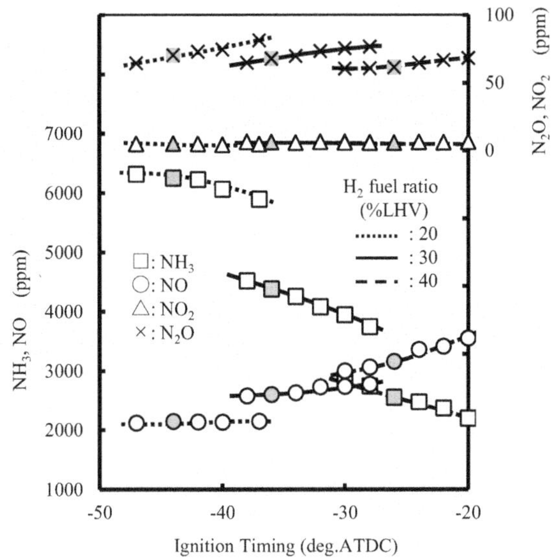

Fig. 37.8 Exhaust gas emissions (colored markers show results at MBT spark timings): supply energy of fuel = 370 J/st

The nitrogen oxides component with the highest concentration was NO, while the concentration of NO_2 was always low regardless of H_2 fuel ratio and spark timing. The N_2O release was a feature of NH_3-H_2 combustion. Although its concentration was not high, N_2O has high global warming potential and should not be ignored.

Nitrogen oxides from NH_3-H_2-fueled engines are considered incomplete combustion products that are different from thermal NO_x from conventional hydrocarbon-fueled engines [11]. Figure 37.9 shows the N_2 production process from NH_3 under engine combustion conditions derived from combustion kinetics [12, 13]. Reaction to N_2 can occur via oxides, which includes production of N_2O as an intermediate. Because N_2O dissociates at high temperatures, N_2O in the exhaust gas may come from flame quenching or post-flame oxidation during expansion and exhaust strokes. The amount of N_2O reduction upon an increase in H_2 fuel ratio mostly corresponded to the reduction in NH_3 fraction in the air–fuel mixture. Thus, N_2O is released in proportion to the NH_3 fraction in the air-fuel mixture.

In contrast, the amount of NO increased with H_2 fuel ratio. Unlike N_2O, NO is present in high concentrations behind a flame front [11]. An increase in H_2 fuel ratio promotes OH radical production that increases reaction via oxides, which in turn promotes NO production.

Another feature was an increase in NO with spark retardation, which also is different from the features of conventional hydrocarbon fuels. Chemical kinetics calculations indicated that NO concentration behind the flame front is several times greater than the equilibrium value, and then decreases gradually during the expansion

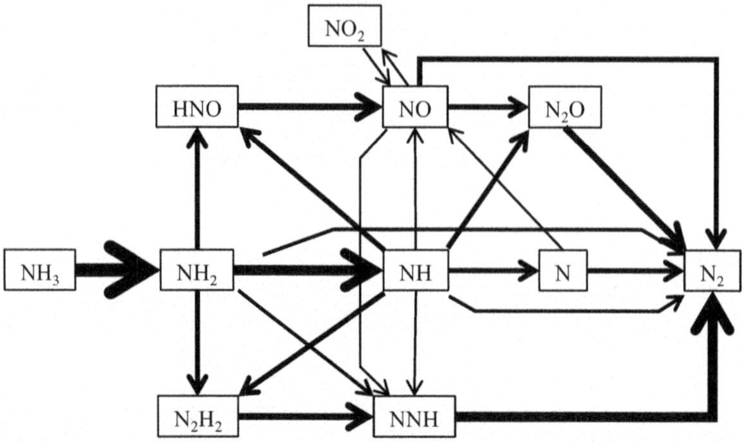

Fig. 37.9 Reaction pathways from NH_3 to N_2 under engine combustion conditions

stroke [11]. The rate-determining step was the Zeldovich mechanism: spark retardation hindered NO decrease by lowering the flame temperature and by shortening the duration that the temperature was high.

Finally, the feasibility of exhaust gas purification was examined. Figure 37.10 shows exhaust gas emissions after a TWC was applied to the test engine. Nearly zero emissions of NH_3 and NO_x were confirmed by maintaining a stoichiometric air-fuel ratio.

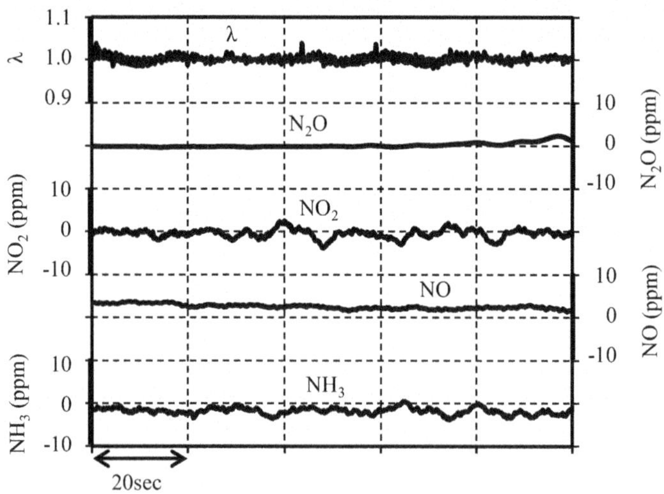

Fig. 37.10 Exhaust gas emissions after TWC with stoichiometric air-fuel mixture: supply energy of fuel = 370 J/st and H_2 fuel ratio = 40%LHV

37.5 Summary

A study of stoichiometric NH_3-H_2 premixed combustion demonstrated the following features. First, laminar burning velocity of NH_3 was approximately the same as that of CH_4 at a dilution of 30% burned gas, while a H_2 fuel fraction of 40% (on LHV basis) was required for the NH_3-H_2 mixture to achieve the same laminar velocity as CH_4. In a study of engine combustion, critical values of MBT spark timing, combustion phasing, and combustion duration of NH_3-H_2 approximately corresponded to those of CH_4-fueled engines at the same laminar burning velocities. The minimal H_2 fuel ratio for stable combustion decreased with load, which allowed an engine fueled solely by NH_3 at high load with small cyclic fluctuations. In a study of thermal efficiency, a smaller H_2 fuel ratio (including fuel without H_2) was advantageous, although a greater H_2 fuel ratio provided more favorable combustion duration, combustion efficiency, and combustion stability. Reducing the H_2 fuel ratio also was vital for mitigating NH_3 dissociation.

Exhaust gas contained N_2O, which is different from conventional engines. The larger the H_2 fuel ratio and the larger the spark retardation, the greater the NO concentration. The amounts of NH_3 and NO_x were reduced to near zero by a TWC under stoichiometric control of the air–fuel ratio.

References

1. Zardi U, Zardi F (2009) History of the birth of the modern synthetic ammonia industry. In: Nitrogen & syngas international conference and exhibition, pp 22–25
2. Kroch E (1945) Ammonia—a fuel for motor buses. J Inst Petrol 213–223
3. Shawn M et al (2008) The fuel mix limits and efficiency of a stoichiometric, ammonia, and gasoline dual fueled spark ignition engine. ASME J Eng Gas Turbines Power 130(2008):042802 1–8
4. Garabedian CG, Johnson JH (1966) The theory of operation of an ammonia burning internal combustion engine. In: Proceedings of the 1966 army science conference, pp 333–348
5. Koike M et al (2013) Ammonia as a hydrogen energy carrier and its application to internal combustion engines. Sustainable Vehicle Technologies (Woodhead Publishing), pp 61–70
6. Frigo S, Gentili R (2013) Analysis of the behavior of a 4-stroke SI engine fueled with ammonia and hydrogen. Int J Hydrogen Energy 38:1607–1615
7. Ogasawara K et al (2014) Effects of residual gas on ammonia engine combustion. In: Proceedings of the 52nd symposium (Japanese) on combustion, pp 282–283 (in Japanese)
8. Kitagawa S et al (2010) Study on turbulent burning velocity of outwardly propagating hydrogen air flames with turbulence Karlovitz Number. J. JSME 76:1426–1432 (in Japanese)
9. Carvalho L et al (2012) Investigation on the fuel and engine parameters that affect the half mass fraction burned (CA50) optimum crank angle. SAE paper 2012-36-0498
10. Zhu GG, Daniels CF, Winkelman J (2003) MBT timing detection and its closed-loop control using in-cylinder pressure signal. SAE Paper 2003-01-3266

11. Koike M et al (2020) NOx emissions from ammonia-hydrogen fueled SI engine. J JSME 86 (in Japanese)
12. Konnov AA, De Ruyck J (2001) A possible new route for NO formation via N_2H_3. Combust Sci Technol 168:1–46
13. Tian Z et al (2009) An experimental and kinetic modeling study of premixed $NH_3/CH_4/O_2/Ar$ flames at low pressure. Combust Flame 156:1413–1426

Chapter 38
Fundamental Study on Application of Ammonia as a Fuel to Marine Diesel Engines

Yoichi Niki, Akira Shimizu, Yoshifuru Nitta, Yasuhisa Ichikawa, and Kazuyoshi Harumi

Abstract In the maritime sector, it has become an urgent issue to be addressed to reduce GHG. Introduction of alternative fuels is expected to promote lower or zero GHG emissions from international shipping. Based on these backgrounds, ammonia (NH_3) has been focused on among alternative fuels, because it has higher energy density and the liquefied NH_3 can be stored at mild conditions compared with hydrogen. It is therefore important to develop NH_3/diesel dual-fuel (ADDF) engines as marine diesel engines. However, it has been reported that emissions of ADDF engines include the unburned NH_3 and nitrous oxide (N_2O) with a high global warming coefficient. To reduce these undesirable emissions from ADDF engines, this chapter reports the results of investigations on ADDF engines. The experiments using an ADDF single-cylinder small-bore engine investigated its efficiency and emission characteristics at the various engine operating conditions and revealed that a pilot injection of diesel fuel reduces both NH_3 and N_2O emissions. The results of numerical simulations of ADDF combustion with a model of an actual marine diesel engine indicated that NH_3 stratification to form a rich NH_3 mixture near the cylinder wall has effects to reduce the unburned NH_3 in combustion. Moreover, it showed that the NH_3 stratification combined with a pilot injection using heavy fuel oil drastically reduced the unburned NH_3.

Keywords Ammonia · Marine diesel engine · Combustion · Dual fuel

---

Originally published in Journal of the Combustion Society of Japan, Vol. 61 (2019), pp. 313-319 (in Japanese);

Y. Niki (✉) · Y. Nitta · Y. Ichikawa · K. Harumi
National Maritime Research Institute, National Institute of Maritime, Port and Aviation Technology, 6-38-1 Shinkawa, Mitaka, Tokyo 181-0004, Japan
e-mail: niki@m.mpat.go.jp

A. Shimizu
JFE Engineering Corporation, 2-1 Suehiro-cho, Tsurumi-ku, Yokohama 230-8611, Japan

K. Aika and H. Kobayashi (eds.), *CO_2 Free Ammonia as an Energy Carrier*,
https://doi.org/10.1007/978-981-19-4767-4_38

38.1 Introduction

Considering climate change, a reduction of greenhouse gas (GHG) is required in the marine sector. Regulations are being tightened to reduce GHG emissions from ships. The regulations regarding GHG emissions for the marine sector are summarized in this section. The regulations for international oceangoing ships are set by the International Maritime Organization (IMO), a specialized agency of the United Nations. Their regulations are universally enforced on any oceangoing ships. IMO has tightened the regulations including NOx emissions from ship engines and sulfur content in heavy fuel oil causing SOx emissions to protect the marine environment. In addition, IMO declared a target to reduce GHG emissions (amount of GHG emission per ton-mile) from the maritime sector by 40% by 2050 compared with that of 2008 so that zero GHG emissions can be achieved as soon as possible in this century [1].

Some strategies have been taken, such as the reduction of hull resistance and the improvement of ship operation to reduce GHG emissions. However, it is difficult to reach zero GHG emissions without shifting to a carbon-free energy source from conventional fuels or the utilization of natural energy. To operate vessels with less or zero GHG emissions, research and development aiming at utilization of various alternative marine fuels are thus conducted. As an example, the shifting to liquefied natural gas has been taken already but it reduces GHG emission by only about 20% compared with the case of using conventional heavy fuel oil.

From the perspective of fuel for oceangoing ships, volumetric energy density is one of the important factors for fuels. In general, ships with larger cargo spaces are beneficial for efficient transportation. On the other hand, lower volumetric energy density fuel requires a larger fuel tank and this causes the suppression of the cargo space. Therefore, the fuel with higher volumetric energy density is desirable, especially for oceangoing vessels which need a large amount of energy for long-distance navigation in weeks. From this point of view, it is clear that liquefied fuels with higher volumetric energy density will be more suitable for oceangoing ships than gaseous fuels. This also means that the condition of storage for the fuel such as pressure and temperature is important. For instance, in the case of hydrogen (H_2) which requires the cryogenic temperature to be liquefied, a particular fuel tank has to be installed on ships to hold it at the required temperature. This increases the cost of ships and consumes space in ships for fuels. In other words, fuel for ships is required to be liquefied near atmospheric temperature and pressure. In addition, the fuel has to be stably stored for the long term. Based on these perspectives, ammonia (NH_3), which does not have a carbon atom and can be liquefied at 1 MPa in 25 °C, has been focused on as carbon-free fuels for ships. However, NH_3 has a lower reactivity compared with hydrocarbon fuels. Several investigations to combust NH_3 with a gas turbine have been conducted. On the other hand, as far as we searched, few researchers have addressed an investigation of the NH_3 combustion with diesel engines, which are widely used in propulsion and electric generation systems of ships.

Fuel supply is also a common concern for alternative fuels. At the moment, an amount of NH_3 production as a fuel for ships is in short supply. Hence, we have thought that utilizing a dual-fuel combustion technology, NH_3 and conventional hydrocarbon fuels, may be one of the solutions in the near future as the first step to GHG emission reduction with NH_3. Since replacing a part of a conventional hydrocarbon fuel with NH_3 in diesel engines reduces GHG emissions by the replaced amount, the dual-fuel combustion technology can meet GHG reduction demands in the future. The aim of this study is therefore to develop the technology of NH_3 combustion which can be applied to the conventional diesel engine and to remove expected technical barriers to use of NH_3 as fuel.

38.2 Technical Barriers of NH_3/diesel Dual-Fuel Engine

In our research, gaseous NH_3 was added to an air intake stream of an experimental diesel engine. Figure 38.1 shows a schematic diagram of NH_3 mixed combustion in a diesel engine. Gaseous NH_3 was injected into the intake, NH_3/air mixture then was drawn in a cylinder. After a compression stroke, diesel fuel was injected to ignite and combust the NH_3/air mixture in the cylinder. Our previous study [2, 3] showed the following results;

(1) An unburned NH_3 in the exhaust gas was emitted about 14% of the amount of NH_3 supplied regardless of the amount of NH_3 supply.
(2) The increase in the amount of supply NH_3 increased concentration levels of unburned NH_3 and nitrous oxide (N_2O) in the exhaust gas.

The results of our previous research suggested a requirement of combustion strategy for NH_3/diesel dual-fuel engine (ADDF) to overcome a low reactivity of NH_3 and reduce N_2O emission which has 265 times higher global warming potential relative to carbon dioxide (CO_2) [4]. To reduce the unburned NH_3 and N_2O from ADDF engines, results obtained from experimental investigations using a single-cylinder engine and 3-D engine combustion simulation are reported in this chapter.

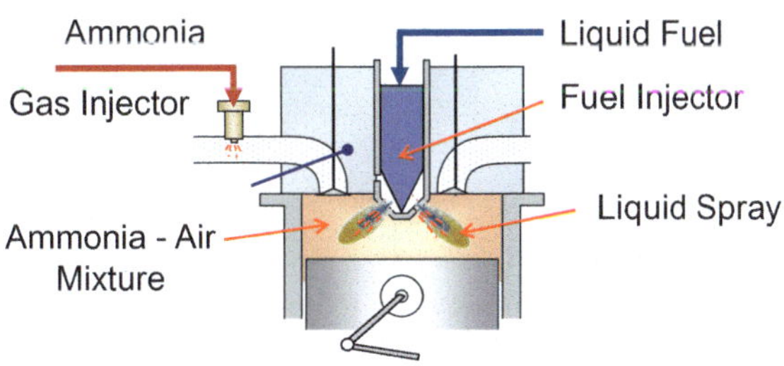

Fig. 38.1 Ammonia mixed combustion in the diesel engine

38.3 Experimental Setup and Method

A schematic layout of an experimental setup used in this study and the photograph of an engine test bench is shown in Figs. 38.2 and 38.3, respectively. An experimental single-cylinder diesel engine (the specification of the engine is listed in Table 38.1) drives to a dynamometer. The dynamometer automatically maintains a specified engine speed by changing a torque. In the experiments, the engine was first started and operated with diesel fuel. Then, gaseous NH_3 was injected in charge air at the intake. The amount of NH_3 supplied and diesel fuel was adjusted to set the engine load. The fraction of lower-heating-value (LHV) from NH_3 to the total LHV of NH_3 and diesel fuel was kept at 20% through the experiments. Engine data were measured after temperatures of coolant and oil at the engine outlet reached a steady state. Cylinder pressure, intake air pressure, and exhaust gas pressure were measured at every 0.5 crank angle degrees by pressure transducers (GH14P, LP12DA, LP21J AVL) and a charge amplifier (FI PIEZO AVL). The in-cylinder pressure was averaged over 50 cycles and corrected to be the same as the value of the averaged pressure of the intake air and the exhaust gas during a valve overlap period.

Emission components in the exhaust gas, which are carbon monoxide (CO), CO_2, nitric oxide (NO), nitrogen dioxide (NO_2), NH_3, N_2O, and water (H_2O), were measured with a Fourier transform infrared (FTIR) gas analyzer (FAST-2200 Iwata Dengyo). The gas analyzer simultaneously samples and measures the multi-gas components in the hot exhaust gas. The sampling line and sampling filter were heated to 191 °C. Thus, NH_3 in the exhaust gas containing H_2O was detected with precision. The total hydrocarbons (THC) emissions and smoke were also measured using a flame ionization detector (600HFID CAI) and a filter smoke meter (GSM-3 Sokken), respectively.

An NH_3 supply system provides gaseous NH_3 from a cylinder storing 50 kg liquefied NH_3. The fraction of heat supplied from NH_3 is calculated by the following equation,

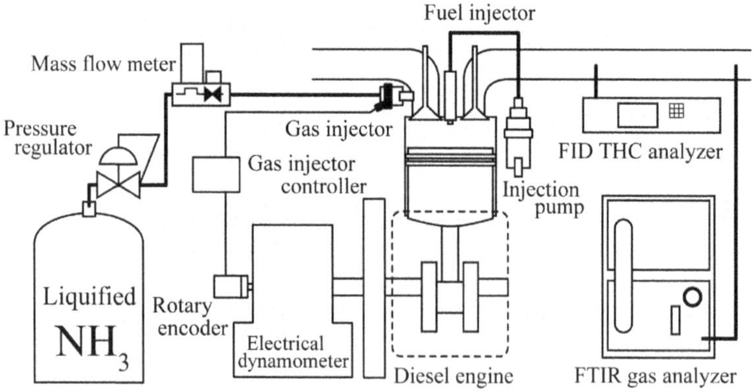

Fig. 38.2 Schematic layout of the experimental setup

Fig. 38.3 Photograph of the engine test bench

Table 38.1 Specifications of test engine

Manufacturer	AVL
Type	4st
Bore/stroke	112/110 mm
Num. cyl	1
Displacement	1.08 L
Comp. ratio	18.5
Aspiration	Naturally aspirated
Rated power	7.7 kW 1500 rpm

$$\text{Fraction of heat supplied from NH}_3 = \frac{\dot{m}_{\text{NH}_3} \times \text{LHV}_{\text{NH}_3}}{\dot{m}_{\text{diesel}} \times \text{LHV}_{\text{diesel}} + \dot{m}_{\text{NH}_3} \times \text{LHV}_{\text{NH}_3}} \tag{38.1}$$

where $\dot{m}$ is the mass flow rate, LHV is lower heating-value, the subscript represents fuel type (diesel or NH_3).

The decrease in the temperature of stored NH_3 in the cylinder causes a decrease in the supply pressure of NH_3. To keep the supply pressure of NH_3, the NH_3 cylinder was warmed by a water jacket using water with the temperature of 40 °C.

38.4 Experimental Results

38.4.1 *Effects of the Engine Speed and Torque on Efficiency and Emissions*

Experimental results are described in this section. The experiments were performed for two fractions of heat supplied from NH_3 of 0 and 20% at various engine speed and torque conditions including two engine speeds of 1000 and 1500 min^{-1} and various torque from 20 to 60 Nm.

Figure 38.4 shows the effects of the engine speed and torque on brake thermal efficiency and emissions. The increase in torque reduced NH_3 and N_2O emissions for the both engine speeds. The higher in-cylinder temperature because of the higher in-cylinder pressure shown in Figs. 38.5 and 38.6 may reduce the NH_3 and N_2O emissions. Comparing the cases for the both engine speeds, the NH_3 emissions are approximately lower for the case of 1000 min^{-1} than for the case of 1500 min^{-1}. The lower engine speed allows NH_3 to combust longer period. The CO, THC, and smoke emissions increased for the case of 1000 min^{-1} in the high torque conditions. This tendency is similar to the results of the diesel fuel only operation shown in Fig. 38.4. It could be affected an air and diesel fuel mixing process in a mixing-controlled combustion phase following premixed combustion in diesel combustion. The air motion is relatively weak in a lower engine speed condition. This gives the insufficient mixing between the injected diesel fuel and the air in the combustion chamber while more amount of diesel fuel is required in the high torque operation.

Comparing the in-cylinder pressures shown in Figs. 38.5 and 38.6, the maximum in-cylinder pressure is higher for the case of 1000 min^{-1} than for the case of 1500 min^{-1}. The higher in-cylinder pressure indicates a higher temperature of bulk gas in the cylinder. The crank angle period of combustion between the heat release rate first rising and return to zero is apparently regardless of the engine speed. The time during one crank angle degree is longer of slow engine speed conditions than that of fast engine speed conditions. Since the decomposition of NH_3 might be promoted by the higher temperature of bulk gas in the cylinder and the longer combustion time, the NH_3 emissions are lower for the case of 1000 min^{-1} than relative to the case of 1500 min^{-1} at the low torque condition.

38.4.2 *Effects of Diesel Fuel Pilot Injection*

The diesel fuel pilot injection was conducted to accelerate the reaction of NH_3. Injected diesel fuel at early timing before ignition spreads into NH_3/air mixture in the cylinder. It can work as a high reactivity fuel of a reaction controlled compression ignition (RCCI) combustion strategy [5]. An electrically controlled injection system was employed for a diesel pilot injection instead of the conventional mechanical injection system.

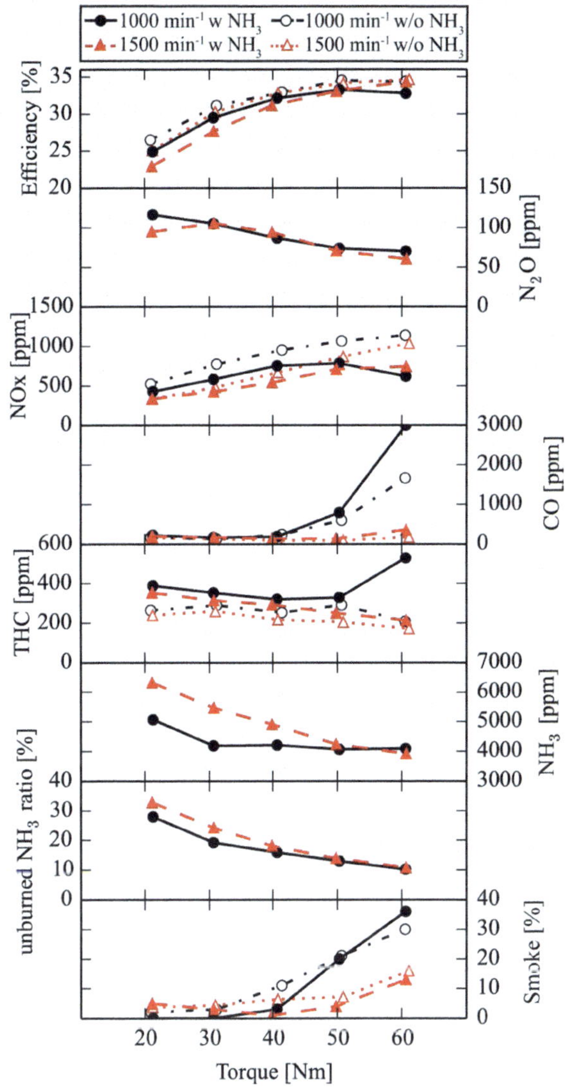

Fig. 38.4 Effect of engine operation condition on brake thermal efficiency and emissions

The fraction of heat supplied from NH_3 and the duration of pilot diesel were respectively held at about 20% and 500 ms through the experiments in which the timing of diesel pilot injection was varied from −30 to −60°. after top dead center (ATDC) in the two engine speeds conditions. The diesel main injection duration was adjusted to maintain the engine torque and the injection timing was held at −5°. ATDC. The effects of the timing of pilot injection on brake thermal efficiency and emissions are shown in Fig. 38.7. The plots of very right in Fig. 38.7 show the results

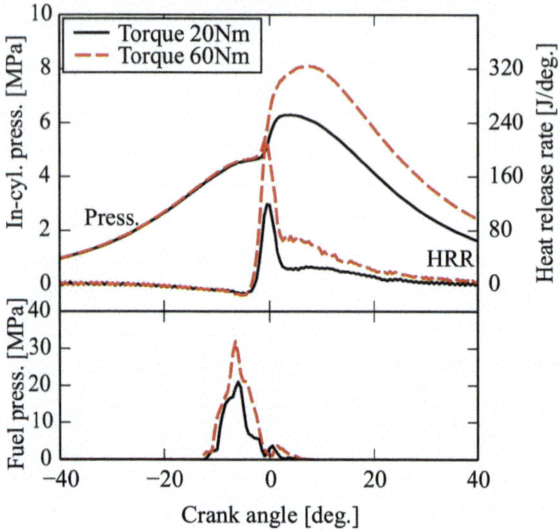

Fig. 38.5 Cylinder pressure, heat release rate, and diesel fuel injection pressure for NH_3 mixed combustion at engine speed of 1000 min^{-1}

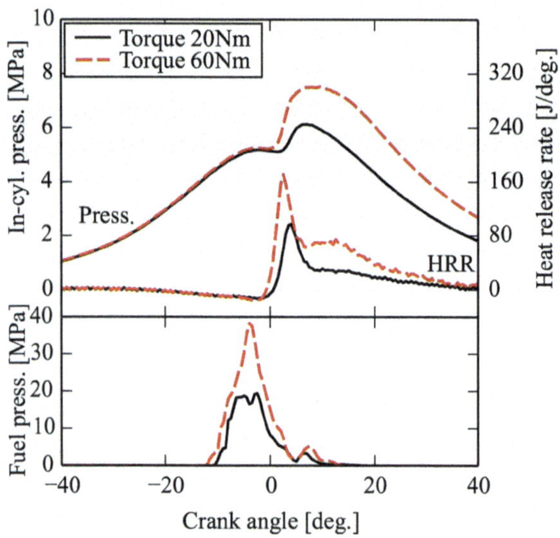

Fig. 38.6 Cylinder pressure, heat release rate, and diesel fuel injection pressure for NH_3 mixed combustion at engine speed of 1500 min^{-1}

without the pilot injection. As the pilot injection timing was advanced beyond −40°. ATDC, NH_3 emissions were reduced. The minimum ratio of unburned NH_3 to the amount of supplied NH_3 was reduced to about 2% for the both engine speeds. The increase in THC and CO was also observed in advancing pilot injection timing beyond −40°. ATDC. The representative in-cylinder pressure histories, heat release rates, and injection signals of the advancing pilot injection cases are shown in Figs. 38.8 and 38.9. Figures also show the in-cylinder pressure data for the cases without the pilot injection. The pilot injections increase maximum cylinder pressures compared with that of all cases without the pilot injection. The heat release rates for the case using the pilot injection are split. Observing the first heat release, the timing of start combustion represented as the first rising heat release rate appeared at approximately −15°. ATDC. These timings are unchanged while the profiles of heat release rate become steep with the advancing in pilot injection timing. Although the NH_3 and N_2O emissions are reduced with the advancing injection timings as shown in Fig. 38.7, a difference of the in-cylinder pressure profiles of the pilot injection case is slightly by the pilot injection timing. These results suggest that the reduction of NH_3 and N_2O is likely attributed to not only the high-temperature state in the cylinder but also an interaction of chemical reaction between diesel fuel and NH_3.

38.5 Numerical Simulation

In the above experiments, a small-bore and no supercharging experimental single-cylinder diesel engine was used. The cylinder bore and output of commercial marine engines are larger than the engine in most cases. Further, an inspection was required to apply the experimental results to actual engines. For this purpose, numerical simulation was performed with a model of a large-scale combustion chamber equivalent to that of an actual engine [6].

38.5.1 Simulation Method

Table 38.2 shows the specifications of the actual engine to be modeled and calculated. As compared with the single-cylinder experimental diesel engine shown in Table 38.1, the cylinder bore and stroke of the actual engine are larger than that of the experimental engine such as 400 mm × 500 mm and the revolution is medium speed level such as 600 min^{-1}. Figure 38.10 shows a 3-D model outline for the combustion chamber and fuel injector of the actual engine. Simulations were performed by transient Reynolds-averaged Navier-Stokes equation (RANS) using an original analysis code. The main fuel is heavy fuel oil, and the NH_3/air mixture is ignited by a heavy fuel oil combustion flame. The fraction of heat supplied from NH_3 to total heat from NH_3 and heavy fuel oil was kept at 20% for all calculations.

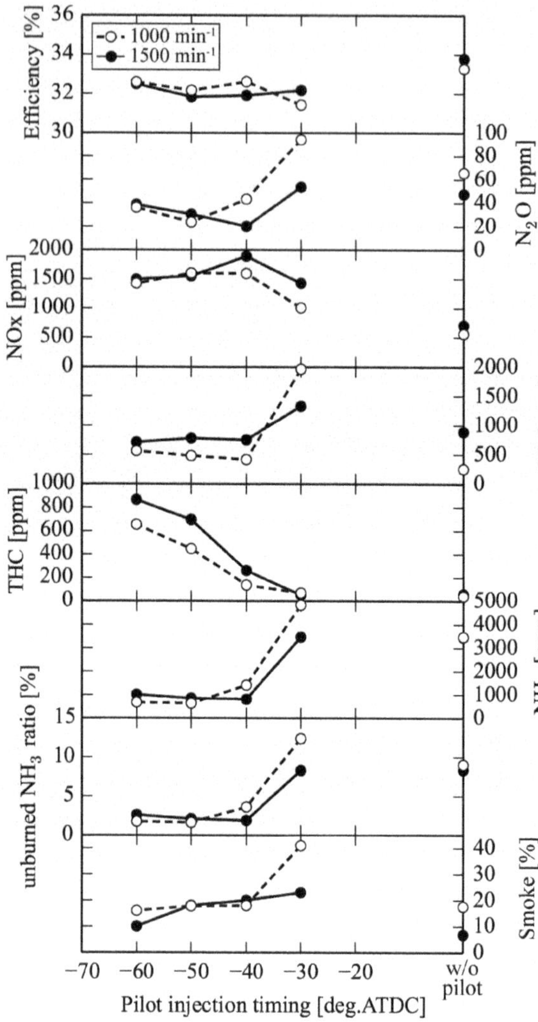

Fig. 38.7 Effect of diesel fuel pilot injection on brake thermal efficiency and emissions for maintaining NH_3 heat contribution at 20%

As the parameters, premixed gaseous NH_3 concentration, stratification (NH_3 distribution), main fuel oil injection timing, and injection patterns during main fuel oil injection were changed and the influence of efficiency, emission, and unburned NH_3 i.e., combustion rate were investigated. Typical calculation results are reported in the next sections.

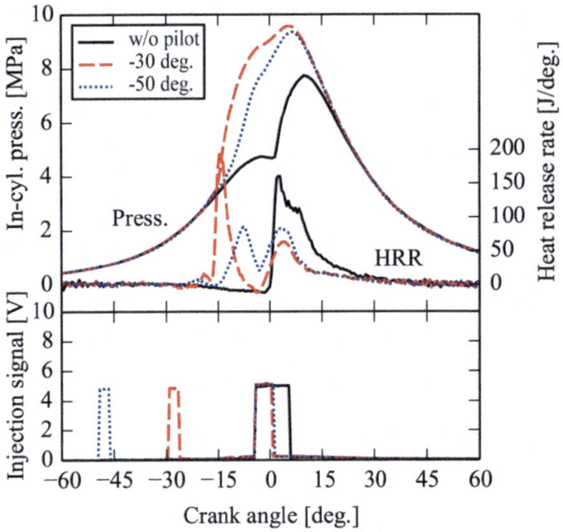

Fig. 38.8 Cylinder pressure, heat release rate, and diesel fuel injection pressure for diesel fuel pilot injection timing sweep at engine speed of 1000 min^{-1}

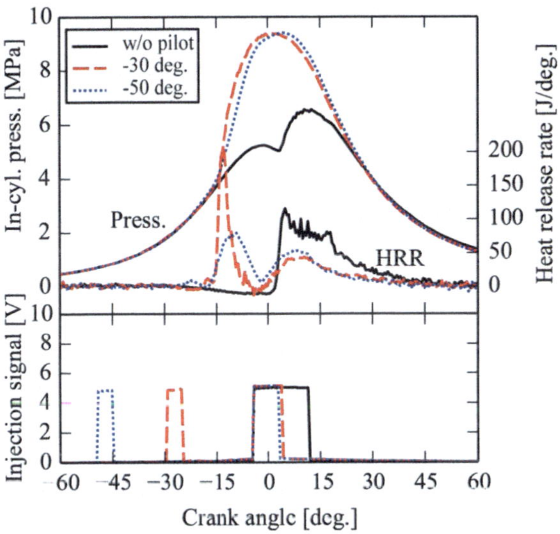

Fig. 38.9 Cylinder pressure, heat release rate, and diesel fuel injection pressure for diesel fuel pilot injection timing sweep at engine speed of 1500 min^{-1}

Table 38.2 Specifications of engine for simulation

Type	4st
Bore/stroke	400/500 mm
Displacement	62.8 L
Comp. Ratio	14.5
Output power (50% load)	375 kW/cyl. 600 min^{-1}

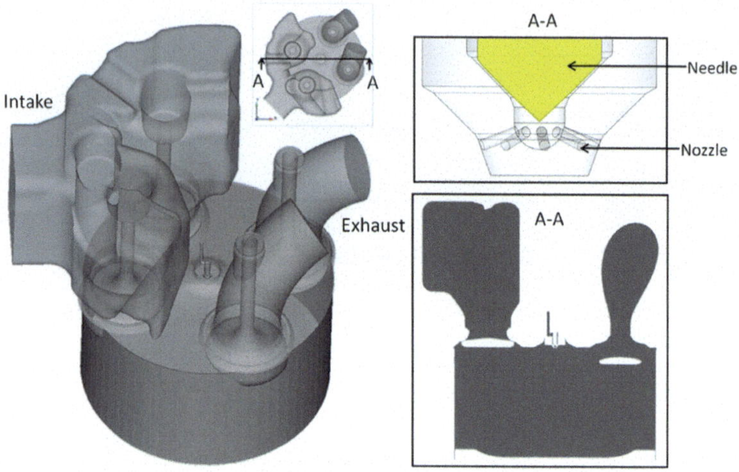

Fig. 38.10 3-D model and the nozzle with 9 holes

38.5.2 Prediction of Effects of Air Excess Ratio

With a no supercharging test engine, research of effects of change in air excess ratio is difficult to perform in experiments, so effects of air excess ratio were predicted by simulation approach. As an initial baseline, the calculation was performed on the condition that air excess ratio $\lambda = 3.2$, which is equivalent to the actual heavy fuel oil operation condition in the actual engine, in adding NH_3, which accounts for 20% of total heat from NH_3 and heavy fuel oil, to the air on homogeneous distribution. The combustion calculation predicted that the ratio of unburned NH_3 to the amount of supplied NH_3 was 37%. In the case of $\lambda = 2.5$ and 2.0 which were adjusted to the fuel-rich side, these cases reduce the ratio of unburned NH_3 to 27 and 21%, respectively. Finally, the calculation results suggest the case of $\lambda = 2.0$ would be optimal due to resulting lower the ratio of unburned NH_3.

According to these calculations, it is necessary to increase the higher temperature region inside of the combustion chamber by decreasing air excess ratio at an optimal rate for improving the combustion of NH_3 having higher ignition temperature characteristics. The temperature distribution inside of the cylinder is shown in Fig. 38.11.

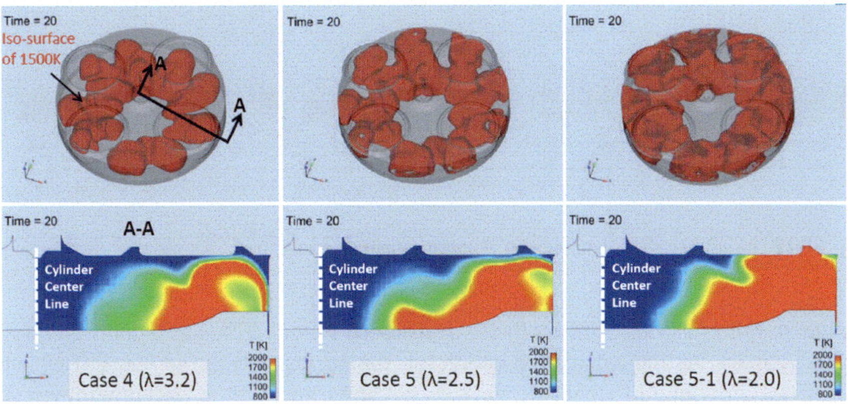

Fig. 38.11 Influence of air excess ratio on temperature profile in the cylinder (left: $\lambda = 3.2$, center: $\lambda = 2.5$, right: $\lambda = 2.0$)

It can also be interpreted as a result corresponding to the decrease in unburned NH_3 in the case that increased torque in the experiments.

38.5.3 *Prediction of Effects of Ammonia Distribution*

The NH_3/air mixture in the cylinder was probably close to homogeneous because the mixture was drawn and compressed after the NH_3 injection at the intake air stream in the experiments. There is the possibility that the unburned NH_3 could be reduced by concentrating NH_3 around the fuel oil main combustion area which is a higher temperature region. It is difficult to manipulate the distribution of NH_3 concentration in experiments, on the other hand, simulation can be performed to search for better distribution. Fuel oil injected by the injector installed at the center of the cylinder mainly combusts around the periphery of the combustion chamber, and the higher temperature region exists near the cylinder wall. To reduce the unburned NH_3 emissions, the combustion strategy with stratified NH_3/air mixture was calculated for various NH_3 stratification states in the cylinder.

Figure 38.12 shows the NH_3 stratification states for each calculation case. Red and blue zones show NH_3 rich and lean areas in Figure, respectively. The left figure shows the ±20% case. The difference in the NH_3 concentration between the center of the cylinder and the cylinder wall was set to ±20%. In the center of the cylinder, the NH_3 concentration difference is −20%. On the other hand, the +20% rich NH_3 mixture is at the cylinder wall. A concentration curve is linear from the center to the wall. The middle figure shows ±70% case and the concentration shape is the same as the ±20% case. The NH_3 concentration difference is −70% at the center and + 70% at the wall position. The right figure shows the ±80% case. In this case, the NH_3 concentration difference is −80% at the center and +80% at the wall, but the

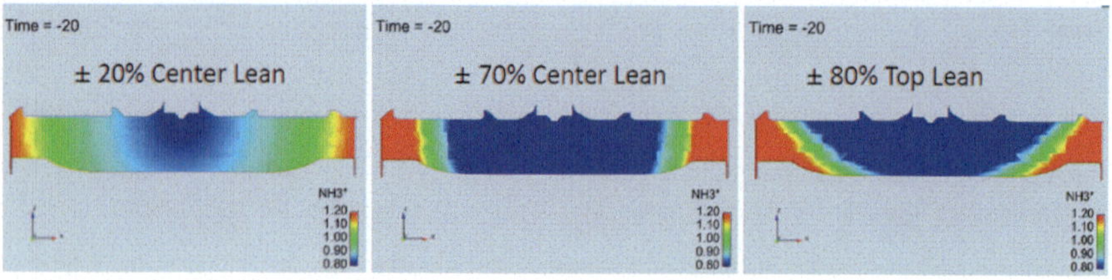

Fig. 38.12 In-cylinder NH_3 contours for three stratification conditions

concentration curve shape is not the same as ±20% and ±70%, and a concentration slope is given. As the calculation condition, the air excess ratio is $\lambda = 2.0$ which is optimal according to λ change calculation results, and fuel oil injection timing is held at −19.5°. ATDC to keep at the same maximum cylinder pressure of original diesel for each case.

In the calculation results, the ratio of unburned NH_3 to the amount of supplied NH_3 for the ±20% case was reduced to 11.5% by the effect of making stratification and an advanced fuel oil injection timing. By increasing the NH_3 stratification to ±70%, the ratio of unburned NH_3 was reduced to 6.2%. In the ±80% case, the ratio of unburned NH_3 was almost the same as the results for the ±70% case and the reduction seemed to be saturated around this condition.

38.5.4 Prediction of Effect of Heavy Fuel Oil Pilot Injection

In the experimental results, the more unburned NH_3 reduction is observed by the effect of diesel fuel pilot injection. To confirm that the same effect could be obtained by simulation, calculations were done for heavy fuel oil pilot injection cases.

Calculations were done under the following condition. The pilot timing is held at −80°. ATDC and the injection amount was 15% of the amount of total fuel oil. The main fuel injection timing was kept at −19.5°. ATDC and the injection amount is 85% of the amount of total fuel oil. The main fuel injection timing was optimal to have the same maximum cylinder pressure as the original diesel operation. The air excess ratio was held at $\lambda = 2.0$ and the NH_3 stratification state is set the ±20% case.

As a result, the ratio of unburned NH_3 to the amount of supplied NH_3 significantly reduced to 6.3%. The ratio of unburned NH_3 was 11.5% for the same λ, stratification state, the main injection timing and the amount of total injected heavy fuel oil without the pilot injection. On the other hand, the concentration of NOx emissions was about 2.4 times (1200 ppm) higher than that of the case without the pilot injection. Figure 38.13 shows the difference between the heavy fuel oil spray and combustion profile by the presence of the pilot injection. As shown in the top right figure, fuel droplets from the pilot injection reached the cylinder wall just after the main fuel oil injection timing. The pilot injection timing was at −80°. ATDC,

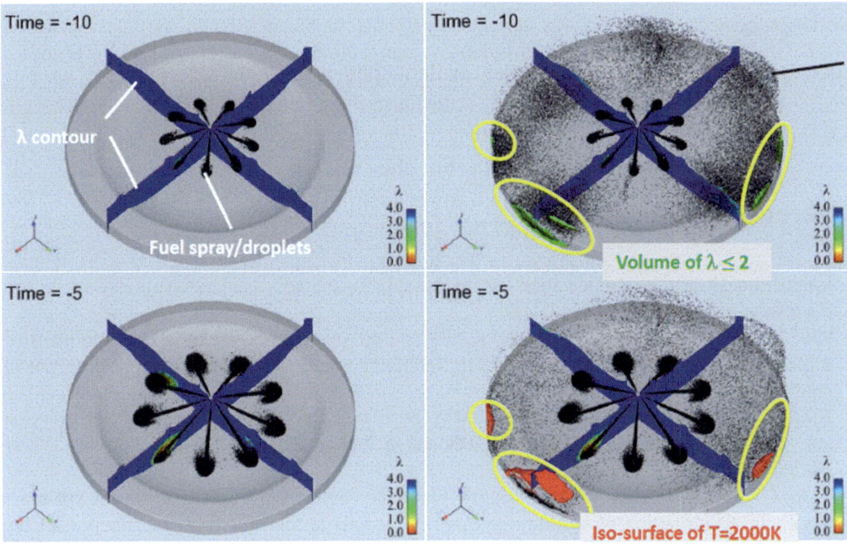

Fig. 38.13 Spray and combustion profile (left: without pilot injection, right: with pilot injection)

during the compression stroke, the gas temperature in the cylinder could not exceed the autoignition temperature of the mixture of pilot injected fuel, air, and NH_3, so the combustible mixture remained without burning on above timing. Due to the pilot injection, some fuel-rich pockets, which are indicated by $\lambda < 2$ (in yellow circle area) at the timing $-10°$. ATDC in the top right figure in Fig. 38.13, are formed near the cylinder wall. This fuel-rich mixture near the cylinder wall combusts relatively early. The earlier combustion near the cylinder wall where the NH_3 concentration is relatively high by the stratification reduces the unburned NH_3 emission.

From these calculation results, the possibility of the reduction of the unburned NH_3 is suggested by devising an optimal operating condition and fuel injection method for actual commercial engines.

38.6 Conclusion

The presented manuscript reports a fundamental study on the application of NH_3 as a fuel to marine diesel engines. A reduction of the unburned NH_3 emission with a diesel fuel pilot injection was performed by experiments. Further reports of the experimental investigation have been published [7, 8]. The Numerical simulations predicted the same trend of the reduction of the unburned NH_3 emission. These results suggested that the diesel fuel pilot injection strategy will be an effective way for a reduction unburned NH_3 from ADDF engines and a numerical simulation will be an effective tool for the development of ADDF engines.

Acknowledgements This work was supported by Council for Science, Technology and Innovation (CSTI), Cross-ministerial Strategic Innovation Promotion Program (SIP), Energy Carriers (Funding agency: JST).

References

1. International Maritime Organization (2018) Initial IMO strategy on reduction of GHG emissions from ships, resolution MEPC. 304(72)
2. Niki Y, Nitta Y, Sekiguchi H, Hirata K (2018) Emission and combustion characteristics of diesel engine fumigated with ammonia. In: Internal combustion engine division fall technical conference: ICEF2018-9634
3. Niki Y, Nitta Y, Sekiguchi H, Hirata K (2019) Diesel fuel multiple injection effects on emission characteristics of diesel engine mixed ammonia gas into intake air. J Eng Gas Turbines Power 141:061020
4. Team CW, Pachauri RK, Meyer LA (2014) Climate change 2014 synthesis report: synthesis report. In: Contribution of working groups I, II and III to the fifth assessment report of the intergovernmental panel on climate change. IPCC: 87
5. Reitz RD, Duraisamy G (2015) Review of high efficiency and clean reactivity controlled compression ignition (RCCI) combustion in internal combustion engines. Prog Energy Combust Sci 46:12–71
6. Shimizu A (2019) Cross-ministerial strategic innovation promotion program (SIP) report. https://www.jst.go.jp/sip/dl/k04/end/team6-17.pdf. Accessed on 10 Mar 2022 (in Japanese)
7. Niki Y (2021) Reduction in unburned ammonia and nitrous oxide emissions from an ammonia-assisted diesel engine with early timing diesel pilot injection. J Eng Gas Turbines Power 143:091014
8. Niki Y (2021) Experimental investigation of effects of split diesel-pilot injection on emission from ammonia-diesel dual fuel engine. In: Internal combustion engine division fall technical conference: ICEF2018-9634

Chapter 39
Ammonia Combustion Chemistry

Hisashi Nakamura

Abstract This chapter summarized studies on chemical kinetic modeling for ammonia combustion in two "Energy Carriers" projects: PREST (Precursory Research for Embryonic Science and Technology) and SIP (Cross-ministry Strategic Innovation Program). Weak flames in a micro flow reactor with a controlled temperature profile (MFR) was applied to investigate ammonia reactivity at low temperatures. Weak flame structures of ammonia/air mixtures were analyzed using a quadrupole mass spectrometer (QMS). Chemical kinetic analysis for ammonia weak flames identified the importance of N_2H_x chemistry for low-temperature reactivity of ammonia. Chemical kinetic model for ammonia combustion was developed and validated using species data of ammonia weak flames and data from the literature. Effects of radiation heat loss on laminar flame speeds and emission characteristics were examined.

Keywords Chemical kinetics · Laminar premixed flames · Autoignition · Microcombustion

39.1 Introduction

The SIP "Energy Carriers" project has demonstrated that ammonia is an important carbon-free fuel for energy systems combined with combustion devices. An accurate and reliable chemical kinetic mechanism of ammonia is necessary to develop ammonia combustors with high thermal efficiency and low NOx emissions. Because ammonia has been regarded as a good reductant in the denitration process [1] and because it is a representative nitrogen compound included in practical coal and biomass for a source of fuel NOx [2], extensive studies of ammonia chemical kinetics have been conducted. Nevertheless, most experiments have been conducted with mixtures at low ammonia concentrations, with validation made using NOx data. Validation of ammonia chemical kinetics for ignition and flame characteristics has not been performed well. Especially for ignition characteristics, ignition data have been

H. Nakamura (✉)
Institute of Fluid Science, Tohoku University, 2-1-1 Katahira, Aoba-ku, Sendai, Miyagi 980-8577, Japan
e-mail: hisashi.nakamura@tohoku.ac.jp

K. Aika and H. Kobayashi (eds.), *CO_2 Free Ammonia as an Energy Carrier*,
https://doi.org/10.1007/978-981-19-4767-4_39

limited at high temperatures (around 2000 K) for highly diluted mixtures [3], probably because of difficulties conducting experiments at low temperatures attributable to the low reactivity of ammonia.

The author was involved in two "Energy Carriers" projects to develop chemical kinetics for ammonia combustion. One is by the Precursory Research for Embryonic Science and Technology (PRESTO, Sakigake in Japanese, FY 2013–2016). The author joined the "Energy Carriers" project by SIP (FY 2014–2018) from FY 2017. Weak flames in a micro flow reactor with a controlled temperature profile (MFR) [4] were used to obtain reactivity of ammonia/air mixtures at low temperatures (around 1200 K). The developed mechanism was validated using measured and literature data. This chapter presents the development and the validation of chemical kinetics for ammonia combustion under "Energy Carriers" projects. Details can be found in a series of publications [5–7]. A summary is presented in this chapter.

39.2 Experimental and Computational Methods

Combustion characteristics of ammonia/air mixtures were observed using MFR [4]. Weak flames in MFR at low inlet flow velocities were used to obtain reactivity of ammonia/air mixtures at low temperatures. A weak flame branch was found on the ignition branch of the Fendell curve in a theoretical study [8]. Reactivity of hydrocarbons has been evaluated using the weak flame position as an index: weak flames of lower/higher reactivity fuels are placed in a higher/lower temperature region of MFR [9–13]. The gas-phase temperature profile in the weak flame regime is nearly equal to the given wall-temperature profile, even in the reaction zone of fuel/air (non-diluted) mixtures [14]. The steady, 1-D numerical model of MFR has enabled the validation of chemical kinetic mechanisms under a well-defined temperature profile [10, 12]. Long residence time under a controlled temperature profile in MFR is suitable for experiments of low reactivity fuels such as ammonia.

Figure 39.1 presents a schematic of MFR. A quartz tube with 2 mm inner diameter was used as a reactor channel. It was heated using a hydrogen/air flat-flame burner to give a stationary temperature ramp along the inner surface of the reactor channel in the flow direction. Hereinafter, "wall temperature" signifies the temperature on the inner surface of the reactor channel. The maximum wall temperature, $T_{w,\, max}$, was set as 1300 or 1400 K. Ammonia/air mixtures were supplied from the low temperature side of the reactor channel at atmospheric pressure. Flow rates of ammonia (purity >99.999%) and synthetic air (O_2:N_2 = 21:79 in moles) were controlled using mass flow controllers.

Two investigations were performed: (1) direct observation of flame dynamics and (2) species measurement for weak flames. For direct observation, a digital still camera (D300; Nikon Corp.) was used without band-pass filters. Flame images were captured in various inlet mean flow velocities. For species measurement, a T-shaped reactor and a quadrupole mass spectrometer (QMS, JMS-Q1500; JEOL) were used (Fig. 39.2). A quartz microprobe (0.1 mm inner diameter; 0.25 mm outer diameter

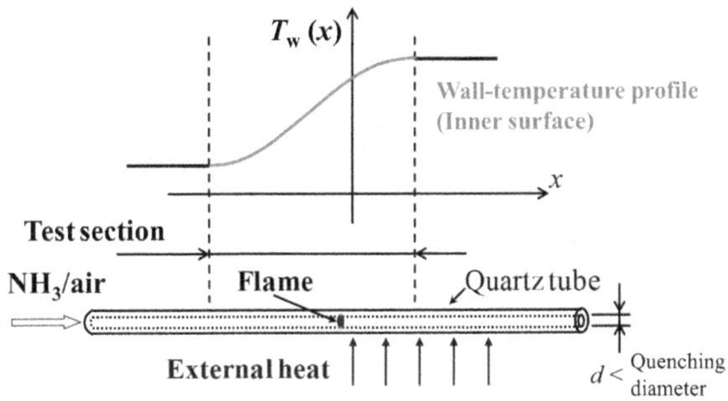

Fig. 39.1 Schematic of MFR [5]

of the edge; 1.6 mm outer diameter of the root) was fused on the reactor channel side wall. A sufficiently small amount of gas in the reactor channel was sampled from the microprobe and was introduced to QMS through a fused-silica capillary. The local temperature at the sampling position was changed by shifting the location of the burner in the x-direction relative to the sampling position. The sampling line temperature was kept at 373 K using electric heaters to prevent water vapor condensation. Using QMS, we identified and quantified NH_3, O_2, H_2O, N_2O, and NO. Electron ionization of 13 eV was chosen to minimize fragmentation effects. The overall error of the measured mole fraction was estimated as 10–15%.

To simulate species profiles of weak flames in MFR, the flow field in the reactor channel in the weak flame regime was modeled as a one-dimensional steady-state reactive flow without a boundary layer. Computations were conducted using PREMIX in ANSYS Chemkin-Pro. A heat convection term between the gas and the reactor wall was added to the gas-phase energy equation [4]. The measured wall temperature profile was used to evaluate the heat convection term. The validity of this

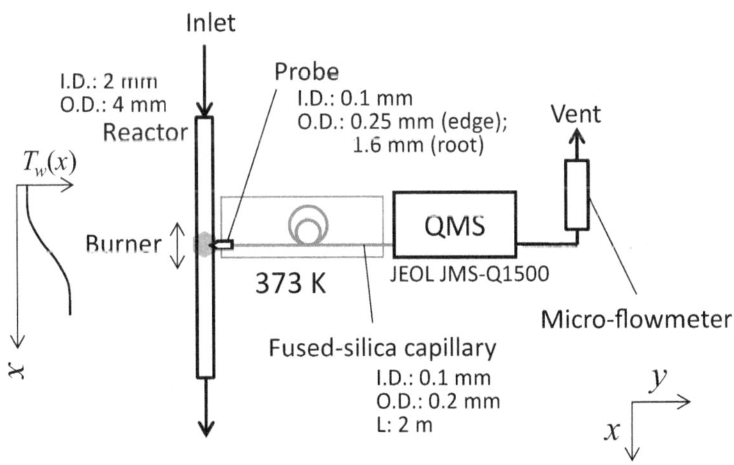

Fig. 39.2 Schematic of species measurement using MFR and QMS [5]

1-D modeling for weak flames has been confirmed for various hydrocarbons [9–13]. Because a radical quenching effect on the quartz surface was confirmed as negligibly small in the heated microchannel [15, 16], no radical quenching mechanism was considered.

39.3 Chemical Kinetic and Radiation Models

The nitrogen chemistry developed by Miller and Bowman [17] was chosen as a starting mechanism. Four steps of improvements were made. The first step of improvements was to adopt the H_2/NHx/N_2O/NO_2/NNH chemistry studied by Mathieu and Petersen [3]. They adopted a series of updates of the H_2/CO chemistry [18], NHx/HCN chemistry [19], N_2O chemistry [20], NO_2 chemistry [21, 22], and NNH chemistry [23]. Results show that their model greatly improved predictions of ignition delay times of highly diluted NH_3/O_2/Ar mixtures at high temperatures [3]. The second step of improvements was to adopt the chemistry of N_2H_x except NNH: reactions related to N_2H_4, N_2H_3, and N_2H_2 studied by Konnov [24]. As discussed later, the N_2H_x chemistry played an important role in ammonia/air reactivity at low temperatures. The third step of improvements was to adopt an accurate and consistent set of thermochemical properties calculated by Bugler et al. [25]. The final step of improvements was to update 30 rate constants, which are mostly related to N_2H_x species, from data available in the literature [26–31]. Details are presented in an earlier report [6]. The developed mechanism comprises 33 species and 232 reactions.

Because of slow laminar flame speeds of ammonia/air mixtures, radiation heat loss was considered for computations of 1-D freely propagating premixed flames. The optically thin model [32] was used to evaluate radiation heat loss. Planck mean absorption coefficients (PMACs) of CO_2, CO, CH_4, H_2O, NO, N_2O, and NH_3 were taken from an earlier report [33], as estimated using HITRAN 2008 database [34]. PMACs expressed by polynomial functions of temperature, which are compatible with the Chemkin-Pro thermodynamic format, are listed in a report of an earlier study [7]. Computations of 1-D freely propagating premixed flames were made in adiabatic and radiative conditions. The computational domain was set to have 10 cm length. The temperature midpoint was fixed at $x = 5$ cm. Computed laminar flame speeds were compared with data reported in literature.

39.4 Results and Discussion

Figure 39.3 shows flame images and a regime diagram of flame dynamics observed in MFR for a stoichiometric ammonia/air mixture. Those for a stoichiometric methane/air mixture [10, 14] are also shown in the figure for the comparison. As observed earlier for methane, ammonia shows three flame dynamics in MFR depending on the inlet mean flow velocities, normal flames, flames with repetitive extinction and ignition (FREI), and weak flames. Notable differences are apparent between ammonia and methane flames. The thickness of ammonia normal flames (approx. 0.7 mm) is much greater than that of methane normal flames (approx. 0.2 mm), as depicted in Fig. 39.3a. Ammonia normal flames are stabilized in higher temperature regions than methane normal flames are. Because of low laminar flame speeds of ammonia, ammonia/air mixtures require high preheating to form normal flames. Ammonia FREI appears to be "thinner" than methane FREI (Fig. 39.3b), which is caused mainly by the extinction of ammonia flames in the higher temperature region than that of methane flames. The thickness of ammonia weak flames (approx. 4 mm) is much wider than that of methane weak flames (approx. 1 mm), as depicted in Fig. 39.3c. Ammonia weak flames are located at 1270 K. Methane weak flames are located at 1230 K, which indicates lower reactivity of ammonia than that of methane. Methane weak flames are visible only at very low flow velocities (less than 3 cm/s) while ammonia weak flames are visible at higher flow velocities (less than 15 cm/s). This trend of a critical velocity condition between the FREI and weak flame regimes is associated with lower heat of formation for ammonia than that for methane.

Figure 39.4 portrays the H_2O mole fraction profiles for an ammonia weak flame obtained from measurements and computations. The weak flame image is included in the figure as a reference. Five reaction mechanisms were used in computations (Konnov [24]; Mathieu [3]; Tian [35]; Lindstedt [36]; and GRI 3.0 [37]). Experiments demonstrate that an ammonia weak flame is located around 1270 K. Also, H_2O, a major product of ammonia combustion, is formed there. However, the mechanisms except Konnov mechanism do not predict the formation of ammonia weak flames within the computational domain. Konnov mechanism predicts H_2O formation at lower temperatures than measurements, indicating that Konnov mechanism overpredicts ammonia reactivity.

A major difference exists in modeling of N_2H_x reactions between Konnov mechanism and other mechanisms. In addition to ammonia decomposition [38], Konnov and De Ruycy investigated hydrazine decomposition and developed N_2H_x chemistry [39]. They also investigated effects of N_2H_x chemistry on NO formation and identified the importance of N_2H_3 reactions in NO formation at low temperatures (below 1500 K) [40]. Konnov mechanism has been developed with incorporation of fundamental findings related to N_2Hx chemistry to improve predictions of NOx formation. Another important fact is that most NOx studies have used low concentrations of ammonia (typically below 0.1%), whereas ammonia/air mixtures contain much higher concentrations of ammonia (approx. 22% at stoichiometry), which indicates the presence of rich NHx radicals in ammonia/air combustion, leading to recombination reactions forming N_2H_x species. Based on these considerations, N_2H_4 signal in the reaction zone of an ammonia weak flame was detected using QMS [6]. Therefore, the present mechanism has markedly updated N_2H_x chemistry, as described in Sect. 39.3.

Figure 39.5 shows validation of the present mechanism using measured species data for ammonia weak flames at equivalence ratios, ϕ, of 0.8, 1.0, and 1.2. Computed

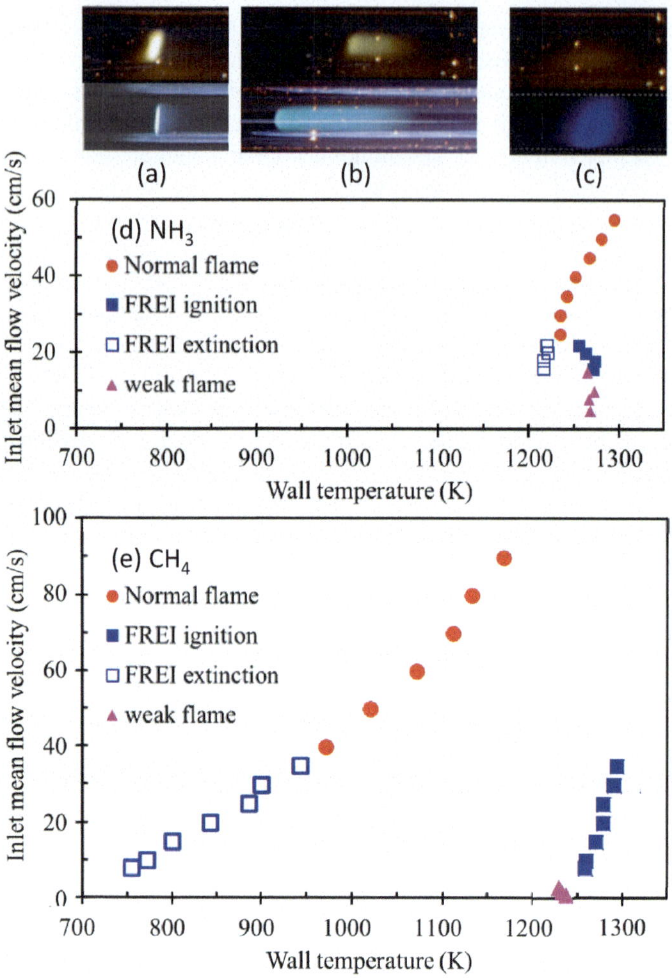

Fig. 39.3 Comparisons of experimentally obtained results between ammonia and methane. Flame images of **a** normal flames, **b** FREI and **c** weak flames (ammonia, upper side; methane, lower side). **d** and **e**: wall temperatures at flame positions (stoichiometric fuel/air mixture; atmospheric pressure; $T_{w,\max} = 1300$ K) [5]

results with Konnov mechanism are also presented in the figure. The present mechanism well predicts mole fraction profiles of the reactants (NH_3 and O_2) and the product (H_2O) at the three equivalence ratios. Measurements show N_2O as an intermediate in the reaction zone of ammonia weak flames. The present mechanism slightly overpredicts the peak values in the measured profiles of the N_2O mole fraction, but it well predicts the final values. Measurements show that NO is formed near the region where N_2O is consumed at $\phi = 0.8$ and 1.0. The present mechanism predicts earlier NO formation than measurements at $\phi = 0.8$ and 1.0.

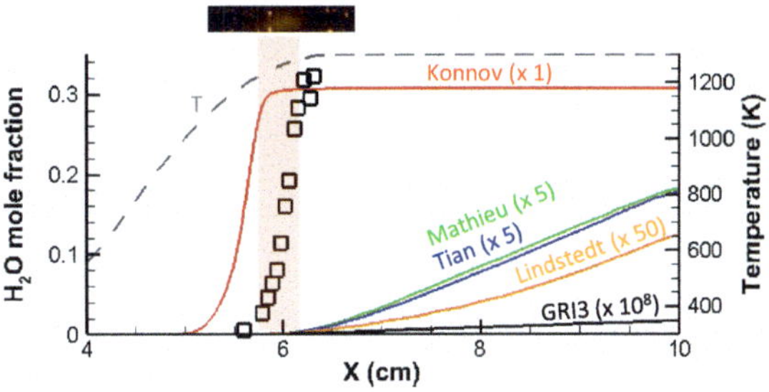

Fig. 39.4 H_2O mole fraction profiles for an ammonia weak flame in experiments and computations (10 cm/s inlet mean flow velocity; atmospheric pressure; $T_{w,\max} = 1300$ K) [5]

However, the present mechanism well predicts the final values of NO at the three equivalence ratios. Compared with the other mechanisms, the present mechanism satisfactorily predicts ammonia weak flame structures. In addition to the validation for ammonia weak flames, the present mechanism was validated for ignition delay times of $H_2/O_2/N_2O/Ar$ mixtures [20], $H_2/O_2/NO_2/Ar$ mixtures [41], and $NH_3/O_2/Ar$ mixtures [3], and for half-life times in N_2H_4 pyrolysis [42], as described in an earlier report [6].

The present mechanism was also validated for laminar flame speeds of ammonia/air mixtures. Figure 39.6 portrays a comparison of adiabatic and radiative laminar flame speeds along with data from the literature for measured laminar flame speeds [43–47]. The ends of the solid line (radiative condition) denote the computational flammability limits of the 1-D planar premixed flames. The computed radiative laminar flame speeds are slower than adiabatic ones for a whole range of equivalence ratios. For most hydrocarbons, radiation heat loss affects laminar flame speeds only at equivalence ratios near the extinction limits [48]. For ammonia, however, results confirmed that radiation heat loss needs to be considered in comparisons between computational and experimentally obtained results because radiation heat loss is inevitable in experiments. The present mechanism well predicts laminar flame speeds of ammonia/air mixtures in fuel-lean conditions, and slightly faster laminar flame speeds than those reported in the literature for stoichiometric and fuel rich conditions.

Figure 39.7 presents emission characteristics of ammonia/air premixed flames in the adiabatic and radiative conditions. Results show that radiative heat loss affects emission characteristics of ammonia/air premixed flames. A low emission window of NH_3 and NO is around $\phi = 1.2$ in the adiabatic condition, but it shifts to around $\phi = 1.1$ in the radiative condition. Because the thermal decomposition of unburned NH_3 is suppressed by the lower radiative flame temperature, NH_3 emissions in the radiative condition become large at a lower equivalence ratio than those in the adiabatic condition. Therefore, NO emissions in the radiative condition approach near-zero at a lower equivalence ratio than those in the adiabatic condition. The emission window has been applied for the successful demonstration of zero-CO_2 and low-NOx emission in power generation with a two-staged, rich-lean combustor [49], namely, low

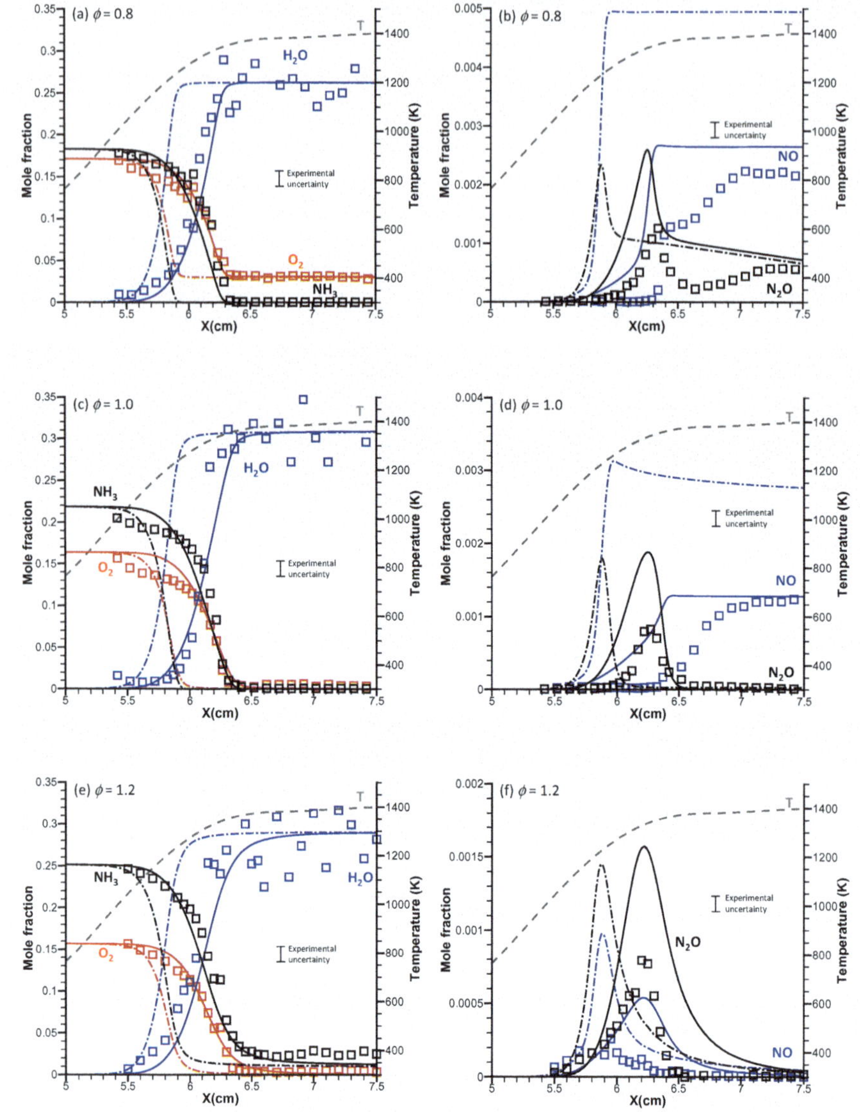

Fig. 39.5 Measured and computational species profiles of weak flames for ammonia/air mixtures at $\phi = 0.8$, 1.0, and 1.2 (symbols, experimentally obtained values; solid lines, computations using the present mechanism; dash-dotted lines, computations using Konnov mechanism; broken lines, wall-temperature profile; 10 cm/s inlet mean flow velocity; atmospheric pressure; $T_{w,\max} = 1400$ K). [6]

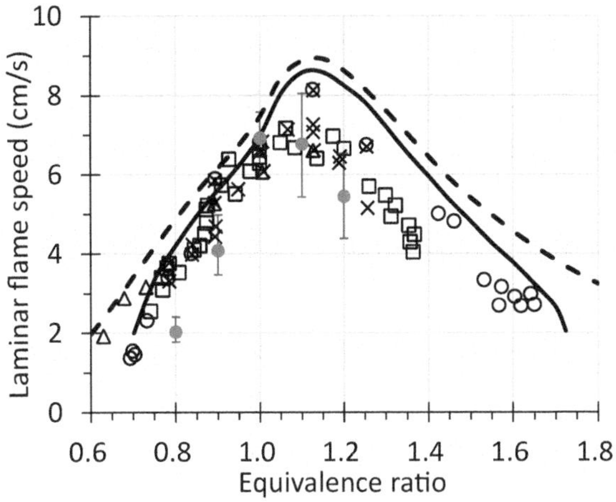

Fig. 39.6 Computed adiabatic (dashed lines) and radiative (solid lines) flame speeds and data obtained from the literature for NH_3/air mixtures [7]. Symbols denote experimentally obtained results from □, Zakaznov et al. [43]; ×, Takizawa et al. [44]; Δ, Pfahl et al. [45]; ○, Ronney [46]; and ●, Hayakawa et al. [47]

emissions of NOx and unburned NH_3 at the emission window in the first stage and combustion with unburned H_2 and secondary air in the second stage. The present computations showed that radiation heat loss affects an equivalence ratio of the emission window, indicating the importance of accurate estimation of flame temperature with consideration of heat loss for improving ammonia combustors.

39.5 Conclusions

The development and validation of chemical kinetics for ammonia combustion under the "Energy Carriers" projects by PRESTO and SIP were summarized in this chapter. Chemical kinetic analysis using ammonia weak flames in MFR highlighted the importance of N_2H_x chemistry for low-temperature reactivity of ammonia. The present mechanism was validated using species data of ammonia weak flames and data from the literature. Regarding validation for ammonia/air laminar premixed flames, the importance of radiation heat loss in laminar flame speeds and emission characteristics was identified. After the "Energy Carriers" projects, extensive experimental studies to assess ignition and flame characteristics have been reported in addition to theoretical studies of chemical kinetics for ammonia and ammonia/hydrocarbon blends. The ammonia mechanism needs to be further improved and validated in light of recently obtained results.

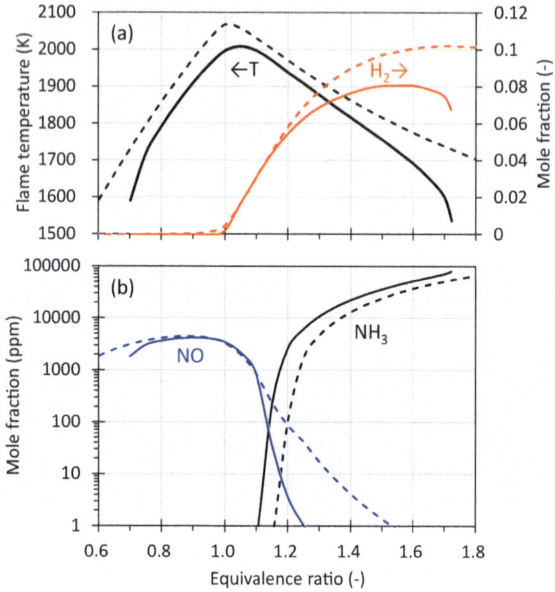

Fig. 39.7 Equivalence ratio dependence of **a** flame temperature and H_2 mole fraction and **b** NH_3 and NO mole fractions at the end of the computational domain in adiabatic (dashed) and radiative (solid) conditions [7]

Acknowledgements Studies introduced into this chapter were supported by Precursory Research for Embryonic Science and Technology (PRESTO), Japan Science and Technology Agency (JPMJPR1343) and Council for Science, Technology and Innovation (CSTI), Cross-ministerial Strategic Innovation Promotion Program (SIP), "Energy Carriers" (Funding agency: the Japan Science and Technology Agency (JST)).

References

1. Smoot LD, Hill SC, Xu H (1998) NO*x* control through reburning. Prog Energy Combust Sci 24:385–408
2. Longwell JP, Rubin ES, Wilson J (1995) Coal: energy for the future. Prog Energy Combust Sci 21:269–360
3. Mathieu O, Petersen EL (2015) Experimental and modeling study on the high-temperature oxidation of ammonia and related NO*x* chemistry. Combust Flame 162:554–570
4. Maruta K, Kataoka T, Kim NI, Minaev S, Fursenko R (2005) Characteristics of combustion in a narrow channel with a temperature gradient. Proc Combust Inst 30:2429–2436
5. Nakamura H, Hasegawa S (2017) Combustion and ignition characteristics of ammonia/air mixtures in a micro flow reactor with a controlled temperature profile. Proc Combust Inst 36:4217–4226
6. Nakamura H, Hasegawa S, Tezuka T (2017) Kinetic modeling of ammonia/air weak flames in a micro flow reactor with a controlled temperature profile. Combust Flame 185:16–27

7. Nakamura H, Shindo M (2019) Effects of radiation heat loss on laminar premixed ammonia/air flames. Proc Combust Inst 37:1741–1748
8. Minaev S, Maruta K, Fursenko R (2007) Nonlinear dynamics of flame in a narrow channel with a temperature gradient. Combust Theory Model 11:187–203
9. Nakamura H, Takahashi H, Tezuka T, Hasegawa S, Maruta K, Abe K (2016) Effects of CO-to-H_2 ratio and diluents on ignition properties of syngas examined by weak flames in a micro flow reactor with a controlled temperature profile. Combust Flame 172:94–104
10. Kamada T, Nakamura H, Tezuka T, Hasegawa S, Maruta K (2014) Study on combustion and ignition characteristics of natural gas components in a micro flow reactor with a controlled temperature profile. Combust Flame 161:37–48
11. Kikui S, Nakamura H, Tezuka T, Hasegawa S, Maruta K (2016) Study on combustion and ignition characteristics of ethylene, propylene, 1-butene and 1-pentene in a micro flow reactor with a controlled temperature profile. Combust Flame 163:209–219
12. Hori M, Yamamoto A, Nakamura H, Tezuka T, Hasegawa S, Maruta K (2012) Study on octane number dependence of PRF/air weak flames at 1–5 atm in a micro flow reactor with a controlled temperature profile. Combust Flame 159:959–967
13. Suzuki S, Hori M, Nakamura H, Tezuka T, Hasegawa S, Maruta K (2013) Study on cetane number dependence of diesel surrogates/air weak flames in a micro flow reactor with a controlled temperature profile. Proc Combust Inst 34:3411–3417
14. Tsuboi Y, Yokomori T, Maruta K (2009) Lower limit of weak flame in a heated channel. Proc Combust Inst 32:3075–3081
15. Kizaki Y, Nakamura H, Tezuka T, Hasegawa S, Maruta K (2015) Effect of radical quenching on CH_4/air flames in a micro flow reactor with a controlled temperature profile. Proc Combust Inst 35:3389–3396
16. Saiki Y, Suzuki Y (2013) Effect of wall surface reaction on a methane-air premixed flame in narrow channels with different wall materials. Proc Combust Inst 34:3395–3402
17. Miller JA, Bowman CT (1989) Mechanism and modeling of nitrogen chemistry in combustion. Prog Energy Combust Sci 15:287–338
18. Kéromnès A, Metcalfe WK, Heufer KA, Donohoe N, Das AK, Sung CJ, Herzler J, Naumann C, Griebel P, Mathieu O, Krejci MC, Petersen EL, Pitz WJ, Curran HJ (2013) An experimental and detailed chemical kinetic modeling study of hydrogen and syngas mixture oxidation at elevated pressures. Combust Flame 160:995–1011
19. Dagaut P, Glarborg P, Alzueta MU (2008) The oxidation of hydrogen cyanide and related chemistry. Prog Energy Combust Sci 34:1–46
20. Mathieu O, Levacque A, Petersen EL (2012) Effects of N_2O addition on the ignition of H_2–O_2 mixtures: experimental and detailed kinetic modeling study. Int J Hydrog Energy 37:15393–15405
21. Dayma G, Dagaut P (2006) Effect of air contamination on the combustion of hydrogen-effect of NO and NO_2 addition on hydrogen ignition and oxidation kinetics. Combust Sci Technol 178:1999–2024
22. Sivaramakrishnan R, Brezinsky K, Dayma G, Dagaut P (2007) High pressure effects on the mutual sensitization of the oxidation of NO and CH_4–C_2H_6 blends. Phys Chem Chem Phys 9:4230–4244
23. Klippenstein SJ, Harding LB, Glarborg P, Miller JA (2011) The role of NNH in NO formation and control. Combust Flame 158:774–789
24. Konnov AA (2009) Implementation of the NCN pathway of prompt-NO formation in the detailed reaction mechanism. Combust Flame 156:2093–2105
25. Bugler J, Somers KP, Simmie JM, Güthe F, Curran HJ (2016) Modeling nitrogen species as pollutants: thermochemical influences. J Phys Chem A 120:7192–7197
26. Dean AM, Bozzelli JW (1999) Combustion chemistry of nitrogen. In: Gardiner WC Jr (ed) Chapter 2 in: gas-phase combustion chemistry, 2nd edn. Springer, pp 125–341
27. Daimon Y, Terashima H, Koshi M (2014) Chemical kinetics of hypergolic ignition in N_2H_4/N_2O_4-NO_2 gas mixtures. J Prop Power 30:707–716

28. Raghunath P, LinYH LMC (2014) Ab initio chemical kinetics for the $N_2H_4 + NOx$ ($x = 1$–3) reactions and related reverse processes. Comput Theor Chem 1046:73–80
29. Chuang YY, Truhlar DG (1997) Improved dual-level direct dynamics method for reaction rate calculations with inclusion of multidimensional tunneling effects and validation for the reaction of H with trans-N_2H_2. J Phys Chem A 101:3808–3814
30. Hanson RK, Salimian S (1984) Survey of rate constants in the N/H/O system. In: Gardiner WC Jr (ed) Chapter 6 in: Combustion chemistry. Springer, pp 361–421
31. Inomata S, Washida N (1999) Rate constants for the reactions of NH_2 and HNO with atomic oxygen at temperatures between 242 and 473 K. J Phys Chem A 103:5023–5031
32. Barlow RS, Karpetis AN, Frank JH, Chen JY (2001) Scalar profiles and NO formation in laminar opposed-flow partially premixed methane/air flames. Combust Flame 127:2102–2118
33. Modest MF (2013) Radiative heat transfer, 3rd edn. Academic Press, pp 367–369
34. Rothman LS, Gordon IE, Barbe A, Benner DC, Bernath PF, Birk M, Boudon V, Brown LR, Campargue A, Champion JP, Chance K, Coudert LH, Dana V, Devi VM, Fally S, Flaud JM, Gamache RR, Goldman A, Jacquemart D, Kleiner I, Lacome N, Lafferty WJ, Mandin JY, Massie ST, Mikhailenko SN, Miller CE, Moazzen-Ahmadi N, Naumenko OV, Nikitin AV, Orphal J, Perevalov VI, Perrin A, Predoi-Cross A, Rinsland CP, Rotger M, Šimečková M, Smith MAH, Sung K, Tashkun SA, Tennyson J, Toth RA, Vandaele AC, VanderAuwera J (2009) The HITRAN 2008 molecular spectroscopic database. J Quant Spectrosc Radiat Transfer 110:533–572
35. Tian Z, Li Y, Zhang L, Glarborg P, Qi F (2009) An experimental and kinetic modeling study of premixed $NH_3/CH_4/O_2$/Ar flames at low pressure. Combust Flame 156:1413–1426
36. Lindstedt RP, Lockwood FC, Selim MA (1994) Detailed kinetic modelling of chemistry and temperature effects on ammonia oxidation. Combust Sci Technol 99:253–276
37. Smith GP, Golden DM, Frenklach M, Moriarty NW, Eiteneer B, Goldenberg M, Bowman CT, Hanson RK, Song S, Gardiner WC Jr, Lissianski VV, Qin Z.http://www.me.berkeley.edu/gri_mech/. Accessed on 13 Mar 2017
38. Konnov AA, Ruyck JD (2000) Kinetic modeling of the thermal decomposition of ammonia. Combust Sci Technol 152:23–37
39. Konnov AA, Ruyck JD (2001) Kinetic modeling of the decomposition and flames of hydrazine. Combust Flame 124:106–126
40. Konnov AA, Ruyck JD (2001) A possible new route for NO formation via N_2H_3. Combust Sci Technol 168:1–46
41. Mathieu O, Levacque A, Petersen EL (2013) Effects of NO_2 addition on hydrogen ignition behind reflected shock waves. Proc Combust Inst 34:633–640
42. Michel KW, Wagner HGG (1965) The pyrolysis and oxidation of hydrazine behind shock waves. Symp Int Combust 10:353–364
43. Zakaznov FZ, Kursheva LA, Felina ZI (1978) Determination of normal flame velocity and critical diameter of flame extinction in ammonia-air mixture. Combust Explos Shock Waves 14:710–713
44. Takizawa K, Takahashi A, Tokuhashi K, Kondo S, Sekiya A (2008) Burning velocity measurements of nitrogen-containing compounds. J Hazard Mater 155:144–152
45. Pfahl UJ, Ross MC, Shepherd JE, Pasamehmetoglu KO, Unal C (2000) Flammability limits, ignition energy, and flame speeds in H_2–CH_4–NH_3–N_2O–O_2–N_2 mixtures. Combust Flame 123:140–158
46. Ronney PD (1988) Effect of chemistry and transport properties on near-limit flames at microgravity. Combust Sci Technol 59:123–141
47. Hayakawa A, Goto T, Mimoto R, Arakawa Y, Kudo T, Kobayashi H (2015) Laminar burning velocity and Markstein length of ammonia/air premixed flames at various pressures. Fuel 159:98–106

48. Yu H, Han W, Santner J, Gou X, Sohn CH, Ju Y, Chen Z (2014) Radiation-induced uncertainty in laminar flame speed measured from propagating spherical flames. Combust Flame 161:2815–2824
49. Kurata O, Iki N, Matsunuma T, Inoue T, Tsujimura T, Furutani H, Kobayashi H, Hayakawa (2017) A performances and emission characteristics of NH_3–air and NH_3 CH_4–air combustion gas-turbine power generations. Proc Combust Inst 36:3351–3359

Part VIII
Ammonia Co-firing Technology Applicable to Pulverized Coal Fired Boilers

Chapter 40
Applicability of Ammonia Co-firing Technology in Pulverized Coal-Fired Boilers

Masayoshi Kimoto, Akira Yamamoto, Yasushi Ozawa, and Saburo Hara

Abstract Ammonia is a potential substitute for fossil fuels because it does not discharge carbon dioxide and is easily handled by liquefaction. One of the possible applications for the direct use of ammonia as a fuel is combustion in thermal power plants. In particular, the co-firing of ammonia in coal-fired power plants seems to have a relatively great advantage in the suppression of greenhouse gases because coal is one of the main emission sources of carbon dioxide. However, the concentration of nitrogen oxides (NO_x), which are typical atmospheric pollutants, in the flue gas would considerably increase owing to the oxidation of ammonia. In this study, we examined the influence of ammonia co-firing with coal on NO_x emissions and investigated methods to reduce NO_x emissions using two coal combustion test furnaces, a single-burner test furnace and a multi-burner test furnace. When the ammonia co-firing rate was increased up to 20% of the low heat value (LHV), there was no significant increase in NO_x concentration at the furnace exit. When ammonia is injected into the pulverized coal combustion flame through the side ports of the furnace wall using a single-burner test furnace, the NO_x concentration could be reduced at the appropriate location of ammonia injection. As a result of examining the effect of changing the burner stage for injecting ammonia using a multi-burner test furnace, it became clear that the NO_x concentration could be reduced by injecting ammonia concentrated in the lower burner rather than injecting ammonia evenly throughout all the burners.

Keywords Ammonia · Co-firing · Pulverized coal-fired boiler

M. Kimoto (✉) · A. Yamamoto · Y. Ozawa · S. Hara
Central Research Institute of Electric Power Industry, 2-6-1 Nagasaka, Yokosuka, Kanagawa 240-0196, Japan
e-mail: mkimoto@criepi.denken.or.jp

K. Aika and H. Kobayashi (eds.), *CO_2 Free Ammonia as an Energy Carrier*,
https://doi.org/10.1007/978-981-19-4767-4_40

40.1 Introduction

One of the means of directly using ammonia as fuel is to use co-firing at an existing thermal power plant. Currently, the use of ammonia in thermal power plants is limited to that of a denitration agent for flue gas denitration equipment. Therefore, the use of ammonia is minimal compared to fuel, and it has not been used as a fuel. The biggest concern when using ammonia as a fuel is that the combustion reaction converts the nitrogen content in ammonia to NO_x, which increases NO_x emissions.

In this study, we used a single-burner furnace with a single pulverized coal burner and a multi-burner furnace with pulverized coal burners, similar to those of a single-burner furnace, arranged in three stages in a vertical shape. An experimental study was conducted on the effect of increasing the co-firing rate of ammonia up to 20% on NO_x emissions, and on the measures to control NO_x emissions applicable to existing pulverized coal-fired boilers. These results are introduced in this study.

40.2 Test Equipment and Methods

40.2.1 Overview of Equipment for Combustion Tests

1. Single-burner furnace

Figure 40.1 shows a schematic diagram of a single-burner furnace. The furnace used in the study was a horizontal cylindrical water-cooled furnace, and the inner wall of the furnace was lined with a refractory material with a thickness of 75 mm. The length was 8.0 m and the inner diameter was 0.85 m. To allow low NO_x combustion, multiple air injection ports for two-stage combustion were installed. In this test, a pair of the ports located approximately 3 m from the burner was used. On the sides of the furnace, 17 probe insertion ports for measuring the temperature and gas concentration inside the furnace were installed at intervals of approximately 0.4 m to the mainstream direction. The rated load of the furnace is 760 kW based on the lower calorific value of coal, which corresponds to a standard coal combustion amount of 100 kg/h. For the pulverized coal burner, we used the CI-α burner [1], which is a low NO_x burner jointly developed by IHI Corporation and our research team.

Ammonia was removed from one of the two 50-kg liquid ammonia containers, decompressed, vaporized by a hot water vaporizer (vapor capacity: maximum 30 kg/h), depressurized again, and then flowed to the upstream side of the furnace. After adjusting the flow rate to a predetermined amount with the flow control valves, it was injected into the furnace. The co-firing rate (injection rate) of ammonia can be set up to 20% (152 kW, 29.6 kg/h) based on the lower calorific value of ammonia for the rated input heat for the furnace (760 kW).

Ammonia injection into the furnace can be performed by two methods: (a) by injecting from the pulverized coal burner, and (b) by injecting into the pulverized

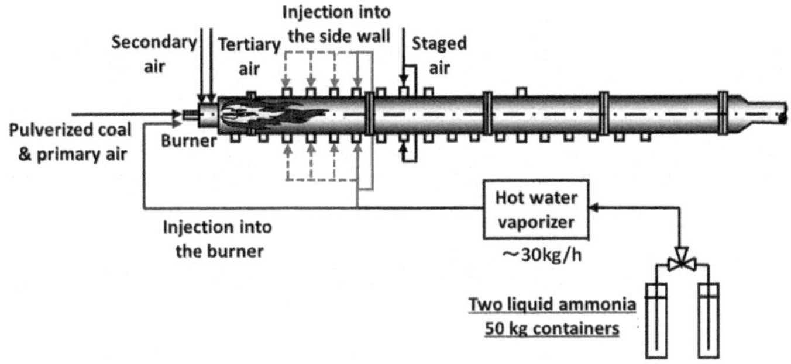

Fig. 40.1 Schematic diagram of a single-burner furnace

coal flame using side ports of the furnace behind the burner (upstream from the location of air injection for two-stage combustion).

2. Multi-burner furnace

Figure 40.2 shows a schematic diagram of the multi-burner furnace. The multi-burner furnace is a test equipment with the same equipment configuration as an actual pulverized coal-fired power plant. The furnace is a steel vertical water-cooled furnace with a length of 1.9 m, width of 0.9 m, and height of 9.7 m. On the front side, pulverized coal burners of the same type as the single-burner furnace are arranged vertically in three stages at 1-m intervals. The rated load of the multi-burner furnace was 2.39 MW, and the amount of coal combustion was approximately 300 kg/h (100 kg/h per burner). To allow low NO_x combustion, air injection ports for two-stage combustion were installed 3 m from the middle burner. Under standard conditions, the two-stage combustion rate was set to 30%.

Ammonia was removed as liquid ammonia from a 500 kg horizontal ammonia container, decompressed, and then vaporized by a hot water vaporizer (vapor capacity: maximum 100 kg/h). After the vaporized ammonia is depressurized again, it is branched into three burners and input through the flow control valves. The

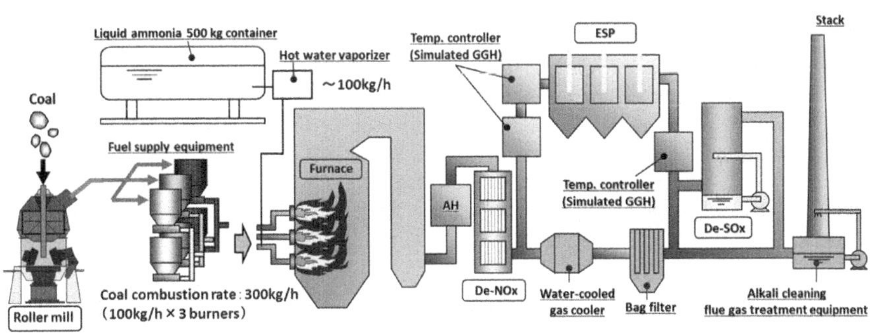

Fig. 40.2 Schematic diagram of the multi-burner furnace

Table 40.1 Fuel properties of test coal

Coal			MW	AL	LM	MO1	MO2	NL
Proximate analysis	Ash	[wt%]	3.2	10.6	4.5	14.0	12.6	15.1
	Volatile matter	[wt%]	51.6	46.7	43.7	36.0	35.4	27.7
	Fixed carbon	[wt%]	45.2	42.7	51.8	50.0	52.0	57.2
	Fuel ratio	–	0.88	0.91	1.19	1.39	1.47	2.06
Ultimate analysis	C	[wt%]	68.4	68.4	75.9	70.1	73.0	71.8
	H	[wt%]	4.89	5.40	5.57	4.82	4.90	4.22
	N	[wt%]	1.33	1.20	1.29	1.64	1.74	1.47
	O	[wt%]	22.1	14.0	12.3	9.0	7.4	7.2
	S	[wt%]	0.19	0.66	0.43	0.49	0.44	0.31
Low heating value		[MJ/kg]	25.4	28.1	30.0	28.4	29.2	27.4

ammonia co-firing rate can be supplied up to 20% (478 kW, 92.8 kg/h) similar to a single-burner furnace. Ammonia gas was blown into the furnace through a supply pipe inserted at the center of the pulverized coal burner.

40.2.2 Methods, Combustion Conditions, and Coal Properties

When co-firing ammonia, the amount of pulverized coal input was reduced by the amount of ammonia input so that the amount of heat input to the furnace would be constant even if the co-firing rate changed. The combustion conditions were an O_2 concentration at the furnace exit of 4% and a two-stage combustion rate of 30%. O_2, CO, CO_2, NO_x, SO_2, and N_2O were measured as exhaust gas components emitted from the furnace exit. Table 40.1 lists the fuel properties of test coal. Among the coal commonly used in pulverized coal-fired power plants, five coal types (three types of MO coal with different lots) with different fuel ratios (weight ratio of fixed carbon to volatile matter) were used.

40.3 Results and Discussion

40.3.1 NO_x Emission Characteristics up to a Maximum Ammonia Co-firing Rate of 20%

When ammonia is co-fired in a pulverized coal combustion plant, an increase in NO_x concentration is expected owing to the oxidation of nitrogen in the ammonia. If a significant increase occurs, removing NO_x concentration from the existing flue gas treatment device becomes difficult. Therefore, we investigated the basic NO_x

emission characteristics when ammonia was injected from a pulverized coal burner to increase the co-firing rate of ammonia. Figure 40.3 shows the relationship between the ammonia co-firing rate in the multi-burner furnace and the NO_x concentration at the furnace exit. The higher the co-firing rate, the higher the NO_x concentration at the furnace exit, but even in the case of 20% ammonia co-firing, the increase was only approximately 1.9 times higher than that of the single combustion of coal. Note that the NO_x concentration in the single-burner furnace did not increase as much as in the multi-burner furnace, and the NO_x concentration in the 20% ammonia co-firing increased by approximately 20% compared to the pulverized coal combustion [2].

Figure 40.4 shows a comparison of the combustion efficiencies of combustible components in a multi-burner furnace. The concentration of unburned matter in the ash was higher when 20% ammonia was co-fired than when coal was exclusively burned. In contrast, the combustion efficiency of combustible components including ammonia was higher in the case of 20% ammonia co-firing than in the case of the single combustion of coal.

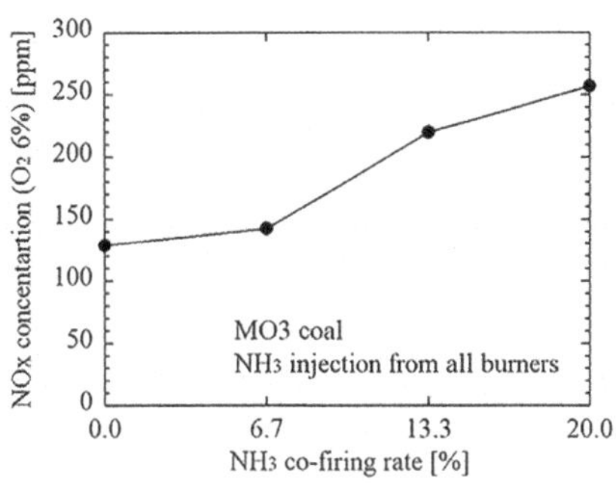

Fig. 40.3 Relationship between the ammonia co-firing rate and NO_x concentration

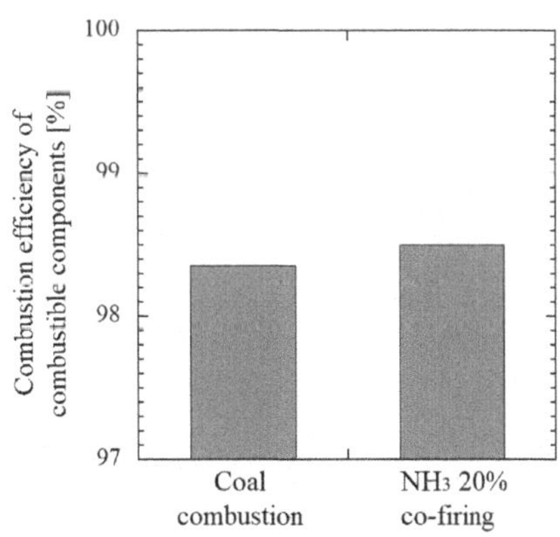

Fig. 40.4 Comparison of the combustion efficiencies of combustible components

40.3.2 Ammonia Injection Method that Enables NO_x Reduction

1. Low NO_x combustion by injecting ammonia into the pulverized coal combustion flame

In our previous research, we developed a method of injecting a secondary fuel for NO_x reduction into a pulverized coal combustion flame as a low NO_x combustion technology [3]. For saturated hydrocarbon gas fuels (methane, propane, and hydrogen) used as secondary fuels and coal with high volatile content, we co-fired up to 25% on a calorific basis. Regarding ammonia, the co-firing rate was as low as 1% or less [4], partly because we aimed to determine the effect of denitration in the flame. The NO_x emission characteristics when large amounts of co-firing are performed are yet to be elucidated.

Figure 40.5 shows the NO_x concentration at the furnace exit when ammonia is injected behind the burner through the side ports of the single-burner furnace with an ammonia co-firing rate of 20%. When the location of ammonia injection was set to 0.6 m or 1.0 m from the burner, the NO_x concentration at the furnace exit became lower compared to when the ammonia was injected through the burner. When the location of the injection was moved further downstream, the NO_x concentration at the furnace exit increased. The characteristics of these changes are common characteristics among multiple coal types [5].

The location 0.6 m or 1.0 m from the burner corresponds to the location where the NO_x concentration in the furnace peaks and the O_2 concentration is the lowest during single combustion of coal. By injecting ammonia at that location, the generation of fuel-NO_x derived from ammonia is suppressed under a strong reducing atmosphere, and there seems to be a local effect of denitration in the flame by ammonia as well.

Figure 40.6 shows the concentrations of unburned carbon in fly ash of the three coal types that were compared under three conditions: when coal was exclusively burned, 20% ammonia was injected from the burner, and 20% ammonia was injected behind the burner. In all coal types, the concentration of unburned carbon in fly ash was slightly higher when injecting ammonia from the burner than in the case of exclusive coal burning, while injecting ammonia behind the burner resulted in the

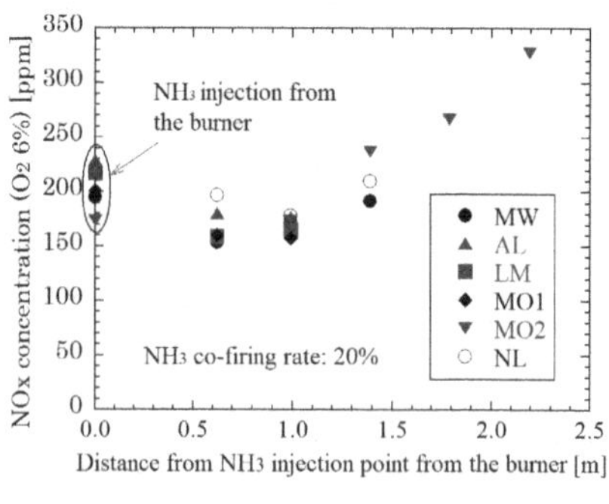

Fig. 40.5 Relationship between the ammonia injection point and NO_x concentration

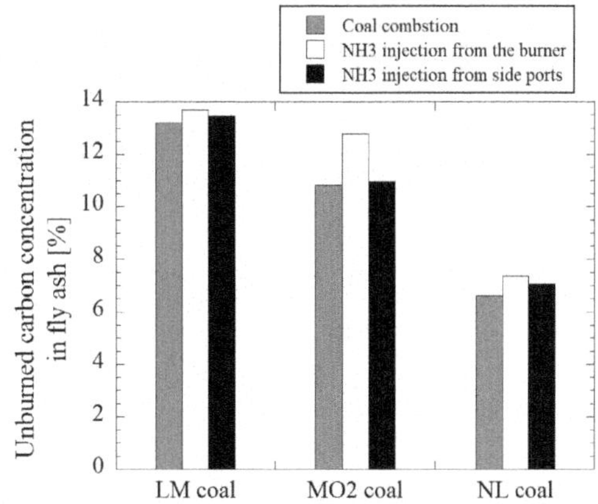

Fig. 40.6 Comparison of the unburned carbon concentration in fly ash

concentration of unburned carbon in fly ash being approximately the same as when coal was exclusively burned. These results suggest that NO_x and unburned carbon in fly ash could be reduced by injecting ammonia at the appropriate location in the pulverized coal combustion flame.

2. Low NO_x combustion by changing the location of the burner to inject ammonia

When ammonia is injected into a furnace with multiple burner stages, injecting only into a certain burner stage is advantageous for the equipment and operation. Therefore, we investigated the effect of changing the burner stage for ammonia injection using a multi-burner furnace. The ammonia injection method is illustrated in Fig. 40.7. As shown in Fig. 40.8, the NO_x concentration was reduced by injecting ammonia concentrated in the lower burner rather than evenly injecting ammonia throughout all the burners. This is because the residence time in the furnace can be lengthened by injecting ammonia into the lower burner stage, and the NO_x reduction effect in the reducing atmosphere is more effective.

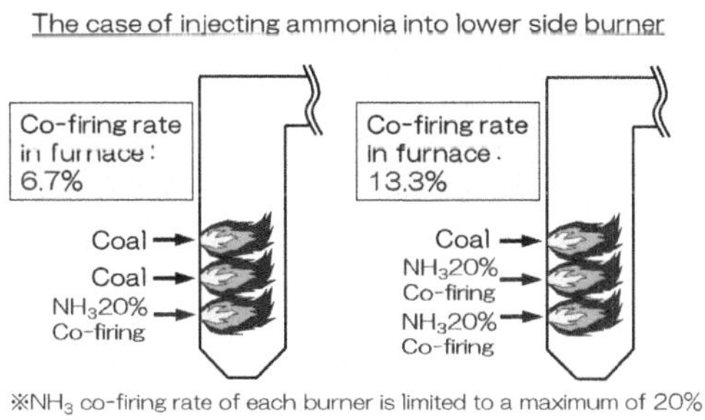

Fig. 40.7 Ammonia injection method in a furnace with multiple burner stages

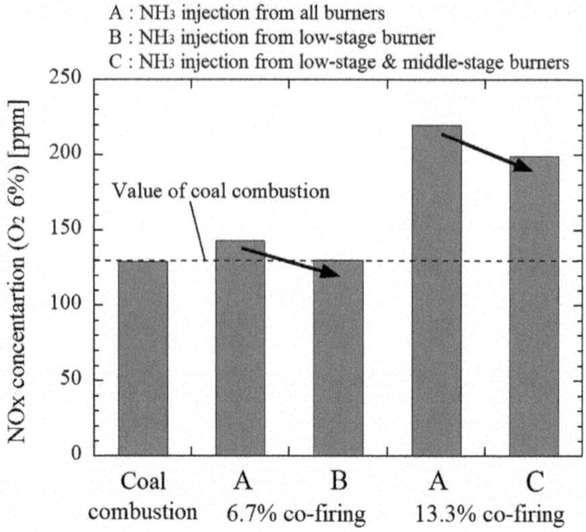

Fig. 40.8 NO_x reduction effect by ammonia injection from lower-stage burner

40.4 Reduction of Greenhouse Gases Accompanied by Ammonia Co-firing

Figure 40.9 shows the results of measuring the CO_2 concentration at the furnace exit. The CO_2 emissions were reduced according to the ammonia co-firing rate. Additionally, we measured the concentration of N_2O, which has a global warming coefficient approximately 300 times that of CO_2, and confirmed that there was no significant increase accompanied by ammonia co-firing.

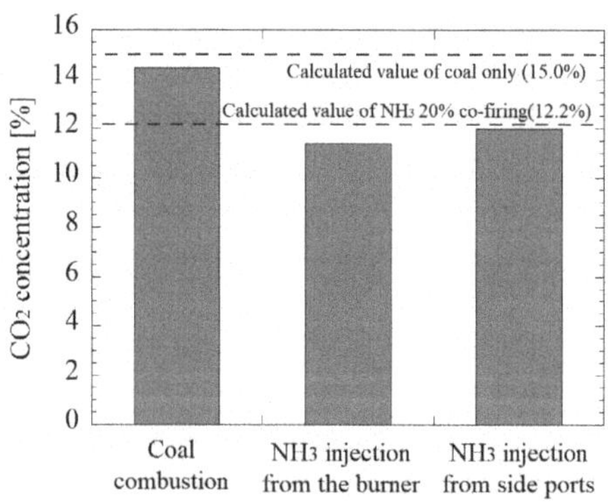

Fig. 40.9 Confirmation of CO_2 reduction effect by ammonia injection

40.5 Conclusions

By conducting an ammonia co-firing test using our equipment for the coal combustion test, we determined the NO_x emission characteristics up to an ammonia co-firing rate of 20%. Furthermore, it was clarified that NO_x may be reduced by injecting ammonia into the pulverized coal combustion flame and injecting ammonia into the lower burner stage. The results suggested that by devising the ammonia injection method, ammonia co-firing is possible without adding a denitration device to the existing pulverized coal-fired boilers.

However, there is still room for reduction of NO_x and optimization of the operation method for the establishment of ammonia co-firing technology in pulverized coal-fired boilers. Hence, we would like to contribute to the early introduction of equipment.

Acknowledgements This study was supported by the first phase of the Strategic Innovation Promotion Program (SIP) "Energy Carrier" of the Japan Science and Technology Agency (JST). We would like to express our deepest gratitude.

References

1. Makino H, Kimoto M, Tsuji H, Kiga T (1999) Development and utilization of advanced low NO_x burner for pulverized coal combustion. In: EPRI-DOE-EPA combined utility air pollutant control symposium
2. Yamamoto A, Kimoto M, Ozawa Y, Hara S (2018) Basic co-firing characteristics of ammonia with pulverized coal in a single burner test furnace. In: 2018 NH_3 fuel conference, Pittsburgh PA, October 31
3. Makino H, Kimoto M, Tanaka T, Nishida S (1993) Development of low NO_x and low ignition loss combustion technology on pulverized coal combustion by secondary fuel injection method for NO_x reduction. In: JSME-ASME international conference on power engineering '93
4. Nishida S, Makino H, Kimoto M, Tanaka T (1990) Development of low NO_x and low ignition loss combustion technology for pulverized coal combustion part 5—NO_x reduction by NH_3 injection. CRIEPI research report: no W90006 (in Japanese)
5. Yamamoto A, Kimoto M, Ozawa Y, Hara S (2018) Effect of coal types on co-firing characteristics of pulverized coal and ammonia in pulverized coal combustion test furnace. In: The 56th Japanese symposium on combustion (in Japanese)

Chapter 41
Development of Pulverized Coal/Ammonia Co-firing Technology

Takamasa Ito, Juwei Zhang, Hiroki Ishii, Toshiyuki Suda, and Toshiro Fujimori

Abstract This paper informs the achievements of the research activity related to the direct combustion of ammonia for the coal-fired power plant. There are some technical issues expected in the ammonia utilization for the coal-fired power plant. One is the increase of the NO concentration in the exhaust gas. Another is the inflection of the local heat flux on the boiler wall. The first issue related to the NO concentration was solved by the experimental approach with the 10 MW thermal input test facility. NO generation was under control by injecting the ammonia into the strong reduction zone, that is created by the coal combustion. Ammonia was well decomposed into the nitrogen and hydrogen in such a strong reduction zone. By this experiment, it was proved that the ammonia can be safely co-fired with coal by 20% in the calorific base. The second issue related to the heat flux on the boiler wall was discussed by the numerical analysis. The size of the boiler in this discussion was 1000 MW which is the biggest size in the present market. Although the gas temperature in the ammonia co-firing is a slightly lower than that in the coal firing, there is not largely different in the local heat flux nor total heat flux between coal firing and ammonia co-firing conditions. For the next step to penetrate this technology into the society, the demonstration in the commercial-scale boiler is demanded.

Keywords Coal · Ammonia · Co-firing · NO

Originally published in Journal of the Combustion Society of Japan, Vol. 61 (2019), pp. 304–08 (in Japanese);

T. Ito (✉) · J. Zhang · T. Fujimori
Technology and Intelligence Integration, IHI Corporation, 1 Shin-nakahara, Isogo-ku, Yokohama, Kanagawa 235-8501, Japan
e-mail: ito9331@ihi-g.com

H. Ishii · T. Suda
Resources, Energy and Environment Business Area, IHI Corporation, 3-1-1 Toyosu, Koto-ku, Tokyo 135-8710, Japan

K. Aika and H. Kobayashi (eds.), *CO_2 Free Ammonia as an Energy Carrier*,
https://doi.org/10.1007/978-981-19-4767-4_41

41.1 Introduction

In this report, the direct use of ammonia in the coal-fired boiler is introduced. In the review of the boiler design, the following three main technical issues should be addressed.

NO_x formation: This is a general issue for the direct use of ammonia as fuel for power generation. Especially for coal combustion, De-NO_x technology is well established. The main discussion to control NO_x formation in the ammonia co-firing will be the diversion of the De-NO_x technology installed in the coal-fired boiler.

Ammonia leakage: It is important that the ammonia is completely consumed in the furnace. If the exhaust gas from the furnace contains residual ammonia, it can become the source of ammonium sulfate. This chemical substance is an adhesive and induces the adhesion of ash particles in the ducts.

Heat absorption in the furnace: Intensity of the radiation from the ammonia flame is lower than that from a coal flame because of lower flame temperature and less soot production. As a result, heat absorption in the boiler may differ largely from the case of pure coal firing. Concerning material durability, it is preferable to have a constant distribution of heat absorption, no matter what the actual co-firing ratio is.

In this study, experimental studies were conducted to develop the ammonia co-firing burner, and numerical studies to evaluate the heat absorption in the boiler.

41.2 Development of Hybrid Burner for Pulverized Coal/Ammonia Co-firing

In this chapter, the development of the ammonia co-firing burner is discussed. Typically, coal contains a small amount of nitrogen that would be a source of NO_x. For environmental protection, boilers are equipped with De-NO_x technology, which is already well established. The main content of this study is to divert the De-NO_x technology from coal firing into ammonia co-firing. The mechanism of the De-NO_x technology for the coal firing is as follows. At first, NO_x is generated due to the combustion of volatile matter (Fig. 41.1). Then, it is retained in a strongly recirculating reduction zone within the flame, where it is reduced to N_2. In some boilers, ammonia is injected into this zone to enhance the reducing reaction. The concept for ammonia co-firing is similar. However, in case of ammonia co-firing, the amount of ammonia injection is much larger than in the case of coal firing. Most of the ammonia injected into the reduction zone is decomposed directly into nitrogen and hydrogen (cf. [1] for further details). Remaining ammonia works as an agent to reduce NO to N_2.

For the proof of this concept, combustion tests were conducted with the test facility shown in Fig. 41.2. Further details are mentioned in [2]. In this furnace, one swirl burner for the pulverized coal is arranged at the front wall. The thermal input into this furnace is around 10 MW. It corresponds to a coal consumption of 1.0 tons per hour.

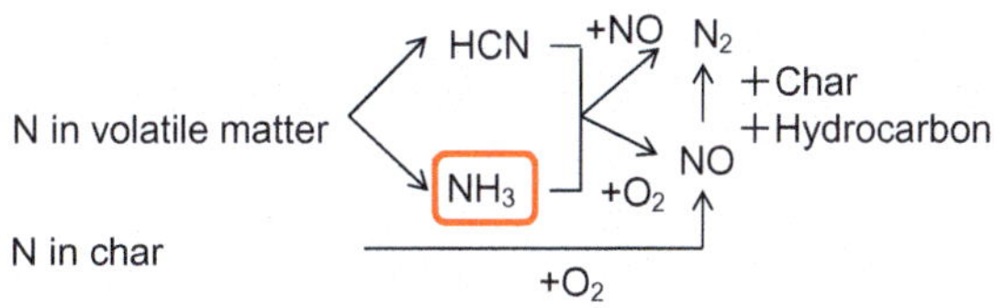

Fig. 41.1 NO formation during coal combustion. © Combustion Society of Japan 2019

Up to 2 MW of ammonia can be supplied, which corresponds to a supply of 0.4 t/h. Main conditions for the combustion tests are summarized in Table 41.1. The excess air ratio (Eq. 41.1) was set to 1.2, and the over-firing ratio (Eq. 41.2) was set to 30%. The excess air ratio is the ratio of the theoretical amount of air necessary to achieve complete oxidation to the total amount of air supplied to the furnace. The over-firing ratio is the ratio of amount of air supplied for over-firing to the total amount of air supplied to the furnace. The composition of the exhaust gas was measured at the "gas sampling" shown in Fig. 41.2.

$$\text{Excess air ratio}\,(-) = \text{Actual supplied air}\,(\text{kg/s})/\text{theoretical amount of air}\,(\text{kg/s}) \quad (41.1)$$

$$\text{Over-firing rate}\,(\%) = \text{Amount of air for over} - \text{firing}\,(\text{kg/s})/\text{total amount of air}\,(\text{kg/s}) \times 100 \quad (41.2)$$

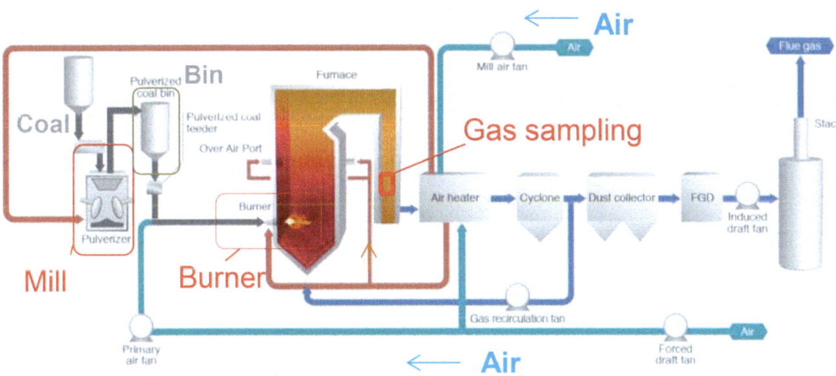

Fig. 41.2 Coal firing test facility

Table 41.1 Main conditions for the coal/ammonia co-firing tests

Parameter	Value
Thermal input (MW)	9.5
Excess air ratio (–)	1.2
Over-firing ratio (%)	31

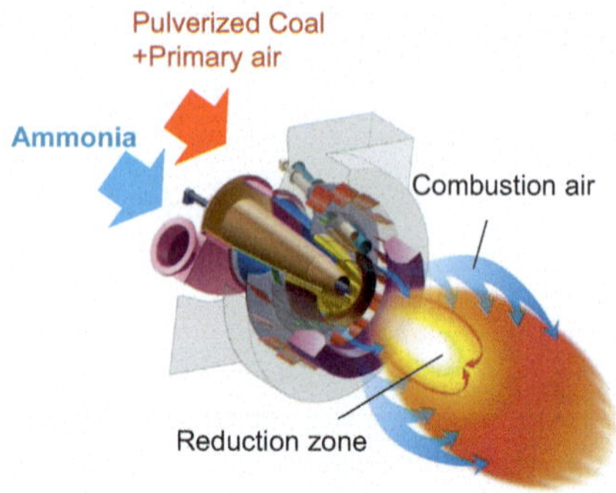

Fig. 41.3 Pulverized coal-fired burner. © Combustion Society of Japan 2019

Figure 41.3 shows an overview of the hybrid burner. This burner can hold the flame at the front of the port and create a strong reduction zone inside the flame. In this study, ammonia was injected into the flame at the appropriate flow speed to enter the reduction zone.

Figure 41.4 shows the state of the flame observed from a window-mounted to the side wall of the furnace. The stability of the flame was confirmed in both cases, coal firing, and ammonia co-firing. Figure 41.5 shows the temperature distribution in the furnace along the burner axis as measured by the monochromatic radiation thermometer. It is readily seen that the temperature is lower in case of ammonia co-firing. As mentioned earlier, this is because the adiabatic flame temperature of ammonia is lower than that of coal.

Figure 41.6 shows the NO_x concentrations in the exhaust gas. The black dashed line represents coal firing, while the red dashed line represents ammonia co-firing; top shows the effect of excess air ratio, while the effect of over-firing ratio is shown at the bottom. In general, NO_x increases when excess air ratio is increased, because the amount of air in the furnace is increased, and nitrogen contained in the fuel is easily oxidized. The effect of over-firing ratio is to increase NO_x since the amount of air around the burner increases as the over-firing ratio is decreased, and fuel-nitrogen is easily oxidized (Fig. 41.6b).

In Fig. 41.6, it is recognized that the trends of NO_x concentration in the case of ammonia co-firing are similar to those of coal firing. According to these combustion tests, it is confirmed that ammonia co-firing can be controlled in such a manner that

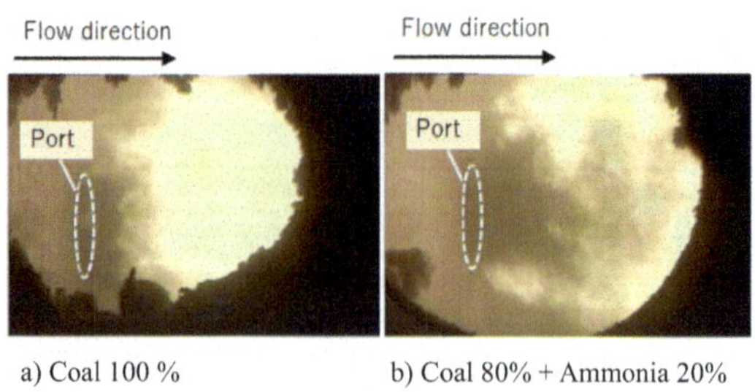

Fig. 41.4 Pictures of flame in the coal/ammonia co-firing test

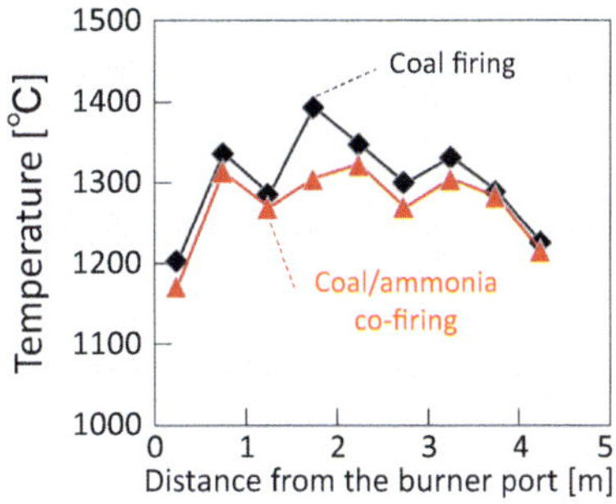

Fig. 41.5 Gas temperature in the furnace. © Combustion Society of Japan 2019

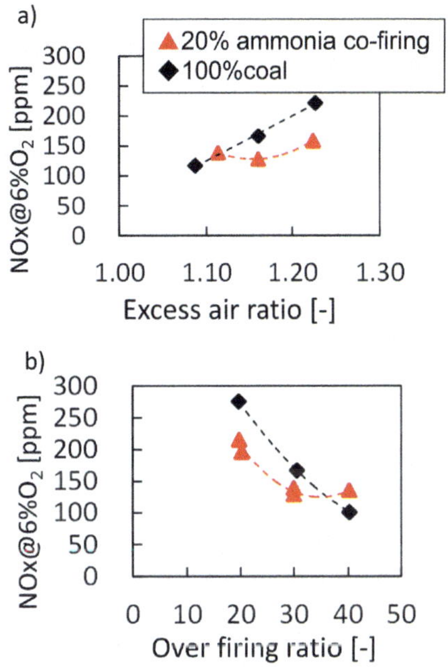

Fig. 41.6 NO_x concentration in the exhaust gas with respect to **a** excess air ratio and **b** over-firing ratio. © Combustion Society of Japan 2019

NO_x concentrations are similar to or lower than in the case of coal firing. Furthermore, no ammonia slip has been detected in the exhaust gas in any of the test runs. As the conclusion, it is clear that no additional denitration facility is necessary; ammonia can be co-fired in existing power plants without any additional measures against NO_x.

41.3 Evaluation of Boiler Performance by Numerical Analysis

Numerical analysis was performed to evaluate the amount of heat absorption in the boiler. Firstly, the numerical accuracy was evaluated by solving the burner test mentioned above. Then, the heat absorption of a 1000 MWe boiler was evaluated.

41.3.1 Ability of the Numerical Code to Analyze Coal/Ammonia Co-firing

Conventionally, some chemical reactions related to ammonia are considered in the numerical analysis of the NO formation in coal combustion. In this study, three reactions are treated as the main reactions for ammonia co-firing. These are ammonia oxidization, NO reduction, and thermal decomposition as shown in Eqs. (41.3) to (41.5), respectively [3].

$$NH_3 + 5/4O_2 \rightarrow NO + 3/2H_2O \quad (41.3)$$

$$NH_3 + NO + 1/4O_2 \rightarrow N_2 + 3/2H_2O \quad (41.4)$$

$$2NH_3 \rightarrow N_2 + 3H_2 \quad (41.5)$$

Figure 41.7 shows the results of simulations concerning the burner test. Shown are distributions of the temperature inside the furnace. It is clearly seen that the temperature is slightly lower in the case of ammonia co-firing. A similar trend has been observed in the experiments (Fig. 41.5). As mentioned before this is the consequence of different adiabatic flame temperatures between coal and ammonia. Table 41.2 presents the composition of the exhaust gas at the furnace outlet. The error in the outlet temperatures is within 10 °C. Errors in the concentration of unburned carbon and CO_2 are within 0.3 points and 0.6 points, respectively. However, the error in NO concentration during ammonia co-firing was approximately 100 ppm. It is a relatively large error compared with the other items. This amount is not critical in the discussion of energy and mass balance inside the furnace. However, for environmental assessments, improvement is necessary.

According to this investigation, it can be said that the energy balance and mass balance are well described in the simulation of both cases, coal firing, and ammonia co-firing. However, some technical issue remains concerning the prediction of NO concentration.

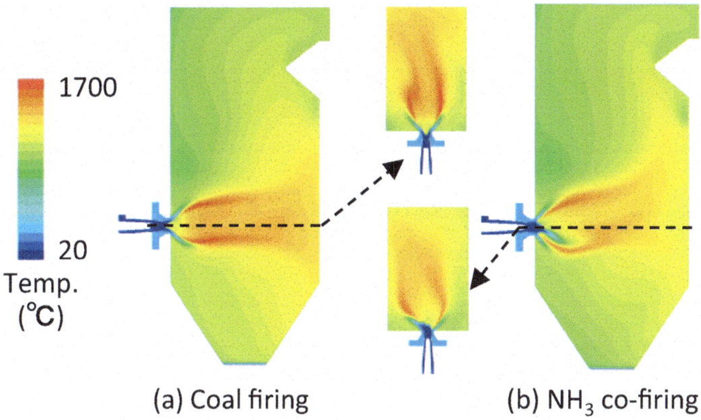

Fig. 41.7 Numerical results: gas temperature in the coal firing test facility (thermal input of 10 MW). © Combustion Society of Japan 2019

Table 41.2 Comparison between experimental and numerical properties of the exhaust gas

	Coal firing		Ammonia co-firing	
	Experiment	Simulation	Experiment	Simulation
Gas temperature (°C)	1045	1051	1033	1043
NO (ppm, 6% O_2)	222	214	160	263
CO_2 (vol%, 6% O_2)	13.0	13.6	11.0	10.9
UC (wt%)	2.4	2.5	3.7	3.4

© Combustion Society of Japan 2019

41.3.2 Evaluation of Heat Absorption for a 1000 MWe Boiler

In this section, we discuss the evaluation of heat absorption in a 1000 MWe boiler [4]. An overview of the boiler is shown in Fig. 41.8. Features of this boiler are multiple burners and over-firing ports compared to the single burner test facility (Fig. 41.2). Objective of the numerical analysis for this boiler is to evaluate the heat absorption over the wall surfaces in both cases, coal firing, and ammonia co-firing.

Figure 41.9 compares the temperature distribution in the central plane of the furnace. As already pointed out before, in case of ammonia co-firing the temperature is slightly lower than in the case of coal firing (Fig. 41.7). Figure 41.10 shows the distribution of heat absorption on front and left walls, respectively. The differences in the distribution of the heat absorption between coal firing and ammonia co-firing are not obvious. Furthermore, total amount of heat absorbed on the wall is not largely changed between coal firing and ammonia co-firing. With respect to the material durability in the boiler, this result is preferable since the ammonia co-firing is feasible in the conventional boilers without affording large retrofit work.

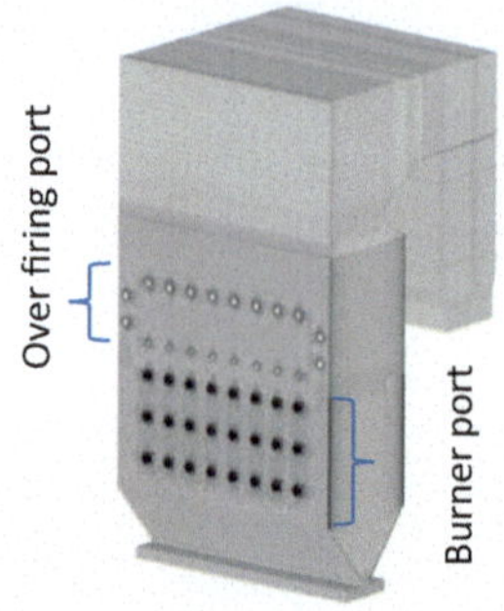

Fig. 41.8 1000 MWe coal-fired boiler. © Combustion Society of Japan 2019

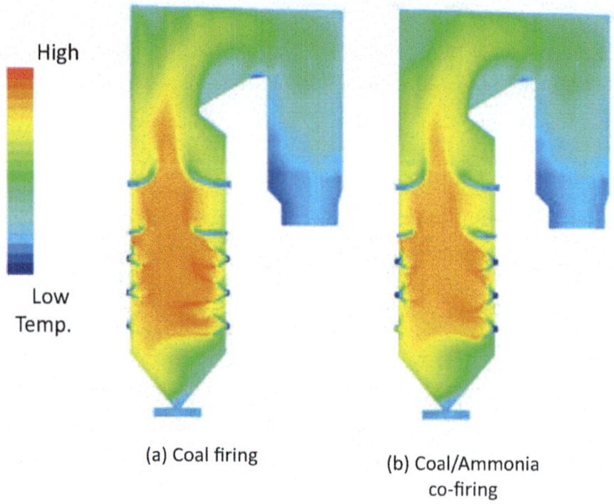

Fig. 41.9 Numerical results: gas temperature in the 1000 MWe coal-fired boiler. © Combustion Society of Japan 2019

41.4 Conclusions

In this study, the direct use of ammonia as a carbonless fuel for the coal-fired boiler is investigated. Firstly, an ammonia co-firing burner was developed, and its performance was confirmed in experiments. NO_x concentrations could be successfully controlled for all conditions considered here (ammonia co-firing rates up to 20%). Secondly, the heat absorption was evaluated by numerical analysis. It was found that neither local nor total heat absorption changed largely when varying the ammonia co-firing ratio. Therefore, no concerns with respect to material selection for the boiler walls are expected. As the next step in the investigation, long-term evaluations with a commercial boiler for the quality of the exhaust gas and heat absorption are recommended.

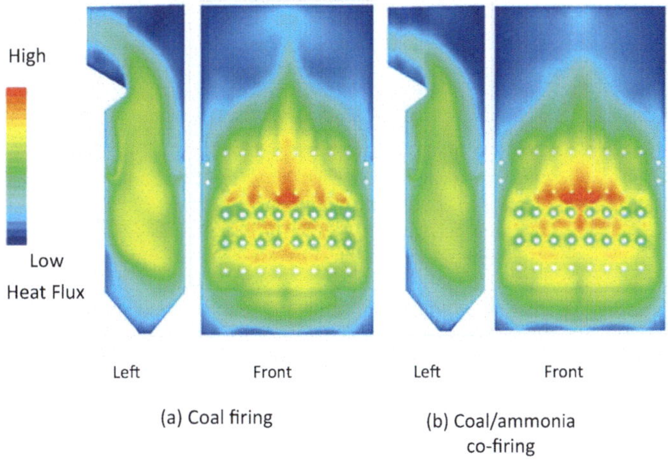

Fig. 41.10 Numerical results: heat absorption on the walls of the 1000 MWe coal-fired boiler. © Combustion Society of Japan 2019

Acknowledgements This research was conducted by the Council for Science, Technology and Innovation Strategic Innovation Promotion Program (SIP) "Energy Carriers" (managing corporation: JST). We would like to thank them for their support.

References

1. Ishihara S, Zhang, J, Ito T, Fujimori T (2018) The effects of operating conditions of ammonia-coal co-firing boilers on the emission gas. In: 56th symposium (Japanese) on combustion (in Japanese)
2. Ishii H, Oono E, Kozaki T, Ito T, Fujimori T (2018) Coal co-firing with ammonia by controlling the NO_x emission. In: 23rd national symposium on power and energy systems
3. Zhang J, Ishihara S, Ito T, Fujimori T (2018) Numerical investigation on co-firing ammonia with pulverized coal in a 10 MWth combustion facility. In: 35th annual international Pittsburgh coal conference, China, Oct 15–18
4. JST (2019) IHI final report of "ammonia direct combustion" in SIP. https://www.jst.go.jp/sip/dl/k04/end/team6-14.pdf

Chapter 42
Test of the Co-firing of Ammonia and Coal at Mizushima Power Station

Tsukasa Yoshizaki

Abstract The ammonia combustion test was conducted at the selected test site, and it was confirmed that the ammonia used as fuel burned completely and that there was no impact on the existing plant due to the combustion. The NO_x value did not differ significantly from that of coal burning, and the test was conducted without any problems within the environmental standard values. The study confirms that ammonia can contribute to power generation as a fuel without significant impact on the environment.

Keywords Co-firing · Coal power plant · Ammonia fuel · NO_x control

42.1 Introduction

Ammonia is attracting attention as a hydrogen energy carrier because of its high hydrogen storage capacity and ease of liquefaction.

It has a boiling point of −33 °C and liquefies at eight atmospheres at room temperature, which makes it much easier to liquefy than hydrogen, and its hydrogen density (amount of hydrogen per volume) is about 1.7 times higher than liquid hydrogen and 2.6 times higher than that of organic hydrides, making it possible to carry a large amount of hydrogen. Another advantage is that the integrated technology is well established from manufacturing to transportation and storage, and existing infrastructure can be used.

Originally published in Journal of the Combustion Society of Japan, Vol. 61 (2019), pp. 309–312 (in Japanese);

T. Yoshizaki (✉)
Energia Economic and Technical Research Institute, The Chugoku Electric Power Co., Inc., 3-9-1 Kagamiyama, Higashihiroshima, Hiroshima 739-0046, Japan
e-mail: 261359@pnet.energia.co.jp

K. Aika and H. Kobayashi (eds.), *CO_2 Free Ammonia as an Energy Carrier*,
https://doi.org/10.1007/978-981-19-4767-4_42

There are two ways to use ammonia: by recomposing it into hydrogen or by direct combustion. Although energy loss during decomposition is a concern for the former, it has the advantage that it can be applied to various hydrogen utilization technologies that are being developed. The latter combusts ammonia directly and thus has no conversion losses and can use energy efficiently.

In the Cross-ministerial Strategic Innovation Promotion Program (SIP) of the Council for Science, Technology and Innovation of the Cabinet Office, energy carriers are featured, and it is expected to promote the development of technologies for the "production, transportation, storage, and utilization" of hydrogen so that Japan can realize an innovative and low-carbon hydrogen energy society by 2030 and become the global market leader in the hydrogen-related industry.

We conducted a feasibility study of ammonia direct combustion in a pulverized coal-fired boiler at a thermal power plant from a user perspective.

An ammonia co-firing test was conducted at our Mizushima power plant Unit 2 (coal-fired power plant, located in Kurashiki City, Okayama Prefecture, rated output: 156,000 kW, Fig. 42.1) to enhance the content and accuracy of the study on the possibility of introducing ammonia.

Fig. 42.1 Mizushima power plant. © Combustion Society of Japan 2019

42.2 Overview

42.2.1 *Selection of a Test Site*

In order to co-fire ammonia with coal, coal-fired power plant boilers need to be equipped with burners capable of burning the gas.

Mizushima Unit 2 is equipped with the facility to burn boil-off gas (BOG), which is generated at the nearby LNG terminal. Supplying ammonia as fuel from an ammonia facility for denitrification to a BOG combustion facility makes it possible to maximize the use of existing facilities and to conduct low-cost co-firing tests.

Therefore, Mizushima Unit 2 was selected as the test site for this test.

42.2.2 *Facilities Design*

To co-fire the ammonia using the existing burners of the Unit 2 boiler, we designed a facility from the existing ammonia facility to the burners.

Several materials are not suitable for using in ammonia facilities in terms of low resistance against the corrosiveness of ammonia, i.e., copper and copper alloys, general structural rolled steel (SS), and carbon steel pipes (SGP) should not be used and pressure vessel steel plates (SPV450 and SPV490), aluminum, and aluminum alloys are not desirable to be used. Rubbers that should not be used are Fluorocarbon Rubber (FKM) and Fluorosilicone Rubber (FVMQ). Nitrile Rubber (NBR) and Fluorinated Ethylene-Propylene Rubber (FEPM) are not desirable to be used.

As a result of confirming the materials used in the existing system, it was found that the above materials were used in some parts of the system. For safety reasons, we decided to install new temporary pipes instead of using the existing BOG pipes to supply ammonia.

The pipe diameter was determined based on the calculation results of temperature and pressure drops, taking into account the burner inlet pressure and flow rate, and the risk of re-liquefaction of the vaporized ammonia gas.

From the above results, about 300 m of piping was installed as shown in Fig. 42.2.

42.2.3 *Test Conditions*

The tests were conducted in an existing ammonia vaporizer (with a nominal capacity of 450 kg/h) under the test conditions shown in Table 42.1. The test coals used were 60% Mount Owen and 40% Boggabri Premium.

In this study, a BOG burner was used, and as shown in Fig. 42.3, the ammonia co-firing area is equivalent to one of the 20 existing BOG burners (approximately 1 MW).

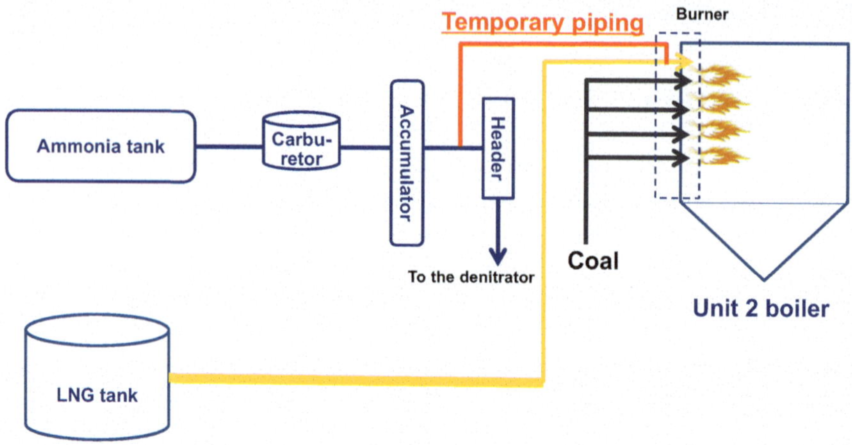

Fig. 42.2 Temporary system

Table 42.1 Test conditions

Testing period	Monday, July 3 to Sunday, July 9, 2017 Only during the daytime (8 am to 6 pm)
Test output	155 MW, 120 MW (July 7)
Type of coal	Mount Owen 60%, Boggabri Premium 40%
Ammonia consumption	Up to 450 kg/h (maximum carburetor flow rate, equivalent to 400 kg of coal)
Pressure	Vaporizer pressure: 0–0.63 MPa Pre-burner pressure: 0–0.52 MPa
Mixed fuel burning ratio	Approximately 0.6%, 0.8% (July 7) (equivalent to 1 MW)
Test content	Voltage boosting test on the system after the carburetor Ammonia burner combustion test, etc.

The ammonia co-firing rate is about 0.6% at 155 MW and about 0.8% at 120 MW.

When the burner valve is opened for ignition, ammonia gas is supplied to the furnace and the burner pressure decreases at the same time.

Considering the pressure drop, we confirmed that the pressure had been raised above 0.4 MPa before blowing ammonia gas into the burner. The burner pressure was then gradually increased and adjusted to the target flow rate (450 kg/h).

The tests were conducted during the daytime only, while the plant's operating conditions were carefully monitored, in order to be prepared for contingencies.

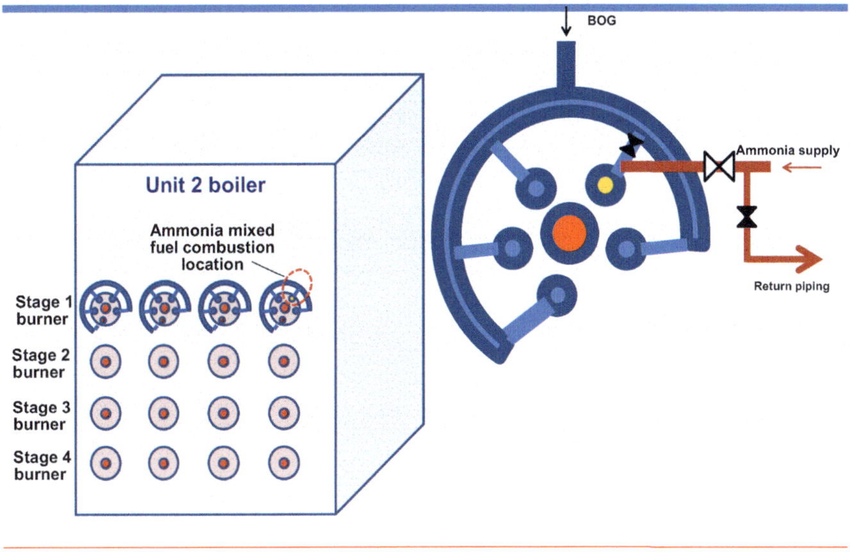

Fig. 42.3 Temporary burner

42.3 Results and Discussion

42.3.1 Burner Characteristics

To confirm that the target ammonia flow rate (450 kg/h) was delivered as planned, the burner characteristics were checked.

The plan was to secure the required pressure at the inlet of the BOG burner from the pressure loss of the pipes and valves to provide the target flow rate.

The actual characteristics of the ammonia flow rate and BOG burner inlet pressure for one BOG burners were shown in Fig. 42.4. The target flow rate (450 kg/h) was achieved, although the burner characteristics were slightly below the planned value due to pressure loss at the modified BOG burner.

42.3.2 Combustibility

To confirm the status of ammonia combustion, we measured the combustion temperature, residual ammonia concentration at the boiler outlet, and monitored the inside of the furnace.

As a result of measuring the temperature of the burner using a radiation thermometer, it was confirmed that the temperature was more than 1210 °C.

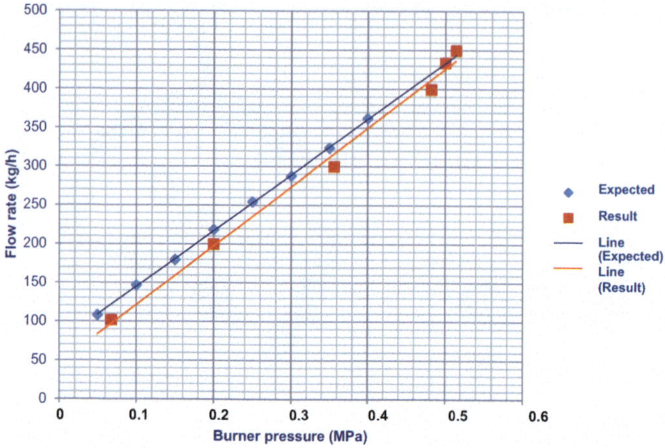

Fig. 42.4 Burner characteristics

Since the ignition temperature of ammonia was 651 °C, it was assumed that the combustion temperature was sufficiently high for complete combustion, but the residual ammonia concentration was measured to confirm that the ammonia was burning reliably.

The concentrations ranged from 0.2 to 0.4 ppm, which was the same as in the absence of ammonia injection. The residual ammonia at the boiler outlet was almost the same as under normal conditions, indicating complete combustion of ammonia in the boiler.

Since the co-firing rate is less than 1%, the impact is very small; however, as shown in Fig. 42.5, it was also confirmed that the amount of coal used was reduced.

During the test period (7 days), about 20 tons of ammonia was injected into the boiler and burned.

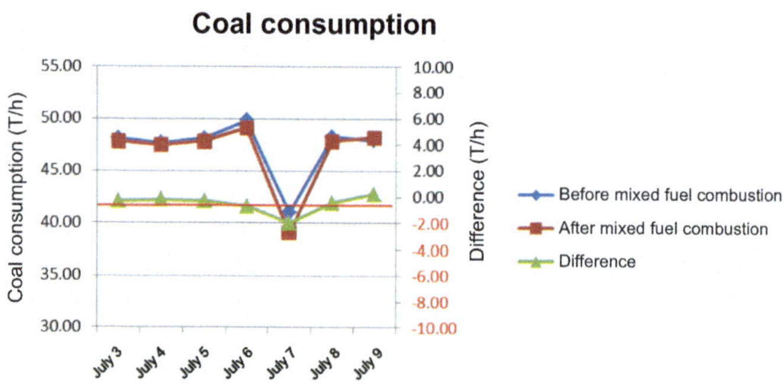

Fig. 42.5 Evaluation of coal amount (dry basis)

42.3.3 Boiler Metal Temperature

One of the important items to monitor for boiler operating conditions is the boiler metal temperature.

The increase in boiler metal temperature can lead to boiler tube blowout due to high-temperature creep, which in the worst case can lead to boiler shutdown, i.e., power generation shutdown.

It is important to avoid overheating due to various factors, and the metal temperature is grasped in thermal power plants to ensure operational reliability.

For this reason, the boiler metal temperature at each load (155 MW and 120 MW) was measured at the measurement position shown in Fig. 42.6 in this test as well.

As shown in Fig. 42.7, no metal temperature problems due to ammonia co-firing occurred at each position.

Other plant data showed no significant changes due to ammonia co-firing.

42.3.4 Change in NO_x Value

While ammonia may produce NO_x due to the nitrogen content of its components, it is also expected to act as a NO_x reducing agent. In this study, the change in NO_x values due to co-firing was confirmed by measuring the NO_x concentration at the boiler outlet.

With a generator output of 155 MW (about 0.6% co-firing), the difference in NO_x concentration before and after co-firing ranged from −8 ppm to +7 ppm (−2.3 to +2.1%), as shown in Fig. 42.8, and the existing denitration system was sufficient to deal with the problem.

On the other hand, the NO_x concentration was reduced by about 14 ppm (5.4%) after co-firing with a generator output of 120 MW (about 0.8% co-firing).

42.3.5 CO_2 Reduction Effect

Ammonia is a CO_2-free fuel that can reduce CO_2 emissions according to the co-firing rate, thus contributing to the reduction of greenhouse gases in thermal power plants. In particular, its application to coal-fired power plants is more effective in reducing CO_2 emissions compared to other fuels.

Table 42.2 shows the estimated annual reduction in CO_2 emissions from the Mizushima Power Plant Unit 2. With an ammonia co-firing rate of approximately 0.6% in the test, if the plant were to operate at 156 MW (70% facility utilization) throughout the year, it would save approximately 4000 tons of CO_2/year. If the ammonia co-firing rate is set at 20%, a reduction of approximately 134000 t-CO_2/year can be expected.

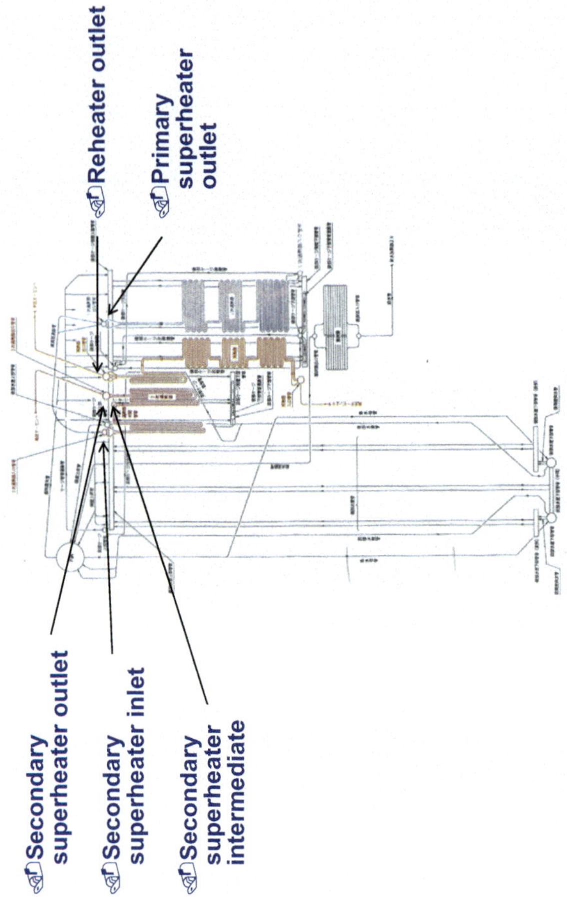

Fig. 42.6 Boiler metal temperature measurement locations

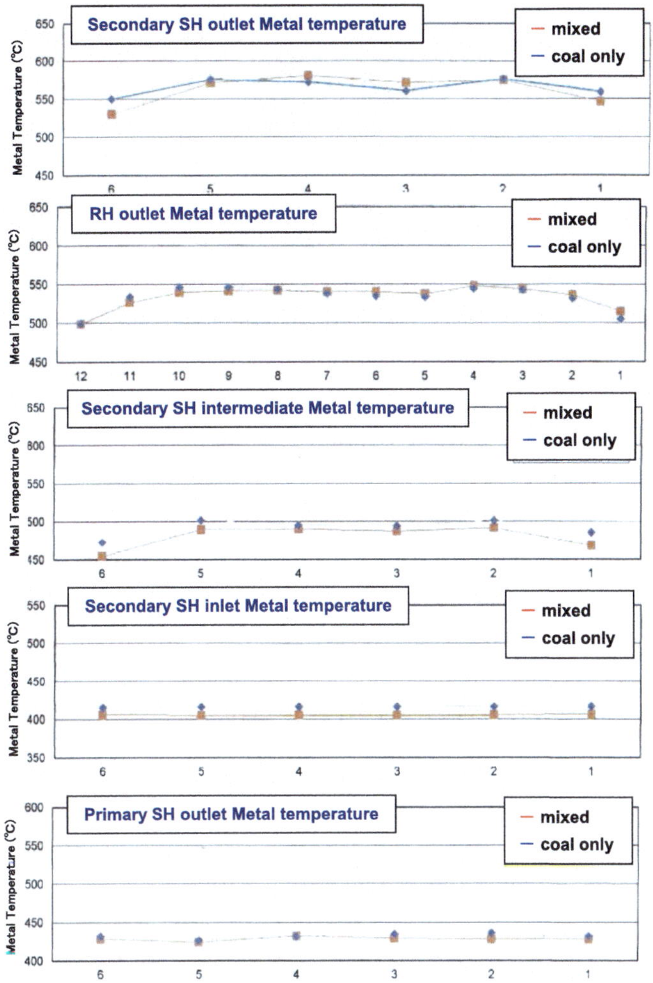

Fig. 42.7 Boiler metal temperature

42.4 Conclusions

The ammonia combustion test was conducted at the selected test site, and it was confirmed that the ammonia used as fuel burned completely and that there was no impact on the existing plant due to the combustion. The NO_x value did not differ significantly from that of coal burning, and the test was conducted without any problems within the environmental standard values.

The study confirms that ammonia can contribute to power generation as a fuel without significant impact on the environment.

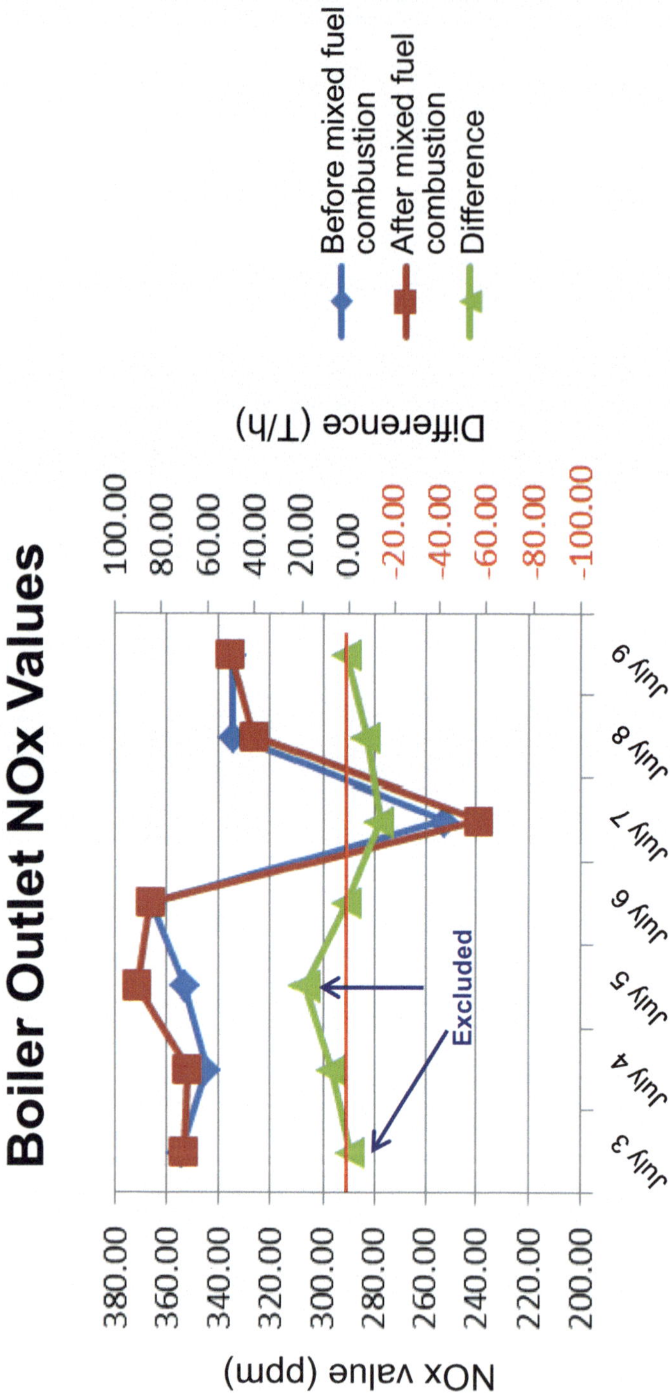

Fig. 42.8 Boiler outlet NO_x values

Table 42.2 Estimated annual reduction in CO_2 emissions from the Mizushima Power Plant Unit 2

Output (MW)	156	
Ammonia co-firing rate (%)	0.6	20
Annual reduction in CO_2 emissions (tons of CO_2/year)	Approximately 4	Approximately 134

Acknowledgements This study was conducted under the contract research project "Ammonia Direct-Combustion" in the Cross-ministerial Strategic Innovation Promotion Program (SIP) of the Council for Science, Technology and Innovation, Cabinet Office. The authors would like to express their deepest gratitude to the Japan Science and Technology Agency (JST) and everyone involved in this study.

Chapter 43
Flame Propagation Characteristics of Coal-Ammonia Co-firing

Nozomu Hashimoto

Abstract Coal-ammonia co-firing in coal-fired power stations is one of ways to reduce the CO_2 emission from the power sector dramatically. For efficient development of the co-firing technology, the fundamental understandings of the flame propagation in the ammonia/coal co-combustion field are necessary. Based on the experimental results for the turbulent flame propagation in the ammonia/coal co-combustion field, the mechanisms of the interaction between ammonia and pulverized coal particles have been clarified. The most important effect of the interaction is the overall equivalence ratio increase effect due to the addition of the volatile matter from coal particles in the preheat zone of the flame. This effect mainly controls the flame propagation velocity in the ammonia/coal co-combustion field.

Keywords Coal ammonia co-firing · Co-combustion · Turbulent flame propagation

43.1 Importance of Research on Flame Propagation Characteristics of Coal-Ammonia Co-combustion

To reduce the CO_2 emission from coal-fired boilers, the development of the technology for the co-combustion of ammonia and pulverized coal particles has been conducted recently [1–4]. For efficient development of this technology, the fundamental understandings are necessary. One of the important phenomena of the ammonia/coal co-combustion is the flame propagation phenomenon, because the flame propagation is an essential phenomenon for the flame stability of the burner. Figure 43.1 shows the schematic of continuous flame propagation in the exit region of pulverized coal burner (The gas temperature distribution with gas velocity vectors was taken from the study by Hashimoto et al. [5]). In most of coal-fired boiler, the stabilized flame is maintained by recirculation flow formed in the burner exit region. In that region, the flame tip continuously propagates toward the flow of fresh air and

N. Hashimoto (✉)
Division of Mechanical and Space Engineering, Hokkaido University, Kita 13 Nishi 8, Kita-ku, Sapporo 060-8628, Japan
e-mail: nozomu.hashimoto@eng.hokudai.ac.jp

K. Aika and H. Kobayashi (eds.), *CO_2 Free Ammonia as an Energy Carrier*,
https://doi.org/10.1007/978-981-19-4767-4_43

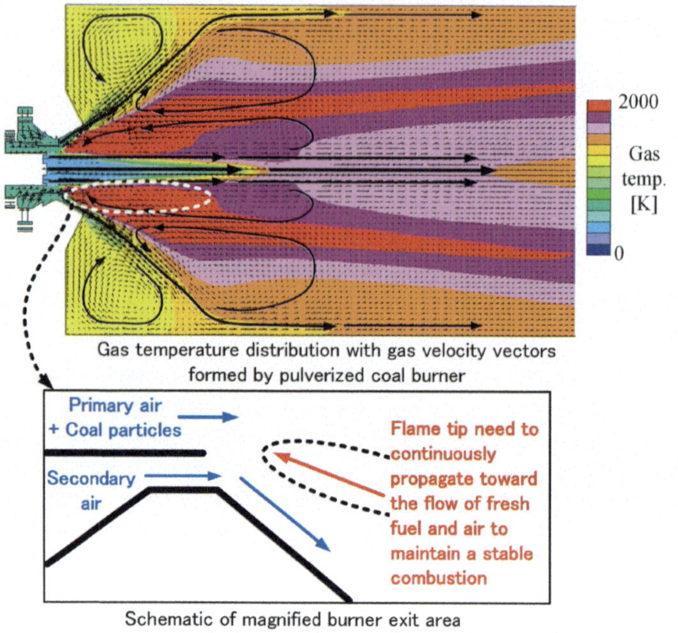

Fig. 43.1 Schematic of continuous flame propagation in exit region of pulverized coal burner. The gas temperature distribution with gas velocity vectors was taken from Hashimoto et al. [5]

fuel. If the flame propagation velocity is not high enough, the flame will be blown off, indicating the lack of residence time for combustion reaction for the mixture of fuel and air. Therefore, the flame propagation velocity is one of the most important factors whether the flame can be maintained stably or not.

43.2 Experimental Apparatus and Procedures for Research on Flame Propagation Characteristics of Coal-Ammonia Co-combustion

To examine this important factor for co-combustion of ammonia and pulverized coal particles, a unique flame propagation experimental apparatus for pulverized solid particles in a turbulent field was developed by the research group of Hokkaido University [6–10]. The schematic of developed apparatus is shown in Fig. 43.2. The apparatus is composed of a combustion chamber, a solid particle dispersion system, an ignition system, gas supply lines, a sequence system, and optical measurement system. The combustion chamber with an inner volume of 6.19 L has two identical counter-rotation seven-bladed fans, which are vertically and symmetrically mounted on the top and bottom covers of the chamber. The chamber has four windows made of quartz glass with a diameter of 50 mm for optical measurement. Pulverized coal particles were stored into four filter cups before the experiment. Coal particle dispersion was made by a sweeping flow of the gas from a dispersion tank, in which the gas with the same composition as the gas in the chamber was filled up to 300 kPa prior to the experiment. Just 0.3 s after the end of dispersion, the ignition was made by the spark ignitor composed of two stainless steel electrodes whose gap was set to 2 mm. After the ignition, the flame was observed by two of three kinds of optical measurements, namely, the direct camera imaging, OH radical imaging and Schlieren photography. Not only the co-combustion experiments of ammonia/pulverized coal but also the experiments of pure ammonia combustion [7, 8] and pure pulverized coal combustion [6] were conducted to compare those flame propagation characteristics. Schlieren photography was employed for pure ammonia combustion experiments because the detailed flame structure can be obtained. For pure pulverized coal combustion or ammonia/pulverized coal co-combustion, however, the Schlieren image could not obtained due to the scattering effect by pulverized particles. Therefore, the direct imaging was used for pure pulverized coal combustion or ammonia/pulverized coal co-combustion experiments. OH radical imaging was employed for all of experiments. Image processing procedures of those three measurements are shown in Fig. 43.3. The raw images were binarized by an image processing software to determine the border of flame front. The value of flame diameter was determined as the distance between the farthest border of flame front. The flame front detected by the OH radical imaging was defined as the reaction front because the OH radical is the essential radical to initiate the combustion reaction when the fuel including H atom in its molecule and oxygen are used as reactants. The flame front detected by the Schlieren photography is defined as the preheat zone front because the Schlieren photography detects a position where the density of gas changes significantly due to the increasing temperature. Therefore, the diameter of the preheat zone front detected by the Schlieren photography is always larger than the reaction front detected by OH radical imaging. The flame front detected by the direct imaging is defined as the luminous flame front because the bright light from the luminous flame, in which there are sufficient amount of soot particles with high temperature, is mainly detected by the direct imaging. Since the soot particles are considered to be formed by the secondary pyrolysis of volatile matter from coal particles after the reaction front passing by, the diameter of the luminous flame front detected by the direct imaging is always smaller than the reaction front detected by the OH radical imaging, as indicated in Fig. 43.3d. Although the relative positions of those three flame fronts are different, the flame propagation velocities calculated by different measurement methods are almost same [8–10]. This means that relative distances between the preheat zone front, the reaction front, and the luminous flame front are almost fixed during the flame propagation. From the consideration of easiness and accurateness of image processing methods, the Schlieren photography was used for the flame propagation measurement of the pure ammonia combustion experiment, and the OH radical imaging was used for that of the ammonia/pulverized coal co-combustion and the pure pulverized coal combustion experiments.

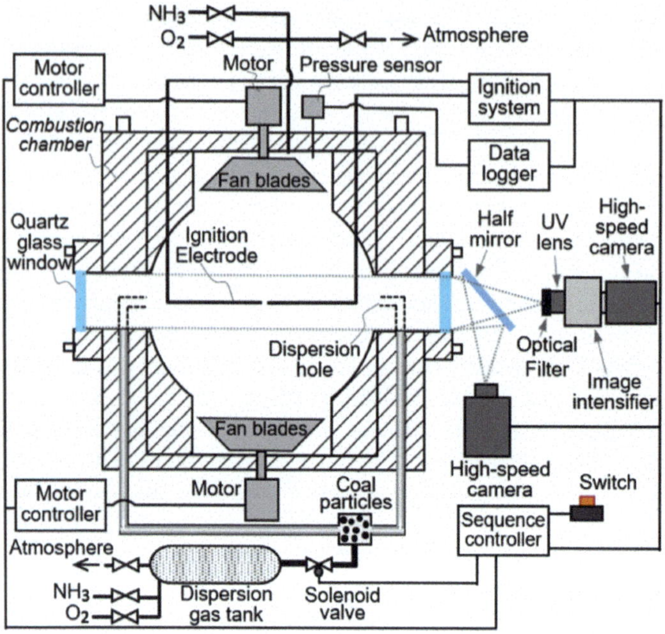

Fig. 43.2 Schematic of flame propagation experiment apparatus for pulverized solid fuel and gaseous fuel in a turbulent environment [9]

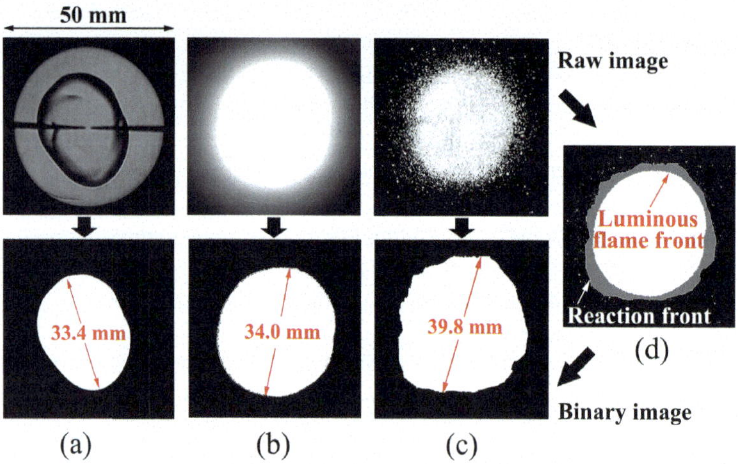

Fig. 43.3 Image processing procedures for **a** Schlieren photography, **b** direct imaging, **c** OH radical imaging [10]

Table 43.1 Properties of coal samples [9]

	Bituminous coal	High-fuel-ratio coal		
	C5	TW	KK	UL
Analysis				
Proximate [wt%]				
Moisture[b]	0.7	2.3	0.3	3.0
Ash[a]	14.2	19.9	12.9	21.2
Volatile matter[a]	33.5	22.9	20.9	12.5
Fixed carbon[a]	52.3	57.2	66.2	66.3
Fuel ratio	1.56	2.5	3.17	5.3
Ultimate [wt%, dry]				
Carbon	70.5	69.2	76.7	71.8
Hydrogen	4.64	3.64	4.09	2.48
Nitrogen	1.66	1.54	1.31	1.52
Oxygen	8.59	5.4	3.6	2.8
Sulfur	0.46	0.41	1.46	0.34
Heating value [MJ/kg]	27.8	27.2	30.1	27.2
Mean particle size [μm]	48	48	33	42

[a] Dry basis
[b] As received

The characteristics of the turbulence formed inside the chamber by the two fans were measured by the particle image velocimetry [7]. It was found that the uniform turbulent field is formed and the turbulence intensity is proportional to the fan rotation speed.

Table 43.1 shows the properties of pulverized coals used for the series of the flame propagation experiments. Pulverized coal samples with various fuel ratios (the ratio of fixed carbon to volatile matter) were employed to examine the effect of fuel ratio on the co-combustion phenomenon.

Since it was difficult to sustain the flame propagation of pulverized coal with air in the small experimental apparatus used in this research, the gas mixture of 40% oxygen + 60% nitrogen was used as oxidizer gas for all experiments.

43.3 Flame Propagation Characteristics of Coal-Ammonia Co-combustion

Figure 43.4 shows the flame radius histories for ammonia/pulverized coal co-combustion and pure ammonia combustion. The plots of "Direct_(coal name)", "OH_(coal name)", and "OH_NH_3" indicate the flame radii measured by the direct imaging for the co-combustion case, the flame radii measured by OH radical imaging for the co-combustion case, and the flame radii measured by OH radical imaging for the pure ammonia combustion case, respectively. It is observed that the flame radius measured by the OH radical imaging is always larger than that measured by the direct imaging. This is caused by the fact that the radius measured by the OH radical imaging is the reaction front, while the radius measured by the direct imaging is luminous flame front, as described previously. From the comparison of Fig. 43.4a, b, the distance between the reaction front and luminous flame front for UL coal co-combustion case is larger than that for C5 coal co-combustion case. This is because the amount of volatile matter content for UL coal is much lower than that for C5 coal. The lower volatile matter amount evolved from UL coal makes the soot particle formation rate low. Consequently, the distance between the reaction front and the luminous flame front for UL coal is longer than that for C5 coal.

Figure 43.5 shows the comparison of the flame propagation velocities as functions of turbulence intensity for ammonia/pulverized coal co-combustion, pure ammonia combustion, and pure pulverized coal combustion under the ammonia/oxidizer equivalence ratio condition of 0.6. For KK coal and UL coal cases (Fig. 43.5c, d), there is no data for pure pulverized coal combustion because the flame propagation could

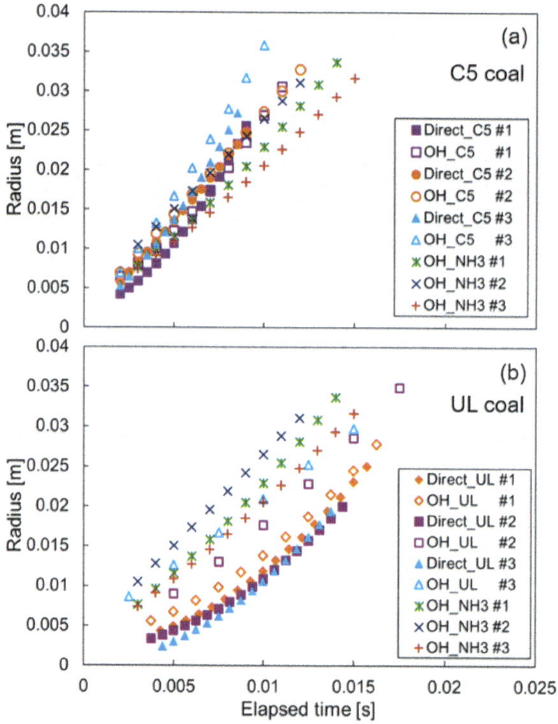

Fig. 43.4 Flame radius as a function of elapsed time for ammonia/pulverized coal co-combustion and pure ammonia combustion. The equivalence ratio of ammonia/oxidizer was 0.6. **a** C5 coal, **b** UL coal [9]

not be sustained without gaseous fuel for these coals. It is observed that the flame propagation velocity for the ammonia/pulverized coal co-combustion case is faster than that for the pure ammonia and pure pulverized coal combustion cases for C5 and TW coals (Fig. 43.5a, b). On the other hand, the flame propagation velocity for the co-combustion case is almost same as that for the pure ammonia combustion case for KK coal (Fig. 43.5c), and the flame propagation velocity for the co-combustion case is slightly slower than that for the pure ammonia combustion case for UL coal (Fig. 43.5d). This is caused by the difference of fuel ratio between four kinds of coals, i.e., the weight ratio of volatile matter in coal particles affects the flame propagation velocity of the ammonia/pulverized coal co-combustion. The mechanism controlling the flame propagation velocity of the ammonia/pulverized coal co-combustion is explained in the following.

Figure 43.6 shows the schematic of the flame propagation mechanism for ammonia/pulverized coal co-combustion. The flame propagation phenomenon is defined as the movement of the reaction front toward the direction of unburned zone. In front of the reaction front, there is a preheat zone where the heat is transferred from the post combustion zone (the soot formation zone and the luminous flame zone in the figure). In this zone, the volatile matter is evolved from coal particles due to the devolatilization reaction caused by the increase in the particle temperature. This means that some amount of heat is absorbed by coal particles in the preheat zone. At

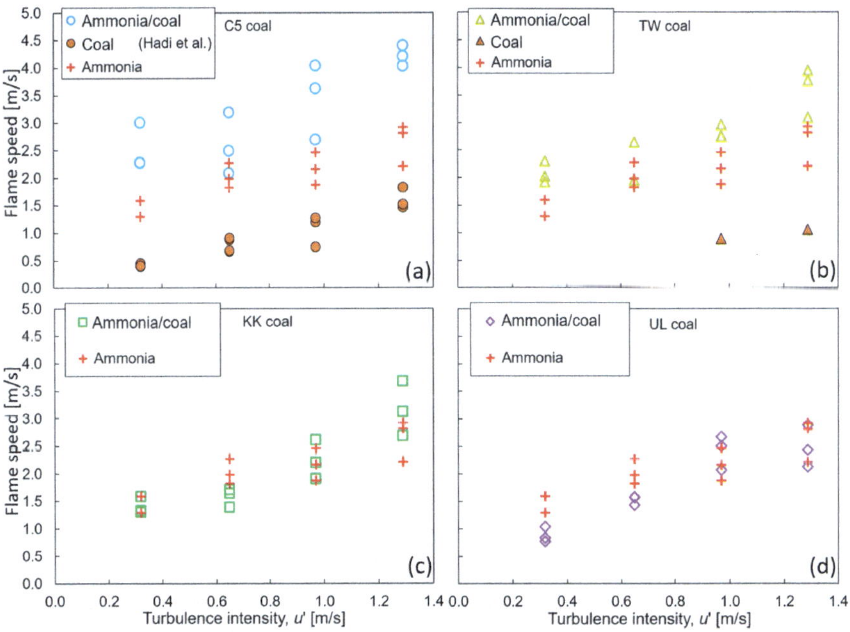

Fig. 43.5 Flame propagation velocity as a function of turbulence intensity for ammonia/pulverized coal co-combustion, pure ammonia combustion, and pure coal combustion. The equivalence ratio of ammonia/oxidizer was 0.6. **a** C5 coal, **b** TW coal, **c** KK coal, **d** UL coal [9]

the reaction front, not only the ammonia but also the volatile matter reacts with the oxidizer gas. Therefore, the overall equivalence ratio at the reaction front is higher than the equivalence ratio calculated by considering just ammonia and oxidizer gases because of the addition of the volatile matter. This is the reason why the flame propagation velocity for the ammonia/pulverized coal co-combustion case is higher than that for the pure ammonia combustion case for C5 and TW coals in Fig. 43.5a, b. Since the equivalence ratio of the ammonia and oxidizer gases was 0.6 (fuel-lean condition) in the experiments for Fig. 43.5, the increase in the overall equivalence ratio due to the addition of the volatile matter resulted in the increase in the flame propagation velocity. Because the heat up and the devolatilization of coal particles require a time, the evolution of volatile matter from middle and large size particles continues even after the reaction front has passed. Due to the secondary pyrolysis reaction of the volatile matter, the soot particles are formed in this zone. After the enough soot particles are formed, the radiation from high temperature soot particles transfers the heat from the luminous flame zone to the pre-heat zone. In short, the controlling factors affecting the flame propagation velocity of ammonia/pulverized coal co-combustion compared to the pure ammonia combustion are summarized as, (1) the overall equivalence ratio increase effect due to the addition of volatile matter from coal particles in the preheat zone, (2) the heat sink effect due to heat absorption by the coal particles in the preheat zone, and (3) the heat transfer effect through the radiation from the luminous flame zone to the unburned coal particles in preheat zone, as indicated in Fig. 43.6. Among the above mentioned three effects, the effect (3) is considered to be weak in the flame propagation inside the experimental apparatus [11]. However, this effect (3) can be stronger in the actual large-scale furnaces because the importance of the radiation heat transfer is significantly affected by the scale of the furnace or the scale of the flame [12]. The previously mentioned effects (1) and (2) are the main effects that affects the flame propagation velocity of the ammonia/pulverized coal co-combustion shown in Fig. 43.5. The effect (1) is stronger than the effect (2) for C5 and TW coal, while the effects (1) and (2) are almost same for KK coal. The effect (1) is weaker than the effect (2) for UL coal because the amount of volatile matter content of UL coal is extremely lower than that of other coals. This is why the flame propagation velocity for the ammonia/pulverized coal co-combustion is higher than that for the pure ammonia combustion for C5 and TW coals, while the flame propagation velocity for the ammonia/pulverized coal co-combustion is slower than that for the pure ammonia combustion for UL coal.

Figure 43.7 shows the flame propagation velocity as a function of equivalence ratio of ammonia/oxidizer mixture for (a) the pure ammonia combustion case, and (b) the ammonia/pulverized coal co-combustion case under various turbulence intensities [10]. It is observed that the flame propagation velocity takes the maximum at the equivalence ratio of 1.0 for the pure ammonia combustion case (Fig. 43.7a), while the flame propagation velocity takes the maximum at the ammonia/oxidizer of 0.8 for the ammonia/pulverized coal co-combustion case (Fig. 43.7b). This is caused by the overall equivalence increase effect as indicated in Fig. 43.6, i.e., the effect (1) is positive for the ammonia-lean case, while the effect (1) is negative for the ammonia-rich case. To visualize this effect clearly, the $\text{Ratio}_{\text{Reaction front}}$ (R_R), which is defined

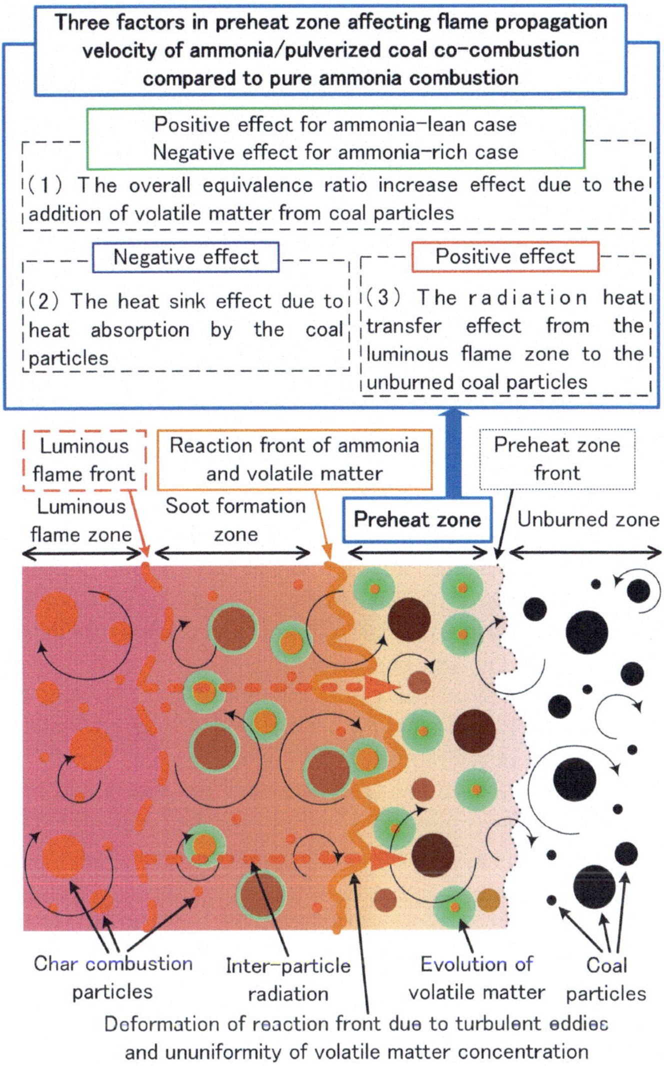

Fig. 43.6 Schematic of flame propagation phenomena for ammonia/pulverized coal co combustion

by the following equation, is introduced:

$$R_R = \frac{S_R - S_A}{S_A}, \tag{43.1}$$

where $S_{\text{Reaction front}}$ (S_R) represents the reaction front propagation velocity for the co-combustion case (Fig. 43.7b), and S_{Ammonia} (S_A) represents the flame propagation velocity for the pure ammonia combustion case (Fig. 43.7a). If the value of R_R is

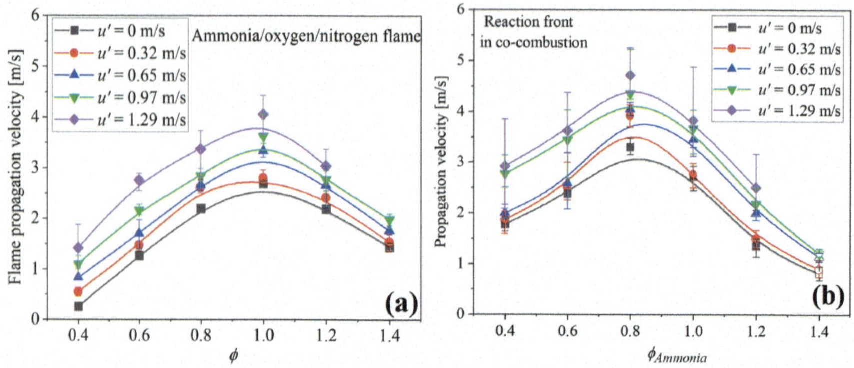

Fig. 43.7 Flame propagation velocity as a function of equivalence ratio of ammonia/oxidizer mixture for **a** pure ammonia combustion case, and **b** ammonia/pulverized coal co-combustion case under various turbulence intensities [10]

positive, it means that the flame propagation velocity of the ammonia/pulverized coal co-combustion is higher than that of the pure ammonia combustion for the same condition. Inversely, if the value of R_R is negative, it means that the flame propagation velocity of the ammonia/pulverized coal co-combustion is lower than that of the pure ammonia combustion for the same condition.

Figure 43.8 shows R_R defined by Eq. (43.1) as a function of the ammonia/oxidizer equivalence ratio. It is observed that R_R decreases monotonically with increasing ammonia/oxidizer equivalence ratio. It should be also noticed that R_R takes positive value in ammonia-lean condition. This is because the effect (1) in Fig. 43.6 has positive effect on the flame propagation velocity for the co-combustion case and overcomes the effect (2). In addition to that, the lower the ammonia/oxidizer equivalence ratio, the stronger the effect of increase of overall equivalence ratio due to the volatile matter addition. This is why R_R increases with decreasing ammonia/oxidizer equivalence ratio. On the other hand, R_R takes negative value in ammonia-rich condition. This is because both effects (1) and (2) in Fig. 43.6 have a negative impact on the flame propagation velocity for the ammonia/pulverized coal co-combustion case compared to the pure ammonia combustion case. As previously described, the effect (3) in Fig. 43.6, which is positive effect regardless of the ammonia/oxidizer equivalence ratio, is limited for the flame with small size in the experimental apparatus. This effect (3) can be stronger in the actual large-scale boiler, which may increase the flame propagation velocity of co-combustion case compared to the pure ammonia combustion case. However, the entire trend of Fig. 43.8, which is that R_R decreases monotonically with increasing ammonia/oxidizer equivalence ratio, is considered to be the same regardless of the flame size.

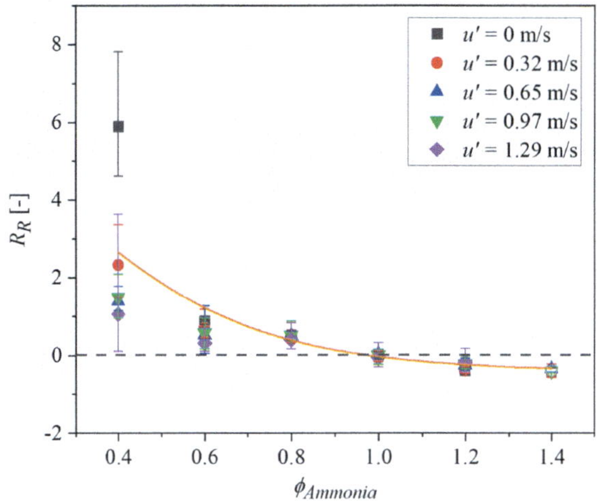

Fig. 43.8 R_R defined by Eq. (43.1) as a function of ammonia/oxidizer equivalence ratio [10]

43.4 Summary for Flame Propagation Characteristics of Coal-Ammonia Co-combustion Mechanism

In this section, the flame propagation characteristics of the ammonia/pulverized coal co-combustion was described based of the recent research results. In summary, there are three effects for the co-combustion compared to the pure ammonia combustion, which are (1) the overall equivalence ratio increase effect due to the addition of volatile matter from coal particles in the preheat zone, (2) the heat sink effect due to heat absorption by the coal particles in the preheat zone, and (3) the heat transfer effect through the radiation from the luminous flame zone to the unburned coal particles in preheat zone. Among them, the effect (1) is the strongest compared to other effects. The effect (1) is positive effect for the flame propagation velocity of co-combustion under the condition that ammonia/oxidizer equivalence ratio is lower than unity, while it is negative effect under the condition that ammonia/oxidizer equivalence ratio is higher than unity. Subsequently, the flame propagation velocity of the ammonia/pulverized coal co-combustion is higher than that of the pure ammonia combustion under the condition that ammonia/oxidizer equivalence ratio is lower than unity, while it is lower under the condition that ammonia/oxidizer equivalence ratio is higher than unity.

Acknowledgements This work was partly supported by JSPS KAKENHI Grant Number JP19180646 and JST Sakigake (PRESTO) Grant Number JPMJPR1542. The authors are indebted to Dr. Suda of IHI Co., Prof. Kitagawa of Kyushu Univ., Prof. Hayakawa of Tohoku Univ. for their helpful advice and discussion. The authors are also indebted to the Central Research Institute of Electric Power Industry for providing coal particle samples.

References

1. Kobayashi H, Hayakawa A, Somarathne KDKA, Okafor EC (2019) Science and technology of ammonia combustion. Proc Combust Inst 37:109–133
2. Zhang J, Ito T, Ishii H, Ishihara S, Fujimori T (2020) Numerical investigation on ammonia co-firing in a pulverized coal combustion facility: effect of ammonia co-firing ratio. Fuel 267:117166
3. Ishihara S, Zhang J, Ito T (2020) Numerical calculation with detailed chemistry of effect of ammonia co-firing on NO emissions in a coal-fired boiler. Fuel 266:116924
4. Tamura M, Gotou T, Ishii H, Riechelmann D (2020) Experimental investigation of ammonia combustion in a bench scale 1.2 MW-thermal pulverised coal firing furnace. Appl Energy 277:115580
5. Hashimoto N, Shirai H (2014) Numerical simulation of sub-bituminous coal and bituminous coal mixed combustion employing TDP model. Energy 71:399–413
6. Hadi K, Ichimura R, Hashimoto N, Fujita O (2019) Spherical turbulent flame propagation of pulverized coal particle clouds in O_2/N_2 atmospheres. Proc Combust Inst 37:2935–2942
7. Ichimura R, Hadi K, Hashimoto N, Hayakawa A, Kobayashi H, Fujita O (2019) Extinction limits of an ammonia/air flame propagating in a turbulent field. Fuel 246:178–186
8. Xia Y, Hashimoto G, Hadi K, Hashimoto N, Hayakawa A, Kobayashi H, Fujita O (2020) Turbulent burning velocity of ammonia/oxygen/nitrogen premixed flame in O_2-enriched air condition. Fuel 268:117383
9. Hadi K, Ichimura R, Hashimoto G, Xia Y, Hashimoto N, Fujita O (2021) Effect of fuel ratio of coal on the turbulent flame speed of ammonia/coal particle cloud co-combustion at atmospheric pressure. Proc Combust Inst 38:4231–4239
10. Xia Y, Hadi K, Hashimoto G, Hashimoto N, Fujita O (2021) Effect of ammonia/oxygen/nitrogen equivalence ratio on spherical turbulent flame propagation of pulverized coal/ammonia co-combustion. Proc Combust Inst 38:4043–4052
11. Xia Y, Hashimoto N, Fujita O (2022) Turbulent flame propagation mechanism of polymethyl-methacrylate particle cloud–ammonia co-combustion. Combust Flame (in press)
12. Hashimoto N, Watanabe H (2016) Numerical analysis on effect of furnace scale on heat transfer mechanism of coal particles in pulverized coal combustion field. Fuel Process Technol 145:20–30

Part IX
Ammonia Combustion for Industrial Furnaces

Chapter 44
Review of Fundamental Study on Ammonia Direct Combustion in Industrial Furnaces

Ryuichi Murai, Noriaki Nakatsuka, Hidetaka Higashino, and Fumiteru Akamatsu

Abstract There are sever issues on increasing amount of CO_2 emission in the world. Many studies are devoted on alternative fuels. One potential candidate is the utilization of hydrogen energy which can realize a low-carbon and hydrogen-based society. Ammonia, which is zero emission of CO_2 and useful for hydrogen energy, might play an important role as a clean energy carrier and can be burned directly as a carbon-free fuel. For direct combustion of ammonia in industrial furnaces, there were three issues which were low burning velocity, weaker radiative heat flux and a huge amount of fuel-NO_x emission compared with the conventional natural gas (methane) combustion. In this review, we introduce those solutions in our studies to overcome these three issues of ammonia direct combustion in an industrial furnace. Firstly, the laminar burning velocity of ammonia, which is about 8 cm/s originally, could be improved with the increase of O_2 concentration in the oxidizer (Takeishi et al. in Measurement of ammonia/N_2/O_2 laminar burning velocity under oxygen-enriched air condition [1]; Takeishi et al. in Combustion characteristics of ammonia/N_2/O_2 mixture in oxygen-enriched air condition INFUB [2]). Secondly, radiative heat flux under the condition of the oxygen-enriched (30 vol%) ammonia non-premixed flame formed in a 10 kW test furnace was improved more than that of the methane-air non-premixed flame (Murai et al. in Energy Procedia 120:325–332 [3]). Thirdly, two-stage combustion, which we proposed in this study, could reduce NO_x emissions due to ammonia generated to be unburned intentionally in the upstream of the furnace which played an important role in chemical reduction process.

Keywords Ammonia · NO_x emissions · Combustion · Enriched oxygen · Industrial furnace

Originally published in Journal of the Combustion Society of Japan, Vol. 61 (2019), pp. 320–325 (in Japanese);

R. Murai (✉) · N. Nakatsuka · H. Higashino · F. Akamatsu
Combustion Laboratory, Department of Mechanical Engineering, Osaka University, 2-1 Yamada-oka, Suita, Osaka 565-0871, Japan
e-mail: murai@combu.mech.eng.osaka-u.ac.jp

K. Aika and H. Kobayashi (eds.), *CO_2 Free Ammonia as an Energy Carrier*,
https://doi.org/10.1007/978-981-19-4767-4_44

44.1 Introduction

It is strongly required to reduce carbon dioxide (hereinafter referred to as CO_2) emission to contribute to the prevention of global warming. With the ratification of the Paris Agreement (COP21), it is necessary to achieve a level of ▲ 26.0% (about 1042 million tons-CO_2) in FY2030 and a level of ▲ 80% in FY2050. Japan's Energy White Paper issued in June 2018 emphasizes the utilization of solar and wind power and hydrogen energy in promoting zero carbon footprint. And this is also a crucial strategy in terms of energy security in Japan to reduce the import ratio of the fossil fuels.

Japanese Prime Minister Suga participated in the G20 Summit held online on November 23th 2020 and declared that Japan will reduce greenhouse gas emissions to virtually zero by 2050 and realize a carbon-free society.

In this way, research and development on hydrogen utilization has become an important item not only in Japan but also all over the world. In addition to being a clean energy that does not emit CO_2, the use of hydrogen can be produced from both fossil fuels and renewable energy, which can contribute to the energy security. Japan, which needs to obtain a large amount of hydrogen from overseas, is planning to convert energy into chemical materials (energy carriers) that contain energy as hydrogen, in addition to the method of storing and transporting hydrogen, and convert this into consumption sectors. The method of transporting and storing to energy and returning it to energy were researched and developed in the Strategic Innovation Promotion Program (SIP) "Energy Carrier" of the Council for Science, Technology, and Innovation of the Cabinet Office. And then ammonia is a strong candidate for a feasible energy carrier. Ammonia has been used in large quantities as a raw material for fertilizers and general chemical raw materials, and its production has been established as the Haber–Bosch method. Regarding transportation and storage, the liquefaction temperature under atmospheric pressure is −33 °C, and at the room temperature it liquefies at 8.5 atm, so liquefaction is easier compared with hydrogen. The amount of hydrogen stored per volume of ammonia is about 18 wt%, which is higher than that of hydrogen storage alloys (<5 wt%) and high-pressure hydrogen, and liquid hydrogen and liquid organic hydrides.

Ammonia can be used directly as fuels and does not emit CO_2. Gas turbines, reciprocating engines, industrial furnaces and industrial burners are fields that can be applied to the direct combustion of ammonia. In addition, since power plants and industrial establishments already have a certain scale of ammonia transportation and storage infrastructure, they are able to quickly apply the results of their respective technological developments to current combustion systems. Among them, various industrial furnaces account for about 25% of the energy consumption of society, and their CO_2 emissions are 350 tons per year, so the impact of reducing CO_2 emissions by combustion of ammonia is similar to that of power generation systems. In addition, there are various scales of industrial furnaces, and this can take a risk management, in which small furnaces are introduced and then gradually expanded to large furnaces.

44.2 Ammonia Combustion in an Industrial Furnace

When ammonia is used as a fuel in an industrial furnace, there are three issues, (1) lower flammability, (2) emissions of high nitrogen oxides (hereinafter referred to as NO_x) and (3) low radiative heat flux compared to conventional hydrocarbon fuels due to the absence of carbon in the fuel. The research for solving three issues is described below.

44.2.1 Improvement of Ammonia Burning Velocity in Oxygen-Enriched Condition

The burning velocity of ammonia and its improvement by oxygen enrichment in the oxidizer are described here. First, the laminar burning velocity of the ammonia flame was measured using a premixed mixture of ammonia/oxygen/nitrogen. The frame stabilized on a slot burner is suitable for measuring the laminar burning velocity, as this flame has a uniform velocity distribution in the long axis direction of the slot burner. Figure 44.1a is the outlet shape of the slot burner and (b) is a photograph of the ammonia flame taken with a single-lens reflex camera (Nikon D7000, exposure time 5 μs) from the long axis direction of the slot burner. In the slot burner, the laminar burning velocity (S_u) is calculated from Eq. (44.1) [4].

$$S_u = U_u \sin \alpha \tag{44.1}$$

In Eq. (44.1), S_u is the laminar burning velocity, U_u is the unburned flow velocity and α is the angle (half apex angle) between the flow of the unburned premixture and the flame surface. In this study, the flame was directly photographed from the long axis direction of the slot burner using a high-speed CMOS camera (Phantom V12.1, Vision Research Inc.) for the purpose of determining the half apex angle (α). (Camera conditions: frame rate 1000 fps, exposure time 400 μs). The observed area of the high-speed camera is 97 mm × 73 mm, and the resolution is 152 μm/pixel. The brightness in each pixel was integrated for 1000 times in each condition. The area where the brightness was maximized in the integrated image was defined as the flame surface. The half apex angle (α) was calculated from the Eq. (44.2) by using the slope of the flame surface from 20 to 80% of the height from the end of the burner to the tip of the flame to eliminate the flame extension due to the buoyancy at the tip of the flame and the flame retention effect at the base of the burner.

$$\alpha = \tan^{-1}(z/h) \tag{44.2}$$

In Eq. (44.2), z is the height from an arbitrary point to the flame surface, and h is the distance from the flame surface in the short axis of the slot burner to an arbitrary point. In the slot burner, the axis is symmetrical at the center in the minor axis of the

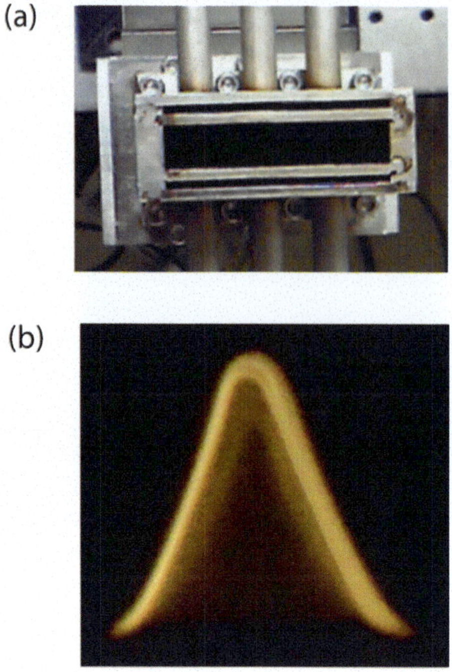

Fig. 44.1 Photograph of the slot burner from the top view (**a**) and the cross-sectional view of ammonia flame enriched oxygen premixed laminar (**b**). © Combustion Society of Japan 2019

burner. Therefore, the half apex angle (α) can be calculated using the left and right flame surfaces formed on the left and right of the central axis. In this experiment, the laminar burning velocity was calculated from the average value of the half apex angle (α) obtained from the left and right flame surfaces. Here, the oxygen concentration in oxidizer was expressed according to Eq. (44.3).

$$\Omega = \frac{Q_{O_2}}{Q_{O_2} + Q_{N_2}} \tag{44.3}$$

where Q_{O_2} is the volumetric flow rate of oxygen and Q_{N_2} is the volumetric flow rate of nitrogen in the oxidizer. In this study, we conducted experiments in the range of $0.21 < \Omega < 0.35$, assuming applications to industrial furnaces.

The laminar burning velocity was measured under the three conditions of $\Omega = 0.28$, $\Omega = 0.30$ and $\Omega = 0.35$ for oxygen-enriched conditions, and the equivalent ratio (ϕ) was changed by 0.10 in the range of 0.80–1.40 and was changed at intervals of 0.05 near the condition of the equivalence ratio where the burning velocity was maximized. Temperature of the premixture was constant at 300 K, and the volumetric flow rate of the supplied premixture was adjusted to the laminar burning velocity so that a laminar flame is stabilized on the slot burner. In this experiment. We changed the volumetric flow rate from 8 to 20 L/min (the velocity at the burner outlet was from

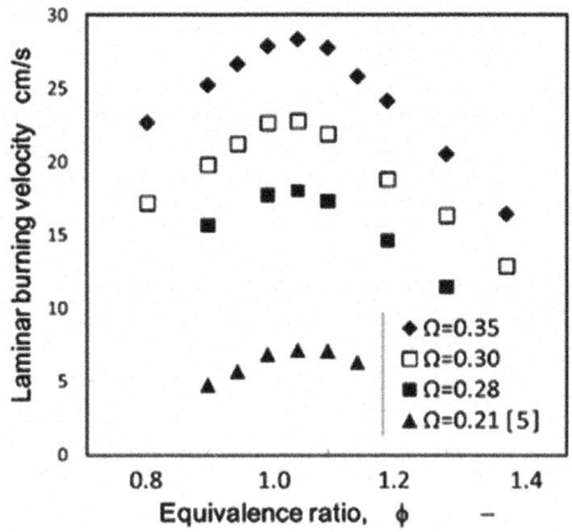

Fig. 44.2 Relationship with laminar burning velocity, degree of enriched oxygen concentration and equivalence ratio (ϕ) [5]. © Combustion Society of Japan 2019

41 to 104 cm/s). Then, the effects of the equivalent ratio ϕ and the oxygen-enriched ratio Ω on the laminar burning velocity were evaluated.

Figure 44.2 shows the measurement results of the laminar burning velocity in the ammonia/N_2/O_2 premixed flame [5]. The horizontal axis is the equivalent ratio and the vertical axis is the laminar burning velocity. From Fig. 44.2, the laminar burning velocity increases as the oxygen concentration increases. In the condition of $\Omega = 0.28$ at $\phi = 1.05$, laminar burning velocity is 18.0 cm/s. In the condition of $\Omega = 0.30$ at $\phi = 1.05$ and $\Omega = 0.35$ at $\phi = 1.05$, laminar burning velocity are 22.8 cm/s and 28.3 cm/s, respectively.

It is considered that this is because the combustion reaction rate increased due to the increase in the adiabatic flame temperature due to the increase in the oxygen-enriched ratio in the supplied premixture. This is 3–4 times value of the laminar burning velocity in the ammonia/air premixed flame, and this results clarify that the burning velocity can be enhanced by oxygen-enriched combustion.

44.2.2 Progress of Radiative Heat Flux by Oxygen-Enriched Ammonia Combustion

It is necessary to increase the radiative heat flux in order to apply the ammonia combustion to boilers or industrial furnaces because the radiative heat flux in ammonia combustion is weaker than that of conventional fossil fuels. The flame temperature can be increased with the oxygen-enriched combustion described in the previous section. If the temperature that affects the amount of radiation can be raised by oxygen-enriched combustion, low radiative heat flux, which is disadvantage in ammonia combustion, can be strengthened. Therefore, we applied oxygen-enriched

Fig. 44.3 Photograph of the 10 kW furnace. © Combustion Society of Japan 2019

combustion to the ammonia combustion, using a 10 kW model furnace. We measured radiative heat flux and light emission spectra of ammonia flame in order to clarify experimentally that oxygen-enriched combustion is effective to increase the radiative heat flux in ammonia combustion.

44.2.2.1 10 kW Experimental Furnace

The experimental equipment consists of a 10 kW model furnace, a radiative heat flux detector and an infrared spectrum meter (FTIR Rocket 2.5–12 μm, ARCoptix SA). Figure 44.3 shows a photograph of the 10 kW model furnace, and the schematic diagram is shown in Fig. 44.4. The total length of the furnace is 1000 mm and the inside wall is surrounded by heat insulating material. On the top of the furnace, there are six ports equipped with thermocouples. On the lower part of the furnace, there are six ports for measuring radiative heat flux and light emission spectra from the inner wall and the flame. Fuels and oxidizers are supplied by using the mass flow controllers (HORIBASTEC N series) to the banner. Burner nozzle had coaxial double tube and fuel flows through the inner tube and oxidizer flows through the outer tube. The coaxial diffusion jet flame are stabilized in test furnace.

Table 44.1 shows the combustion conditions in this experiment. The volumetric fuel flow rate for each condition was determined as the lower heating value was 10 kW in each condition.

44.2.2.2 Measurement of Radiative Heat Flux

The radiative heat flux detector in this experiment (Fig. 44.5) consists of a radiation sensor (CAPTHERM-R) and the cooling unit. Blackbody plates and reflectors are alternately deposited on the endothermic film on the surface of the radiation sensor, and only the radiative heat flux generate temperature difference at each contact point of the thermopile, generating an electromotive force in the sensor. The heat flux

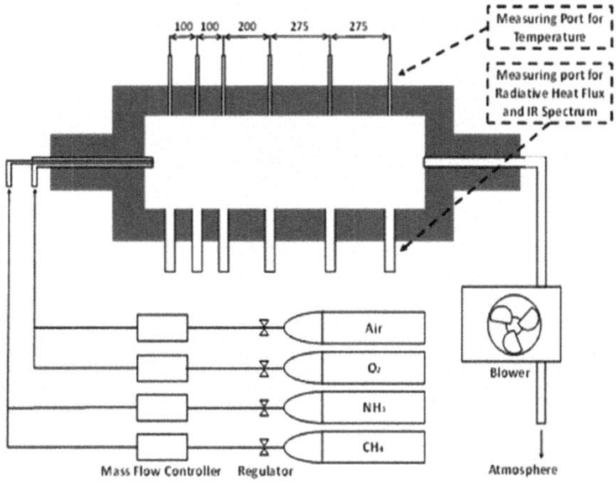

Fig. 44.4 Schematic diagram of the 10 kW model furnace, the gas supply lines and exhaust lines. © Combustion Society of Japan 2019

Table 44.1 Experimental conditions on combustion

	Fuel	Flow rate [Nm^3/h]	Lower heating value [kW]	O_2 concentration [Ω]	Air ratio [α]
Case1	NH_3	2.55	10	0.21	1.05
Case2	CH_4	1.01	10	0.21	1.05
Case3	NH_3	2.55	10	0.30	1.05

that reaches the sensor due to convection heat transfer is uniformly transferred to the inside of the sensor, and that generate no temperature difference at each contact point of the thermocouple. Before measuring the radiative heat flux in ammonia combustion, a preliminary experiment was conducted using a blackbody furnace and an infrared spectrometer to confirm the wavelength sensitivity dependence of the radiative heat flux detector. As a result, it was confirmed that there was almost no difference in wavelength sensitivity for each absorption band of this detector. For example, the difference between the sensitivity in 2.5–3.1 μm (strong absorption band of H_2O and CO_2) and the sensitivity in 4.2–5.0 μm (strong absorption band of CO_2) is about 0.1%, that is sufficient to measure precisely radiative heat flux in the test furnace. Figure 44.6a shows a photograph of the radiative heat flux measurement system with a water-cooling system on the radiation sensor and Fig. 44.6b shows its schematic diagram. A water-cooled chiller was connected to the radiative heat flux detector to keep the temperature of the measuring unit at 287 K. In addition, nitrogen was supplied to the inside of the water-cooling tube. The flow rate for each radiation sensor was kept at 1 L/min to prevent the condensation of water vapor on the surface and contact with high temperature combustion gases on the surface. The

Fig. 44.5 Photograph of measurement sensor for radiative heat flux. © Combustion Society of Japan 2019

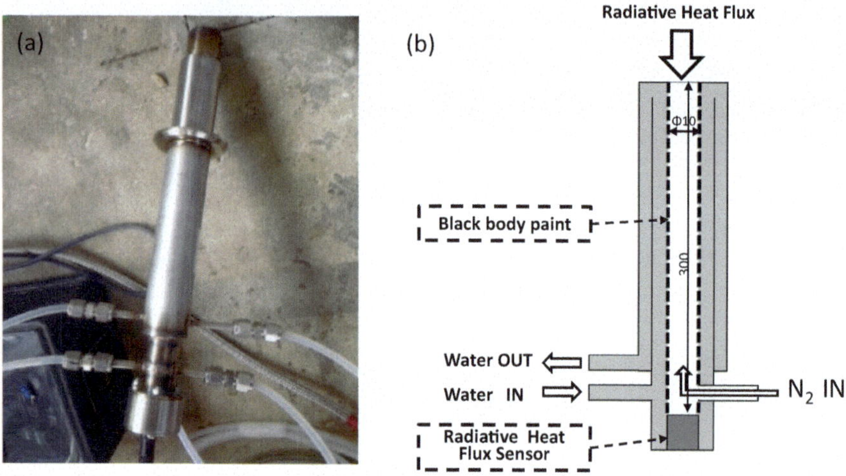

Fig. 44.6 Overview photograph of measurement detector system for radiative heat flux (left) and its schematic structure (right). © Combustion Society of Japan 2019

nitrogen supply affected only 2% value difference for both the furnace temperature and the radiative heat flux. The output of the sensor was recorded at 5 Hz, and the average value of 900 measurements was used to analyze the results. There are four measurement points in the axial direction from the burner side: $X = 100$ mm, 500 mm, 775 mm and 1050 mm. The furnace wall temperature was measured 6000 times at 5 Hz at the same time as the radiative heat flux measurement and the average value was used to analyze the results.

44.2.2.3 Results of Radiative Heat Flux and Oxygen-Enriched Combustion

Figure 44.7 shows the results of the radiative heat flux measurement. The measured values were normalized with the maximum value q_{r1} (the measured value at the port of 1000 mm) of the radiative heat flux under the conditions of ammonia/air combustion (Case 3 of Table 44.1). Figure 44.8 shows the results of furnace wall temperature measurement. Comparing Figs. 44.7 and 44.8, measurement results of the radiative heat flux and the wall temperature shows the same trend. From Fig. 44.7, the radiative heat flux in the methane/air combustion (Case 2 of Table 44.1) showed convex curve and the maximum value was at the position of $X = 500$ mm to 750 mm. On the other hand, the radiative heat flux measured in the ammonia/air combustion (Case 1 of Table 44.1) showed the maximum value at the port position of $X =$ 1000 mm, and the radiative heat flux increases from 500 mm to the downstream as the distance from the burner increases. This result indicates that the distribution of the radiative heat flux in ammonia combustion appears in the downstream as compared with that in methane combustion. Comparing the maximum values of the radiative heat flux, ammonia combustion showed the maximum value about 22% lower than that of methane combustion. For the combustion condition of the oxygen-enriched (30 vol%) ammonia (Case of Table 44.1), the position of the maximum value of the radiative heat flux was at the position of $X = 500$ mm. From this, distribution of radiative heat flux moved upstream when ammonia was burned under the oxygen-enriched conditions. The maximum value of the radiative heat flux under the oxygen-enriched 30% condition (Case 3 of Table 44.1) at $X = 500$ mm was about 46% higher than the maximum value of the radiative heat flux of ammonia combustion without oxygen enrichment (Case 1 of Table 44.1) at $X = 1000$ mm. From these results, it was confirmed that by increasing the oxygen concentration Ω to 0.30, stronger radiation was obtained than on the case of methane/air combustion (Case 2 of Table 44.1).

The result of the above-mentioned radiative heat flux measurement was the total of the radiative heat flux from the combustion gas and the furnace wall.

When we aim to clarify the radiative heat flux characteristics due to combustion, it is necessary to consider the radiative heat flux from the combustion gas separately from the furnace wall. Therefore, in this study, when the radiative heat flux measured as the radiative heat flux from the furnace wall, we terminated the fuel supply to turn

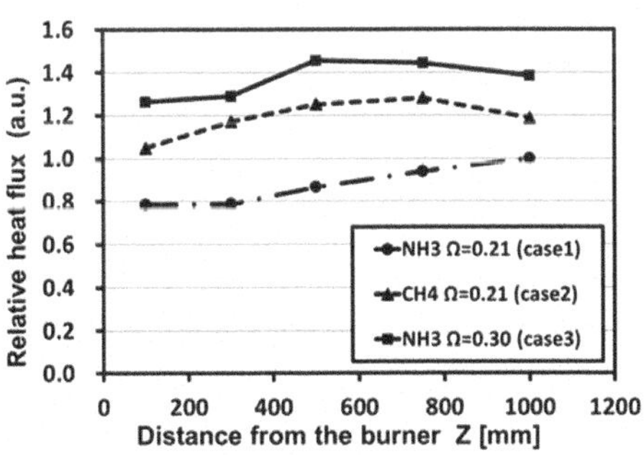

Fig. 44.7 Experimental results of radiative heat flux measurements. Circle dot and dot-and-dash line, NH_3, $\Omega = 0.21$, square dot and dot-line, CH_4, $\Omega = 0.21$, triangle dot and black line, NH_3, $\Omega = 0.30$ (enriched O_2). © Combustion Society of Japan 2019

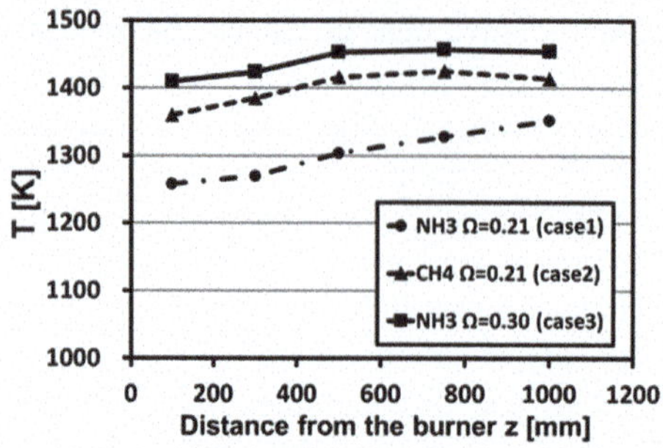

Fig. 44.8 Experimental results of temperature measurements. Circle dot and dot-and-dash line, NH_3, $\Omega = 0.21$, square dot and dot-line, CH_4, $\Omega = 0.21$, triangle dot and black line, NH_3, $\Omega = 0.30$ (enriched O_2). © Combustion Society of Japan 2019

off the combustion and purged the combustion gases inside the furnace with air. In addition, the furnace wall temperature was also measured when measuring the radiative heat flux from the furnace wall. Figure 44.9 shows these results. The blue line in the figure is the result of radiative heat flux after turning off the combustion that corresponds to the radiative heat flux from the furnace wall. It was found that about 95% of the radiative heat flux during the combustion was the radiative heat flux from the furnace wall. From these results, it was found that for heating in industrial furnaces, it is sufficient to use the radiation from the furnace wall without using the radiation emitted directly from flames and soot in the furnaces.

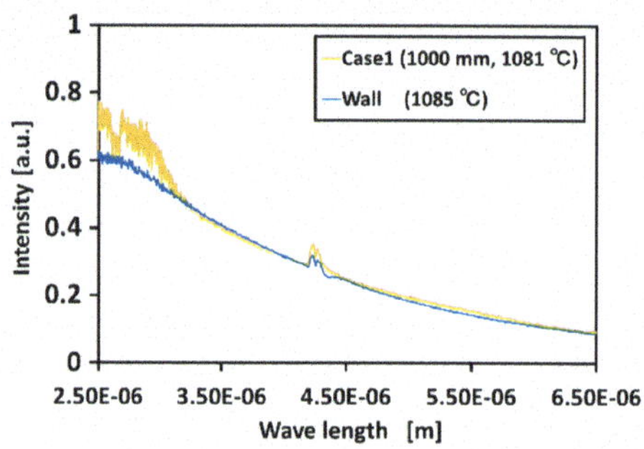

Fig. 44.9 Experimental results of the radiative heat flux comparison with radiative heat flux of the case 1 and turn off combustion at $Z = 1000$ mm. Yellow line; the radiative heat flux from the total heat flux within the measured solid angle. Blue line; the radiative heat flux only from the furnace wall in case without combustion. © Combustion Society of Japan 2019

44.2.3 Reduction of NO_x Emissions by Two-Stage Combustion

In hydrocarbon fuels, the production of NO_x is dominated by the following chemical reaction (44.4) so called Thermal-NO_x in the high temperature range (1500 °C or higher). On the other hand, the Fuel-NO_x, which comes from nitrogen derived from the fuel itself, are produced when ammonia is used as fuel. This chemical reaction is shown by Eq. (44.5) [6].

$$N_2 + O \rightleftarrows NO + N \tag{44.4}$$

$$NH_i \rightleftarrows N, N + OH \rightleftarrows NO + H \tag{44.5}$$

The operating temperature range in many industrial furnaces is 1000–1500 °C, which is relatively low compared to that in gas turbines. When ammonia is burned in an industrial furnace designed for conventional hydrocarbon-based fuels, the Fuel-NO_x shown in Eq. (44.5) is dominant, so a large amount of NO_x is emitted regardless of the low temperature. Under these circumstances, we applied the two-stage combustion to ammonia combustion, obtained a significant NO_x reduction effect, and realized ammonia combustion that achieved the environmental regulation for current industrial furnaces in Japan as shown in Fig. 44.10.

In general, the two-stage combustion aims at reducing NO_x emissions by supplying oxidizers from multiple positions to divide the combustion region into

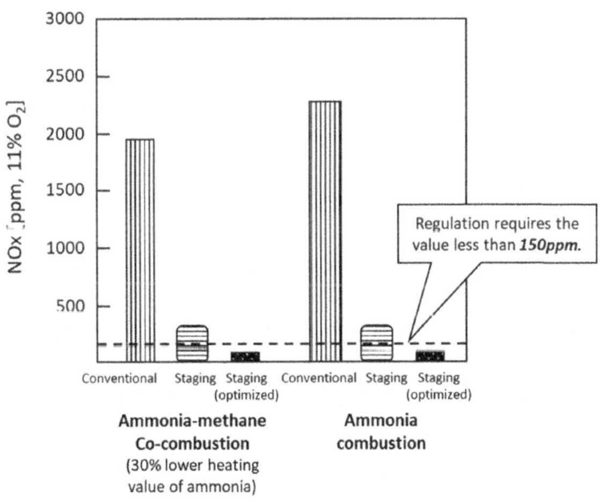

Fig. 44.10 Experimental results of two-stage combustion in a 10 kW test furnace. Combustion conditions were ammonia methane co-combustion (30% lower heating value of ammonia) and ammonia combustion

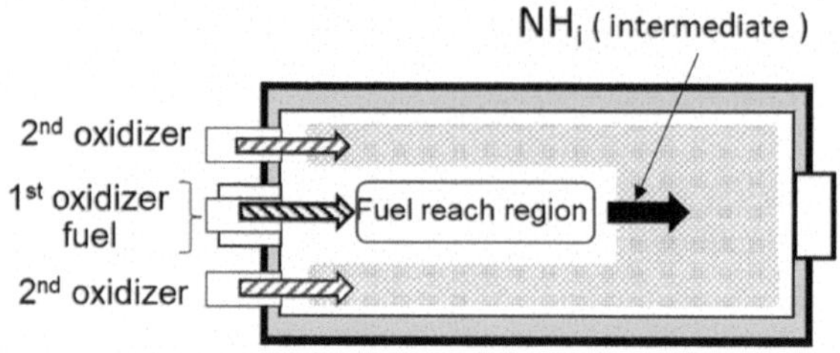

Fig. 44.11 Conceptual diagram of the two-stage combustion in a 10 kW model furnace

two regions, a fuel-rich region, and a fuel-lean region, to avoid the formation of local high temperature regions.

The 10 kW experimental furnace (Fig. 44.3) introduced in Sect. 44.2.2 was used in this study. As shown in Fig. 44.11, this burner had two portions. One of them is the coaxial double tubes for introducing the fuel and the primary oxidizer, and the other consists of two nozzles for introducing the secondary oxidizer separately on the left and right sides of the burner to rearise the two-stage combustion.

It is revealed that NO_x emissions (O_2 11% conversion) change significantly depending on the burner structure, fuel, and the conditions of secondary oxidizer supply. When methane is used as the hydrocarbon fuel, NO_x emissions were reduced from 120 to 50 ppm by the two-stage combustion. On the other hand, under the combustion conditions in which ammonia was mixed with methane at a volume fraction of 30% (hereinafter referred to as 30% ammonia mixed combustion), NO_x emissions could be significantly reduced from 2000 to 120 ppm. Similarly, under 100% ammonia combustion, the amount of NO_x emissions could be reduced from 2350 to 120 ppm.

Compared with methane combustion, NO_x can be significantly reduced by the two-stage combustion developed in this study for ammonia combustion. The temperature inside the 10 kW test furnace in this study is only about 50–100 °C lower than that in the case of methane combustion. This significant reduction of NO_x emissions in the ammonia combustion cannot be explained only by avoiding the existence of the high temperature region.

The mechanism of NO_x reduction in ammonia combustion is considered as follows. In the upstream area of the combustion furnace, the supply of oxidizer is divided into the primary oxidizer and the secondary oxidizer. The combustion region where the fuel and the primary oxidizer were supplied was in a fuel-rich condition, and it was possible to intentionally generate unburned ammonia or intermediate products of ammonia. The produced unburned ammonia or intermediate product flowed to the downstream of the furnace, and NO_x was reduced by the chemical reactions of Eqs. (44.6) and (44.7).

$$NH + NO \rightleftarrows N_2O + H \tag{44.6}$$

$$N_2O + H \rightleftarrows N_2 + OH \tag{44.7}$$

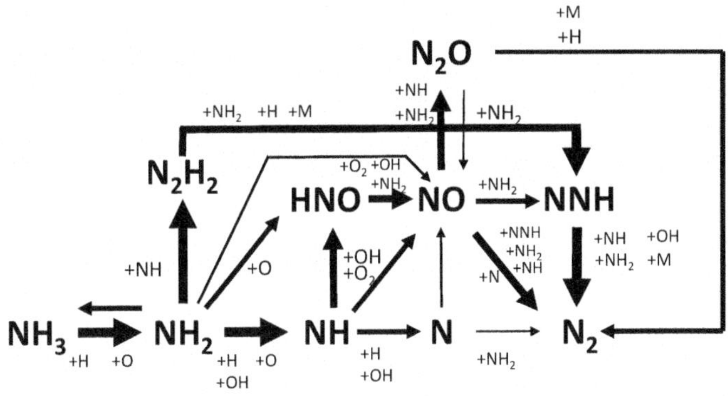

Fig. 44.12 Reaction path of the combustion condition with NH_3, equivalence ratio = 0.8, $\Omega = 0.28$ (enriched O_2) calculated using the Lindstedt mechanism. © Combustion Society of Japan 2019

A detailed elementary reaction pathway diagram was derived by applying a one-dimensional freely propagation flame model to the intermediate product based on the Lindstedt mechanism [7]. Figure 44.12 shows the result. Intermediate products of ammonia here include NH_2, NH and HNO. As shown in Fig. 44.12, NO has a reaction pathway that is reduced to N_2 by the Eq. (44.7). Although details are not given here, the emission spectrum of N_2O, which is an intermediate product in Eqs. (44.6) and (44.7), is measured by optical measurement using the Fourier Transform Infrared Spectroscopy (FTIR) method, and the gas-sampling probe inserted into the furnace is used in these measurements. We could detect unburned ammonia in the furnace by the gas-sampling probe and FTIR.

It was confirmed that an equal results demonstrated in the 10 kW test furnace was reproduced in in the 100 kW test furnace, which is close to the burner scale of actually working furnaces in the industrial field.

44.3 Conclusions

Issues in direct combustion of ammonia in an industrial furnace are (1) instability of combustion due to low burning velocity, (2) decrease in radiative heat flux, (3) large amount of NO_x emissions. It was clarified that (1) and (2) can be solved by oxygen-enriched combustion and (3) can be solved by applying the two-stage combustion. The strategy of two-stage combustion applied to ammonia combustion is different from the case of conventional hydrocarbon fuels, which aims to avoid local high temperature region. It was clarified that applying the two-stage combustion, unburned ammonia or intermediate products of ammonia was generated and flowed to the downstream of the furnace where the NO_x is reduce by the reduction chemical reaction.

Acknowledgements This work was supported by Council for Science, Technology, and Innovation (CSTI), Cross-ministerial Strategic Innovation Promotion Program (SIP), "Energy Carrier" (Funding agency: JST).

We thank TAIYO NIPPON SANSO Corporation, Dr. Hiroyuki Takeishi, Mr. Wataru Arita, Mr. Ryohei Omori, Mr. Ryuki Kano, and Mr. Yuji Tada in the combustion engineering laboratory of Osaka University, and for kind help in accomplishing this study.

References

1. Takeishi H, Hayashi J, Suzuki M, Iino K, Akamatsu F (2014) Measurement of ammonia/N_2/O_2 laminar burning velocity under oxygen-enriched air condition. In: Proceedings of grand renewable energy 2014
2. Takeishi H, Hayashi J, Iino K, Akamatsu F (2015) Combustion characteristics of ammonia/N_2/O_2 mixture in oxygen-enriched air condition INFUB 2015
3. Murai R, Omori R, Kano R, Tada Y, Higashino H, Nakatsuka N, Hayashi J, Akamatsu F, Iino K, Yamamoto Y (2017) The radiative characteristics of NH_3/N_2/O_2 non-premixed flame on a 10kW test furnace. Energy Procedia 120:325–332
4. Mizutani Y (2002) Combustion engineering version 3 82. Morikita Publishing (in Japanese)
5. Takizawa K et al (2008) Burning velocity measurements of nitrogen-containing compounds. J Hazard Mater 155:144–155
6. Takeishi H (2016) Basic study on the combustion of laminar premixed flame using ammonia/N_2/O_2 as a fuel. Doctoral dissertations of Osaka University (in Japanese)
7. Niioka T et al (2001) Basic combustion phenomena. Ohm Sha, pp 260–271 (in Japanese)

Chapter 45
Development of Impinging Jet Burner Using Ammonia Fuel for Degreasing Steel Sheets

Yoshiyuki Hagihara, Yasuyuki Yamamoto, Masaaki Numata, and Taizo Matsuda

Abstract Ammonia has been expected as a hydrogen energy carrier to prevent global warming because hydrogen has no carbon and it means CO_2 free combustion. However, there are some challenges to burn ammonia. The reasons are its low combustion intensity, low radiation intensity, and high NO_x emission. This paper shows the potential of a burner using ammonia fuel in an industrial furnace. Burner itself should be developed to burn ammonia fuel which is mixed in natural gas. Several types of burner design and several ratios of ammonia volume in natural gas were evaluated. Arrangement of burners in an industrial furnace is also important to get uniform heating on a steel sheet because dirt and oil on the steel sheet should be burned for degreasing. Distance of burners in one line and distance of lines were decided with evaluating uniformity of heating. Minimal capacity of furnace which has a developed impinge jet burner with optimal arrangement was installed in a continuous galvanizing line which produces galvanized steel sheets being supplied to automotive industry, appliance industry, and construction industry. The developed impinge jet burners performed as expected ability in the industrial furnace.

Keywords Ammonia · CO_2 free · Impinging jet burner · Industrial furnace · Galvanizing

45.1 Introduction

In the Strategic Innovation Creation Program (SIP), which is one of science and technology policies of the Cabinet Office, we have focused on the use of ammonia as

Originally published in Journal of the Combustion Society of Japan, Vol. 61 (2019), pp. 326–330 (in Japanese); Copyright © Combustion Society of Japan. All Rights Reserved.

Y. Hagihara (✉) · Y. Yamamoto
Taiyo Nippon Sanso Co Ltd, 3054-3 Shimokurosawa, Takane-cho, Hokuto, Yamanashi 408-0015, Japan
e-mail: hagiharay.qxr@tn-sanso.co.jp

M. Numata · T. Matsuda
Nippon Steel Nisshin Co Ltd, 5 Ishizunishimachi, Nishiku, Sakai, Osaka 592-8332, Japan

K. Aika and H. Kobayashi (eds.), *CO_2 Free Ammonia as an Energy Carrier*,
https://doi.org/10.1007/978-981-19-4767-4_45

a “Hydrogen energy carrier” and have been conducting research on transportation, stockpiling, conversion as energy, and utilization. Ammonia is hard to burn and has a low combustion rate and flame temperature. In addition, the amount of radiant heat transfer is low because CO_2, which is a radiant gas, is not generated. Therefore, there are many technical problems in using industrial furnaces.

On the other hand, it is possible to suppress the generation of CO_2 during combustion, and it is possible to remove the generated NO_x with the current general equipment such as a denitration device. In addition, since technologies for suppressing the generation of NO_x such as low NO_x burners have been developed, ammonia can be expected to be used as a fuel for industrial furnaces in the future.

We aim to introduce ammonia combustion technology into the industrial furnace. Therefore, we participated in SIP and developed an ammonia co-firing furnace of impinging jet burner using Ammonia fuel for degreasing steel sheets and conducted its evaluation test.

In this program, we have developed the following technologies.

1. Performance evaluation of the ammonia co-firing imping jet burner for degreasing steel sheet.
2. Optimization of ammonia co-firing impinging jet burners placement in steel sheet degreasing furnaces.
3. Performance evaluation of the ammonia co-firing furnace with impinging jet burners in a continuous hot-dip galvanizing process.

45.2 Outline of Galvanized Steel Sheet Manufacturing Process and Evaluation Test Furnace

45.2.1 Galvanized Steel Sheet Manufacturing Process

Figure 45.1 shows the manufacturing process of a galvanized steel sheet by a general continuous hot-dip galvanizing line. In the cold rolling process, a thin steel sheet (called Raw sheet) is manufactured and then hot-dip galvanized continuously. Oil and iron powder generated in the rolling process adhere to the surface of the raw sheet. (called dirt component) The degreasing furnace removes this dirt component. In order to keep the appearance of the product good and to coat zinc uniformly to the surface of the raw sheet, it is important to remove the dirt component.

45.2.2 Evaluation Test Furnace

Figure 45.2 shows a degreasing furnace in a general continuous hot-dip galvanizing line. In the degreasing furnace, a flame ejected in parallel with the steel sheet heats the steel and removes dirt components at the same time. In this method, if the dirt is

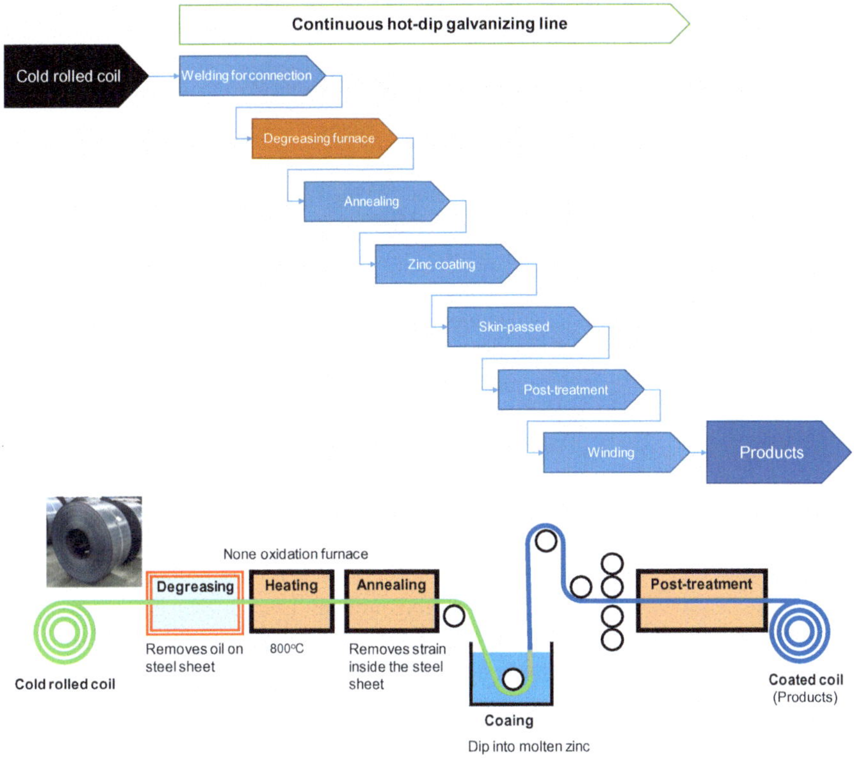

Fig. 45.1 Overview of continuous hot-dip galvanizing line

unevenly distributed on a part of the raw sheet to be plated, the dirt component cannot be sufficiently removed, which causes quality deterioration. On the other hand, in the evaluation test furnace, the flame of the burner hits the steel sheet directly in the vertical direction. In this method the heat transfer efficiency from the burner flame to steel sheet is improved, and the installation space of the burner can be reduced. In addition, the kinetic energy of the burner flame can be expected to effectively remove local dirt components. In this furnace, impinging jet burner using oxygen combustion is used. Therefore, even if ammonia is used as the fuel, a high-speed flame can be formed at a high temperature. Therefore, it is effective in a degreasing furnace that directly heats a steel sheet (Table 45.1).

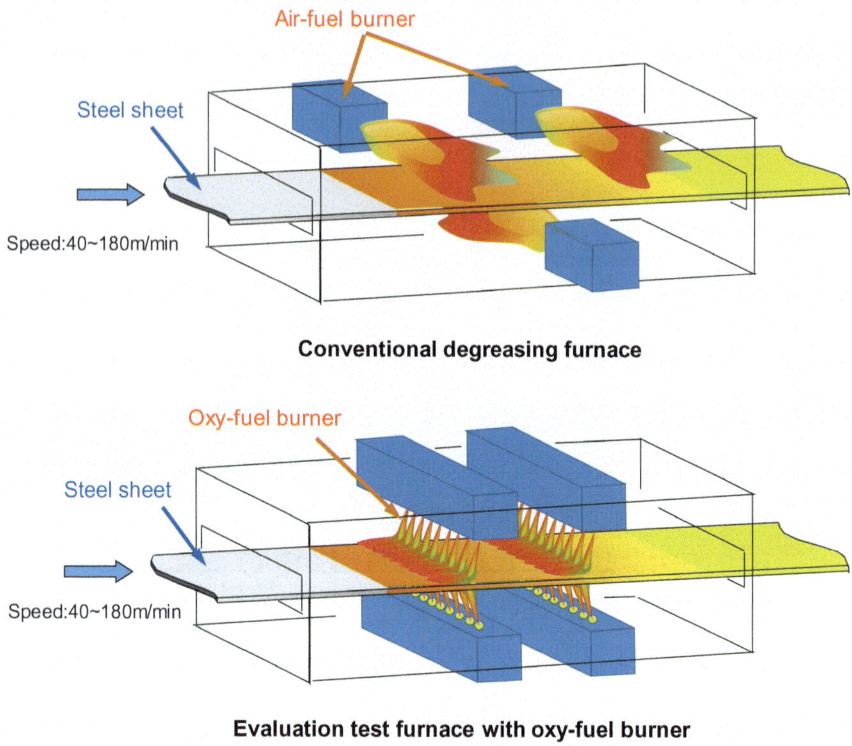

Fig. 45.2 Schematic diagram comparing degreasing furnace

Table 45.1 Comparison of existing degreasing furnace and evaluation test furnace

Item		Conventional furnace	Evaluation test furnace
Burner	Heat transfer	Radiative	Impinging jet
	Burner type	Diffuse combustion	Nozzle mix
Fuel		Natural gas	Natural gas + Ammonia
Oxidant		Air	Oxygen
Heat transfer efficiency		40–60%	80%

45.3 Performance Evaluation of the Ammonia Co-firing Impinging Jet Burner for Degreasing Steel Sheet

Ammonia is a fuel with a slow combustion rate. Therefore, in order to increase the combustion rate of the fuel, pure oxygen was used as the oxidant. In addition, the nozzle shape was optimized to improve heat transfer efficiency. The burner for

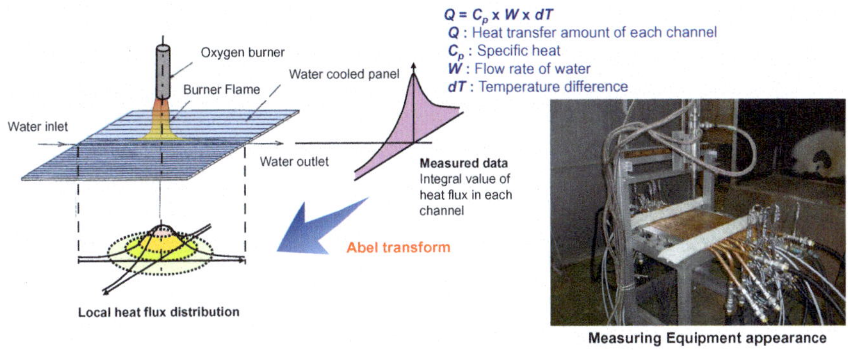

Fig. 45.3 Collision convection heat flux measuring equipment

degreasing the steel sheet is required to have the ability to locally heat with high speed flames. Therefore, the developed burner has a combustion chamber at the nozzle tip. In addition, local heating is possible by ejecting the fuel gas diagonally toward the central axis of the burner so as to focus. To evaluate the performance of the burner, the heat flux distribution was measured using the collision convection heat flux measuring device shown in Fig. 45.3. The measurement was performed at the Akamatsu laboratory of Osaka University.

A water-cooled panel is installed in the measuring equipment. A thermocouple is installed at the inlet and outlet of the water cooling panel to measure the temperature difference between the inlet and outlet of water. At the same time, the flow rate of the cooling water flowing through the panel is also measured. The flame of the burner collides with this water-cooled panel to heat the water flowing inside. As a result, the amount of heat transferred from the flame to the water cooling panel can be obtained, and the heat flux distribution can be obtained.

As a result of measurement using the developed burner, it was confirmed that the heat flux was concentrated in the center of the flame when it collided with the water cooling panel.

45.4 Optimization of Ammonia Co-firing Impinging Jet Burners Placement in Steel Sheet Degreasing Furnaces

To optimize the placement of the degreasing burner, the steel sheet heating test equipment shown in Fig. 45.4 was used. With this device, a test piece of steel sheet can be placed on a trolley and moved. Also, multiple degreasing burners can be installed to heat the test piece. The moving speed of the test piece placed on the trolley was adjusted so that the heating temperature reached the target temperature. The heated test piece was analyzed for oil residue adhering to the surface. As for the degreasing burners, the burners were arranged in two rows. As a result of an evaluation test using this device, it was confirmed that the heating efficiency was maximized by arranging the burner spacing at 60 mm, the distance from the test piece to the burner tip at 200 mm, and the two rows of burners in a staggered arrangement. This evaluation test was conducted with Taiyo Nippon Sanso Co. For the evaluation of degreasing performance, test pieces were analyzed using a glow discharge light emitting surface analyzer. Analysis was performed in the depth direction from the surface of the test piece. The oil residue on the surface of the test piece was evaluated by the integrated value of the carbon content obtained in the sputtering time of 10 s. (Figs. 45.5, 45.6 and 45.7).

Generally, the rolling oil used in the rolling process can be removed by heating to about 400 °C. In the degreasing performance evaluation test using a degreasing burner, the effect of removing oil by the burner was observed when the temperature of the test piece reached 200 °C, and the same effects as when removing with alcohol were observed at 400 °C or higher. Methane was used as the fuel for the degreasing burner. The degreasing performance did not change even when ammonia was mixed by 30% in the fuel.

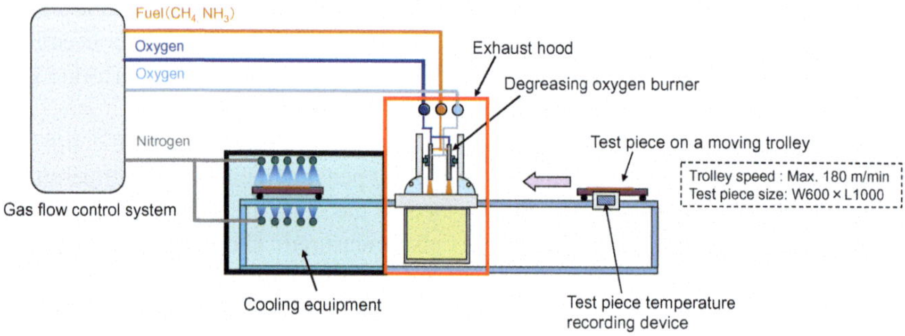

Fig. 45.4 Schematic diagram of steel sheet heating test equipment

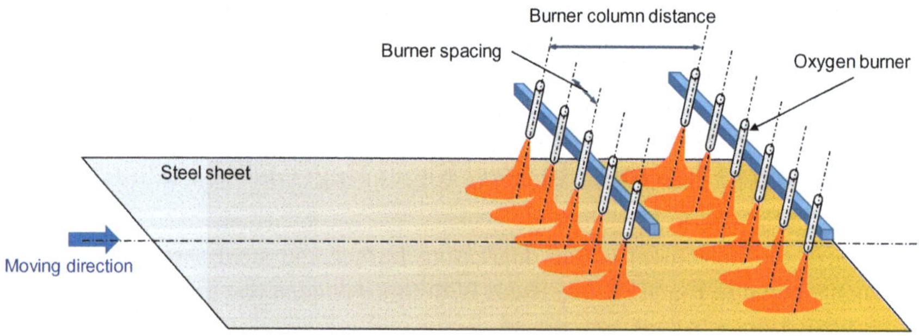

Fig. 45.5 Schematic diagram of the arrangement of the degreasing oxygen burner

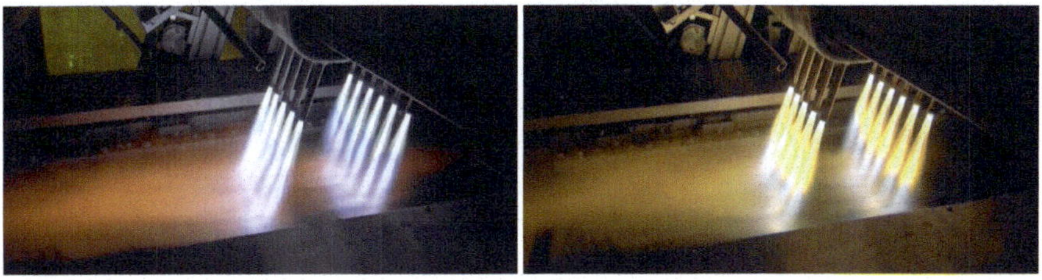

Methane combustion | Methane-ammonia(30%) combustion

Fig. 45.6 Flame condition of degreasing burner. © Combustion Society of Japan 2019

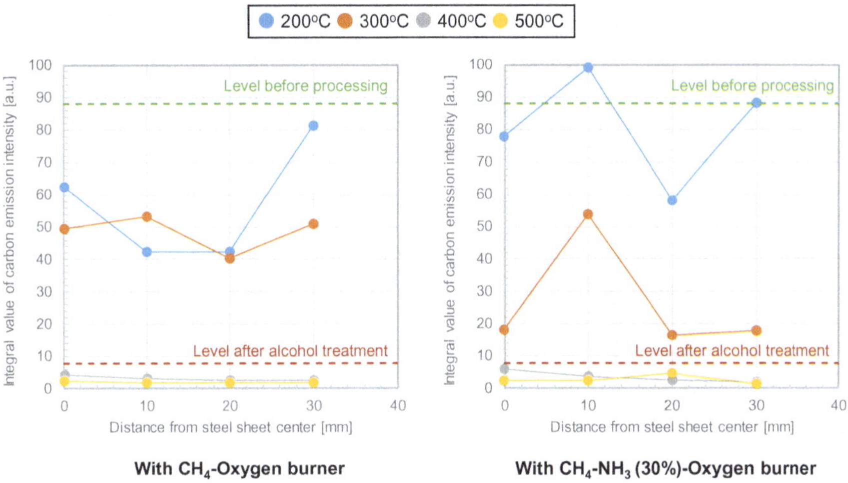

Fig. 45.7 Measurement results of carbon residue of test pieces heat-treated with a degreasing burner

45.5 Performance Evaluation of the Ammonia Co-firing Furnace with Impinging Jet Burners in a Continuous Hot-Dip Galvanizing Process

To evaluate the degreasing performance, an evaluation test furnace was installed in the continuous hot-dip galvanizing line of Nippon Steel Nisshin Sakai works. Based on the development test results of the degreasing burner, the evaluation test furnace has the specifications and details shown in Tables 45.2, 45.3, 45.4, Figs. 45.8, 45.9, and 45.10. In the actual manufacturing process, oil adheres to both sides of the steel sheet, so it is necessary to degreasing both sides. However, in the evaluation test furnace, the burner was placed on only one side because the purpose was to evaluate the degreasing performance of the burner. Thirty-one burners were installed in the evaluation test furnace to enable degreasing of steel sheet with a width of 900 mm. Burner staggered 16 in the first row and 15 in the second row. The burner spacing

Table 45.2 Specification of degreasing burner

Capacity	34.8 kW per burner
Number	31 (16 in 1st row and 15 in 2nd row)

Table 45.3 Specification of the evaluation test furnace

Input	1078 kW
Natural gas	91.0 Nm^3/h
Ammonia	13.3 Nm^3/h
Oxygen	241 Nm^3/h

Table 45.4 Production condition of steel sheet

Thickness	0.20–0.40 mm
Width	900 mm
Speed	50 m/min
Temperature	200–400 °C

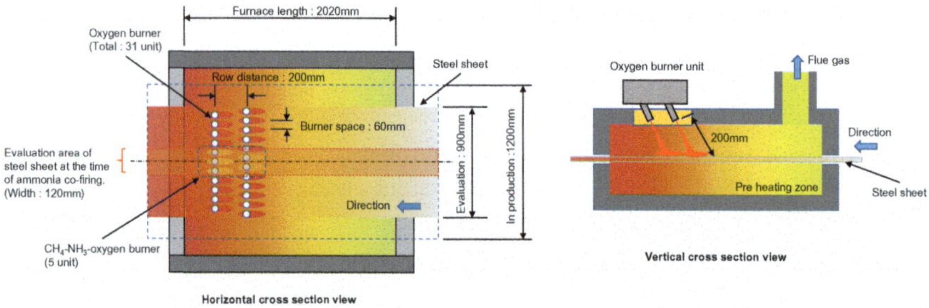

Fig. 45.8 Schematic diagram of evaluation test furnace

(60 mm) and the distance from the steel sheet (200 mm) were determined based on the results of the development test. By supplying ammonia to the five burners placed in the center, the degreasing performance during ammonia co-firing was evaluated.

45.6 Evaluation Result

In the evaluation test, the speed of the steel sheet was kept constant, and the output of the degreasing burner was adjusted so that the temperature of the steel sheet at the exit of the evaluation test furnace became a target temperature. By using the steel sheet temperature measured at the exit of the evaluation test furnace, we investigated the relationship between amount of heat transfer to the steel sheet and heat transfer rate, as well as the relationship between the amount of heat transfer and burner output (Fig. 45.11).

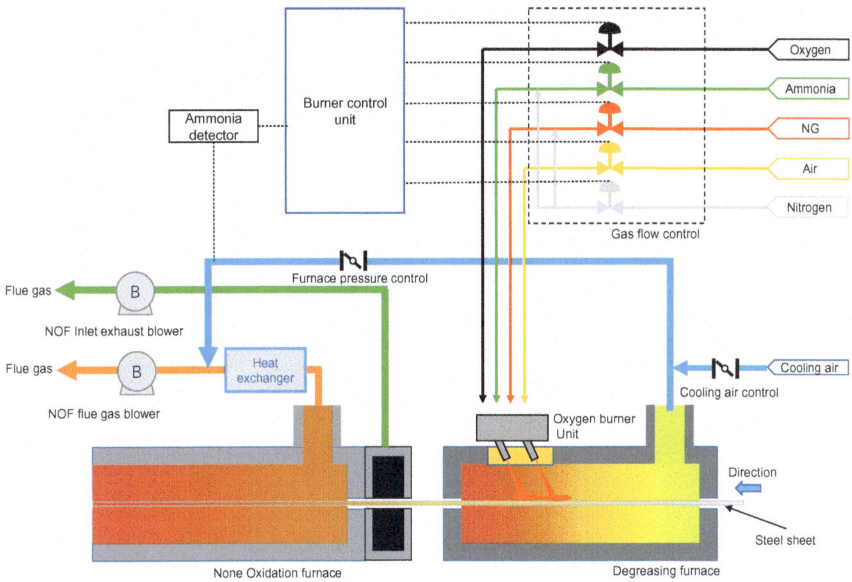

Fig. 45.9 Schematic diagram of gas flow control of evaluation test furnace

Fig. 45.10 Evaluation test furnace (**a**) and degreasing oxygen burner control unit (**b**). © Combustion Society of Japan 2019

The thermal transfer efficiency is as shown in Fig. 45.12. The thermal efficiency reached the target value of 30% at any of the burner output of 60, 70, and 80%.

Fig. 45.11 Combustion flame of degreasing oxygen burner in the evaluation test furnace. © Combustion Society of Japan 2019

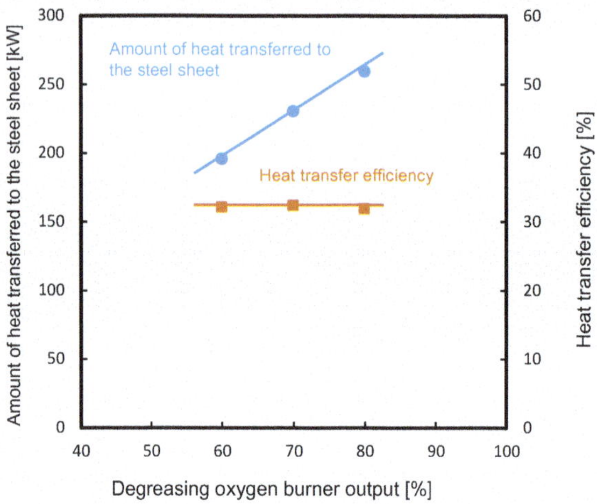

Fig. 45.12 Relationship between burner output and heat transfer efficiency to steel sheet

45.7 Conclusions

1. **Challenges for practical use**

In the continuous hot-dip galvanizing process, in a test using an evaluation test furnace, the degreasing burner showed the performance as planned. However, considering the economic efficiency in actual operation, the degreasing burner needs a heat transfer efficiency of 80% or more. In order to achieve this thermal efficiency, it is necessary to install a preheating furnace before the degreasing furnace to optimize the preheating condition to improve the thermal efficiency.

2. **Evaluation of economic efficiency**

In the test using the evaluation test furnace, it was confirmed that the process using the degreasing burner had a smaller energy consumption than the conventional process. However, considering the newly required oxygen and ammonia costs, the operating cost exceeds the conventional method. However, in this project, the effectiveness of oxygen combustion and the degreasing performance by ammonia co-firing were confirmed for the continuous hot-dip galvanizing process. These are effective means for reducing CO_2. In the future, incentives to satisfy economic rationally and establishment of an ammonia supply system are desired.

Acknowledgements For this research, we asked the Akamatsu Laboratory of Osaka University to cooperate in collecting the basic data necessary for the design of the evaluation furnace. Thank you for writing here.

Part X
Evaluation of Green Ammonia Fuel in View of LCA and Economics

Chapter 46
Life Cycle Carbon Dioxide Emissions from Ammonia-Based Power Generation Technology

Yuki Kudoh and Akito Ozawa

Abstract Hydrogen energy carriers are expected to play vital roles in realising hydrogen economy and carbon neutrality. However, their low-carbon potential should be evaluated from a life cycle perspective. Here, life cycle inventory (LCI) analyses were conducted to understand the low-carbon potential of using ammonia (NH_3) in the power generation sector. The NH_3 was produced overseas from natural gas, both with and without carbon capture and storage (CCS), or from renewable energy, was imported to Japan and used in NH_3 co-firing with coal and NH_3 mono-firing power plants. The CO_2 emissions both from the operations and the capital goods related to the NH_3 value chain and power generation were calculated. Foreground data related to the supply chain were obtained from literature, and the Japanese LCI database was used to calculate the emissions. The results confirmed that the life cycle CO_2 emissions from the NH_3 mono-firing power plant, which had a low-carbon NH_3 value chain, were lower than those from coal power generation conducted with CCS. The results indicate that CO_2 emissions attributed to the capital goods of the fuel value chain may have an enormous impact on life cycle emissions, especially when utilising a variable renewable energy source with a low capacity factor.

Keywords Life cycle assessment · Ammonia · Power generation · Natural gas with carbon capture and storage · Renewable electricity

46.1 Introduction

Hydrogen has environmental merit as an energy carrier as it includes no carbon molecules, and because the only emission released by obtaining energy from hydrogen is water; no other environmental emissions such as carbon dioxide (CO_2) are produced. Therefore, hydrogen is recognised as a 'carbon-free' energy carrier, and is expected to play a vital role in realising the future hydrogen economy and achieving worldwide carbon neutrality. For example, in Japan's Basic Hydrogen

Y. Kudoh (✉) · A. Ozawa
Global Zero Emission Research Center, National Institute of Advanced Industrial Science and Technology, Onogawa, Tsukuba, Ibaraki 305-8569, Japan
e-mail: kudoh.yuki@aist.go.jp

K. Aika and H. Kobayashi (eds.), *CO_2 Free Ammonia as an Energy Carrier*,
https://doi.org/10.1007/978-981-19-4767-4_46

Strategy [1], which was formulated in December 2017, the vision and goals to be pursued in the future (with a view to 2050) involved transitioning to a hydrogen society ahead of the rest of the world; this strategy also outlined an action plan to achieve this by 2030. The strategy also outlined the possibility of using a variety of hydrogen energy carriers, including ammonia (NH_3), to lower the cost of hydrogen storage and transport, and to establish an international hydrogen supply chain. The 'Green Growth Strategy Through Achieving Carbon Neutrality in 2050' [2], which was formulated in December 2020, referred to NH_3 as a 'decarbonised fuel that will be the mainstay in the transition to a hydrogen society'. It also highlighted that both the development of NH_3 combustion technology to decarbonise thermal power generation, and the establishment of an international supply chain, are necessary to expand the use of NH_3 as a fuel.

To utilise these hydrogen energy carriers in society, it is necessary to establish hydrogen supply chains that cover hydrogen production from various primary energy sources, through energy conversion, storage and transportation, to their final use. As large amounts of energy and resources would be needed to establish these supply chains, it can be said that the use of 'carbon-free' hydrogen energy carriers may induce a certain amount of CO_2 emissions from the viewpoint of life cycle assessment (LCA). Hence, establishing 'low-carbon' hydrogen supply chains is crucial to utilising hydrogen energy carriers to achieve carbon neutrality.

The National Institute of Advanced Industrial Science and Technology (AIST) has been conducting LCA studies of hydrogen supply chains to identify the research and development challenges for a low-carbon supply chain for hydrogen energy carriers [3, 4]. These studies analysed literature to collect process data regarding the supply chains, and then linked them with CO_2 emission intensity data from available LCA databases to calculate life cycle CO_2 emissions through life cycle inventory (LCI) analysis.

Amongst the sectors where hydrogen energy carriers can be used, the power generation sector is not only expected to become a future low-carbon driver, but also has the potential to drastically increase demand for hydrogen energy carriers, thus reducing costs by establishing a large hydrogen supply chain to meet this demand. To understand the low-carbon potential of using hydrogen energy carriers in the power generation sector, this study evaluated the life cycle CO_2 emissions of power generation technologies that use imported low-carbon NH_3. It then compared these emissions with those of conventional power generation technologies in Japan.

46.2 Methods and Assumptions

46.2.1 LCI Analysis

Amongst the four steps of LCA stipulated by ISO 14040 [5], LCI analysis entails collecting and estimating the relevant inputs and outputs of a product system

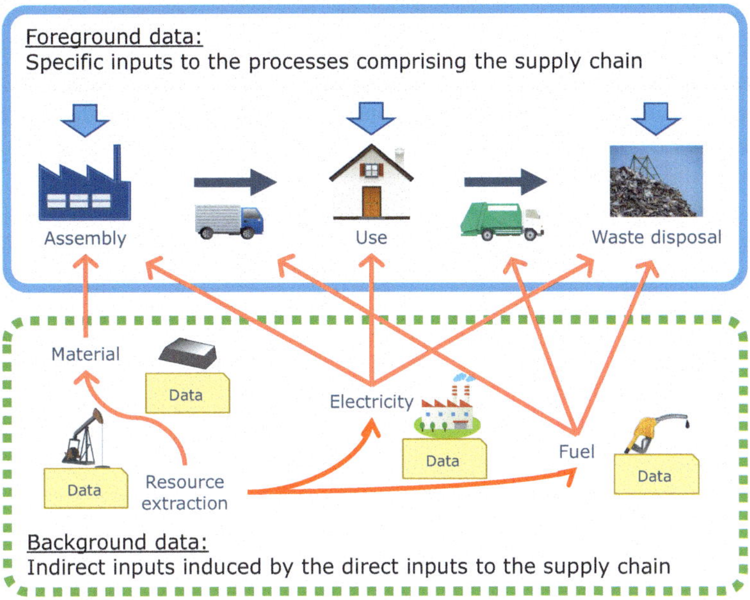

Fig. 46.1 Foreground and background data for LCI analysis

throughout its life cycle. As shown in Fig. 46.1, both foreground and background data are necessary for LCI analysis. Foreground data comprise direct inputs, such as energy and materials specific to the target product system; they are usually collected by those who conduct LCI analysis. Background data are not directly related to the product system, but instead relate to all of the indirect inputs induced by the direct inputs into the product system. It is difficult for those who are conducting LCI analysis to collect of all the indirect input data, including data from upstream processes. Thus, average industry data from LCA databases are often used as background data.

The life cycle CO_2 emissions, E, from a product system can be calculated using Eq. (46.1), where $x_{i,j}$ is the foreground data of input i to process j, and e_i represents the background embodied CO_2 emissions data of input i.

$$E = \sum_{i,j} \left(x_{i,j} \times e_i \right) \tag{46.1}$$

46.2.2 Background Data Used in This Study

To conduct an LCI analysis in this study, various foreground data related to the target fuel value chain were obtained from the Cross-ministerial Strategic Innovation Promotion Programme (SIP) 'Energy Carriers' project, previous LCI studies and demonstration project reports, as outlined in the following subsections. To calculate the life cycle CO_2 emissions using Eq. (46.1), the background data outlined below were used, depending on the characteristics of the collected foreground data.

When the foreground data were expressed in physical units, the Inventory Database for Environmental Analysis (IDEA) was used as one of the background data sources for the inputs to Japanese processes. IDEA provides environmental emission intensity data for approximately 3800 processes, covering almost all of the economic activities of Japanese industries [6]. This study used IDEA version 2.2, which was released in December 2017. As the CO_2 intensity data available for the overseas processes related to the NH_3 supply chain are limited, the following data were used for term e_i in Eq. (46.1) to calculate emissions related to overseas processes: (a) IDEA data were used for non-energy inputs, assuming that overseas countries had the same emission intensities as Japan; (b) International Energy Agency (IEA) statistics on CO_2 emissions from electricity [7] were used for electricity inputs and (c) direct emission factors were used for fuel inputs.

If the material components of a target process were unknown, and only the cost data of the process were available, the emission intensity data (given in monetary units) were used. The 'Embodied Energy and Emission Intensity Data for Japan Using Input–Output Tables' (referred to as '3EID' hereafter) [8] was used to obtain embodied CO_2 emission intensity data expressed in monetary units for the sectors of the input–output table. Again, due to the limited availability of similar background data in other countries, 3EID data were used as background data when the foreground data were only available in monetary units.

46.2.3 System Boundary and Assumptions

Figure 46.2 shows the system boundary of this study. The life cycle CO_2 emissions from power plants using NH_3 as fuel were estimated, and the emissions from coal power plants, both with and without post-combustion carbon capture and storage (CCS) facilities (assuming a 90% CO_2 capture rate), were also calculated to provide a conventional counterpart for comparison. Both the emissions attributed to capital goods and those arising from the operation of the power plant and its fuel value chain were included in this calculation.

This study assumed that NH_3 was produced from natural gas and renewable energy in the United Arab Emirates (UAE). The produced NH_3 was assumed to have been stored in stationary tanks at the loading port and shipped to Japan by NH_3 tankers fuelled with heavy fuel oil (HFO). The CO_2 emissions arising from round trips between Japan and the UAE were considered. After this international transportation, the unloaded NH_3 was assumed to have been stored in stationary tanks at the port, before being transported to the power plants by pipelines and then being finally re-evaporated and combusted at said power plants.

Figure 46.3 shows the process flow diagram of the NH_3 production process from natural gas designed under the SIP 'Energy Carriers' project; it has a daily production

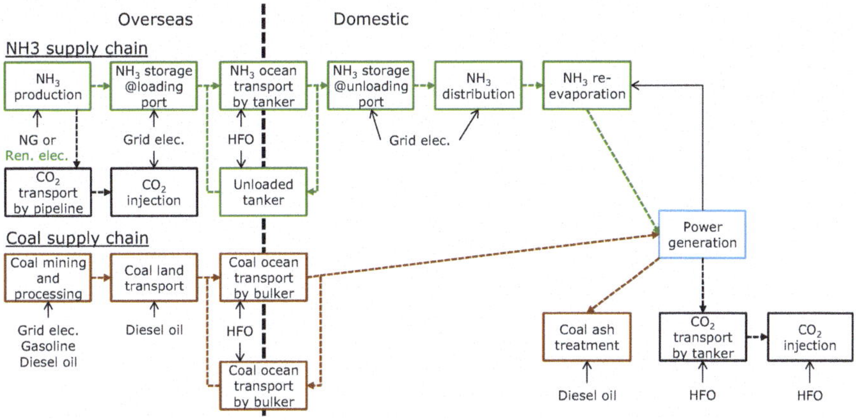

Fig. 46.2 System boundary of this study

capacity of 1850 t-NH_3·day^{-1}. In this process, NH_3 was produced from natural gas by the Haber–Bosch process. Natural gas was also used as a fuel for the reformer and the auxiliary steam boiler. Steam from the reformer and boiler was used to supply electricity for the process. The two cases shown in Fig. 46.3 (NG-1 and NG-2) were configured in this study. Case NG-1 was the base case, in which CO_2 was not captured from the NH_3 production plant. In case NG-2, CCS technology was implemented to capture all of the CO_2 emissions originating from the feedstock natural gas (referred to as 'process CO_2' hereafter) and 90% of CO_2 emissions from fuel gas (referred to as 'utility CO_2' hereafter). For case NG-2, it was assumed that no CO_2 separation process was required for the process CO_2 (due to its high CO_2 density), whereas chemical absorption using monoethanolamine (MEA) was applied to capture the utility CO_2. The captured CO_2 was transported by pipelines and injected into a storage reservoir [9]. The transport distance between the NH_3 production plant and the storage reservoir was set to be 50 km.

Figure 46.4 shows the process flow diagram for the NH_3 production process from renewable electricity that was assumed in this study, with a daily production capacity of 460 t-NH_3·day^{-1}. Renewable hydrogen (H_2) was produced by water electrolysis using solar photovoltaic (PV) power or concentrated solar power (CSP), the specifications and calculated life cycle CO_2 emissions of which are shown in Table 46.2. References [10] and [11] were used as foreground data for the solar PV power and CSP facilities. It was assumed that these renewable power plants were newly constructed, and that they were dedicated to producing NH_3 from renewable energy sources. It was also assumed that the building materials required for the foundations of the renewable power plants were procured in the UAE, whereas the other core components for renewable power generation were produced in, and shipped from, Japan. The produced H_2 was then compressed and reacted with nitrogen supplied from an air separation unit to produce NH_3, via the Haber–Bosch method. The air separation unit and compressor were both assumed to run on grid electricity.

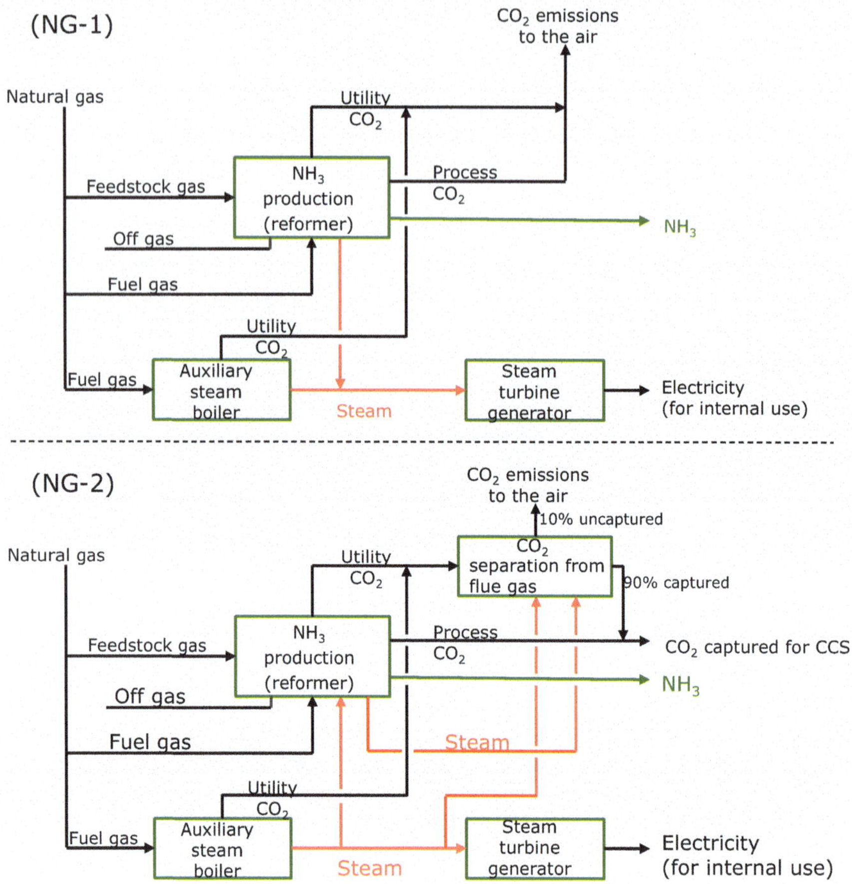

Fig. 46.3 NH_3 production from natural gas without CCS (NG-1) and with CCS (NG-2)

Natural gas was used to compress H_2 for the NH_3 reactor and refrigerant for the NH_3 condenser. Industrial water was used to supply steam to the compressor.

The CO_2 emissions from the coal value chain were calculated using data estimated from report [10] and the 2016 Japanese statistics on coal imports. Coal was extracted from an overseas coal mine. Methane released from this coal mine was also taken into account. The processed coal was then assumed to have been exported to Japan by a bulker, using HFO as fuel. The emissions from both the laden and ballast voyages were included in the LCA estimation.

Table 46.2 shows the configured specifications of the target power plants considered in this study. The coal and NH_3 mono-firing power plant parameters were set the same as the ultra-supercritical coal (600 °C class) and liquid natural gas (LNG)-fired combined cycle (1600 °C class) power plants as outlined in report [10], respectively. For the coal power plant, emissions arising from the treatment of coal ash treatment were also calculated. The CCS option was considered for a coal power plant

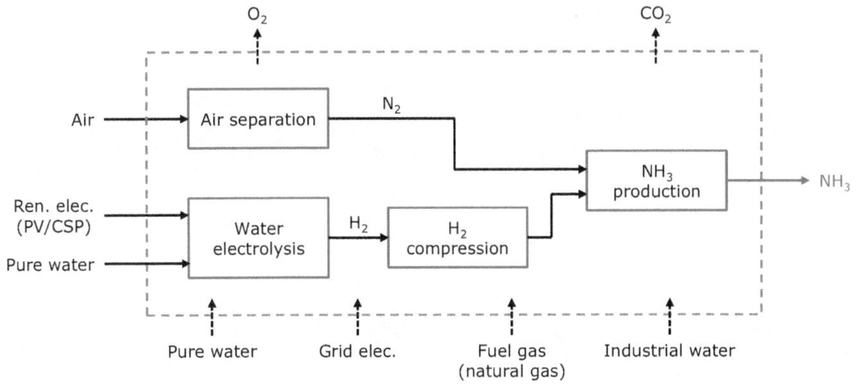

Fig. 46.4 NH_3 production from renewable electricity

comprising post-combustion CO_2 capture achieved by chemical absorption using MEA, CO_2 liquefaction, transportation by tanker to the offshore storage reservoir and finally injection [9].

In most of the processes shown in Fig. 46.2, the foreground data were obtained in physical units ($x_{i,j}$ of Eq. (46.1)). Hence, to calculate CO_2 emissions, either IDEA CO_2 emission intensity data, IEA data on electricity or direct emission factors of fuels were used as the background data (e_i of Eq. (46.1)) according to the application rule expressed in Sect. 46.2.2. However, only the capital expenditures calculated for the processes (electrolyser, compressor, H_2 storage tank, NH_3 production plant, loading port, NH_3 tanker, unloading port, NH_3 distribution and NH_3 re-evaporation) designed under the SIP project [12] were used as foreground data to calculate the capital goods of the NH_3 value chain, except for renewable power plant construction and CCS facilities for producing NH_3 from natural gas. To calculate the life cycle CO_2 emissions in these cases, the estimated cost data were first adjusted to the 2011 price using the gross domestic product (GDP) deflator and then multiplied by the embodied CO_2 emissions intensity data of the 3EID estimated from the 2011 input–output table.

Table 46.1 Assumed values for NH_3 and coal power plants

	NH_3-coal co-firing	NH_3 mono-firing	Coal mono-firing
NH_3 co-firing rate (on HHV basis) (%)	20		
Capacity (MW)	1000	1000	1000
Capacity factor (%)	70	70	70
Efficiency (HHV) (%)	39.6	53	39.6
Auxiliary power ratio (%)	5.06	2.08	5.06 (36.9% with CCS)

Table 46.2 Specifications of renewable power plants

Parameters	Solar PV	CSP with thermal storage
Capacity (MW)	10	15
Capacity factor (%)	20	71
Auxiliary power ratio (%)	3	0
Lifetime (years)	30	30
Life cycle CO_2 emissions (g-CO_2·kWh^{-1})	39.2	35.4
Rate of capital goods in the life cycle CO_2 emissions (%)	82.2	76.9

46.3 Results and Discussion

Figure 46.5 shows the CO_2 emissions attributed to the NH_3 value chains (NH_3 production from natural gas and renewables to its re-evaporation at the power plant). CO_2 reductions compared to the NG-1 case (without CCS) were calculated as 85%, 63% and 64% for the NG-2, solar PV and CSP cases, respectively. The CO_2 emissions in the NG-2 case were lower than those of the solar PV and CSP cases because natural gas was used as the power source for the H_2 compressor before the NH_3 reactor for the latter two cases. It was also used for refrigerant compression for the NH_3 condenser, whose CO_2 emissions accounted for 69% and 67% of the total emissions from the solar PV and CSP cases, respectively. The emissions from the NH_3 production process from renewable electricity could possibly be reduced by using low-carbon compression and condensation methods.

Figure 46.6 shows the life cycle CO_2 emissions attributed to the operation and capital goods of the NH_3 value chains. In terms of the contribution of capital goods

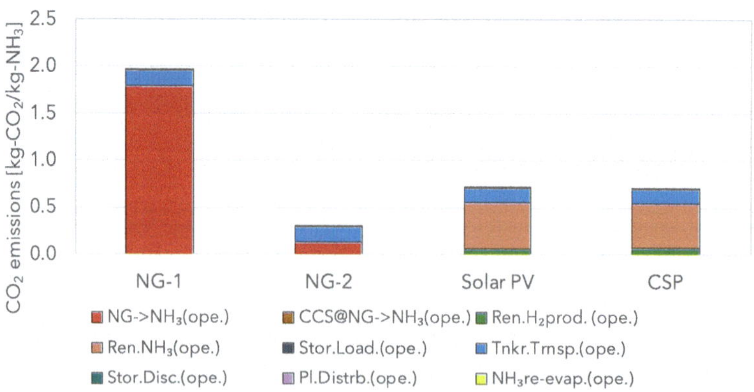

Fig. 46.5 CO_2 emissions attributed to the NH_3 value chains operation. NG → NH_3: NH_3 production from natural gas, CCS@NG → NH_3: CCS for NH_3 production from natural gas, ren. H_2 prod.: renewable H_2 production, ren. NH_3: renewable NH_3 production, stor. load.: storage at loading port, tnkr. trnsp.: tanker transport, stor. disc.: storage at discharging port, pl. distrb.: pipeline distribution, NH_3 re-evap.: NH_3 re-evaporation, ope.: operation

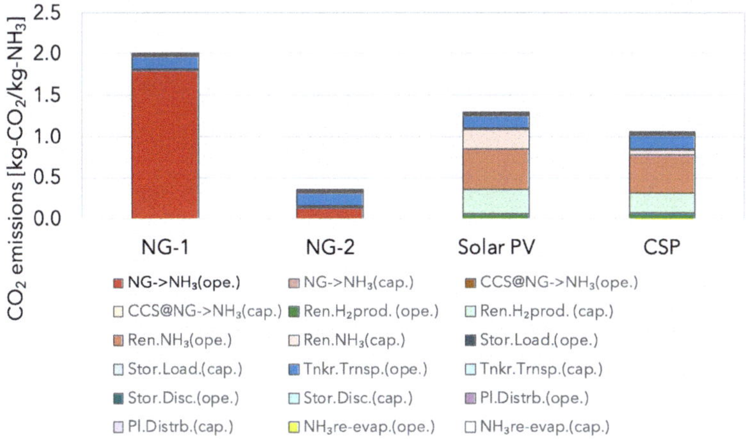

Fig. 46.6 Life cycle CO_2 emissions from the NH_3 value chains. Cap.: capital goods. Other legends are the same as those in Fig. 46.5

to the total CO_2 emissions, for cases NG-1 and NG-2 capital goods accounted for 2 and 13% of emissions, respectively, but for the solar PV and CSP cases capital goods accounted for 44 and 33% of emissions, respectively. The main reason for this high contribution was that capital goods emissions were included during the construction of renewable power plants. In addition, the capacity factor of solar PV was set to 20%, as shown in Table 46.2; this was lower than that of CSP with thermal storage. Given the solar PV specifications, both a large water electrolyser and H_2 storage tanks were needed to produce and store H_2 within the limited sunlight time to operate a 460 t-NH_3·day^{-1} scale NH_3 production plant. This resulted in the CO_2 emissions attributed to the capital goods for the renewable production of H_2 and NH_3 being relatively high.

Figure 46.7 shows the life cycle CO_2 emissions (both the operational and capital goods emissions) from coal mono-firing, NH_3-coal co-firing and NH_3 mono-firing power plants. For coal mono-firing, the emissions released by applying post-combustion CCS were also calculated; this process was found to reduce the life cycle CO_2 emissions by 80% compared to conventional coal mono-firing without CCS. For a NH_3-coal co-firing power plant with a 20% NH_3 co-firing rate, the emissions were predicted to decrease by 1, 16, 8 and 10% when using NH_3 produced by the NG-1, NG-2, solar PV and CSP cases, respectively, compared with conventional coal power generation without post-combustion CCS. If large-scale NH_3 mono-firing technology for power generation becomes available in the future, then emissions could be reduced by 35, 89, 58 and 66% for the NG-1, NG-2, solar PV and CSP cases, respectively. This indicates that the life cycle CO_2 emissions of the NH_3 mono-firing power generation could be smaller than those of the coal power plant with post-combustion CCS were a low-carbon NH_3 value chain to be established.

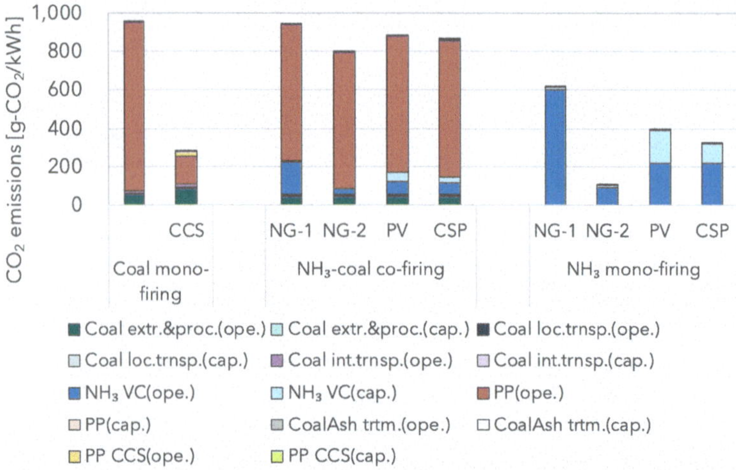

Fig. 46.7 Life cycle CO_2 emissions from coal and NH_3 power plants. Coal extr. & proc.: coal extraction and processing, coal loc. trnsp.: coal local transport, coal int. trnsp.: coal international transport, NH_3 VC: NH_3 value chain, PP: power plant, Coal Ash trtm.: coal ash treatment, PP CCS: post-combustion CCS at power plant

46.4 Conclusion

To understand the CO_2 emission profile of a power plant that uses NH_3 as fuel, here LCI analysis was conducted for NH_3-coal co-firing and NH_3 mono-firing power plants. During this analysis, NH_3 was assumed to be produced from natural gas or renewables in the UAE. The application of CCS to the process of producing NH_3 from natural gas was also considered. The foreground data were collected through literature surveys, and the background data were chosen according to the properties of the foreground data.

The results indicate that CO_2 emissions from the NH_3 value chain could be significantly reduced by applying CCS technology to the process of producing NH_3 from natural gas. Producing NH_3 from renewables was also found to have a lower CO_2 profile than the conventional process of producing NH_3 from natural gas, but low-carbon compression and condensation processes are needed to further reduce these emissions. It was also suggested that emissions from the capital goods required to establish the NH_3 value chain originating from renewables could have an enormous impact on total emissions. A NH_3-coal co-firing power plant with 20% NH_3 co-firing was confirmed to reduce life cycle CO_2 emissions compared to a conventional coal mono-firing power plant, but the degree of reduction depended on the CO_2 emissions profile assumed for the NH_3 value chain. NH_3 mono-firing was predicted to reduce life cycle CO_2 emissions significantly; emissions lower than those of a coal mono-firing power plant with post-combustion CCS could be achieved if the NH_3 were to be produced from natural gas with the combination of CCS technology (NG-2 case).

Conducting LCI studies based on the energy flow of the fuel value chain is essential to determining the CO_2 reduction effects of new energy technologies, compared to conventional technologies. However, this study's results indicate that the CO_2 emissions attributed to the capital goods of the fuel value chain may have an enormous impact on the whole life cycle emissions, especially when utilising a variable renewable energy source with a low capacity factor.

Acknowledgements This work was supported by the Council for Science, Technology and Innovation (CSTI), the Cross-ministerial Strategic Innovation Promotion Program (SIP), 'Energy Carriers' (management agency: Japan Science and Technology Agency (JST)).

References

1. Ministerial council on renewable energy, hydrogen and related issues. Basic hydrogen strategy. https://www.meti.go.jp/english/press/2017/pdf/1226_003b.pdf. Accessed 1 Mar 2022
2. Ministry of Economy, Trade and Industry. Green growth strategy through achieving carbon neutrality in 2050. https://www.meti.go.jp/english/press/2020/pdf/1225_001b.pdf. Accessed 1 Mar 2022
3. Ozawa A, Inoue M, Kitagawa N, Muramatsu R, Anzai Y, Genchi Y, Kudoh Y (2017) Assessing uncertainties of well-to-tank greenhouse gas emissions from hydrogen supply chains. Sustainability 9(7):1101. https://doi.org/10.3390/su9071101
4. Ozawa A, Kudoh Y, Kigatawa N, Muramatsu R (2019) Life cycle CO_2 emissions from power generation using hydrogen energy carriers. Int J Hydrogen Energy 44(21):11212–11232. https://doi.org/10.1016/j.ijhydene.2019.02.230
5. International Organization of Standardization (2006) ISO 14040: 2006 environmental management—life cycle assessment—principles and framework. ISO, Geneva
6. National Institute of Advanced Industrial Science and Technology. IDEA. http://idea-lca.jp/index.html. Accessed 1 Mar 2022
7. International Energy Agency (2017) CO_2 emissions from fuel combustion 2017. OECD, Paris. https://doi.org/10.1787/co2_fuel-2017-en
8. Nansai K. Embodied energy and emission intensity data for Japan using input–output tables (3EID). National Institute for Environmental Studies, Japan. http://www.cger.nies.go.jp/publications/report/d031/index.html. Accessed 1 Mar 2022
9. Tang L, Yokoyama T, Kubota H, Shimoda A (2014) Life cycle assessment of a pulverized coal power generation with CCS technology in Japan (Part 2)—comparative study on environmental impact of photovoltaic and geothermal power generation (in Japanese). Tokyo
10. Imamura E, Iuchi M, Bando S (2016) Comprehensive assessment of life cycle CO_2 emissions from power generation technologies in Japan (in Japanese). Tokyo
11. Sullivan JL, Clark CC, Han J, Wang M (2010) Life-cycle analysis results of geothermal systems in comparison to other power systems. Argonne National Lab, Argonne, IL, USA
12. Hiraoka K, Fujimura Y, Kai M, Sakata K, Ishimoto Y, Mizuno Y, Kudoh Y, Ozawa A, Kitagawa N, Muramatsu R (2019) Potential of low carbon ammonia as fuel for power generation. In: Proceedings of Nitrogen + Syngas

Chapter 47
Techno-economic Considerations for Ammonia Production, Storage, and Transportation

Kevin H. R. Rouwenhorst and Trevor Brown

Abstract Ammonia is a feedstock for the fertilizer industry and the chemical industry, currently mainly produced from fossil feedstocks such as natural gas and coal. Renewable ammonia production accounts for only 0.01% of the current global ammonia production. However, decarbonization of the global economy also requires decarbonization of ammonia production, for current uses and for new applications. Future markets for low-carbon ammonia include its use as a decarbonized fuel and hydrogen carrier. In the current Chapter, the ammonia market is briefly highlighted. Furthermore, the production cost of ammonia production from fossil sources, as well as carbon capture and storage (CCS) is discussed, after which the production cost of renewable ammonia production is discussed. The storage and transportation of ammonia are key for ammonia as a decarbonized energy vector in the hydrogen economy and is briefly discussed thereafter. Various case studies are presented regarding recently announced low-carbon ammonia projects. The Chapter is concluded with some final remarks regarding the techno-economics of ammonia production, storage, and transportation. **Highlights**: • Renewable ammonia has almost achieved cost parity with fossil-based ammonia production, especially due to increasing CO_2 taxation and decreasing electricity cost from solar and wind. • Size still matters for renewable ammonia plants, due to economies of scale. Small-scale ammonia can work well if constrained within a local energy system, but not within the global commodity market. • Renewable ammonia will be a commodity for current uses, such as fertilizers (nitrogen), as well as for new energy markets (hydrogen). Large-scale renewable ammonia projects have recently been announced, up to 10 million tons of ammonia per year capacity. • The ammonia market is expected to

K. H. R. Rouwenhorst (✉) · T. Brown
Ammonia Energy Association, 77 Sands Street, 6th Floor, Brooklyn, NY 11201, USA
e-mail: krouwenhorst@ammoniaenergy.org

T. Brown
e-mail: tbrown@ammoniaenergy.org

K. H. R. Rouwenhorst
Catalytic Processes and Materials, MESA+ Institute for Nanotechnology, University of Twente, P.O. Box 217, 7500 AE Enschede, The Netherlands

Proton Ventures, Karel Doormanweg 5, 3115 JD Schiedam, The Netherlands

K. Aika and H. Kobayashi (eds.), *CO_2 Free Ammonia as an Energy Carrier*,
https://doi.org/10.1007/978-981-19-4767-4_47

grow to at least three times the current demand in 30 years, to roughly 600 million tons per year by 2050.

Keywords Ammonia · Carbon capture and storage (CCS) · Renewable ammonia · CapEx · OpEx4

47.1 Background

Ammonia is an important feedstock for the fertilizer industry and the chemical industry. In fact, ammonia is the second-most manufactured chemical in the world, with a global ammonia production of 183 Mt in 2020 [1]. Ammonia is currently almost exclusively produced from fossil feedstocks, sometimes termed *brown ammonia* or *gray ammonia.*

Most ammonia is further processed, mainly for fertilizer applications [1]. However, about 25–30 Mt ammonia is transported around the world annually, by ship, road, train, or pipeline [2]. Since 2000, the market price of ammonia in the Caribbean, Europe, and the Middle East has fluctuated between 100 and 600 USD/t [1], while the market price of ammonia has been about 200–300 USD/t over the past five years [3]. In 2021, the market price of ammonia has increased again, due to lower production rates in 2020, and shortages in natural gas in Europe and Asia.

Hereafter, the global ammonia market is discussed, with a focus on fertilizer applications. Thereafter, the economic build-up of ammonia production from fossil fuels and from renewable feedstock is discussed. Furthermore, the cost of carbon emission and carbon capture and storage (CCS) are discussed.

47.2 The Global Ammonia Market

The global ammonia production was about 183 Mt in 2020 [1]. As ammonia is mainly produced from fossil fuels, primarily natural gas and coal, current ammonia production emits about 0.5 Gt CO_2 annually [4], e.g., about 1% of the global CO_2 equivalent emissions.

A schematic overview of various pathways for ammonia synthesis is shown in Fig. 47.1. The majority of ammonia production is derived from natural gas (72%), while Chinese ammonia production mainly relies on coal (22%) [5]. Other feedstocks (6%) include naphtha, heavy fuel oil, and byproduct hydrogen from other industrial processes. Renewable ammonia production currently accounts for about 0.01% of global ammonia production [2]. Ammonia is a major hydrogen consumer, accounting for about 40–45% of global pure hydrogen demand [6].

Ammonia is mainly used for fertilizer applications (about 85%) [1]. Urea ($CO(NH_2)_2$) production accounts for 55% of the global ammonia utilization [1]. Other fertilizer applications include ammonium nitrate, diammonium phosphate,

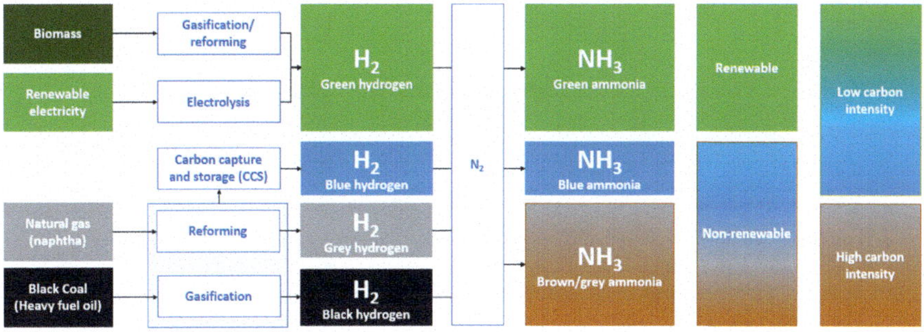

Fig. 47.1 Schematic overview of various pathways for ammonia synthesis. Adapted and modified from IRENA [2]

ammonium sulfate, and monoammonium phosphate [7]. Ammonia is also used in the textile industry, for explosives and mining, for pharmaceuticals, for refrigeration, plastics manufacturing, and waste treatment [3].

Current markets for ammonia production grow at about 2% annually [8], resulting in about 250–350 Mt ammonia production by 2050. On top of this, new markets for ammonia are developing, such as its use as a decarbonized fuel and hydrogen carrier. Ammonia may be used as hydrogen feedstock after ammonia decomposition, or it can be directly used for stationary power, and as shipping fuel (see Fig. 47.2). This could result in an ammonia demand of 600 Mt by 2050 [9], giving ammonia a prominent role in the hydrogen economy. These novel markets rely on low-carbon ammonia, e.g., fossil-based ammonia production with carbon capture and storage (CCS) or renewable ammonia production. Fossil-based ammonia production without carbon mitigation has little benefits over conventional fossil fuels in terms of greenhouse gas emissions [10]. However, fossil-based ammonia may allow for bridging the fuel security and cost gap to low-carbon ammonia. Furthermore, this may accelerate retrofitting coal-based and natural gas-based technology to be ammonia-ready.

47.3 Natural Gas-Based Ammonia and Coal-Based Ammonia Production

Ammonia is a globally traded commodity. The global ammonia production capacity is about 243 Mt [3], while the demand was 183 Mt in 2020 [1]. Thus, ammonia production is a highly competitive market. This results in ammonia production mainly produced at large-scale and at locations with low-cost feedstock.

The cost of ammonia production can be divided into capital investment (CapEx) and operational expenses (OpEx). The CapEx consists of all costs associated with building the ammonia production plant. The OpEx consists of feedstock cost and expenses during operation and maintenance. The cost of ammonia production mainly depends on the plant capacity and the choice of feedstock.

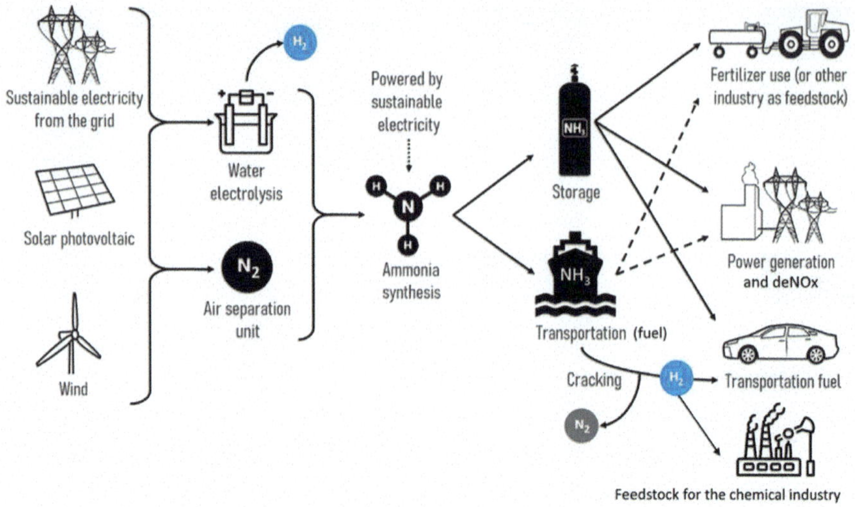

Fig. 47.2 Schematic of the ammonia economy. Adapted and modified from [11]

Upon increasing the ammonia plant size, the relative capital investment per amount ammonia produced decreases due to the economy of scale. Thus, modern ammonia plants have a capacity as high as 1000–3300 t/d ammonia [12].

The choice of feedstock also has a profound impact on the capital investment. The relative capital investment for a 1150 stpd (standard ton per day) ammonia plant is as follows: natural gas (1.0) > naphtha (1.18), heavy fuel oil (1.5), coal (2.0) [13]. Furthermore, the choice of feedstock determines the energy consumption for ammonia synthesis. A modern natural gas-based ammonia plant consumes about 29 GJ/t, while a coal-based ammonia plant consumes 42 GJ/t [12]. This means more coal feedstock must be processed, resulting in a higher capital investment and operational costs next to the feedstock cost. It should be noted, however, that coal is typically less expensive on an energy basis (0.5–2.5 USD/MMBtu) as compared to natural gas (2–10 USD/MMBtu).

The feedstock cost, e.g., the natural gas price or coal price, determines the profitability of ammonia production to a large extent. For locations with high natural gas prices such as western Europe (usually >4 USD/MMBtu), it is not profitable to build new ammonia plants. In such locations, ammonia production can only be profitable with current assets, as the capital investment has already been paid off. Newly built factories are mainly located in other areas with very low natural gas prices (<3 USD/MMBtu), such as Russia, Trinidad & Tobago, and the United States.

Large-scale ammonia plants with capacities in the range of 1000–3300 t/d ammonia require substantial investments of multiple billion USD. Typical capital intensities reported for large-scale, natural gas-based ammonia plants are about 1000 USD/t/y [14, 15]. However, various recent plants in the United States have reported a higher cost, with figures around 1300–2000 USD/t/y [15].

47.4 Carbon Taxation and Blue Ammonia Production

In order to account for the environmental impact of fossil-based ammonia production, carbon penalties are introduced in various regions, e.g., with a carbon tax or with emission trading schemes. In order to determine the cost associated with CO_2 equivalent emissions, the carbon footprint is required.

Ammonia production from natural gas and coal results in about 1.6 tons of CO_2 and 3.8 tons of CO_2 emitted per ton of ammonia produced during the ammonia synthesis process [12]. However, CO_2 equivalent emissions also occur during the feedstock extraction. For example, methane slip during natural gas extraction can cause significant CO_2 equivalent emissions [16]. This should be accounted for when determining the carbon footprint of the ammonia product.

The average resulting CO_2 footprint of ammonia derived from natural gas is about 2.2 tons of CO_2 per ton of ammonia, while that of coal-based ammonia has a total footprint of 4.0 tons of CO_2 per ton of ammonia [2]. Upon assuming a CO_2 penalty of about 60 USD/t CO_2, e.g., current levels in the European Union and the United Kingdom, this results in an added cost of about 135 USD/t ammonia from natural gas and 240 USD/t ammonia from coal.

Alternatively, the CO_2 emitted during ammonia synthesis can be captured and stored. This is sometimes termed *blue ammonia*. Carbon capture and storage (CCS) costs approximately 100–150 USD/t ammonia [3]. It should be noted that CO_2 is still emitted during feedstock extraction, which can result in a CO_2 footprint of about 0.7 tons CO_2 per ton of ammonia [2].

Based on Lifecycle Assessment (LCA) studies, it is found that a CO_2 penalty of about 60–90 USD/t CO_2 is required for *blue ammonia* to become competitive with *brown ammonia* [2]. For reference, the CO_2 emission penalty in the European Union and in the United Kingdom is currently about 60 USD/t CO_2. Thus, CCS can already be competitive in some locations, especially when the CO_2 is not utilized for production of ammonia derivatives such as urea.

Case Study: Barents Blue Ammonia Project

Various ammonia plants with CCS have recently been announced with a combined capacity of a few Mt [2]. This lower carbon ammonia may find applications for fuel use, such as for stationary power or for the shipping sector, or as a hydrogen feedstock.

An example of this is the Barents Blue ammonia project of Horisont Energi. Low-cost, abundant natural gas in Finnmark, Northern Norway is converted to ammonia with auto thermal reforming technology, which allows for a high CO_2 capture rate of 98% as compared to conventional steam methane reforming technology with a CO_2 capture rate of 90% [16]. Also, methane slip from Norwegian natural gas is relatively low, resulting in a CO_2 footprint as low as 0.2 tons CO_2 per ton of ammonia [16]. It is expected that the plant will produce about 1 Mt blue ammonia per annum from 2025 onward, although there are already plans to expand this capacity to 3 Mt blue ammonia per annum [17]. Recently, a memorandum of understanding was signed between Horisont Energi and the Port of Rotterdam to transport blue ammonia from Finnmark to Rotterdam [18], where the ammonia may be utilized as shipping fuel or be converted to hydrogen for the European hydrogen grid.

47.5 Renewable Ammonia Production

Ammonia can also be produced from renewable energy sources, e.g., from renewable electricity coupled with electrolysis-based hydrogen and subsequent ammonia production (see Fig. 47.3). Historically, various hydroelectric power plants have been combined with alkaline electrolysis plants to produce renewable ammonia in locations, such as Egypt, Italy, and Norway [19]. In fact, the first renewable ammonia plant became operational in Terni, Italy by 1921 [19, 20]. Most renewable ammonia plants have been closed down or revamped to natural gas-based ammonia plants, due to more favorable economics of fossil-based ammonia production. Currently, one renewable ammonia plant is still in operation, located in Cusco, Peru [21].

Recent decreases in renewable electricity costs from solar and wind, as well as decreasing costs in electrolyzers, have caused renewed interest in renewable ammonia production. Furthermore, carbon taxes and emission trading schemes bridge the cost gap to renewable technologies. Renewable electricity from solar and wind can already be produced at a cost below 20 USD/MWh at locations with the highest wind load hours and high solar irradiation, such as in Saudi Arabia [23]. The cost of electrolysis is expected to decrease due to scale-up of electrolyzer production and a cost-learning curve upon installing electrolysis capacity [24].

A main difference between renewable ammonia production and fossil-based ammonia production is the cost build-up. In case of renewable ammonia production, the energy input in the form of renewable electricity is part of the capital investment, rather than an operational expense. For reference, a fossil-based ammonia plant mainly requires investment for the ammonia plant, including hydrogen production and nitrogen purification. The natural gas or coal feedstock is bought during operation, causing less investment risk. Thus, the initial investment for renewable ammonia is typically multiple times that of fossil-based ammonia production. This means that

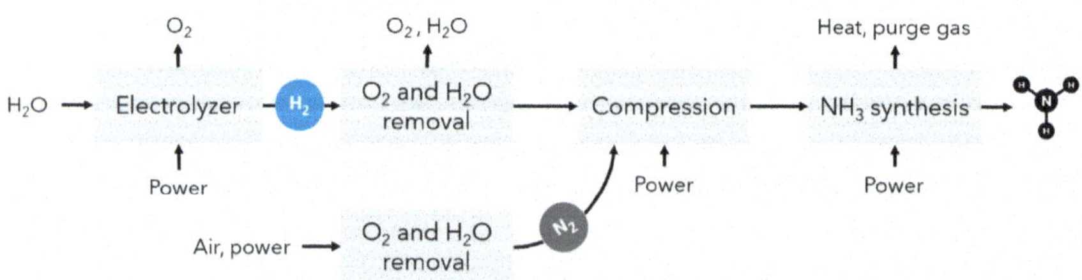

Fig. 47.3 Schematic overview of steps involved in ammonia synthesis from water and air. Reproduced from Sousa Cardoso et al. [11] and Brown [22]

the economic models used for determining profitability, e.g., with value depreciation over time, determine the profitability of renewable ammonia to a great extent.

Electricity production typically accounts for more than half of the renewable ammonia production cost. As a rule of thumb, every 10 USD/MWh adds about 100 USD/t to the ammonia cost [25]. Furthermore, the hydrogen production typically accounts for the majority of the investment for the ammonia production facility, followed by the ammonia synthesis loop. Process parts, such as water desalination, nitrogen purification, and ammonia storage, only account for typically 5–10 USD/t ammonia [26, 27]. Economy of scale still applies for renewable ammonia production, implying plants with a capacity above 1000 t/d ammonia are most economical (see the case study below) [2].

An indicative cost comparison of fossil-based ammonia production and renewable ammonia production is shown in Fig. 47.4. Newly built renewable ammonia can be competitive with newly built fossil-based ammonia production if electricity prices are low (≤20 USD/MWh), and if carbon penalties and/or fossil feedstock prices are high. Furthermore, the cost of renewable ammonia is expected to decrease over the upcoming decades, while carbon penalties are expected to increase [28, 29].

The cost of newly built fossil-based ammonia production and renewable ammonia production is higher than the market value of 200–300 USD/t ammonia (cost range in 2015–2020) in most cases for Fig. 47.4. It should be noted that the ammonia

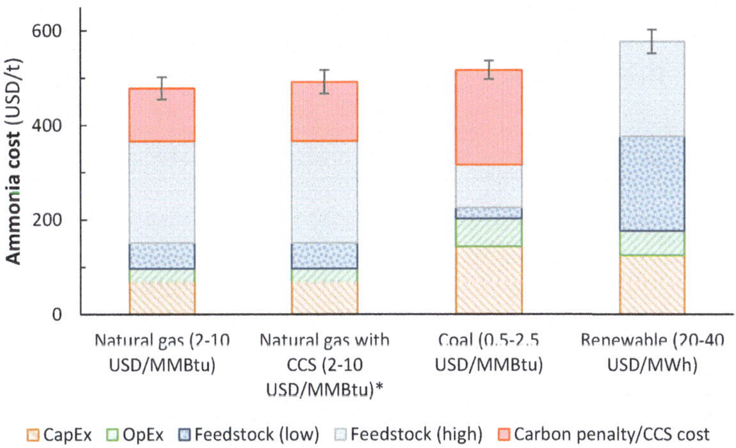

Fig. 47.4 Indicative cost comparison for large-scale fossil-based ammonia production and renewable ammonia production. Estimates for CapEx and OpEx for natural gas-based ammonia production and coal-based ammonia production reproduced from Appl [31]. A carbon penalty of 50 USD/t CO_2 is assumed, as well as a CO_2 footprint of 2.2 t-CO_2/t-NH_3 for natural gas-based ammonia production and 4.0 t-CO_2/t-NH_3 for coal-based ammonia production. The estimates for the CapEx for renewable ammonia production is based on Armijo et al. [32], and the OpEx is scaled accordingly. The cost of CCS is assumed to be 125 USD/t-NH_3, based on the AmmonFuel report [3]. Solar and wind energy is shown as 'feedstock'

production cost in Fig. 47.4 does not include transportation, which typically adds 30–100 USD/t ammonia [30].

The current market price for ammonia mainly relies on factories that have already fully depreciated the capital investment, implying the ammonia cost is only determined by the feedstock cost, operational expenses, and transportation. Thus, a comparison of renewable ammonia production with the current market price is not entirely fair. In fact, a comparison for added capacity should be made between newly built projects for fossil-based ammonia production plants (including a carbon penalty or CCS), and renewable ammonia production plants, as shown in Fig. 47.4.

Case Study: Renewable Ammonia Demand

Various large-scale renewable ammonia plants have been announced in recent years [2]. This is in line with hydrogen strategies and increasing carbon penalties in various economic areas. The announced renewable ammonia production capacity totals about 17 Mt/y ammonia production by 2030 [2]. Saygin and Gielen [9] estimated 330 Mt/y renewable ammonia production from electrolysis-based hydrogen by 2050 for the 1.5 °C scenarios, in accordance with the Paris Agreement (see Fig. 47.5).

Additionally, Saygin and Gielen [9] estimated 162 Mt/y renewable ammonia production from biomass-based hydrogen by 2050 for the 1.5 °C scenarios [9]. Biomass-based hydrogen production results in circular CO_2 as byproduct, which may be reacted with ammonia to form urea ($CO(NH_2)_2$). Urea production accounts for about 55% of current ammonia production [1]. It should be noted, however, that biomass-based ammonia plants have not materialized so far.

Case Study: Renewable Energy Hubs

Due to the modular design of solar panels, wind turbines, and electrolyzers, it has been argued that renewable ammonia may be produced at the MW scale, e.g., multiple

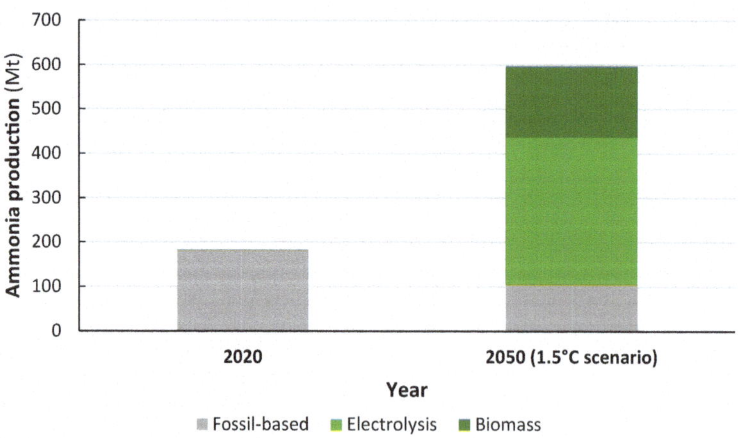

Fig. 47.5 Current and future ammonia production by source. Future scenario based on 1.5 °C scenarios, in accordance with the Paris Agreement. Reproduced from [9]

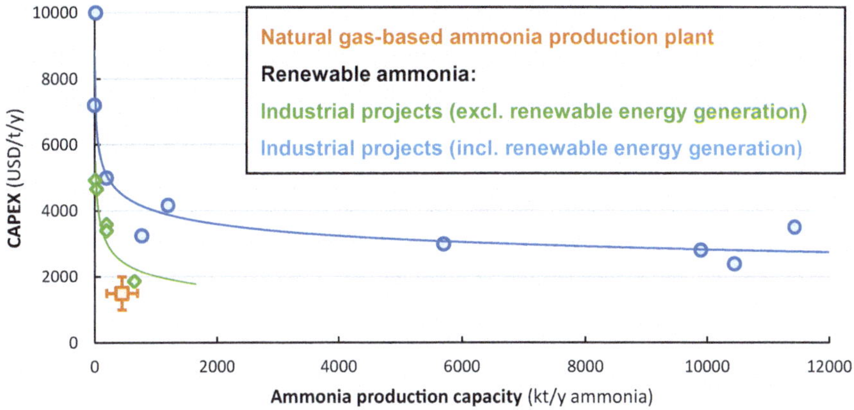

Fig. 47.6 Capital intensity of ammonia synthesis plants versus ammonia production capacity. Adapted and modified from IRENA [2]

kt ammonia per annum [15, 33]. More recently, however, the consensus seems to be that the future of renewable ammonia as hydrogen feedstock and renewable fuel will be based mainly on large-scale, renewable energy hubs. Ammonia capacities of multiple Mt ammonia per annum have been announced for these GW scale renewable energy hubs in for example Australia, Mauritania, and Oman [2]. Economies of scale still apply for the hydrogen production, nitrogen purification, and ammonia synthesis loop [34]. Furthermore, the required port infrastructure for ammonia export is only affordable upon producing renewable ammonia in large-scale plants [35]. Thus, the ammonia plant size matters. To illustrate this, the projected capital intensity of the announced renewable ammonia plants is shown in Fig. 47.6.

The current electrolyzer manufacturing capacity is currently a bottleneck to the development of large-scale, renewable energy hubs [2]. Furthermore, coupling intermittent solar and wind electricity with ammonia synthesis is not yet demonstrated technology at commercial scale, although various small-scale renewable ammonia demonstrators are operational in Japan, the United Kingdom, and the United States [36–38]. Thus, the focus in the upcoming few years will be on demonstrating renewable ammonia synthesis at the MW scale, in order to reduce the risk for scale-up to the renewable energy hubs at the GW scale.

47.6 Transportation Cost

After ammonia production, ammonia is converted or stored and transported. Ammonia storage is performed in tanks with a capacity of up to about 50 kt ammonia [31, 39]. Such large-scale refrigerated tanks operate at atmospheric pressure and −33 °C.

About 25–30 Mt ammonia is transported annually, by pipeline, rail, road, and ship [2, 6]. International ammonia trade is facilitated by shipping, totaling about 18–20 Mt annually [1, 3]. This is equivalent to about one-tenth of the current global ammonia production. Various authors have estimated the cost of ammonia transportation by ship [30]. Typical values reported are in the range of 30–100 USD/t ammonia, depending on the distance and fuel cost. Ammonia is currently shipped in vessels fueled with heavy fuel oil. However, these vessels may run on ammonia as a fuel in the future.

In the United States, and between Ukraine and Russia, ammonia is mainly transported by pipeline, spanning thousands of kilometers [3, 40]. Ammonia transportation by pipeline costs about 0.07 USD/t/km [6]. Transportation of natural gas and hydrogen is more expensive than transportation of ammonia by pipeline [6, 41]. In Europe, ammonia pipelines typically only span a few kilometers [42]. Ammonia is mainly transported by train in Europe [3]. Ammonia transportation by train has a cost similar to ammonia transportation by pipeline [43].

47.7 Final Remarks

Ammonia is an important feedstock for the fertilizer industry and the chemical industry. Currently, ammonia is mainly produced from fossil feedstocks such as natural gas and coal, resulting in about 1% of global CO_2 equivalent emissions. Renewable ammonia production currently accounts for only 0.01% of the global ammonia production [2]. However, various large-scale projects for renewable ammonia production have recently been announced [2]. Future markets for such low-carbon ammonia include its use as a decarbonized fuel and as a hydrogen carrier.

Fossil-based ammonia production currently has the lowest cost among alternatives. However, at carbon emission penalties of 60–90 USD/t CO_2, carbon capture and storage (CCS) becomes cost-competitive [2]. Furthermore, renewable ammonia production may become cost-competitive on the long term [28, 29], due to cost reductions in renewable electricity and electrolyzers. The main hurdle for renewable ammonia plants is the high capital investment. Afterward, the operational costs of renewable ammonia are very competitive with fossil-based ammonia production.

Ammonia can be transported across continents at a cost typically below 100 USD/t. This means ammonia is a commodity traded across the globe [3]. Newly built ammonia plants are mainly located near low-cost feedstock, as this is the only way to stay competitive on the global market. When incentives are provided to decarbonize, e.g., carbon penalties, newly built ammonia production will probably move to locations with high wind load hours and high solar irradiation.

References

1. Hatfield O (2020) A review of global ammonia supply. In: NH_3 energy conference
2. IRENA, Ammonia Energy Association (2022) Innovation outlook: renewable ammonia. IRENA. https://www.irena.org/-/media/Files/IRENA/Agency/Publication/2022/May/IRENA_Innovation_Outlook_Ammonia_2022.pdf
3. Laval A, Hafnia HT et al (2020) Ammonfuel—an industrial view of ammonia as a marine fuel. https://www.topsoe.com/hubfs/DOWNLOADS/DOWNLOADS%20-%20White%20papers/Ammonfuel%20Report%20Version%2009.9%20August%203_update.pdf
4. David B et al (2020) Ammonia: zero-carbon fertiliser, fuel and energy store. The Royal Society. https://royalsociety.org/-/media/policy/projects/green-ammonia/green-ammonia-policy-briefing.pdf
5. Bicer Y, Dincer I, Zamfirescu C et al (2016) Comparative life cycle assessment of various ammonia production methods. J Clean Prod 135:1379–1395. https://doi.org/10.1016/j.jclepro.2016.07.023
6. IEA (2019) The future of hydrogen: seizing today's opportunities. International Energy Agency. https://www.iea.org/reports/the-future-of-hydrogen
7. Yara (2018) Yara fertilizer industry handbook. https://www.yara.com/siteassets/investors/057-reports-and-presentations/other/2018/fertilizer-industry-handbook-2018-with-notes.pdf/
8. de Pee A, Pinner D, Roelofsen O et al (2018) Decarbonization of industrial sectors: the next frontier. https://www.mckinsey.com/~/media/mckinsey/business%20functions/sustainability/our%20insights/how%20industry%20can%20move%20toward%20a%20low%20carbon%20future/decarbonization-of-industrial-sectors-the-next-frontier.pdf
9. Saygin D, Gielen D (2021) Zero-emission pathway for the global chemical and petrochemical sector. Energies 14:3772. https://doi.org/10.3390/en14133772
10. Al-Aboosi FY, El-Halwagi MM, Moore M, Nielsen RB (2021) Renewable ammonia as an alternative fuel for the shipping industry. Curr Opin Chem Eng 31.https://doi.org/10.1016/j.coche.2021.100670
11. Sousa Cardoso J, Silva V, Rocha RC, et al (2021) Ammonia as an energy vector: current and future prospects for low-carbon fuel applications in internal combustion engines. J Clean Prod 296.https://doi.org/10.1016/j.jclepro.2021.126562
12. Brightling JR (2018) Ammonia and the fertiliser industry: the development of ammonia at Billingham. Johnson Matthey Technol Rev 62:32–47. https://doi.org/10.1595/205651318X696341
13. Czuppon TA, Buividas LJ (1980) Hydrogen for ammonia production and the economics of alternate feedstocks. In: Hydrogen: production and marketing, pp 47–66
14. Argus Media (2020) Green shift to create 1 billion tonne "green ammonia" market? https://view.argusmedia.com/rs/584-BUW-606/images/Argus%20White%20Paper%20-%20Green%20Ammonia.pdf
15. Brown T (2018) The capital intensity of small-scale ammonia plants. In: Ammon. Ind. https://ammoniaindustry.com/the-capital-intensity-of-small-scale-ammonia-plants/. Accessed 14 Mar 2021
16. Hydrogen Council (2021) Hydrogen decarbonization pathways: a life-cycle assessment. https://hydrogencouncil.com/wp-content/uploads/2021/01/Hydrogen-Council-Report_Decarbonization-Pathways_Part-1-Lifecycle-Assessment.pdf
17. Horisont Energi (2021) Horisont Energi, Equinor and Vår Energi enter cooperation agreement for Barents Blue ammonia plant. https://www.horisontenergi.no/horisont-energi-equinor-and-var-energi-enter-cooperation-agreement-for-barents-blue-ammonia-plant/. Accessed 20 Sep 2021
18. Energi H (2021) Horisont Energi and Port of Rotterdam sign memorandum of understanding regarding blue ammonia. https://www.horisontenergi.no/horisont-energi-and-port-of-rotterdam-sign-memorandum-of-understanding-regarding-blue-ammonia/. Accessed 20 Sep 2021

19. Travis AS (2018) Nitrogen capture: the growth of an international industry (1900–1940). Springer International Publishing
20. Ernst FA, Sherman MS (1927) The world's inorganic nitrogen industry. Ind Eng Chem 19:196–204. https://doi.org/10.1021/ie50206a003
21. Brown T (2019) Green ammonia plants in Chile, Australia, New Zealand. In: Ammon. Energy. https://www.ammoniaenergy.org/articles/green-ammonia-plants-in-chile-australia-new-zealand/. Accessed 13 Apr 2021
22. Rouwenhorst KHR, Krzywda PM, Benes NE et al (2020) Ammonia. 4. Green ammonia production. Ullmann's Encycl Ind Chem
23. IRENA (2021) Renewable power generation costs in 2020. https://www.irena.org/-/media/Files/IRENA/Agency/Publication/2021/Jun/IRENA_Power_Generation_Costs_2020.pdf
24. IRENA (2020) Green hydrogen cost reduction: scaling up electrolysers to meet the 1.5 °C climate goal. https://irena.org/-/media/Files/IRENA/Agency/Publication/2020/Dec/IRENA_Green_hydrogen_cost_2020.pdf
25. Grundt T, Christiansen K (1982) Hydrogen by water electrolysis as basis for small scale ammonia production: a comparison with hydrocarbon based technologies. Int J Hydrogen Energy 7:247–257. https://doi.org/10.1016/0360-3199(82)90088-X
26. Batool M, Wetzels W (2019) Decarbonisation options for the Dutch fertilizer industry. https://www.pbl.nl/sites/default/files/downloads/pbl-2019-decarbonisation-options-for-the-dutch-fertiliser-industry_3657.pdf
27. Morgan ER (2013) Techno-economic feasibility study of ammonia plants powered by offshore wind. University of Massachusetts Amherst
28. Fasihi M, Weiss R, Savolainen J, Breyer C (2021) Global potential of green ammonia based on hybrid PV-wind power plants. Appl Energy 294:116170. https://doi.org/10.1016/j.apenergy.2020.116170
29. Nayak-Luke RM, Bañares-Alcántara R (2020) Techno-economic viability of islanded green ammonia as a carbon-free energy vector and as a substitute for conventional production. Energy Environ Sci 13:2957–2966. https://doi.org/10.1039/d0ee01707h
30. Salmon N, Bañares-Alcántara R (2021) Green ammonia as a spatial energy vector: a review. Sustain Energy Fuels. https://doi.org/10.1039/d1se00345c
31. Appl M (1999) Ammonia: principles and industrial practice, 1st edn. Wiley-VCH Verlag GmbH, Weinheim (Germany)
32. Armijo J, Philibert C (2020) Flexible production of green hydrogen and ammonia from variable solar and wind energy: case study of Chile and Argentina. Int J Hydrogen Energy 45:1541–1558. https://doi.org/10.1016/j.ijhydene.2019.11.028
33. Brown T (2018) Small-scale ammonia production is the next big thing. In: Ammon. Eng. https://www.ammoniaenergy.org/articles/small-scale-ammonia-production-is-the-next-big-thing/
34. Nayak-Luke RM, Forbes C, Cesaro Z, et al (2020) Techno-economic aspects of production, storage and distribution of ammonia. In: Bañares-Alcántara R, Valera-Medina A (eds) Techno-economic challenges of green ammonia as energy vector. Elsevier Science Publishing Co Inc., pp 191–208
35. Salmon N, Bañares-Alcántara R, Nayak-Luke R (2021) Optimization of green ammonia distribution systems for intercontinental energy transport. iScience 24:102903. https://doi.org/10.1016/j.isci.2021.102903
36. Reese M, Marquart C, Malmali M et al (2016) Performance of a small-scale Haber process. Ind Eng Chem Res 55:3742–3750. https://doi.org/10.1021/acs.iecr.5b04909
37. Cesaro Z, Bañares-Alcántara R (2019) Siemens demonstrator update & investigation into ammonia for energy infrastructure in developing countries. In: NH_3 event, Rotterdam (the Netherlands)
38. Brown T (2018) Green ammonia demonstration plants now operational, in Oxford and Fukushima. In: Ammon. Eng. https://www.ammoniaenergy.org/articles/green-ammonia-demonstration-plants-now-operational-in-oxford-and-fukushima/
39. Nielsen A (1995) Ammonia: catalysis and manufacture, 1st edn. Springer-Verlag, Berlin Heidelberg

40. Papavinasam S (2014) Oil and gas industry network. In: Corrosion control in the oil and gas industry, pp 41–131
41. Green L (1982) An ammonia energy vector for the hydrogen economy. Int J Hydrogen Energy 7:355–359. https://doi.org/10.1016/0360-3199(82)90128-8
42. Fertilizers Europe (2012) Guidance for inspection of and leak detection in liquid ammonia pipeline. https://www.fertilizerseurope.com/wp-content/uploads/2019/08/Guidance_for_inspection_of_and_leak_detection_in_liquid_ammonia_pipelines_FINAL_01.pdf
43. (1955). Ammonia transportation. J Agric Food Chem 3:461–462. https://doi.org/10.1021/jf60052a602

Chapter 48
Environmental Concerns About Fuel Ammonia

Ken-ichi Aika

Abstract The Haber–Bosch process saved the world from starvation by creating an artificial fertilizer supply. However, plants accept only about 50% of reactive nitrogen from fertilizer, and the rest is wasted into soil and water. Part of it remains on the earth because the deposition rate has exceeded the decomposition rate by bacteria. It has been said that N-impact is a more serious environmental concern than global warming. Despite this, fuel ammonia has started to replace fossil fuels in order to decrease CO_2 emissions. The environmental concerns about fuel ammonia are discussed from three viewpoints: N_2O formation, NO formation, and NH_3 leakage. Although N_2O and NO are intermediates of ammonia combustion, these molecules can be controlled if the reaction progresses to completion. Thermal NO_x is still an issue in fossil fuel combustion. The technology and mechanism of the de-NO_x process is reviewed. Since the ammonia industry has a long history, safety systems and regulations are well developed. However, further technology is expected such as NH_4^+ and NO_3^- decomposition in water. It is discussed that ammonia fuel shall not have a worse impact on the N-issue and there is still a chance to mitigate the issue.

Keywords Fuel ammonia · Reactive nitrogen · Eutrophication · Fertilizer · NO_x · N_2O warming

48.1 Global Nitrogen Issue

Proteins and amino acids play important roles as information media to translate DNA in the body structures of plants and animals. Nitrogen is one of their central elements; however, it is supplied from di-nitrogen through enzyme (nitrogenase) metabolism. Seeds gather proteins within crop plants, which in turn serve as animal food. Humans have learned how to grow crops and are at the top of the hierarchy among the biospheres. Our ancestors learned that organic fertilizers, beans of Rhizobia-symbiosis, and thunderstorms were effective for crop growth. Scientists

K. Aika (✉)
Professor Emeritus, Tokyo Institute of Technology, 2-12-1 Ohokayama, Meguro-ku, Tokyo 152-8550, Japan
e-mail: kenaika@c01.itscom.net

K. Aika and H. Kobayashi (eds.), *CO2 Free Ammonia as an Energy Carrier*,
https://doi.org/10.1007/978-981-19-4767-4_48

have shown that reactive nitrogen (*Nr*), with ammonia as one of these, is an essential fertilizer for plants. We had understood that the nitrogen fertilizer feed rate was nearly proportional to the wheat production rate. European countries imported Chile saltpeter, Chilean nitrate, as a fertilizer (and explosives) for some decades. When resources declined, society required an artificial ammonia synthesis process. It was at just such a time when Haber and Bosch invented the industrial ammonia production process. This saved human beings from worldwide starvation in the twentieth century.

Advanced countries have enjoyed a stable food supply over the last century, however, an excess supply of *Nr* has created N-pollution (eutrophication of soil or water, and NO_x, N_2O emission) on the earth. At present, about 180 Mt-NH_3 is produced annually, 80% of which is utilized for fertilizer. Let's consider urea as an example of a fertilizer. Urea, $(NH_2)_2CO$, is decomposed gradually to give *Nr* (NH_3 in this case) in accordance with the growth rate of plants. However, just 94% of *Nr* reaches the plant root area, with 47% said to be absorbed into the plant [1]. The rest is oxidized slowly by the bacteria system ($NH_3 \rightarrow N_2O \rightarrow NO_x$, $NH_3 \rightarrow N_2$, $NH_4^+ \rightarrow NO_3^-$, etc.) in soil or water. Currently, the amount of useless ammonia exceeds the digestion ability of earth's bacteria. Research on the annual rates of these processes is challenging, which will be discussed later. For a crop that has absorbed 47% of *Nr*, only 31% is harvested and just 26% goes to food, and surprisingly, only 14% is eventually absorbed into organisms. This situation is illustrated in Fig. 48.1. If we consume meat, the situation becomes worse. Animals hold just 4% of *Nr* supplied, while we digest just 0.6% of the original *Nr*, fertilizer [1]. The wasted plants and animals emit *Nr* in addition to di-nitrogen into the environment. Ideally, all the *Nr* should return to di-nitrogen.

Galloway et al. [2] estimated the *Nr* transport rate in the year before the industrial revolution (1860), the year after Haber–Bosch (1995) and a future day (2050) from 249 references. Here, we summarize the data for 1995 and 1860 by referring to Lin's expression [3]. These authors stated that denitrification to N_2 and deposition rate (into the earth) are difficult to determine since material balance is not attained. Nevertheless, we can learn much from this work. They speculated that there had been

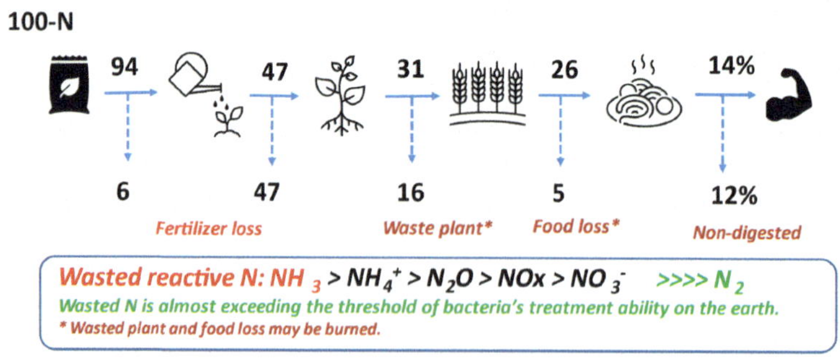

Fig. 48.1 Nitrogen pathway starting from ammonia fertilizer. The numbers express the percentage of fertilizer N

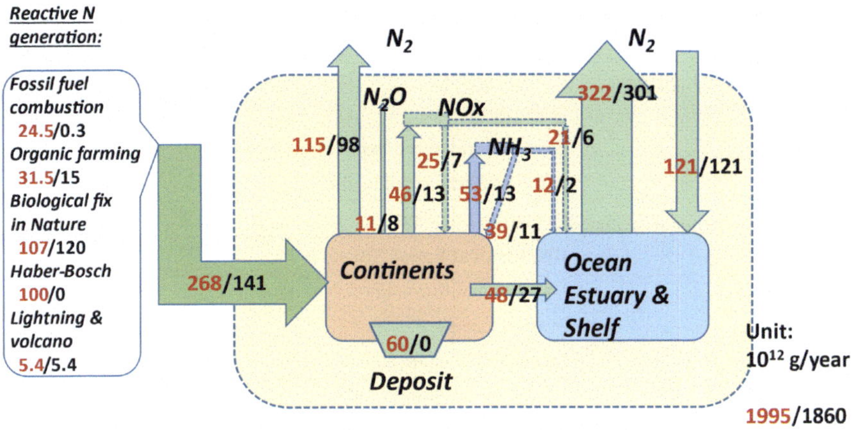

Fig. 48.2 Components of the global nitrogen cycle for 1995 versus 1860 based on Galloway [2] and Lin [3]. Unit: Tg-N/year. Reactive nitrogen (*Nr*) is surrounded by a dotted square. Total mass is balanced for continents region, but it is not so for the ocean region because of the uncertainty. T means 10^{12}

no *Nr* deposition (meaning the steady-state was maintained) before the industrial revolution (1860). This hypothesis is supported by the constancy of the atmospheric N_2O recorded in ice bubbles back to 1000 AD [4].

Here we compare the data for 1995 (red numbers) versus that for 1860 (black numbers) as shown in Fig. 48.2.

1. In 1860, *Nr* input to earth occurred mostly through biological fixation in nature (120 Tg-N/year) with some through organic agriculture (15 Tg-N/year).
2. In 1995, the Haber–Bosch process (100 Tg-N/year as NH_3) made a strong impact on *Nr* input flow. Fossil fuel combustion (24.5 Tg-N/year as NO_x) also added a new risk.
3. As a result, transportation of *Nr* to rivers, oceans, and the troposphere increased.
4. Although the rate of denitrification and deposition is difficult to quantify, they concluded some amount (~60 Tg-N/year) of *Nr* deposition occurs every year.

The amount of nitrogen deposition on the earth (~60 Tg-N/year) is integrated every year, and it might exceed the limit that bacteria can control. Steffen et al. have warned that the present nitrogen input might exceed planetary boundaries (safe operation space for humanity to regulate the earth system) as shown in Table 48.1 [5]. Their analysis shows that "biochemical flows of nitrogen" is ranked as one of the highest risks, even higher than "climate change".

Under these troubling circumstances, we are starting to use ammonia as a fuel. How can this situation be managed? We shall discuss this issue from a scientific viewpoint in the next section. Of course, population control, agriculture revolution, and meal style changes are the first task.

Table 48.1 Risk in 11 fields of planetary boundaries (safe operation space to regulate the earth system) [5]

Planetary boundaries	Risk	
Climate change	3	Increasing
Nobel entities		Un-quantified
Stratospheric ozone depletion	3	Increasing
Atmospheric aerosol loading		Un-quantified
Ocean acidification	2	Safe
Biochemical flows; Nitrogen	**7**	**High risk**
Biochemical flows; Phosphorus	5	Increasing
Fresh water use	1	Safe
Land-system change	3	Increasing
Biosphere integrity; Functional diversity		Un-quantified
Biosphere integrity; Genetic diversity	7	High risk

48.2 Global Warming with N_2O

We first take a look at the generation of reactive N (*Nr*) shown on the left side of Fig. 48.2. We understand that "Fossil fuel combustion" means NO_x formation, which shall be discussed in the next Sect. (48.3). "Lightning" makes NO from the air (N_2–O_2), which is turned to HNO_3 in the rain. We skip this case here. "Organic farming", "Biological fixation", and "Haber–Bosch" are related to ammonia synthesis and fertilizer. About half of ammonia is not absorbed by plants and goes into soil or water, then is gradually oxidized with bacteria to N_2O, NO_x, NO_3^-, etc. In this section, we pick up N_2O in relation to NH_3 combustion. Leaked ammonium ions in water will be discussed in Sect. 48.4.

Escaped N_2O is known to destroy the ozone layer and absorb infrared rays. The latter action has 298 times higher global warming potential than that of CO_2. In agricultural and industrial areas, there are many spots where N_2O is emitted. The Japanese Ministry of Environment is asking for those spots to calculate their N_2O emissions and report them annually. The Ministry is disclosing benchmarks of N_2O emission examples [6]. Figure 48.3 illustrates some of these. Chicken droppings emit about 3% of *Nr* as N_2O when they are dried (blue column at first left). Fertilizer emits about 1% of *Nr* as N_2O in crop fields (blue column at 2nd left). These correspond to 338 and 106 g-CO_2/MJ-ammonia (brown column at first and 2nd left), respectively (*Nr* weight is transformed to NH_3 mol and further to combustion heat of NH_3). These values are much higher than gasoline combustion (73 g-CO_2/MJ, Aika K Chap. 1). Farmland can be considered a warming gas emission site. However, if N-containing materials are burned, N_2O emissions are 38, 17, and 0.6 ppm of *Nr*, shown on the right three blue columns in Fig. 48.3 (wheat waste open burning, solid waste furnace burning, and solid fuel furnace burning). Thus, the fictitious emission of CO_2 is negligible (0.4, 0.2, 0.006 g-CO_2/MJ-ammonia for these three cases; right three brown columns). This means that there is a significant amount of N_2O produced from just biochemical processes, but that it is negligible from combustion processes.

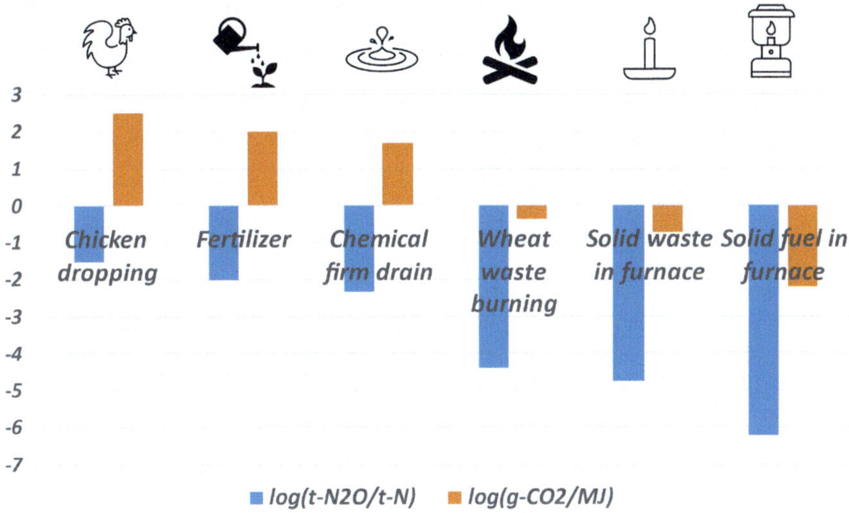

Fig. 48.3 Annual N_2O emissions expressed as log of t-N_2O/t-N (blue columns), and calculated carbon footprint expressed as log of g-CO_2/MJ-ammonia (orange columns) [6]

As discussed in Chaps. 32 [7], 39 [8], and 44 [9], N_2O can be an intermediate of ammonia combustion, however, the concentration is below the observation limit in ammonia combustion experiments. Before publishing the pilot experiment of coal-ammonia mix combustion, the central electric power institute was very careful to check that the concentration of N_2O was null [10]. We understand that chemical equilibrium governs the reaction at "high temperature" such as in combustion. We compare three reactions of ammonia combustion and their standard Gibbs free energy changes as follows:

$$NH_3 + NO + 0.25O_2 = N_2 + 1.5H_2O \quad -413\,\text{kJ/mol} \tag{48.1}$$

$$NH_3 + N_2O + 0.25O_2 = 1.5N_2 + 1.5H_2O \quad -430\,\text{kJ/mol} \tag{48.2}$$

$$NH_3 + 0.5O_2 + 0.25O_2 = 0.5N_2 + 1.5H_2O \quad -447\,\text{kJ/mol} \tag{48.3}$$

A negative Gibbs free energy change moves the reaction more to the right side under the thermodynamic equilibrium. Judging from the Gibbs free energy change, either NO or N_2O can achieve a negligible concentration in complete combustion of ammonia (with excess O_2 concentration), with N_2O far less than NO under equilibrium. Although N_2O is a potential global warming gas, we can control it under careful combustion where the reaction is almost under equilibrium.

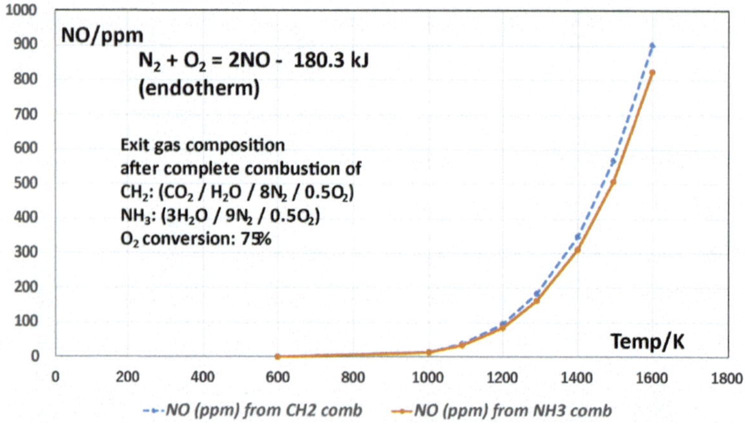

Fig. 48.4 Equilibrium NO concentration (ppm) as a function of temperature (K) when excess oxygen (1.33 times) is used for CH_2 or NH_3 complete combustion

48.3 NO from Combustion

Annual ammonia production (Haber–Bosch) was recorded as 180 Mt/y in 2016, i.e., 148 Tg-*Nr*/y (Mt = Tg). This number is 1.48 times higher than that in 1995 (Fig. 48.2). Clearly, the ammonia fertilizer supply is increasing, with a concomitant increase in nitrogen risk (e.g., NO_x emission, *Nr* deposition). NO is known to form cancer-triggering compounds through reactions with polyaromatic hydrocarbons [11, 12] in the urban atmosphere. This is one of the reasons why NO and hydrocarbon emissions from automobiles and power plants are strictly controlled. NO is produced from "fossil fuel" combustion sites. Here, we compare NO from "ammonia fuel" combustion sites.

We have proposed a simple model to estimate NO emissions from fossil fuel burning [13]. If petroleum is expressed as CH_2 and 75% of oxygen in air is consumed for combustion, it generates about 10.5 product molecules.[1]

$$CH_2 + 1.5(O_2 + 4N_2) + 0.5(O_2 + 4N_2) \rightarrow CO_2 + H_2O + 6N_2 + 0.5(O_2 + 4N_2) \quad (48.4)$$

From this, we get 0.048 bar O_2 and 0.762 bar N_2 at the furnace exit, which allows us to calculate NO concentration under equilibrium as a function of temperature (Fig. 48.4). NO is produced by a reaction between O_2 and N_2 in air at high temperatures. Since the reaction is endothermic, hundred ppm level NO production is remarkable at temperatures higher than 1000 K as shown in Fig. 48.4.

[1] Petroleum is assumed as a fictitious molecule CH_2. BP (Statistical Review of World Energy 2020) data, where total energy of natural gas (G), petroleum (P) and coal (C) is turned to petroleum weight, is used. This also means CH_2 as an average chemical formula of G-P-C mixture.

Returning to the world scale, we assume the average NO concentration in the furnace exit to be 300 ppm.[2] In 1995, 7740 Tg of fossil fuel (as CH_2) was utilized (burned) in the world. This gives NO production of 1.74 (7740/14 * 10.5 * 0.0003) Tmol; 24.4 Tg of *Nr*. Interestingly, this number happens to be almost identical to the data (24.5 Tg) in Fig. 48.2. However, we must use caution because NO emissions are closely related to the burning condition (temperature, exit O_2 concentration, and so on).

Once the equilibrium is attained at high temperature, NO (thermal NO_x) produced in a furnace is carried to the exit where the gas is quenched in the heat exchanger. At low temperatures, NO is difficult to decompose (no reverse reaction) and the molecules retain their memory at high temperatures. This is the reason why combustion gas carries NO into the exit, which has been a difficult issue for many years. Around the 1970s, a new technology (de-NO_x system) was invented by several Japanese companies. This system contains a catalyst and reductant such as ammonia. Some SO_x-tolerant oxides such as V_2O_5 are used as the catalysts. The unique function of the catalyst is that the intermediate containing NH_3 reacts preferentially with (low concentration such as 0.03%) NO rather than (high concentration such as 4.8%) O_2 at 250–410 °C, which is called selective catalytic reduction (SCR). If we think of the Gibbs free energy change, the O_2–NH_3 reaction (Eq. 48.3) is more feasible than the NO–NH_3 reaction (Eq. 48.1). However, the catalyst directs it down a different path. The mechanism is shown in Eqs. 48.5–48.7.

$$\mathrm{M{-}OH + NH_3 \rightarrow M{-}OH{-}NH_3} \tag{48.5}$$

$$\mathrm{M{-}OH{-}NH_3 + NO \rightarrow M{-}OH{-}NH_3{-}NO} \tag{48.6}$$

$$\mathrm{M{-}OH{-}NH_3{-}NO + 0.25O_2 \rightarrow M{-}OH + N_2 + 1.5H_2O} \tag{48.7}$$

Next, it is shown how the de-NO_x system is useful in Japan (Fig. 48.5). In accordance with economic development after the 2nd World War, fossil fuel was used increasingly, and the NO_x level increased around the 1960s. However, the new de-NO_x catalyst systems were introduced to the boiler systems (mainly at power stations) in the 1970s. The technology depressed NO_x emissions strikingly in Japan as shown in Fig. 48.5. If this system were applied worldwide, NO_x emissions might be reduced significantly.

What about NO_x emissions during ammonia combustion? It has been widely believed that ammonia emits NO_x easily because NO_x is one of the intermediates of NH_3 oxidation ($NH_3 \rightarrow NH_x, NO_x \rightarrow N_2 + H_2O$). This is the challenge of balancing kinetics, mechanism, and flow control in ammonia combustion. Indeed, these were

[2] This number (300 ppm) is an equilibrium concentration of NO when 0.048 bar O_2 and 0.64bar of N_2 react at 1365K (1092C). We assumed this example as an average, although each combustion site over the world has different condition from this.

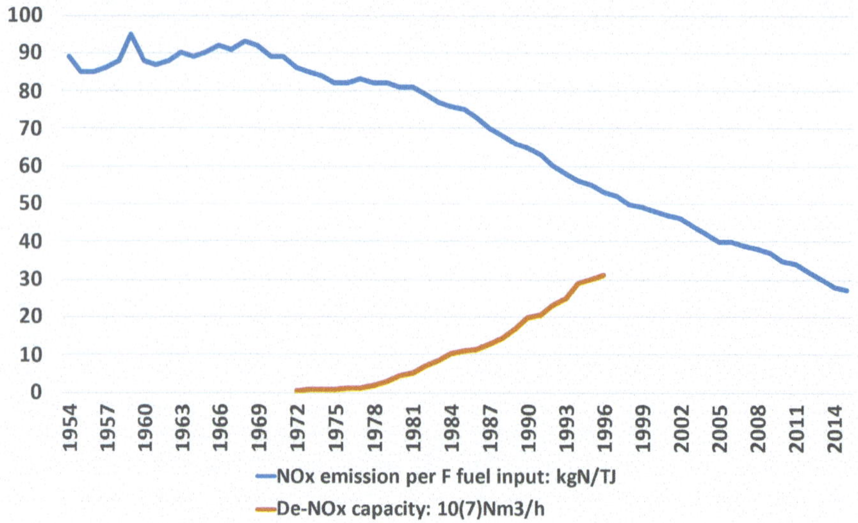

Fig. 48.5 Annual change of NO_x emission per fossil fuel input energy in Japan (kg-N/TJ) [14] and the accumulation of de-NO_x facilities expressed as the treated exhaust gas volume rate (100 Mm^3/h)

the research targets of the SIP program (Aika K in Chap. 1). The combustion team succeeded in controlling the NO_x to the level of fossil fuels [7–9].

Of course, we cannot escape from the equilibrium of the N_2–O_2 reaction. If ammonia is burned in 33.3% excess O_2 condition, $2NH_3$ molecules give about 12.5 molecules (Eq. 48.8).

$$2NH_3 + 1.5(O_2 + 4N_2) + 0.5(O_2 + 4N_2) \rightarrow 3H_2O + 7N_2 + 0.5(O_2 + 4N_2) \tag{48.8}$$

We know that after ammonia combustion 0.04 bar O_2 and 0.72 bar N_2 stay at the exit, which leads to the NO concentration under equilibrium as a function of temperature (Fig. 48.4). Neither CH_2 nor NH_3 itself gives NO under combustion with excess oxygen condition (NO_x intermediates from NH_3 oxidation under equilibrium); however, the O_2–N_2 reaction occurs at high temperature. This situation is common for both fossil fuels and ammonia fuels (only thermal NO_x, no fuel NO_x). Ammonia does not increase world NO because it is used as a substitute for fossil fuel (for the purpose of de-C). One advantage may be that ammonia combustion does not emit any carbonaceous intermediates (such as polyaromatic hydrocarbons) to react with NO to form carcinogens.

48.4 Ammonia Leakage to Environment

As is shown in Fig. 48.2, there are three independent *Nr* pathways: N_2O, NO, and NH_3. N_2O travels in the air (inner space), absorbs infrared rays, and reacts with ozone. It mainly comes from fertilizer. NO goes into the air (near land), turns to smog, and is oxidized to acid rain. It comes mainly from fuel (both fossil and ammonia) combustion. Since ammonia fuel is used as a substitute for fossil fuel, the total NO would not increase if the combustion is controlled properly as discussed above.

The NH_3 pathway may spread into air, land, and water. To discuss the effect of fuel ammonia on the environmental ammonia pathway is beyond our ability at present. Probably the most difficult task is to evaluate fuel ammonia leakage. Ammonia is a poisonous chemical and causes eutrophication. However, ammonia is widely used in industry, and there are many safety and environmental regulations. We believe that industry and the government will expand those rules to include fuel ammonia. At the same time, we should acknowledge that the earth has been overwhelmed with reactive nitrogen. Thus, we should continue to study the situation and develop new technologies to mitigate it.

Here, we briefly highlight two such technologies of denitrification of wasted ammonia in water. In industry, gaseous ammonia is easily decomposed to N_2 and H_2 over catalysts such as supported-Ru or Ni at 400–500 °C as is shown by Muroyama H in Chap. 27 [15]. Once ammonia is dissolved in water, it is difficult to decompose it to N_2. In this case, supported metal (oxide) can work as a catalyst for oxidation of ammonia in water at around 200 °C or under several bars of air (wet air oxidation; WAO). The inorganic catalyst works like an enzyme that decomposes ammonia to NO_x and finally to N_2. NH_3 is oxidized through N_2O, NO, NO_2^-, NO_3^- and finally into N_2. The product control (reaction selectivity) is a challenge for both inorganic catalysis and enzyme reactions. However, reaction rates with inorganic catalysts are much faster than those with enzymes. In this sense, wet air oxidation would be a possible technology to treat leaked ammonia in water if it can be controlled in an industrial apparatus.

Here two examples are discussed: WAO of ammonia solution [16] and decomposition of ammonium nitrate solution [17]. RuO_2/Al_2O_3 catalyst showed remarkable activity (99% conversion) in WAO of 1500 ppm ammonia aqueous solution at 230 °C under 1.5 MPa. The main product was N_2 with a trace amount of NO_3^-. However, NaOH addition was necessary to keep the pH above 12, where NH_4^+ (less reactive) turns to NH_3 (more reactive) [16].

Ammonium nitrate is also difficult to decompose in water, but an active carbon (AC) supported Pt catalyst was found to be effective [17]. 0.1 mol/L of NH_4NO_3 solution was treated with 3% Pt/AC in an autoclave (0.1–3.5 MPa) at 180 °C for 2 h. The results are shown in Fig. 48.6 together with the reaction mechanism. Ammonium ion and nitrate ion react on the Pt catalyst giving N_2O (adsorbed intermediate) and water. N_2O (adsorbed intermediate) has two reaction pathways: desorption (to N_2O) and reaction with AC support (to N_2). Under 0.5 MPa of air feed pressure, both ions are almost 100% decomposed with N_2O selectivity of just 3%. We found that

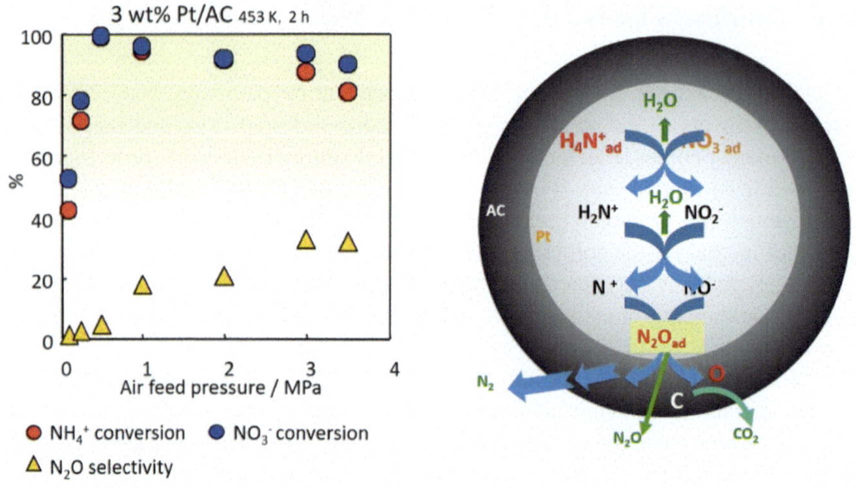

Fig. 48.6 Simultaneous wet air decomposition of ammonium and nitrate with active carbon-supported platinum catalyst at 453 K

some reductant (AC in this case) was necessary to give complete denitrification [17]. These are just two examples of ammonia solution reactions, where we can harness the potential of inorganic catalysis.

48.5 Final Remarks: Effect of Ammonia Fuel Input on Reactive Nitrogen World

Figure 48.7 illustrates the reactive N pathway and the effect on the environment. Ammonia is produced through the Haber–Bosch process and utilized mainly for fertilizer. However, almost half of it escapes to the environment and turns to reactive nitrogen (*Nr*). The environmental impact of this route may be N_2O and ionic forms of nitrogen (NH_4^+, NO_3^-). The former has a very high global warming potential and the latter causes eutrophication. NO_x comes mostly from fossil fuel combustion (thermal NO_x). Ammonia combustion does not contribute to N_2O and NO formation through intermediate formation (fuel NO_x) if the process is well controlled. However, it emits a similar level of thermal NO_x as fossil fuel does. Since ammonia fuel replaces fossil fuel, the total NO_x would not be increased. The better news is that NO_x from ammonia is not accompanied by organic exhaust such as polyaromatic hydrocarbons, which is a cause of cancer when reacted with NO_x.

The author described in Sect. 48.2 that farmland can be considered a warming gas emission site. This is true if *Nr* in fertilizer is deemed as NH_3 fuel. However, once fertilizer is acknowledged to assist in growing plants, it is also deemed to be a helper for CO_2 fixation. The function of fertilizer to accelerate CO_2 fixation is several times

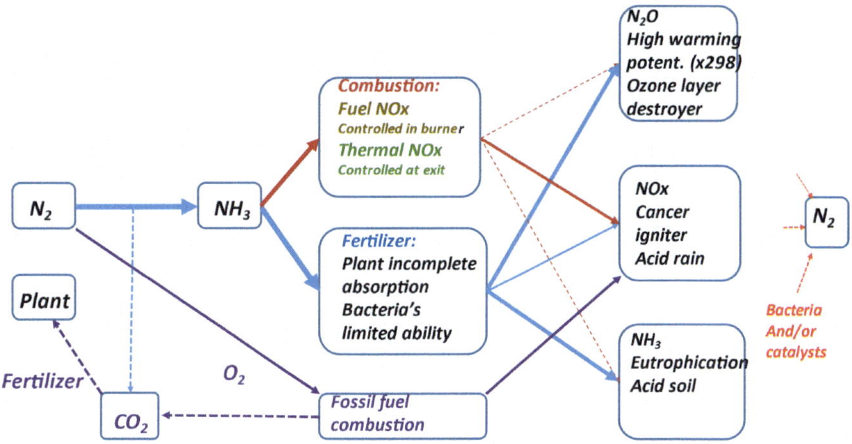

Fig. 48.7 Formation and transformation of reactive nitrogen and impact on the environment: effect of ammonia fuel on ammonia fertilizer

higher than the carbon footprint of brown ammonia. Fertilizer is a risk in global warming (through N_2O production) on one hand; however, it mitigates C-impact on the other hand.

Finally, we summarize how NO_x is produced through NH_3 oxidation (Eq. 48.3).

1. Since NH_3 oxidation (Eq. 48.3) is a steep downhill reaction, the intermediates such as N_2O or NO are negligible products under excess oxygen conditions (with well-mixed combustion at high temperatures around 1000 °C). The complete reaction occurs in a furnace with a swirl burner.
2. If the reaction gases are not well mixed, both oxygen-rich and -poor areas might occur. Through the oxygen-poor area, the intermediates (N_2O and/or NO) might be emitted even from a furnace.
3. In an open fire, the flame (reaction zone) is connected to open air where the intermediates might pass through the boundary and be quenched.
4. Biochemical reactions proceed quite slowly at ambient temperature with bacteria. The reaction is controlled with enzyme catalysis and need not always be controlled with thermodynamic equilibrium. The intermediates (e.g., N_2O, NO_x, NO_3^-) are produced.
5. As for the *Nr* transition, inorganic catalysis processes such as in Eqs. 48.5–48.7 and Fig. 48.6 occur at medium temperature (200–500 °C). A well-developed catalyst can control the reaction path.

References

1. Pais H et al (2020) The two faces of nitrogen. https://www.youtube.com/watch?v=PTtU8ZzKu18. Accessed 12 Aug 2020
2. Galloway JN, Dentener FJ, Capone DG, Boyer EW, Howarth RW, Seitzinger SP, Asner GP, Cleveland CC, Green PA, Holland EA, Karl DM, Michaels AF, Porter JH, Townsend AR, Vorosmarty CJ (2004) Nitrogen cycles: past, present, and future. Biogeochemistry 70:153–226
3. Lin B-L (2011) Expectation and anxiety in nitrogen in terms of bio-mass energy exploration. Petrotech 34(8):559–566
4. Ayres RU, Schlesinger WH, Socolow RH (1994) Human impacts on the carbon and nitrogen cycles. In: Socolow RH, Andrews C, Berkhout R, Thomas V (eds) Industrial ecology and global change. Cambridge University Press, New York, pp 121–155
5. Steffen W, Richardson K, Rockström J, Cornell SE, Fetzer I, Bennett EM, Biggs R, Carpenter SR, de Vries W, de Wit CA, Folke C, Gerten D, Heinke J, Mace GM, Persson LM, Ramanathan V, Reyers B, Sörlin S (2015) Planetary boundaries: guiding human development on a changing planet. Research 347(6223):1258855/1-10
6. Ministry of Environment (2020) Calculation and reporting of global warming gas emission. https://ghg-santeikohyo.env.go.jp/files/calc/itiran_2020_rev.pdf. Accessed 4 Sep 2021, http://ghg-santeikohyo.env.go.jp/files/manual/chpt2_4-0.pdf. Accessed 12 May 2017
7. Kobayashi H (2022) Chapter 32 Significance and impact of ammonia combustion research (in this book)
8. Nakamura H (2022) Chapter 39 Ammonia combustion chemistry (in this book)
9. Murai R et al (2022) Chapter 44 Review of fundamental study of ammonia direct combustion in industrial furnaces (in this book)
10. Kimoto M (2022) Chapter 40 Applicability of ammonia co-firing technology in pulverized coal-fired boilers (in this book)
11. Ishii S, Hisamatsu Y, Inazu K, Kadoi M, Aika K (2000) Ambient measurement of nitrotriphenylenes and possibility of nitrotriphenylenes formation by atmospheric reaction. Environ Sci Technol 34:1893–1899
12. Ishii S, Hisamatsu Y, Inazu K, Kobayashi T, Aika K (2000) Mutagenic nitrated benzo[a]pyrene derivatives in the reaction product of benzo[a]pyrene in NO_2-air in the presence of O_3 or under photoirradiation. Chemosphere 41:1809–1819
13. Gosho A, Aika K (2000) Zero emission possibility with nitrogen cycle analysis: Present status and measure in view of chemical reaction principle. Chem Econ (Kagaku Keizai) Sept:84–94
14. Hayashi K (2020) Toward sustainable nitrogen utilization. Petrotech 43:37
15. Muroyama H (2022) Chapter 27 Ammonia decomposition catalysts for fuel cell application (in this book)
16. Qin J, Aika K (1998) Catalytic wet air oxidation of ammonia over alumina supported metals. Appl Catal B: Environ 16:261–268
17. Kitahara M, Inazu K, Aika K (2002) Simultaneous wet air decomposition of ammonium and nitrate over supported platinum catalyst. In: 81st spring meeting, Japan. Chem Soc, 22 Mar

Chapter 49
Clean Ammonia Fuel; Practical Use and the Road Map

Shigeru Muraki

Abstract The governmental SIP energy carrier program started in 2014 was successfully ended focusing on ammonia fuel technology in 2019. The program was succeeded to "Clean Fuel Ammonia Association" supported by both Japanese and overseas industries. Japanese government decided to develop ammonia fuel technology as one of the de-carbon energy policies. The author describes such development in an international view.

Keywords Ammonia fuel · Value chain · Blue ammonia · Green ammonia · Road map

Ammonia has been attracting attention as a clean fuel that does not emit CO_2 when burned. Research, development and demonstration for the use of ammonia as a fuel were conducted for five years from April 2014 under the Japanese government's Cross-ministerial Strategic Innovation Promotion Program (SIP), and significant progress in the technology of direct ammonia combustion as a fuel in the energy market was realized. As a result, ammonia is expected to play a major role as a promising hydrogen carrier for future decarbonization and carbon neutrality. Since April 2019, when the SIP ended, Clean Fuel Ammonia Association (CFAA), a cross-industry organization, has been taking the lead in promoting further technological development, implementing fuel ammonia utilization technologies, and building a value chain from supply to use.

Among the direct use technologies of clean fuel ammonia in the energy market, as for the power generation sector, mixed combustion in coal power boilers is ready for commercialization, and mixed and single-fuel combustion in gas turbines and solid oxide fuel cells are underway, and those technologies are expected to be commercialized in the next five years. In the industrial sector, mixed combustion technology with natural gas is being developed and is expected to be commercialized in the late 2020s (Figs. 49.1 and 49.2).

S. Muraki (✉)
SIP, Tokyo Gas, Kaigan, Minato-ku, Tokyo 105-8527, Japan
e-mail: s-muraki@tokyo-gas.co.jp

K. Aika and H. Kobayashi (eds.), *CO_2 Free Ammonia as an Energy Carrier*,
https://doi.org/10.1007/978-981-19-4767-4_49

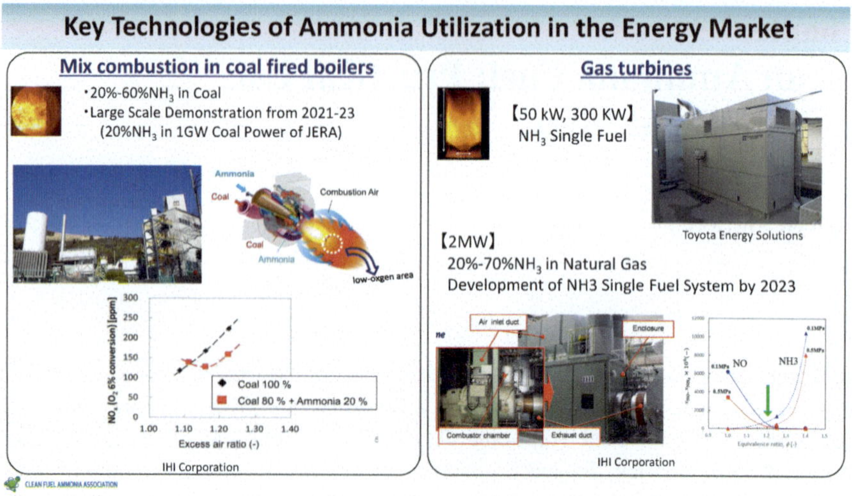

Fig. 49.1 Key technologies of ammonia utilization: mixed combustion with coal and gas turbines

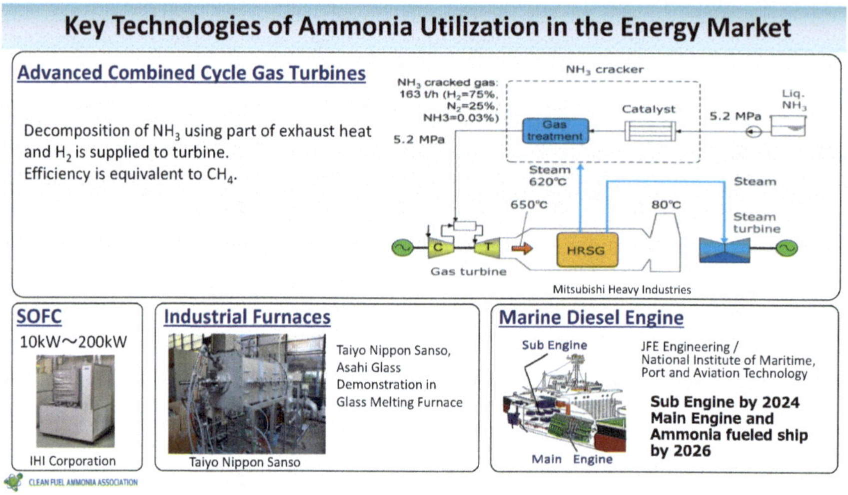

Fig. 49.2 Key technologies of ammonia utilization: combined cycle gas turbine, SOFC, industrial furnaces, and marine diesel

After the power generation market, utilization as a marine fuel is the next promising application. The development of ammonia-fueled diesel engines is already underway. It is expected to be completed and put into practical use by 2026. Ammonia is considered to be the most promising candidate among CO_2-free fuels for deep-sea shipping in the future. The International Energy Agency (IEA) predicts that 46% of marine fuel will be ammonia by 2050.

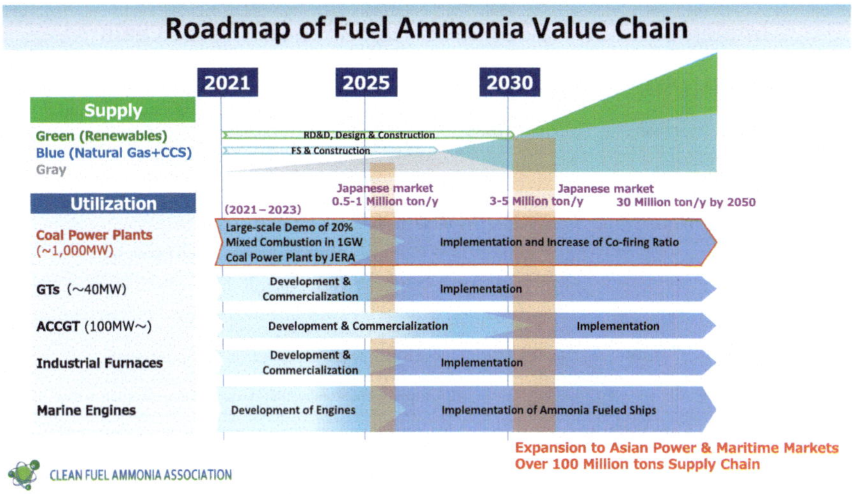

Fig. 49.3 Road map of fuel ammonia value chain

As for the supply of clean fuel ammonia, studies and discussions are underway for blue and green ammonia. Blue ammonia is produced from natural gas as a feedstock using the Haber–Bosch process while treating CO_2 generated during the process by CCS or EOR. Green ammonia uses hydrogen produced by electrolysis of water using electricity generated from renewable energy sources. As for the supply cost of blue ammonia to Japan and other Asian countries, it is estimated to realize $2/kg of hydrogen equivalent from $3/mmbtu Natural Gas, which is the necessary cost to be used in the power generation market, and it is expected that early market introduction will be possible. On the other hand, green ammonia still needs significant cost reduction, including water electrolysis equipment. Therefore, the introduction of green ammonia will be in the 2030s as the market for fuel ammonia expands and cost reduction progresses (Fig. 49.3).

For the supply of clean fuel ammonia, a different scheme is being considered from the current main market of fertilizer. It is assumed that dedicated ammonia plants for the fuel market will be built, and sales contracts with a term agreement that sets the price, supply volume, and supply period will be introduced. It is expected that the market for clean fuel ammonia, including green ammonia, will be steadily realized through this type of contracts. Australia, New Zealand, the United States, Canada, the Middle East region, and Chile are the leading candidates for clean fuel ammonia supply countries and regions in the Asia–Pacific region (Fig. 49.4).

As for the supply infrastructure in Japan, the concept of a fuel ammonia hub terminal, a large-scale import terminal with loading function to smaller coastal vessels for secondary domestic delivery, is being considered in cooperation with Japan's Carbon Neutral Port Project in order to efficiently supply ammonia to a wide range of users. This project is expected to contribute greatly to the decarbonization of Japan by making fuel ammonia widely available.

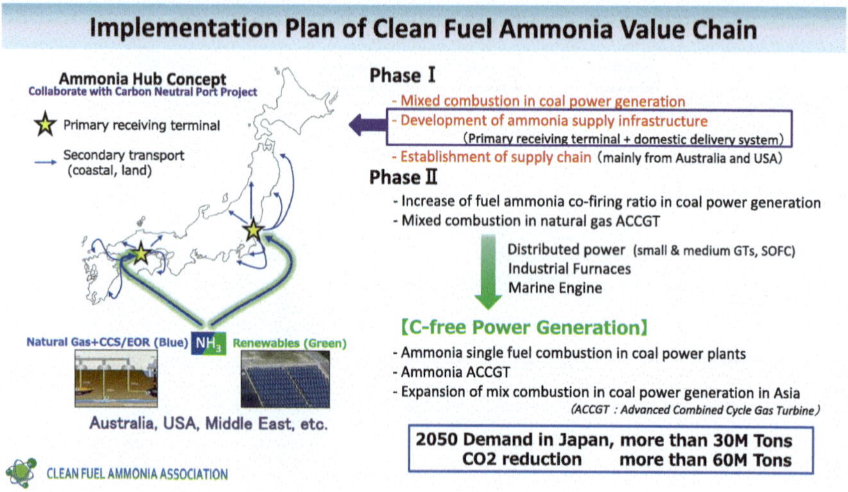

Fig. 49.4 Implementation plan of clean fuel ammonia value chain

The formation of such infrastructure could be applied to other Asian countries as well.

In order to establish a fuel ammonia value chain, considerations on technical standards for equipment, operational and safety standards, and fuel ammonia specifications are underway based on standards already in operation in the industry. In addition, discussions on standards and certification of clean fuel ammonia are also underway, and by continuing to discuss with supplier countries, Japan aims to establish the international standards (Fig. 49.5).

The large-scale introduction of fuel ammonia in the power generation market is expected to start in 2027 with mixed combustion in coal-fired power plants. According to the introduction plan jointly formulated by the government and the private sector, expected demands in Japan are 3 million tons per year in 2030 and 30 million tons in 2050, respectively. The introduction plan also aims to build a 100 million tons supply chain by 2050, mainly in Asia, with Japanese companies taking the lead. Nevertheless, given the ambitious efforts toward decarbonization and carbon neutrality in the future, demand for clean fuel ammonia, a realistic and viable option, is likely to increase significantly beyond this plan.

This clean fuel, ammonia, will undoubtedly play an important role in the decarbonization of Japan, Asia, and the world, and has the potential to be a game-changer in the energy market of the twenty-first century.

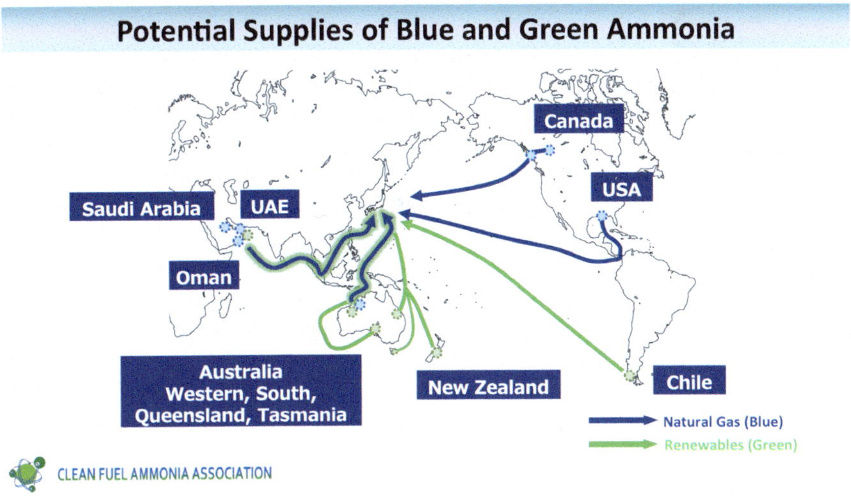

Fig. 49.5 Potential supplies of blue and green ammonia

Acknowledgements The following organizations are acknowledged for support of research, development, and industrialization of hydrogen energy carrier technology and ammonia fuel technology.

Advanced Low Carbon Research and Technology and Development Program (ALCA 2013, program officer: Ken-ichi Aika).

Cross-ministerial Strategic Innovation Program (SIP) "Energy Carrier" (2014–2019, director: Shigeru Muraki), the Cabinet Office (2014–2019, CAO SIP program representative: Kazuo Kyuma), and Japan Science and Technology Institute (JSP 2014–2019, logistics members: A Koga).

Green Ammonia Energy Consortium (2019–2021, representative: Shigeru Muraki, Osamu Ishitobi).

Clean Fuel Ammonia Association (2021-present, representative: Osamu Ishitobi).

The manufacturer's authorised representative in the EU is Springer Nature Customer Service Centre GmbH, Europaplatz 3, 69115 Heidelberg, Germany. If you have any concerns regarding our products, please contact ProductSafety@springernature.com

Printed and bound by CPI Group (UK) Ltd, Croydon, CR0 4YY
30/09/2026
02225315-0018